IET TELECOMMUNICATIONS SERIES 68

Advanced Relay Technologies in Next Generation Wireless Communications

Other volumes in this series:

Volume 9	**Phase noise in signal sources** W.P. Robins
Volume 12	**Spread spectrum in communications** R. Skaug and J.F. Hjelmstad
Volume 13	**Advanced signal processing** D.J. Creasey (Editor)
Volume 19	**Telecommunications traffic, tariffs and costs** R.E. Farr
Volume 20	**An introduction to satellite communications** D.I. Dalgleish
Volume 26	**Common-channel signalling** R.J. Manterfield
Volume 28	**Very small aperture terminals & (VSATs)** J.L. Everett (Editor)
Volume 29	**ATM: the broadband telecommunications solution** L.G. Cuthbert and J.C. Sapanel
Volume 31	**Data communications and networks, 3rd edition** R.L. Brewster (Editor)
Volume 32	**Analogue optical fibre communications** B. Wilson, Z. Ghassemlooy and I.Z. Darwazeh (Editors)
Volume 33	**Modern personal radio systems** R.C.V. Macario (Editor)
Volume 34	**Digital broadcasting** P. Dambacher
Volume 35	**Principles of performance engineering for telecommunication and information systems** M. Ghanbari, C.J. Hughes, M.C. Sinclair and J.P. Eade
Volume 36	**Telecommunication networks, 2nd edition** J.E. Flood (Editor)
Volume 37	**Optical communication receiver design** S.B. Alexander
Volume 38	**Satellite communication systems, 3rd edition** B.G. Evans (Editor)
Volume 40	**Spread spectrum in mobile communication** O. Berg, T. Berg, J.F. Hjelmstad, S. Haavik and R. Skaug
Volume 41	**World telecommunications economics** J.J. Wheatley
Volume 43	**Telecommunications signalling** R.J. Manterfield
Volume 44	**Digital signal filtering, analysis and restoration** J. Jan
Volume 45	**Radio spectrum management, 2nd edition** D.J. Withers
Volume 46	**Intelligent networks: principles and applications** J.R. Anderson
Volume 47	**Local access network technologies** P. France
Volume 48	**Telecommunications quality of service management** A.P. Oodan (Editor)
Volume 49	**Standard codecs: image compression to advanced video coding** M. Ghanbari
Volume 50	**Telecommunications regulation** J. Buckley
Volume 51	**Security for mobility** C. Mitchell (Editor)
Volume 52	**Understanding telecommunications networks** A. Valdar
Volume 53	**Video compression systems: from first principles to concatenated codecs** A. Bock
Volume 54	**Standard Codecs: image compression to advanced video coding, 3rd edition** M. Ghanbari
Volume 59	**Dynamic Ad Hoc Networks** H. Rashvand and H. Chao (Editors)
Volume 60	**Understanding Telecommunications Business** A Valdar and I Morfett
Volume 905	**ISDN applications in education and training** R. Mason and P.D. Bacsich

Advanced Relay Technologies in Next Generation Wireless Communications

Edited by
Ioannis Krikidis and Gan Zheng

The Institution of Engineering and Technology

Published by The Institution of Engineering and Technology, London, United Kingdom

The Institution of Engineering and Technology is registered as a Charity in England & Wales (no. 211014) and Scotland (no. SC038698).

First published 2016

The Institution of Engineering and Technology
Michael Faraday House
Six Hills Way, Stevenage
Herts, SG1 2AY, United Kingdom

www.theiet.org

British Library Cataloguing in Publication Data
A catalogue record for this product is available from the British Library

ISBN 978-1-78561-003-5 (hardback)
ISBN 978-1-78561-004-2 (PDF)

Typeset in India by MPS Limited
Printed in the UK by CPI Group (UK) Ltd, Croydon

To my parents, Apostolos and Sophia,
for their love and encouragement.
— Ioannis Krikidis

To my parents, my wife and my lovely children,
for their generous support.
— Gan Zheng

Contents

Preface

A basic relay channel is a fundamental network structure consisting of one source, one destination, and an intermediate device, i.e., "the relay," which facilitates the communication between the source and the destination. From the information theory standpoint, it is decomposed into one broadcast channel (from the source to the relay and the destination) and one multi-access channel (from the source and the relay to the destination). The main research questions refer to the encoding of the broadcast channel at the source, the management of the multi-user interference at the destination, and the signal processing at the relay node. These questions still lack answers, specifically in the fundamental level. Despite the recent advances on the information theoretical aspects of the relay channel, the Shannon capacity of the general relay channel is still unknown.

The relay channel has received significant interest from both industry and academia over the last decades. A tremendous number of studies have appeared in the literature, dealing with the exploitation of the relay channel in practical network structures such as cellular networks, mesh networks, and ad-hoc networks. Researchers have focused on two basic relaying architectures: the first one is the multi-hop transmission where several intermediate nodes assist the source to transmit its data to the destination; the main goal of this architecture is to combat large-scale fading and path-loss effects. The second architecture, named cooperative diversity, provides multi-antenna benefits to single-antenna devices, i.e., terminals with single antennas create virtual multiple antenna configurations to achieve spatial diversity and mitigate small-scale fading effects.

Although relaying was originally introduced as an effective means to extend coverage and repair broken links in wireless networks, in the current networks it is considered as a core element that assists the network to achieve different objectives such as reliability, spectral efficiency, energy savings, and secrecy. However, relaying technology is also facing major challenges when applied to the evolving wireless networks, including complexity, energy consumption, interference management, spectrum sharing, non-idealities, and signaling overhead. These challenges need to be addressed in order to reap the benefits of relaying technology, calling for joint efforts from wireless industry and academic researchers. Therefore, it is necessary to have a complete understanding of the relaying protocols, performance gains, practical limitations and potential applications.

This book provides a thorough treatment of relaying technology in future wireless networks, including the fifth-generation (5G) wireless communications and beyond. The book features contributions from internationally leading researchers in relaying technology, giving cutting-edge tutorials on fundamental relaying principles, rigorous analysis, practical considerations, and most recent application scenarios. It covers a wide range of hot topics such as distributed antennas, full-duplex radios, relay selection, green communications, wireless security, wireless power transfer, smart grid, optical wireless networks, and multi-beam satellite communications.

The book is organized into 15 chapters. Chapter 1 focuses on the integration of the spatial modulation (SM) concept in relay-aided and cooperative communications. By using appropriate encoding mechanism, the activation of the relay nodes (spatial dimension) is exploited as an efficient means to convey information. Two SM-based distributed cooperative protocols, with different complexities and design goals, are analyzed. This chapter covers the design of efficient demodulators at the relay nodes, which are robust to demodulation errors and ensure diversity.

Chapter 2 presents the combination of massive multiple-input multiple-output (MIMO) with relay technologies. Massive MIMO has been identified as one of the main technologies in the upcoming 5G mobile communication systems, and its combination with relaying can be exploited for improving spectral efficiency versus coverage. This chapter discusses fundamental issues of massive MIMO relaying such as system configurations, transmission frame structure, channel estimation, and signal processing techniques.

Chapter 3 deals with the fundamental concept of distributed/virtual mmWave MIMO relaying in the context of shared user equipment-side distributed antenna system (SUDAS) for 5G outdoor-to-indoor communications. The SUDAS principle exploits both licensed and unlicensed frequency bands and translates spatial multiplexing into frequency multiplexing through relaying. Fundamental characteristics of SUDAS as well as implementation aspects, potential applications, and design challenges are discussed.

Chapter 4 re-examines the MIMO relaying architecture for multi-user networks with co-channel interference (CCI). This chapter covers the design of low-complexity linear processing techniques that mitigate and/or exploit CCI at the relay node. Performance analysis studies demonstrate the benefits of linear processing as well as the fundamental performance limitations. Wireless power transfer scenarios, where the relay node uses a portion of the received signal in order to power the relaying links, are also discussed.

Full-duplex (FD) relay technology, where relay nodes can simultaneously transmit/receive data in the same frequency band, is presented in Chapter 5. Until recently, FD radio was considered to be unrealistic mainly due to the harmful effects of the loopback self-interference (LI) from the transmit antennas to the receive antennas. This chapter highlights the core elements of the FD relaying technology and presents transmit power adaptation as an efficient technique to mitigate LI. An opportunistic switching between FD and conventional half-duplex relaying is also analyzed.

Relay selection is a well-known technique in the cooperative literature to ensure spatial diversity benefits with low computational demands. Chapter 6 discusses its application into three different cooperative scenarios, namely (a) FD relaying, (b) buffer-aided cooperative networks, and (c) wireless powered cooperative networks with batteries. By addressing the fundamental technical challenges of relay selection in these three communication scenarios, the authors show that relay selection is a promising physical layer design tool in cooperative networks.

Energy efficiency is a critical design goal for future sustainable communications. Relaying provides a unique solution to improve energy efficiency, because it effectively decreases the distances between communications nodes. We have two chapters dedicated to this topic. Chapter 7 provides a detailed energy evaluation of relaying protocols. It emphasizes the importance of using practical models to account for the total energy consumption including not only transmit power, but also signal processing and circuit power consumption. This chapter also discusses the benefits in energy efficiency of using multiple relays to form virtual MIMO networks.

Chapter 8 focuses on the optimization of energy efficiency of amplify-and-forward relaying networks, motivated by the fact that in many practical systems this simple relaying protocol outperforms more complex protocols such as decode-and-forward relaying. Energy-efficient optimization of power control, precoding matrix and receive filters are developed for single-stream and multiple-streams MIMO systems. Distributed resource allocation algorithms are also discussed.

Relaying has been mostly used for cooperation of information transfer. Chapter 9 introduces energy cooperation through the concept of wireless power transfer via relaying. This is particularly important to create strong incentives for cooperation in cognitive radio networks, where the primary and the secondary users normally compete for spectrum access. This chapter describes information and energy cooperation that allows the primary transmitters to supply energy to the secondary transmitters in exchange of data relaying assistance.

Chapter 10 provides an in-depth study on the often ignored non-ideal channel and signal conditions in wireless relaying, including feedback delay, link correlation, non-Gaussian interference, and limited relay power. For each non-ideal condition, it presents rigorous analysis on its effect and shows that traditionally optimal designs are no longer suitable. The author introduces new solutions including amplitude-based relay selection, adopting practical interference model with approximation techniques, and energy harvesting to combat these effects.

Relays not only can help improve efficiency and reliability of wireless communications, but also enhance security. Chapter 11 investigates the problem of physical-layer security in wireless networks with the help of cooperative relays. The authors study two main approaches to deaf cooperation, namely, the cooperative jamming and the noise forwarding approach, to reinforce security of transmission in wireless relay networks. A novel combination of these two approaches is discussed by decomposing the multiple spatial dimensions. It then investigates active relay cooperation in decode-and-forward networks.

Modern communication technology is revolutionizing the smart grid to improve power delivery efficiency and reduce operating costs. Chapter 12 aims to identify relaying strategies that can achieve greater spectral efficiency, extended transmission range, and improved reliability for smart grid applications. The authors first present an overview of communication technologies for the smart grid applications and analyze the potential challenges. Then they focus on wireless communications for smart girds, and study advanced relaying strategies including two-relay strategy, bidirectional relaying, and virtual MIMO strategy.

Chapter 13 looks into relay-assisted interference channel, where multiple-antenna relays use harvested radio frequency (RF) energy from the source to forward the message to the destination, thus providing a sustainable approach to prolonging the network's lifetime. The authors develop a distributed framework to maximize the sum rate by formulating a non-cooperative game and prove that the formulated game always admits a unique Nash equilibrium. They show that in general the proposed game-theoretic approach can achieve a near-optimal network-wide performance.

Optical wireless communication has recently attracted a renewed interest. Chapter 14 investigates relaying strategies to overcome atmospheric degradations in optical wireless networks. It first provides an overview of the optical wireless channel, and then evaluates the performance of relay-assisted optical wireless communications systems. It also demonstrates the advantage of using a free space optical relaying link to extend the reach of the RF link. It offers an in-depth description of more robust optical orthogonal frequency division multiplexing techniques and discusses schemes that are robust to non-linear distortions of the transmitted signal.

Satellite communications are one of the most widely used relaying systems to retransmit radio, TV, telephone signals and data around the earth, working mostly in an amplify-and-forward fashion. Chapter 15 introduces state-of-the-art research developments on multi-beam satellite communications. It provides an overview of three main practical limitations, i.e., framed-based precoding, non-idealities of amplifiers, and receiver design. This chapter gives a promising perspective that cooperative satellite relaying is a mature technology in research and has reached the doorstep of practical implementation.

Ioannis Krikidis
Gan Zheng

Acknowledgements

First we would like to thank our publisher team, the IET and MPS Limited for their professional support throughout this project, special thanks to Paul Deards, Jennifer Grace, and Kate Carveth for their encouragement that motivated this book, and Priya Devaraj and Yassar Arafat for typesetting and keeping us on track.

We would like to express our sincere thanks to chapter authors, who have made every endeavour to provide the draft and corrections in time. Without their efforts, commitment, and hard work, the timely completion of this book would not have been possible.

We would also like to thank many of our collaborators around the world who have directly or indirectly contributed to this book and we appreciate the long-lasting rewarding cooperation.

Containing contributions by some of the most prominent researchers in the field, this book offers probably the most comprehensive presentation of advanced technologies for wireless relaying that is available to date. With a clear emphasis on emerging techniques, ranging from massive MIMO, full-duplex and mmWave relaying to cognitive and optical wireless relaying, including green networking considerations and addressing applications such as satellite communications, physical layer security, and power systems, this impressive collection of works is well-placed to become a key reference for next-generation wireless communication networks of all types, in which relaying is poised to be a key component.

Constantinos B. Papadias, Fellow, IEEE
Dean, Athens Information Technology

This is a compelling anthology on the topical subject of relaying. The amalgam of an exquisite author-team and the influential editorial touch resulted in an excellent state-of-the-art anthology.

To elaborate a little further, relaying has attracted wide-ranging research efforts, especially since the conceptual linkage between co-located and distributed multiple antenna elements has been explored by the community. A limitation of the co-located antenna elements is that unless they are sufficiently far apart, the correlation of their signals erodes the achievable capacity. This impediment may be mitigated with the aid of intelligently sharing the antennas of the cooperating nodes, albeit in some applications this is only feasible at the cost of setting aside extra time-slots for the cooperative exchange of information, whilst tolerating the associated extra interference.

This anthology provides a comprehensive tour of the associated solutions in a system-oriented context, touching upon the beneficial combination of relaying with a whole host of sophisticated enabling techniques, such as spatial modulation, large-scale antenna systems, in full-duplex solutions, cognitive radio techniques, whilst touching upon the achievable energy-reduction, on the novel facets of joint information and power transfer, just to name a few. Since relaying is capable of improving the signal-to-interference-plus-noise ratio (SINR) of the desired link, it also improves the capacity of the desired link, which hence allows the legitimate users to communicate at a higher rate than the attainable rate of the eavesdropper, hence enhancing the physical-layer security. Given the recent advances in optical wireless communications, relaying has also found beneficial applications in the optical domain, as well as in the emerging smart grid and in the satellite-systems above the clouds.

Clearly, a comprehensive anthology – wishing you intellectual fun.

Lajos Hanzo
FREng, FIEEE, FIEE, DSc
Chair of Telecomms.
University of Southampton

List of contributors

Chapter no.	Authors affiliation
1	**Sandeep Narayanan and Mark Flanagan** School of Electrical, Electronic and Communications Engineering, University College Dublin, Dublin, Ireland **Marco Di Renzo** Laboratoire des Signaux et Systémes, CNRS, CentraleSupélec, Univ Paris Sud, Université Paris-Saclay, France **Harald Haas** Institute for Digital Communications, University of Edinburgh, Edinburgh, UK **Fabio Graziosi** Department of Information Engineering, Computer Science and Mathematics, University of L'Aquila, Coppito, L'Aquila, Italy
2	**Gayan Amarasuriya Aruma Baduge** Department of Electrical and Computer Engineering, Southern Illinois University, Carbondale, IL, USA **H. Vincent Poor** School of Engineering and Applied Science, Princeton University, Princeton, NJ, USA
3	**Derrick Wing Kwan Ng** School of Electrical Engineering and Telecommunications, The University of New South Wales, Sydney, Australia **Marco Breiling, Christian Rohde and Frank Burkhardt** Fraunhofer Institute for Integrated Circuits (IIS), Erlangen, Germany **Robert Schober** Institute for Digital Communications, Friedrich-Alexander-University Erlangen-Nurnberg (FAU), Germany

Chapter no.	Authors affiliation
4	**Caijun Zhong** Institute of Information and Communication Engineering, Zhejiang University, Hangzhou, China **Himal A. Suraweera** Department of Electrical and Electronic Engineering, University of Peradeniya, Peradeniya, Sri Lanka **George K. Karagiannidis** Department of Electrical & Computer Engineering, Aristotle University of Thessaloniki, Thessaloniki, Greece
5	**Taneli Riihonen** Department of Signal Processing and Acoustics, Aalto University School of Electrical Engineering, Helsinki, Finland **Xiaodong Wang** Department of Electrical Engineering, Columbia University, New York, NY, USA
6	**Constantinos Psomas, Christodoulos Skouroumounis and Ioannis Krikidis** Department of Electrical and Computer Engineering, Faculty of Engineering, University of Cyprus, Nicosia, Cyprus
7	**Pan Cao, Miryam Gonzalez-Perez and John S. Thompson** School of Engineering, Institute for Digital Communications, University of Edinburgh, Edinburgh, UK
8	**Alessio Zappone and Eduard A. Jorswieck** Technische Universität Dresden, Communication Theory, Communications Laboratory, Dresden, Germany
9	**Gan Zheng** Wolfson School of Mechanical, Electrical and Manufacturing Engineering, Loughborough University, Loughborough, Leicestershire, UK
10	**Yunfei Chen** School of Engineering, University of Warwick, Coventry, UK

Chapter no.	Authors affiliation
11	**Raef Bassily** Department of Computer Science & Engineering, Center for Information Theory and Applications, University of California San Diego, La Jolla, CA, USA **Sennur Ulukus** Department of Electrical and Computer Engineering, Institute for Systems Research, University of Maryland, MD, USA
12	**Jing Jiang and Hongjian Sun** School of Engineering and Computing Sciences, Durham University, Durham, UK
13	**He (Henry) Chen and Yonghui Li** School of Electrical and Information Engineering, The University of Sydney, Sydney, New South Wales, Australia
14	**Majid Safari and Wasiu Popoola** Institute for Digital Communications, University of Edinburgh, Edinburgh, UK **Sinan Sinanovic** School of Engineering and Built Environment, Glasgow Caledonian University, Glasgow, UK
15	**Dimitrios Christopoulos, Stefano Andrenacci, Danilo Spano, Symeon Chatzinotas and Björn Ottersten** Interdisciplinary Centre for Security, Reliability and Trust (SnT), University of Luxembourg, Luxembourg **Jens Krause** SES, Château de Betzdorf, Betzdorf, Luxembourg

Chapter 1

Spatial modulation for cooperative networks

Sandeep Narayanan, Marco Di Renzo, Fabio Graziosi, Harald Haas and Mark Flanagan

1.1 Introduction

In this chapter, Distributed Spatial Modulation (DSM) and Distributed Spatially Modulated Space-Time Block Code (DSM-STBC) are elaborately described. DSM and DSM-STBC are two new cooperative wireless protocols for multi-relay networks, which are based on the principle of Spatial Modulation (SM). The distinguishable feature of DSM lies in improving the reliability of the source via distributed diversity and by increasing the aggregate throughput of the cooperative network, since new data is transmitted during each transmission phase. This is achieved by encoding the data transmitted from the source into the spatial positions of the available relays and by exploiting the signal domain for transmitting the data of the relays. In DSM-STBC, SM and Space-Time-Block-Codes (STBCs) are synergistically combined for a distributed scenario. The distinguishable feature of DSM-STBC lies in offering throughput enhancement, by achieving the same order of diversity and having to activate the same number of relay nodes as conventional distributed STBC schemes. At the destination, demodulators robust to demodulation errors at the relays are developed, and their end-to-end error probability and achievable diversity are studied. With the aid of Monte Carlo simulations, DSM and DSM-STBC are compared against state-of-the-art cooperative protocols, and they are shown to provide better performances.

It is well known that Multiple-Input-Multiple-Output (MIMO) systems have significant advantages over single-antenna wireless systems [1–3]. The multiple-antenna in MIMO systems can be exploited in different ways to achieve multiplexing, diversity or antenna gains. However, regardless of the use as spatial multiplexing, diversity or smart antenna system, one of the main drawbacks of conventional MIMO schemes is its high energy consumption. This is because conventional MIMO schemes, such as spatial-multiplexing MIMO, need multiple active Radio Frequency (RF) front-ends to simultaneously transmit multiple data streams from each available transmit-antenna [4, 5]. Furthermore, this issue will be exacerbated for the implementation of potential 5G technologies, such as Large-Scale Multiple-Antenna Systems (LSMAS) [6, 7]. This consideration has encouraged researchers to shift the design paradigm to the single-RF-based MIMO principle [8].

SM is an emerging single-RF MIMO concept that can outperform many conventional MIMO transmission schemes with practical implementation constraints [9–17]. In SM, in addition to the conventional signal-constellation diagram given by Quadrature Amplitude Modulation (QAM)/Phase Shift Keying (PSK) modulation, data is also encoded onto the so-called spatial-constellation diagram. More specifically, instead of simultaneously transmitting multiple data streams from the available antennas, SM encodes part of the information data to be transmitted onto the spatial (physical) position of each transmit-antenna in the antenna-array. Thus, in addition to the usual signal-constellation diagram, the antenna-array plays the role of a second-constellation diagram. Since two information carrying units are used to transmit data, SM can increase the data rate with respect to single-antenna wireless systems. Furthermore, this increase in data rate is achieved by activating a single transmit-antenna and with optimum single-stream decoding. Thus, unlike conventional spatial-multiplexing MIMO techniques, SM is implicitly a single-RF MIMO scheme and requires only a single active RF front-end. A comprehensive state-of-the-art survey of SM research is available in References 18 and 19, to which the reader is referred for further information.

In general, SM requires a large number of transmit-antenna elements to be effective from the point of view of spectral-efficiency. Unfortunately, there are practical constraints in equipping the mobile terminals with multiple antennas. The exploitation of relaying and distributed cooperation can provide effective solutions in this context. In fact, in the recent years, cooperative protocols have established themselves as a promising technology for future wireless networks [20, 21]. They enable single-antenna mobile terminals to harvest the benefits of MIMO systems by being part of a distributed antenna-array. Recent results have shown that cooperation is an effective means for combating multi-path fading and shadowing, for improving the network coverage, as well as for reducing the transmit power. Also, several authors have investigated the potential benefits of exploiting MIMO-aided transmission and relay-aided cooperation, e.g., References 22–24.

More recently, several researchers have leveraged the SM principle for application to relay-aided and cooperative communications. Notable examples related to the present chapter include References 25–33. These protocols have shown the potential benefits of using SM in the context of relay-aided transmission compared to conventional modulation and MIMO schemes. Many of these relaying protocols, however, rely on certain simplifying assumptions and do not fully study, the true potential of SM in the context of relay-aided cellular networks. For example, the protocols in References 25–30 require multiple-antenna source and/or relays. The protocol in Reference 31 does not account for demodulation errors at the relays and does not investigate the achievable diversity order. The protocol in Reference 32 relies on ideal Cyclic Redundancy Check (CRC) at the relays. In Reference 33, the optimal demodulator at the destination is developed by considering only a simple two-relay network topology. Furthermore, most of the previously proposed protocols treat SM as a conventional modulation scheme, since they aim either to increase the coverage, e.g., Reference 25 or to enhance link reliability of SM transmission via distributed diversity, e.g., Reference 27. A summary of SM-based relay-aided protocols is available in Section V.B of Reference 18, to which the interested reader is referred for further information. *In the present chapter, we describe two new protocols for relay-aided communications based on the principle of SM.*

The first protocol that we describe is referred to as DSM [34, 35], and is aimed at overcoming some important design challenges relay-aided cooperation faces, which include [36, 37]: (i) the half-duplex constraint, i.e., the relays cannot transmit and receive data on the same frequency at the same time, which results in a loss of system throughput. Although full-duplex relays may solve this problem and promising results have recently been reported [38, 39], half-duplex radios are widespread used and will remain so in the near future due to their maturity and low implementation cost. In the present chapter, for this reason, we focus our attention on half-duplex relaying; (ii) the relays are forced to use their own resources for forwarding the data of the source, usually without receiving any rewards except for the fact that the whole system may become more efficient; (iii) in classical cooperative protocols, the relays that perform a re-transmission on behalf of the sources must delay the transmission of their own data frames, which has an impact on the network's latency.

Recently, many attempts for overcoming the aforementioned limitations of relaying and distributed cooperation have been made. Notable examples include: (i) non-orthogonal relaying [40–42]; (ii) successive relaying [43, 44]; (iii) shifted successive relaying [45]; (iv) two-way relaying (also known as physical-layer or analog Network Coding, NC) [46–48]; (v) cognitive cooperation [49] and (vi) complex-field NC [50]. A comprehensive survey and assessment of advantages/disadvantages of these solutions is available in Figure 6 of Reference 37, to which the reader is referred for further information. According to Reference 36, a common assumption and limitation of these protocols is that the relays receive no direct reward from cooperation. They help the sources by performing re-transmissions on their behalf, but they do not pursue any personal interests. In particular, *the relays must, in general, delay the transmission of their own data frames to first help the sources*. This operating principle is realistic when the relays are dedicated network elements with no data to be transmitted, but it might be a disincentive to distributed cooperation when the relays have data to be transmitted.

Motivated by these considerations, two cooperative diversity protocols have recently been introduced, which allow the relays to help the sources while simultaneously pursuing their own interests, i.e., to avoid delaying the transmission of their data frames while re-transmitting those of the source. The first solution is referred to as MIMO-NC and was introduced in References 51 and 52. MIMO-NC is a transmission protocol that allows the relays to encode the data received from the source with their own data by leveraging the principle of digital NC. This approach avoids throughput reductions that are typical of half-duplex relaying. Recent results, however, have highlighted that the source receives no distributed diversity gain regardless of the Galois field being used for NC [53–55]. The second solution is referred to as Super-Position Modulation (SPM) and was introduced in Reference 56. SPM can be thought as a NC scheme that is implemented in the modulation domain rather than in the code domain. Its performance and achievable diversity critically depend on the weighting factors being used for superimposing the modulated symbols of source and relay, which should be appropriately optimized for reducing the impact of self-interference and for avoiding a non-negligible performance degradation [57]. Compared with MIMO-NC, SPM is capable of increasing the diversity order of the source without throughput degradation. The choice of the weight factors is, however, a non-trivial optimization problem.

The DSM, similar to MIMO-NC and SPM, allows the relays to transmit their own data frames while re-transmitting the data frames of the source. It takes advantage of the unique encoding mechanism of SM in a distributed fashion. More specifically, each symbol transmitted from the source is univocally encoded into a unique identifier associated to each available relay. During the broadcasting phase, each relay decodes the received symbol independently of the others. During the relaying phase, only that particular relay whose identifier coincides with the demodulated symbol is allowed to transmit, while the other relays are kept silent. With the aid of this encoding mechanism, the activation of a given relay (spatial dimension) is exploited as an implicit means for relaying the symbol of the source instead of using conventional amplitude/phase modulations. Thus, this latter signal dimension can be used by the activated relay for transmitting its own data, instead of that of the source. With the aid of DSM, as a result, the relays are capable of helping the source without delaying their own data transmission. Also, the average aggregate throughput is higher compared to half-duplex relaying and, in particular, it is the same as that of single-hop transmission if the modulation order of source and relay are the same.

Although SM was originally proposed as a low-complexity spatial-multiplexing technique, many attempts have been made recently to design transmit-diversity-achieving SM schemes [15, 58–60]. In particular, in these papers, the authors combine SM with STBCs and/or time-orthogonal shaping filters for the sake of achieving transmit-diversity gains. The second cooperative protocol that we describe in this chapter applies the STBC-SM concept of Reference 15 to a distributed scenario, and throughout this chapter, this protocol is referred to as DSM-STBC [61]. The distinguishable feature of DSM-STBC is that it offers throughput enhancement, by achieving the same order of diversity and having to activate the same number of relay nodes as conventional distributed STBC schemes [62, 63].

The remainder of the present chapter is organized as follows. In Section 1.2, the DSM protocol is introduced and is studied extensively. In Section 1.3, the DSM-STBC protocol is introduced and is studied extensively. In Section 1.4, numerical results to substantiate the analytical findings and to compare DSM against state-of-the-art cooperative protocols are shown. Finally, Section 1.5 concludes this chapter.

Notation: $P\{\cdot\}$ denotes probability. $(\cdot)^*$ and $|\cdot|$ denote complex conjugate and absolute value operators, respectively. $\mathbb{E}_X\{\cdot\}$ denotes the expectation computed with respect to the Random Variable (RV) X. $\mathrm{Re}\{\cdot\}$ and $\mathrm{Im}\{\cdot\}$ denote real and imaginary part operators, respectively. $j=\sqrt{-1}$ denotes the imaginary unit. $Q(x)=(1/\sqrt{2\pi})\int_x^{+\infty}\exp(-t^2/2)dt$ denotes the Q-function. $\mathrm{card}\{\cdot\}$ denotes the cardinality of a set. Bold symbols denote strings of bits with fixed length. $\propto$ denotes the proportionality operator. $X\sim\mathcal{E}(\sigma_X^2)$ denotes an exponential RV with Probability Density Function (PDF) and Moment Generating Function (MGF) equal to $f_X(x)=(1/\sigma_X^2)\exp\{-x/\sigma_X^2\}$ and $\Phi_X(s)=\mathbb{E}_X\{\exp\{-sX\}\}=(1+s\sigma_X^2)^{-1}$, respectively. The function $\delta(a,b)$ is defined as $\delta(a,b)=0$ if $a\neq b$ and $\delta(a,b)=1$ if $a=b$. The function $\overline{\delta}(a,b)$ is defined as $\overline{\delta}(a,b)=1-\delta(a,b)$.

1.2 Protocol I: DSM

1.2.1 Transmission model

A multi-relay network topology with one source (S), N_R half-duplex relays (R_r, for $1, 2, \ldots, N_R$) and one destination (D) is considered. All nodes are equipped with a single antenna. Both source and relays have independent data to be delivered to the destination. The source transmits complex modulated symbols, by using either PSK or QAM of constellation size M. The M complex modulated symbols are denoted by p_m for $m = 1, 2, \ldots, M$. Likewise, the relays use PSK/QAM modulation of constellation size N for transmitting their own data symbols. These N complex modulated symbols are denoted by q_n for $n = 1, 2, \ldots, N$. In general, the constellation size of source and relays is different, i.e., $N \neq M$. As it will become apparent in the sequel, DSM requires that the number of relays N_R is no less than the constellation size of the source, i.e., $N_R \geq M$. For ease of illustration, $N_R = M$ is assumed throughout this chapter. The case study $N_R > M$, in fact, would require some criteria for choosing the M relays, e.g., based on the minimization of the end-to-end error probability. This usually requires a feedback channel from the destination to the source and to the relays. The mathematical analysis of this case study is postponed to future research. A numerical example is, however, illustrated in Section 1.4. Thus, without ambiguity, N_R is replaced by M throughout this chapter. Each relay is assigned a unique digital identifier $\mathbf{ID}_{R_r}$ for $1, 2, \ldots, M$, which is a string of $\log_2(M)$ bits. Without loss of generality, a lexicographic labelling is used throughout this chapter. If $M = 4$, for example, the four relays are univocally identified by the strings of bits $\mathbf{ID}_{R_1} = 00$, $\mathbf{ID}_{R_2} = 01$, $\mathbf{ID}_{R_3} = 10$ and $\mathbf{ID}_{R_4} = 11$.

DSM consists of two transmission phases, i.e., broadcasting and relaying, that occur in two non-overlapping time-slots. In the first phase, the source is in transmission mode, and both relays and destination are in reception mode. In the second phase, the source is off while the relays and the destination are in transmission and reception modes, respectively. Both phases are described in what follows and are illustrated in Figure 1.1.

1.2.1.1 Broadcasting phase

In this phase, the source broadcasts its data to the M relays and to the destination. The signals received at relays and destination can be written as follows ($r = 1, 2, \ldots, M$):

$$y_{SR_r} = \sqrt{E_S} h_{SR_r} s_{tx} + n_{SR_r}, \quad y_{SD} = \sqrt{E_S} h_{SD} s_{tx} + n_{SD} \tag{1.1}$$

where (i) E_S is the source's average transmit energy per symbol; (ii) h_{XY} is the fading gain of the link from node X to node Y, which is a circular symmetric complex Gaussian RV with zero mean and variance $\sigma_{XY}^2/2$ per real dimension. The fading gains over different links are independent but non-identically distributed (i.n.i.d.) RVs, in order to account for different propagation distances and shadowing effects; (iii) n_{XY} is the noise at the input of node Y and related to the transmission from node X, which is a complex Additive White Gaussian (AWG) RV with variance $N_0/2$ per real dimension. n_{XY} in different time-slots or at the input of different nodes are independent

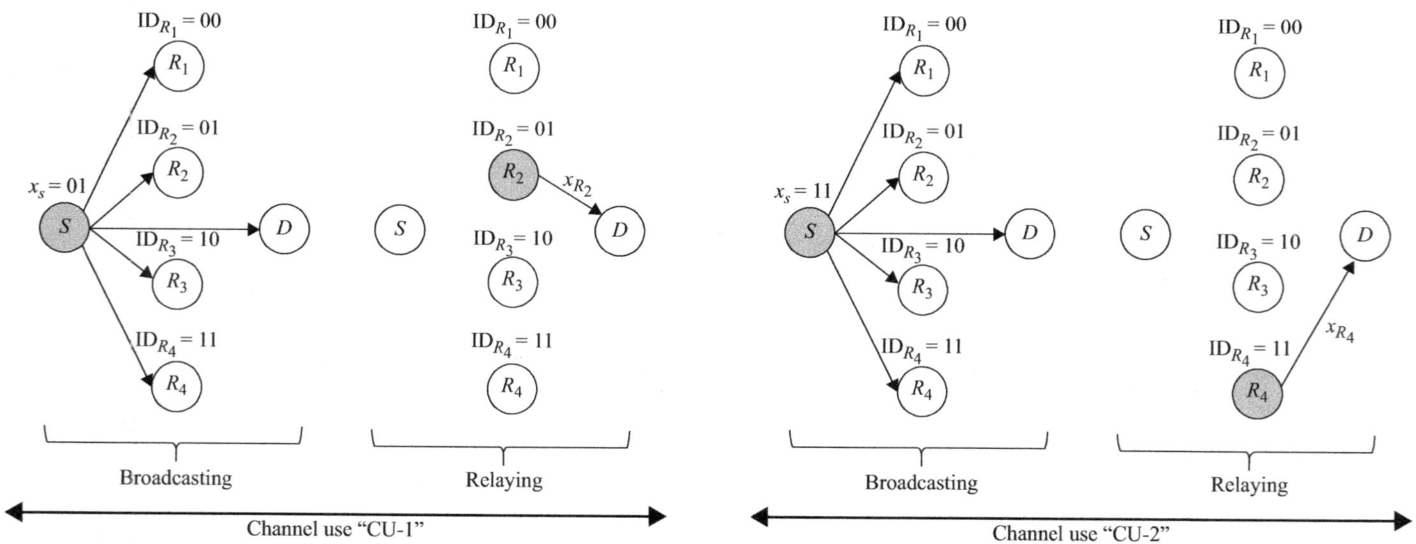

Figure 1.1 Illustration of the DSM protocol for $N_R = M = 4$ and in the absence of demodulation errors at the relays. The shaded nodes are in transmission mode. The others are in reception mode, and the source (S) is in idle mode during the relaying phase. Reproduced with permission from Reference 35. ©2015 IEEE

and identically distributed (i.i.d.) RVs; and (iv) $s_{tx} = \mathcal{M}_S(\mathbf{x}_S) \in \{p_1, p_2, \ldots, p_M\}$ is the complex modulated symbol transmitted by the source, where $\mathbf{x}_S$ are the $\log_2(M)$ bits emitted by the source and $\mathcal{M}_S(\cdot)$ is the bits-to-symbol mapping function at the source.

The destination keeps the received signal for further processing. Each relay, on the other hand, demodulates the signal received from the source independently of the others. With the aid of the Maximum-Likelihood (ML) criterion, the demodulated symbol can be formulated as follows:

$$\hat{s}_{tx}^{(R_r)} = \underset{p_m \in \{p_1, p_2, \ldots, p_M\}}{\arg\min} \left\{ \left| y_{SR_r} - \sqrt{E_S} h_{SR_r} p_m \right|^2 \right\}, \quad \hat{\mathbf{x}}_S^{(R_r)} = \mathcal{M}_S^{-1}\left(\hat{s}_{tx}^{(R_r)}\right) \tag{1.2}$$

where: (i) $\hat{s}_{tx}^{(R_r)}$ is the demodulated symbol at R_r for $r = 1, 2, \ldots, M$; (ii) $\mathcal{M}_S^{-1}(\cdot)$ is the inverse bits-to-symbol mapping function; and (iii) $\hat{\mathbf{x}}_S^{(R_r)}$ is the estimate of $\mathbf{x}_S$ at R_r for $r = 1, 2, \ldots, M$.

1.2.1.2 Relaying phase

In this phase, the SM encoding principle applied to the set of M distributed relays is leveraged in order to *implicitly* forward the estimate of the source's data and to *explicitly* transmit the data of the activated relay. In particular, each relay, independently of the others, compares $\hat{\mathbf{x}}_S^{(R_r)}$ with its own unique digital identifier $\mathbf{ID}_{R_r}$. If they coincide, the relay is activated. If not, the relay is kept inactive. As such, the activation of a given relay is interpreted at the destination as the transmission of a specific symbol by the source. The activated relay exploits the unused signal dimension for modulating and for transmitting its own data symbol to the destination, instead of re-modulating and transmitting the estimate of the source's data. As a result, relayed and new data are simultaneously transmitted in this phase.

If the source-to-relay links are error-free, no demodulation errors occur at the relays. Due to the uniqueness of the digital identifiers associated to each relay, a single relay is activated during the relaying phase. In practical operating conditions, however, demodulation errors may occur at the relays. Since each relay operates independently of the others, multiple or no relays may be activated during the relaying phase. In the presence of demodulation errors at the relays, the signal received at the destination can be formulated as follows:

$$y_{RD} = \sum_{r=1}^{M} \left(\sqrt{E_{R_r}} h_{R_r D} r_{tx}^{(R_r)} \right) + n_{RD}, \quad r_{tx}^{(R_r)} = \begin{cases} \mathcal{M}_R\left(\mathbf{x}_{R_r}\right) & \text{if } \mathbf{ID}_{R_r} = \hat{\mathbf{x}}_S^{(R_r)} \\ 0 & \text{if } \mathbf{ID}_{R_r} \neq \hat{\mathbf{x}}_S^{(R_r)} \end{cases} \tag{1.3}$$

where (i) E_{R_r} is the average transmit energy per symbol of R_r; (ii) $\mathbf{x}_{R_r}$ are the $\log_2(N)$ bits emitted by the relay if activated. These are relay's own bits that are independent of the source's bits $\mathbf{x}_S$; and (iii) $\mathcal{M}_R(\cdot)$ is the bits-to-symbol mapping function at the relays. Since $M \neq N$, $\mathcal{M}_R(\cdot)$ and $\mathcal{M}_S(\cdot)$ are, in general, different. It is worth mentioning that in (1.3) the source's data is implicitly encoded into the channel impulse responses $h_{R_r D}$ that correspond to non–zero values of $r_{tx}^{(R_r)}$. To better understand, let us assume that no demodulation errors occur at the relays and that $R_r = R_*$ is the only relay for which $\mathbf{ID}_{R_*} = \mathbf{x}_S$. Thus, (1.3) reduces to $y_{RD} = \sqrt{E_{R_*}} h_{R_* D} \mathcal{M}_R(\mathbf{x}_{R_*}) + n_{RD}$, which shows that the source's data $\mathbf{x}_S$ is univocally encoded into $h_{R_* D}$.

The optimal demodulator in the absence of demodulation errors at the relays is formulated and studied in Reference 42. In Section 1.2.2, a diversity combiner at the destination based on (1.3), i.e., in the presence of demodulation errors at the relays, is rigorously formulated. Furthermore, its error performance and achievable diversity are studied in Section 1.2.3.

1.2.1.3 Comments

Some comments about practical implementation issues of the DSM protocol are as follows:

- In (1.1)–(1.3), a symbol-by-symbol signal model is considered. This is assumed for ease of description, since DSM is compliant with packet-based transmission protocols in both broadcasting and relaying phases. In the relaying phase, in particular, "sparse" packets are transmitted by each relay, whose symbols are non-zero only if they are activated according to the DSM principle. The packets transmitted in broadcasting and relaying phases by source and relays, respectively, have the same length. Only the number of non-zero elements is different. This number may be added to the header of the packet. The zero symbols are discarded at the destination to reconstruct the bitstreams emitted by source and relays.
- The performance of the DSM protocol may be enhanced by reducing the impact of relay-induced error propagation with the aid of a CRC mechanism. The adoption of a CRC usually increases the demodulation complexity at the relays and reduces the spectral efficiency, since entire packets are discarded even though a few symbols are not demodulated correctly. CRC-based implementations are not described in the present chapter.
- In this chapter, for ease of description, channel coding is not considered. It may, however, be applied to the DSM protocol by using the line of thought recently introduced in References 64 and 65 for application to a broad class of relay-aided networks. It is worth noting, in particular, that the application of channel coding at the relays may relax the need of taking the reliability of the source-to-relay links into account at the destination, since relay-induced errors are less probable. The price to be paid for reducing the demodulation complexity at the destination is, however, the need of increasing the demodulation complexity at the relays.
- The signal formulation in (1.3) highlights that DSM requires, for appropriate demodulation at the destination, symbol-level synchronization among the relays. This requirement is needed for many cooperative transmission protocols, e.g., distributed space–time coding [66]. The analysis of the impact of synchronization errors on the performance of DSM is out of the scope of the present chapter and is left to future research.
- In DSM, the source transmits only in the broadcasting phase. Thus, it uses 1/2 of the available degrees of freedom for its transmission. Likewise, the relays *collectively* exploit 1/2 of the available degrees for their transmissions. The transmission of the relays, however, is discontinuous, since they transmit only if they are activated by the source. Under the assumption that the source emits equally likely symbols, the M relays equally share the remaining degrees of freedom,

i.e., on average each of them uses $1/2M$ of the available degrees of freedom for their transmissions.

- Due to demodulation errors, multiple relays may be activated in the relaying phase. Nevertheless, the destination must be capable of associating, without ambiguity, each transmitted symbol to the active relay that has transmitted it. This is guaranteed by the fact that the channel impulse responses of all relay-to-destination links are different. This concept is behind the correct operation of multiple-access channels, whose signal model is similar to (1.3). The difference in DSM is that the active relays have to be identified at the destination. To this end, we develop appropriate demodulators in the next section.

1.2.2 Demodulation at the destination

In this section, we develop a diversity-achieving demodulator at the destination for DSM, which is capable of retrieving the symbols transmitted from source and relays in the presence of demodulation errors at the relays. The demodulator is based on the Maximum A Posteriori (MAP) criterion of optimality [67]. Using y_{SD} and y_{RD} in (1.1) and (1.3), respectively, the diversity combiner can be formulated as follows:

$$\begin{aligned}\left[\hat{s}_{tx}^{(D)}, \hat{r}_{tx}^{(R_1,D)}, \ldots, \hat{r}_{tx}^{(R_M,D)}\right] &= \underset{\substack{\mathcal{M}_S\left(\tilde{\mathbf{x}}_S^{(D)}\right)\in\{p_1,p_2,\ldots,p_M\} \\ \widetilde{\mathcal{M}}_R\left(\tilde{\mathbf{x}}_{R_r}^{(D)}\right)\in\{0,q_1,q_2,\ldots,q_N\} \text{ for } r=1,2,\ldots,M}}{\arg\max} \left\{P_1\left(y_{SD}, y_{RD}\middle|\tilde{\mathbf{x}}^{(D)}\right) P_2\left(\tilde{\mathbf{x}}^{(D)}\right)\right\} \\ \hat{\mathbf{x}}_S^{(D)} = \mathcal{M}_S^{-1}\left(\hat{s}_{tx}^{(D)}\right);\quad \hat{\mathbf{x}}_{R_r}^{(D)} &= \widetilde{\mathcal{M}}_R^{-1}\left(\hat{r}_{tx}^{(R_r,D)}\right) \text{ for } r=1,2,\ldots,M\end{aligned} \tag{1.4}$$

$$\begin{aligned}P_1\left(y_{SD}, y_{RD}\middle|\tilde{\mathbf{x}}^{(D)}\right) &= \mathcal{P}\left\{y_{SD}, y_{RD}\middle|\tilde{\mathbf{x}}_S^{(D)}, \tilde{\mathbf{x}}_{R_1}^{(D)}, \tilde{\mathbf{x}}_{R_2}^{(D)}, \ldots, \tilde{\mathbf{x}}_{R_M}^{(D)}\right\} \\ &= \exp\left(-\frac{\left|y_{SD} - \sqrt{E_S}h_{SD}\mathcal{M}_S\left(\tilde{\mathbf{x}}_S^{(D)}\right)\right|^2}{N_0}\right)\exp\left(-\frac{\left|y_{RD} - \sum_{r=1}^{M}\sqrt{E_{R_r}}h_{R_rD}\widetilde{\mathcal{M}}_R\left(\tilde{\mathbf{x}}_{R_r}^{(D)}\right)\right|^2}{N_0}\right)\end{aligned} \tag{1.5}$$

$$\begin{aligned}P_2\left(\tilde{\mathbf{x}}^{(D)}\right) &= \mathcal{P}\left\{\tilde{\mathbf{x}}_S^{(D)}, \tilde{\mathbf{x}}_{R_1}^{(D)}, \tilde{\mathbf{x}}_{R_2}^{(D)}, \ldots, \tilde{\mathbf{x}}_{R_M}^{(D)}\right\} \\ &= \frac{1}{M}\prod\nolimits_{r\in\Omega_R^{(\mathrm{ON})}}\frac{1}{N}\mathcal{P}\left\{\hat{\mathbf{x}}_S^{(R_r)} = \mathbf{ID}_{R_r}\middle|\tilde{\mathbf{x}}_S^{(D)}\right\} \times \prod\nolimits_{r\in\Omega_R^{(\mathrm{OFF})}}\left(1 - \mathcal{P}\left\{\hat{\mathbf{x}}_S^{(R_r)} = \mathbf{ID}_{R_r}\middle|\tilde{\mathbf{x}}_S^{(D)}\right\}\right)\end{aligned} \tag{1.6}$$

where (i) $\hat{s}_{tx}^{(D)}$ and $\hat{\mathbf{x}}_S^{(D)}$ denote the estimates at the destination of information symbol, s_{tx}, and information bits, $\mathbf{x}_S$, respectively, transmitted by the source; (ii) $\hat{r}_{tx}^{(R_r,D)}$ and $\hat{\mathbf{x}}_{R_r}^{(D)}$ denote the estimates at the destination of information symbol, $r_{tx}^{(R_r)}$, and information bits, $\mathbf{x}_{R_r}$, respectively, transmitted by the relay R_r for $r = 1, 2, \ldots, M$; (iii)

$\tilde{\mathbf{x}}_S^{(D)}$ and $\tilde{\mathbf{x}}_{R_r}^{(D)}$ for $r=1,2,\ldots,M$ are the trial bits used in the hypothesis detection problem, where $\tilde{\mathbf{x}}^{(D)}=(\tilde{\mathbf{x}}_S^{(D)},\tilde{\mathbf{x}}_{R_1}^{(D)},\tilde{\mathbf{x}}_{R_2}^{(D)},\ldots,\tilde{\mathbf{x}}_{R_M}^{(D)})$ is a short-hand introduced for notational simplicity; (iv) $\mathcal{M}_R^{-1}(\cdot)$ is the inverse function of $\mathcal{M}_R(\cdot)$, i.e., the symbol-to-bits mapping function; (v) $\widetilde{\mathcal{M}}_R(\cdot)$ is the generalized mapping function used at the relays, which account for the no transmission state of the inactive relays. The no transmission state is identified by symbol "0". Likewise, $\widetilde{\mathcal{M}}_R^{-1}(\cdot)$ is the inverse of $\widetilde{\mathcal{M}}_R(\cdot)$, where $\widetilde{\mathcal{M}}_R^{-1}(0)=\mathcal{N}$ and $\mathcal{N}$ denotes the "null" (no transmission) state in the bits domain; (vi) $P_1(y_{SD},y_{RD}|\tilde{\mathbf{x}}^{(D)})$ denotes the joint probability density function of the received signals conditioned upon the hypothesis $\tilde{\mathbf{x}}^{(D)}$; and (vii) $P_2(\tilde{\mathbf{x}}^{(D)})$ is the a priori probability that the M relays are active/inactive given the hypothesis of the data transmitted by the source $\tilde{\mathbf{x}}_S^{(D)}$. In (1.6), in particular, $\widetilde{\Omega}_R^{(\mathrm{ON})}$ and $\widetilde{\Omega}_R^{(\mathrm{OFF})}$ denote the sets of active and inactive relays, respectively. They are defined as $\widetilde{\Omega}_R^{(\mathrm{ON})}=\{r=1,2,\ldots,M|\widetilde{\mathcal{M}}_R(\tilde{\mathbf{x}}_{R_r}^{(D)})\neq 0\}$ and $\widetilde{\Omega}_R^{(\mathrm{OFF})}=\{r=1,2,\ldots,M|\widetilde{\mathcal{M}}_R(\tilde{\mathbf{x}}_{R_r}^{(D)})=0\}$. Furthermore, $\mathcal{P}\{\hat{\mathbf{x}}_S^{(R_r)}=\mathbf{ID}_{R_r}|\tilde{\mathbf{x}}_S^{(D)}\}$ denotes the probability that the relay R_r for $r=1,2,\ldots,M$ is activated conditioned upon $\tilde{\mathbf{x}}_S^{(D)}$, i.e., the probability that the symbol demodulated at R_r is $\mathbf{ID}_{R_r}$ by assuming that the source has transmitted $\tilde{\mathbf{x}}_S^{(D)}$. The terms $1/M$ and $1/N$ take into account that the symbols transmitted by source and relays are equally likely.

The computation of (1.6) requires a closed-form expression of $\mathcal{P}\{\hat{\mathbf{x}}_S^{(R_r)}=\mathbf{ID}_{R_r}|\tilde{\mathbf{x}}_S^{(D)}\}$. The exact computation of this probability is a non-trivial problem for general modulation schemes, as recently remarked in References 54 and 55. Fortunately, recent results related to the development of diversity-achieving demodulators in the context of relay-aided communications have highlighted that approximated expressions of these probabilities based on, e.g., the union-bound method are sufficiently accurate from the diversity order standpoint. The interested reader is referred to References 55 and 68 for further information. By using a similar line of thought and with the aid of the union-bound [67], $\mathcal{P}\{\hat{\mathbf{x}}_S^{(R_r)}=\mathbf{ID}_{R_r}|\tilde{\mathbf{x}}_S^{(D)}\}$ can be approximated as follows:

$$\begin{aligned}
&\mathcal{P}\left\{\hat{\mathbf{x}}_S^{(R_r)}=\mathbf{ID}_{R_r}|\tilde{\mathbf{x}}_S^{(D)}\right\}\\
&\approx\begin{cases}Q\left(\sqrt{\dfrac{E_S}{2N_0}\left|\mathcal{M}_S\left(\tilde{\mathbf{x}}_S^{(D)}\right)-\mathcal{M}_S(\mathbf{ID}_{R_r})\right|^2|h_{SR_r}|^2}\right) & \text{if}\quad \tilde{\mathbf{x}}_S^{(D)}\neq\mathbf{ID}_{R_r}\\
1-\sum\limits_{z\neq r=1}^{M}Q\left(\sqrt{\dfrac{E_S}{2N_0}\left|\mathcal{M}_S\left(\tilde{\mathbf{x}}_S^{(D)}\right)-\mathcal{M}_S(\mathbf{ID}_{R_z})\right|^2|h_{SR_z}|^2}\right) & \text{if}\quad \tilde{\mathbf{x}}_S^{(D)}=\mathbf{ID}_{R_r}\end{cases}\\
&\overset{(a)}{\approx}\begin{cases}Q\left(\sqrt{\dfrac{E_S}{2N_0}\left|\mathcal{M}_S\left(\tilde{\mathbf{x}}_S^{(D)}\right)-\mathcal{M}_S(\mathbf{ID}_{R_r})\right|^2|h_{SR_r}|^2}\right) & \text{if}\quad \tilde{\mathbf{x}}_S^{(D)}\neq\mathbf{ID}_{R_r}\\
1-\beta Q\left(\sqrt{2\alpha\dfrac{E_S}{N_0}|h_{SR_r}|^2}\right) & \text{if}\quad \tilde{\mathbf{x}}_S^{(D)}=\mathbf{ID}_{R_r}\end{cases}
\end{aligned}\tag{1.7}$$

where (a) further simplifies the union-bound by avoiding the summation with respect to the modulated symbols. The approximation in (a) is widespread used in the literature, as discussed in Reference 69. The parameters α and β depend on the modulation scheme being used. For example, $\alpha=3/(2(M-1))$ and $\beta=4(1-1/\sqrt{M})$ for QAM. If $M=2$, the union-bound and the approximation in (a) coincide and provide an exact expression of $P\{\hat{\mathbf{x}}_S^{(R_r)}=\mathbf{ID}_{R_r}|\tilde{\mathbf{x}}_S^{(D)}\}$ for $\alpha=\beta=1$.

1.2.2.1 Remarks

Some comments about the demodulator in (1.4) are as follows:

- The demodulator in (1.4) is designed to jointly demodulate the source's and relays' data, so that the relays are allowed to transmit their own data during the relaying phase. The active/inactive state of the relays is modelled as an additional symbol "0" that is added to the constellation diagram. Thus, a symbol error occurs if a relay is estimated as active instead of inactive and vice versa, e.g., the symbol "0" is erroneously demodulated as any other symbols of the signal-constellation diagram.
- Demodulation errors at the relays are taken into account via the a priori probabilities $P\{\hat{\mathbf{x}}_S^{(R_r)} = \mathbf{ID}_{R_r} | \tilde{\mathbf{x}}_S^{(D)}\}$ which depend on the channel state information of the source-to-relay links. This is a common requirement for diversity-achieving demodulators that do not rely upon CRC-based error mitigation methods [55]. This channel state information may be estimated at the relays and then forwarded to the destination, as discussed in Reference 64.
- The demodulator in (1.4) is optimal, however, the approximations used in (1.7) for computing $P\{\hat{\mathbf{x}}_S^{(R_r)} = \mathbf{ID}_{R_r} | \tilde{\mathbf{x}}_S^{(D)}\}$ make it sub-optimal. This sub-optimality refers to the resulting coding gain of the error probability [70]. The demodulator is optimal only if $M = 2$.
- In the absence of demodulation errors at the relays, e.g., if CRC–based methods are used, the demodulator in (1.4) can be simplified by letting $Q(\cdot) \to 0$ for $r = 1, 2, \ldots, M$ in (1.7) and by taking into account that only one relay is activated during the relaying phase. As mentioned in Section 1.2.1.2, this demodulator is available in Table I of Reference 35 (c.f. DSM – sub-optimal demodulator) and whose performance is studied in Section 1.4.
- The demodulator in (1.4) can be thought as a generalized multi-user detector, since the data of different users is jointly demodulated. It is "generalized" because the reliability of the source-to-relay links is taken into account by design.

1.2.2.2 High-signal-to-noise-ratio approximation

Although the demodulator in (1.4) is near-optimal, the associated signal processing complexity is non-negligible, especially because of the computation of several exponential and $Q(\cdot)$ functions. Furthermore, the presence of these functions does not facilitate its mathematical analysis in terms of error probability and diversity order. Motivated by these considerations, in this section, we propose a high-SNR approximation for (1.4).

In the high-SNR regime, $1 - Q\left(\sqrt{(E_S/N_0)x}\right) \approx 1$. Thus, $P_2(\tilde{\mathbf{x}}^{(D)})$ can be approximated as follows:

$$P_2\left(\tilde{\mathbf{x}}^{(D)}\right) \approx \widetilde{\mathcal{K}} \left(\prod_{r \in \widetilde{\Omega}_{R,S}^{(\mathrm{ON})}} Q\left(\sqrt{\frac{E_S}{2N_0} \left| \mathcal{M}_S\left(\tilde{\mathbf{x}}_S^{(D)}\right) - \mathcal{M}_S\left(\mathbf{ID}_{R_r}\right) \right|^2 \left| h_{SR_r} \right|^2} \right) \right) \times \left(\prod_{r \in \widetilde{\Omega}_{R,S}^{(\mathrm{OFF})}} Q\left(\sqrt{2\alpha \frac{E_S}{N_0} \left| h_{SR_r} \right|^2} \right) \right) \quad (1.8)$$

where $\widetilde{\mathcal{K}} = (1/M)(1/N)^{\text{card}\{\widetilde{\Omega}_{R,S}^{(\text{ON})}\}}\beta^{\text{card}\{\widetilde{\Omega}_{R,S}^{(\text{OFF})}\}}\widetilde{\Omega}_{R,S}^{(\text{ON})}\widetilde{\Omega}_{R,S}^{(\text{OFF})}$, and $\widetilde{\Omega}_{R,S}^{(\text{ON})}$ and $\widetilde{\Omega}_{R,S}^{(\text{OFF})}$ are defined as follows:

$$\begin{cases} \widetilde{\Omega}_{R,S}^{(\text{ON})} = \left\{ r = 1,2,\ldots,M \left| \widetilde{\mathcal{M}}_R\left(\tilde{\mathbf{x}}_{R_r}^{(D)}\right) \neq 0 \quad \text{and} \quad \tilde{\mathbf{x}}_S^{(D)} \neq \mathbf{ID}_{R_r} \right. \right\} \\ \widetilde{\Omega}_{R,S}^{(\text{OFF})} = \left\{ r = 1,2,\ldots,M \left| \widetilde{\mathcal{M}}_R\left(\tilde{\mathbf{x}}_{R_r}^{(D)}\right) = 0 \quad \text{and} \quad \tilde{\mathbf{x}}_S^{(D)} = \mathbf{ID}_{R_r} \right. \right\} \end{cases} \tag{1.9}$$

By applying the Chernoff bound, which is accurate in the high-SNR regime, i.e., $Q\left(\sqrt{(E_S/N_0)x}\right) \leq (1/2)\exp\left(-(1/2)(E_S/N_0)x\right)$, (1.4) can be re-formulated, with the aid of some algebra and by neglecting some irrelevant constants, as follows:

$$\begin{aligned} &\left[\hat{s}_{tx}^{(D)}, \hat{r}_{tx}^{(R_1,D)}, \ldots, \hat{r}_{tx}^{(R_M,D)}\right] \\ &= \underset{\substack{\mathcal{M}_S\left(\tilde{\mathbf{x}}_S^{(D)}\right) \in \{p_1,p_2,\ldots,p_M\} \\ \mathcal{M}_R\left(\tilde{\mathbf{x}}_{R_r}^{(D)}\right) \in \{0,q_1,q_2,\ldots,q_M\}, r=1,2,\ldots,M}}{\arg\min} \left\{ \Lambda\left(\tilde{\mathbf{x}}_S^{(D)}, \tilde{\mathbf{x}}_{R_1}^{(D)}, \ldots, \tilde{\mathbf{x}}_{R_M}^{(D)}\right) \right\} \end{aligned} \tag{1.10}$$

$$\begin{aligned} &\Lambda\left(\tilde{\mathbf{x}}_S^{(D)}, \tilde{\mathbf{x}}_{R_1}^{(D)}, \ldots, \tilde{\mathbf{x}}_{R_M}^{(D)}\right) \\ &= \left(\frac{\left| y_{SD} - \sqrt{E_S} h_{SD} \mathcal{M}_S\left(\tilde{\mathbf{x}}_S^{(D)}\right) \right|^2}{N_0} \right) + \left(\frac{\left| y_{RD} - \left(\sum_{r=1}^{M} \sqrt{E_{R_r}} h_{R_r D} \widetilde{\mathcal{M}}_R\left(\tilde{\mathbf{x}}_{R_r}^{(D)}\right) \right) \right|^2}{N_0} \right) \\ &\quad + \frac{E_S}{4N_0} \sum_{r \in \widetilde{\Omega}_{R,S}^{(\text{ON})}} \left| \mathcal{M}_S\left(\tilde{\mathbf{x}}_S^{(D)}\right) - \mathcal{M}_S\left(\mathbf{ID}_{R_r}\right) \right|^2 \left| h_{SR_r} \right|^2 + \alpha \frac{E_S}{N_0} \sum_{r \in \widetilde{\Omega}_{R,S}^{(\text{OFF})}} \left| h_{SR_r} \right|^2 - \widetilde{\mathcal{H}} \end{aligned} \tag{1.11}$$

where $\widetilde{\mathcal{H}} = \ln(\widetilde{\mathcal{K}}) + (\text{card}\{\widetilde{\Omega}_{R,S}^{(\text{ON})}\} + \text{card}\{\widetilde{\Omega}_{R,S}^{(\text{OFF})}\})\ln(1/2)$. It is worth noting that $\widetilde{\mathcal{H}}$ is independent of the SNR and that it can usually be neglected for high-SNR without a significant performance degradation. As such, it is neglected in the remainder of the present chapter.

1.2.3 Error probability and diversity order analysis

The methodology to the computation of the error probability of DSM consists of two steps: first, the Average Pairwise Error Probability (APEP) is computed, and then the Average Symbol Error Probability (ASEP) is obtained from it with the aid of the union-bound [67]. In particular, the APEP is related to the codeword that is composed of the data of the source and the relays. Thus, it can be interpreted as a codeword error probability. The ASEP of the source is subsequently obtained by considering only the codewords that correspond to an error for the source. This approach is usually used for the analysis of unequal error protection codes, which provide different diversity orders to different information bits of a codeword [71]. The approach for the performance evaluation relies upon some high-SNR approximations, in order to make the mathematical derivation tractable. This is sufficient, since we are mainly interested in characterizing the diversity order of the protocol. In this section, in particular, we focus our attention only on the computation of ASEP and diversity order of the source. In fact, the destination receives a single copy of the packets transmitted by each relay and, as a consequence, the diversity order of the relays is equal to one. The same mathematical development as for the analysis of the source may be applied, however, to the analysis of the relays. It is worth mentioning that the fact that the relays have diversity order equal to one is not a limitation of DSM. This is, in fact, the usual assumption of relay-aided transmission [20].

1.2.3.1 Error probability

Let $\mathbf{x}^{(D)} = (\mathbf{x}_S^{(D)}, \mathbf{x}_{R_1}^{(D)}, \mathbf{x}_{R_2}^{(D)}, \ldots, \mathbf{x}_{R_M}^{(D)})$ and $\tilde{\mathbf{x}}^{(D)} = (\tilde{\mathbf{x}}_S^{(D)}, \tilde{\mathbf{x}}_{R_1}^{(D)}, \tilde{\mathbf{x}}_{R_2}^{(D)}, \ldots, \tilde{\mathbf{x}}_{R_M}^{(D)})$ be transmitted and hypothesis symbols' vectors, respectively. The Pairwise Error Probability (PEP) is denoted by $\text{PEP}(\mathbf{x}^{(D)} \to \tilde{\mathbf{x}}^{(D)}|\mathbf{n}, \mathbf{h})$, and it is defined as the probability of demodulating $\tilde{\mathbf{x}}^{(D)}$ in lieu of $\mathbf{x}^{(D)}$, by assuming that they are the only two available codewords and that AWGN ($\mathbf{n}$) and fading channels ($\mathbf{h}$) are fixed. In particular, $\mathbf{n}$ and $\mathbf{h}$ are short-hands used to denote all noises and fading channels available in the decision variable, respectively. The APEP is obtained from the PEP by removing the conditioning with respect to AWGN and fading gains. In mathematical terms, they can be formulated as follows:

$$\begin{aligned}
\text{PEP}\left(\mathbf{x}^{(D)} \to \tilde{\mathbf{x}}^{(D)}|\mathbf{n}, \mathbf{h}\right) &= \mathcal{P}\left\{\Lambda\left(\tilde{\mathbf{x}}^{(D)}\right) < \Lambda\left(\mathbf{x}^{(D)}\right)|\mathbf{n}, \mathbf{h}\right\} \mathcal{P}\{\mathbf{x}^{(D)}|\mathbf{h}\} \\
&= \mathcal{P}\left\{\Delta\left(\mathbf{x}^{(D)}, \tilde{\mathbf{x}}^{(D)}\right) < 0|\mathbf{n}, \mathbf{h}\right\} \mathcal{P}\{\mathbf{x}^{(D)}|\mathbf{h}\} \\
\text{APEP}\left(\mathbf{x}^{(D)} \to \tilde{\mathbf{x}}^{(D)}\right) &= \mathbb{E}_{\mathbf{h}}\left\{\mathbb{E}_{\mathbf{n}}\left\{\text{PEP}\left(\mathbf{x}^{(D)} \to \tilde{\mathbf{x}}^{(D)}|\mathbf{n}, \mathbf{h}\right)\right\}\right\} \\
&= \mathbb{E}_{\mathbf{h}}\left\{\mathcal{P}\left\{\Delta\left(\mathbf{x}^{(D)}, \tilde{\mathbf{x}}^{(D)}\right) < 0|\mathbf{h}\right\} \mathcal{P}\{\mathbf{x}^{(D)}|\mathbf{h}\}\right\}
\end{aligned} \tag{1.12}$$

where $\mathcal{P}\{\Delta(\mathbf{x}^{(D)}, \tilde{\mathbf{x}}^{(D)}) < 0|\mathbf{h}\} = \mathbb{E}_{\mathbf{n}}\{\mathcal{P}\{\Delta(\mathbf{x}^{(D)}, \tilde{\mathbf{x}}^{(D)}) < 0|\mathbf{n}, \mathbf{h}\}\}$, $\Delta(\mathbf{x}^{(D)}, \tilde{\mathbf{x}}^{(D)}) = \Lambda(\tilde{\mathbf{x}}^{(D)}) - \Lambda(\mathbf{x}^{(D)})$ and $\mathcal{P}\{\mathbf{x}^{(D)}|\mathbf{h}\} = P_2(\mathbf{x}^{(D)})$ is defined in (1.6). In the high-SNR regime, $\mathcal{P}\{\mathbf{x}^{(D)}|\mathbf{h}\}$ can be formulated similar to (1.8) and by applying the Chernoff bound.

Based on (1.12), the ASEP can be formulated as follows:

$$\text{ASEP} \approx \sum_{\mathbf{x}^{(D)}} \sum_{\tilde{\mathbf{x}}^{(D)}} \bar{\delta}\left(\mathbf{x}_S^{(D)}, \tilde{\mathbf{x}}_S^{(D)}\right) \text{APEP}\left(\mathbf{x}^{(D)} \to \tilde{\mathbf{x}}^{(D)}\right) \tag{1.13}$$

where the function $\delta(\cdot, \cdot)$ takes into account that, for the source, an error occurs only if $\mathbf{x}_S^{(D)} \neq \tilde{\mathbf{x}}_S^{(D)}$.

The computation of the APEP in (1.12) is not an easy task. To proceed further, we introduce an approximation that is accurate in the high-SNR regime. For ease of description, the approximated expression of the ASEP is first provided and then it is discussed.

Let Ω_{exact} be the set of codewords $\overline{\mathbf{x}}^{(D)}$ corresponding to the ideal scenario in the absence of demodulation errors at the relays. Accordingly, Ω_{exact} is a subset of all possible codewords $\mathbf{x}^{(D)}$, which can be formulated as follows:

$$\begin{aligned}
\Omega_{\text{exact}} &= \left\{\overline{\mathbf{x}}^{(D)}|\overline{\mathbf{x}}^{(D)} = \left(\mathbf{x}_S^{(D)}, \mathbf{x}_{R_1}^{(D)}, \mathcal{N}, \ldots, \mathcal{N}\right), \overline{\mathbf{x}}^{(D)}\right. \\
&\left.= \left(\mathbf{x}_S^{(D)}, \mathcal{N}, \mathbf{x}_{R_2}^{(D)}, \mathcal{N}, \ldots, \mathcal{N}\right), \ldots, \overline{\mathbf{x}}^{(D)} = \left(\mathbf{x}_S^{(D)}, \mathcal{N}, \ldots, \mathcal{N}, \mathbf{x}_{R_M}^{(D)}\right)\right\}
\end{aligned} \tag{1.14}$$

where the position of the single active relay in each codeword satisfies the condition $\mathbf{x}_S^{(D)} = \mathbf{ID}_{R_r}$ for $r = 1, 2, \ldots, M$.

The following high-SNR approximation for the ASEP is used:

$$\begin{aligned}\text{ASEP} &\approx \sum_{\overline{\mathbf{x}}^{(D)}} \sum_{\tilde{\mathbf{x}}^{(D)}} \delta\left(\overline{\mathbf{x}}^{(D)}, \tilde{\mathbf{x}}_S^{(D)}\right) \text{APEP}\left(\overline{\mathbf{x}}^{(D)} \to \tilde{\mathbf{x}}^{(D)}\right) \\ &\overset{(a)}{\approx} \frac{1}{MN} \sum_{\overline{\mathbf{x}}^{(D)}} \sum_{\tilde{\mathbf{x}}^{(D)}} \delta\left(\overline{\mathbf{x}}^{(D)}, \tilde{\mathbf{x}}_S^{(D)}\right) \mathbb{E}_{\mathbf{h}}\left\{\mathcal{P}\left\{\Delta\left(\overline{\mathbf{x}}^{(D)}, \tilde{\mathbf{x}}^{(D)}\right) < 0 \middle| \mathbf{h}\right\}\right\}\end{aligned} \tag{1.15}$$

where $\Delta(\overline{\mathbf{x}}^{(D)}, \tilde{\mathbf{x}}^{(D)}) = \Lambda(\tilde{\mathbf{x}}^{(D)}) - \Lambda(\overline{\mathbf{x}}^{(D)})$ and (a) follows by taking into account that, for high-SNR, $\mathcal{P}\{\overline{\mathbf{x}}^{(D)}|\mathbf{h}\} \approx 1/MN$. $\Lambda(\overline{\mathbf{x}}^{(D)})$ can be formulated as in (1.11), by replacing $\tilde{\mathbf{x}}^{(D)}$ with, $\overline{\mathbf{x}}^{(D)}$ as follows:

$$\begin{aligned}\Lambda\left(\overline{\mathbf{x}}^{(D)}\right) &= \frac{\left|y_{SD} - \sqrt{E_S} h_{SD} \mathcal{M}_S\left(\mathbf{x}_S^{(D)}\right)\right|^2}{N_0} + \frac{\left|y_{RD} - \sum_{r=1}^{M} \sqrt{E_{R_r}} h_{R_r D} \widetilde{\mathcal{M}}_R\left(\mathbf{x}_{R_r}^{(D)}\right)\right|^2}{N_0} \\ &\quad + \frac{E_S}{4N_0} \sum_{r \in \Omega_{R,S}^{(\text{ON})}} \left|\mathcal{M}_S\left(\mathbf{x}_S^{(D)}\right) - \mathcal{M}_S(\mathbf{ID}_{R_r})\right|^2 \left|h_{SR_r}\right|^2 + \alpha \frac{E_S}{N_0} \sum_{r \in \Omega_{R,S}^{(\text{OFF})}} \left|h_{SR_r}\right|^2 - \mathcal{H} \\ &\overset{(a)}{\approx} \frac{\left|y_{SD} - \sqrt{E_S} h_{SD} \mathcal{M}_S\left(\mathbf{x}_S^{(D)}\right)\right|^2}{N_0} + \frac{\left|y_{RD} - \sum_{r=1}^{M} \sqrt{E_{R_r}} h_{R_r D} \widetilde{\mathcal{M}}_R\left(\mathbf{x}_{R_r}^{(D)}\right)\right|^2}{N_0} - \mathcal{H}_{\text{exact}}\end{aligned} \tag{1.16}$$

where $\mathcal{K} = (1/M)(1/N)^{\text{card}\{\Omega_{R,S}^{(\text{ON})}\}} \beta^{\text{card}\{\Omega_{R,S}^{(\text{OFF})}\}}$, $\mathcal{H} = (\text{card}\{\Omega_{R,S}^{(\text{ON})}\} + \text{card}\{\Omega_{R,S}^{(\text{OFF})}\}) \ln(1/2) + \ln(\mathcal{K})$, and $\Omega_{R,S}^{(\text{ON})}$ and $\Omega_{R,S}^{(\text{OFF})}$ are defined, similar to (1.9), as follows:

$$\begin{cases} \Omega_{R,S}^{(\text{ON})} = \left\{r = 1, 2, \ldots, M \middle| \widetilde{\mathcal{M}}_R\left(\mathbf{x}_{R_r}^{(D)}\right) \neq 0 \quad \text{and} \quad \mathbf{x}_S^{(D)} \neq \mathbf{ID}_{R_r}\right\} \\ \Omega_{R,S}^{(\text{OFF})} = \left\{r = 1, 2, \ldots, M \middle| \widetilde{\mathcal{M}}_R\left(\mathbf{x}_{R_r}^{(D)}\right) = 0 \quad \text{and} \quad \mathbf{x}_S^{(D)} = \mathbf{ID}_{R_r}\right\} \end{cases} \tag{1.17}$$

The approximation in (a) follows by noting that $\Omega_{R,S}^{(\text{ON})}$ and $\Omega_{R,S}^{(\text{OFF})}$ are simply empty sets if $\mathbf{x}^{(D)} = \overline{\mathbf{x}}^{(D)}$, and thus, $\mathcal{H} = \mathcal{H}_{\text{exact}} = \ln(1/M)$. Since $\mathcal{H}_{\text{exact}}$ is independent of $\overline{\mathbf{x}}^{(D)}$, it can be neglected when solving the hypothesis-testing problem. It is worth noting, finally, that a single addend of $\sum_{r=1}^{M} \sqrt{E_{R_r}} h_{R_r D} \widetilde{\mathcal{M}}_R(\mathbf{x}_{R_r}^{(D)})$ in (1.16) is non-zero if $\mathbf{x}^{(D)} = \overline{\mathbf{x}}^{(D)}$.

The rationale behind (1.15) lies in realizing that, for high-SNR, the a priori probabilities $\mathcal{P}\{\mathbf{x}^{(D)}|\mathbf{h}\}$ that mainly contribute to the ASEP in (1.13) are those corresponding to $\mathbf{x}^{(D)} = \overline{\mathbf{x}}^{(D)}$. In (1.15), in particular, it is assumed that $\mathcal{P}\{\mathbf{x}^{(D)} = \overline{\mathbf{x}}^{(D)}|\mathbf{h}\} \approx 1/MN$ and $\mathcal{P}\{\mathbf{x}^{(D)} \neq \overline{\mathbf{x}}^{(D)}|\mathbf{h}\} \approx 0$. The proposed approximation is substantiated with the aid of Monte Carlo simulations in Section VI. The obtained tightness justifies its use for evaluating the diversity order of DSM. It is worth noting that the approximations in (1.15) and (1.16) do not boil down to a system with error-free relays. The proposed approximation just takes into account the events that, for high-SNR, contribute the most to the ASEP. The reliability of the source-to-relay links, in fact, affects the APEP, since $\Lambda(\tilde{\mathbf{x}}^{(D)})$ still depends on these links. In other words, some codewords (those for which $\mathbf{x}^{(D)} \neq \bar{\mathbf{x}}^{(D)}$ hold) are excluded from the computation of the ASEP, but they are taken into account in the decision metric $\Lambda(\tilde{\mathbf{x}}^{(D)})$ and, thus, in the implementation of the demodulator.

A detailed step-by-step explanation for the computation of APEP is available in Reference 35 and is not presented here due to space constraints. The final formula is provided below:

$$\mathrm{ASEP} \approx \frac{1}{MN} \sum_{\bar{\mathbf{x}}^{(D)}} \sum_{\tilde{\mathbf{x}}^{(D)}} \bar{\delta}\left(\mathbf{x}_S^{(D)}, \tilde{\mathbf{x}}_S^{(D)}\right) \gamma_0^{-\mathcal{G}_d\left(\bar{\mathbf{x}}^{(D)}, \tilde{\mathbf{x}}^{(D)}\right)} \mathcal{G}_c\left(\bar{\mathbf{x}}^{(D)}, \tilde{\mathbf{x}}^{(D)}\right) \mathcal{I}\left(\bar{\mathbf{x}}^{(D)}, \tilde{\mathbf{x}}^{(D)}\right) \tag{1.18}$$

where the following functions are introduced:

$$\begin{cases} \mathcal{G}_d\left(\bar{\mathbf{x}}^{(D)}, \tilde{\mathbf{x}}^{(D)}\right) = \omega_{SD}\left(\bar{\mathbf{x}}^{(D)}, \tilde{\mathbf{x}}^{(D)}\right) + \omega_{RD}\left(\bar{\mathbf{x}}^{(D)}, \tilde{\mathbf{x}}^{(D)}\right) + \omega_{SR}^{(\mathrm{ON})}\left(\tilde{\mathbf{x}}^{(D)}\right) + \omega_{SR}^{(\mathrm{OFF})}\left(\tilde{\mathbf{x}}^{(D)}\right) \\ \mathcal{G}_c\left(\bar{\mathbf{x}}^{(D)}, \tilde{\mathbf{x}}^{(D)}\right) = \left(\kappa_S \sigma_{SD}^2 \left|d_S\right|^2\right)^{-\omega_{SD}\left(\bar{\mathbf{x}}^{(D)}, \tilde{\mathbf{x}}^{(D)}\right)} \left(\sum_{r=1}^{M} \kappa_{R_r} \sigma_{R_r D}^2 \left|d_{R_r}\right|^2\right)^{-\omega_{RD}\left(\bar{\mathbf{x}}^{(D)}, \tilde{\mathbf{x}}^{(D)}\right)} \\ \quad \times \prod_{r \in \tilde{\Omega}_{R,S}^{(\mathrm{ON})}} \left((1/4)\kappa_S \sigma_{SR_r}^2 \left|d_S(\mathbf{ID}_{R_r})\right|^2\right)^{-\delta\left(\left|d_S(\mathbf{ID}_{R_r})\right|^2, 0\right)} \prod_{r \in \tilde{\Omega}_{R,S}^{(\mathrm{OFF})}} \left(\alpha \kappa_S \sigma_{SR_r}^2\right)^{-1} \\ \mathcal{I} = \mathcal{I}\left(s; \bar{\mathbf{x}}^{(D)}, \tilde{\mathbf{x}}^{(D)}\right) = s^{-\mathcal{G}_d\left(\bar{\mathbf{x}}^{(D)}, \tilde{\mathbf{x}}^{(D)}\right)} (1-s)^{-\omega_{SD}\left(\bar{\mathbf{x}}^{(D)}, \tilde{\mathbf{x}}^{(D)}\right) - \omega_{RD}\left(\bar{\mathbf{x}}^{(D)}, \tilde{\mathbf{x}}^{(D)}\right)} \\ \mathcal{I}\left(\bar{\mathbf{x}}^{(D)}, \tilde{\mathbf{x}}^{(D)}\right) = \frac{1}{2\pi j} \int_{\mathcal{C}-j\infty}^{\mathcal{C}+j\infty} \mathcal{I}\left(s; \bar{\mathbf{x}}^{(D)}, \tilde{\mathbf{x}}^{(D)}\right) \frac{ds}{s} \\ \quad \overset{(a)}{\approx} \frac{1}{2\pi} \int_{-1}^{+1} \mathrm{Re}\left\{\mathcal{I}\left(\mathcal{C} + j\mathcal{C}\frac{\sqrt{1-x^2}}{x}\right)\right\} \\ \quad + \frac{1}{2\pi} \int_{-1}^{+1} \mathrm{Im}\left\{\mathcal{I}\left(\mathcal{C} + j\mathcal{C}\frac{\sqrt{1-x^2}}{x}\right)\right\} \frac{dx}{x} \end{cases} \tag{1.19}$$

and, the approximation in (a) of (1.19) follows from page 17 in Reference 72. The complex integral $\mathcal{I}(\cdot,\cdot)$ may be solved in closed-form by using the residues theorem [72], but the result would depend on the pair $(\bar{\mathbf{x}}^{(D)}, \tilde{\mathbf{x}}^{(D)})$ and, thus, for brevity it is not provided here. From the expression of $\mathcal{I}(\cdot,\cdot)$, the integration pole $\mathcal{C}$ can be chosen equal to $\mathcal{C} = 1/2$, based on the guidelines reported in Reference 72. As highlighted in the next sub-section, in fact, $\omega_{SD}(\bar{\mathbf{x}}^{(D)}, \tilde{\mathbf{x}}^{(D)}) = 1$ for every pair $\tilde{\mathbf{x}}^{(D)} \neq \bar{\mathbf{x}}^{(D)}$. As a consequence, the pole $s = 1$ always exists in $\mathcal{I}(\cdot,\cdot)$.

Finally, we close this section with a comment about the novelty of the methodological approach that is used for performance analysis in the present chapter. From Reference 18, it is apparent that the error probability performance of SM has been studied extensively. The most general frameworks for computing its error probability, in particular, are available in References 12 and 15. These mathematical frameworks, however, are applicable to classical combining methods at the destination, which do not account for the reliability of the source-to-relay links. For this reason, in the present chapter, we have used a methodology that is different from most of the other papers available in the literature. The proposed approach, in particular, is based on complex analysis and on residues theory [72], which has never been applied in the context of SM.

1.2.3.2 Diversity order

In this section, we focus our attention on the diversity order of the source. Based on the operating principle DSM, the diversity order of the source may only be 0, 1 or 2. In fact, the destination receives two copies of the data of the source, one explicitly from the source itself and one implicitly from the relay activation process. In this section, we prove that the diversity order is equal to 2 for every pair $\tilde{\mathbf{x}}^{(D)} \neq \overline{\mathbf{x}}^{(D)}$.

According to Reference 70, the diversity order of the source can be formulated as follows:

$$\begin{aligned}\mathcal{G}_d &= \min_{\overline{\mathbf{x}}^{(D)},\tilde{\mathbf{x}}^{(D)}\neq\overline{\mathbf{x}}^{(D)}}\left\{\mathcal{G}_d\left(\overline{\mathbf{x}}^{(D)},\tilde{\mathbf{x}}^{(D)}\right)\right\}\\ &\overset{(a)}{\approx} 1+\min_{\overline{\mathbf{x}}^{(D)},\tilde{\mathbf{x}}^{(D)}\neq\overline{\mathbf{x}}^{(D)}}\left\{\omega_{RD}\left(\overline{\mathbf{x}}^{(D)},\tilde{\mathbf{x}}^{(D)}\right)+\omega_{SR}^{(\mathrm{ON})}\left(\tilde{\mathbf{x}}^{(D)}\right)+\omega_{SR}^{(\mathrm{OFF})}\left(\tilde{\mathbf{x}}^{(D)}\right)\right\}\end{aligned} \tag{1.20}$$

where (a) follows from (1.18) by definition of ASEP of the source, i.e., $\mathbf{x}_S^{(D)} \neq \tilde{\mathbf{x}}_S^{(D)}$ which implies $|d_S|^2 \neq 0$ and then $\omega_{SD}(\overline{\mathbf{x}}^{(D)},\tilde{\mathbf{x}}^{(D)}) = 1$.

From (1.20), we conclude that $\mathcal{G}_d \geq 1$. In order to prove second-order diversity, it is necessary to prove that $\omega_{RD}(\overline{\mathbf{x}}^{(D)},\tilde{\mathbf{x}}^{(D)}) + \omega_{SR}^{(\mathrm{ON})}(\tilde{\mathbf{x}}^{(D)}) + \omega_{SR}^{(\mathrm{OFF})}(\tilde{\mathbf{x}}^{(D)}) \geq 1$ for every pair $\tilde{\mathbf{x}}^{(D)} \neq \overline{\mathbf{x}}^{(D)}$. To this end, it is sufficient to prove that $\omega_{RD}(\cdot,\cdot)$, $\omega_{SR}^{(\mathrm{ON})}(\cdot)$ and $\omega_{SR}^{(\mathrm{OFF})}(\cdot)$ are not equal to zero simultaneously. This can be proved as follows:

- If $\omega_{RD}(\overline{\mathbf{x}}^{(D)},\tilde{\mathbf{x}}^{(D)})=0$, then $\omega_{SR}^{(\mathrm{ON})}(\tilde{\mathbf{x}}^{(D)})=1$. Based on (1.14), in fact, the condition $\omega_{RD}(\overline{\mathbf{x}}^{(D)},\tilde{\mathbf{x}}^{(D)})=0$ implies that a single relay is active in the hypothesis $\tilde{\mathbf{x}}^{(D)}$, and it is the correct[1] one. If so, the set $\widetilde{\Omega}_{R,S}^{(\mathrm{ON})}$ cannot be empty, since $\mathbf{x}_S^{(D)} \neq \tilde{\mathbf{x}}_S^{(D)}$ by assumption.
- If $\omega_{SR}^{(\mathrm{OFF})}(\tilde{\mathbf{x}}^{(D)})=0$ or $\omega_{SR}^{(\mathrm{ON})}(\tilde{\mathbf{x}}^{(D)})=0$, then $\omega_{RD}(\overline{\mathbf{x}}^{(D)},\tilde{\mathbf{x}}^{(D)})=1$. If $\omega_{SR}^{(\mathrm{OFF})}(\tilde{\mathbf{x}}^{(D)})=0$, then $\widetilde{\Omega}_{R,S}^{(\mathrm{OFF})}$ is empty. Thus, (1.9) implies that the single relay of the hypothesis $\tilde{\mathbf{x}}^{(D)}$ for which $\tilde{\mathbf{x}}_S^{(D)}=\mathbf{ID}_{R_r}$ is active. Since $\mathbf{x}_S^{(D)} \neq \tilde{\mathbf{x}}_S^{(D)}$ by assumption, this active relay cannot be the correct one. If $\omega_{SR}^{(\mathrm{ON})}(\tilde{\mathbf{x}}^{(D)})=0$, the $\widetilde{\Omega}_{R,S}^{(\mathrm{ON})}$ is empty. Thus, (1.9) implies that the correct relay is inactive in the hypothesis $\tilde{\mathbf{x}}^{(D)}$. In conclusion, if $\omega_{SR}^{(\mathrm{OFF})}(\tilde{\mathbf{x}}^{(D)}) = 0$ or $\omega_{SR}^{(\mathrm{ON})}(\tilde{\mathbf{x}}^{(D)}) = 0$, based on (14), this implies that $\sum_{r=1}^{M}|d_{R_r}|^2 \neq 0$ and, then, $\omega_{RD}(\overline{\mathbf{x}}^{(D)},\tilde{\mathbf{x}}^{(D)})=1$.

In the light of these considerations, we conclude that the diversity order of the source is two, i.e., $\mathcal{G}_d = 2$.

1.2.4 Average energy consumption and average rate of DSM

In DSM, the number of activated relays during the relaying phase depends on the demodulation outcome at the relays, as mathematically formulated in (1.2) and (1.3),

[1] In this section, correct relay is always referred to $\overline{\mathbf{x}}^{(D)}$, i.e., to the relay R_* for which $\mathbf{x}_S^{(D)}=\mathbf{ID}_{R_*} \neq \tilde{\mathbf{x}}_S^{(D)}$.

which, in turn, depends on the quality of the source-to-relay links. During the relaying phase, in particular, the transmitted energy is equal to $\sum_{r\in\Omega_R^{(\mathrm{ON})}} E_{R_r} = E_R \mathrm{card}\{\Omega_R^{(\mathrm{ON})}\}$, where, for ease of description, it is assumed that $E_R = E_{R_r}$ for $r = 1, 2, \ldots, M$. In other words, the larger the number of activated relay is, the higher the energy consumption is. This is in stark contrast compared to conventional relaying protocols, where the set of activated relays is fixed and the net effect of an incorrect demodulation at the relays results only in forwarding erroneous data to the destination. In DSM, on the other hand, the activation of multiple relays may result in a large increase of the energy consumption. The objective of this section is to show that the average (i.e., with respect to all possible activation patterns of the relays) energy consumption of DSM in the presence of demodulation errors at the relays is the same as the energy consumption of DSM in the absence of demodulation errors at the relays, i.e., $E_S + E_R$. This result is important for guaranteeing a fair comparison with conventional relaying protocols and for ensuring that the power consumption does not increase in the presence of unreliable source-to-relay links.

Similar comments apply to the average (i.e., with respect to all possible activation patterns of the relays) rate of DSM. In particular, the rate of DSM during the relaying phase is equal $\log_2(N)\mathrm{card}\{\Omega_R^{(\mathrm{ON})}\}$. Similar to the energy consumption, it is possible to show that the average rate of DSM in the presence of demodulation errors at the relays is the same as the average rate of DSM in the absence of demodulation errors at the relays, i.e., $\log_2(M) + \log_2(N)$ bits. Since the mathematical development is the same, i.e., it is sufficient to replace E_R with $\log_2(N)$ and E_S with $\log_2(M)$, the details are omitted. This result implies that the average rate of DSM is $\mathcal{R}_{\mathrm{av}} = (1/2)(\log_2(M) + \log_2(N))$ bits per channel use (bpcu), since $\log_2(M) + \log_2(N)$ bits are transmitted, on average, in two time-slots.

Let $\mathbf{x}_S$ be the equiprobable $\log_2(M)$-long bitstream emitted by the source. The average energy consumption of DSM can be formulated as follows:

$$\begin{cases} E_{\mathrm{av}} = (1/M)\sum_{\tilde{\mathbf{x}}_S^{(D)}} E_{\mathrm{av}}(\mathbf{x}_S) \\ E_{\mathrm{av}}(\mathbf{x}_S) = E_S + E_R \sum_{\mathbf{m}} \Big(\mathrm{card}\left\{\Omega_R^{(\mathrm{ON})}(\mathbf{m})\right\} \prod_{r\in\Omega_R^{(\mathrm{ON})}(\mathbf{m})} \mathcal{Q}_r(\mathbf{x}_S) \\ \qquad\qquad \times \prod_{r\in\Omega_R^{(\mathrm{OFF})}(\mathbf{m})} (1 - \mathcal{Q}_r(\mathbf{x}_S))\Big) \end{cases} \tag{1.21}$$

where $\mathbf{m}$ is a M-long binary vector whose rth entry is $\mathbf{m}[r] \in \{0, 1\}$ for $r = 1, 2, \ldots, M$. Thus, the number of binary vectors $\mathbf{m}$ is 2^M, and each of them identifies one possible active/inactive state of the M available relays. If $M = 2$, for example, $\mathbf{m} = (0, 0)$ denotes that both relays are inactive; $\mathbf{m} = (0, 1)$ denotes that R_1 is inactive and R_2 is active; $\mathbf{m} = (1, 0)$ denotes that R_1 is active and R_2 is inactive; and $\mathbf{m} = (1; 1)$ denotes that both relays are active. The sets $\Omega_R^{(\mathrm{ON})}(\cdot)$ and $\Omega_R^{(\mathrm{OFF})}(\cdot)$ denote the active and inactive relays, respectively, and are defined as follows:

$$\begin{cases} \Omega_R^{(\mathrm{ON})}(\mathbf{m}) = \{r = 1, 2, \ldots, M \,|\, \mathbf{m}[r] = 1\} \\ \Omega_R^{(\mathrm{OFF})}(\mathbf{m}) = \{r = 1, 2, \ldots, M \,|\, \mathbf{m}[r] = 0\} \end{cases} \tag{1.22}$$

and similar to (1.7), $\mathcal{Q}_r(\cdot)$ accounts for the demodulation reliability at the relays according to (1.2), and it can be formulated as follows:

$$\begin{aligned}
\mathcal{Q}_r(\mathbf{x}_S) &= \mathbb{E}\left\{\mathcal{P}\left\{\hat{\mathbf{x}}_S^{(R_r)} = \mathbf{ID}_{R_r} \middle| \mathbf{x}_S\right\}\right\} \\
&\approx \begin{cases} \mathbb{E}_{|h_{SR_r}|^2}\left\{Q\left(\sqrt{\frac{E_S}{2N_0}\left|\mathcal{M}_S(\mathbf{x}_S) - \mathcal{M}_S(\mathbf{ID}_{R_r})\right|^2 |h_{SR_r}|^2}\right)\right\} & \text{if } \mathbf{x}_S \neq \mathbf{ID}_{R_r} \\ 1 - \sum\limits_{z \neq r = 1}^{M} \mathbb{E}_{|h_{SR_z}|^2}\left\{Q\left(\sqrt{\frac{E_S}{2N_0}\left|\mathcal{M}_S(\mathbf{x}_S) - \mathcal{M}_S(\mathbf{ID}_{R_z})\right|^2 |h_{SR_z}|^2}\right)\right\} & \text{if } \mathbf{x}_S = \mathbf{ID}_{R_r} \end{cases} \\
&\overset{(a)}{=} \begin{cases} 1/2 - 1/2\sqrt{\dfrac{(E_S/N_0)\sigma_{SR_r}^2\left|\mathcal{M}_S(\mathbf{x}_S) - \mathcal{M}_S(\mathbf{ID}_{R_r})\right|^2}{2 + (E_S/N_0)\sigma_{SR_r}^2\left|\mathcal{M}_S(\mathbf{x}_S) - \mathcal{M}_S(\mathbf{ID}_{R_r})\right|^2}} & \text{if } \mathbf{x}_S \neq \mathbf{ID}_{R_r} \\ 1 - \sum\limits_{z \neq r = 1}^{M}\left(1/2 - 1/2\sqrt{\dfrac{(E_S/N_0)\sigma_{SR_z}^2\left|\mathcal{M}_S(\mathbf{x}_S) - \mathcal{M}_S(\mathbf{ID}_{R_z})\right|^2}{2 + (E_S/N_0)\sigma_{SR_z}^2\left|\mathcal{M}_S(\mathbf{x}_S) - \mathcal{M}_S(\mathbf{ID}_{R_z})\right|^2}}\right) & \text{if } \mathbf{x}_S = \mathbf{ID}_{R_r} \end{cases}
\end{aligned} \tag{1.23}$$

where (a) follows from the identity $\mathbb{E}_X\{Q(\sqrt{kX})\} = 1/2 - 1/2\sqrt{k\sigma_X^2(2 + k\sigma_X^2)^{-1}}$, which holds for $X \sim \mathcal{E}(\sigma_X^2)$. It is worth mentioning that the mathematical formulation of $E_{\text{av}}(\mathbf{x}_S)$ in (1.21) is an approximation because the union-bound is used in (1.23). Furthermore, we note that the modulation scheme of the relays does not affect (1.21), since modulation schemes with average unit-energy are assumed throughout this chapter.

For any number of relays, M, the average energy consumption of DSM can be computed by using (1.21). In what follows, we provide explicit results for two case studies: $M = 2$ and $M = 4$. The mathematical development for arbitrary M is, in fact, quite tedious, and numerical examples are reported for ease of illustration instead.

- $M = 2$: If $M = 2$, $E_{\text{av}}(\mathbf{x}_S)$ in (1.21) can be explicitly formulated as follows:

$$\begin{cases} E_{\text{av}}(\mathbf{x}_S = 0) = E_S + E_R + (E_R/2)(\chi_{SR_1}(1) - \chi_{SR_2}(1)) \\ E_{\text{av}}(\mathbf{x}_S = 1) = E_S + E_R + (E_R/2)(-\chi_{SR_1}(1) + \chi_{SR_2}(1)) \end{cases} \tag{1.24}$$

  where $\chi_{SR_r}(\upsilon) = \sqrt{(1/\upsilon)(E_s/N_0)\sigma_{SR_r}^2(1 + (1/\upsilon)(E_S/N_0)\sigma_{SR_r}^2)^{-1}}$. From (1.21) and (1.24), it follows, by direct inspection, that $E_{\text{av}} = E_S + E_R$.
- $M = 4$: If $M = 4$, $E_{\text{av}}(\mathbf{x}_S)$ in (1.21) can be explicitly formulated as follows:

$$\begin{cases} E_{\text{av}}(\mathbf{x}_S = 00) = E_S + E_R + (E_R/2) \\ \qquad \times\left(2\chi_{SR_1}(2) + \chi_{SR_1}(1) - \chi_{SR_2}(2) - \chi_{SR_3}(2) - \chi_{SR_4}(1)\right) \\ E_{\text{av}}(\mathbf{x}_S = 01) = E_S + E_R + (E_R/2) \\ \qquad \times\left(-\chi_{SR_1}(2) + 2\chi_{SR_2}(2) + \chi_{SR_2}(1) - \chi_{SR_3}(1) - \chi_{SR_4}(2)\right) \\ E_{\text{av}}(\mathbf{x}_S = 10) = E_S + E_R + (E_R/2) \\ \qquad \times\left(-\chi_{SR_1}(2) - \chi_{SR_2}(1) + 2\chi_{SR_3}(2) + \chi_{SR_3}(1) - \chi_{SR_4}(2)\right) \\ E_{\text{av}}(\mathbf{x}_S = 11) = E_S + E_R + (E_R/2) \\ \qquad \times\left(-\chi_{SR_1}(1) - \chi_{SR_2}(2) - \chi_{SR_3}(2) + 2\chi_{SR_4}(2) + \chi_{SR_4}(1)\right) \end{cases} \tag{1.25}$$

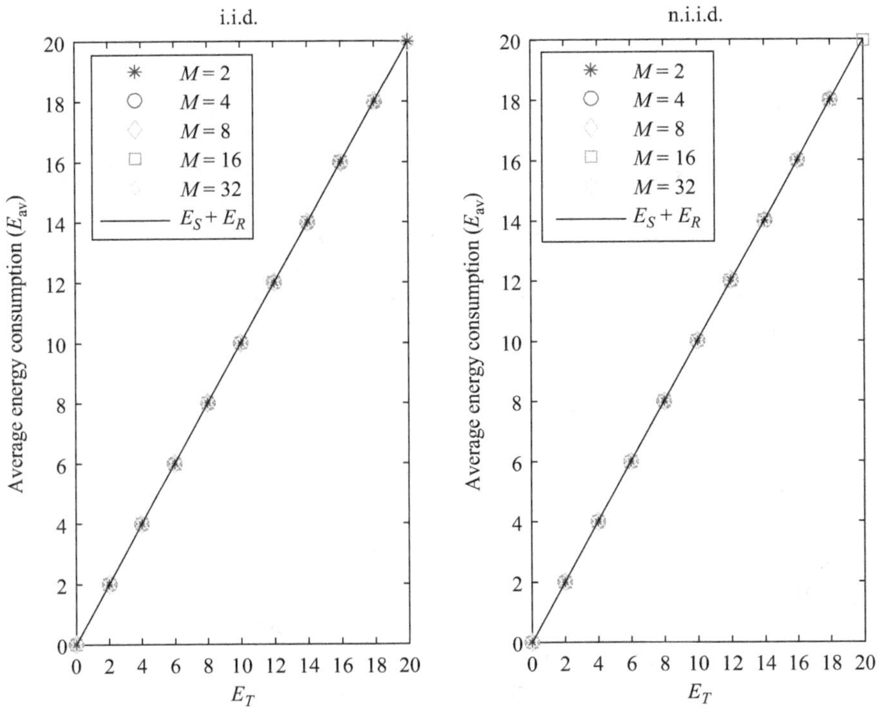

Figure 1.2 Energy consumption of the DSM protocol for $N_R = M$ and in the presence of demodulation errors at the relays. Setup: (i) PSK modulation; (ii) $E_T = E_S + E_R$ with $E_S = E_T/2$ and $E_R = E_T/2$; (iii) $N_0 = 1$; (iv) MC = 100,000; (v) $\sigma^2_{SR_r} = 1$ for the i.i.d. setup and (vi) $\sigma^2_{SR_r}$ are generated uniformly at random in [0,10] for the i.n.i.d. setup. Reproduced with permission from Reference 35. ©2015 IEEE

Again, from (1.21) and (1.25), it follows, by direct inspection, that $E_{\text{av}} = E_S + E_R$.

- *Generic M*: Since the computation of (1.21) is quite tedious for arbitrary M, we show some numerical results in Figure 1.2. These results are obtained by simulating the DSM protocol based on (1.1)–(1.3) and by computing the energy consumption as $\frac{1}{\text{MC}} \sum_{i=1}^{\text{MC}} (E_S + E_R \text{card}\{\Omega_R^{(\text{ON})}\})_i$, where MC is the number of Monte Carlo trials. The figure confirms that regardless of the number of relays and the i.i.d./i.n.i.d. channel conditions, the average energy consumption of DSM is equal to $E_{\text{av}} = E_S + E_R$, similar to conventional relaying protocols. This result is important for the performance comparison of DSM against other relaying protocols discussed in Section 1.4. A detailed analysis of the energy consumption of state-of-the-art relaying protocols is summarized in Table I of Reference 35, to which the reader is referred for further information.

1.3 Protocol II: DSM-STBC

In this section, we introduce the DSM-STBC protocol and develop an error-aware MAP demodulator, which is robust to demodulation errors at the relays. Due to its low implementation constraints when applied to distributed relay networks, in this chapter, we make use of the transmit-diversity-achieving SM-STBC scheme proposed in Reference 15. In particular, the so-called "set partitioning" transmission mode is used. Set partitioning mode implies that the symbols of the spatial-constellation diagram of SM are chosen such that the active antennas in each symbol result in a partition of the antenna-array. Interested readers are encouraged to consult Reference 15 for further information.

1.3.1 Transmission model

We consider a similar network model as illustrated in Section 1.2.1, with one source (S), N_R relays (R_r, with $r=1,2,\ldots,N_R$) and one destination (D), all of which are equipped with a single antenna. The relays are assumed to be dedicated network elements and do not have their own data to transmit. In order to better highlight the advantage of DSM-STBC, the channel between the source and the destination is not considered, as it is assumed to be very poor. Similar to DSM, in DSM-STBC, the transmission of the data from the source to the destination takes place in two phases[2]: (1) broadcasting phase, where the source broadcasts its data sequentially to the N_R cooperating relays. The relays demodulate the received data using the ML criterion [67]; and (2) relaying phase, where the relays apply the proposed DSM-STBC protocol, in order to forward the received data of the source to the destination. During this phase, the source is kept silent in order to avoid receive collisions. An illustration of the two transmission phases is provided in Figure 1.3.

The N_R cooperating relays are partitioned into virtual sets, with each set consisting of two distinct relays. These sets are denoted as V_s, with $s=1,2,\ldots,N_R/2$. Without loss of generality, R_1 and R_2 belong to V_1, R_3 and R_4 belong to V_2, and R_{N_R-1} and R_{N_R} belong to $V_{N_R/2}$. Although the relays are partitioned into virtual sets, there is no coordination between the relays. Similar to DSM, each relay is assigned a digital identifier ($\mathbf{ID}_{R_r}$, $r=1,2,\ldots,N_R$), of length $\log_2(N_R/2)$ bits. The relays in the same set have the same digital identifier, whereas the relays in different sets have different identifiers. For example, if $N_R=4$, the identifiers of the four relays are, $\mathbf{ID}_{R_1}=\mathbf{ID}_{R_2}=0$ and $\mathbf{ID}_{R_3}=\mathbf{ID}_{R_4}=1$.

1.3.1.1 Broadcasting phase

Let $\mathbf{x}_S^{(1)}$, $\mathbf{x}_S^{(2)}$ and $\mathbf{x}_S^{(3)}$ be the three data symbols that the source needs to transmit to the destination. In general, the number of bits in $\mathbf{x}_S^{(1)}$ can be different from that of $\mathbf{x}_S^{(2)}$ and $\mathbf{x}_S^{(3)}$. As we will see in detail later, $\mathbf{x}_S^{(1)}$ is forwarded from the relays to the destination via

[2] Unless otherwise specified, the notation used is similar as that in Section 1.2.

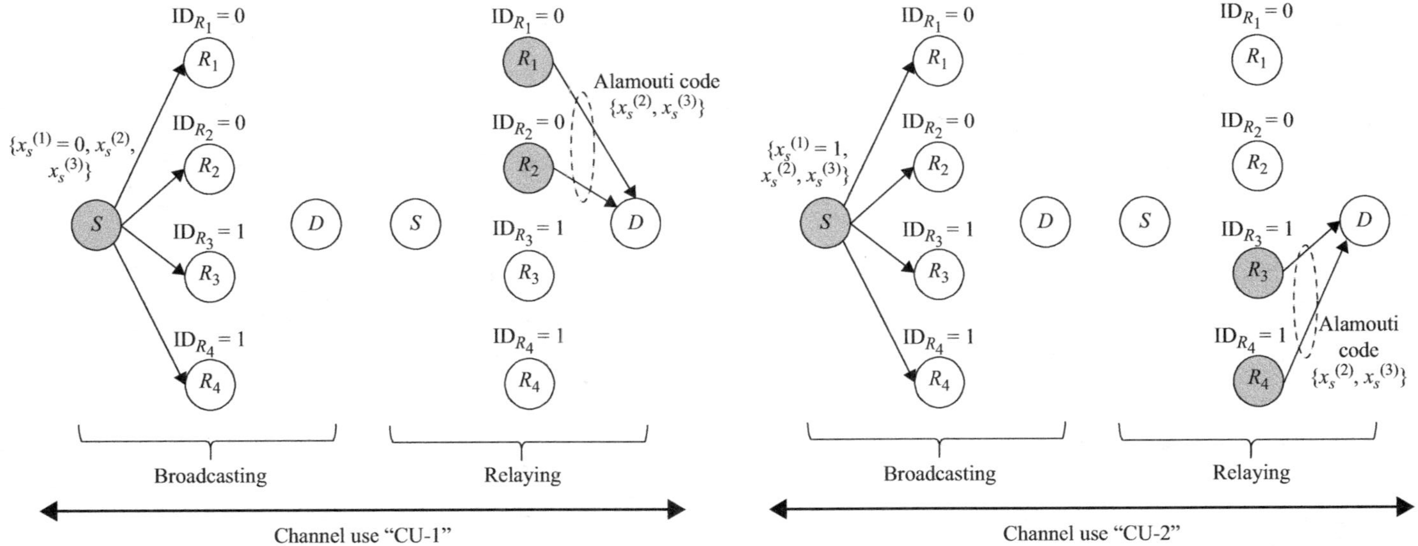

Figure 1.3 Illustration of the DSM-STBC protocol for $N_R = 4$ and in the absence of demodulation errors at the relays. The N_R relays are partitioned into $M = 2$ virtual sets. The shaded nodes are in transmission mode. The others are in reception mode, and the source (S) is in idle mode during the relaying phase

a relay activation process, and $\mathbf{x}_S^{(2)}$ and $\mathbf{x}_S^{(3)}$ are forwarded via the Alamouti scheme. Therefore, if there is a large number of cooperating relay nodes, it is possible to achieve a higher data throughput. However, for the sake of simplicity of exposition, in this chapter, we assume $\mathbf{x}_S^{(1)}$, $\mathbf{x}_S^{(2)}$ and $\mathbf{x}_S^{(3)}$ are of $\log_2(M)$ bits each. Let $s_{tx}^{(1)}$, $s_{tx}^{(2)}$ and $s_{tx}^{(3)}$ be the PSK/QAM modulated symbol transmitted from the source, corresponding to $\mathbf{x}_S^{(1)}, \mathbf{x}_S^{(2)}$ and $\mathbf{x}_S^{(3)}$, respectively. In the broadcasting phase, the source sequentially transmits $s_{tx}^{(1)}$, $s_{tx}^{(2)}$ and $s_{tx}^{(3)}$, to all the cooperating relays. Thus, the signal received at the relay, R_r, for $r = 1, 2, \ldots, N_R$, at time-slots, $t = 1$, 2 and 3 can be formulated as follows:

$$y_{SR_r}^{(t)} = \sqrt{E_S} h_{SR_r} \mathcal{M}_S\left(\mathbf{x}_S^{(t)}\right) + n_{SR_r}^{(t)} \tag{1.26}$$

where the channel h_{SR_r}, for $r = 1, 2, \ldots, N_R$, remains constant over the transmission of $\mathbf{x}_S^{(1)}$, $\mathbf{x}_S^{(2)}$ and $\mathbf{x}_S^{(3)}$.

The signal $y_{SR_r}^{(t)}$ is demodulated using the ML criterion as follows:

$$\begin{cases} \hat{s}_{tx}^{(R_r,t)} = \underset{p_m \in \{p_1, p_2, \ldots, p_M\}}{\arg\min} \left\{ \left| y_{SR_r}^{(t)} - \sqrt{E_S} h_{SR_r} p_m \right|^2 \right\} \\ \hat{\mathbf{x}}_S^{(R_r,t)} = \mathcal{M}_S^{-1}\left(\hat{s}_{tx}^{(R_r,t)}\right) \end{cases} \tag{1.27}$$

where (i) $\hat{s}_{tx}^{(R_r,t)}$ and $\hat{\mathbf{x}}_S^{(R_r,t)}$ are the estimates of $s_{tx}^{(t)}$ and $\mathbf{x}_S^{(t)}$, respectively, at R_r for $r = 1, 2, \ldots, N_R$, at $t = 1$, 2 and 3.

1.3.1.2 Relaying phase

After demodulating the signal received from the source by using (1.27), each cooperating relay, R_r, for $r = 1, 2, \ldots, N_R$, applies the SM-STBC principle [15], to forward the three data symbols which are received during the broadcasting phase.

The Alamouti STBC matrix for co-located antennas is given as [1]:

$$\begin{bmatrix} s_{tx}^{(1)} & s_{tx}^{(2)} \\ -s_{tx}^{(2)*} & s_{tx}^{(1)*} \end{bmatrix} \tag{1.28}$$

In an Alamouti STBC scheme with two co-located transmit-antennas, each transmit-antenna transmits a distinct column of (1.28) in two consecutive time-slots. In DSM-STBC, the Alamouti code[3] is formed using the demodulated symbols, $\hat{s}_{tx}^{(R_r,2)}$ and $\hat{s}_{tx}^{(R_r,3)}$, and they are transmitted from a pair of active relays belonging to a particular set, V_s. More specifically, $\hat{\mathbf{x}}_S^{(R_r,1)} = \mathcal{M}_S^{-1}(\hat{s}_{tx}^{(R_r,1)})$ is used to select the relays that should be active in the relaying phase, and through these active relays, $\hat{s}_{tx}^{(R_r,2)}$ and $\hat{s}_{tx}^{(R_r,3)}$ are forwarded to the destination by using the Alamouti code.

[3] Higher-order STBCs can also be used as "mother STBC" in DSM-STBC. Alamouti STBC has been considered here due to its simplicity and full-rate property.

Thus, the signal received at the destination in two consecutive time-slots during the relaying phase can be written as follows:

$$
\begin{cases}
y_{RD}^{(1)} = \sum\limits_{r\in\Theta} \left(\begin{array}{l} \sqrt{\frac{E_{R_r}}{2}} h_{R_rD} \hat{s}_{tx}^{(R_r,2)} \delta\left(\hat{\mathbf{x}}_S^{(R_r,1)}, \mathbf{ID}_{R_r}\right) \\ + \sqrt{\frac{E_{R_r}}{2}} h_{R_{r+1}D} \hat{s}_{tx}^{(R_{r+1},3)} \delta\left(\hat{\mathbf{x}}_S^{(R_{r+1},1)}, \mathbf{ID}_{R_{r+1}}\right) \end{array} \right) + n_{RD}^{(1)} \\
y_{RD}^{(2)} = \sum\limits_{r\in\Theta} \left(\begin{array}{l} \sqrt{\frac{E_{R_r}}{2}} h_{R_rD} \left(-\hat{s}_{tx}^{(R_r,3)}\right)^{*} \delta\left(\hat{\mathbf{x}}_S^{(R_r,1)}, \mathbf{ID}_{R_r}\right) \\ + \sqrt{\frac{E_{R_{r+1}}}{2}} h_{R_{r+1}D} (\hat{s}_{tx}^{(R_r,2)})^{*} \delta\left(\hat{\mathbf{x}}_S^{(R_{r+1},1)}, \mathbf{ID}_{R_{r+1}}\right) \end{array} \right) + n_{RD}^{(2)}
\end{cases}
\tag{1.29}
$$

where $\Theta = \{1, 3, 5, \ldots N_R - 1\}$, and $\delta(\cdot, \cdot)$ is defined in Section 1.1.

From (1.29), it follows that: (i) the estimate of the source's data, $\hat{s}_{tx}^{(R_r,1)}$ is encoded into the activation process of the cooperating relays. More specifically, R_r, belonging to the set V_s, transmits during the relaying phase if and only if the demodulated data, $\hat{s}_{tx}^{(R_r,1)}$, obtained from (1.27), coincides with its own digital identifier. If this is not the case, that particular relay is silent and will have no active role to play during the relaying phase. This activation process is represented using the $\delta(\cdot, \cdot)$ function in (1.29). In other words, $\hat{s}_{tx}^{(R_r,1)}$ is *implicitly* forwarded using the active relays. On the other hand, $\hat{s}_{tx}^{(R_r,2)}$ and $\hat{s}_{tx}^{(R_r,3)}$ are forwarded using the Alamouti code from the active relays. This results in a throughput enhancement for the Alamouti-based DSM-STBC compared to the conventional distributed Alamouti code; (ii) since each relay demodulates the data independently and without any coordination, some relays may decode the data correctly and some others incorrectly. Hence, similar to DSM, the relays that are supposed to be active during the relaying phase might be silent, and vice versa. The demodulation at the destination must be robust to these errors at the relays in order to achieve the diversity-gain provided by the Alamouti code; (iii) if there are no errors at the relays, only two relays will be active during the relaying phase. In this case, the conventional SM-STBC demodulator, given in Reference 15, can be used for DSM-STBC.

For convenience, let us re-write (1.29) in the matrix form as follows:

$$
\mathbf{y}_{RD} = \sum_{r\in\Theta} \mathbf{G}_{RD}^{(r)} \mathbf{X}_R^{(r)} + \mathbf{n}_{RD} \tag{1.30}
$$

where (i) $\mathbf{y}_{RD} = [y_{RD}^{(1)}, y_{RD}^{(2)}]$ is the received data vector at the destination; (ii) $\mathbf{G}_{RD}^{(r)} = [G_{R_rD}, G_{R_{r+1}D}]$, with $G_{R_mD} = \sqrt{\frac{E_{R_m}}{2}} h_{R_mD} \delta(\hat{\mathbf{x}}_S^{(R_m,1)}, \mathbf{ID}_{R_m})$, for $m = r,\ r+1$ can be interpreted as the effective channel matrix; and (iii) $\mathbf{X}_R^{(r)}$ is the Alamouti code transmitted by the active relays in a distributed fashion, and it is defined as follows:

$$
\mathbf{X}_R^{(r)} = \begin{bmatrix} \hat{s}_{tx}^{(R_r,2)} & \hat{s}_{tx}^{(R_{r+1},3)} \\ \left(-\hat{s}_{tx}^{(R_{r+1},3)}\right)^{*} & \left(\hat{s}_{tx}^{(R_{r+1},2)}\right)^{*} \end{bmatrix} \tag{1.31}
$$

It can readily be seen from (1.31) that if $\hat{s}_{tx}^{(R_r,t')} \neq \hat{s}_{tx}^{(R_{r+1},t')}$, for $t' \in \{2, 3\}$, then, $\mathbf{X}_R(\mathbf{X}_R)^H$ is not necessarily an orthogonal STBC, which is in fact the case when there are decoding errors at the relays. Hence, single-stream decoding might not be possible for DSM-STBC, which is also the case for conventional distributed Alamouti scheme [62].

1.3.2 Demodulation at the destination

In this section, we develop a MAP demodulator at the destination for the DSM-STBC scheme. Similar to DSM, the MAP demodulator in DSM-STBC is also developed by taking advantage of the knowledge of the instantaneous channel impulse responses of the source-to-relay links. By using the MAP criterion [67], the three data symbols transmitted by the source can be jointly demodulated at the destination as follows:

$$
\begin{aligned}
\left[\hat{\mathbf{s}}_{tx}^{(D)}\right] &= \arg\max_{\mathcal{M}_S\left(\tilde{\mathbf{x}}_S^{(t,D)}\right)\in\mathcal{A},\, t=1,2,3} \left\{ \sum_{\mathcal{M}_S\left(\hat{\mathbf{x}}_S^{(R_{1,1})}\right)\in\mathcal{A}} \sum_{\mathcal{M}_S\left(\hat{\mathbf{x}}_S^{(R_{1,2})}\right)\in\mathcal{A}} \sum_{\mathcal{M}_S\left(\hat{\mathbf{x}}_S^{(R_{1,3})}\right)\in\mathcal{A}} \cdots \sum_{\mathcal{M}_S\left(\hat{\mathbf{x}}_S^{(R_{N_R},1)}\right)\in\mathcal{A}} \right. \\
&\left. \sum_{\mathcal{M}_S\left(\hat{\mathbf{x}}_S^{(R_{N_R},2)}\right)\in\mathcal{A}} \sum_{\mathcal{M}_S\left(\hat{\mathbf{x}}_S^{(R_{N_R},3)}\right)\in\mathcal{A}} \begin{pmatrix} \overline{P}_1\left(y_{RD}^{(1)}, y_{RD}^{(2)} | \tilde{\mathbf{x}}^{(D)}, \hat{\mathbf{x}}^{(R)}\right) \\ \times \overline{P}_2\left(\tilde{\mathbf{x}}^{(D)}, \hat{\mathbf{x}}^{(R)}\right) \end{pmatrix} \right\} \\
\hat{\mathbf{x}}_S^{(D,1)} &= \mathcal{M}_S^{-1}\left(\hat{s}_{tx}^{(D,1)}\right);\ \hat{\mathbf{x}}_S^{(D,2)} = \mathcal{M}_S^{-1}\left(\hat{s}_{tx}^{(D,2)}\right);\ \hat{\mathbf{x}}_S^{(D,3)} = \mathcal{M}_S^{-1}\left(\hat{s}_{tx}^{(D,3)}\right)
\end{aligned}
\quad (1.32)
$$

where (i) $\hat{\mathbf{s}}_{tx}^{(D)} = [\hat{s}_{tx}^{(D,1)}, \hat{s}_{tx}^{(D,2)}, \hat{s}_{tx}^{(D,3)}]$ and $\hat{\mathbf{x}}_S^{(D)} = [\hat{\mathbf{x}}_S^{(D,1)}, \hat{\mathbf{x}}_S^{(D,2)}, \hat{\mathbf{x}}_S^{(D,3)}]$ denote the estimates of the information symbol and information bits, sequentially transmitted by the source during the broadcasting phase; (ii) $\mathcal{A} = \{p_1, p_2, \ldots, p_M\}$, $\hat{\mathbf{x}}^{(R)} = \{\hat{\mathbf{x}}_S^{(R_1,1)}, \hat{\mathbf{x}}_S^{(R_1,2)}, \hat{\mathbf{x}}_S^{(R_1,3)}, \ldots, \hat{\mathbf{x}}_S^{(R_{N_R},1)}, \hat{\mathbf{x}}_S^{(R_{N_R},2)}, \hat{\mathbf{x}}_S^{(R_{N_R},3)}\}$, and $\tilde{\mathbf{x}}^{(D)} = \{\tilde{\mathbf{x}}_S^{(D,1)}, \tilde{\mathbf{x}}_S^{(D,2)}, \tilde{\mathbf{x}}_S^{(D,3)}\}$ are short-hands introduced for notational simplicity; (iii) $\tilde{\mathbf{x}}_S^{(D,t)}$, for $t = 1, 2$ and 3, are trial bits used in the hypothesis detection problem; and (iv) $\overline{P}_1(y_{RD}^{(1)}, y_{RD}^{(2)} | \tilde{\mathbf{x}}^{(D)}, \hat{\mathbf{x}}^{(R)})$ denotes the joint probability density function of the received signals conditioned on $\tilde{\mathbf{x}}^{(D)}$ and $\hat{\mathbf{x}}^{(R)}$, and $\overline{P}_2(\tilde{\mathbf{x}}^{(D)}, \hat{\mathbf{x}}^{(R)})$ denotes the a priori joint probability of the demodulation events at the relays. $\overline{P}_1(\cdot|\cdot)$ and $\overline{P}_2(\cdot)$ can be formulated as follows:

$$
\begin{aligned}
&\overline{P}_1\left(y_{RD}^{(1)}, y_{RD}^{(2)} | \tilde{\mathbf{x}}^{(D)}, \hat{\mathbf{x}}^{(R)}\right) \\
&= \mathcal{P}\left\{y_{RD}^{(1)}, y_{RD}^{(2)} | \tilde{\mathbf{x}}_S^{(D,1)}, \tilde{\mathbf{x}}_S^{(D,2)}, \tilde{\mathbf{x}}_S^{(D,3)}, \hat{\mathbf{x}}_S^{(R_1,1)}, \hat{\mathbf{x}}_S^{(R_1,2)}, \hat{\mathbf{x}}_S^{(R_1,3)}, \ldots, \hat{\mathbf{x}}_S^{(R_{N_R},1)}, \hat{\mathbf{x}}_S^{(R_{N_R},2)}, \hat{\mathbf{x}}_S^{(R_{N_R},3)}\right\} \\
&= \exp\left(\frac{\left| y_1^{(RD)} - \sum_{r\in\Theta} G_{R_rD}\mathcal{M}_S\left(\hat{\mathbf{x}}_S^{(R_r,2)}\right) + G_{R_{r+1}D}\mathcal{M}_S\left(\hat{\mathbf{x}}_S^{(R_{r+1},3)}\right)\right|^2}{N_0} \right) \\
&\times \exp\left(\frac{\left| y_2^{(RD)} - \sum_{r\in\Theta} G_{R_rD}\left(-\mathcal{M}_S(\hat{\mathbf{x}}_S^{(R_r,3)})\right)^* + G_{R_{r+1}D}\left(\mathcal{M}_S\left(\hat{\mathbf{x}}_S^{(R_{r+1},2)}\right)\right)^*\right|^2}{N_0} \right)
\end{aligned}
\quad (1.33)
$$

$$\begin{aligned}
&\overline{P}_2\left(\tilde{\mathbf{x}}^{(D)}, \hat{\mathbf{x}}^{(R)}\right)\\
&= \mathcal{P}\left\{\tilde{\mathbf{x}}_S^{(D,1)}, \tilde{\mathbf{x}}_S^{(D,2)}, \tilde{\mathbf{x}}_S^{(D,3)}, \hat{\mathbf{x}}_S^{(R_1,1)}, \hat{\mathbf{x}}_S^{(R_1,2)}, \hat{\mathbf{x}}_S^{(R_1,3)}, \ldots, \hat{\mathbf{x}}_S^{(R_{N_R},1)}, \hat{\mathbf{x}}_S^{(R_{N_R},2)}, \hat{\mathbf{x}}_S^{(R_{N_R},3)}\right\}\\
&= \prod_{r=1}^{N_R} \prod_{t=1}^{3} \frac{1}{M} \mathcal{P}\left\{\tilde{\mathbf{x}}_S^{(D,t)}, \hat{\mathbf{x}}_S^{(R_r,t)}\right\}
\end{aligned} \tag{1.34}$$

In (1.34), $\mathcal{P}\{\tilde{\mathbf{x}}_S^{(t,D)}, \hat{\mathbf{x}}_S^{(R_r,t)}\}$ denotes the probability that the relay R_r, for $r = 1, 2, \ldots, N_R$ and $t = 1$, 2 and 3, demodulates the received signal as $\hat{\mathbf{x}}_S^{(R_r,t)}$, when the source transmits $\tilde{\mathbf{x}}_S^{(t,D)}$ during the broadcasting phase. By using similar arguments as in Section 2.2, an approximation for $\mathcal{P}\{\tilde{\mathbf{x}}_S^{(t,D)}, \hat{\mathbf{x}}_S^{(R_r,t)}\}$ can be formulated as follows:

$$\mathcal{P}\left\{\tilde{\mathbf{x}}_S^{(D,t)}, \hat{\mathbf{x}}_S^{(R_r,t)}\right\} \approx \begin{cases} Q\left(\sqrt{\frac{E_S}{2N_0}\left|\mathcal{M}_S\left(\tilde{\mathbf{x}}_S^{(D,t)}\right) - \mathcal{M}_S\left(\hat{\mathbf{x}}_S^{(R_r,t)}\right)\right| \left|h_{SR_r}\right|^2}\right) \\ 1 - \beta Q\left(\sqrt{2\alpha\frac{E_S}{N_0}\left|h_{SR_r}\right|^2}\right) \end{cases} \tag{1.35}$$

1.3.2.1 High-SNR approximation

A high-SNR approximation for the demodulator in (1.32) can be developed by following a similar line of thought as that used in Section 2.2.2. The final result is as follows:

$$\left[\hat{\mathbf{s}}_{tx}^{(D)}, \hat{\mathbf{s}}_{tx}^{(R)}\right] = \underset{\mathcal{M}_S\left(\tilde{\mathbf{x}}_S^{(t,D)}\right) \in \mathcal{A}, t=1,2,3}{\arg\min} \left\{\Lambda'\left(y_{RD}^{(1)}, y_{RD}^{(2)} \middle| \tilde{\mathbf{x}}^{(D)}, \hat{\mathbf{x}}^{(R)}\right)\right\} \tag{1.36}$$

$$\begin{aligned}
&\Lambda'\left(y_{RD}^{(1)}, y_{RD}^{(2)} \middle| \tilde{\mathbf{x}}^{(D)}, \hat{\mathbf{x}}^{(R)}\right)\\
&= \left| y_1^{(RD)} - \sum_{\Theta_r} G_{R_rD}\mathcal{M}_S\left(\hat{\mathbf{x}}_S^{(R_r,2)}\right) + G_{R_{r+1}D}\mathcal{M}_S\left(\hat{\mathbf{x}}_S^{(R_{r+1},3)}\right)\right|^2\\
&\quad + \left| y_2^{(RD)} - \sum_{\Theta_r} G_{R_rD}\left(-\mathcal{M}_S\left(\hat{\mathbf{x}}_S^{(R_r,3)}\right)\right)^* + G_{R_{r+1}D}\left(\mathcal{M}_S\left(\hat{\mathbf{x}}_S^{(R_{r+1},2)}\right)\right)^*\right|^2\\
&\quad + \sum_{r=1}^{N_r}\sum_{t=1}^{3}\left(E_S\left|h_{SR_r}\right|^2\left|\mathcal{M}_S\left(\mathbf{x}_S^{(D,t)}\right) - \mathcal{M}_S\left(\hat{\mathbf{x}}_S^{(R_r,t)}\right)\right|^2\right)
\end{aligned} \tag{1.37}$$

where $\hat{\mathbf{s}}_{tx}^{(R)} = [\mathcal{M}_S(\hat{\mathbf{x}}_S^{(R_1,1)}), \mathcal{M}_S(\hat{\mathbf{x}}_S^{(R_1,2)}), \mathcal{M}_S(\hat{\mathbf{x}}_S^{(R_1,3)}), \ldots, \mathcal{M}_S(\hat{\mathbf{x}}_S^{(R_{N_R},1)}), \mathcal{M}_S(\hat{\mathbf{x}}_S^{(R_{N_R},2)}), \mathcal{M}_S(\hat{\mathbf{x}}_S^{(R_{N_R},3)})]$.

From (1.36) and (1.37), it can be seen that the general structure of the demodulator is now a two-dimensional array, and hence searching the code set is computationally less complex. However, the total search space has not been reduced. For large signal-constellation size and number of relays, the implementation of (1.36) and (1.37) still requires a large number of computations.

It is also worth noting that, in (1.32), the need for the multiple summation on the right-hand side is circumvented by considering all the possible demodulation events on the left-hand side. This form is more suitable for use in the mathematical analysis of the DSM-STBC protocol. Let us note that we are not interested in the estimates $\hat{\mathbf{s}}_{tx}^{(R)}$ in (1.36). In fact, an error is detected for the data transmitted from the source, if and only if $\hat{s}_{tx}^{(D,t)} \neq s_{tx}^{(t)}$, for $t = 1$, 2 and 3.

1.3.3 Remarks

Some additional comments on the proposed DSM-STBC protocol are as follows:

- *Error probability and diversity order*: The analytical error-performance and the achievable diversity order of DSM-STBC have been are omitted due to space limitations. However, they can be derived by using a similar mathematical development as that used in Section 1.2.3. Extensive Monte Carlo simulations show that, using the demodulators in (1.32) and (1.36), DSM-STBC is capable of achieving second-order diversity.
- *Average energy consumption*: By using the mathematical consideration introduced in Section 2.4, the average energy consumption of DSM-STBC is found to be $E'_{\text{av}} = E_S + 2E_R$, which is similar to the conventional distributed Alamouti scheme.
- *Average rate*: In general, the average rate of DSM-STBC is $\mathcal{R}'_{\text{av}} = (2/5) \times (\log_2(M)) + (1/5)\,(\log_2([N_R/2]))$ bpcu, where [.] represents the floor function. Therefore, when there is a large number of cooperating relay nodes, DSM-STBC achieves a significantly higher rate compared to the distributed Alamouti scheme, for which the average rate is only $(1/2)(\log_2(M))$ bpcu. For the particular scenario considered in this chapter, where $\mathbf{x}_S^{(1)}$, $\mathbf{x}_S^{(2)}$ and $\mathbf{x}_S^{(3)}$ are of $\log_2(M)$ bits each, the average rate of DSM-STBC is $\mathcal{R}'_{\text{av}} = (3/5)(\log_2(M))$.

1.4 Numerical and simulation results

In this section, we present numerical results for assessing the performance of DSM and DSM-STBC, and for comparing their ASEPs against state-of-the-art relaying protocols. For ease of reproducibility, i.i.d. fading channels are assumed, i.e., $\sigma^2 = \sigma_{SD}^2 = \sigma_{SR_r}^2 = \sigma_{R_rD}^2 = 1$ for $r = 1, 2, \ldots, M$. Also, the relays are assumed to transmit the same average energy during the relaying phase, i.e., $E_{R_r} = E_R$ for $r = 1, 2, \ldots, M$. PSK modulation is used. Other simulation parameters are provided in the caption of the figures.

For ensuring a fair comparison among different relaying protocols (with which DSM and DSM-STBC are compared), which may provide a different rate (in bpcu) and may rely upon a different number of active relays, the following two conditions are enforced to all considered relaying protocols: (i) the average energy consumption (E_{av} for DSM and E'_{av} for DSM-STBC) is the same and (ii) the average rate in bpcu ($\mathcal{R}_{\text{av}}$ for DSM and $\mathcal{R}'_{\text{av}}$ for DSM-STBC) is the same. As an example, let us consider DSM. Similar calculations for state-of-the-art relaying protocols analysed in this section

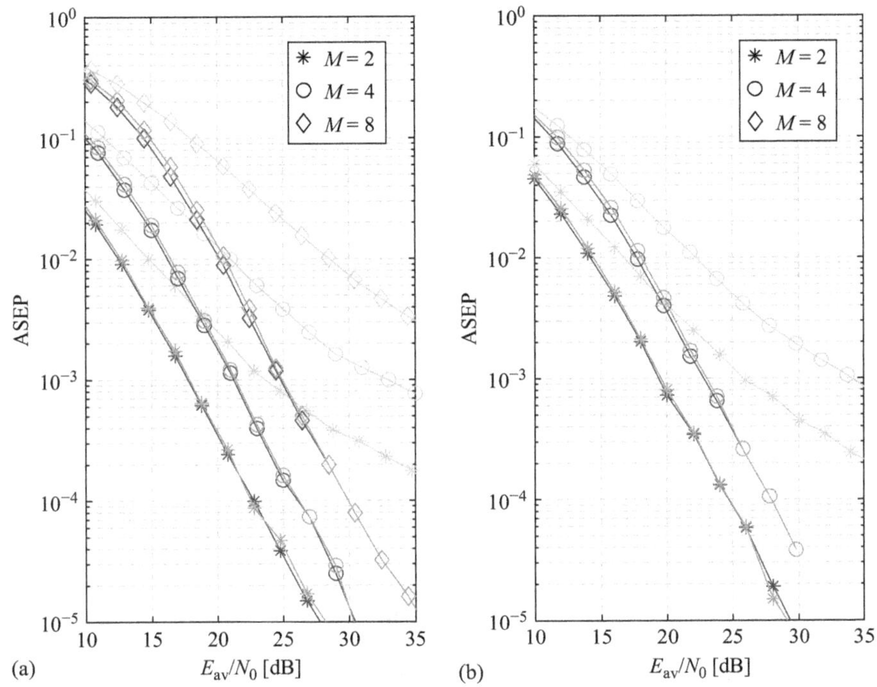

Figure 1.4 Comparison of different demodulators for DSM. (a) $N=2$ and (b) $N=4$. The solid lines are obtained using (1.4), and the dashed lines are obtained using (1.10). The dotted lines denote the DSM-sub-optimal demodulator, and are obtained from Table I of Reference 35. As for the demodulator in (1.10), $\tilde{\mathcal{H}}$ is neglected. Reproduced with permission from Reference 35. ©2015 IEEE

are summarized in Table I of Reference 35. Given $\mathcal{R}_{av}$, the possible pairs (M,N) are obtained such that the equality $\mathcal{R}_{av}=(1/2)(\log_2(M)+\log_2(N))$ is satisfied (see Section 1.2.4). Let $E_{av}=E_S+E_R$ (see Section 1.2.4) and (M,N), the average energy per bit (E_b) is computed such that the transmit energies for broadcasting and relaying phases are set equal to $E_S=E_b\log_2(M)$ and $E_R=E_b(\log_2(M)+\log_2(N))$. Similar considerations are also applied to the figures related to DSM-STBC. The proposed DSM and DSM-STBC protocols are not compared with each other, as they are applicable to different scenarios. More specifically: in DSM, the relays have their own data to transmit, whereas in DSM-STBC, the relays are dedicated network elements and do not have their own data.

- *Diversity-achieving demodulation for DSM:* In Figure 1.4, the ASEP of the three diversity combiners for DSM is compared. The results confirm that second-order diversity cannot be achieved if relay-induced demodulation errors are not taken into account at the destination, i.e., the sub-optimal demodulator (referred to as "DSM – sub-optimal demodulator" [35, Table I]) is used. Furthermore, it is shown

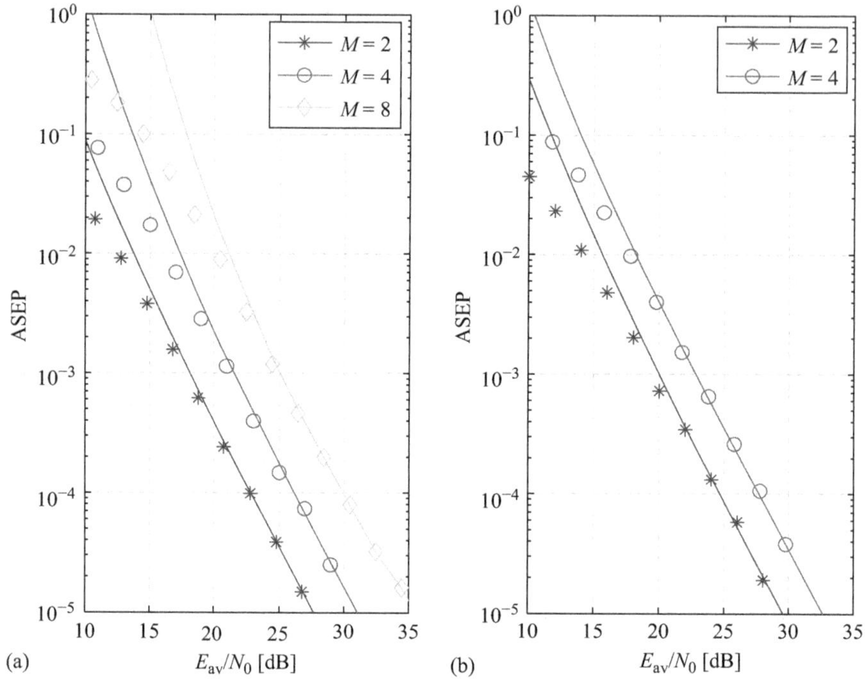

Figure 1.5 Comparison of Monte Carlo simulations (markers) and the mathematical framework in (1.18) and (1.19) (solid lines). (a) $N = 2$ and (b) $N = 4$. Reproduced with permission from Reference 35. ©2015 IEEE

that the performance difference between the demodulators in (1.4) and (1.10) is negligible. In the rest of this section, as a consequence, only the demodulator in (1.10) is considered for DSM.

- *Validation of mathematical framework and diversity analysis of DSM:* In Figure 1.5, Monte Carlo simulations are compared against the mathematical framework of Section 1.2.3. The results confirm the tightness of the proposed approach for performance evaluation. This provides a sound justification of the approximation used in (1.15) for computing the ASEP and for using the resulting framework for studying the achievable diversity order of DSM. Indeed, Figure 1.5 confirms that the demodulators introduced in Section 1.2.2 are capable of achieving second-order diversity.
- *Performance comparison of DSM with state-of-the-art relaying protocols:* In Figures 1.6 and 1.7, DSM is compared against relevant state-of-the-art relaying protocols [35, Table I]. In particular, Figure 1.6 confirms that DSM is, for the analysed range of SNRs, always better than Demodulate-and-Forward (DemF) relaying. On the other hand, both DemF and DSM may be worse than Single-Hop (SH) transmission for low-SNR values. This originates from the i.i.d. assumption for the fading channels and from the need of using two time-slots for DemF and

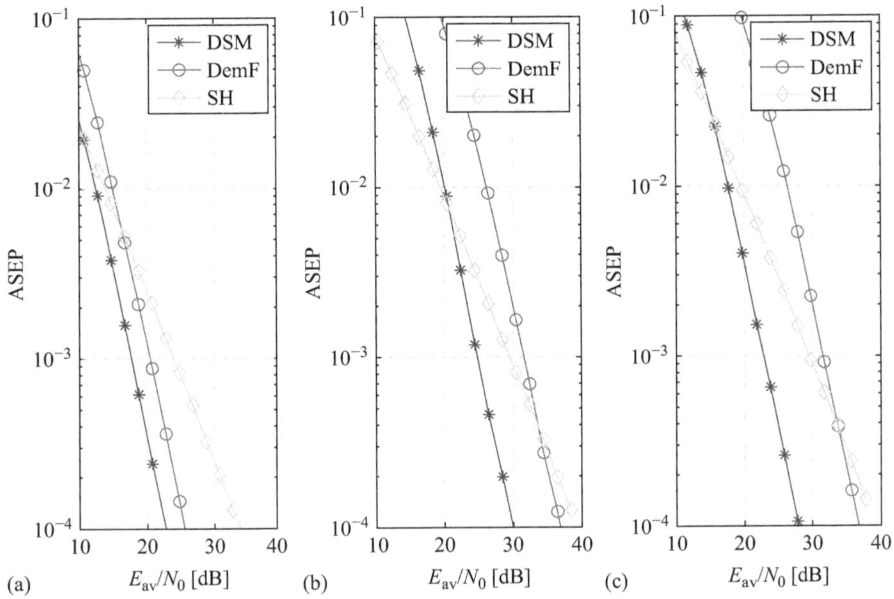

Figure 1.6 Comparison of DSM (demodulator in (1.10) where $\widetilde{\mathcal{H}}$ is neglected) against DemF relaying and SH transmission [35, Table I]. (a) $\mathcal{R}_{av} = 1$ bpcu: $M = 2$ and $N = 2$ for DSM; $M = 4$ for DemF and $M = 2$ for SH. (b) $\mathcal{R}_{av} = 2$ bpcu: $M = 8$ and $N = 2$ for DSM; $M = 16$ for DemF and $M = 4$ for SH. (c) $\mathcal{R}_{av} = 2$ bpcu: $M = 4$ and $N = 4$ for DSM; $M = 16$ for DemF and $M = 4$ for SH. Reproduced with permission from Reference 35. ©2015 IEEE

DSM. As expected, however, DemF and DSM are superior for high-SNR values, since they provide second-order diversity. Furthermore, Figure 1.7 confirms that DSM provides a better ASEP than SPM and Network-Coded Cooperative (NCC) relaying. Compared with NCC, the gain is significant since NCC is not capable of achieving any diversity gains for the considered setup [55]. Compared with SPM, the gain is at least equal to 1 dB for the considered setup, and it depends on the parameter used for SPM. The values of γ^2 considered in Figure 1.7 provide near-optimal performance for the analysed setup. This gain comes, however, at the cost of a higher demodulation complexity for DSM. Possible solutions for reducing the demodulation complexity of DSM are currently under investigation by the authors. Potential options encompass, e.g., the application of sphere decoding [73]. More specifically, the demodulator in (1.4) can be interpreted as a multi-user detector and, thus, several algorithms for reducing its signal processing complexity exist. The difference compared to conventional sphere decoders and multi-user demodulators lies in the need of generalising them for taking into account the reliability of the source-to-relay links.

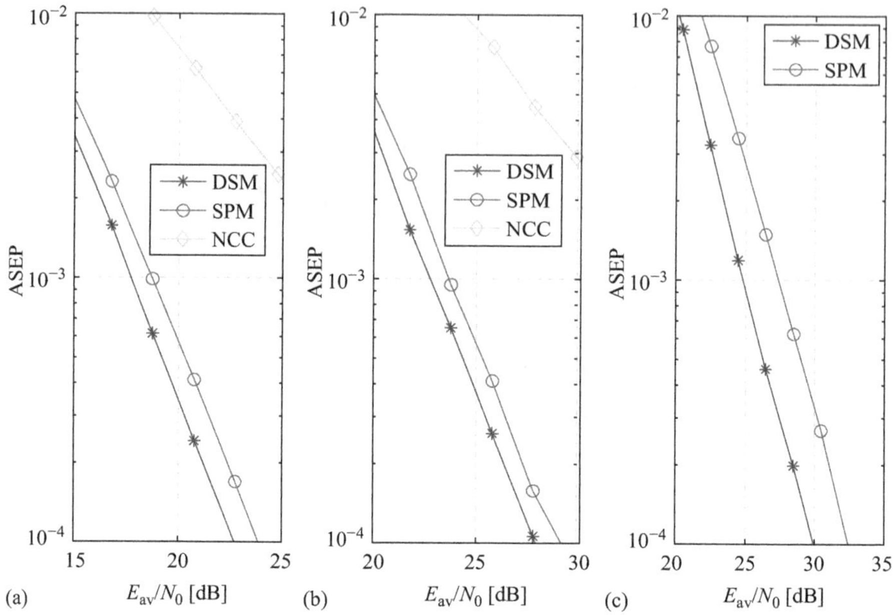

Figure 1.7 Comparison of DSM (demodulator in (1.10) where $\widetilde{\mathcal{H}}$ is neglected) against SPM and NCC relaying [35, Table I]. (a) $\mathcal{R}_{av} = 1$ bpcu: $M = 2$ and $N = 2$ for DSM, SPM ($\gamma^2 = 0.2$) and NCC. (b) $\mathcal{R}_{av} = 2$ bpcu: $M = 4$ and $N = 4$ for DSM, SPM ($\gamma^2 = 0.25$) and NCC. (c) $\mathcal{R}_{av} = 2$ bpcu: $M = 8$ and $N = 2$ for DSM and SPM ($\gamma^2 = 0.4$). A "zoom in" view is provided for enhancing the readability of the figure. Reproduced with permission from Reference 35. ©2015 IEEE

- *Performance enhancement of DSM with relay selection:* Figure 1.8 investigates the potential gain of combining DSM with relay selection for those scenarios where $N_R > M$. The criterion used for relay selection is based on choosing the relays that provide the lowest (instantaneous) SEP at the destination. For simplicity, the relays are selected by neglecting demodulation errors at the relays. Accordingly, the system reduces to a MIMO scheme with co-located antennas, and a (simplified) closed-form expression of the instantaneous SEP is obtained from Reference 12. The demodulation errors, however, are accounted for after relay selection and for computing the ASEP shown in Figure 1.8. The figure highlights the potential gain of combining DSM with relay selection, both in terms of coding gain and diversity order.
- *Diversity-achieving demodulation for DSM-STBC:* In Figure 1.9, the ASEP of the three demodulators for DSM-STBC is compared. As illustrated in Section 1.3, no direct link is considered in this scenario. The results further confirm the necessity of demodulators that have the potential to mitigate the effects of relay-induced error propagation, in order to obtain distributed diversity gains.

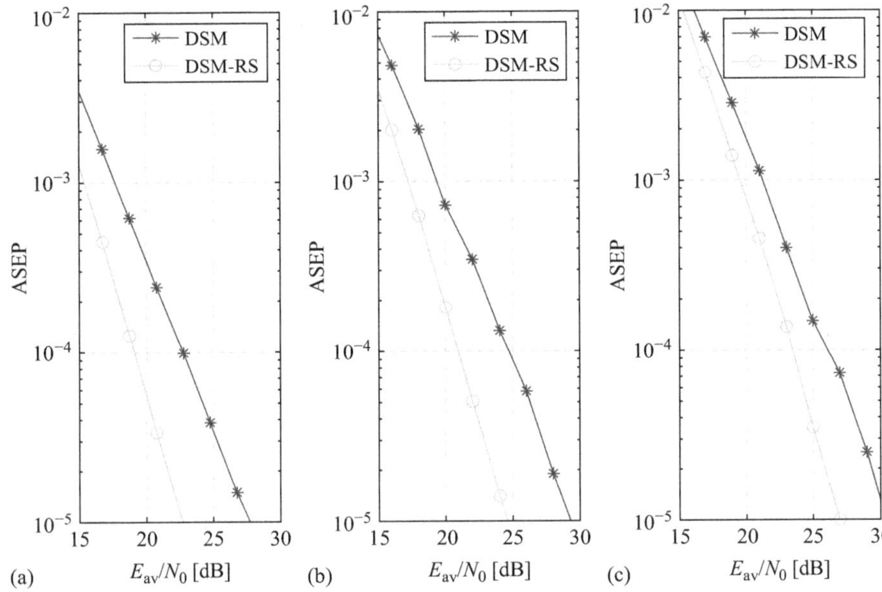

Figure 1.8 *Comparison of DSM with Relay Selection (RS) and without (demodulator in (1.10) where $\widetilde{\mathcal{H}}$ is neglected). (a) $\mathcal{R}_{av} = 1$ bpcu: $N_R = M = 2$ and $N = 2$ for DSM; $N_R = 3$, $M = 2$ and $N = 2$ for DSM-RS. (b) $\mathcal{R}_{av} = 1.5$ bpcu: $N_R = M = 2$ and $N = 4$ for DSM; $N_R = 3$, $M = 2$ and $N = 4$ for DSM-RS. (c) $\mathcal{R}_{av} = 1.5$ bpcu: $N_R = M = 4$ and $N = 2$ for DSM; $N_R = 5$, $M = 4$ and $N = 2$ for DSM-RS. Reproduced with permission from Reference 35. ©2015 IEEE*

More specifically, the error-aware demodulators in (1.32) and (1.36) achieve second-order diversity, whereas the sub-optimal demodulator (referred to as "DSM-STBC – sub-optimal demodulator") [15] achieves only first-order diversity.

1.5 Conclusion

In this chapter, DSM and DSM-STBC, which are two new cooperative wireless protocols based on the concept of SM, have been introduced. The distinguishable feature of DSM lies in allowing the network nodes acting as relays to forward the data of the source, while transmitting their own data. This increases the aggregate throughput. The distinguishable feature of DSM-STBC lies in offering throughput enhancement, by achieving the same order of diversity and having to activate the same number of relay nodes as the conventional distributed Alamouti scheme. Demodulators robust to demodulation errors at the relays have been introduced for both DSM and DSM-STBC, and their achievable performances have been studied. In particular, it has been mathematically proved that, with the aid of the introduced demodulator, DSM

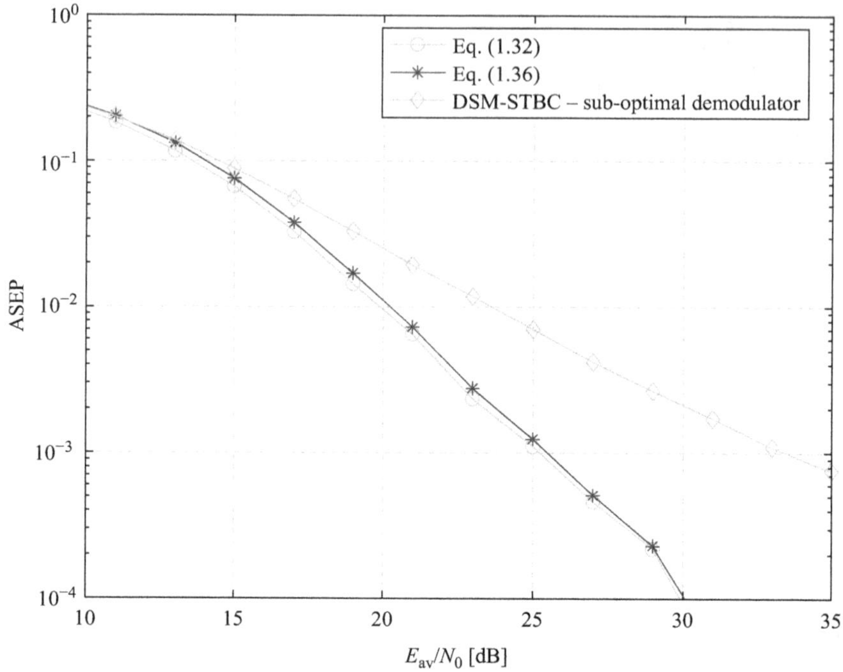

Figure 1.9 Comparison of different demodulators for DSM-STBC (without direct link). $\mathcal{R}'_{av} = 0.6$ bpcu, $N_R = 4$ and $M = 2$

is capable of providing second-order diversity to the data transmitted by the source. Furthermore, with the aid of the introduced demodulator, DSM-STBC, even with no direct link, is also capable of achieving second-order diversity. This is not possible, on the other hand, by using the conventional ML-optimum demodulators for SM and SM-STBC. With the aid of Monte Carlo simulations, the performances of DSM and DSM-STBC have been compared against those of several state-of-the-art relaying protocols, and it has been shown that DSM and DSM-STBC are capable of providing better performances. The price to pay for this performance improvement is an increase of the demodulation complexity at the destination. Hence, the development of low-complexity but diversity-achieving demodulators is an interesting future research direction, in order to leverage the potential of DSM and DSM-STBC.

Acknowledgements

This work was supported in part by the European Commission under the auspices of the FP7-PEOPLE MITN-GREENET project (grant number 264759) and by the Science Foundation of Ireland (grant number 13/CDA/2199).

References

[1] S. M. Alamouti, "A simple transmit diversity technique for wireless communications," *IEEE J. Select. Areas Commun.*, vol. 16, no. 8, pp. 1451–1458, October 1998.

[2] V. Tarokh, H. Jafarkhani, and R. Calderbank, "Space–time block codes from orthogonal designs," *IEEE Trans. Inform. Theory*, vol. 45, no. 5, pp. 1456–1467, July 1999.

[3] J. Mietzner, R. Schober, L. Lampe, W. H. Gerstacker, and P. A. Hoeher, "Multiple-antenna techniques for wireless communications – A comprehensive literature survey," *IEEE Commun. Surveys Tuts.*, vol. 11, no. 2, pp. 87–105, 2nd quarter 2009.

[4] G. Auer, V. Giannini, I. Godor, *et al.*, "Cellular energy efficiency evaluation framework," *IEEE Veh. Technol. Conf. – Spring*, pp. 1–6, May 2011.

[5] Z. Hasan, H. Boostanimehr, and V. K. Bhargava, "Green cellular networks: A survey, some research issues and challenges," *IEEE Commun. Surveys Tuts.*, vol. 13, no. 4, pp. 524–540, November 2011.

[6] T. L. Marzetta, "Noncooperative cellular wireless with unlimited numbers of base station antennas," *IEEE Trans. Wireless Commun.*, vol. 9, no. 11, pp. 3590–3600, November 2010.

[7] E. G. Larsson, F. Tufvesson, O. Edfors, and T. L. Marzetta, "Massive MIMO for next generation wireless systems," *IEEE Commun. Mag.*, vol. 52, no. 2, pp. 186–195, February 2014.

[8] A. Mohammadi and F. M. Ghannouchi, "Single RF frontend MIMO transceivers," *IEEE Commun. Mag.*, vol. 49, no. 12, pp. 104–109, December 2011.

[9] R. Y. Mesleh, H. Haas, S. Sinanovic, C. W. Ahn, and S. Yun, "Spatial modulation," *IEEE Trans. Veh. Technol.*, vol. 57, no. 4, pp. 2228–2241, July 2008.

[10] M. Di Renzo, H. Haas, and P. Grant, "Spatial modulation for multiple-antenna wireless systems – A survey," *IEEE Commun. Mag.*, vol. 49, no. 12, pp. 182–191, December 2011.

[11] M. Di Renzo and H. Haas, "Bit error probability of space shift keying MIMO over multiple-access independent fading channels," *IEEE Trans. Veh. Technol.*, vol. 60, no. 8, pp. 3694–3711, October 2011.

[12] M. Di Renzo and H. Haas, "Bit error probability of SM-MIMO over generalized fading channels," *IEEE Trans. Veh. Technol.*, vol. 61, no. 3, pp. 1124–1144, March 2012.

[13] M. Di Renzo, D. De Leonardis, F. Graziosi, and H. Haas, "Space shift keying (SSK-) MIMO with practical channel estimates," *IEEE Trans. Commun.*, vol. 60, no. 4, pp. 998–1012, April 2012.

[14] A. Stavridis, S. Sinanovic, M. Di Renzo, H. Haas, and P. M. Grant, "An energy saving base station employing spatial modulation," *IEEE Int. Workshop on Computer-Aided Modeling Analysis and Design of Communication Links and Networks*, pp. 1–6, September 2012.

[15] M. Di Renzo and H. Haas, "On transmit-diversity for spatial modulation MIMO: Impact of spatial-constellation diagram and shaping filters at the transmitter," *IEEE Trans. Veh. Technol.*, vol. 62, no. 6, pp. 2507–2531, July 2013.

[16] A. Stavridis, S. Sinanovic, M. Di Renzo, and H. Haas, "Energy evaluation of spatial modulation at a multi-antenna base station," *IEEE Veh. Technol. Conf. – Fall*, pp. 1–5, September 2013.

[17] N. Serafimovski, A. Younis, R. Mesleh, *et al.*, "Practical implementation of spatial modulation," *IEEE Trans. Veh. Technol.*, vol. 62, no. 9, pp. 4511–4523, November 2013.

[18] M. Di Renzo, H. Haas, A. Ghrayeb, S. Sugiura, and L. Hanzo, "Spatial modulation for generalized MIMO: Challenges, opportunities and implementation," *Proc. IEEE*, vol. 102, no. 1, pp. 56–103, January 2014.

[19] P. Yang, M. Di Renzo, Y. Xiao, S. Li, and L. Hanzo, "Design guidelines for spatial modulation," *IEEE Commun. Surveys Tuts.*, vol. 17, no. 1, pp. 6–26, March 2015.

[20] A. Nosratinia, T. E. Hunter, and A. Hedayat, "Cooperative communications in wireless networks," *IEEE Commun. Mag.*, vol. 42, no. 10, pp. 74–80, October 2004.

[21] J. N. Laneman, D. Tse, and G. W. Wornell, "Cooperative diversity in wireless networks: Efficient protocols and outage behavior," *IEEE Trans. Inform. Theory*, vol. 50, no. 12, pp. 3062–3080, December 2004.

[22] Y. Rong, X. Tang, and Y. Hua, "A unified framework for optimizing linear non-regenerative multicarrier MIMO relay communication systems," *IEEE Trans. Sig. Proc.*, vol. 57, no. 12, pp. 4837–4851, December 2009.

[23] W. Xu, X. Dong, and H. Yongming, "Asymptotic achievable rate analysis for selection strategies in amplify-and-forward MIMO two-hop networks with feedback," *IEEE Trans. Veh. Technol.*, vol. 59, no. 7, pp. 3662–3668, September 2010.

[24] W. Xu, X. Dong, and W.-S. Lu, "Joint precoding optimization for multiuser multiantenna relaying downlinks using quadratic programming," *IEEE Trans. Commun.*, vol. 59, no. 5, pp. 1228–1235, May 2011.

[25] N. Serafimovski, S. Sinanovic, M. Di Renzo, and H. Haas, "Dual-hop spatial modulation (Dh-SM)," *IEEE Veh. Technol. Conf. – Spring*, pp. 1–5, May 2011.

[26] P. Som and A. Chockalingam, "Performance analysis of space shift keying in decode-and-forward multi-hop MIMO networks," *IEEE Trans. Veh. Technol.*, vol. 64, no. 1, pp. 132–146, January 2015.

[27] R. Y. Mesleh, S. Ikki, H. Aggoune, and A. Mansour, "Performance analysis of space shift keying (SSK) modulation with multiple cooperative relays," *EURASIP J. Adv. Signal Process.*, vol. 2012, 10 p., September 2012.

[28] P. Yang, B. Zhang, Y. Xiao, *et al.*, "Detect-and-forward relaying aided cooperative spatial modulation for wireless networks," *IEEE Trans. Commun.*, vol. 61, no. 11, pp. 4500–4511, November 2013.

[29] E. Zhou and L. Hao, "Optimal power allocation for space shift keying modulation in a distributed antenna system," *IEEE Commun. Lett.*, vol. 17, no. 9, pp. 1734–1737, September 2013.

[30] A. Stavridis, D. Basnayaka, S. Sinanovic, M. Di Renzo, and H. Haas, "A virtual MIMO dual-hop architecture based on hybrid spatial modulation," *IEEE Trans. Commun.*, vol. 62, no. 9, September 2014.

[31] Y. Yang and S. Aissa, "Information-guided transmission in decode-and-forward relaying systems: Spatial exploitation and throughput enhancement," *IEEE Trans. Wireless Commun.*, vol. 10, no. 7, pp. 2341–2351, July 2011.

[32] S. Sugiura, S. Chen, H. Haas, P. M. Grant, and L. Hanzo, "Coherent versus non-coherent decode-and-forward relaying aided cooperative space–time shift keying," *IEEE Trans. Commun.*, vol. 59, no. 6, pp. 1707–1719, June 2011.

[33] S. Narayanan, M. Di Renzo, F. Graziosi, and H. Haas, "Distributed space shift keying for the uplink of relay-aided cellular networks," *IEEE International Workshop on Computer-Aided Modeling Analysis and Design of Communication Links and Networks*, pp. 130–134, September 2012.

[34] S. Narayanan, M. Di Renzo, F. Graziosi, and H. Haas, "Distributed spatial modulation for relay networks," *IEEE Veh. Technol. Conf. – Fall*, pp. 1–6, September 2013.

[35] S. Narayanan, M. Di Renzo, F. Graziosi, and H. Haas, "Distributed spatial modulation: A cooperative diversity protocol for half-duplex relay-aided wireless networks," *IEEE Trans. Veh. Technol.*, to appear [Online]. doi: 10.1109/TVT.2015.2442754, http://ieeexplore.ieee.org/xpl/login.jsp?tp=&arnumber=7120187&url=http%3A%2F%2Fieeexplore.ieee.org%2Fiel7%2F25%2F4356907%2F07120187.pdf%3Farnumber%3D7120187.

[36] A. Munari, F. Rossetto, and M. Zorzi, "Phoenix: Making cooperation more efficient through network coding in wireless networks," *IEEE Trans. Wireless Commun.*, vol. 8, no. 10, pp. 5248–5258, October 2009.

[37] Z. Ding, I. Krikidis, B. Rong, J. S. Thompson, C. Wang, and S. Yang, "On combating the half-duplex constraint in modern cooperative networks: Protocols and techniques," *IEEE Wireless Commun. Mag.*, vol. 19, no. 6, pp. 20–27, June 2012.

[38] I. Krikidis, H. Suraweera, S. Yang, and K. Berberidis, "Full-duplex relaying over block fading channel: A diversity perspective," *IEEE Trans. Wireless Commun.*, vol. 11, no. 12, pp. 4524–4535, December 2012.

[39] D. Bharadia and S. Katti, "Full duplex MIMO radios," *ACM UNISEX*, pp. 1–13, April 2014.

[40] R. U. Nabar, H. Bolcskei, and F. W. Kneubuhler, "Fading relay channels: Performance limits and space–time signal designs," *IEEE J. Select. Areas Commun.*, vol. 22, no. 6, pp. 1099–1109, August 2004.

[41] K. Azarian, H. El Gamal, and P. Schniter, "On the achievable diversity-multiplexing tradeoff in half-duplex cooperative channels," *IEEE Trans. Inf. Theory*, vol. 51, no. 12, pp. 4152–4172, December 2005.

[42] S. Yang and J.-C. Belfiore, "Optimal space–time codes for the MIMO amplify-and-forward cooperative channel," *IEEE Trans. Inf. Theory*, vol. 53, no. 2, pp. 647–663, February 2007.

[43] S. Yang and J.-C. Belfiore, "Towards the optimal amplify-and-forward cooperative diversity scheme," *IEEE Trans. Inf. Theory*, vol. 53, no. 9, pp. 3114–3126, September 2007.

[44] Y. Fan, C. Wang, J. S. Thompson, and H. V. Poor, "Recovering multiplexing loss through successive relaying using repetition coding," *IEEE Trans. Wireless Commun.*, vol. 6, no. 12, pp. 4484–4493, December 2007.

[45] C. Wang, Y. Fan, J. S. Thompson, Skoglund M., and Poor H. V., "Approaching the optimal diversity-multiplexing tradeoff in a four-node cooperative network," *IEEE Trans. Wireless Commun.*, vol. 9, no. 12, pp. 3690–3700, December 2010.

[46] S. Zhang, S. Liew, and P. Lam, "Physical layer network coding," *ACM MobiCom*, pp. 358–365, September 2006.

[47] S. Katti, S. Gollakota, and D. Katabi, "Embracing wireless interference: Analog network coding," *ACM SIGCOMM*, pp. 397–408, October 2007.

[48] A. Yang, Z. Fei, C. Xing, M. Xiao, J. Yuan, and J. Kuang, "Design of binary network codes for multiuser multiway relay networks," *IEEE Trans. Veh. Technol.*, vol. 62, no. 8, pp. 3786–3799, October 2013.

[49] A. K. Sadek, K. J. R. Liu, and A. Ephremides, "Cognitive multiple access via cooperation: Protocol design and performance analysis," *IEEE Trans. Inf. Theory*, vol. 53, no. 10, pp. 3677–3696, October 2007.

[50] T. Wang and G. B. Giannakis, "Complex field network coding for multiuser cooperative communications," *IEEE J. Select. Areas Commun.*, vol. 26, no. 3, pp. 561–571, April 2008.

[51] E. Fasolo, F. Rossetto, and M. Zorzi, "Network coding meets MIMO," *IEEE Work. Network Coding, Th., App.*, pp. 1–6, January 2008.

[52] F. Rossetto and M. Zorzi, "Mixing network coding and cooperation for reliable wireless communications," *IEEE Wireless Commun. Mag.*, vol. 18, no. 2, pp. 15–21, February 2011.

[53] M. Di Renzo, M. Iezzi, and F. Graziosi, "On diversity order and coding gain of multisource multirelay cooperative wireless networks with binary network coding," *IEEE Trans. Veh. Techol.*, vol. 62, no. 3, pp. 1138–1157, March 2013.

[54] M. Di Renzo, M. Iezzi, and F. Graziosi, "Error performance and diversity analysis of multi-source multi-relay wireless networks with binary network coding and cooperative MRC," *IEEE Trans. Wireless Commun.*, vol. 12, no. 6, pp. 2883–2903, June 2013.

[55] M. Di Renzo, "On the achievable diversity of repetition-based and relay selection network-coded cooperation," *IEEE Trans. Commun.*, Vol. 62, No. 7, pp. 2296–2313, July 2014.

[56] E. G. Larsson and B. R. Vojcic, "Cooperative transmit diversity based on superposition modulation," *IEEE Commun. Lett.*, vol. 9, pp. 778–780, September 2005.

[57] I. Krikidis, "Analysis and optimization issues for superposition modulation in cooperative networks," *IEEE Trans. Veh. Tech.*, vol. 58, pp. 4837–4847, November 2009.

[58] E. Basar, U. Aygolu, E. Panayirci, and H. V. Poor, "Space-time block coded spatial modulation," *IEEE Trans. Commun.*, vol. 59, no. 3, pp. 823–832, March 2011.

[59] D. Yang, C. Xu, L. L. Yang, and L. Hanzo, "Transmit-diversity-assisted space-shift keying for colocated and distributed/cooperative MIMO elements," *IEEE Trans. Veh. Technol.*, vol. 60, no. 6, pp. 2864–2869, July 2011.

[60] R. Rajashekar and K. V. S. Hari, "Modulation diversity for spatial modulation using complex interleaved orthogonal design," *TENCON 2012 – 2012 IEEE Region 10 Conference*, vol. 1, no. 6, pp. 19–22, November 2012.

[61] S. Narayanan, A. Stavridis, M. Di Renzo, F. Graziosi, and H. Haas, "Distributed spatially-modulated space-time-block-codes," *IEEE International Workshop on Computer-Aided Modeling Analysis and Design of Communication Links and Networks*, pp. 159–163, September 2013.

[62] A. Bansal, M. R. Bhatnagar, and A. Hjorungnes, "Decoding and performance bound of demodulate-and-forward based distributed Alamouti STBC," *IEEE Trans. Wireless Commun.*, vol. 12, no. 2, pp. 702–713, February 2013.

[63] S. Yiu, R. Schober, and L. Lampe, "Distributed space-time block coding," *IEEE Trans. Commun.*, vol. 54, no. 07, pp. 1195–1206, July 2006.

[64] X.-T. Vu, M. Di Renzo, and P. Duhamel, "Multiple-access relaying with network coding: Iterative network/channel decoding with imperfect CSI," *EURASIP J. Adv. Signal Process.*, vol. 2013, no. 170, 18 p., November 2013.

[65] X.-T. Vu, P. Duhamel, and M. Di Renzo, "Performance analysis of network coded cooperation with channel coding and adaptive DF-based relaying in Rayleigh fading channels," *IEEE Sig. Process. Lett.*, vol. 22, no. 9, pp. 1354–1358, September 2015.

[66] J. N. Laneman and G. W. Wornell, "Distributed space–time coded protocols for exploiting cooperative diversity in wireless networks," *IEEE Trans. Inform. Theory*, vol. 49, no. 10, pp. 2415–2425, October 2003.

[67] M. K. Simon and M.-S. Alouini, Digital Communication over Fading Channels, John Wiley & Sons, Inc., Hoboken, New Jersey, 2nd ed., 2005.

[68] A. Nasri, R. Schober, and M. Uysal, "Performance and optimization of network-coded cooperative diversity systems," *IEEE Trans. Commun.*, vol. 61, no. 3, pp. 1111–1122, March 2013.

[69] H. Kim and C. Tepedelenlioglu, "Performance bounds on average error rates using the AM–GM inequality and their applications in relay networks," *IEEE Trans. Wireless Commun.*, vol. 11, no. 8, pp. 2986–2995, August 2012.

[70] Z. Wang and G. B. Giannakis, "A simple and general parameterization quantifying performance in fading channels," *IEEE Trans. Commun.*, vol. 51, no. 8, pp. 1389–1398, August 2003.

[71] B. Masnick and J. Wolf, "On linear unequal error protection codes," *IEEE Trans. Inform. Theory*, vol. IT-3, no. 4, pp. 600–607, October 1967.

[72] E. Biglieri, C. Caire, G. Taricco, and J. Ventura-Traveset, "Computing error probabilities over fading channels: A unified approach," *Eur. Trans. Telecommun.*, vol. 9, no. 1, pp. 15–25, February 1998.

[73] A. Younis, S. Sinanovic, M. Di Renzo, R. Y. Mesleh, and H. Haas, "Generalised sphere decoding for spatial modulation," *IEEE Trans. Commun.*, vol. 61, no. 7, pp. 2805–2815, July 2013.

Chapter 2

Relaying for massive MIMO

Gayan Amarasuriya Aruma Baduge and H. Vincent Poor

2.1 Introduction

Wireless technologies have undergone tremendous growth in the past decades. For instance, the data rate of mobile communication services has increased by about 100 times every six to seven years, and this growth translates into about a million-fold capacity increase since the 1950s [1]. Further, wireless data traffic demand in the future is forecast to increase dramatically mainly due to the recent proliferation of data-centric portable wireless devices. It has been predicted that mobile data traffic demand will grow up to about 10.8 exabyte/month by 2016, an 18-fold increase over 2011 [2]. In particular, this unprecedented demand for wireless data traffic is expected to account for a 16–20% increase in energy consumption, and consequently, the global CO_2 emissions are predicted to increase by another 2% [3]. In conventional techniques, the wireless channels are exploited in time and frequency dimensions for increasing data rate, enhancing link-reliability, and reducing energy consumption. Nevertheless, these techniques are approaching their fundamental limits or provide only marginal performance improvements. Hence, innovative wireless network architectures should be designed to satisfy the aforementioned future wireless requirements.

Conventional single-hop multiple-antenna technology enhances the achievable diversity and/or multiplexing gains [4, 5]; however, it may not be capable of extending coverage. The traditional techniques of extending coverage and improving the data rates include network densification, in which more base-stations are deployed to serve in smaller cells. Nevertheless, this approach in effect drives up deployment cost significantly. As a remedy to this problem as well as to circumvent the challenge in next-generation wireless network architecture, cooperative relaying, in which multiple spatially distributed terminals operate cooperatively, has recently emerged [6–8]. Wireless terminals act as relays whenever they forward the signal from one terminal to another by enabling end-to-end data transmission through multiple-hops. Although a terminal sacrifices some of its resources (e.g., bandwidth and battery power) for cooperation, the overall throughput, coverage and reliability metrics can be substantially improved as a result. Cooperative diversity is one of the key benefits of cooperative relay networks [6, 7]. In these networks, spatially distributed terminals engaging in cooperative communication share their antennas and other wireless resources in order to assist a transmission of a particular terminal. Cooperative diversity arises when the aforementioned sharing is used primarily to leverage the spatial diversity available among distributed terminals. Thus, by exploiting the benefits of multiple-hops and cooperative diversity, relay networks enable reliable end-to-end communication between two terminals that are far apart, which is not possible with traditional single-hop networks. Cooperative relaying techniques have already been standardized in IEEE 802.16j – multi-hop relay specification [9].

Although single-antenna relay networks are capable of expanding coverage and enhancing the link reliability, they may solely not be able to increase the data rates. Multiple-antenna or MIMO relay networks are the natural generalization of single-antenna cooperative relaying for multiple-antenna terminals [10–12]. In MIMO relaying, the spatial dimension of the wireless channels is exploited by sending multiple-streams of data signals via multiple-antennas over multiple-relayed hops. Thus, MIMO relay networks can simultaneously reap both the benefits of cooperative relaying and MIMO technology, however, at the expense of increased signal processing complexity. Specifically, MIMO relaying can provide some unique benefits, which would not necessarily be achieved by using either relay or MIMO technologies separately. For instance, MIMO relay networks can provide diversity against large-scale fading or shadowing, which cannot typically be mitigated with collocated antennas. Due to these advantages, MIMO relaying techniques have recently gained significant research interest and are being investigated for emerging next-generation wireless standards [13, 14].

Very large antenna arrays or massive MIMO has been identified as one of the disruptive technology directions for fifth generation (5G) wireless standards [15], as it provides substantial performance gains in terms of spectral efficiency improvements and transmit power reductions [16–21]. The massive MIMO technology utilizes a very large number of antennas (e.g., orders of magnitude more than conventional MIMO) for steering wireless signals among multiple spatially distributed user nodes [17, 21]. Specifically, very large antenna arrays can be exploited to focus radio frequency (RF) signals towards intended directions by using transmit beamforming techniques and thereby minimizing the intra-cell and inter-cell interferences. One of the most prominent benefits of massive MIMO is the unprecedented gains in spectral and energy efficiencies [16–19, 21]. Specifically, the spectral efficiency gain is due to the viability of enabling aggressive spatial multiplexing techniques, while the energy efficiency improvement is obtained by focusing the radiated energy on smaller regions of space with extreme sharpness [17, 21]. In particular, the substantial energy efficiency gains rendered by massive MIMO indeed enable wireless networks operate with drastically lesser output RF power than that of the current wireless technology [21]. Thus, not only the total energy consumption but also the CO_2 emissions can be reduced, thereby paving the way to greener wireless technologies. The other promising advantages of massive MIMO include (i) significant reduction of latency in the air interface, (ii) simplified frequency/time domain scheduling in the medium access control layer, and (iii) improved robustness against unintended interference and intended jamming [21].

One of the key architectural design challenges of massive MIMO systems is to accommodate a very large number of antennas, transceivers and associated power amplifiers in a space-limited base-station. Nevertheless, as a result of the recent research advances in millimetre-wave (mmWave) wireless communication techniques, massive MIMO has increasingly become practically viable. In this context, extremely short wavelengths associated with the mmWave frequency bands (e.g., 28 and 38 GHz) can promisingly be exploited to design massive MIMO transceivers with very large antenna arrays and corresponding RF elements having much smaller physical dimensions [22, 23].

Interestingly, cooperative relaying can be further generalized by employing massive MIMO technology, and the corresponding new wireless architectural design is expected to play a key role in furthering the emerging next-generation wireless standards, which are primarily being developed to meet the unprecedented increasing demand for faster, reliable and seamless wireless connectivity. In particular, the coverage area of massive MIMO systems is expected to be relatively smaller than that of conventional MIMO systems due to the greater path loss incurred by the usage of higher frequency bands. Thus, the relaying techniques for massive MIMO systems can potentially extend the coverage area while retaining its inherent unprecedented spectral and energy efficiency gains. Thus, in this chapter, the system configurations, physical-layer transmission strategies, important performance metrics and practical implementation aspects of relaying for massive MIMO systems are reviewed.

In this chapter, wireless relaying techniques for massive multiple-input multiple-output (MIMO) systems are investigated by presenting the underlying theoretical and practical aspects. One of the key motivational design aspects of future-generation wireless communication standards is to investigate efficient physical-layer techniques for increasing the spectral efficiency of wireless data transmission. To this end, massive MIMO has been identified as one of the distributive technology directions towards achieving the goal of a 5 to 15 times spectral efficiency improvement compared to the existing technology which has been proposed by the International Telecommunication Union as one of the 5G requirements. Furthermore, wireless relaying has widely been investigated as one of the pragmatic and cost-effective approaches for extending the coverage area of wireless networks. In this context, the combination of massive MIMO and relay technologies can potentially be exploited for improving one of the fundamental trade-offs of wireless systems, namely spectral efficiency versus coverage. Thus, in this chapter, system configurations, channel models and performance metrics for multi-user massive MIMO relay networks are described for the primary objective of increasing the achievable area spectral efficiency.

The contributions of this chapter can be summarized as follows: In the introductory section, the key motivation for the emerging concept of massive MIMO relay networks is discussed. Then, several important contributions to the development of both single-hop and multi-hop massive MIMO systems are summarized. The key theoretical as well as practical aspects of realizing massive MIMO relay networks are discussed. To begin with, system configurations for massive MIMO relays are presented by reviewing the underlying techniques for antenna configurations, channel state information acquisition methods and transmission frame structures. Next, the practical channel models including the favourable propagation concepts in massive MIMO are presented. Subsequently, practical channel estimation techniques are investigated. In particular, signal processing aspects for massive MIMO relay configurations are investigated by analysing the corresponding transmission strategies including the signal precoding/detection techniques employed at the base-stations, relay and user nodes. Next, a general end-to-end signal model is developed by considering three specific system configurations for massive MIMO relay networks. Thereby, two fundamental metrics used for characterizing the system performance are presented, and new mathematical tools for investigating the asymptotic performance metrics are discussed. To strengthen the understanding of the aforementioned key technical aspects, a case study involving multi-user massive MIMO relay networks is next presented by developing the corresponding signal models, investigating transmit power scaling laws, and deriving the signal-to-interference-plus-noise ratio (SINR) and achievable sum rate expressions. Finally, by using the analysis and numerical results, several key insights, which can potentially be exploited in designing practical multi-user massive MIMO relay networks, are presented, and their implications for system performance is investigated.

2.1.1 Single-hop massive MIMO systems

In this subsection, several important research contributions to the development of single-hop massive MIMO are summarized. In the seminal work of Marzetta [16], the concept of massive MIMO is introduced for non-cooperative cellular systems by proposing the use of very large antenna arrays in base-stations to serve multiple spatially distributed user nodes. It has been shown in Reference 16 that in the limit of infinitely many base-station antennas, the simple linear precoders and decoders become nearly optimal, and the performance of multi-user MIMO is limited by only the pilot contamination, which is the residual interference caused by the reuse of pilot sequences in adjacent cells. Further, in Reference 16, the asymptotic SINR expressions corresponding to maximal ratio transmission and maximal ratio combining (MRC) for the forward and reverse links, respectively, are derived whenever the number of base-station antennas is allowed to grow without bound. The other key findings of Reference 16 include (i) the throughput and the number of terminals are independent of the cell size, (ii) the spectral efficiency is independent of the bandwidth utilized, and (iii) the transmit power can be made arbitrary small without incurring performance penalties.

Moreover, in References 19 and 24, the transmit power scaling laws and corresponding asymptotic sum rate expressions for the uplink of massive MIMO systems are derived. In Reference 18, the relative energy versus spectral efficiency trade-off of the downlink of massive MIMO systems with linear precoding is studied. Specifically, Ngo *et al.* [19] show that the transmit power at the user nodes of multi-user massive MIMO systems can be scaled down inversely proportional to the number of antennas at the base-station for perfect channel state information (CSI) cases, whereas for imperfect (estimated) CSI cases, the transmit power can only be scaled down inversely proportional to the square-root of the number of base-station antennas. In Reference 20, the performance of multi-cell MIMO systems is investigated for the case in which the number of base-station antennas is not significantly large compared to the number of user nodes. In particular, in Reference 20, a unified performance analysis is presented for multi-cell MIMO systems with linear precoders/detectors for the case in which the number of antennas at the base-station is not extremely large compared to the number of user nodes. Thereby, Hoydis *et al.* [20] derive how many antennas per user node are required to obtain a certain percentage of the ultimate performance limits, which are achieved when there are infinitely many antennas at the base-stations. In Reference 25, the fundamental problem of pilot contamination in multi-cell multi-user massive MIMO systems is investigated, and thereby, the analytical tools for quantifying the corresponding performance degradation are presented. Further, Jose *et al.* [25] propose novel minimum mean square error (MMSE)-based precoding techniques for mitigating the performance degradation due to pilot contamination. In Reference 26, by first revisiting the asymptotic SINR expressions derived in Reference 25, efficient transmission techniques are developed to avoid simultaneous interference from adjacent cells due to reuse of non-orthogonal pilot sequences. In Reference 27, the uplink performance of time-reversal MRC in massive MIMO systems subject to phase noise is investigated, and thereby, it is concluded that the phase noise effectively limits the length of the interval used for data transmission and the number of scheduled users. In References 28 and 29, the detrimental effects of channel ageing (delayed-CSI) are investigated, and thereby, the amount of performance degradation is quantified. In Reference 30, the uplink achievable rates are investigated for massive MIMO systems in Rician fading channels having an arbitrary-rank deterministic component as well as a Rayleigh-distributed random component. Further, it has been shown in Reference 30 that for an increasing Rician K-factor, the uplink rates readily converge to fixed values for both MRC and zero-forcing (ZF) receivers. Choi *et al.* [31] study open-loop and closed-loop downlink training techniques for massive MIMO systems with frequency-division duplexing (FDD).

2.1.2 Dual-hop massive MIMO systems

In this subsection, several key contributions to the development of relaying techniques for massive MIMO are summarized. In References 32 and 33, efficient topologies and transmission strategies are proposed and investigated for multi-user massive MIMO relay networks, in which multiple spatially distributed user nodes transmit their signals to massive MIMO base-stations via intermediate multiple-antenna relay nodes. Specifically, in Reference 32, the uplink performance of multi-user massive MIMO relay networks is investigated by employing ZF-based linear detectors constructed by using the CSI of the cascaded relay channel. Moreover, in Reference 33, the uplink performance is studied by utilizing linear detectors at both relay nodes and base-stations. In Reference 34, system models and antenna configurations are developed for multi-way relay networks with massive MIMO in which spatially distributed user nodes fully/mutually exchange their data through intermediate relay nodes equipped with very large antenna arrays. In particular, in References 32–34, the asymptotic performance is investigated by deriving the asymptotic SINR and spectral efficiency expressions whenever the numbers of antennas at the relay or/and base-station are allowed to grow without bound for both perfect and imperfect CSI cases. In Reference 35, a generic performance analysis framework is developed for investigating the cumulative performance degradation due to practical transmission impairments including co-channel interference, pilot contamination and channel ageing. Notably, in References 32–35, a set of generic analytical tools which can be utilized for efficient evaluation of asymptotic performance of massive MIMO relay networks is developed.

In the context of dual-hop multi-pair relay networks, Suraweera *et al.* [36] investigate the asymptotic sum rate performance of multi-pair one-way relay networks (OWRNs) with very large relay antenna arrays by using three transmit power scaling laws. In Reference 37, massive MIMO techniques for multi-pair full-duplex one-way relaying are developed, and thereby, it has been shown that the loop interference can be effectively cancelled out by using a very large antenna array at the relay for both perfect and imperfect CSI cases. Further, in Reference 38, the results of Reference 36 are extended to multi-pair two-way massive MIMO relay networks by deriving the corresponding asymptotic SINR and spectral efficiency expressions. In particular, Ngo and Larsson [39] investigate the performance of two-way relay networks (TWRNs) consisting of single-antenna user and relay nodes by employing distributed beamforming. It is shown in Reference 39 that for infinitely many relays, distributed beamforming achieves the capacity scaling laws provided by the cut-set upper bound.

2.2 System configurations

In this section, several important relaying configurations for massive MIMO systems are investigated. A schematic diagram of the uplink of a single-cell massive MIMO relay network is depicted in Figure 2.1. All user terminals are single-antenna nodes, whereas both relay and destination nodes are multiple-antenna terminals. Here, multiple spatially distributed user nodes are categorized into two groups as cell-edge user nodes and direct user nodes. Specifically, M cell-edge user nodes communicate with the destination via an intermediate relay, while L direct users communicate with the destination without any relaying. The usability of the direct channel between the user nodes and the destination depends heavily on the propagation characteristics including the path loss and shadowing. The aforementioned system set-up can readily be expanded into multi-cell wireless systems. For example, in Figure 2.2, the corresponding schematic diagram of a multi-cell massive MIMO relay network is depicted.

In particular, the concept of massive MIMO is a natural generalization of the conventional multi-user MIMO. Nevertheless, the former technology is fundamentally different from the latter in several ways. Correspondingly, massive MIMO relaying also departs from conventional MIMO relaying in terms of the same fundamental aspects. These key differences can be further reviewed by means of the antenna configuration, CSI acquisition and signal processing at the various nodes in the network.

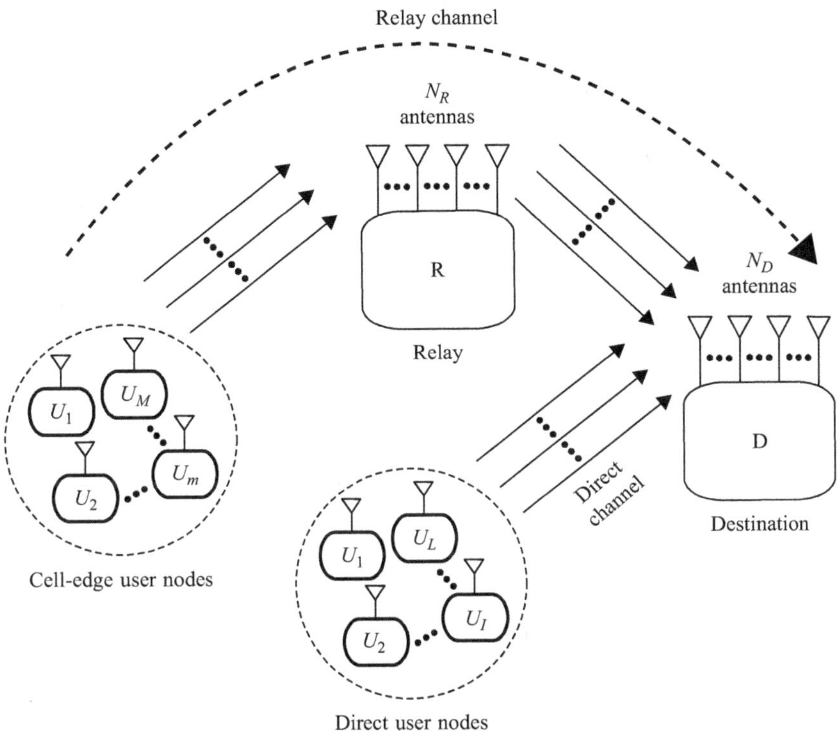

Figure 2.1 A schematic diagram of a single-cell massive MIMO relay network

2.2.1 Antenna configuration

Massive MIMO is fully scalable with respect to the number of antennas at the destination (N_D) which is typically orders of magnitude more than the number of user nodes [16, 21]. In the context of massive MIMO relay networks, the number of antennas at the relay nodes (N_R) stays in between the number of user nodes (M) and N_D. Whenever, both relay and destination nodes are equipped with very large (massive) antenna arrays, N_D can be typically chosen as an integer multiple (α) of N_R, i.e., $N_D = \alpha N_R$ [32]. However, N_R need not be significantly larger than M always. The benefits of using a massive number of antennas at the relay and destination nodes are two-fold. First, a large number of antennas enables not only a higher overall spectral efficiency but also provides a guaranteed quality-of service (QoS) uniformly across spatially distributed user nodes. Second, whenever N_D and N_R are significantly higher than M, the physical phenomenon known as channel hardening comes into play resulting in simpler signal processing at relay and destination nodes. Specifically, the small-scale fading and the underlying frequency dependencies of wireless channels disappear due to channel hardening, and the corresponding performance metrics depend only on the large-scale fading effects. Consequently, simple linear precoders/detectors become nearly optimal. Further, wireless resource allocation and

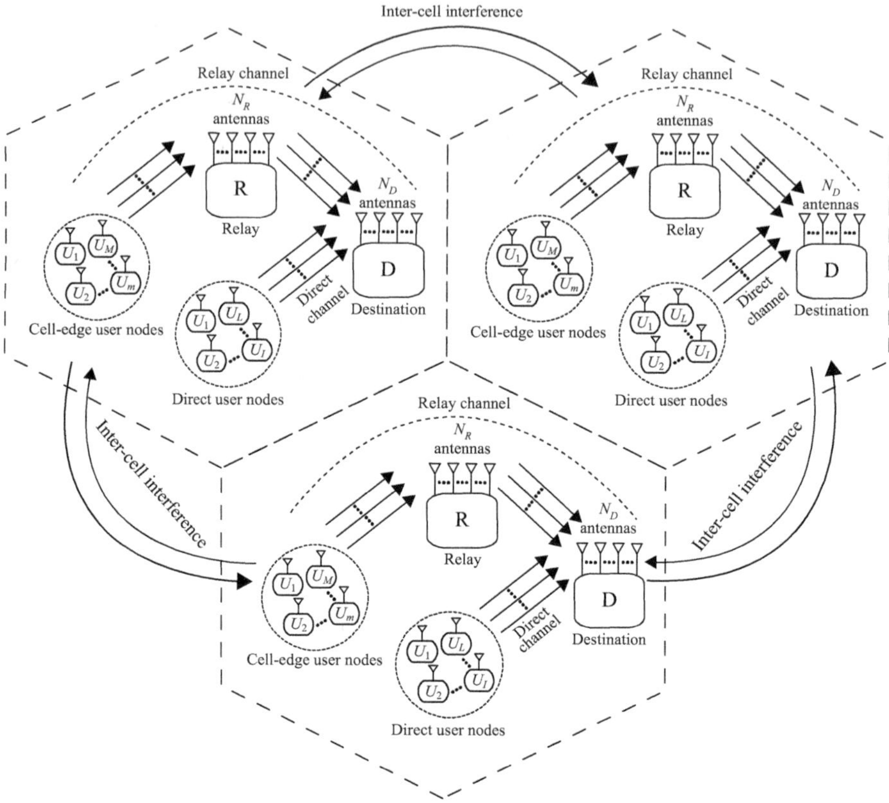

Figure 2.2 A schematic diagram of a multi-cell massive MIMO relay network

scheduling techniques become significantly less complicated, hence, resulting a simpler medium-access control layer. Consequently, the latency in the air interface can be considerably reduced rendering enhanced overall QoS.

2.2.2 Acquisition of channel state information and duplexing modes

The CSI acquisition techniques of massive MIMO strictly depend on the duplexing mode. Let us first consider the single-cell/single-hop massive MIMO sub-network in Figure 2.1. Next, the channel acquisition techniques employed by massive MIMO networks operating in FDD and time-division duplexing (TDD) modes are discussed by emphasizing the lengths of pilot sequences and scalability aspects.

2.2.2.1 CSI acquisition in FDD mode

In FDD mode, the user nodes can estimate the downlink channel by receiving the pilot sequences transmitted by the base-station. Nevertheless, in order to acquire this channel at the base-station, the user nodes typically transmit the estimated channel coefficients back to the base-station through a reliable feedback (control) channel.

Moreover, the uplink channel at the base-station can be estimated by using the pilots transmitted by the user nodes. Thus, in order to estimate the downlink channel, at least N_D pilot symbols are required. Further, L pilots need to be transmitted by the user nodes for estimating the uplink channel. An additional set of N_D channel coefficients are required to be transmitted by the user nodes via the feedback channel for learning the downlink channel at the base-station. In this context, massive MIMO operating in FDD mode is clearly not scalable with respect to the number of base-station antennas [16, 17, 21].

2.2.2.2 CSI acquisition in TDD mode

In TDD mode, the base-station estimates the uplink channel by using the pilots transmitted by the user nodes. Due to the TDD mode of operation, by exploiting the channel reciprocity property, the base-station can use the same uplink channel coefficients for constructing the transmit precoders in the downlink. In this context, the base-station can learn both uplink and downlink channels by using at least L pilots. Thus, the amount of pilot symbols required is indeed independent of the number of base-station antennas, and hence, massive MIMO operating in TDD mode is fully scalable with respect to N_D. According to the aforementioned facts, TDD mode is more practically viable than FDD mode for implementing massive MIMO. It is worth noting that the channel reciprocity assumption heavily depends on the TDD mode of operation. Although the actual wireless channel segments in the uplink and downlink are reciprocal, the overall end-to-end channel responses may not always be reciprocal due to the hardware differences in the transmitter and receiver. Nevertheless, a simple calibration in the hardware chains can be used in practice to make the overall channel responses of downlink and uplink reciprocal.

2.2.2.3 CSI acquisition for relaying in TDD mode

In the context of massive MIMO relay networks, either two individual channel segments (i.e., users-to-relay and relay-to-destination channels) or the cascaded relay (users-to-relay-to-destination) channel need to be estimated for constructing precoders/detectors. Since scalability is one of the fundamental motivations for massive MIMO, only the channel acquisition techniques for massive MIMO relaying operating in TDD mode are discussed. Whenever linear processing is required at both the relay and destination nodes, the users-to-relay and relay-to-destination channels can be separately estimated by using the pilots transmitted by the user and relay nodes, respectively. Moreover, the cascaded relay channel needs to be estimated by using the user pilots for massive MIMO relay networks with linear processing solely at the destination.

2.2.2.4 CSI acquisition for multi-cell massive MIMO

Channel estimation for multi-cell massive MIMO can be more challenging than the single-antenna counterpart. The number of possible orthogonal pilot sequences depends on the coherence time and coherence bandwidth of the corresponding wireless channels, and hence, is limited in practice [16]. Thus, the same orthogonal pilot set must be reused in multiple cells. If these cells with same pilot sequences happen to operate in the same frequency band as well, then the channel estimation at the base-stations becomes imperfect. To be more specific, in the event of estimating its own uplink channel, a base-station inadvertently learns the uplink channels of the adjacent cells, which share the same pilot sequence [16, 25]. Consequently, whenever the base-station transmits data to its own user nodes, it also selectively transmits the same data to user nodes belonging to adjacent cells having the same pilot sequence. Moreover, when combining signals received by its own user nodes, the base-station also coherently combines signals transmitted by the user nodes in the adjacent cells. This phenomenon is typically termed pilot contamination in the massive MIMO literature [16]. It can be shown that the resulting inter-cell interference does not disappear even in the limit of infinitely many base-station antennas, and hence, the corresponding residual interference limits the ultimate performance of massive MIMO systems [16, 25].

2.2.3 Transmission frame structure

In this subsection, the transmission frame structure is defined for massive relay networks. To begin with, several important channel parameters, namely coherence time, coherence bandwidth and coherence interval, are defined as follows: The time duration in which the channel can be approximated by a time-invariant system is termed the coherence time (T_c). The frequency interval in which the channel frequency response can be approximated by a constant is called the coherence bandwidth (B_c). Then the coherence interval can be defined as the length of the time–frequency space over which the channel can be reasonably approximated by a constant in both time and frequency dimensions. Thus, the coherence interval can be quantified to be $\tau_c = T_c B_c$ [16] in sampling instances.

Let us first consider a single-hop massive MIMO system. In TDD mode, the channel coherence interval can be divided into three segments as the pilot phase, uplink phase and downlink phase. The corresponding transmission frame structure for single-hop massive MIMO is depicted in Figure 2.3. Here, τ_p symbols per coherence interval are used as uplink pilots which should be at least the number of user nodes served by the base-station. The next τ_d symbols per coherence interval are used for transmitting downlink payload data by the base-station. The remaining τ_u symbols are used for the transmission of uplink payload data by the user nodes, thereby constituting the following relationship $\tau_c = \tau_p + \tau_d + \tau_u$.

Next, we investigate the transmission frame structure of a massive MIMO relay network described in Figure 2.1. Here the relay is assumed to be a half-duplex terminal, and hence, the spectral efficiency of the relay channel is significantly lower than that

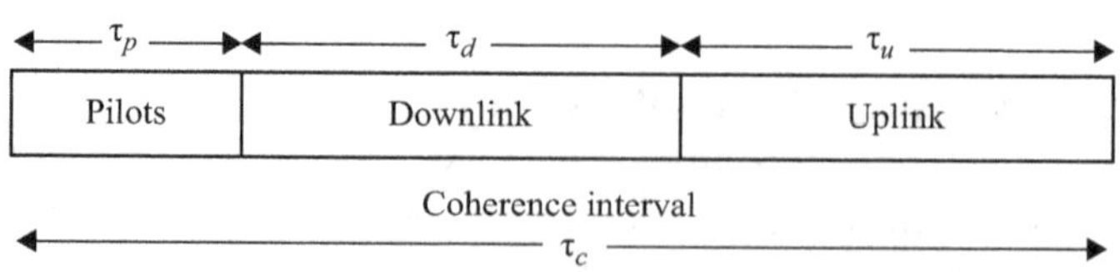

Figure 2.3 Transmission frame structure for single-hop massive MIMO systems

of the direct channel. Specifically, the users-to-relay and relay-to-destination channels are treated as two separate channels having their individual coherence intervals. The transmission sequences and the corresponding frame structure of a massive MIMO relay network are depicted in Figure 2.4. During τ_{p_1} and τ_{u_1} symbol periods in the coherence interval of the first hop, the user nodes transmit pilots and uplink data to the relay, respectively. Similarly, during τ_{p_2} and τ_{u_2} symbol periods in the coherence interval of the second hop, the relay transmits its own pilots and an amplified version of its received uplink data to the destination. Next, the destination uses the next τ_{d_2} symbol periods in the coherence interval of the second hop for transmitting its own downlink data to the relay. Finally, during τ_{d_1} symbol periods in the coherence interval of the first hop, the relay forwards an amplified version of its received downlink data towards the user nodes.

Again, it should be emphasized that the loss of spectral efficiency is due to the half-duplex operation of the relay, and this fact can be clearly seen from the idle symbol periods in the coherence intervals of the first and second hops. It is worth noting that during the idle symbol periods in the coherence interval of the first hop, both the relay and destination fully engage in either reception or transmission. Nevertheless, during the idle symbol periods in the coherence interval of the second hop, the destination does not transmit or receive, and hence, it may remain idle too. By exploiting this void, the overall spectral efficiency of the entire cell can indeed be improved by serving direct user nodes during the idle period in the coherence interval of the second hop. This fact can be clearly understood by looking at the transmission sequence and coherence interval of the direct channel in Figure 2.4.

The aforementioned loss of spectral efficiency can be avoided by using full-duplex relays. In Figure 2.5, the transmission sequences and the corresponding frame

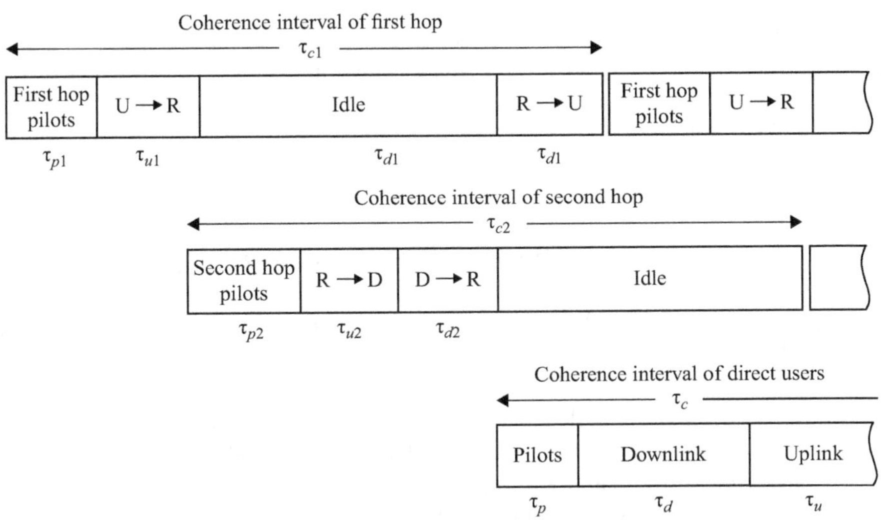

Figure 2.4 Transmission frames for massive MIMO half-duplex relay networks

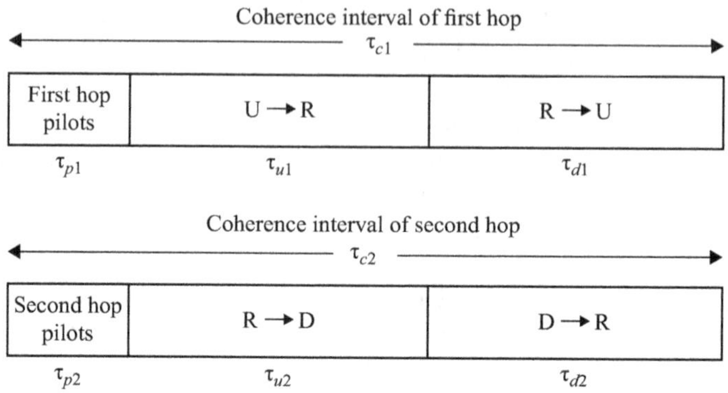

Figure 2.5 Transmission frames for massive MIMO full-duplex relay networks

structure for a massive MIMO full-duplex relay network are depicted. In this context, during τ_{p_1} and τ_{p_2} symbol periods in the coherence intervals of the first and second hops, the user nodes and relay transmit their individual pilots, respectively. The relay is now capable of simultaneously receiving the pilots sent by user nodes as well as transmitting its own pilots towards the destination thanks to its full-duplex mode of operation. During next τ_{u_1} symbol periods in the coherence interval of the first hop, the relay receives its uplink payload data from user nodes. At the same time, the relay forwards an amplified version of its received signal to the destination by spending τ_{u_2} symbol periods in the coherence interval of the second hop. During the next τ_{d_2} symbol periods in the coherence interval of the second hop, the destination transmits its own downlink payload data towards the relay. At the same moment, the relay forwards an amplified version of its received downlink data back to the user nodes by spending the remaining τ_{d_1} symbol periods in the coherence interval of the first hop. It can clearly be seen from Figure 2.5 that by using a full-duplex relay, the coherence intervals of both the first and second hops can be utilized without any idling, and consequently, the spectral efficiency loss incurred due to the half-duplex relay can be completely circumvented.

2.3 Favourable propagation in massive MIMO

In conventional multi-user MIMO, non-linear precoders and detectors have been shown to be optimal in the sense of achievable spectral efficiency [40], whereas the simple linear counterparts are strictly suboptimal. Nevertheless, one of the key design aspects of massive MIMO is to achieve ultra-high spectral efficiencies by simultaneously serving multiple spatially distributed user nodes by exploiting large antenna arrays at the base-stations employing simple linear spatial multiplexing and de-multiplexing transmission techniques [16].

The propagation condition that the massive MIMO systems with linear precoders and detectors asymptotically operate optimally as the number of base-station antennas grows without bound is known as favourable propagation [17]. Let us denote the $1 \times N_D$ channel vector from the kth user node to the base-station by $\mathbf{h}_k$ for $k \in \{1, \cdots K\}$. It can be shown that the channels, $\mathbf{h}_k$s, maximize the information rate whenever $\mathbf{h}_k$s are as different as possible or mutually orthogonal subject to a specific constraint on the channel norms $||\mathbf{h}_k||$, and hence, the ideal favourable propagation conditions can be written as follows [17]:

$$\mathbf{h}_k^H \mathbf{h}_{k'} = 0 \quad \text{for} \quad k \in \{1, \cdots, K\}, \quad k' \in \{1, \cdots, K\} \quad \text{and} \quad k \neq k', \tag{2.1}$$

$$\mathbf{h}_k^H \mathbf{h}_k \neq 0 \quad \text{for} \quad k \in \{1, \cdots, K\}. \tag{2.2}$$

However, the ideal conditions in (2.1) and (2.2) are extremely unlikely to be achieved in practice. Nevertheless, it has been shown that an approximate favourable propagation condition can indeed be achieved in the limit of infinitely many base-station antennas, and the corresponding asymptotically favourable propagation condition can be written as [16, 17, 21, 41]

$$\lim_{N_D \to \infty} \frac{\mathbf{h}_k^H \mathbf{h}_{k'}}{N_D} \to 0 \text{ for } k \in \{1, \cdots, K\}, \ k' \in \{1, \cdots, K\} \text{ and } k \neq k', \tag{2.3}$$

$$\lim_{N_D \to \infty} \frac{\mathbf{h}_k^H \mathbf{h}_k}{N_D} \neq 0 \quad \text{for } k \in \{1, \cdots, K\}. \tag{2.4}$$

Interestingly, the asymptotically favourable propagation conditions in (2.3) and (2.4) seem to be independent of the probability distributions of the underlying channel vectors $\mathbf{h}_k$. In theory, massive MIMO systems operating over practical wireless channels are capable of achieving the aforementioned asymptotically favourable propagation conditions, and consequently, simple linear precoders and decoders used for spatial multiplexing and de-multiplexing become nearly optimal whenever the number of base-station antennas is allowed to grow without bound [16].

2.3.1 Channel modelling

In this subsection, practical propagation channel models are discussed in the context of massive MIMO relay networks. Let us consider the multi-user massive MIMO relay network depicted in Figure 2.6. Here K spatially distributed single-antenna user nodes communicate with an N_D-antenna destination (or a base-station) via an intermediate N_R-antenna relay.

The antenna arrays at the relay and the base-station are assumed to be located in a dense, isotropic scattering environment. The users-to-relay and the relay-to-destination channel matrices are denoted by $\mathbf{F}$ and $\mathbf{G}$, respectively. By assuming that each antenna receives a superimposition of a large number of waveforms originating from independent scatterers, $\mathbf{F}$ can be modelled as a independently distributed random matrix as follows:

$$\mathbf{F} = \tilde{\mathbf{F}} \mathbf{D}_F^{1/2}, \tag{2.5}$$

where $\tilde{\mathbf{F}} \sim \mathcal{CN}_{N_R \times K}(\mathbf{0}_{N_R \times K}, \mathbf{I}_{N_R} \otimes \mathbf{I}_K)$ represents the independently distributed small-scale Rayleigh fading. Further, the $K \times K$ diagonal matrix, $\mathbf{D}_F$, captures the large-scale fading effects such as path loss and shadowing. In this context, the channel

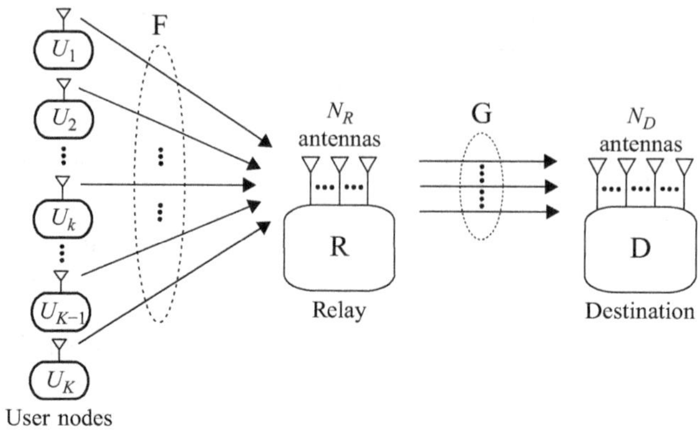

Figure 2.6 A dual-hop multi-user massive MIMO relay network

coefficient between the kth user node and the mth antenna at the relay can be explicitly written as

$$[\mathbf{F}]_{m,k} = \sqrt{\zeta_k}\big[\tilde{\mathbf{F}}\big]_{m,k} \quad \text{for} \quad m \in \{1, \cdots, N_R\}, \tag{2.6}$$

where ζ_k is the kth diagonal element of $\mathbf{D}_F$ and assumed to be independent over $m \in \{1, \cdots, N_R\}$ because the distance between the kth user node and the relay is much larger than the antenna spacing [17, 19]. Further, ζ_k is assumed to be constant over many channel coherence intervals and to be known a priori as it changes very slowly with time [19].

The channel matrix between the relay and destination, $\mathbf{G}$, can also be modelled as an independently distributed random matrix as follows:

$$\mathbf{G} = \tilde{\mathbf{G}}\mathbf{D}_G^{1/2}, \tag{2.7}$$

where $\tilde{\mathbf{G}} \sim CN_{N_D \times N_R}(\mathbf{0}_{N_D \times N_R}, \mathbf{I}_{N_D} \otimes \mathbf{I}_{N_R})$ accounts for the small-scale Rayleigh fading. Again, the diagonal matrix, $\mathbf{D}_G$, captures the large-scale fading including the path loss and shadowing. It is assumed that the antennas at both relay and destination are co-located, and hence, the distance between any antenna pair belonging to the relay and destination is the same. This fact can further be justified by noting that the distance between the relay and destination is far greater than the antenna spacing. Consequently, $\mathbf{D}_G$ can be defined as $\mathbf{D}_G = \eta \mathbf{I}_{N_R}$. Thus, the channel coefficient between the mth relay antenna and nth destination antenna is denoted by

$$[\mathbf{G}]_{n,m} = \sqrt{\eta}\big[\tilde{\mathbf{G}}\big]_{n,m}, \tag{2.8}$$

where $m \in \{1, \cdots, N_R\}$ and $n \in \{1, \cdots, N_D\}$.

Let us investigate the asymptotically favourable propagation conditions for the aforementioned two channel models in (2.5) and (2.7). As the number of relay antennas grows without bound ($N_R \to \infty$), the column vectors of $\tilde{\mathbf{F}}$ become mutually orthogonal, and hence, the favourable propagation condition for $\mathbf{F}$ can be written as [41]

$$\lim_{N_R\to\infty} \frac{[\tilde{\mathbf{F}}]_i^H[\tilde{\mathbf{F}}]_j}{N_R} = \begin{cases} 1, & i=j \\ 0, & i\neq j \end{cases} \quad \text{and} \quad \lim_{N_R\to\infty} \frac{\mathbf{F}^H\mathbf{F}}{N_R} = \mathbf{D}_F. \tag{2.9}$$

Similarly, whenever the number of antennas at the destination grows without bound ($N_D \to \infty$), the column vectors of $\tilde{\mathbf{G}}$ become orthogonal, and consequently, the corresponding favourable propagation condition can be written as

$$\lim_{N_D\to\infty} \frac{[\tilde{\mathbf{G}}]_i^H[\tilde{\mathbf{G}}]_j}{N_D} = \begin{cases} 1, & i=j \\ 0, & i\neq j \end{cases} \quad \text{and} \quad \lim_{N_D\to\infty} \frac{\mathbf{G}^H\mathbf{G}}{N_D} = \eta\mathbf{I}_{N_R}. \tag{2.10}$$

It follows from (2.9) and (2.10) that the pairwise inner products of column vectors of $\mathbf{F}$ and $\mathbf{G}$ asymptotically converge to deterministic constants as $\mathbf{f}^H\mathbf{f}_k/N_R \to \zeta_k$ and $\mathbf{g}_l^H\mathbf{g}_l/N_D \to \eta$ as $N_R \to \infty$ and $N_D \to \infty$, respectively, for $k \in \{1,\cdots,K\}$ and $l \in \{1,\cdots,N_R\}$. This phenomenon is typically known as channel hardening in the massive MIMO literature [17].

For massive MIMO relay networks operating in TDD mode, the destination-to-relay and relay-to-users channels can be defined by using the reciprocity property of wireless channels as $\mathbf{G}^T$ and $\mathbf{F}^T$, respectively.

2.4 Channel estimation

In this section, the MMSE channel estimation of the users-to-relay and relay-to-destination channels is presented. In the massive MIMO relay network in Figure 2.6, the users-to-relay channel is estimated by using the pilots transmitted by K user nodes. The length of the coherence interval of the first hop is τ_{c_1} in symbol durations. Further, the length of the pilot sequence is assumed to be τ_{p_1} in symbol durations. In this context, the received pilot signal at the relay can be written as

$$\mathbf{Y}_p = \sqrt{P_U}\mathbf{F}\mathbf{X}_p + \mathbf{N}_p, \tag{2.11}$$

where $\mathbf{X}_p$ is a $K \times \tau_{p_1}$ pilot symbol matrix and can be defined as $\mathbf{X}_p = \sqrt{\tau_{p_1}}\mathbf{\Psi}^H$. Here, $\mathbf{\Psi}$ is a unitary matrix satisfying $\mathbf{\Psi}^H\mathbf{\Psi} = \mathbf{I}_K$. Moreover, $\mathbf{W}_p$ is the noise matrix at the relay with independent and identically distributed (i.i.d.) Gaussian elements, i.e., $[\mathbf{W}_p]_{i,j} \sim CN(0,1)$. The despreading of the signal received at the relay can be written as

$$\tilde{\mathbf{Y}}_p = \mathbf{Y}_p\mathbf{\Psi} = \sqrt{\tau_{p_1}P_U}\mathbf{F} + \tilde{\mathbf{N}}_p, \tag{2.12}$$

where $\tilde{\mathbf{N}}_p = \mathbf{N}_p\mathbf{\Psi}$ and retains its i.i.d. Gaussian properties as $\mathbf{\Psi}$ is unitary. Next, the MMSE estimate of $\mathbf{F}$ can be derived as [19, 20]

$$\hat{\mathbf{F}} = \frac{1}{\sqrt{P_p}}\mathbf{Y}_p\mathbf{\Psi}\left(\mathbf{D}_F + \mathbf{I}_K/P_{p_1}\right)^{-1}\mathbf{D}_F = \left(\mathbf{F} + \tilde{\mathbf{N}}_F/P_{p_1}\right)\left(\mathbf{D}_F + \mathbf{I}_K/P_{p_1}\right)^{-1}\mathbf{D}_F, \tag{2.13}$$

where P_{p1} is defined as $P_{p1} = \tau_{p1} P_U$. In (2.13), $\tilde{\mathbf{N}}_F \sim \mathcal{CN}_{M\times K}(\mathbf{0}_{M\times K}, \mathbf{I}_M \otimes \mathbf{I}_K)$. Further in (2.12), $\mathbf{F}$ and $\tilde{\mathbf{N}}_F$ are statistically independent. Next, the estimated channel can be decomposed by using the MMSE properties as follows:

$$\hat{\mathbf{F}} = \mathbf{F} + \mathbf{E}_F, \tag{2.14}$$

where $\mathbf{E}_F$ is the zero-mean i.i.d. Gaussian estimation error matrix. Here, $\hat{\mathbf{F}}$ and $\mathbf{E}_F$ are uncorrelated. The variances of the (i,j)th component of $\hat{\mathbf{F}}$ and $\mathbf{E}_F$ can be written as follows:

$$\sigma^2_{[\hat{F}]_{i,j}} = \frac{\tau_{p1} P_U \zeta_k^2}{1 + \tau_{p1} P_U \zeta_k} \quad \text{and} \quad \sigma^2_{[E]_{i,j}} = \frac{\zeta_k}{1 + \tau_{p1} P_U \zeta_k}. \tag{2.15}$$

Next, the relay-to-users channel can be obtained by using the reciprocity properties of the wireless channel as $\hat{\mathbf{F}}^T$.

The relay-to-destination channel can be estimated at the destination by using the pilots transmitted by the relay antennas. To this end, by using techniques similar to those in (2.11)–(2.13), the MMSE estimate of the relay-to-destination channel can be derived as follows:

$$\hat{\mathbf{G}} = \left(\mathbf{G} + \tilde{\mathbf{N}}_G/P_{p2}\right)\left(\mathbf{D}_G + \mathbf{I}_K/P_{p2}\right)^{-1}\mathbf{D}_G \quad \text{and} \quad \hat{\mathbf{G}} = \mathbf{G} + \mathbf{E}_G, \tag{2.16}$$

where $P_{p2} = \tau_{p2} P_R$ and $\mathbf{E}_G$ is the estimation error matrix, distributed as zero-mean i.i.d. Gaussian. Further, $\mathbf{G}$ and $\mathbf{E}_G$ are statistically independent. Similarly, the variances of the (i,j)th element of $\hat{\mathbf{G}}$ and $\mathbf{E}_G$ can be written as

$$\sigma^2_{[\hat{G}]_{i,j}} = \frac{\tau_{p2} P_U \eta^2}{1 + \tau_{p2} P_U \eta} \quad \text{and} \quad \sigma^2_{[E]_{i,j}} = \frac{\eta}{1 + \tau_{p2} P_U \eta}. \tag{2.17}$$

Again, the destination-to-relay channel can be obtained by using the reciprocity property as $\hat{\mathbf{G}}^T$.

In particular, the cascaded relay channel (or users-to-relay-to-destination channel) can directly be estimated by using the pilots transmitted by the user nodes [42].

2.5 Massive MIMO configurations and signal processing aspects

In this section, massive MIMO configurations for multi-user relay networks are discussed by reviewing the signal processing aspects at the relay and destination. To this end, a general end-to-end signalling model is presented by explicitly developing expressions for received and transmitted signals at the relay and destination over fading channels. Next, the signal processing steps at the relay and destination are discussed by presenting relay amplification techniques and corresponding precoder/decoder designs at the relay and destination.

2.5.1 General end-to-end signal model

In this subsection, a general signal model for the uplink of multi-user massive MIMO relay networks operating over fading channels is presented. In this context, we revisit the schematic diagram of the multi-user relay network depicted in Figure 2.6 for developing the corresponding signal model for the users-to-destination or uplink data transmission. Here, K spatially distributed user nodes transmit their data to the destination via an intermediate relay. The processing matrices at the relay and destination are denoted by $\mathbf{W}_R$ and $\mathbf{W}_D$, respectively. Further, the amplification factor at the relay is denoted by β_R.

First, the user nodes transmit their signal vector, $\mathbf{x}_U$, towards the relay. Then the received signal at the relay can be written as

$$\mathbf{y}_R = \sqrt{P_U}\mathbf{F}\mathbf{x}_U + \mathbf{n}_R, \tag{2.18}$$

where P_U is the transmit power at each user node and the user signal vector $\mathbf{x}_U$ satisfies $\mathcal{E}[\mathbf{x}_U\mathbf{x}_U^H] = \mathbf{I}_k$, where $\mathcal{E}[\cdot]$ is the expectation operator. Further, $\mathbf{F}$ is the channel matrix between the user nodes and relay, and $\mathbf{n}_R$ is an additive white Gaussian noise (AWGN) vector at the relay satisfying $\mathcal{E}[\mathbf{n}_R\mathbf{n}_R^H] = \sigma_R^2\mathbf{I}_K$.

Next, the relay processes its received signal by applying the signal processing matrix $\mathbf{W}_R$. Then, the relay amplifies its received signal by using an amplification factor (β_R) designed to constrain the instantaneous transmit power. In this context, the transmitted signal at the relay can be written as

$$\mathbf{y}_{R^{tx}} = \beta_R\mathbf{W}_R\mathbf{y}_R, \tag{2.19}$$

where $\mathbf{y}_R$ is defined in (2.18). Here, the relay amplification factor, β_R, can be written as follows:

$$\beta_R = \sqrt{\frac{P_R}{P_U\mathbf{Tr}(\mathbf{W}_R\mathbf{F}\mathbf{F}^H\mathbf{W}_R^H) + \sigma_R^2\mathbf{Tr}(\mathbf{W}_R\mathbf{W}_R^H)}}, \tag{2.20}$$

where P_R is the transmit power at the relay.

Finally, the relay forwards the processed-and-amplified signal towards the destination. The post-processing received signal at the destination can then be written as follows:

$$\mathbf{y}_D = \mathbf{W}_D(G\mathbf{y}_{R^{tx}} + \mathbf{n}_D), \tag{2.21}$$

where the relay transmitted signal $\mathbf{y}_{R^{tx}}$ is defined in (2.19). Here, $\mathbf{W}_D$ is the signal processing matrix at the destination, and $\mathbf{G}$ is the channel matrix between the relay and destination. Further, $\mathbf{n}_D$ is AWGN at the destination satisfying $\mathcal{E}[\mathbf{n}_D\mathbf{n}_D^H] = \sigma_D^2\mathbf{I}_K$. By substituting (2.19) into (2.21), the received signal vector at the destination can be expanded as follows:

$$\mathbf{y}_D = \sqrt{P_U}\beta_R\mathbf{W}_D\mathbf{G}\mathbf{W}_R\mathbf{F}\mathbf{x}_U + \beta_R\mathbf{W}_D\mathbf{G}\mathbf{W}_R\mathbf{n}_R + \mathbf{W}_D\mathbf{n}_D. \tag{2.22}$$

2.5.2 Signal processing aspects

Let us next review the signal processing aspects at the relay and destination. Specifically, the signal processing matrices $\mathbf{W}_R$ and $\mathbf{W}_D$ need to be designed to optimize the fundamental trade-off between the achievable performance and implementation complexity. Nevertheless, simple linear precoders and detectors have been shown to

Table 2.1 Signal processing matrices for case-1

Case-1	$\mathbf{W}_R$	$\mathbf{W}_D$	β_R
MRC	$\mathbf{I}_{N_R}$	$(\mathbf{GF})^H$	$\sqrt{\dfrac{P_R}{P_U \mathrm{Tr}(\mathbf{F}^H\mathbf{F}) + N_R\sigma_R^2}}$
ZF	$\mathbf{I}_{N_R}$	$((\mathbf{GF})^H\mathbf{GF})^{-1}(\mathbf{GF})^H$	$\sqrt{\dfrac{P_R}{P_U \mathrm{Tr}(\mathbf{F}^H\mathbf{F}) + N_R\sigma_R^2}}$

Table 2.2 Signal processing matrices for case-2

Case-2	$\mathbf{W}_R$	$\mathbf{W}_D$	β_R
MRC	$\mathbf{G}^H\mathbf{F}^H$	$\mathbf{I}_{N_D}$	$\sqrt{\dfrac{P_R}{P_U \mathrm{Tr}((\mathbf{F}^H\mathbf{F})^2\mathbf{GG}^H) + \sigma_R^2 \mathrm{Tr}(\mathbf{F}^H\mathbf{FGG}^H)}}$
ZF	$\mathbf{G}^H(\mathbf{GG}^H)^{-1}(\mathbf{F}^H\mathbf{F})^{-1}\mathbf{F}^H$	$\mathbf{I}_{N_D}$	$\sqrt{\dfrac{P_R}{P_U \mathrm{Tr}((\mathbf{GG}^H)^{-1}) + \sigma_R^2 \mathrm{Tr}((\mathbf{F}^H\mathbf{F})^{-1}(\mathbf{GG}^H)^{-1})}}$

be nearly optimal for massive MIMO [16, 17, 19, 21]. In this context, three specific precoder/detector configurations are presented for the multi-user massive MIMO relay network shown in Figure 2.6.

2.5.2.1 Case-1: Signal processing at destination

In the first case, linear detectors are utilized only at the destination. In particular, the received signal at the relay is amplified transparently by using the amplification factor derived in (2.20). It is worth noting that neither a linear detector nor a linear precoder is employed for signal processing at the relay. In this context, the corresponding amplification factor and signal processing matrices at the relay and destination can be tabulated as in Table 2.1. Here, the linear detectors at the destination can be constructed for the cascaded users-to-relay channel by following MRC and ZF strategies. In order to employ these linear detectors, the number of user nodes and numbers of relay and destination antennas should satisfy the constraint $N_D > N_R > K$.

2.5.2.2 Case-2: Signal processing at relay

In the second case, signal processing is performed only at the relay. Specifically, the relay utilizes a cascaded linear detector–precoder. Again, the amplification factor at the relay is designed to constrain its instantaneous transmit power. In this case, it is not required to apply any linear detector at the destination. These signal processing matrices at the relay can again be constructed by using MRC and ZF strategies as shown in Table 2.2. In this case, the constraint that the number of user nodes

Table 2.3 Signal processing matrices for case-3

Case-3	$\mathbf{W}_R$	$\mathbf{W}_D$	β_R
MRC	$\mathbf{F}^H$	$\mathbf{G}^H$	$\sqrt{\dfrac{P_R}{P_U \text{Tr}((\mathbf{F}^H\mathbf{F})^2) + \sigma_R^2 \text{Tr}(\mathbf{F}^H\mathbf{F})}}$
ZF	$(\mathbf{F}^H\mathbf{F})^{-1}\mathbf{F}^H$	$(\mathbf{G}^H\mathbf{G})^{-1}\mathbf{G}^H$	$\sqrt{\dfrac{P_R}{P_U K + \sigma_R^2 \text{Tr}((\mathbf{F}^H\mathbf{F})^{-1})}}$

and the numbers of relay and destination antennas need to satisfy can be written as $N_R > N_D > K$.

2.5.2.3 Case-3: Signal processing at relay and destination

In the third case, the signal processing is performed at both the relay and the destination. The relay applies a linear detector and an amplification factor constructed to constrain its instantaneous transmit power. Further, a linear detector is utilized at the destination. These linear detectors at the relay and destination can again be constructed by using MRC and ZF strategies as shown in Table 2.3. Specifically, the constraint that the number of user nodes and numbers of antennas at the relay and destination need to satisfy can be written as $N_D > N_R > K$.

2.6 Fundamental performance metrics

In this section, two important metrics for quantifying the achievable performance of multi-user massive MIMO relay networks are presented by using the generalized end-to-end signal model in Section 2.5.1.

One of the fundamental performance metrics that can be employed for evaluating the multi-user MIMO systems is the SINR. In particular, the SINR is defined as the ratio between the desired signal power and the total of effective interference and noise powers.

By using the received signal vector at the destination in (2.22), the kth data substream can be extracted as follows:

$$[\mathbf{y}_D]_k = \sqrt{P_U}\beta_R \left([\mathbf{W}_D^T]_k\right)^T \mathbf{P}_k x_k + \sqrt{P_U}\beta_R \sum_{i=1, i\neq k}^{K} \left([\mathbf{W}_D^T]_k\right)^T \mathbf{P}_i x_i + \beta_R \left([\mathbf{W}_D^T]_k\right)^T \mathbf{G}\mathbf{W}_R\mathbf{n}_R + \left([\mathbf{W}_D^T]_k\right)^T \mathbf{n}_D, \quad (2.23)$$

where $[\mathbf{W}_D^T]_k$ is the kth row of $\mathbf{W}_D$, $[\mathbf{P}]_k$ is the kth column of the matrix product $\mathbf{P} = \mathbf{G}\mathbf{W}_R\mathbf{F}$, and x_k is the kth signal component of the vector $\mathbf{x}_U$. In (2.23), the first term represents the desired signal component, while the second term accounts for the inter-stream interference. Moreover, the third and fourth terms of (2.23) represent the effective noise components.

By using (2.23), the end-to-end SINR of the kth data substream can be derived by constructing the ratio between the desired signal power and the total of interference and noise powers of (2.23) as follows:

$$\gamma_k = \frac{P_R\beta_R^2 \left| \left([\mathbf{W}_D^T]_k \right)^T \mathbf{P}_k \right|^2}{P_R\beta_R^2 \sum_{i=1, i\neq k}^{K} \left| \left([\mathbf{W}_D^T]_k \right)^T \mathbf{P}_i \right|^2 + \sigma_R^2\beta_R^2 \left\| \left([\mathbf{W}_D^T]_k \right)^T \mathbf{G}\mathbf{W}_R \right\|^2 + \sigma_D^2 \left\| \left([\mathbf{W}_D^T]_k \right)^T \right\|^2}. \tag{2.24}$$

The sum rate is another fundamental metric used for characterizing the achievable performance of multi-user wireless networks. The sum rate can be defined as the sum of the achievable rate of each data substream at the destination as follows:

$$R = \frac{1}{2} \sum_{k=1}^{K} \log_2 \left(1 + \gamma_k \right), \tag{2.25}$$

where γ_k is the end-to-end SINR of the kth data substream and is defined in (2.24).

In the subsequent sections, a case study of a multi-user massive MIMO relay network is presented to strengthen our understanding of the aforementioned fundamental performance metrics in (2.24) and (2.25).

2.7 Case study

In this section, a multi-user massive MIMO deployment scenario is investigated by using the basic concepts and techniques developed in the previous sections. In particular, our objective is to characterize the achievable performance by deriving several important measures, including the end-to-end SINR and achievable sum rate.

Again, we revisit the multi-user relay network depicted in Figure 2.6. In order to strengthen the understanding of signal processing aspects, we adopt case-1 discussed in Section 2.5.2.1. In this context, the MRC-based linear detector given in Table 2.1 is utilized at the destination. Hence, the corresponding relay amplification factor and signal processing matrices can be explicitly written as follows:

$$\beta_R = \sqrt{\frac{P_R}{P_U \mathrm{Tr}(\mathbf{F}^H\mathbf{F}) + N_R\sigma_R^2}}, \tag{2.26}$$

$$\mathbf{W}_D = (\mathbf{G}\mathbf{F})^H \quad \text{and} \quad \mathbf{W}_R = \mathbf{I}_{N_R}. \tag{2.27}$$

By substituting (2.26) and (2.27) into (2.22), the received signal vector at the destination can be written as

$$\mathbf{y}_D = \sqrt{P_U}\beta_R(\mathbf{G}\mathbf{F})^H\mathbf{G}\mathbf{F}\mathbf{x}_U + \beta_R(\mathbf{G}\mathbf{F})^H\mathbf{G}\mathbf{n}_R + (\mathbf{G}\mathbf{F})^H\mathbf{n}_D. \tag{2.28}$$

Next, the kth signal component or data substream at the destination can be extracted by using (2.28) as

$$[\mathbf{y}_D]_k = \sqrt{P_U}\beta_R \mathbf{f}_k^H \mathbf{G}^H \mathbf{G}\mathbf{f}_k x_k + \sqrt{P_U}\beta_R \sum_{i=1,i\neq k}^{K} \mathbf{f}_k^H \mathbf{G}^H \mathbf{G}\mathbf{f}_i x_i + \beta_R \mathbf{f}_k^H \mathbf{G}^H \mathbf{G}\mathbf{n}_R + \mathbf{f}_k^H \mathbf{G}^H \mathbf{n}_D, \tag{2.29}$$

where $\mathbf{f}_m$ is the mth column of the matrix $\mathbf{F}$, and x_m is the mth element of the vector $\mathbf{x}_U$. The first term of (2.29) represents the desired signal component, while the second term accounts for the inter-stream interference terms. Further, the last two terms of (2.29) represent the effective noise components.

By using (2.29), the end-to-end SINR of the kth data substream at the destination can be derived as follows:

$$\gamma_k = \frac{P_U \beta_R^2 \left|\mathbf{f}_k^H \mathbf{G}^H \mathbf{G}\mathbf{f}_k\right|^2}{P_U \beta_R^2 \sum_{i=1,i\neq k}^{K} \left|\left|\mathbf{f}_k^H \mathbf{G}^H \mathbf{G}\mathbf{f}_i\right|\right|^2 + \sigma_R^2 \beta_R^2 \left|\left|\mathbf{f}_k^H \mathbf{G}^H \mathbf{G}\right|\right|^2 + \sigma_D^2 \left|\left|\mathbf{f}_k^H \mathbf{G}^H\right|\right|^2}. \tag{2.30}$$

The term in the numerator of (2.30) accounts for desired signal power. The first term in the denominator represents the inter-stream interference power. The second and third terms in the denominator account for the effective noise powers.

In multi-user massive MIMO relay networks, the system performance needs to be characterized for very large numbers of relay and destination antennas. To this end, we first let the number of antennas grow without bound while allowing the transmit power at the nodes to scale down according to a certain proportion to the number of antennas. In the next three subsections, our objective is to derive three asymptotic SINR expressions corresponding to three specific transmit power scaling laws at the user and relay nodes.

2.7.1 Transmit power scaling at user nodes

In this subsection, an asymptotic SINR expression is derived for transmit power scaling at the user nodes whenever the numbers of antennas at the relay and destination grow without bound while keeping a fixed ratio. To be more specific, the transmit power at the user nodes is scaled inversely proportional to number of relay antennas as $P_U = E_U/N_R$ where E_U is a power constant defined to maintain the product $P_U N_R$ fixed. Next, by letting the numbers of antennas at the relay and destination grow as $N_R \to \infty$, $N_D \to \infty$ and $\alpha = N_D/N_R$, an asymptotic SINR expression can be derived as follows:

To begin with, by substituting β_R in (2.26) and $P_U = E_U/N_R$ into (2.30), the desired signal power can be written in an alternative form as

$$P_S = \frac{\frac{P_R E_U}{N_R}\left(\mathbf{f}_k^H \mathbf{G}^H \mathbf{G}\mathbf{f}_k \mathbf{f}_k^H \mathbf{G}^H \mathbf{G}\mathbf{f}_k\right)}{\frac{E_U}{N_R}\mathrm{Tr}\left(\mathbf{F}^H \mathbf{F}\right) + \sigma_R^2 N_R}. \tag{2.31}$$

The desired signal power in (2.31) can be written in an alternative form as follows:

$$\frac{P_S}{N_D^2} = \frac{P_R E_U \left(\dfrac{\mathbf{f}_k^H \mathbf{G}^H \mathbf{G} \mathbf{f}_k \mathbf{f}_k^H \mathbf{G}^H \mathbf{G} \mathbf{f}_k}{N_R^2 N_D^2} \right)}{\dfrac{E_U}{N_R} \mathrm{Tr} \left(\dfrac{\mathbf{F}^H \mathbf{F}}{N_R} \right) + \sigma_R^2}. \tag{2.32}$$

Next, by using the identities in (2.9) and (2.10), whenever the numbers of relay and destination antennas are allowed to grow without bound, it can readily be shown that

$$\lim_{\substack{N_R \to \infty,\ N_D \to \infty \\ N_D/N_R = \alpha}} \left(\frac{\mathbf{f}_k^H \mathbf{G}^H \mathbf{G} \mathbf{f}_k \mathbf{f}_k^H \mathbf{G}^H \mathbf{G} \mathbf{f}_k}{N_R^2 N_D^2} \right) = \eta^2 \zeta_k^2 \quad \text{and} \quad \lim_{N_R \to \infty} \mathrm{Tr} \left(\frac{\mathbf{F}^H \mathbf{F}}{N_R} \right) = \mathrm{Tr}\left(\mathbf{D}_F \right). \tag{2.33}$$

By substituting (2.33) into (2.32), the asymptotic value of P_S/N_D^2 can be derived as

$$\lim_{\substack{N_R \to \infty,\ N_D \to \infty \\ N_D/N_R = \alpha}} \left(\frac{P_S}{N_D^2} \right) = \frac{P_R E_U \eta^2 \zeta_k^2}{\sigma_R^2}. \tag{2.34}$$

By substituting $P_U = E_U/N_R$ into (2.30), the inter-stream interference power term can be written as follows:

$$P_{I_1} = \frac{\dfrac{P_R E_U}{N_R} \displaystyle\sum_{i=1,1 \neq k}^{K} \left(\mathbf{f}_k^H \mathbf{G}^H \mathbf{G} \mathbf{f}_k \mathbf{f}_k^H \mathbf{G}^H \mathbf{G} \mathbf{f}_k \right)}{\dfrac{E_U}{N_R} \mathrm{Tr} \left(\mathbf{F}^H \mathbf{F} \right) + \sigma_R^2 N_R}. \tag{2.35}$$

By using steps similar to those in (2.32), the term P_{I_1}/N_D^2 can be written as

$$\frac{P_{I_1}}{N_D^2} = \frac{P_R E_U \displaystyle\sum_{i=1,1 \neq k}^{K} \left(\frac{\mathbf{f}_k^H \mathbf{G}^H \mathbf{G} \mathbf{f}_i \mathbf{f}_i^H \mathbf{G}^H \mathbf{G} \mathbf{f}_k}{N_R^2 N_D^2} \right)}{\dfrac{E_U}{N_R} \mathrm{Tr} \left(\dfrac{\mathbf{F}^H \mathbf{F}}{N_R} \right) + \sigma_R^2}. \tag{2.36}$$

Again, by using the identities in (2.9) and (2.10), it can be shown that

$$\lim_{\substack{N_R \to \infty,\ N_D \to \infty \\ N_D/N_R = \alpha}} \sum_{i=1,1 \neq k}^{K} \left(\frac{\mathbf{f}_k^H \mathbf{G}^H \mathbf{G} \mathbf{f}_i \mathbf{f}_i^H \mathbf{G}^H \mathbf{G} \mathbf{f}_k}{N_R^2 N_D^2} \right) = 0. \tag{2.37}$$

By substituting (2.37), one can readily show that the contribution of the term P_{I_1}/N_D^2 vanishes in the limit of infinitely many relay and destination antennas, and this fact can be formally written as

$$\lim_{\substack{N_R \to \infty,\ N_D \to \infty \\ N_D/N_R = \alpha}} \left(\frac{P_{I_1}}{N_D^2} \right) = 0. \tag{2.38}$$

Next, by substituting β_R in (2.26) and $P_U = E_U/N_R$ into (2.30), the second power term in the denominator in (2.30) can be written as

$$P_{I_2} = \frac{P_R\sigma_R^2\left(\mathbf{f}_k^H\mathbf{G}^H\mathbf{G}\mathbf{G}^H\mathbf{G}\mathbf{f}_k\right)}{\dfrac{E_U}{N_R}\mathrm{Tr}\left(\mathbf{F}^H\mathbf{F}\right)+\sigma_R^2N_R}. \tag{2.39}$$

The expression in (2.39) can then be re-arranged as follows:

$$\frac{P_{I_2}}{N_D^2} = \frac{P_R\sigma_R^2\left(\dfrac{\mathbf{f}_k^H\mathbf{G}^H\mathbf{G}\mathbf{G}^H\mathbf{G}\mathbf{f}_k}{N_RN_D^2}\right)}{\dfrac{E_U}{N_R}\mathrm{Tr}\left(\dfrac{\mathbf{F}^H\mathbf{F}}{N_R}\right)+\sigma_R^2}. \tag{2.40}$$

By using the identities in (2.9) and (2.10), one can readily show that

$$\lim_{\substack{N_R\to\infty,\ N_D\to\infty\\ N_D/N_R=\alpha}}\left(\frac{\mathbf{f}_k^H\mathbf{G}^H\mathbf{G}\mathbf{G}^H\mathbf{G}\mathbf{f}_k}{N_RN_D^2}\right) = \eta^2\zeta_k. \tag{2.41}$$

By substituting (2.41) into (2.40), the asymptotic expression for P_{I_2}/N_D^2 can be derived as

$$\lim_{\substack{N_R\to\infty,\ N_D\to\infty\\ N_D/N_R=\alpha}}\left(\frac{P_{I_2}}{N_D^2}\right) = P_R\eta^2\zeta_k. \tag{2.42}$$

The third power term in the denominator in (2.30) can be explicitly written as

$$P_{I_3} = \sigma_D^2\left(\mathbf{f}_k^H\mathbf{G}^H\mathbf{G}\mathbf{f}_k\right). \tag{2.43}$$

Next, the quantity P_{I_3}/N_D^2 can be written as

$$\frac{P_{I_3}}{N_D^2} = \sigma_D^2\left(\frac{N_R}{N_D}\right)\left(\frac{\mathbf{f}_k^H\mathbf{G}^H\mathbf{G}\mathbf{f}_k}{N_RN_D}\right). \tag{2.44}$$

By using the identities in (2.9) and (2.10), it can be shown that

$$\lim_{\substack{N_R\to\infty,\ N_D\to\infty\\ N_D/N_R=\alpha}}\left(\frac{\mathbf{f}_k^H\mathbf{G}^H\mathbf{G}\mathbf{f}_k}{N_RN_D}\right) = \eta\zeta_k. \tag{2.45}$$

By substituting (2.45) into (2.44), the asymptotic expression for P_{I_3}/N_D^2 can be derived as follows:

$$\lim_{\substack{N_R\to\infty,\ N_D\to\infty\\ N_D/N_R=\alpha}}\left(\frac{P_{I_3}}{N_D^2}\right) = \frac{\sigma_D^2\eta\zeta_k}{\alpha}. \tag{2.46}$$

Now, by using (2.31), (2.35), (2.39) and (2.43), the end-to-end SINR in (2.30) can be written alternatively as

$$\gamma_k = \frac{P_S}{N_D^2}\left(\frac{P_{I_1}}{N_D^2}+\frac{P_{I_2}}{N_D^2}+\frac{P_{I_3}}{N_D^2}\right)^{-1}, \tag{2.47}$$

where the terms P_S/N_D^2 and P_{I_1}/N_D for $i \in \{1,2,3\}$ are defined in (2.32), (2.36), (2.40) and (2.44), respectively. By first letting $N_R \to \infty$ and $N_D \to \infty$ while keeping a fixed ratio $\alpha = N_D/N_R$, and then by using (2.34), (2.38), (2.46) and (2.46), the asymptotic SINR expression for transmit power scaling at the user nodes can be derived as follows:

$$\gamma_k^{\infty} = \lim_{\substack{N_R\to\infty,\ N_D\to\infty \\ N_D/N_R=\alpha}} \left[\frac{P_S}{N_D^2}\left(\frac{P_{I_1}}{N_D^2}+\frac{P_{I_2}}{N_D^2}+\frac{P_{I_3}}{N_D^2}\right)^{-1}\right] = \frac{E_U\eta\zeta_k}{\eta\sigma_R^2+\frac{\sigma_R^2\sigma_D^2}{\alpha P_R}}. \tag{2.48}$$

The achievable asymptotic sum rate corresponding to the transmit power scaling at the user nodes can be derived by substituting (2.48) into (2.25) as follows:

$$R^{\infty} = \frac{1}{2}\sum_{k=1}^{K}\log_2\left(1+\frac{E_U\eta\zeta_k}{\eta\sigma_R^2+\frac{\sigma_R^2\sigma_D^2}{\alpha P_R}}\right). \tag{2.49}$$

2.7.2 Transmit power scaling at relay nodes

In this subsection, our objective is to develop an asymptotic SINR expression for transmit power scaling at the relay node. In this context, the transmit power at the relay node is scaled inversely proportional to number of antennas at the destination as $P_R = E_R/N_D$. Again, the constant E_R is defined to keep the product P_RN_D fixed. Here, we derive an asymptotic SINR expression by letting the numbers of antennas at the relay and destination grow without bound while keeping a fixed ratio $\alpha = N_D/N_R$ for transmit power scaling at the relay as follows: To begin with, by substituting $P_R = E_R/N_D$ into (2.30), the end-to-end SINR can be re-written as

$$\gamma_k = \frac{P_S}{N_D^2}\left(\frac{P_{I_1}}{N_D^2}+\frac{P_{I_2}}{N_D^2}+\frac{P_{I_3}}{N_D^2}\right)^{-1}, \tag{2.50}$$

where the term P_S/N_D^2 can be defined as

$$\frac{P_S}{N_D^2} = \frac{P_UE_R\left(\frac{N_R}{N_D}\right)\left(\frac{\mathbf{f}_k^H\mathbf{G}^H\mathbf{G}\mathbf{f}_k\mathbf{f}_k^H\mathbf{G}^H\mathbf{G}\mathbf{f}_k}{N_R^2N_D^2}\right)}{P_U\mathrm{Tr}\left(\frac{\mathbf{F}^H\mathbf{F}}{N_R}\right)+\sigma_R^2}. \tag{2.51}$$

Moreover, the terms P_{I_i}/N_D^2 for $i \in \{1,2,3\}$ can be defined as

$$\frac{P_{I_1}}{N_D^2} = \frac{P_UE_R\left(\frac{N_R}{N_D}\right)\sum_{i=1,i\neq k}^{K}\left(\frac{\mathbf{f}_k^H\mathbf{G}^H\mathbf{G}\mathbf{f}_i\mathbf{f}_i^H\mathbf{G}^H\mathbf{G}\mathbf{f}_k}{N_R^2N_D^2}\right)}{P_U\mathrm{Tr}\left(\frac{\mathbf{F}^H\mathbf{F}}{N_R}\right)+\sigma_R^2}, \tag{2.52a}$$

$$\frac{P_{I_2}}{N_D^2} = \frac{\frac{E_R\sigma_R^2}{N_D}\left(\frac{\mathbf{f}_k^H\mathbf{G}^H\mathbf{G}\mathbf{G}^H\mathbf{G}\mathbf{f}_k}{N_RN_D^2}\right)}{P_U\mathrm{Tr}\left(\frac{\mathbf{F}^H\mathbf{F}}{N_R}\right)+\sigma_R^2}, \tag{2.52b}$$

$$\frac{P_{I_3}}{N_D^2} = \sigma_D^2\left(\frac{N_R}{N_D}\right)\left(\frac{\mathbf{f}_k^H\mathbf{G}^H\mathbf{G}\mathbf{f}_k}{N_RN_D}\right). \tag{2.52c}$$

By using the identity in (2.33), an asymptotic expression for P_S/N_D^2 can be derived as follows:

$$\lim_{\substack{N_R\to\infty,\ N_D\to\infty\\ N_D/N_R=\alpha}} \left(\frac{P_S}{N_D^2}\right) = \frac{P_U E_R \eta^2 \zeta_k^2}{\alpha\left(P_U \mathrm{Tr}\left(\mathbf{D}_F\right) + \sigma_R^2\right)}. \tag{2.53}$$

Similarly, by using the identities in (2.37), (2.41) and (2.45), respectively, the asymptotic expressions for P_{I_i}/N_D^2 for $i \in \{1, 2, 3\}$ can be derived as follows:

$$\lim_{\substack{N_R\to\infty,\ N_D\to\infty\\ N_D/N_R=\alpha}} \left(\frac{P_{I_1}}{N_D^2}\right) = 0, \tag{2.54a}$$

$$\lim_{\substack{N_R\to\infty,\ N_D\to\infty\\ N_D/N_R=\alpha}} \left(\frac{P_{I_2}}{N_D^2}\right) = 0, \tag{2.54b}$$

$$\lim_{\substack{N_R\to\infty,\ N_D\to\infty\\ N_D/N_R=\alpha}} \left(\frac{P_{I_2}}{N_D^2}\right) = \frac{\sigma_D^2 \eta \zeta_k}{\alpha}. \tag{2.54c}$$

Finally, by using the asymptotic expressions in (2.53), (2.54a), (2.54b) and (2.54c), the asymptotic SINR expression for transmit power scaling at the relay can be derived as follows:

$$\gamma_k^\infty = \lim_{\substack{N_R\to\infty,\ N_D\to\infty\\ N_D/N_R=\alpha}} (\gamma_k) = \frac{P_U E_R \eta \zeta_k}{\sigma_D^2 \left(P_U \mathrm{Tr}\left(\mathbf{D}_F\right) + \sigma_R^2\right)}. \tag{2.55}$$

Then, the achievable asymptotic sum rate for the case of transmit power scaling at the relay nodes can be derived by substituting (2.55) into (2.25) as follows:

$$R^\infty = \frac{1}{2}\sum_{k=1}^{K} \log_2\left(1 + \frac{P_U E_R \eta \zeta_k}{\sigma_D^2 \left(P_U \mathrm{Tr}\left(\mathbf{D}_F\right) + \sigma_R^2\right)}\right). \tag{2.56}$$

2.7.3 *Transmit power scaling at user and relay nodes*

In this subsection, the transmit power scaling laws at the user and relay nodes are investigated. Thereby, an asymptotic SINR expression is derived whenever the numbers of antennas grow without bound as $N_R \to \infty$, $N_D \to \infty$ and $\alpha = N_D/N_R$. To be more specific, the transmit powers at the user and relay nodes are scaled inversely proportional to the square-root of the number of antennas at the relay and destination, respectively, i.e., $P_U = E_U/\sqrt{N_R}$ and $P_R = E_R/\sqrt{N_D}$ where E_U and E_R are two constants defined to keep the products $P_U\sqrt{N_R}$ and $P_R\sqrt{N_D}$ fixed.

To begin with, by scaling the transmit power at the user and relay nodes inversely proportional to $\sqrt{N_R}$ and $\sqrt{N_D}$, respectively, the end-to-end SINR can be written in an alternative form as follows:

$$\gamma_k = \frac{P_S}{\sqrt[3]{N_D}\sqrt{N_R}} \left(\frac{P_{I_1}}{\sqrt[3]{N_D}\sqrt{N_R}} + \frac{P_{I_2}}{\sqrt[3]{N_D}\sqrt{N_R}} + \frac{P_{I_3}}{\sqrt[3]{N_D}\sqrt{N_R}}\right)^{-1}, \tag{2.57}$$

where the desired signal component in (2.57) can be defined as

$$\frac{P_S}{\sqrt[3]{N_D}\sqrt{N_R}} = \frac{E_U E_R \sum_{i=1,1\neq k}^{K} \left(\frac{\mathbf{f}_k^H \mathbf{G}^H \mathbf{G} \mathbf{f}_i \mathbf{f}_i^H \mathbf{G}^H \mathbf{G} \mathbf{f}_k}{N_R^2 N_D^2} \right)}{\frac{E_U}{\sqrt{N_R}} \mathrm{Tr}\left(\frac{\mathbf{F}^H \mathbf{F}}{N_R} \right) + \sigma_R^2}. \tag{2.58}$$

Furthermore, the interference terms $P_{I_i}/\sqrt[3]{N_D}\sqrt{N_R}$ for $i \in \{1,2,3\}$ can be defined as

$$\frac{P_{I_1}}{\sqrt[3]{N_D}\sqrt{N_R}} = \frac{E_U E_R \sum_{i=1,i\neq k}^{K} \left(\frac{\mathbf{f}_k^H \mathbf{G}^H \mathbf{G} \mathbf{f}_i \mathbf{f}_i^H \mathbf{G}^H \mathbf{G} \mathbf{f}_k}{N_R^2 N_D^2} \right)}{\frac{E_U}{\sqrt{N_R}} \mathrm{Tr}\left(\frac{\mathbf{F}^H \mathbf{F}}{N_R} \right) + \sigma_R^2}, \tag{2.59a}$$

$$\frac{P_{I_2}}{\sqrt[3]{N_D}\sqrt{N_R}} = \frac{\frac{E_R \sigma_R^2}{\sqrt{N_R}} \left(\frac{\mathbf{f}_k^H \mathbf{G}^H \mathbf{G} \mathbf{G}^H \mathbf{G} \mathbf{f}_k}{N_R N_D^2} \right)}{\frac{E_U}{\sqrt{N_R}} \mathrm{Tr}\left(\frac{\mathbf{F}^H \mathbf{F}}{N_R} \right) + \sigma_R^2}, \tag{2.59b}$$

$$\frac{P_{I_3}}{\sqrt[3]{N_D}\sqrt{N_R}} = \sigma_D^2 \sqrt{\frac{N_R}{N_D}} \left(\frac{\mathbf{f}_k^H \mathbf{G}^H \mathbf{G} \mathbf{f}_k}{N_R N_D} \right). \tag{2.59c}$$

Again, by using the identity in (2.33), an asymptotic expression for $P_S/\sqrt[3]{N_D}\sqrt{N_R}$ can be next derived as

$$\lim_{\substack{N_R\to\infty,\ N_D\to\infty \\ N_D/N_R=\alpha}} \left(\frac{P_S}{\sqrt[3]{N_D}\sqrt{N_R}} \right) = \frac{E_U E_R \eta^2 \zeta_k^2}{\sigma_R^2}. \tag{2.60}$$

By using the identities in (2.37), (2.41) and (2.45), respectively, the asymptotic expressions for $P_{I_i}/\sqrt[3]{N_D}\sqrt{N_R}$ for $i \in \{1,2,3\}$ can be derived as follows:

$$\lim_{\substack{N_R\to\infty,\ N_D\to\infty \\ N_D/N_R=\alpha}} \left(\frac{P_{I_1}}{\sqrt[3]{N_D}\sqrt{N_R}} \right) = 0, \tag{2.61a}$$

$$\lim_{\substack{N_R\to\infty,\ N_D\to\infty \\ N_D/N_R=\alpha}} \left(\frac{P_{I_2}}{\sqrt[3]{N_D}\sqrt{N_R}} \right) = 0, \tag{2.61b}$$

$$\lim_{\substack{N_R\to\infty,\ N_D\to\infty \\ N_D/N_R=\alpha}} \left(\frac{P_{I_2}}{\sqrt[3]{N_D}\sqrt{N_R}} \right) = \frac{\sigma_D^2 \eta \zeta_k}{\sqrt{\alpha}}. \tag{2.61c}$$

By employing (2.60), (2.61a), (2.61b) and (2.61c), the asymptotic SINR for transmit power scaling at both user and relay nodes can next be derived as

$$\gamma_k^\infty = \lim_{\substack{N_R\to\infty,\ N_D\to\infty \\ N_D/N_R=\alpha}} (\gamma_k) = \frac{\sqrt{\alpha} E_U E_R \eta \zeta_k}{\sigma_R^2 \sigma_D^2}. \tag{2.62}$$

Again, by substituting (2.62) into (2.25), the achievable asymptotic sum rate for transmit power scaling at both user and relay nodes can be derived as

$$R^{\infty} = \frac{1}{2}\sum_{k=1}^{K}\log_2\left(1 + \frac{\sqrt{\alpha}E_U E_R \eta \zeta_k}{\sigma_R^2 \sigma_D^2}\right). \tag{2.63}$$

2.7.4 Numerical results

In this subsection, the achievable sum rate is plotted as a function of the number of antennas at the relay node. In this context, the system parameters used for Monte Carlo simulations can be defined as follows:

The users-to-relay (**F**) and relay-to-destination (**G**) channels are modelled as independently distributed Rayleigh fading. The path loss between any two nodes is modelled as $\mathrm{PL} = \mathrm{PL}_0 + 10n\log_{10}\left(d_0/d_{i,j}\right)$, where d_0 is the reference distance, $d_{i,j}$ is the distance between node i and node j, and n is the path loss exponent. Further, PL_0 is the path loss at the reference distance d_0. The number of user nodes is set to $K = 6$, and the ratio between the number of destination and relay antennas is defined as $\alpha = N_D/N_R = 4$. The distance vector between user nodes and the relay is given by $d_{U,R} = [10\,\mathrm{m}, 15\,\mathrm{m}, 20\,\mathrm{m}, 25\,\mathrm{m}, 30\,\mathrm{m}, 35\,\mathrm{m}]$, and the reference distance is set to $d_0 = 1\,\mathrm{m}$. Further, the distance between R and D is $d_{R,D} = 15\,\mathrm{m}$. The path loss exponent to be defined as $n = 2.1$. The diagonal matrix accounting for path loss is defined as $\mathbf{D}_F = \mathrm{diag}(\zeta_1, \cdots, \zeta_k, \cdots, \zeta_K)$, where its diagonal elements are obtained by using the aforementioned path loss model. The noise variances at the relay and destination are defined as $\sigma_R^2 = -5$ dBm and $\sigma_D^2 = -5$ dBm, respectively. Further, the power constants at the user nodes and relay are defined as $E_U = 31.5$ dBm and $E_R = 31.5$ dBm, respectively.

In Figure 2.7, the achievable sum rate curves are plotted for the three specific transmit power scaling laws presented in Sections 2.7.1–2.7.3. For the sake of brevity, these three transmit power scaling laws are denoted as follows: (i) case-1: transmit power scaling at user nodes only, (ii) case-2: transmit power scaling at relay nodes only, and (iii) case-3: transmit power scaling at both user and relay nodes. The asymptotic sum rate curves are plotted by using the analytical expressions given in (2.49), (2.56) and (2.63). Further, the Monte Carlo simulated sum rate curves for MRC- and ZF-based precoders/detectors are plotted by using the system parameters presented in Tables 2.1–2.3. Figure 2.7 clearly shows that the sum rate curves corresponding to both MRC- and ZF-based precoders/detectors converge to their pertinent asymptotic curves whenever the numbers of antennas at the relay and destination are allowed to grow without bound. Nevertheless, the rate of convergence of ZF-based system is faster than that of MRC-based system. In particular, case-3 outperforms both case-1 and case-2 in terms of sum rate. Moreover, case-1 outperforms case-2, but still its performance is inferior to case-3. This behaviour is not surprising because in case-3, the transmit powers at both user and relay nodes are scaled inversely proportional to the square-root of the number of relay and destination antennas, respectively. Whereas in case-1 and case-2, the transmit powers at user and relay nodes are scaled inversely proportional to the number of antennas at the relay and user nodes, respectively.

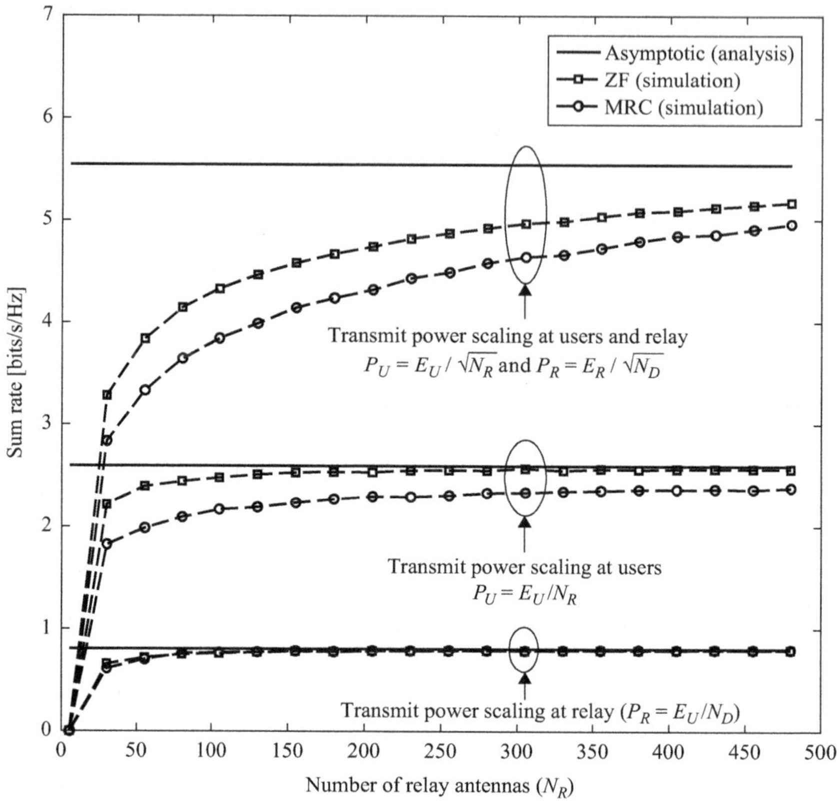

Figure 2.7 The sum rate versus the number of relay antennas

Thus, the case-3 indeed possesses an advantage in terms of the amount of transmitted power over the other two cases, and this advantage is transformed into a significant sum rate gain. Nevertheless, the user nodes in case-1 are more power efficient than the other two cases as their transmit power can become infinitesimal at a much faster rate whenever the numbers of antennas at the relay and destination nodes are allowed to grow without bound.

2.7.5 System design insights

Our objective of this subsection is to draw important system design insights through the asymptotic performance analysis presented in Sections 2.7.1–2.7.3. To this end, valuable insights about transmit power scaling laws, asymptotic SINR and achievable sum rate can be enumerated as follows:

1. Whenever the numbers of antennas at the relay and destination grow without bound, the asymptotic SINR expressions in (2.48), (2.55) and (2.62) become independent of the fast fading components of the corresponding wireless channels. Consequently, the sum rate expressions in (2.56) and (2.63) also become independent of fast fading.

2. Consequently, the medium-access control layer of the massive MIMO relay networks becomes significantly less complicated due to the much simpler radio resource allocation and scheduling techniques compared to those of conventional MIMO relay networks.
3. As the SINR expressions are independent of the fast fading, the latency in the air interface can be significantly reduced, and hence, the overall system QoS can be considerably improved.
4. The transmit powers at the user and relay nodes can become infinitesimal for infinitely many relay and destination antennas. To this end, the transit powers only at the user or relay nodes and both user and relay nodes can be allowed to scale down inversely proportional to the number of antennas at either the relay or destination.
5. The transmit power at the user nodes can be scaled down inversely proportional to the number of relay antennas without degrading the achievable sum rate.
6. The transmit power at the relay node can also be scaled down inversely proportional to the number of destination antennas.
7. The transmit powers at the user and relay nodes can only be scaled down inversely proportional to the square-root of the number of antennas at the relay and destination nodes, respectively.

2.8 Summary

Multi-user massive MIMO relaying is a cost-effective, pragmatic solution to one of the fundamental bottlenecks of the wireless physical layer, namely, the low-area spectral efficiency. Specifically, a combination of massive MIMO and wireless relaying techniques can potentially be exploited for improving the trade-off between spectral efficiency and coverage, thereby substantially improving the achievable area spectral efficiency. To this end, in this chapter, system configurations, transmission frame structures, CSI acquisition techniques and signal processing aspects for multi-user massive MIMO relay networks have been presented. Moreover, the corresponding channel models and favourable propagation conditions have been reviewed, and thereby, a generalized end-to-end signal model has been developed. First, the performance analysis techniques and the underlying mathematical tools have been presented, and then, they have been used for deriving basic performance metrics including asymptotic SINR and achievable sum rate expressions. Finally, important insights have been drawn by using these performance metrics, and thereby, the validity and viability of the proposed system, channel and signal models can be ascertained.

References

[1] M. Dohler, R. Heath, A. Lozano, C. Papadias, and R. Valenzuela, "Is the PHY layer dead?" *IEEE Commun. Mag.*, vol. 49, no. 4, pp. 159–165, 2011.

[2] *Global Mobile Data Traffic Forecast Update, 2012–2017*, Cisco, Cisco Visual Networking Index, Feb. 2013.

[3] M. Ismail and W. Zhuang, "Network cooperation for energy saving in green radio communications," *IEEE Wireless Commun. Mag.*, vol. 18, no. 5, pp. 76–81, 2011.

[4] G. J. Foschini, "Layered space-time architecture for wireless communication in a fading environment when using multi-element antennas," *Bell Labs Tech. J.*, vol. 5, no. 2, pp. 41–59, Autumn 1996.

[5] A. Goldsmith, S. A. Jafar, N. Jindal, and S. Vishwanath, "Capacity limits of MIMO channels," *IEEE J. Sel. Areas Commun.*, vol. 21, no. 5, pp. 684–702, Jun. 2003.

[6] J. N. Laneman, D. N. C. Tse, and G. W. Wornell, "Cooperative diversity in wireless networks: Efficient protocols and outage behavior," *IEEE Trans. Inf. Theory*, vol. 50, no. 12, pp. 3062–3080, Dec. 2004.

[7] A. Sendonaris, E. Erkip, and B. Aazhang, "User cooperation diversity – Part I: System description," *IEEE Trans. Commun.*, vol. 51, no. 11, pp. 1927–1938, Nov. 2003.

[8] A. Nosratinia, T. E. Hunter, and A. Hedayat, "Cooperative communication in wireless networks," *IEEE Commun. Mag.*, vol. 42, no. 10, pp. 74–80, Oct. 2004.

[9] *Air Interface for Broadband Wireless Access Systems – Amendment 1: Multihop Relay Specification*, IEEE Std 802.16j-2009, Jun. 2009.

[10] R. U. Nabar, O. Oyman, H. Bölcskei, and A. J. Paulraj, "Capacity scaling laws in MIMO wireless networks," in *Proc. 41st Allerton Conference on Communication, Control, and Computing*, Monticello, IL, Oct. 2003.

[11] Y. Fan and J. Thompson, "MIMO configurations for relay channels: Theory and practice," *IEEE Trans. Wireless Commun.*, vol. 6, no. 5, pp. 1774–1786, May 2007.

[12] M. Yuksel and E. Erkip, "Multiple-antenna cooperative wireless systems: A diversity-multiplexing tradeoff perspective," *IEEE Trans. Inf. Theory*, vol. 53, no. 10, pp. 3371–3393, Oct. 2007.

[13] Y. Yang, H. Hu, J. Xu, and G. Mao, "Relay technologies for WiMAX and LTE-advanced mobile systems," *IEEE Commun. Mag.*, vol. 47, no. 10, pp. 100–105, Oct. 2009.

[14] K. Loa, C.-C. Wu, S.-T. Sheu, *et al.*, "IMT-advanced relay standards," *IEEE Commun. Mag.*, vol. 48, no. 8, pp. 40–48, Aug. 2010.

[15] F. Boccardi, R. Heath, A. Lozano, T. Marzetta, and P. Popovski, "Five disruptive technology directions for 5G," *IEEE Commun. Mag.*, vol. 52, no. 2, pp. 74–80, Feb. 2014.

[16] T. Marzetta, "Noncooperative cellular wireless with unlimited numbers of base station antennas," *IEEE Trans. Wireless Commun.*, vol. 9, no. 11, pp. 3590–3600, Nov. 2010.

[17] F. Rusek, D. Persson, B. K. Lau, *et al.*, "Scaling up MIMO: Opportunities and challenges with very large arrays," *IEEE Signal Process. Mag.*, vol. 30, no. 1, pp. 40–60, Jan. 2013.

[18] H. Yang and T. Marzetta, "Performance of conjugate and zero-forcing beamforming in large-scale antenna systems," *IEEE J. Sel. Areas Commun.*, vol. 31, no. 2, pp. 172–179, Feb. 2013.

[19] H. Q. Ngo, E. Larsson, and T. Marzetta, "Energy and spectral efficiency of very large multiuser MIMO systems," *IEEE Trans. Commun.*, vol. 61, no. 4, pp. 1436–1449, Apr. 2013.

[20] J. Hoydis, S. ten Brink, and M. Debbah, "Massive MIMO in the UL/DL of cellular networks: How many antennas do we need?" *IEEE J. Sel. Areas Commun.*, vol. 31, no. 2, pp. 160–171, Feb. 2013.

[21] E. Larsson, O. Edfors, F. Tufvesson, and T. Marzetta, "Massive MIMO for next generation wireless systems," *IEEE Commun. Mag.*, vol. 52, no. 2, pp. 186–195, Feb. 2014.

[22] A. Sulyman, A. Nassar, M. Samimi, G. Maccartney, T. Rappaport, and A. Alsanie, "Radio propagation path loss models for 5G cellular networks in the 28 GHz and 38 GHz millimeter-wave bands," *IEEE Commun. Mag.*, vol. 52, no. 9, pp. 78–86, Sep. 2014.

[23] A. Swindlehurst, E. Ayanoglu, P. Heydari, and F. Capolino, "Millimeter-wave massive MIMO: The next wireless revolution?" *IEEE Commun. Mag.*, vol. 52, no. 9, pp. 56–62, Sep. 2014.

[24] Q. Zhang, S. Jin, K.-K. Wong, H. Zhu, and M. Matthaiou, "Power scaling of uplink massive MIMO systems with arbitrary-rank channel means," *IEEE J. Sel. Areas Signal Process.*, vol. 8, no. 5, pp. 966–981, Oct. 2014.

[25] J. Jose, A. Ashikhmin, T. Marzetta, and S. Vishwanath, "Pilot contamination and precoding in multi-cell TDD systems," *IEEE Trans. Wireless Commun.*, vol. 10, no. 8, pp. 2640–2651, Aug. 2011.

[26] F. Fernandes, A. Ashikhmin, and T. Marzetta, "Inter-cell interference in noncooperative TDD large scale antenna systems," *IEEE J. Sel. Areas Commun.*, vol. 31, no. 2, pp. 192–201, Feb. 2013.

[27] A. Pitarokoilis, S. Mohammed, and E. Larsson, "Uplink performance of time-reversal MRC in massive MIMO systems subject to phase noise," *IEEE Trans. Wireless Commun.*, vol. 14, no. 2, pp. 711–723, Feb. 2015.

[28] K. Truong and R. Heath, "Effects of channel aging in massive MIMO systems," *J. Commun. Networks*, vol. 15, no. 4, pp. 338–351, Aug. 2013.

[29] A. Papazafeiropoulos and T. Ratnarajah, "Uplink performance of massive MIMO subject to delayed CSIT and anticipated channel prediction," in *Proc. IEEE Int. Conf. Acoustics, Speech Signal Process. (ICASSP)*, May 2014, pp. 3162–3165.

[30] Q. Zhang, S. Jin, K.-K. Wong, H. Zhu, and M. Matthaiou, "Power scaling of uplink massive MIMO systems with arbitrary-rank channel means," *IEEE J. Sel. Top. Signal Process.*, vol. 8, no. 5, pp. 966–981, Oct. 2014.

[31] J. Choi, D. Love, and P. Bidigare, "Downlink training techniques for FDD massive MIMO systems: Open-loop and closed-loop training with memory," *IEEE J. Sel. Top. Signal Process.*, vol. 8, no. 5, pp. 802–814, Oct. 2014.

[32] G. Amarasuriya and H. V. Poor, "Multi-user relay networks with massive MIMO," in *Proc. IEEE Int. Conf. Commun. (ICC)*, Jun. 2015, pp. 3620–3626.

[33] G. Amarasuriya, "Sum rate analysis for multi-user massive MIMO relay networks," *2015 IEEE Global Commun. Conf. (GLOBECOM)*, San Diego, CA, 2015, pp. 1–7.

[34] G. Amarasuriya and H. V. Poor, "Multi-way amplify-and-forward relay networks with massive MIMO," in *Proc. IEEE Int. Symp. Personal, Indoor, and Mobile Radio Commun. (PIMRC)*, Sep. 2014, pp. 595–600.

[35] G. Amarasuriya and H. V. Poor, "Impact of channel aging in multi-way relay networks with massive MIMO," in *Proc. IEEE Int. Conf. Commun. (ICC)*, Jun. 2015, pp. 3554–3560.

[36] H. Suraweera, H. Q. Ngo, T. Duong, C. Yuen, and E. Larsson, "Multi-pair amplify-and-forward relaying with very large antenna arrays," in *Proc. IEEE Int. Conf. Commun. (ICC), 2013*, Jun. 2013, pp. 4635–4640.

[37] H. Q. Ngo, H. Suraweera, M. Matthaiou, and E. Larsson, "Multipair full-duplex relaying with massive arrays and linear processing," *IEEE J. Sel. Areas Commun.*, vol. 32, no. 9, pp. 1721–1737, Sep. 2014.

[38] H. Cui, L. Song, and B. Jiao, "Multi-pair two-way amplify-and-forward relaying with very large number of relay antennas," *IEEE Trans. Wireless Commun.*, 2014, available at IEEE Explore – early access.

[39] H. Q. Ngo and E. Larsson, "Large-scale multipair two-way relay networks with distributed AF beamforming," *IEEE Commun. Lett.*, vol. 17, no. 12, pp. 1–4, Dec. 2013.

[40] A. Paulraj, R. Nabar, and D. Gore, *Introduction to Space-Time Wireless Communications*, 1st ed. Cambridge University Press, UK, 2003.

[41] H. Cramer, *Random Variables and Probability Distributions*. Cambridge University Press, UK, 1970.

[42] J. Ma, P. Orlik, J. Zhang, and G. Li, "Pilot matrix design for estimating cascaded channels in two-hop MIMO amplify-and-forward relay systems," *IEEE Trans. Wireless Commun.*, vol. 10, no. 6, pp. 1956–1965, Jun. 2011.

Chapter 3

SUDAS: mmWave relaying for 5G outdoor-to-indoor communications

Marco Breiling, Derrick Wing Kwan Ng, Christian Rohde, Frank Burkhardt and Robert Schober

3.1 Introduction

The spatial degrees of freedom offered by multiple antennas at transmitters and receivers of wireless communication systems facilitate a trade-off between multiplexing gains and diversity gains [1–3], which provides a high flexibility in resource allocation. A wireless point-to-point link with M transmit and N receive antennas constitutes an M-by-N multiple-input multiple-output (MIMO) communication system. In particular, under certain mild conditions [1–3], the ergodic capacity of an M-by-N MIMO fading channel increases practically linearly with min $\{M, N\}$ for a fixed transmit power and bandwidth. Therefore, MIMO has attracted a lot of research interest over the past decades, as it enables significant performance and throughput gains. Unfortunately, the signal processing complexity required in MIMO receivers and the small number of receive antennas that a wireless mobile device can support limit the potential MIMO gain in practice. As an alterative, the concept of cooperative communication has been proposed for wireless networks as an effective technique for realizing the MIMO performance gains. The basic idea of cooperative communication is to enable multiple single-antenna terminals of a multiuser (MU) system to share their antennas to create a virtual MIMO (VMIMO) system [4, 5]. In this context, user cooperation, base station (BS) cooperation, and relaying have been proposed in the literature for implementation of cooperative communications. Theoretically, BS cooperation and user cooperation are able to provide huge performance gains, when compared to non-cooperative networks. However, the required information exchange between BSs via backhaul links [6] and the information exchange between users via wireless channels make these options less attractive in practice. In contrast, cooperative relaying with dedicated relays requires significantly less signaling overhead and allows for low-cost implementations while achieving significant coverage extension, diversity gains, and throughput gains compared to non-cooperative transmission. Therefore, cooperative relaying has attracted significant interest from both academia and industry in the last few years.

In practice, relays can transmit and receive signals at the same time and over the same frequencies, which is known as full-duplex (FD) relaying [7–9]. However, implementing such FD relays requires precise and expensive hardware components which may not be desirable and cost effective. Alternatively, relays may operate in a half-duplex manner [10–27], i.e., transmission and reception at relays are separated in either time or frequency. Such relays are also known as cheap relays in the literature due to their low hardware complexity. Different relaying strategies such as amplify-and-forward (AF), compress-and-forward (CF), and decode-and-forward (DF) protocols have been proposed in the literature [10–27]. There is no uniformly optimal relaying protocol, and each protocol can outperform the others, depending on the system configuration. Nevertheless, AF is particularly appealing for practical implementation, as the relays only amplify and linearly process the received signal which leads to low-complexity transceiver designs. More importantly, AF relays are transparent to the adaptive modulation techniques that are typically employed at the BSs. For these reasons, AF was selected as one of the possible relaying modes in IEEE 802.16j (mobile multihop relay) and is being proposed for the fourth generation (4G) Long-Term Evolution (LTE) Advanced systems. Despite the significant advances in communication researches in the past decades, the data rate of current 4G systems is still limited, especially for outdoor-to-indoor communication. Besides, it is foreseen that existing network architectures may become a bottleneck in the development of next generation communication. Thus, an effective system architecture to assist outdoor-to-indoor communication is needed.

The remainder of this chapter is organized as follows. In Section 3.2, we provide a brief introduction to next generation communication systems. In particular, we discuss three existing potential technologies for realizing the goals of next generation communication systems. In Section 3.2.3, we propose a novel shared user-equipment side distributed antenna system (SUDAS) to assist outdoor-to-indoor communication which can effectively improve the system data rate. Then, in Section 3.3, practical implementation details of the proposed system are discussed. Sections 3.4 and 3.5 present the mathematical model and numerical performance results for the proposed system, respectively. Finally, in Section 3.6, we discuss some future research and implementation challenges for the designed system.

3.2 5G communication systems

High energy efficiency, low communication latency, high spectral efficiency, and cost effectiveness are the primary goals of the fifth generation (5G) mobile communication systems [28–32]. In particular, a target data rate of 10 Gbit/s per device is expected for both uplink and downlink communications. The data rate originally envisaged for 4G mobile communication standards was 1 Gbit/s, and the LTE-A Release 10 standard extended this target to 3 Gbit/s. The ambitious goal of 5G will enable the download of high-definition (HD) movies within seconds and will offer very high data rates to many users simultaneously. The tenfold increase in the required data rate from 4G to 5G is one of the key performance indicators specified by the European Commission for 5G research within the Horizon 2020 programme. However, with existing technologies, it appears very difficult to reach such high data rates even under favorable conditions. Currently, three technologies are considered as top candidates for realizing

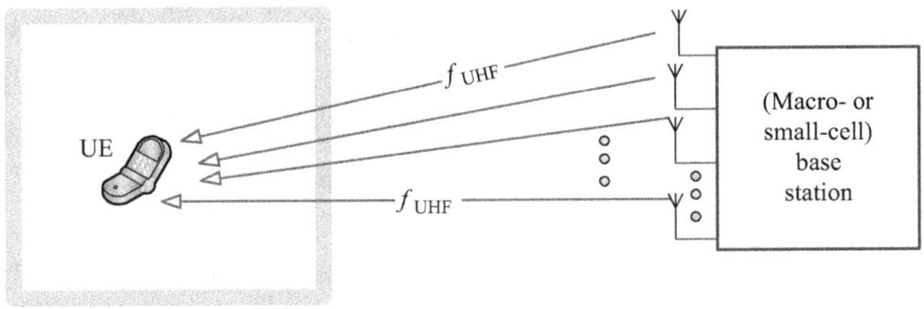

Figure 3.1 A massive MIMO BS serving a UE inside a building in the ultra-high-frequency (UHF) band, f_{UHF}

the 5G goals: massive MIMO, small cells, and the use of unlicensed millimeter wave (mmWave) bands. We will show in the sequel that it is highly questionable whether these technologies alone are able to reach the elusive goal of 10 Gbit/s.

3.2.1 Massive MIMO

In the past decades, MIMO has been implemented in modern communication systems for providing high-speed data rate communication services due to its potential to improve spectral efficiency [1–3]. Especially, massive MIMO technology, where a BS is equipped with a very large number of antennas to serve a comparatively small number of users [33–36], is considered as a potential candidate for achieving the 10 Gbit/s data rate target in 5G, cf. Figure 3.1. However, state-of-the-art smartphones are typically equipped with only two receive antennas which limit the spatial multiplexing gain offered by MIMO. In practice, some expensive user-equipments (UEs) might possess even up to four internal antennas. Yet, deploying more than four receive antennas at each UE may not be realistic due to the limited physical size of mobile receivers. Thus, despite the use of massive MIMO, at most four spatial streams can be exploited simultaneously for increasing the data rate of a UE. On the other hand, with carrier aggregation, a maximum of 200 MHz of licensed spectrum [37] can be created for a single mobile network operator (MNO) with today's technology. In other words, a total of 50 bit/s/Hz spectral efficiency is required to meet the 10 Gbit/s requirement. With a maximum of four spatial streams per UE offered by massive MIMO, 12.5 bit/s/Hz per spatial stream is needed which leads to a minimum required signal-to-noise ratio (SNR) of 38 dB. Besides, the use of at least 16384-quadrature amplitude modulation (QAM) in each stream is needed for achieving the required spectral efficiency. However, the computational complexity for information decoding at the receiver grows exponentially with the size of the constellation, and it is challenging to implement such decoders for real-time applications at a reasonable cost.

3.2.2 Small cells and mmWaves

The concept of small cells, e.g., femtocells, has been proposed as a network architecture for reducing the network power consumption and extending service

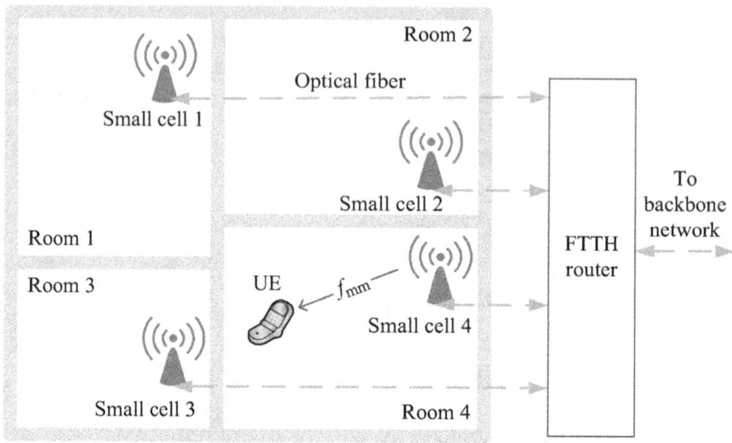

Figure 3.2 An indoor environment equipped with multiple small cells in every room. The small-cell BSs operate in the mmW frequency band, f_{mm}

coverage, cf. Figure 3.2. Thanks to the very large available bandwidth for wireless communication in the mmWave range, e.g., ~7 GHz of bandwidth in the 60 GHz bands, it appears realistic to deliver 10 Gbit/s from a small-cell BS indoors to a UE over a short distance. However, the bottleneck now becomes the backhauling of the data from a service provider to the indoor small-cell BS. In general, the last mile connection from a backbone network to the terminal users at homes cannot support such high data rates, except if optical fibers are deployed, which is known as fiber-to-the-home (FTTH). However, the cost of deploying FTTH for all home users is prohibitive. For instance, the cost of equipping every building (not home!) with FTTH in Germany is estimated to be around 67 billion € [38]. Moreover, most non-business buildings do not possess an internal optical fiber infrastructure or a wired ethernet for further distribution of the data received from FTTH to the UEs. Furthermore, as mmWave signals are heavily blocked by walls, a small-cell BS transmitting in the mmWave bands would be required in all rooms of a building, cf. Figure 3.2. Hence, extra costs in connecting each room of each building must be added on top of the 67 billion €, and these tremendous costs make this a less attractive option for service providers for realizing the goals of 5G.

Another potential difficulty in utilizing the mmWave bands for communication is finding a suitable location for installing a small-cell BS such that a line-of-sight (LOS) to all UEs is available. Without such a LOS connection, exploiting mmWave links for wireless communication is hardly possible. Observe that if a small-cell BS belongs to a single MNO, the envisaged high data rate is only available to the UEs in the small cell which are associated with this MNO. In order to allow different UEs associated with different MNOs to enjoy high data rate services, the small-cell BS must be neutral with regard to the MNO which is not the case for the small-cell BSs on the market.

3.2.3 Combinations of massive MIMO and mmWave

Recently, different approaches for utilizing a combination of massive MIMO and mmWave have been proposed. Basically, these approaches can be divided into two categories, and we summarize the corresponding disadvantages in the following:

- Backhauling of small cells by massive MIMO over mmWave
  - The problem of this scheme is the high penetration loss of building walls from outdoor to indoor. Specifically, high propagation losses are expected over longer distances at mmWave frequencies.
  - A small-cell BS is required in each room and cannot be shared by different MNOs simultaneously.
- Exploiting massive MIMO for communication to UEs in mmWave band
  - The high penetration loss of walls renders this approach for outdoor use only, i.e., outdoor-to-indoor communication via mmWave may not be supported.
  - A large number of backhauling BSs are needed in order to provide sufficient service coverage for small-cell BSs. Specifically, a LOS connection is required from any UE location to at least one BS at any time.

The combination of massive MIMO and small-cell schemes has limited potential to improve the system performance as long as their operation is limited to the high-frequency unlicensed mmWave bands. Since both licensed and unlicensed bands are available to the service providers, it is reasonable to utilize both spectra to meet the stringent data rate requirements of 5G which motivates our proposed design in the next section. In the last decade, most of the data rate improvement in mobile communication systems was due to the use of MIMO with an increasing number of antennas [4–42]. However, only a small number of UHF antennas can be integrated into a UE which limits the MIMO gain for improving the spectral efficiency. On the other hand, mmWave band communication facilitates small antenna sizes and provides a huge bandwidth for achieving high data rate communication. Yet, mmWave communication suffers from very difficult propagation conditions [43] compared to UHF. Therefore, we propose a system architecture, which efficiently combines the advantages of MIMO over a mobile UHF band channel with its favorable propagation conditions and mmWave with its large bandwidth. Our novel architecture is derived from the idea of VMIMO [4, 39, 40]. It extends the VMIMO concept to the use of mmWave for relaying the MIMO signals transmitted in the UHF band, and to sharing an infrastructure between multiple UEs and multiple MNOs. VMIMO is also known as cooperative MIMO, distributed MIMO, and virtual antenna arrays [4, 39]. A variant of VMIMO, coordinated multi-point (CoMP), has already been implemented in LTE-A at the BS-side, where the received signals are relayed or forwarded to one entity (e.g., a joint signal processing unit) over the backhauling network [40–42].

3.2.4 SUDAS – overview

The ambitious data rate target of 10 Gbit/s for the 5G standard is elusive despite the use of advanced air interface technologies such as massive MIMO, small cells, and mmWave communication links. In order to achieve this high target data rate for UEs

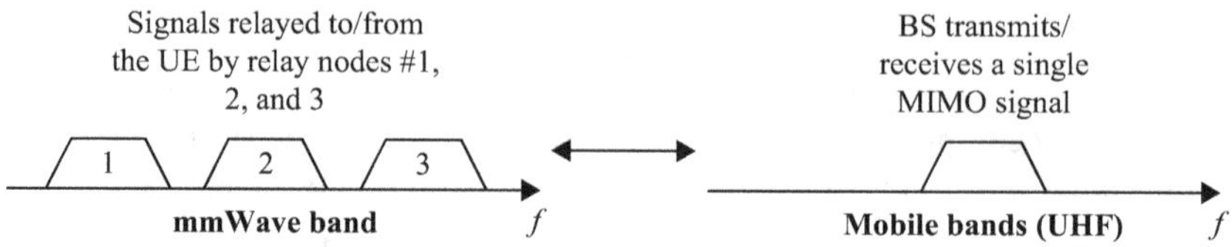

Figure 3.3 Spectrum occupation in the mmWave band and mobile bands (UHF): a single VMIMO carrier in the mobile bands is relayed to multiple non-overlapping signals in the mmWave band

inside a building, we propose in this section for 5G to employ an infrastructure of many low-price relay nodes installed in fixed indoor locations. These nodes relay signals received from the BS in a mobile UHF band via non-overlapping mmWave links to the UE, and vice versa, cf. Figure 3.3. We refer to this scattered infrastructure as a SUDAS [44]. The proposed relaying scheme is limited to only two hops, namely between the BS and the relay nodes, and between the relay nodes and the UE. This simplification allows a low end-to-end latency, and it avoids the significant overhead needed for exchanging routing information. In particular, each relay is called a shared UE-side distributed antenna component (SUDAC) in the proposed SUDAS. In practice, such SUDACs could be integrated into many devices with continuous power supply, including electrical outlets (cf. Figure 3.4), lamps, and light outlets or other devices that carry machine-type-communication (MTC) circuits in the future, such as TV sets and fridges. Besides, the SUDACs are dedicated low-cost devices and scattered in a room. In the following sections, we discuss:

- Working principle of SUDAS
- Potential application scenarios
- Implementation details
- Challenges in SUDAS

3.2.5 SUDAS – working principle

Figure 3.5 illustrates the working principle of the proposed SUDAS. All links in the mobile bands from the BS of MNO "A" utilize the same frequency resource $f_{\mathrm{UHF,A}}$ in the UHF range. Besides, the mmWave links employ the orthogonal frequency bands $f_{\mathrm{mm},i}$, $\forall i \in \{1, \ldots, M\}$, where M is the number of SUDACs. The SUDAS forms a scattered infrastructure which is in contrast to the centralized infrastructure realized by deploying one small-cell BS per room. As shown in Figure 3.5, the SUDAS is not only shared among all UEs in a room but also possibly shared between multiple BSs associated with different MNOs. Because of the high attenuation of stone walls in the mmWave bands, multiple SUDASs located in different rooms do hardly interfere with each other. Therefore, extensive frequency reuse can be deployed in a network with different SUDASs.

The SUDAS concept is particularly promising because it translates spatial multiplexing in the mobile bands (referred to as backend) into frequency multiplexing in the

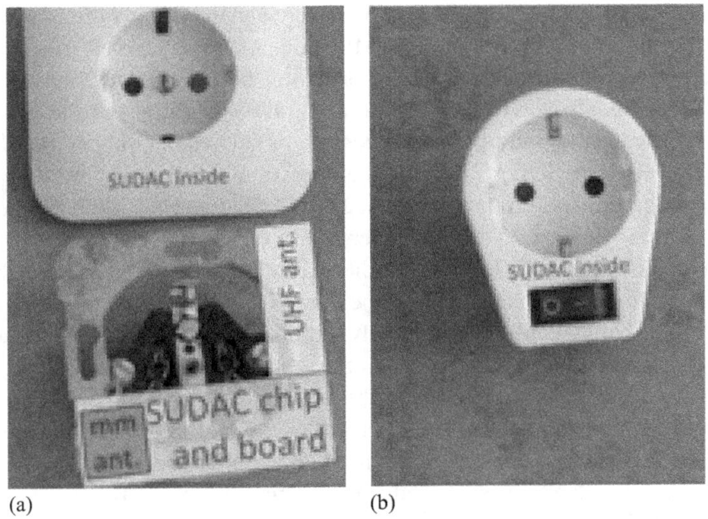

Figure 3.4 Two possible implementations of SUDACs for electrical outlets: (a) integrated into the socket and (b) integrated into an adapter which is to be plugged into a power outlet

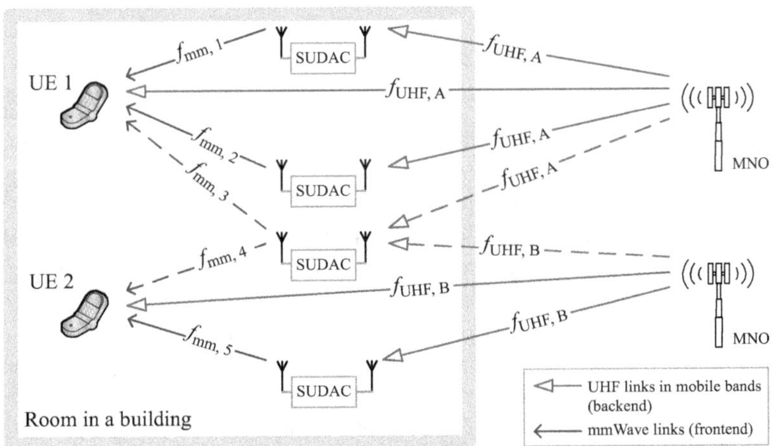

Figure 3.5 Basic principle of a SUDAS for downlink communication. Multiple UEs of different MNOs share the scattered infrastructure represented by the SUDACs via different mmWave frequency bands. Direct links from different MNO BSs to the UEs may exist via different UHF bands

mmWave bands (referred to as frontend), since spatial multiplexing works perfectly in the mobile bands and frequency multiplexing is highly desirable in the mmWave frequency bands. The end-to-end connection between a BS and the UEs is hence split into two segments, and the SUDAS acts as a translator between them. Besides, this does not preclude the existence of direct links between the BS and the UEs which can be further exploited to improve the VMIMO gain. An advantageous of the SUDACs is that they do not need to carry out information decoding. They simply forward the received signal which will be decoded at the UE (downlink) or the BS (uplink). Similarly, MIMO spatial multiplexing precoding is carried out only at the BS and the UEs, respectively. The SUDACs can hence operate in either an AF or a CF manner. Since signal reception and transmission at each SUDAC have a large frequency separation (one in the UHF band and the other one in the mmWave band), both can take place simultaneously.[1] Different from traditional relaying systems, a SUDAS utilizes both a licensed and an unlicensed frequency band in parallel. As the spectrum in the licensed mobile bands is scarce and costly, its efficient use is a major concern and should be maximized. On the contrary, in the very wide unlicensed mmWave band, high spectral efficiency is not needed as the unlicensed band is virtually free of cost. As a result, SUDAS is able to exploit these properties for improving the system performance. Furthermore, a SUDAS offers markedly higher degrees of spatial diversity than a small-cell BS due to its scattered nature. This is particularly pronounced in the mmWave band. For instance, a human body may completely shadow a mmWave link between a small-cell BS and a UE. However, it is rather unlikely that the links to all SUDACs in a room will be shadowed simultaneously. Also, the SUDAS infrastructure can be shared between multiple UEs and multiple MNOs, which is not possible for traditional relaying networks.

3.2.6 SUDAS – application scenarios

The main application scenario for the utilization of a SUDAS is outdoor-to-indoor communication, i.e., the transmitter is located outdoors while the receivers are inside buildings, such as a living room or a large open-plan office. In fact, since most mobile traffic is consumed indoors (especially at home in the evening [45]) and while commuting, such SUDAS installations would cover the most important mobile use cases. SUDACs could be integrated into electrical outlets and many electrical devices, e.g., Figure 3.4. Therefore, having eight SUDACs or more per room appears to be a realistic scenario. Besides providing very high data rates for one UE, a SUDAS is also able to the improve MU-MIMO performance to support high-speed communication for multiple simultaneous users. In the following, we discuss three different practical application scenarios for which SUDAS is able to improve the system performance.

3.2.6.1 Scenario 1

A SUDAS could enable the data rates and spectral efficiencies that future broadcast systems such as LTE-multimedia broadcast multicast services strive to achieve in order to deliver ultra-HD television content to fixed TV sets (which act as a receive-only UE) with indoor antennas (Figure 3.6). To achieve these goals, the use of MIMO

[1] In practice, the time delay incurred by frequency up-/down-conversion at each SUDAC is much shorter than the delay spread of the channel. Thus, the time delay introduced by the frequency conversion does not have a large impact on the system performance.

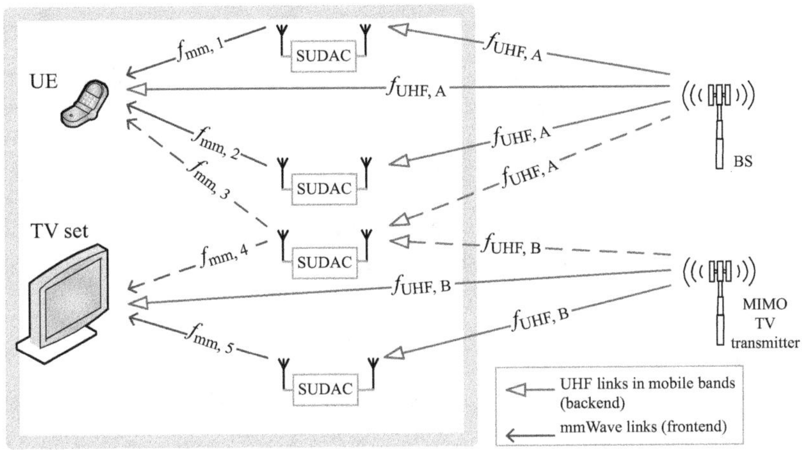

Figure 3.6 A SUDAS used for TV broadcast signal reception with a MIMO TV transmitter and shared by a UE for mobile reception. The TV set can use internal or external indoor antennas

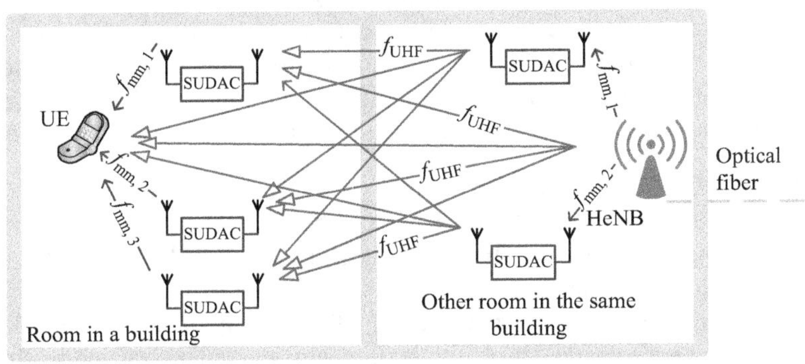

Figure 3.7 SUDAS in two rooms used for connecting a UE to an HeNB in another room

is indispensable and the multiplexing gain brought to the TV set could be increased by the proposed SUDAS.

3.2.6.2 Scenario 2

Figure 3.7 shows another application of the proposed SUDAS. Specifically, a private Home-eNodeB (HeNB) is connected to the public network via optical fiber in one room of the building and can transmit high data rates to UEs in other rooms in the building by first transmitting to the SUDACs in the first room over mmWave. These SUDACs then relay the signal to a sub-6 GHz or UHF band such as the 2.4 GHz industrial, scientific, and medical (ISM) band, where they all transmit concurrently over the same VMIMO channel to another SUDAS in a second room using spatial multiplexing. The second SUDAS finally relays these MIMO signals to the destination

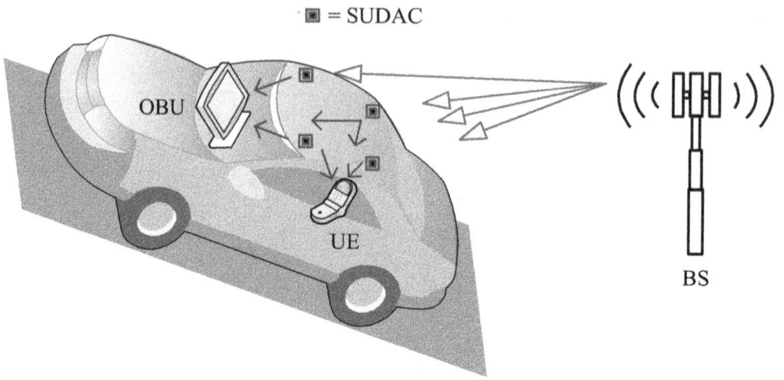

Figure 3.8 BS to vehicle passengers communication via SUDAS

UE once again in the mmWave band. The mmWave transmissions in the first and second rooms do not interfere with each other because of the high attenuation of the walls. In other words, the same mmWave band can be reused extensively. The two new aspects in this scenario are that the HeNB is assisted by several virtual antennas, which can be used in addition to its built-in antennas, and that the UE extends its receiving antennas by the surrounding SUDACs in the same way. The advantage of SUDAS over the existing HeNB concept is hence that rather small devices (HeNB and UE) can share the antennas with SUDACs to achieve higher VMIMO gains.

3.2.6.3 Scenario 3

Besides the installation in indoor devices, another important usage scenario for SUDAS is transportation systems. In particular, SUDACs could be mounted in buses, trains, and cars, too, cf. Figure 3.8. Nowadays, vehicles manufacturers equip cars with multiple antennas to achieve receive diversity for the on-board units (OBU). However, these antennas do not provide data multiplexing gains to the UEs carried by the passengers. As an alternative, the installed antennas could be replaced by SUDACs, and the communication between the antennas/SUDACs and the OBU could thereby be replaced by a standardized wireless air interface operating over mmWave. Then, the UEs of all passengers in the car could benefit from the improved car infrastructure. Possibly, the passengers could employ not only their own car's SUDAS, but even the SUDAS infrastructure of nearby cars could be used when needed, as long as such a resource sharing can be made fair and reasonable. Using an approach similar to the one described above for HeNBs, a device-to-device (D2D) or car-to-car communication between the OBUs of two cars could be established over such a VMIMO link between the two SUDAS, such that higher data rates and/or increased diversity can be achieved.

3.2.7 Comparisons with VMIMO

Existing work on VMIMO covers among others the aspects of channel modeling, analysis of channel capacity and energy efficiency, resource allocation, and routing algorithms. In some system models, all network nodes have identical functionalities, i.e., any node can be source, destination, and relay, while in other models, there are dedicated source and destination nodes (also dubbed "data-gathering nodes") as well as sheer relaying nodes [10–27]. Moreover, some models are hierarchical and allocate the nodes to groups or layers/stages according to their location or distance to the source node, while simpler models represent a plain uniform network [46]. Most of the systems studied in the literature are partly using in-band relaying, e.g., by using time-division-multiplexing with a first time slot reserved for the hop from a node X to a node Y and a second slot for the hop from Y to Z, or they use short-range links for connecting nearby nodes within one of the aforementioned groups or layers. The model in Reference 4 uses a combination of both, i.e., in-band relaying between the groups and short-range links within a group. Our new proposal uses a simpler system model compared to most of the aforementioned approaches: a SUDAS is a relaying system with only two hops in order to ensure the low latencies required by 5G. As a consequence, routing is not an issue here (it is clear that a SUDAC shall forward the signals to the UE or BS, respectively), and the routing algorithm and associated signaling overhead is omitted. An important property of SUDAS is its use of mmWave for the short-range links to connect each individual "virtual antenna" to a UE. For many devices in a room, a wide spectrum is needed for supporting the links established by the devices. As a result, the mmWave band is perfectly suited for this purpose as the links are almost completely insulated from those in the neighboring rooms. It should be also noted that 5G research already considers D2D links and (multihop) relaying by UEs for other UEs at the cell edge. In contrast to the proposed scheme, 5G does not yet envisage a scattered infrastructure realized by SUDACs that serve not only one mobile network but are shared between multiple MNOs. When relaying is realized by other UEs (as suggested for 5G), then this depends strongly on the goodwill of their owners, who might not be willing to spend their battery energy for relaying the signals of other users. In contrast, since SUDACs are fixed and installed in places with electric power supply (no batteries), these problems do not exist. The infrastructure simply belongs to the owner of the apartment who can make it available to his guests.

3.3 SUDAC – implementation

In practice, a SUDAC can perform AF or CF relaying. In this section, we discuss two different options for the implementation of the out-of-band relaying SUDAC: AF and CF[2].

[2] We note that LTE-A has already enabled both in-band and out-of-band relaying. However, both types of relaying take place in licensed bands. Besides, such relays are (high-cost) regenerative (i.e., DF) relays and represent a centralized infrastructure which is in contrast to the scattered structure of the proposed SUDAS.

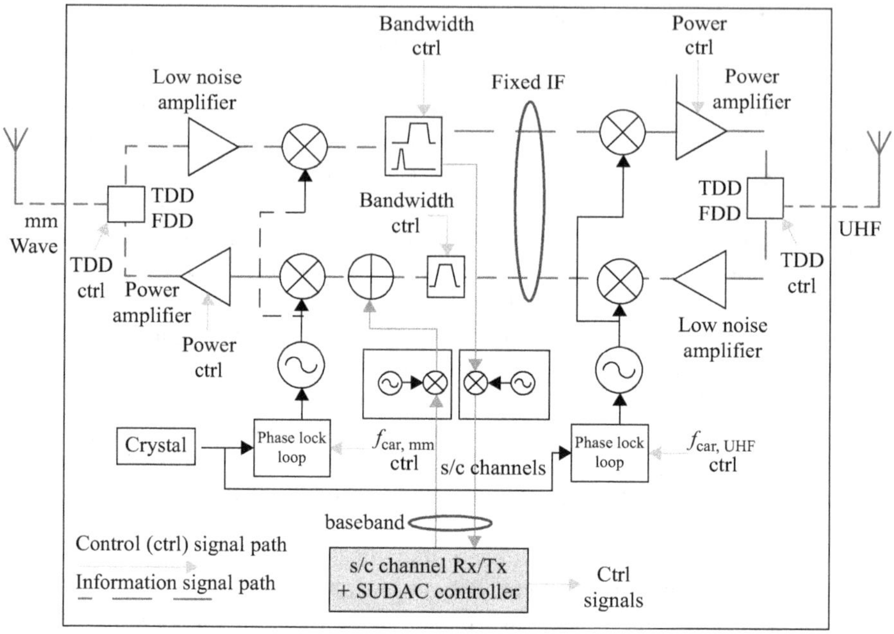

Figure 3.9 A simplified block diagram of an AF SUDAC, where the payload is processed only in the analog domain. Band selection filters in the UHF and mmWave bands have been omitted for specificity

3.3.1 Amplify-and-forward

The main advantage of an AF SUDAC is its potentially low cost and can be implemented with both TDD mode and FDD mode. Figure 3.9 shows a simplified block diagram of such an AF device. The dashed lines denote payload signals which are only processed in the analog domain. In an alternative realization, digital processing can be used, which requires down-conversion of the payload signals to the baseband. The processing (analog or digital) in the SUDAC comprises bandpass filtering in order to avoid the noise amplification and/or relaying of interference surrounding the useful signal. Besides, if the signals can have a flexible bandwidth as required in LTE, then the passband of the bandpass filter has to be dynamically adjusted accordingly. Besides, a SUDAC needs to be configured to know the carrier frequency and bandwidth of the relayed signal. This information can be provided by the UE over a dedicated status/control (s/c) channel in the mmWave band that is frequency-multiplexed with the payload signal.

The s/c channels should carry pilots that enable channel estimation in the mmWave links. Thanks to the high path loss in the mmWave band, the impulse response of an indoor channel is rather short (10–20 ns [47]). This leads to little frequency selectivity over typical bandwidths of up to 20 MHz and simple channel estimation. In the FDD mode, a UE needs to transmit and receive in the mmWave

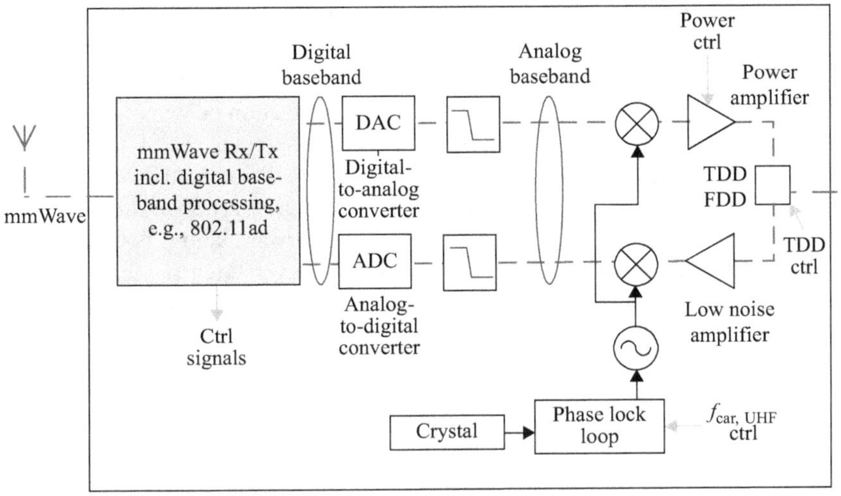

Figure 3.10 Block diagram of a CF SUDAC (either using IEEE 802.11ad or a dedicated air interface). Band selection filters in UHF have been omitted for simplicity

band simultaneously. In order to avoid self-interference due to leakage, the transmit and receive frequency bands have to be separated sufficiently. For applications such as TV reception, unidirectional transmission is used. However, also in this case there needs to be an s/c channel in the return direction to allow handshaking between the UE (i.e., TV set) and the SUDAC for control information exchange. This solitary s/c channel without accompanying payload channel must be sufficiently separated in frequency from the signal in the forward direction to avoid self-interference, of course. Such a solitary s/c channel might also be needed to realize a return link for exchanging control information continuously in the mmWave band, when TDD transmission is used by the BS. Additionally, special rendezvous channels may be needed for the discovery procedure.

Besides the cost, further arguments in favor of an AF SUDAC could be reduced power consumption and a shorter end-to-end latency. However, the price to pay is a lower flexibility than for the solutions presented hereafter. For instance, AF SUDACs are limited to use frequency division multiple access for the mmWave links, while CF SUDACs, as will be discussed in the next section, can employ further multiplexing and multiple access schemes. Moreover, for purely analog AF SUDACs (Figure 3.9), it is difficult to realize fully flexible bandwidths (e.g., from 1.4 to 20 MHz as in LTE) and to equalize the mmWave channel, e.g., compensate the Doppler shift and the slight frequency selectivity. These issues will be discussed later in this chapter.

3.3.2 Compress-and-forward

If the SUDACs are closer to the UEs compared to the BS, the CF protocol provides superior performance compared to AF and DF in relaying networks. Figure 3.10 shows the block diagram of a CF SUDAC. In the downlink, the signal from the BS received in the mobile (UHF) band is sampled. The corresponding binary representation of the in-phase and quadrature-phase (IQ) samples is then compressed and transmitted over the mmWave bands to the UE. For instance, vector quantization or Wyner-Ziv coding can be applied to CF-SUDAS to compress the observed signal at the SUDAS before sending it to the UEs. In the uplink, the payload from the UE received by the SUDAC contains the binary representation of the IQ samples that are transmitted over the mobile (UHF) band to the BS. This scheme is straightforward, and it is flexible due to the adopted digital baseband signal processing. However, such flexibility comes at the expense of higher cost and/or higher power consumption than other alternative SUDAC realizations.

3.4 Mathematical system model

In this section, we propose a mathematical system model for the AF-SUDAS and evaluate the performance of the proposed SUDAS via simulation.

3.4.1 SUDAS downlink communication model

We consider a mathematical model for a SUDAS assisted orthogonal frequency division multiple access downlink transmission network which consists of one BS, one SUDAS, and one UE, cf. Figure 3.11. The SUDAS consists of M SUDACs, and we assume that the SUDACs are installed in electrical wall outlets and can cooperate with each other by sharing channel state information and received signals via low data rate power line communication links. In other words, joint processing between SUDACs is possible such that the SUDACs can fully utilize their antennas.

The BS is equipped with N_{T} transmit antennas transmitting signals in a licensed frequency band. The UEs are equipped with two set of antennas. The first set consists of R antennas used for receiving signal in the licensed band and the other set consists of one antenna used for receiving signal in the unlicensed frequency band. We focus on a wideband multicarrier communication system with n_{F} subcarriers. The communication channel is time-invariant within a scheduling slot. The BS performs spatial multiplexing in the licensed band. The data symbol vector $\mathbf{d}^{[i]} \in \mathbb{C}^{N_{\mathrm{S}}\times 1}$ on subcarrier $i \in \{1, \ldots, n_F\}$ for the UE is precoded at the BS as

$$\mathbf{x}^{[i]} = \mathbf{P}^{[i]}\mathbf{d}^{[i]}, \tag{3.1}$$

where $\mathbf{P}^{[i]} \in \mathbb{C}^{N_{\mathrm{T}}\times N_{\mathrm{S}}}$ is the precoding matrix adopted by the BS on subcarrier i. The signals received on subcarrier i at the M SUDACs for the UE are given by

$$\mathbf{y}_{\mathrm{S}}^{[i]} = \mathbf{H}_{\mathrm{B}\to\mathrm{S}}^{[i]}\mathbf{x}^{[i]} + \mathbf{z}^{[i]}, \tag{3.2}$$

where $\mathbf{y}_{\mathrm{S}}^{[i]} = [y_{\mathrm{S}_1}^{[i]} \ldots y_{\mathrm{S}_m}^{[i]} \ldots y_{\mathrm{S}_M}^{[i]}]^T$ and $y_{\mathrm{S}_m}^{[i]} \in \{1, \ldots, M\}$ denotes the received signal at SUDAC m. $\mathbf{H}_{\mathrm{B}\to\mathrm{S}}^{[i]}$ is the $M \times N_{\mathrm{T}}$ MIMO channel matrix between the BS and the M SUDACs on subcarrier i and captures the joint effects of path loss, shadowing, and multi-path fading. $\mathbf{z}^{[i]}$ is the additive white Gaussian noise (AWGN) vectors with

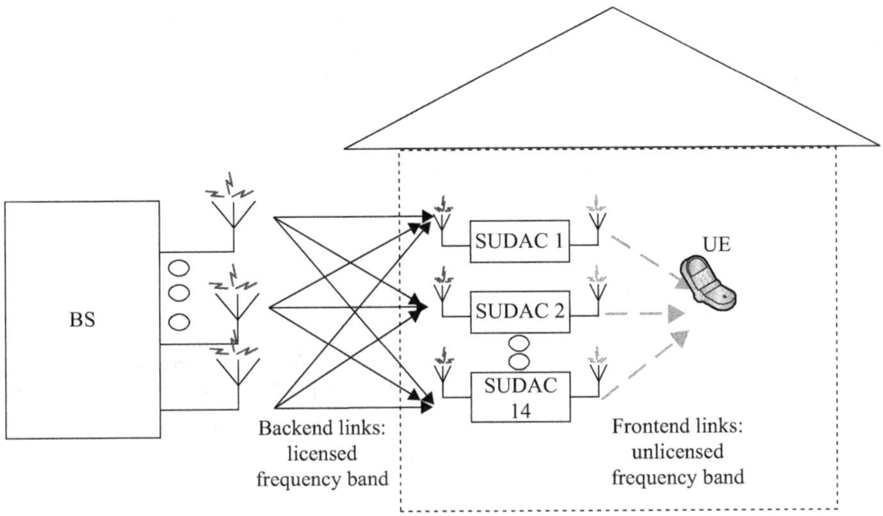

Figure 3.11 SUDAS simulation topology with a BS, M = 14 SUDACs, and one UE

distribution $\mathcal{CN}(\mathbf{0}, \boldsymbol{\Sigma})$ on subcarrier i impairing the M SUDACs where $\boldsymbol{\Sigma}$ is an $M \times M$ diagonal covariance matrix with each main diagonal element given by N_0.

Then, each SUDAC performs frequency repetition in the unlicensed band. In particular, the M SUDACs multiply the received signal vector on subcarrier i by $\mathbf{F}^{[i]} \in \mathbb{C}^{M\times M}$ and forward the processed signal vector to the UE on subcarrier i in M different independent frequency sub-bands in the unlicensed spectrum, cf. Figure 3.3. In other words, each SUDAC forwards its received signal in a different sub-band and thereby avoids further multiple access interference in the unlicensed spectrum.

Then, the signal received at the UE on subcarrier i from the SUDACs in the M frequency bands, $\mathbf{y}^{[i]}_{\text{S}\to\text{UE}} \in \mathbb{C}^{M\times 1}$, and the BS, $\mathbf{y}^{[i]}_{\text{B}\to\text{UE}} \in \mathbb{C}^{R\times 1}$, can be expressed as

$$\begin{aligned}\mathbf{y}^{[i]}_{\text{S}\to\text{UE}} &= \mathbf{H}^{[i]}_{\text{S}\to\text{UE}}\mathbf{F}^{[i]}\left(\mathbf{H}^{[i]}_{\text{B}\to\text{S}}\mathbf{x}^{[i]} + \mathbf{z}^{[i]}\right) + \mathbf{n}^{[i]} \\ &= \underbrace{\mathbf{H}^{[i]}_{\text{S}\to\text{UE}}\mathbf{F}^{[i]}\mathbf{H}^{[i]}_{\text{B}\to\text{S}}\mathbf{P}^{[i]}\mathbf{s}^{[i]}}_{\text{desired signal}} + \underbrace{\mathbf{H}^{[i]}_{\text{S}\to\text{UE}}\mathbf{F}^{[i]}\mathbf{z}^{[i]}}_{\text{amplified noise}} + \mathbf{n}^{[i]} \quad \text{and} \\ \mathbf{y}^{[i]}_{\text{B}\to\text{UE}} &= \mathbf{H}^{[i]}_{\text{B}\to\text{UE}}\mathbf{x}^{[i]} + \mathbf{u}^{[i]},\end{aligned} \tag{3.3}$$

respectively. The m-th element of vector $\mathbf{y}^{[i]}_{\text{S}\to\text{UE}}$ represents the received signal at the UE in the m-th unlicensed frequency sub-band. Besides, since the SUDACs forward the received signals in different frequency bands, $\mathbf{H}^{[i]}_{\text{S}\to\text{UE}}$ is a diagonal matrix with the diagonal elements representing the channel gain between the SUDACs and the UE on subcarrier i in the unlicensed sub-band m. Matrix $\mathbf{H}^{[i]}_{\text{B}\to\text{UE}} \in \mathbb{C}^{R\times N_\text{T}}$ is the direct channel between the BS and the UE. $\mathbf{n}^{[i]} \in \mathbb{C}^{M\times 1}$ and $\mathbf{u}^{[i]} \in \mathbb{C}^{R\times 1}$ are the AWGN vector at the UE on subcarrier i in the unlicensed and licensed bands with distribution $\mathcal{CN}(\mathbf{0}, \boldsymbol{\Sigma})$ and $\mathcal{CN}(\mathbf{0}, \boldsymbol{\Lambda})$, respectively. $\boldsymbol{\Sigma}$ and $\boldsymbol{\Lambda}$ are $M \times M$ and $R \times R$ diagonal matrices, and each main diagonal element is equal to N_0. In order to simplify the subsequent mathematical expressions and without loss of generality, we assume in the following a normalized noise variance of $N_0 = 1$ at all receive antennas of the SUDACs and the UEs. Besides, $M + R \geq N_\mathrm{S}$ and the UE employs a linear receiver for estimating the data vector symbol received in the M different frequency bands in the unlicensed band. The estimated data vector symbols, $\hat{\mathbf{d}}^{[i]} \in \mathbb{C}^{N_\mathrm{S} \times 1}$, on subcarrier i are given by:

$$\hat{\mathbf{d}}^{[i]} = (\mathbf{W}^{[i]})^H \begin{bmatrix} \mathbf{y}^{[i]}_{\mathrm{S}\rightarrow\mathrm{UE}} \\ \mathbf{y}^{[i]}_{\mathrm{B}\rightarrow\mathrm{UE}} \end{bmatrix}, \tag{3.4}$$

where $\mathbf{W}^{[i]} \in \mathbb{C}^{(M+R) \times N_\mathrm{S}}$ is a post-processing matrix used for subcarrier i at the UE. Without loss of generality, we assume that $\mathcal{E}\{\mathbf{d}^{[i]}(\mathbf{d}^{[i]})^H\} = \mathbf{I}_{N_\mathrm{S}}$ where $\mathcal{E}\{\cdot\}$ denotes statistical expectation. As a result, the mean square error matrix for the data transmission on subcarrier i for the UE via the SUDAS and the optimal post-processing matrix are given by

$$\begin{aligned} \mathbf{E}^{[i]} &= \mathcal{E}\{(\hat{\mathbf{d}}^{[i]} - \mathbf{d}^{[i]})(\hat{\mathbf{d}}^{[i]} - \mathbf{d}^{[i]})^H\} \\ &= \left[\mathbf{I}_{N_\mathrm{S}} + (\mathbf{\Gamma}^{[i]})^H(\mathbf{\Theta}^{[i]})^{-1}(\mathbf{\Gamma}^{[i]})\right]^{-1} && (3.5) \\ \text{and } \mathbf{W}^{[i]} &= (\mathbf{\Gamma}^{[i]}(\mathbf{\Gamma}^{[i]})^H + \mathbf{\Theta}^{[i]})^{-1}\mathbf{\Gamma}^{[i]}, && (3.6) \end{aligned}$$

respectively, where $\mathbf{\Gamma}^{[i]}$ is the effective MIMO channel matrix between the BS and the UE via the SUDAS on subcarrier i, and $\mathbf{\Theta}^{[i]}$ is the corresponding equivalent noise covariance matrix. These matrices are given by

$$\begin{aligned} \mathbf{\Gamma}^{[i]} &= \mathbf{H}^{[i]}_{\mathrm{S}\rightarrow\mathrm{UE}}\mathbf{F}^{[i]}\mathbf{H}^{[i]}_{\mathrm{B}\rightarrow\mathrm{S}}\mathbf{P}^{[i]} \quad \text{and} \\ \mathbf{\Theta}^{[i]} &= \left(\mathbf{H}^{[i]}_{\mathrm{S}\rightarrow\mathrm{UE}}\mathbf{F}^{[i]}\right)\left(\mathbf{H}^{[i]}_{\mathrm{S}\rightarrow\mathrm{UE}}\mathbf{F}^{[i]}\right)^H + \mathbf{I}_M. \end{aligned} \tag{3.7}$$

In this following, we first introduce the adopted system performance measure. Then, the resource allocation and scheduling design is formulated as an optimization problem.

3.4.2 System throughput

The end-to-end achievable data rate $R^{[i]}$ on subcarrier i between the BS and UE via the SUDAS is given by

$$R^{[i]} = -\log_2\left(\det\left[\mathbf{E}^{[i]}\right]\right), \tag{3.8}$$

where $\det(\cdot)$ represents the determinant of a matrix. The data rate (bit/s) delivered to the UE can be expressed as

$$\rho = \sum_{i=1}^{n_F} R^{[i]}. \tag{3.9}$$

This system model can be easily extended to a CF SUDAS by following a similar approach as in Reference 48.

Table 3.1 System parameters

Mobile band carrier center frequency	**800 MHz**
mmWave band carrier center frequency	60 GHz
Useful signal bandwidth	200 MHz
Channel model for mobile band to/from BS	Winner+ outdoor-to-indoor
Channel model for mmWave links	IEEE 802.11ad LOS living room scenario
BS antenna height	25 m
Height of SUDAC and UE	1.5 m
Distance BS ↔ SUDAC	96 m
Distance BS ↔ UE	100 m
Distance SUDAC ↔ UE	4 m
Antenna (mobile band) gain of SUDAC and UE	0 dBi
Antenna (mmWave band) gain of SUDAC and UE	3 dBi (semi-isotropic)
Noise figure of SUDAC and UE	8 dB
Maximum transmit power per SUDAC in mmWave	13 dBm

3.5 Numerical results

We focus on a single-cell downlink communication scenario where an LTE-type BS equipped with 16 antennas serves a UE via a SUDAS with 14 SUDACs. The UE is equipped with two mobile band antennas and one mmWave band antenna, while each SUDAC has one mobile band antenna and one mmWave antenna. The important simulation parameters are summarized in Table 3.1. Carrier aggregation is performed such that a total bandwidth of 200 MHz is available in the mobile band for downlink communication. We assume that the downlink channels from the BS to the SUDAS are mutually statistically independent for any pair of BS antennas. Also, in order to reveal the potential of SUDAS, we assume perfect synchronization and channel estimation. Furthermore, we adopt equal power allocation at both the BS and each SUDAC. The system achievable data rate is computed by (3.9).

3.5.1 Average system throughput versus transmit power

Figure 3.12 shows the average system throughput versus the equivalent isotropically radiated power (EIRP) (dBm) for different downlink transmission schemes. As can be seen, both SUDAS with AF and CF relaying realize an extraordinarily high data rate bringing 10 Gbit/s to the indoor UEs over a distance of 100 m. For comparison, we also show the system performance of a benchmark MIMO system and a baseline system. For the benchmark MIMO system, we assume that the UE is equipped with 16 receive antennas without the help of the SUDAS, and optimal resource allocation is performed to maximize the system throughput. In other words, the average system throughput of the benchmark system serves as a performance upper bound for the proposed SUDAS. As for the baseline system, the BS performs optimal resource allocation

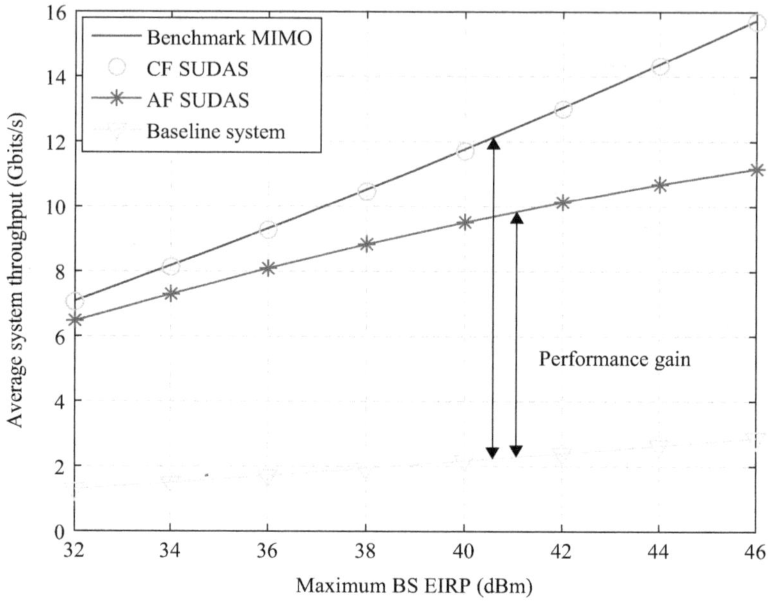

Figure 3.12 Average system throughput (bit/s) versus the EIRP (dBm) for different downlink transmission schemes

and utilizes only the licensed frequency band without the help of the SUDAS and the UE. As can be observed, the proposed SUDAS is able to exploit most of the spatial multiplexing gain, even though the UE is equipped with a small number of receive antennas compared to the BS. Besides, a huge performance gain is achieved by the SUDAS compared to the baseline system as the SUDAS utilizes both the licensed and the unlicensed bands. On the other hand, the performance of the proposed scheme and the benchmark system improves rapidly with increasing BS transmit power budget alleviating the system bottleneck, i.e., the BS-to-SUDAS link.

3.5.2 Average system throughput versus number of SUDACs

Figure 3.13 depicts the average system throughput versus the number of SUDACs for $N_T = 16$. The maximum EIRP of the BS is 46 dBm. It can be observed that the system throughput grows with the number of SUDACs for the proposed SUDAS. In particular, a higher spatial multiplexing gain can be achieved when we increase the number of SUDACs M for $N_T \geq M + 2$. On the other hand, for $M + 2 > N_T$, increasing the number of SUDACs in the system leads to more spatial diversity which also improves the system throughput. Besides, for a large enough number of SUDACs, the performance of the proposed SUDAS closely approaches that of the benchmark system which demonstrates the excellent potential of the proposed scheme in exploiting the spatial multiplexing gain.

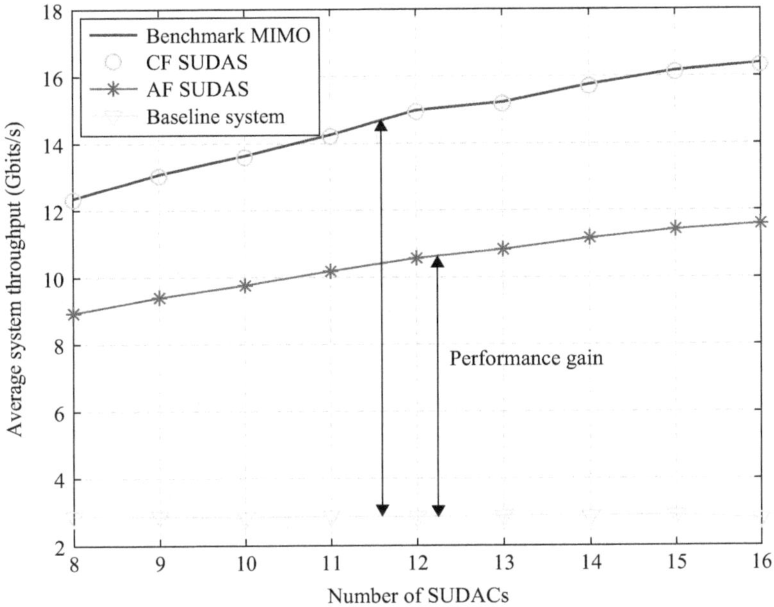

Figure 3.13 Average system throughput (Mbits/s) versus the number of SUDACs for a maximum EIRP at the BS of 46 dBm for different downlink transmission schemes

3.6 SUDAS – challenges

Although the proposed SUDAS has a strong potential to reach the data rate requirements set by 5G, there are many questions and challenges that need to be addressed individually. In this section, we discuss some research challenges arising in the practical implementation of SUDAS for outdoor-to-indoor communication.

3.6.1 Keyhole effect

The SUDAS performance requires suitable channel conditions for creating a VMIMO system from an outdoor BS to an indoor SUDAC. However, it is known that the so-called keyhole effect can lead to performance degradation of the VMIMO channel because of the following three phenomena [49, 50]:

1. Spatial keyholes
2. Diffraction-induced keyholes
3. Modal keyholes

Fortunately, Almers *et al.* [51] show that there is only a negligible spatial keyhole effect if an opening of at least 30 cm × 30 cm exists in a metal wall. As a result, whenever the SUDAS is used in an environment that meets this condition (e.g., rooms with typical window sizes), we can safely assume that spatial keyholes will not affect the

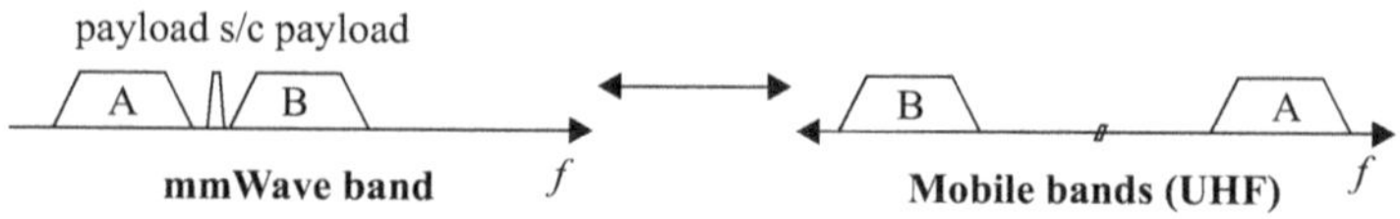

Figure 3.14 Spectrum occupation in mmWave band and mobile bands (UHF): aggregation of multiple mobile carriers in the mmWave band

performance of a SUDAS. On the other hand, Chizhik *et al.* [50] relate diffraction-induced keyholes to outdoor scenarios and also propose a solution for the placement of the BS antennas to counter this effect. Besides, modal keyholes will occur when the waves propagate along hallway-like structures. This could be the case in an indoor scenario or a street canyon. The authors in Reference 49 proposed a relaying system for MIMO which could mitigate such effects. Therefore, it appears that SUDAS will not severely suffer from the above three described keyhole effects. However, further channel measurements and studies should be conducted to verify the channel conditions needed for the successful implementation of SUDAS.

3.6.2 Carrier aggregation

As carrier aggregation will probably be used in 5G, a SUDAS has to allow the relaying of several carriers simultaneously. A simple solution is to assign one SUDAC to one mobile carrier frequency only and to assign further carriers to other SUDACs. Alternatively, several independent RF chains could be bundled in a single SUDAC device. Similarly, several mobile downlink channels could be aggregated inside a SUDAC. For instance, a "double" AF SUDAC could bundle two carriers that are separated in frequency in the mobile bands to adjacent channels in the mmWave band, see Figure 3.14.

3.6.3 Resource allocation for multiple MNOs

As stated before, a SUDAS should allow any UE to connect to the network of their associated MNOs. Thereby, one SUDAS can host different UEs associated with different MNOs concurrently. This must be taken into account by the resource allocation. The resources allocated to a SUDAS are not only the time, frequency, code, and space occupied in the mmWave band but also the relaying (i.e., hardware) resources represented by the pool of SUDACs in the indoor environment. A fair resource allocation must ensure that the available data rates are assigned to each UE according to the quality of service requirements of its services, the number of carriers that can be relayed, and to the UE's and SUDAC's capabilities to support certain carrier frequencies, carrier spacings, bandwidths, channel switching times, transmit powers, etc. Moreover, policies defined by the MNO concerning the data rates granted to certain UEs or services should be taken into account. Besides, the resource allocation must ensure that interference between multiple mmWave links or other users of the mmWave bands (such as IEEE 802.11ad) is avoided.

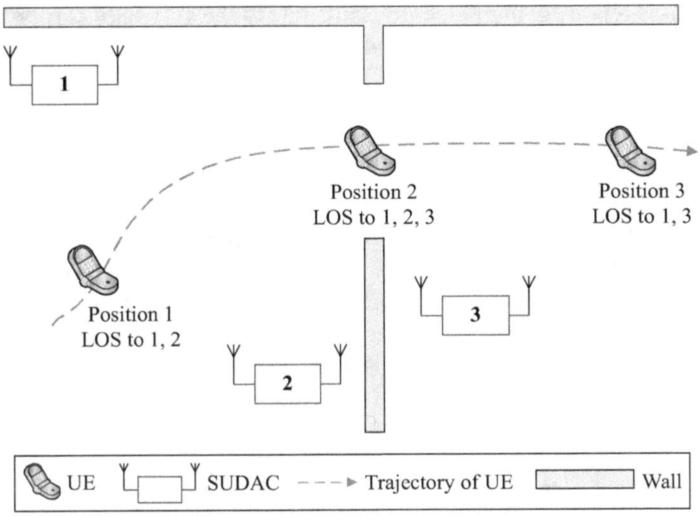

Figure 3.15 New SUDACs discovery scenario

The resource allocation can be implemented in a centralized (e.g., by an entity in the core network cooperating with the other MNOs) or in a distributed manner, cf. e.g., Reference 52. According to our investigations, the best solution appears to be a distributed resource allocation carried out in the UEs but supported by the BSs. The motivation for a distributed resource allocation is among others a lower latency and a lower control plane overhead compared to a centralized approach. In the distributed approach, the UEs can interact with the BSs in order to assist the radio resource management, e.g., in the carrier selection for the mobile bands.

3.6.4 Mobility

Figure 3.15 shows a typical scenario where a UE moves from one room equipped with two SUDACs to another room equipped with one SUDAC. One might wonder whether this system contains two individual infrastructures (one SUDAS per room) or a single larger SUDAS with three SUDACs. The shown trajectory of the UE suggests that that both rooms cannot simply be treated as two separated SUDASs. In position 2, all three SUDACs have a LOS connection to the UE, and the resource allocation algorithm has to cover such cases. Hence, a SUDAS can become a quite extended network of SUDACs, and any UE sees only its local network environment at any time. As the UEs move, the SUDACs must constantly scan passively for approaching UEs and contact them. Then, the resources must be re-allocated locally in the immediate neighborhood of the UE and potentially a hand-over to another SUDAC has to be initiated. Passive scanning should be chosen for the SUDACs in this discovery procedure in order to reduce the electromagnetic pollution and to improve the energy efficiency of the system. With regard to the channel characteristics, the user mobility causes relatively high Doppler shifts due to the very high carrier frequency in the mmWave band. For instance, moving the UE at 10 km/h when turning around will cause an instantaneous Doppler shift of 550 Hz.

3.6.5 Synchronization and channel estimation

The aforementioned momentary Doppler shift is high but it is still within the range supported by LTE, where the maximum tolerable Doppler frequency is 840 Hz. Therefore, for an AF SUDAC, the UEs or the BS are in principle able to compensate the channel variations over the complete end-to-end link. However, the time-derivative of the Doppler shift can be very high in the mmWave band and might exceed the capabilities of the BS or the UEs. Moreover, indoor SUDACs are fortunately not mobile but installed in a more or less fixed location. Therefore, the channel between the BS and the SUDACs in the mobile bands is frequency selective but quite static, while the channel between the SUDACs and the UEs in the mmWave band is time-variant but affected only by relatively flat fading (over up to 20 MHz bandwidth). Sophisticated AF SUDACs could exploit these properties and compensate the channel variations over the mmWave link, e.g., remove the Doppler shift if required. The situation changes when it comes to the usage of SUDACs in trains, buses, and cars. Now, also the channel between the BS and the SUDAC is time-variant with possibly high Doppler shift. This would add to the Doppler shift originating from the mmWave link, e.g., when the user moves in a train. Hence, the compensation of the Doppler shift by the SUDAC as described above becomes indispensable in the case of AF relaying. On the other hand, for fixed indoor SUDACs installed in an industrial environment such as a factory building, the delay spread may be much longer than the aforementioned 10–20 ns [47]. The longer delays are due to the larger spaces lacking absorbing walls and including metallic reflectors. For such scenarios, the mmWave channel can be quite frequency selective, and an AF SUDAC is unable to compensate these effects. Instead, equalization must be performed at the BS or the UEs. Additionally, the phase noise mask of a mmWave carrier is significantly worse than that typically encountered in mobile UHF bands. An AF SUDAC could compensate part of this phase noise by utilizing the pilots embedded in the s/c channel. Without such compensation, it is questionable whether a BS or a UE would be able to estimate and compensate the phase noise.

3.6.6 Power consumption

With regard to power consumption, according to Reference 53, the total power consumption of a transceiver chip (that processes similar data rates in receive and transmit directions) appears to be governed largely by the power dissipated by the power amplifier. In contrast, the power required for the digital baseband processing (e.g., forward error correction decoding) is probably a minor part compared to the consumption in the power amplifier. As a result, AF and CF SUDASs are expected to consume similar amounts of power, and the choice of the SUDAS relaying protocol may depend more on the hardware complexity.

3.7 Conclusions

In this chapter, we have proposed a novel SUDAS with the objective of achieving the 10 Gbit/s data rate goal set by 5G for indoor UEs. The proposed SUDAS exploits the benefits of the licensed and unlicensed frequency bands simultaneously. In particular, it translates the spatial multiplexing in the licensed bands into frequency multiplexing in the unlicensed bands for boosting the end-to-end data rate via VMIMO. It is expected that the proposed SUDAS can further enhance the system performance when advanced resource allocation technique is employed. Besides, we have also discussed some potential application scenarios where deployment of SUDAS appears to be beneficial. Also, we have investigated different potential realizations of SUDAS and the corresponding implementation challenges. It is expected that the proposed SUDAS is able to bridge the gap between the current technology and the high data rate requirement of the next generation communication systems.

Acknowledgements

This work was supported in part by the AvH Professorship Program of the Alexander von Humboldt Foundation.

References

[1] D. Tse and P. Viswanath, *Fundamentals of Wireless Communication*, 1st ed. Cambridge University Press, United Kingdom, 2005.

[2] A. Goldsmith, *Wireless Communications*. Cambridge University Press, United Kingdom, 2005.

[3] V. K. N. Lau and Y. K. Kwok, *Channel Adaptation Technologies and Cross Layer Design for Wireless Systems with Multiple Antennas – Theory and Applications*, 1st ed. Wiley John Proakis Telecom Series, New York, 2005.

[4] M. Dohler, "Virtual Antenna Arrays," Ph.D. dissertation, Kings College London, University of London, Nov. 2003.

[5] D. W. K. Ng and R. Schober, "Cross-Layer Scheduling Design for Cooperative Wireless Two-Way Relay Networks," *Cooperative Cellular Wireless Networks*, Cambridge University Press, United Kingdom, p. 259, 2011.

[6] D. W. K. Ng, E. S. Lo, and R. Schober, "Energy-Efficient Resource Allocation in Multi-Cell OFDMA Systems with Limited Backhaul Capacity," *IEEE Trans. Wireless Commun.*, vol. 11, pp. 3618–3631, Oct. 2012.

[7] D. Ng and R. Schober, "Dynamic Resource Allocation in OFDMA Systems with Full-Duplex and Hybrid Relaying," in *Proc. IEEE Intern. Commun. Conf.*, Jun. 2011, pp. 1–6.

[8] T. Riihonen, S. Werner, and R. Wichman, "Hybrid Full-Duplex/Half-Duplex Relaying with Transmit Power Adaptation," *IEEE Trans. Wireless Commun.*, vol. 10, pp. 3074–3085, Sep. 2011.

[9] D. W. K. Ng, E. S. Lo, and R. Schober, "Dynamic Resource Allocation in MIMO-OFDMA Systems with Full-Duplex and Hybrid Relaying," *IEEE Trans. Commun.*, vol. 60, pp. 1291–1304, May 2012.

[10] I. Hammerstrom and A. Wittneben, "Power Allocation Schemes for Amplify-and-Forward MIMO-OFDM Relay Links," *IEEE Trans. Wireless Commun.*, vol. 6, pp. 2798–2802, Aug. 2007.

[11] N. Zlatanov, R. Schober, and P. Popovski, "Buffer-Aided Relaying with Adaptive Link Selection," *IEEE J. Select. Areas Commun.*, vol. 31, pp. 1530–1542, Aug. 2013.

[12] N. Zlatanov and R. Schober, "Buffer-Aided Relaying with Adaptive Link Selection-Fixed and Mixed Rate Transmission," *IEEE Trans. Inf. Theory*, vol. 59, pp. 2816–2840, May 2013.

[13] Y. Liu, M. Tao, B. Li, and H. Shen, "Optimization Framework and Graph-Based Approach for Relay-Assisted Bidirectional OFDMA Cellular Networks," *IEEE Trans. Wireless Commun.*, vol. 9, no. 11, pp. 3490–3500, Nov. 2010.

[14] Y. Liu and M. Tao, "Optimal Channel and Relay Assignment in OFDM-Based Multi-Relay Multi-Pair Two-Way Communication Networks," *IEEE Trans. Commun.*, vol. 60, no. 2, pp. 317–321, Feb. 2012.

[15] Y. Liu, J. Mo, and M. Tao, "QoS-Aware Transmission Policies for OFDM Bidirectional Decode-and-Forward Relaying," *IEEE Trans. Wireless Commun.*, vol. 12, pp. 2206–2216, May 2013.

[16] E. Calvo, J. Vidal, and J. R. Fonollosa, "Optimal Resource Allocation in Relay-Assisted Cellular Networks with Partial CSI," *IEEE Trans. Signal Process.*, vol. 57, pp. 2809–2823, Jul. 2009.

[17] Y. Yu and Y. Hua, "Power Allocation for a MIMO Relay System with Multiple-Antenna Users," *IEEE Trans. Signal Process.*, vol. 58, pp. 2823–2835, May 2010.

[18] C. Wang, H. Wang, D. Ng, X. Xia, and C. Liu, "Joint Beamforming and Power Allocation for Secrecy in Peer-to-Peer Relay Networks," *IEEE Trans.* Wireless Commun., vol. 14, no. 6, pp. 3280–3293, June 2015.

[19] C. Li, X. Wang, L. Yang, and W.-P. Zhu, "A Joint Source and Relay Power Allocation Scheme for a Class of MIMO Relay Systems," *IEEE Trans. Signal Process.*, vol. 57, pp. 4852–4860, Dec. 2009.

[20] I. Ahmed, A. Ikhlef, D. Ng, and R. Schober, "Power Allocation for a Hybrid Energy Harvesting Relay System with Imperfect Channel and Energy State Information," in *Proc. IEEE Wireless Commun. Netw. Conf.*, Apr. 2014, pp. 990–995.

[21] Y. Fu, L. Yang, and W.-P. Zhu, "A Nearly Optimal Amplify-and-Forward Relaying Scheme for Two-Hop MIMO Multi-Relay Networks," *IEEE Commun. Lett.*, vol. 14, pp. 229–231, Mar. 2010.

[22] D. Ng and R. Schober, "Resource Allocation for Secure OFDMA Decode-and-Forward Relay Networks," in *12th Canadian Workshop Inf. Theory (CWIT)*, May 2011, pp. 202–205.

[23] D. Ng and R. Schober, "Resource Allocation and Scheduling in Multi-Cell OFDMA Decode-and-Forward Relaying Networks," in *Proc. IEEE Global Telecommun. Conf.*, Dec. 2010, pp. 1–6.

[24] I. Ahmed, A. Ikhlef, D. Ng, and R. Schober, "Optimal Resource Allocation for Energy Harvesting Two-Way Relay Systems with Channel Uncertainty," in *IEEE Global Conf. Signal Inf. Process.*, Dec. 2013, pp. 345–348.

[25] D. Ng and R. Schober, "Cross-Layer Scheduling Design for OFDMA Two-Way Amplify-and-Forward Relay Networks," in *Proc. IEEE Wireless Commun. Netw. Conf.*, Apr. 2010, pp. 1–6.

[26] D. Ng and R. Schober, "Cross-Layer Scheduling for OFDMA Amplify-and-Forward Relay Networks," in *Proc. IEEE Veh. Techn. Conf.*, Sep. 2009, pp. 1–5.

[27] X. Chen, L. Lei, H. Zhang, and C. Yuen, "Large-Scale MIMO Relaying Techniques for Physical Layer Security: AF or DF?" *IEEE Trans. Wireless Commun.*, vol. PP, no. 99, 2015.

[28] J. Zeng and H. Minn, "A Novel OFDMA Ranging Method Exploiting Multiuser Diversity," *IEEE Trans. Commun.*, vol. 58, pp. 945–955, Mar. 2010.

[29] P. Chan and R. Cheng, "Capacity Maximization for Zero-Forcing MIMO-OFDMA Downlink Systems with Multiuser Diversity," *IEEE Trans. Wireless Commun.*, vol. 6, pp. 1880–1889, May 2007.

[30] D. Ng, E. Lo, and R. Schober, "Energy-Efficient Resource Allocation in OFDMA Systems with Hybrid Energy Harvesting Base Station," *IEEE Trans. Wireless Commun.*, vol. 12, pp. 3412–3427, Jul. 2013.

[31] H. Bogucka and A. Conti, "Degrees of Freedom for Energy Savings in Practical Adaptive Wireless Systems," *IEEE Commun. Mag.*, vol. 49, pp. 38–45, Jun. 2011.

[32] A. Chowdhery, W. Yu, and J. M. Cioffi, "Cooperative Wireless Multicell OFDMA Network with Backhaul Capacity Constraints," in *Proc. IEEE Intern. Commun. Conf.*, Jun. 2011, pp. 1–6.

[33] T. Marzetta, "Noncooperative Cellular Wireless with Unlimited Numbers of Base Station Antennas," *IEEE Trans. Wireless Commun.*, vol. 9, pp. 3590–3600, Nov. 2010.

[34] Y. Wu, R. Schober, D. W. K. Ng, C. Xiao, and G. Caire, "Secure Massive MIMO Transmission in the Presence of an Active Eavesdropper," in *Proc. IEEE Intern. Commun. Conf.*, Jun. 2015.

[35] D. W. K. Ng, E. Lo, and R. Schober, "Energy-Efficient Resource Allocation in OFDMA Systems with Large Numbers of Base Station Antennas," *IEEE Trans. Wireless Commun.*, vol. 11, pp. 3292–3304, Sep. 2012.

[36] K. Kumar, G. Caire, and A. Moustakas, "Asymptotic Performance of Linear Receivers in MIMO Fading Channels," *IEEE Trans. Inf. Theory*, vol. 55, pp. 4398–4418, Oct. 2009.

[37] Z. Qi, LTE Throughput Leader Nokia Sets World Record with SK Telecom of Close to 4 Gbps Using TDD-FDD Carrier Aggregation, 2014. [Online] Available: http://nsn.com/news-events/press-room/press-releases/lte-throughput-leader-nokia-sets-world-record-with-sk-telecom-of-close-to-4-gbps-using-tdd

[38] L. D. Heyn, A. Grooten, C. Holden, *et al.*, Cost Model – Country Analysis Report (CAR) for Germany, 2013. [Online] Available: http://www.ftthcouncil.eu/documents/Reports/2013/Cost_Model_CAR_Germany_August2013.pdf

[39] R. Pabst, B. Walke, D. Schultz, *et al.*, "Relay-Based Deployment Concepts for Wireless and Mobile Broadband Radio," *IEEE Commun. Mag.*, vol. 42, pp. 80–89, Sep 2004.

[40] D. Ng, E. Lo, and R. Schober, "Energy-Efficient Resource Allocation in Multi-Cell OFDMA Systems with Limited Backhaul Capacity," in *Proc. IEEE Wireless Commun. Netw. Conf.*, Apr. 2012, pp. 1146–1151.

[41] D. W. K. Ng and R. Schober, "Secure and Green SWIPT in Distributed Antenna Networks with Limited Backhaul Capacity," *IEEE Trans. Wireless Commun.*, vol. 14, no. 9, pp. 5082–5097, Sep. 2015.

[42] D. Ng and R. Schober, "Resource Allocation for Coordinated Multipoint Networks with Wireless Information and Power Transfer," in *Proc. IEEE Global Telecommun. Conf.*, Dec. 2014, pp. 4281–4287.

[43] T. Rappaport, S. Sun, R. Mayzus, *et al.*, "Millimeter Wave Mobile Communications for 5G Cellular: It Will Work!" *IEEE Access*, vol. 1, pp. 335–349, 2013.

[44] M. Breiling, D. W. K. Ng, C. Rohde, F. Burkhardt, and R. Schober, "Resource Allocation for Outdoor-to-Indoor Multicarrier Transmission with Shared UE-Side Distributed Antenna Systems," in *Proc. IEEE Veh. Techn. Conf.*, 2015.

[45] H. Andersson, M. Björn, S. Carson, *et al.*, Ericsson Mobility Report, 2013. [Online] Available: http://ec.europa.eu/digital-agenda/en/news/2014-report-implementation-eu-regulatory-framework-electronic-communications

[46] H. Ding, D. B. da Costa, and J. Ge, "A Centralized Role Selection Scheme for Two-User Amplify-and-Forward Relaying Systems," in *Proc. IEEE Wireless Commun. Netw. Conf.*, Apr. 2013, pp. 3476–3481.

[47] T. Rappaport, J. Murdock, and F. Gutierrez, "State of the Art in 60-GHz Integrated Circuits and Systems for Wireless Communications," *Proc. IEEE*, vol. 99, pp. 1390–1436, Aug. 2011.

[48] J. Jiang, M. Dianati, M. Imran, and Y. Chen, "Energy Efficiency and Optimal Power Allocation in Virtual-MIMO Systems," in *Proc. IEEE Veh. Techn. Conf.*, Sep. 2012, pp. 1–6.

[49] O. Souihli and T. Ohtsuki, "Cooperative Diversity Can Mitigate Keyhole Effects in Wireless MIMO Systems," in *Proc. IEEE Global Telecommun. Conf.*, Nov. 2009, pp. 1–6.

[50] D. Chizhik, G. Foschini, M. Gans, and R. Valenzuela, "Keyholes, Correlations, and Capacities of Multielement Transmit and Receive Antennas," *IEEE Trans. Wireless Commun.*, vol. 1, pp. 361–368, Apr. 2002.

[51] P. Almers, F. Tufvesson, and A. Molisch, "Keyhole Effect in MIMO Wireless Channels: Measurements and Theory," *IEEE Trans. Wireless Commun.*, vol. 5, pp. 3596–3604, Dec. 2006.

[52] R. A. Ramirez, E. Altman, J. S. Thompson, and R. M. Ramos, "A Stable Marriage Framework for Distributed Virtual MIMO Coalition Formation," in *Proc. IEEE Personal, Indoor and Mobile Radio Commun. Symp.*, Sep. 2013, pp. 2707–2712.

[53] A. Jensen, M. Lauridsen, P. Mogensen, T. Sorensen, and P. Jensen, "LTE UE Power Consumption Model: For System Level Energy and Performance Optimization," in *Proc. IEEE Veh. Techn. Conf.*, Sep. 2012, pp. 1–5.

Chapter 4

Linear processing techniques for multi-antenna relaying systems with interference

Caijun Zhong, Himal A. Suraweera and George K. Karagiannidis

4.1 Introduction

The advances in communication theory and practice during the last few decades have vastly improved the performance of current cellular systems. However, providing satisfactory throughput in the cell edge region is still a major challenge [1, 2]. Toward this end, one effective solution, that has received enormous, attention is the deployment of relaying nodes to assist the communication between the source and destination [3, 4]. Two relaying protocols that have been extensively studied in the literature are: amplify-and-forward (AF) and decode-and-forward (DF) [5, 6]. According to the AF protocol, the relay amplifies the received signal from the source before forwarding it to the destination (without decoding the received signal). Hence, AF relaying has a low implementation complexity and can potentially find applications in a variety of wireless systems.

Due to the spectrum scarcity along with dense deployment of access nodes [7], future commercial wireless systems are likely to adopt aggressive frequency reuse policies in order to meet the increasing demand for high-quality wireless services. However, the use of the same frequency in the same cell site inevitably results in an interference-limited communication environment.

When relaying technology is adopted in cellular systems, the co-channel interference (CCI) environment becomes increasingly complex [2, 8]. Standard relaying approaches ignore CCI and therefore exhibit an inferior performance in cellular networks. Therefore, understanding the fundamental performance limitations in the presence of CCI, and devising efficient methods to handle the CCI in relaying systems is of paramount importance in future wireless systems.

4.1.1 Interference mitigation in relaying systems

In the design of conventional cellular networks, CCI has always been regarded as a detrimental factor [9, 10]. The presence of CCI corrupts the desired signal, hence, it can severely degrade the system performance, as demonstrated in a rich body of publications on the performance of dual-/multi-hop relaying systems with CCI. For example, the ergodic capacity, outage probability, and error probability are some of the common performance measures considered to investigate the decremental effects of CCI on relaying systems [11–14]. The influence of CCI on the performance degradation tends to be different for AF and DF relaying protocols. In DF relaying, the relay would decode the signals in the presence of CCI and hence would forward erroneous information to the destination. On the other hand, an AF relay would simply forward the accumulated interference to the destination, and it must employ some form of suppression before decoding. Hence to aid the end node, the relay could also apply some form of CCI suppression before forwarding.

The effect of CCI on the outage probability of single-antenna AF relaying systems has been investigated assuming various fading models and communication scenarios, including Rayleigh and Nakagami-*m* fading [12, 13, 15], and single/multiple interferer with different cases of interference at the relay and/or the destination [16, 17]. The performance of relay selection in the presence of CCI has been investigated in References 11 and 18. Due to the distributed positions of relaying nodes, relay selection can help to alleviate the harmful effects of CCI to some extent. For example, performance degradation can be minimized, if relays experiencing low levels of CCI at a given time are chosen to forward signals than the ones with high levels of CCI. In Reference 11, it has been shown that the conventional *max–min* relay selection criterion for multiple-relay scenarios becomes inefficient in the presence of CCI. Furthermore, the outage analysis reported in Reference 11 has demonstrated that the diversity slope is bounded by a static point in the high signal-to-noise ratio (SNR) regime. In Reference 18, the outage probability of dual-hop DF systems with opportunistic relay selection has been investigated, with CCI both at the relays and destination.

It is well known that multiple-input multiple-output (MIMO) technology provides extra spatial degrees of freedom, which can be efficiently utilized for CCI cancellation [19]. MIMO has been identified as one of the key enabling physical layer technologies in contemporary wireless standards. Toward this end, the performance of multi-antenna relaying systems with CCI has been investigated in References 20–26. In Reference 20, the outage probability of a dual-hop fixed-gain AF multi-antenna source, and destination, single-antenna relay ($N-1-N$) system has been investigated. In spite of the use of multiple antennas, the results in Reference 20 show that CCI can degrade the outage performance by reducing the array gain. In Reference 21, outage probability of three different multi-antenna relaying systems, namely, (i) multi-antenna source, single-antenna relay, and destination ($N-1-1$), (ii) multi-antenna destination, single-antenna source, and relay ($1-N-1$), and (iii) multi-antenna relay, single-antenna source, and destination ($1-1-N$) systems with fixed-gain AF and variable-gain AF relaying have been investigated. The results in Reference 21 show that variable-gain AF relaying outperforms fixed-gain relaying in all three systems studied. In these multi-antenna systems, fixed-gain AF relaying can only achieve a diversity order of one, while in the $1-N-1$ system, variable-gain AF relaying can achieve a diversity order of N. Therefore, increasing the number of antennas at the relay is beneficial to mitigate the effects of CCI. In order to perform CCI suppression, optimum combining can be used [24, 25]. Assuming an arbitrary number of interferers and noise at the relay and destination, an AF relaying scheme with optimum combining has been analyzed in Reference 24 for end-to-end signal-to-interference-plus-noise ratio (SINR) maximization. Such a multi-antenna relay can trade diversity for interference-cancellation gains and is useful to deploy in situations where CCI significantly limits the performance. In Reference 25, the performance of a multi-antenna DF relaying system using optimum combining with joint relay and antenna selection has been studied. If at least one relay has more number of antennas than the number of interferers, the scheme in Reference 25 can achieve diversity gains. The combined effect of outdated channel state information (CSI) and interference at the relay and the destination have been analyzed for a dual-hop MIMO system in Reference 26. In particular, transmit antenna selection and maximum ratio combining (MRC) for the first and second hops, respectively have been assumed. The observation in Reference 26 reveals that the system's outage probability highly depends on the number of antenna at each node, the strength of interference signals, and the correlation between the outdated CSI.

Assuming three linear CCI suppression schemes at the multi-antenna relay, the authors of Reference 22 have considered a $1-N-1$ dual-hop system and derived exact and simple high-SNR expressions for the outage probability. The main motivation for the use of linear combining schemes is to simplify the MIMO operation [27–30]. To this end, a heuristic two-stage relay processing scheme, where the relay first applies a linear combining method to suppress the CCI and then forwards the transformed signal to the destination by using the maximal ratio transmission (MRT), has been proposed. The ergodic capacity of these schemes has been investigated in Reference 23. The results in References 22 and 23 reveal that some of the studied schemes are capable of outperforming each other, in terms of the outage probability or ergodic capacity, depending on the SNR regime of operation. Furthermore, it has been demonstrated that increasing the number of antennas at the relay helps to improve the system's performance.

4.1.2 Interference exploitation for wireless-powered relaying systems

Traditionally, interference has been regarded as a deleterious factor which should be limited in the design of wireless communications systems. As such, sophisticated techniques have been proposed in the literature to avoid, suppress, or cancel interference. However, recent advances in communication theory have demonstrated that interference can be potentially exploited to enhance the system performance [31]. For instance, in the multiuser downlink system, the base station can intelligently make use of the known interference data symbols of users to design efficient precoders (beamformers), and consequently, improve the quality of the useful signal [32–34].

To exploit CCI in conventional cellular systems, inherent structure and prior knowledge of the interference are required. Another system capable of exploiting

CCI to improve performance are the simultaneous wireless information and power transfer (SWIPT) systems [35–37]. For example, radio frequency (RF) signals are capable of carrying both information and energy. As a consequence, they are useful to remotely transfer power to sensor-like communication nodes. Hence in newly emerging wireless-powered communication system paradigms, CCI (another source of RF signals) could be potentially exploited to enhance the system performance. In such systems, the key is whether to suppress or exploit the CCI. The RF energy harvesting techniques also find important applications in relaying systems, where an energy constrained relay with limited battery reserves relies on some external charging mechanism to assist the transmission of source information to the destination [38–40]. Moreover, in wireless-powered relaying systems, the use of multiple antennas at the relay offers several benefits. For example, multiple antennas help to accumulate more energy, as well as it allows the deployment of MIMO techniques to enhance the performance [41].

The benefits of using multiple antennas for wireless-powered relaying systems have been presented in References 39 and 40. To be more specific, two cases, namely, a benchmark case without CCI and a case with CCI have been studied in References 39 and 40 to highlight some interesting observations for the design of such systems. In wireless-powered relay transmission, although CCI provides additional energy, it corrupts the desired signal. Therefore, in order to exploit CCI as a beneficial prospect one must carefully design spatial processing schemes at the multi-antenna relay. Specifically in Reference 39, a linear zero-forcing (ZF) scheme has been applied at the relay for interference removal and low complexity, while in [40], the performance with three linear combining schemes has been analyzed. These results demonstrate that CCI could be potentially exploited to significantly improve the system performance. However, performance gains due to CCI in practice depend heavily on the choice of the linear processing scheme used. Further, the benefits due to a specific linear processing scheme depend on the CCI conditions, and some processing schemes are able to exploit strong CCI to favor the performance at the end destination, while others show significant performance degradation under the same CCI conditions.

4.1.3 Organization of the chapter

The main purpose of this chapter is to introduce linear processing techniques, useful for interference suppression and exploitation in multi-antenna relaying systems. First we introduce three commonly used linear processing techniques in Section 4.2 and apply them for interference suppression in dual-hop systems with a multi-antenna relay in Section 4.3. Specifically, we investigate the outage probability of the system when these three schemes are applied and compare the performance. Next, we consider a wireless-powered multi-antenna relaying system in Section 4.4 and discuss how interference can be exploited to improve the performance with simple linear processing techniques. We conclude the chapter in Section 4.5 with a summary of results and present possible future directions for the topic of linear processing techniques for multi-antenna relaying systems with interference.

4.2 Linear processing techniques

The goal of this section is to provide the reader with a basic understanding of some concepts on popular linear processing techniques discussed in the MIMO literature [27–30]. Linear processing techniques have received wide attention, due to low implementation complexity and satisfactory performance. However, these schemes are not optimal. While simple linear processing schemes are nearly optimal in some cases such as when the number of base station antennas is large [42].

To illustrate the concept of linear processing techniques, we consider an uplink system with $K+1$ users. The received signal $\mathbf{y}$ at the base station is given by

$$\mathbf{y} = \mathbf{h}_0 s_0 + \sum_{i=1}^{K} \mathbf{h}_i s_i + \mathbf{n}_0, \tag{4.1}$$

where the $N \times 1$ vector $\mathbf{h}_i$, $i = 0, 1, \ldots, K$ denotes the channel between the ith user and the base station. In (4.1) s_i is the information symbol satisfying $\mathrm{E}\left\{s_i s_i^*\right\} = P_i$, and $\mathbf{n}_0$ is an $N \times 1$ vector that denotes the additive white Gaussian noise (AWGN) at the base station with $\mathrm{E}\{\mathbf{n}_0 \mathbf{n}_0^\dagger\} = N_0 \mathbf{I}$. Finally, $\mathrm{E}\{x\}$ stands for the expectation of the random variable x.

Without loss of generality, assuming single user detection, we focus on "User 0". To detect the desired signal, s_0, the receiver multiplies the received signal $\mathbf{y}$ with a linear processing vector $\mathbf{w}$ with $\|\mathbf{w}\|_F = 1$ to obtain an estimate of the s_0, i.e.,

$$\hat{s}_0 = \mathbf{w}\mathbf{y} = \mathbf{w}\mathbf{h}_0 s_0 + \sum_{i=1}^{K} \mathbf{w}\mathbf{h}_i s_i + \mathbf{w}\mathbf{n}_0. \tag{4.2}$$

Therefore, the resultant SINR can be expressed as

$$\rho = \frac{|\mathbf{w}\mathbf{h}_0|^2 P_0}{\sum_{i=1}^{K} |\mathbf{w}\mathbf{h}_i|^2 P_i + N_0}. \tag{4.3}$$

Depending on the requirement of the CSI and the consideration of the performance and complexity trade-off, three popular linear processing techniques, i.e., MRC, ZF, and minimum mean square error (MMSE) have been proposed in the literature. We now discuss these techniques below.

4.2.1 Maximum ratio combining

For the MRC scheme, the receiver simply applies the principle of matched filtering to maximize the strength of the desired signal, ignoring the effect of interference. Hence, the processing vector $\mathbf{w}$ is given by

$$\mathbf{w} = \frac{\mathbf{h}_0^\dagger}{\|\mathbf{h}_0\|_F}. \tag{4.4}$$

Obviously, to implement the MRC scheme, only the CSI of the desired signal, i.e., $\mathbf{h}_0$, is required. However, the MRC scheme performs poorly in interference-limited conditions since it simply neglects the effect of interference.

4.2.2 Zero-forcing

For the ZF scheme, the base station utilizes the available multiple antennas to completely eliminate the interference. Define an $N \times M$ matrix $\mathbf{H}_I = [\mathbf{h}_1, \mathbf{h}_2 \cdots \mathbf{h}_K]$ as the interference channel matrix, it can be shown that the receiver processing vector is the solution of the following optimization problem:

$$\begin{aligned} &\mathbf{w}_{\text{opt}} = \arg\max_{\mathbf{w}} \ |\mathbf{w}\mathbf{h}_0|^2 \\ &\text{s.t. } \mathbf{w}\mathbf{H}_I = \mathbf{0} \ \& \ \|\mathbf{w}\|_F = 1. \end{aligned} \tag{4.5}$$

A closed-form expression for the optimal processing vector $\mathbf{w}_{\text{opt}}$ is

$$\mathbf{w}_{\text{opt}} = \frac{\mathbf{h}_0^\dagger \mathbf{P}}{\sqrt{\mathbf{h}_0^\dagger \mathbf{P} \mathbf{h}_0}}, \tag{4.6}$$

where $\mathbf{P} = \mathbf{I}_N - \mathbf{H}_I(\mathbf{H}_I^\dagger \mathbf{H}_I)^{-1}\mathbf{H}_I^\dagger$. Note that with ZF, the interference is completely nulled out by projecting the desired stream (User 0) onto the orthogonal complement of the interuser interference subspace.

In order to implement the ZF scheme, the base station receiver requires the knowledge of interference CSI, $\mathbf{H}_I$, in addition to $\mathbf{h}_0$. The ZF scheme works well in interference-limited scenarios. The ZF scheme has a higher implementation complexity compared to the MRC scheme since it involves calculating the matrix inverse $(\mathbf{H}_I^\dagger \mathbf{H}_I)^{-1}$.

4.2.3 Minimum mean square error

The ZF receiver completely eliminates the interference, which however may cause an elevated noise level (if the channel matrix is not well conditioned and any of the eigenvalues of $(\mathbf{H}_I^\dagger \mathbf{H}_I)^{-1}$ is small, inversion will amplify the noise significantly), thereby substantially degrading the performance. In contrast, the MMSE scheme does not fully eliminate the interference, instead, it provides the optimum trade-off between interference suppression and noise enhancement.

According to the principle of MMSE [43], $\mathbf{w}$ is given by

$$\mathbf{w} = \mathbf{h}_0^\dagger \left(\mathbf{h}_0 \mathbf{h}_0^\dagger + \mathbf{H}_I \mathbf{H}_I^\dagger + N_0 \mathbf{I} \right)^{-1}. \tag{4.7}$$

Note that compared to the ZF scheme, the implementation of the MMSE receiver requires the additional knowledge of the noise variance N_0. However, it is worth pointing out that $\mathbf{w}$ is the optimal solution, which maximizes the SINR given in (4.3). Therefore, among the MRC, ZF, and MMSE schemes, MMSE achieves the best performance.

4.3 Linear processing for interference suppression in relaying systems

In this section, we apply linear processing techniques to suppress CCI in multi-antenna AF relaying systems. The performance of these techniques is compared in terms of the outage probability. We also discuss the performance under large antenna regime and implementation complexity in terms of the CSI requirement.

Consider a dual-hop multi-antenna AF relaying system as shown in Figure 4.1, where both source and destination are equipped with a single antenna, while the relay is equipped with N antennas. We consider the scenario where the relay is subjected to M independently but not necessarily identically distributed co-channel interferers and AWGN, while the destination is corrupted by AWGN only. The considered scenario is relevant to frequency-division relaying systems, where the relay and the destination experience different interference patterns.

We assume half-duplex relaying, and hence transmission occurs in two phases. During the first phase, the source transmits the signal to the relay, and the signal received at the relay is given by

$$\mathbf{y}_r = \mathbf{h}_1 x + \sum_{i=1}^{M} \mathbf{h}_{Ii} s_{Ii} + \mathbf{n}_1, \tag{4.8}$$

where the $N \times 1$ vector $\mathbf{h}_1$ denotes the channel for the source–relay link. The entries of $\mathbf{h}_1$ are identically and independently distributed (i.i.d.) random variables which follow $\mathcal{CN}(0,1)$. The $N \times 1$ vector $\mathbf{h}_{Ii}$ denotes the channel for the ith interference–relay link, and its entries follow i.i.d. $\mathcal{CN}(0,1)$, and x is the source symbol satisfying $\mathrm{E}\{xx^*\} = P_s$. The ith interference symbol is s_{Ii} with $\mathrm{E}\{s_{Ii}s_{Ii}^*\} = P_{Ii}$, $\mathbf{n}_1$ is an $N \times 1$ vector and denotes the AWGN at the relay with $\mathrm{E}\{\mathbf{n}_1\mathbf{n}_1^\dagger\} = N_0\mathbf{I}$.

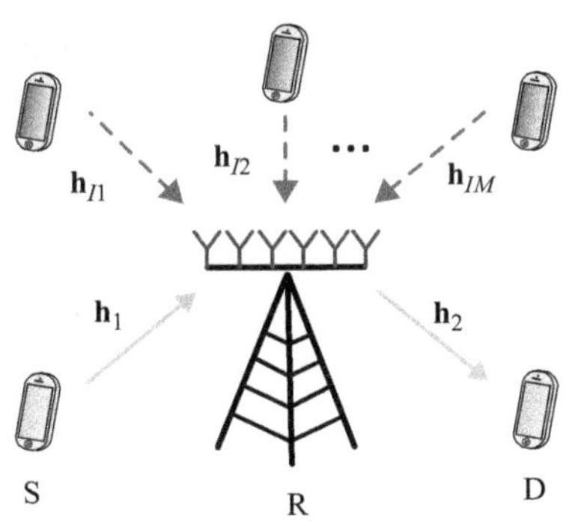

Figure 4.1 The relaying system model with multiple co-channel interferers: S, R, and D denote the source, relay, and the destination, respectively

In the second phase, the relay transmits a transformed version of the received signal to the destination, and the signal at the destination can be written as

$$y_d = \mathbf{h}_2 \mathbf{W} \mathbf{y}_r + n_2, \tag{4.9}$$

where $\mathbf{h}_2$ is a $1 \times N$ vector and denotes the channel for the relay–destination link, and its entries follow i.i.d. $\mathcal{CN}(0, 1)$, n_2 is the AWGN with $\mathrm{E}\{n_2^* n_2\} = N_0$, $\mathbf{W}$ is the transformation matrix at the relay with $\mathrm{E}\{\|\mathbf{W}\mathbf{y}_r\|_F^2\} = P_r$.

Combining (4.8) and (4.9), the end-to-end SINR can be computed as

$$\gamma = \frac{|\mathbf{h}_2 \mathbf{W} \mathbf{h}_1|^2 P_s}{\sum_{i=1}^{M} |\mathbf{h}_2 \mathbf{W} \mathbf{h}_{Ii}|^2 P_{Ii} + \|\mathbf{h}_2 \mathbf{W}\|_F^2 N_0 + N_0}. \tag{4.10}$$

The optimal relay transformation matrix, $\mathbf{W}$, which maximizes the end-to-end SINR, γ, does not seem to be analytically tractable, due to the non-convex nature of the problem. In order to overcome this problem, linear combining methods are used to suppress the CCI in the first hop and the MRT scheme in the second hop since it maximizes the SNR of the relay–destination link. Therefore, the heuristic relay transformation matrix, $\mathbf{W}$, is a rank-1 matrix, i.e.,

$$\mathbf{W} = \omega \frac{\mathbf{h}_2^\dagger}{\|\mathbf{h}_2\|_F} \mathbf{w}_1, \tag{4.11}$$

where ω is the power constraint factor, $\frac{\mathbf{h}_2^\dagger}{\|\mathbf{h}_2\|_F}$ is used for matching the second-hop relay–destination link, and $\mathbf{w}_1$ is a $1 \times N$ linear combining vector, which depends on the linear combining scheme employed by the relay.

4.3.1 MRC/MRT scheme

For the MRC/MRT scheme, $\mathbf{w}_1$ is set to match the first-hop source–relay link, hence,

$$\mathbf{w}_1 = \frac{\mathbf{h}_1^\dagger}{\|\mathbf{h}_1\|_F}. \tag{4.12}$$

To meet the transmit power constraint at the relay, the constant ω^2 can be computed as

$$\omega^2 = \frac{P_r}{\mathbf{h}_1^\dagger \mathbf{h}_1 P_s + \frac{\sum_{i=1}^{M} \left|\mathbf{h}_1^\dagger \mathbf{h}_{Ii}\right|^2 P_{Ii}}{\|\mathbf{h}_1\|_F^2} + N_0}. \tag{4.13}$$

Therefore, the end-to-end SINR for the MRC/MRT scheme, γ_{MRC}, can be expressed as

$$\gamma_{\text{MRC}} = \frac{|\mathbf{h}_2|^2|\mathbf{h}_1|^2 P_s}{|\mathbf{h}_2|^2 \frac{\sum_{i=1}^{M}\left|\mathbf{h}_1^{\dagger}\mathbf{h}_{Ii}\right|^2 P_{Ii}}{|\mathbf{h}_1|^2} + \left|\mathbf{h}_2^{\dagger}\right|^2 N_0 + \frac{N_0}{P_r}\left(|\mathbf{h}_1|^2 P_s + \frac{\sum_{i=1}^{M}\left|\mathbf{h}_1^{\dagger}\mathbf{h}_{Ii}\right|^2 P_{Ii}}{|\mathbf{h}_1|^2} + N_0\right)}. \quad (4.14)$$

4.3.1.1 Outage probability

The outage probability is an important performance metric for wireless systems. It is defined as the probability that the instantaneous SINR falls below a predefined, threshold γ_{th}. Therefore, outage probability is useful to understand the percentage of time the performance is viewed as unsuitable. Depending on the application, various values of outage probability are acceptable. Mathematically, the outage probability can be expressed as

$$P_{\text{out}} = \text{Prob}(\gamma < \gamma_{\text{th}}). \quad (4.15)$$

With the help of the definition in (4.15), outage probability of the MRC/MRT scheme can be written as

$$P_{\text{out}}^{\text{MRC}} = \text{Prob}(\gamma_{\text{MRC}} < \gamma_{\text{th}}). \quad (4.16)$$

Theorem 4.1. *The outage probability of the dual-hop AF relaying system with the MRC/MRT scheme can be expressed as*

$$P_{\text{out}}^{\text{MRC}} = 1 - \frac{2e^{-\frac{\gamma_{\text{th}}}{\rho_1}-\frac{\gamma_{\text{th}}}{\rho_2}}}{\Gamma(N)} \sum_{m=0}^{N-1}\left(\frac{\gamma_{\text{th}}}{\rho_1}\right)^m \frac{1}{m!}\sum_{j=0}^{m}\binom{m}{j}\left(\frac{1}{\rho_2}\right)^{m-j}\sum_{k=0}^{N+j-1}\binom{N+j-1}{k}$$
$$\times\left(\frac{\gamma_{\text{th}}}{\rho_2}\right)^{N+j-1-k}\left(\frac{(\gamma_{\text{th}}+1)\gamma_{\text{th}}}{\rho_1\rho_2}\right)^{\frac{k-m+1}{2}} \mathcal{I}_1(\gamma_{\text{th}}), \quad (4.17)$$

where $\rho_1 = \frac{P_s}{N_0}$ and $\rho_2 = \frac{P_r}{N_0}$ denote the per hop SNRs. Moreover, the integral $\mathcal{I}_1(\gamma_{\text{th}})$, is given by

$$\mathcal{I}_1(\gamma_{\text{th}}) = \sum_{p=1}^{\rho(\mathbf{D})}\sum_{q=1}^{\tau_p(\mathbf{D})} \chi_{p,q}(\mathbf{D})\frac{\rho_{I\langle p\rangle}^{-q}}{(q-1)!}\int_0^{\infty} K_{k-m+1}\left(2\sqrt{\frac{(\gamma_{\text{th}}+1)\gamma_{\text{th}}}{\rho_1\rho_2}(x+1)}\right)$$
$$\times(x+1)^{\frac{k+m+1}{2}}x^{q-1}e^{-\left(\frac{\gamma_{\text{th}}}{\rho_1 N_0}+\frac{1}{\rho_{I\langle p\rangle}}\right)x}dx, \quad (4.18)$$

where $\rho_{Ii} = \frac{P_{Ii}}{N_0}$, $i = 1,\ldots,M$, $\mathbf{D} = \text{diag}(\rho_{I1},\rho_{I2},\cdots,\rho_{IM})$, $\rho(\mathbf{D})$ is the number of distinct diagonal elements of $\mathbf{D}$, $\rho_{I\langle 1\rangle} > \rho_{I\langle 2\rangle} > \cdots > \rho_{I\langle\rho(\mathbf{D})\rangle}$ are the distinct diagonal elements in decreasing order, $\tau_i(\mathbf{D})$ is the multiplicity of $\rho_{I\langle i\rangle}$, $\chi_{i,j}(\mathbf{D})$ is the (i,j)th characteristic coefficient of $\mathbf{D}$ and $K_\nu(x)$ is the νth order modified Bessel function of the second kind [44, Eq. (8.407.1)].

Proof. A detailed proof can be found in Appendix I-A in Reference 22. □

Although Theorem 4.1 presents an analytical formula for the outage probability, it is complicated and does not give any useful insight for the system design. Motivated by this, we present a simple approximation for the outage probability in the high SNR regime, which enables the characterization of the achievable diversity order of the MRC/MRT scheme. The diversity order corresponds to the slope in a plot of log outage versus SNR in dB [45]. Therefore, a higher value for the diversity order indicates better high SNR performance.

Theorem 4.2. *In the high SNR regime, i.e., $\rho_2 = \mu\rho_1$, $\rho_1 \to \infty$, with μ being a finite constant, the outage probability of the MRC/MRT scheme can be approximated as*[1]

$$P_{\text{out}}^{\text{MRC}} \approx \left(\frac{1}{\mu^N} + \sum_{k=0}^{N}\binom{N}{k}\sum_{i=1}^{\rho(\mathbf{D})}\sum_{j=1}^{\tau_i(\mathbf{D})}\chi_{i,j}(\mathbf{D})\frac{\Gamma(k+j)}{\Gamma(j)}\rho_{I\langle i\rangle}^{k}\right)\frac{\left(\frac{\gamma_{\text{th}}}{\rho_1}\right)^N}{\Gamma(N+1)} + O\left(\left(\frac{\gamma_{\text{th}}}{\rho_1}\right)^{N+1}\right), \tag{4.19}$$

where $\Gamma(x)$ is the gamma function [44, Eq. (8.310.1)].

Proof. A detailed proof can be found in Appendix I-C in Reference 22. □

Theorem 4.2 shows that the MRC/MRT scheme achieves a diversity order of N. Moreover, it implies that the number of interferers does not affect the achievable diversity order, however, it causes a detrimental effect on the array gain. This key observation suggests that, in the presence of strong CCI, the outage performance of the MRC/MRT scheme will be significantly affected. Hence, in such a scenario, the MRC/MRT scheme may not be suitable.

4.3.2 ZF/MRT scheme

In the ZF/MRT scheme, the relay utilizes the available multiple antennas to completely eliminate the CCI. To ensure that, the number of the antennas equipped at the relay should be greater than the number of interferers. Hence, for the ZF/MRT scheme, it is assumed that $N > M$.

Define an $N \times M$ matrix, $\mathbf{H}_I = [\mathbf{h}_{I1}, \mathbf{h}_{I2} \cdots \mathbf{h}_{IM}]$, as the interference channel matrix, then the SINR expression in (4.10) can be alternatively expressed as

$$\gamma_{\text{ZF}} = \frac{\omega^2\|\mathbf{h}_2\|_F^2|\mathbf{w}_1\mathbf{h}_1|^2\rho_1}{\omega^2\|\mathbf{h}_2\|_F^2\left(\mathbf{w}_1\mathbf{H}_I\mathbf{D}\mathbf{H}_I^\dagger\mathbf{w}_1^\dagger\right) + \omega^2\|\mathbf{h}_2\|_F^2\|\mathbf{w}_1\|_F^2 + 1}. \tag{4.20}$$

[1] If $f(x) = \sum_{n=0}^{\infty} a_n x^n$ as $x \to 0^+$, we write $f(x) = \sum_{n=0}^{N} a_n x^n + O\left(x^{N+1}\right)$ as $x \to 0^+$ to imply that $|f(x) - \sum_{n=0}^{N} a_n x^n|$ is smaller than $C|x^{N+1}|$ for some constant C as $x \to 0^+$.

Hence, the optimal combining vector $\mathbf{w}_1$ should be the solution of the following maximization problem:

$$\begin{aligned} \mathbf{w}_1 &= \arg\max_{\mathbf{w}_1} \ \gamma_{\mathrm{ZF}} \\ \text{s.t.} \ \ \mathbf{w}_1 \mathbf{H}_I &= \mathbf{0} \ \ \& \ \ \|\mathbf{w}_1\|_F = 1. \end{aligned} \tag{4.21}$$

The optimal combining vector, $\mathbf{w}_1$, in (4.21) can be obtained as

$$\mathbf{w}_1 = \frac{\mathbf{h}_1^\dagger \mathbf{P}}{\sqrt{\mathbf{h}_1^\dagger \mathbf{P} \mathbf{h}_1}}, \tag{4.22}$$

where $\mathbf{P} = \mathbf{I}_N - \mathbf{H}_I \left(\mathbf{H}_I^\dagger \mathbf{H}_I\right)^{-1} \mathbf{H}_I^\dagger$.

4.3.2.1 Outage probability

Using the optimal $\mathbf{w}_1$, as in (4.22), we now establish the outage probability of the ZF/MRT scheme given by

$$P_{\mathrm{out}}^{\mathrm{ZF}} = \mathrm{Prob}(\gamma_{\mathrm{ZF}} < \gamma_{\mathrm{th}}), \tag{4.23}$$

in Theorem 4.3 below.

Theorem 4.3. *The outage probability of the dual-hop AF relaying system with the ZF/MRT scheme can be expressed as*

$$\begin{aligned} P_{\mathrm{out}}^{\mathrm{ZF}} &= 1 - \frac{2e^{-\frac{\gamma_{\mathrm{th}}}{\rho_1} - \frac{\gamma_{\mathrm{th}}}{\rho_2}}}{\Gamma(N)} \sum_{m=0}^{N-M-1} \left(\frac{\gamma_{\mathrm{th}}}{\rho_1}\right)^m \frac{1}{m!} \sum_{j=0}^{m} \binom{m}{j} \\ &\times \left(\frac{1}{\rho_2}\right)^{m-j} \sum_{k=0}^{N+j-1} \binom{N+j-1}{k} \left(\frac{\gamma_{\mathrm{th}}}{\rho_2}\right)^{N+j-k-1} \\ &\times \left(\frac{(1+\gamma_{\mathrm{th}})\gamma_{\mathrm{th}}}{\rho_1 \rho_2}\right)^{\frac{k-m+1}{2}} K_{k-m+1}\left(2\sqrt{\frac{(1+\gamma_{\mathrm{th}})\gamma_{\mathrm{th}}}{\rho_1 \rho_2}}\right). \end{aligned} \tag{4.24}$$

Proof. A detailed proof can be found in Reference 22. □

To gain further insights, we now look into the high SNR regime and present a simple and informative approximation for the outage probability.

Theorem 4.4. *In the high SNR regime, i.e.,* $\rho_2 = \mu \rho_1$, $\rho_1 \to \infty$, *the outage probability with the ZF/MRT scheme can be approximated as*

$$P_{\mathrm{out}}^{\mathrm{ZF}} = \frac{1}{\Gamma(N-M+1)} \left(\frac{\gamma_{\mathrm{th}}}{\rho_1}\right)^{N-M} + o\left(\left(\frac{\gamma_{\mathrm{th}}}{\rho_1}\right)^{N-M+1}\right). \tag{4.25}$$

Proof. A detailed proof can be found in Appendix II-B in Reference 22. □

We observe that μ does not affect the outage probability at high SNR. In addition, Theorem 4.4 shows that the achievable diversity order of the ZF/MRT scheme is $N - M$. Hence, compared with the MRC/MRT scheme, the ZF/MRT scheme incurs a diversity loss of M. This important observation suggests that complete elimination of CCI may not be the best option in terms of the outage performance.

4.3.3 MMSE/MRT scheme

The ZF scheme completely eliminates the CCI at the relay, which however may cause an elevated noise level. In contrast, the MMSE scheme does not fully eliminate the CCI, instead, it provides the optimum trade-off between interference suppression and noise enhancement. In the following, we study the outage performance of the MMSE/MRT scheme.

To make the analysis tractable, an equal interference power assumption, i.e., $\rho_{Ii} \equiv \rho_I, \forall i = 1, 2, \ldots, M$ can be adopted. Although a non-i.i.d. interference power assumption is more general, the equal interference power assumption is also of practical interest. For example, such a situation can occur when the interference sources are clustered together or when the interference originates from a multi-antenna transmitter implementing an uniform power allocation policy. Furthermore, in Reference 22 numerical examples have been presented to show that the equal power assumption can also provide accurate approximations to the outage probability for scenarios with distinct interference power.

Using the principle of MMSE, $\mathbf{w}_1$ can be written as

$$\mathbf{w}_1 = \mathbf{h}_1^\dagger \left(\mathbf{h}_1 \mathbf{h}_1^\dagger + \mathbf{H}_I \mathbf{H}_I^\dagger + \frac{1}{\rho_I} \mathbf{I} \right)^{-1}. \tag{4.26}$$

Also, in order to meet the power constraint at the relay, we have

$$\omega^2 = \frac{\rho_2}{|\mathbf{w}_1 \mathbf{h}_1|^2 \rho_1 + \sum_{i=1}^{M} |\mathbf{w}_1 \mathbf{h}_{Ii}|^2 \rho_I + \|\mathbf{w}_1\|_F^2}. \tag{4.27}$$

Therefore, the end-to-end SINR can be expressed as

$$\gamma_{\text{MMSE}} = \frac{|\mathbf{w}_1 \mathbf{h}_1|^2 \rho_1}{\left(1 + \frac{1}{\rho_2 \|\mathbf{h}_2\|_F^2}\right) \left(\sum_{i=1}^{M} |\mathbf{w}_1 \mathbf{h}_{Ii}|^2 \rho_I + \|\mathbf{w}_1\|_F^2 \right) + \frac{|\mathbf{w}_1 \mathbf{h}_1|^2 \rho_1}{\rho_2 \|\mathbf{h}_2\|_F^2}}. \tag{4.28}$$

4.3.3.1 Outage probability

In order to characterize the outage probability of the MMSE/MRT scheme given by

$$P_{\text{out}}^{\text{MMSE}} = \text{Prob}(\gamma_{\text{MMSE}} < \gamma_{\text{th}}), \tag{4.29}$$

it is necessary to find the distribution of γ_{MMSE}. In Reference 22, the cumulative distribution function of γ_{MMSE} has been derived. The result shown in Theorem 4.5 for the outage probability requires the numerical evaluation of a single integral.

Theorem 4.5. *The outage probability of the dual-hop AF relaying system with the MMSE/MRT scheme can be expressed as*

$$\begin{aligned}P_{\text{out}}^{\text{MMSE}} &= 1 - \frac{2e^{-\frac{\gamma_{\text{th}}}{\rho_1}-\frac{\gamma_{\text{th}}}{\rho_2}}}{\Gamma(N)}\sum_{m=0}^{N-1}\left(\frac{\gamma_{\text{th}}}{\rho_1}\right)^m\frac{1}{m!}\sum_{j=0}^{m}\binom{m}{j}\left(\frac{1}{\rho_2}\right)^{m-j}\sum_{k=0}^{N+j-1}\binom{N+j-1}{k}\\
&\quad\times\left(\frac{\gamma_{\text{th}}}{\rho_2}\right)^{N+j-k-1}\left(\frac{(1+\gamma_{\text{th}})\gamma_{\text{th}}}{\rho_1\rho_2}\right)^{\frac{k-m+1}{2}}K_{k-m+1}\left(2\sqrt{\frac{(1+\gamma_{\text{th}})\gamma_{\text{th}}}{\rho_1\rho_2}}\right)+e^{-\frac{\gamma_{\text{th}}}{\rho_1}-\frac{\gamma_{\text{th}}}{\rho_2}}\\
&\quad\times\left(\frac{\gamma_{\text{th}}}{\rho_1}\right)^N\frac{\Gamma(M+1)}{\Gamma(N)}\sum_{m=m_1}^{N}\frac{\rho_I^{N-m+1}}{\Gamma(m)\,\Gamma(N-m+2)\,\Gamma(m-N+M)}\\
&\quad\times\sum_{j=0}^{N}\binom{N}{j}\left(\frac{1}{\rho_2}\right)^{N-j}\sum_{k=0}^{N+j-1}\binom{N+j-1}{k}\left(\frac{\gamma_{\text{th}}}{\rho_2}\right)^{N+j-1-k}\mathcal{I}_2(\gamma_{\text{th}}),\end{aligned}\tag{4.30}$$

where $m_1 = \max(0, N-M)+1$. *Moreover the integral* $\mathcal{I}_2(\gamma_{\text{th}})$ *is given by*

$$\begin{aligned}\mathcal{I}_2(\gamma_{\text{th}}) &= \int_0^\infty e^{-\frac{(1+\gamma_{\text{th}})\gamma_{\text{th}}}{\rho_1\rho_2 x}}e^{-x}x^{k-N}\\
&\quad\times{}_2F_1\left(M+1, N-m+1; N-m+2; -\frac{\rho_I\gamma_{\text{th}}}{\rho_1}\left(1+\frac{\gamma_{\text{th}}+1}{\rho_2 x}\right)\right)dx,\end{aligned}\tag{4.31}$$

where ${}_2F_1(a,b;c,z)$ *is the Gauss hyper-geometric function [44, Eq. (9.100)].*

Proof. A detailed proof can be found in Appendix III-B in Reference 22. □

It is also useful to look into the high SNR regime and present an outage probability expression in order to characterize the diversity order of the MMSE/MRT scheme.

Theorem 4.6. *In the high SNR regime, i.e.,* $\rho_2 = \mu\rho_1$, $\rho_1 \to \infty$, *the outage probability of the dual-hop AF relaying system with the MMSE/MRT scheme can be approximated as*

$$\begin{aligned}P_{\text{out}}^{\text{MMSE}} &= \left(\Gamma(M+1)\sum_{m=m_1}^{N}\frac{\rho_I^{N-m+1}}{\Gamma(m)\,\Gamma(N-m+2)\,\Gamma(m-N+M)}\right.\\
&\quad\left.+\left(1+\left(\frac{1}{\mu}\right)^N\right)\frac{1}{\Gamma(N+1)}\right)\left(\frac{\gamma_{\text{th}}}{\rho_1}\right)^N+o\left(\left(\frac{\gamma_{\text{th}}}{\rho_1}\right)^{N+1}\right).\end{aligned}\tag{4.32}$$

Proof. A detailed proof can be found in Reference 22. □

Theorem 4.6 indicates that the MMSE/MRT scheme achieves a diversity order of N, the same as that of MRC/MRT scheme. MMSE/MRT in general requires more CSI, compared with the MRC/MRT scheme, however it always achieves a strictly better outage performance compared to MRC/MRT in the high SNR regime.

4.3.4 Large N analysis

We now focus our attention into the large N regime with fixed M and investigate the asymptotic behavior of the three linear combining schemes. The use of antenna arrays with a very large number of elements at communication terminals is increasingly attracting a lot of research interest due to massive MIMO technology (also known as "Large-Scale Antenna Systems") [42]. Massive MIMO brings significant improvements in throughput and energy efficiency for wireless communications. The outage probability of the ZF/MRT and MMSE/MRT schemes as $N \to \infty$ can be studied using the following theorem.

Theorem 4.7. *When $N \to \infty$, the end-to-end SINR of the ZF/MRT and the MMSE/MRT schemes converges to*

$$\gamma^{\infty} = \frac{\rho_2 \|\mathbf{h}_2\|_F^2 \rho_1 \|\mathbf{h}_1\|_F^2}{\rho_2 \|\mathbf{h}_2\|_F^2 + \rho_1 \|\mathbf{h}_1\|_F^2 + 1}, \tag{4.33}$$

and the corresponding outage probability is given by

$$P_{\text{out}}^{\infty} = 1 - \frac{2e^{-\frac{\gamma_{\text{th}}}{\rho_1} - \frac{\gamma_{\text{th}}}{\rho_2}}}{\Gamma(N)} \sum_{m=0}^{N-1} \left(\frac{\gamma_{\text{th}}}{\rho_1}\right)^m \frac{1}{m!} \sum_{j=0}^{m} \binom{m}{j} \left(\frac{1}{\rho_2}\right)^{m-j} \sum_{k=0}^{N+j-1} \binom{N+j-1}{k}$$
$$\times \left(\frac{\gamma_{\text{th}}}{\rho_2}\right)^{N+j-k-1} \left(\frac{(1+\gamma_{\text{th}})\gamma_{\text{th}}}{\rho_1\rho_2}\right)^{\frac{k-m+1}{2}} K_{k-m+1}\left(2\sqrt{\frac{(1+\gamma_{\text{th}})\gamma_{\text{th}}}{\rho_1\rho_2}}\right). \tag{4.34}$$

Proof. A detailed proof can be found in Appendix IV in Reference 22. □

We see that the asymptotic SINR γ^{∞} presented in Theorem 4.7 is equivalent to the end-to-end SNR of the same dual-hop AF relaying system without CCI at the relay. In other words, when N is large, interference has no impact on the ZF/MRT and the MMSE/MRT schemes. Unfortunately, this appealing result does not hold true in the case of the MRC/MRT scheme. Let us analyse the interference term, $U_1 = \frac{\sum_{i=1}^{M} \left|\mathbf{h}_1^{\dagger}\mathbf{h}_{Ii}\right|^2 \rho_{Ii}}{\|\mathbf{h}_1\|_F^2}$ in (4.14). It can be seen that U_1 is a hyper-exponential random variable which is independent of N. Hence, the effect of CCI does not vanish even in the large N regime for the MRC/MRT scheme. This analysis implies that it is always desirable to have a large number of antennas at the relay. Moreover, linear combining schemes with superior interference suppression capability such as ZF/MRT and MMSE/MRT should be adopted over the MRC/MRT scheme.

4.3.5 Comparison of the schemes

A detailed comparison in terms of the required CSI for the MRC/MRT, ZF/MRT, and MMSE/MRT schemes is shown in Table 4.1. In general, if a large amount of CSI is available at the relay node, more sophisticated transmission schemes could be designed to improve the transmission efficiency and to achieve a better performance. Such performance improvement, however, comes at a cost. The acquisition of CSI

Table 4.1 Comparison of the MRC/MRT, ZF/MRT, and MMSE/MRT schemes

	MRC/MRT	ZF/MRT	MMSE/MRT
CSI requirement	$\mathbf{h}_1$ and $\mathbf{h}_2$	$\mathbf{h}_1$, $\mathbf{h}_2$, and $\mathbf{H}_I$	$\mathbf{h}_1$, $\mathbf{h}_2$, $\mathbf{H}_I$, and N_0
Antenna number requirement	None	$N > M$	None
Diversity order	N	$N - M$	N
Impact of interference power	Reduces the array gain	No impact	Reduces the array gain
Decay rate of outage probability vs. N	Slow	Fast	Fast

involves additional feedback overhead. Therefore, in order to make a fair comparison among the three schemes, the CSI requirement of each scheme must be characterized.

The MRC/MRT scheme only requires the channel knowledge of the source–relay link, $\mathbf{h}_1$, and the relay–destination link, $\mathbf{h}_2$. On the other hand, the ZF/MRT scheme requires the channel knowledge of $\mathbf{h}_1$, $\mathbf{h}_2$, and in addition to the interference link, $\mathbf{H}_I$. Among the three schemes we analyzed, the MMSE/MRT scheme has the highest CSI requirement, since the noise variance N_0 at the relay is also required besides the channel knowledge of $\mathbf{h}_1$, $\mathbf{h}_2$, and $\mathbf{H}_I$.

The CSI of $\mathbf{h}_1$ can be directly acquired through a channel estimation process at the relay while knowledge of $\mathbf{h}_2$ can be acquired using a destination–relay feedback link. In a practical system with a limited feedback rate, full CSI of the relay–destination link is difficult to acquire. However, if channel reciprocity condition is assumed, the destination could send out a brief broadcast signal that will allow the relay to determine $\mathbf{h}_2$. The channel knowledge of the interference link, $\mathbf{H}_I$, can be obtained by utilizing the methods given in the literature, see e.g., References 46 and 47.

4.3.6 Numerical results

Now we show some numerical results to confirm the correctness of the analysis and reveal the effect of various system parameters on the outage probability of the considered relay system. In Figure 4.2, the outage performance of the proposed linear processing schemes are compared under two scenarios with different level of interference power, i.e., weak interference, $\rho_I = 0$ dB, and strong interference, $\rho_I = 10$ dB. In these examples, we set $\gamma_{\text{th}} = 1$, and equal per hop SNR ($\mu = 1$) is assumed. It can be readily observed that, when the interference power increases, the outage performance of MRC/MRT and MMSE/MRT schemes degrades, while the outage performance of the ZF/MRT scheme remains the same, which is intuitive since the ZF scheme completely eliminates interference. Furthermore, it is evident that the MMSE/MRT scheme always achieves the best outage performance, while the ZF/MRT scheme is superior than the MRC/MRT scheme in the low SNR regime, especially when the interference power is large.

The impact of the number of relay antennas on the outage performance is illustrated in Figure 4.3, assuming fixed source and relay transmit power. Once again, we see that the MMSE/MRT scheme always achieves the best outage performance. On the

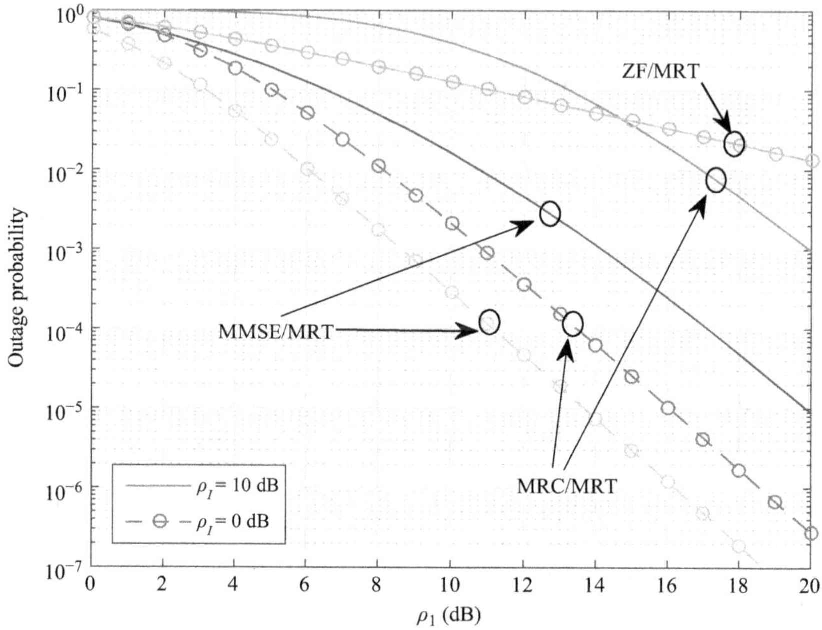

Figure 4.2 Outage probability of MRC/MRT, ZF/MRT, and MMSE/MRT schemes with $N = 4$, $M = 3$ and different ρ_I

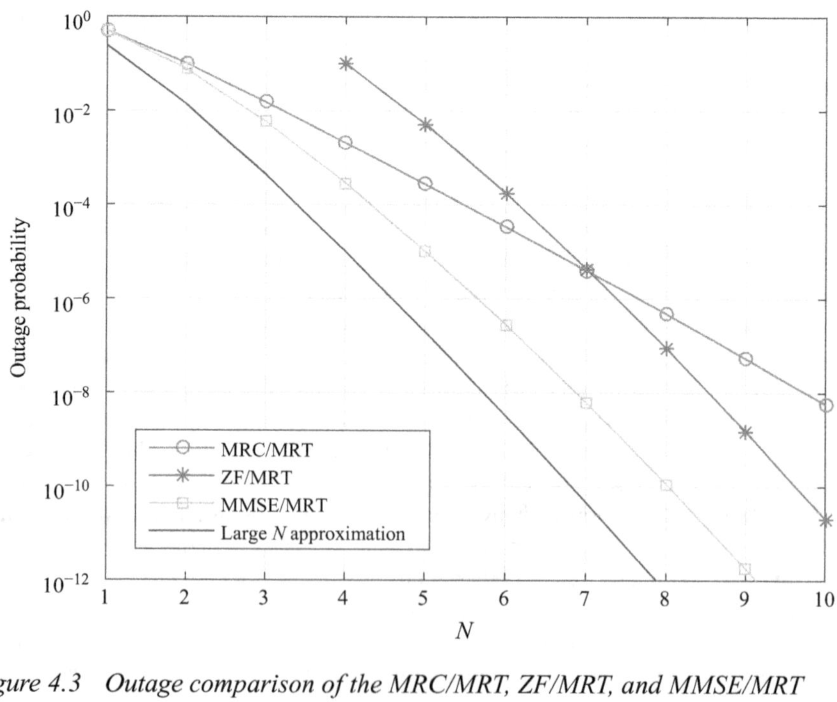

Figure 4.3 Outage comparison of the MRC/MRT, ZF/MRT, and MMSE/MRT schemes with $\rho_1 = \rho_2 = 10\,dB$, $\rho_I = 0\,dB$, $M = 3$, and different N

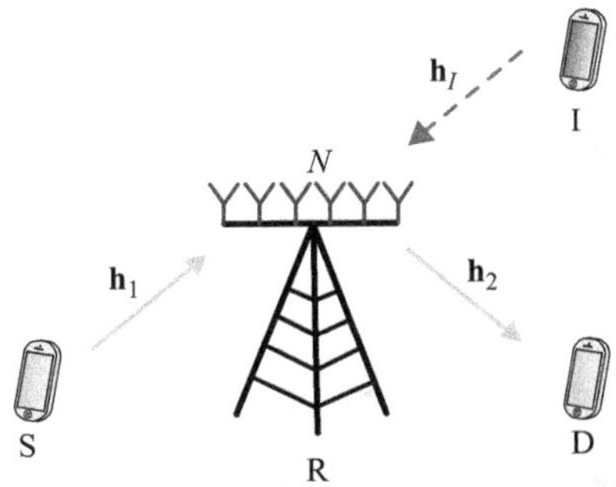

Figure 4.4 System model: S, R, and D denote the source, relay, and destination nodes, respectively

other hand, the MRC/MRT scheme is superior to the ZF/MRT scheme with small N, ($N \leq 7$), and becomes inferior to the ZF/MRT scheme when N is large. Moreover, the slope of the outage curves of the ZF/MRT and MMSE/MRT schemes is almost identical, which is steeper than that of the MRC/MRT scheme, hence implying that, to attain some specified outage probability, the minimum required antenna number N is less for the ZF/MRT and MMSE/MRT schemes compared with the MRC/MRT scheme. Finally, as expected, the large N approximation shown in Figure 4.3 is not tight under the considered small values of N, but becomes accurate in the asymptotic regime of N.

4.4 Linear processing for interference exploitation in wireless-powered relaying systems

In this section, we investigate the role of linear processing schemes for interference exploitation in a wireless-powered multi-antenna relaying system. In particular, we consider the outage probability of these three schemes under the power splitting protocol and present an analysis for the optimal splitting ratio that minimizes the outage probability.

Consider a dual-hop multi-antenna SWIPT system as shown in Figure 4.4, where both the source and the destination are equipped with a single antenna, while the relay is equipped with N antennas. The source sends information to the destination through an energy constrained relay node. It is assumed that direct link between the source and the destination does not exist due to obstacles and/or severe fading, and the channel remains constant over the block time T and varies independently and identically from one block to the other and has a Rayleigh distributed magnitude. Also, no CSI is assumed at the source, full CSI is assumed at the relay, and local CSI is assumed at the destination.

We focus on the power splitting receiver architecture based protocol studied in References 38 and 48. Specifically, the entire communication consists of two time slots with duration of $\frac{T}{2}$ each. At the end of the first phase, each antenna at the relay splits the received source signal into two streams, one for energy harvesting from RF radiation and the other for information processing as depicted in Figure 4.5.

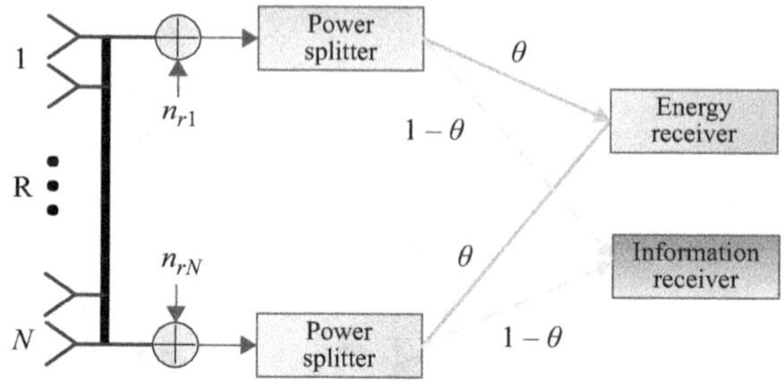

Figure 4.5 Block diagram of the relay receiver with the power splitting protocol

We assume that the relay is subjected to a single dominant interferer and noise, while the destination is corrupted by noise only. This system model enables us to gain key insights on the role of interference and multiple antennas in the considered wireless-powered relaying system.

In such a case, the signal at the input of the information receiver at the relay is given by

$$\mathbf{y}_r = \sqrt{(1-\theta)P_s/d_1^\tau}\mathbf{h}_1 x + \sqrt{(1-\theta)P_I/d_I^\tau}\mathbf{h}_I s_I + \mathbf{n}_r, \tag{4.35}$$

where P_I is the interference power, d_I denotes the distance between the interferer and the relay, s_I is the interference symbol with unit power, and $\mathbf{h}_I$ is an $N \times 1$ vector with entries following i.i.d. $\mathcal{CN}(0, 1)$.

At the end of the first phase, the overall energy harvested during half of the block time $\frac{T}{2}$ is given by

$$E_h = \eta\theta\left(\frac{P_s}{d_1^\tau}\|\mathbf{h}_1\|_F^2 + \frac{P_I}{d_I^\tau}\|\mathbf{h}_I\|_F^2\right)\frac{T}{2}. \tag{4.36}$$

During the second phase, the relay transmits a transformed version of the received signal to the destination using the harvested energy. Hence, the signal at the destination can be expressed as

$$y_d = \sqrt{1/d_2^\tau}\mathbf{h}_2\mathbf{W}\mathbf{y}_r + n_d, \tag{4.37}$$

where $\mathbf{h}_2$ is a $1 \times N$ vector and denotes the relay–destination channel and its entries follow i.i.d. $\mathcal{CN}(0, 1)$, d_2 denotes the distance between the relay and the destination, n_d is the AWGN at the destination with $\mathrm{E}\{n_d^* n_d\} = N_0$, $\mathbf{W}$ is the transformation matrix applied at the information receiver at the relay with $\mathrm{E}\{\|\mathbf{W}\mathbf{y}_r\|_F^2\} = P_r$. Obviously, the performance of the system depends on the choice of $\mathbf{W}$. In the presence of CCI, the optimal relay processing matrix $\mathbf{W}$ maximizing the end-to-end SINR is the solution of the following optimization problem:

$$\max_{\mathbf{W}} \gamma = \frac{\frac{(1-\theta)P_s}{d_1^\tau d_2^\tau}|\mathbf{h}_2\mathbf{W}\mathbf{h}_1|^2}{\frac{(1-\theta)P_I}{d_I^\tau d_2^\tau}|\mathbf{h}_2\mathbf{W}\mathbf{h}_I|^2 + \frac{\|\mathbf{h}_2\mathbf{W}\|_F^2}{d_2^\tau}N_0 + N_0}$$

$$\text{s.t. } \mathbb{E}\left\{\|\mathbf{W}\mathbf{y}_r\|_F^2\right\} = P_r = \eta\theta\left(\frac{P_s}{d_1^\tau}\|\mathbf{h}_1\|_F^2 + \frac{P_I}{d_I^\tau}\|\mathbf{h}_I\|_F^2\right).$$

Due to the non-convex nature of the above problem, a closed-form solution for $\mathbf{W}$ is hard to derive. Hence, we now consider three heuristic two-stage relay processing techniques, proposed in Reference 22, i.e., the matrix $\mathbf{W}$ admits a rank-1 structure given by

$$\mathbf{W} = \omega\frac{\mathbf{h}_2^\dagger}{\|\mathbf{h}_2\|_F}\mathbf{w}_1, \tag{4.38}$$

where $\mathbf{w}_1$ is a $1 \times N$ linear combining vector, which depends on the specific linear combining scheme employed at the relay.

4.4.1 MRC/MRT scheme

For the MRC/MRT scheme, $\mathbf{w}_1$ is set to match the first-hop (source-relay) channel. Also, to meet the transmit power constraint at the relay, the power constraint factor ω^2 should be given by

$$\omega^2 = \frac{P_r}{\frac{(1-\theta)P_s}{d_1^\tau}\|\mathbf{h}_1\|_F^2 + \frac{(1-\theta)P_I}{d_I^\tau}\frac{\left|\mathbf{h}_1^\dagger\mathbf{h}_I\right|^2}{\|\mathbf{h}_1\|_F^2} + N_0}, \tag{4.39}$$

where $P_r = \frac{E_h}{T/2} = \eta\theta\left(\frac{P_s}{d_1^\tau}\|\mathbf{h}_1\|_F^2 + \frac{P_I}{d_I^\tau}\|\mathbf{h}_I\|_F^2\right)$. Therefore, the end-to-end SINR of the MRC/MRT scheme can be expressed as

$$\gamma_I^{\text{MRC}} = \frac{\gamma_{I1}^{\text{MRC}}\gamma_{I2}^{\text{MRC}}}{\gamma_{I1}^{\text{MRC}} + \gamma_{I2}^{\text{MRC}} + 1}, \tag{4.40}$$

where

$$\gamma_{I1}^{\text{MRC}} = \frac{(1-\theta)\rho_1}{d_1^\tau}\|\mathbf{h}_1\|_F^2\left(\frac{(1-\theta)\rho_I}{d_I^\tau}\frac{\left|\mathbf{h}_1^\dagger\mathbf{h}_I\right|^2}{\|\mathbf{h}_1\|_F^2} + 1\right)^{-1}, \tag{4.41}$$

and

$$\gamma_{I2}^{\text{MRC}} = \frac{\eta\theta}{d_2^\tau}\left(\frac{\rho_1}{d_1^\tau}\|\mathbf{h}_1\|_F^2 + \frac{\rho_I}{d_I^\tau}\|\mathbf{h}_I\|_F^2\right)\|\mathbf{h}_2\|_F^2, \tag{4.42}$$

and $\rho_I = \frac{P_I}{N_0}$.

4.4.1.1 Outage probability

The exact analysis of the outage probability is difficult, and in Reference 40 an outage lower bound and a simple high SNR approximation have been presented.

According to References 21 and 22, the end-to-end SINR in (4.40) can be tightly upper bounded by

$$\gamma_I^{\mathrm{MRC}} \leq \gamma_I^{\mathrm{up}} = \min\left(\gamma_{I1}^{\mathrm{MRC}}, \gamma_{I2}^{\mathrm{MRC}}\right), \tag{4.43}$$

the outage probability of the MRC/MRT scheme is lower bounded by

$$P_{I\mathrm{out}}^{\mathrm{LMRC}} = \mathrm{Prob}\left(\gamma_I^{\mathrm{up}} < \gamma_{\mathrm{th}}\right). \tag{4.44}$$

We present the following theorem to lower bound the outage probability of the MRC/MRT scheme.

Theorem 4.8. *If $\rho_1 \neq \rho_I$, the outage probability of the MRC/MRT scheme can be lower bounded as*

$$P_{I\mathrm{out}}^{\mathrm{LMRC}} = 1 - F_1^{\mathrm{MRC}} F_2^{\mathrm{MRC}}, \tag{4.45}$$

with

$$F_1^{\mathrm{MRC}} = \frac{d_I^{\tau} e^{-\frac{d_1^{\tau}\gamma_{\mathrm{th}}}{(1-\theta)\rho_1}}}{(1-\theta)\rho_I} \sum_{m=0}^{N-1} \left(\frac{d_1^{\tau}\gamma_{\mathrm{th}}}{(1-\theta)\rho_1}\right)^m \sum_{n=0}^{m} \frac{1}{(m-n)!} \left(\frac{(1-\theta)\rho_1\rho_I}{d_I^{\tau}\rho_1 + d_1^{\tau}\rho_I\gamma_{\mathrm{th}}}\right)^{n+1},$$

and

$$\begin{aligned} F_2^{\mathrm{MRC}} &= \frac{2 d_1^{N\tau} d_I^{N\tau}}{\rho_1^N \rho_I^N} \sum_{s=1}^{N} \frac{\prod_{j=1}^{s-1}(1-N-j)}{(N-s)!\,(s-1)!} \left(\frac{d_I^{\tau}}{\rho_I} - \frac{d_1^{\tau}}{\rho_1}\right)^{1-N-s} \\ &\quad \times \sum_{m=0}^{N-1} \frac{1}{m!} \left(\frac{d_2^{\tau}\gamma_{\mathrm{th}}}{\eta\theta}\right)^{N+1-s} \left(\frac{d_1^{\tau} d_2^{\tau}\gamma_{\mathrm{th}}}{\eta\theta\rho_1}\right)^{\frac{m+s-N-1}{2}} K_{m+s-N-1}\left(2\sqrt{\frac{d_1^{\tau} d_2^{\tau}\gamma_{\mathrm{th}}}{\eta\theta\rho_1}}\right) \\ &\quad + \frac{2 d_1^{N\tau} d_I^{N\tau}}{\rho_1^N \rho_I^N} \sum_{s=1}^{N} \frac{\prod_{j=1}^{s-1}(1-N-j)}{(N-s)!\,(s-1)!} \left(\frac{d_1^{\tau}}{\rho_1} - \frac{d_I^{\tau}}{\rho_I}\right)^{1-N-s} \sum_{m=0}^{N-1} \frac{1}{m!} \left(\frac{d_2^{\tau}\gamma_{\mathrm{th}}}{\eta\theta}\right)^{N+1-s} \\ &\quad \times \left(\frac{d_2^{\tau} d_I^{\tau}\gamma_{\mathrm{th}}}{\eta\theta\rho_I}\right)^{\frac{m+s-N-1}{2}} K_{m+s-N-1}\left(2\sqrt{\frac{d_2^{\tau} d_I^{\tau}\gamma_{\mathrm{th}}}{\eta\theta\rho_I}}\right). \end{aligned} \tag{4.46}$$

Proof. A detailed proof can be found in Appendix IV in Reference 40. □

While Theorem 4.8 is useful for the evaluation of the system's outage probability, the expression is complex to yield much useful insights. Motivated by this, we look into the high SNR region and derive a simple approximation for the outage probability, which enables the characterization of the achievable diversity order of the system.

Theorem 4.9. *In the high SNR region, i.e., $\rho_1 \to \infty$, the outage probability of the MRC/MRT scheme can be approximated as*[2]

$$P_{I\text{out}}^{\text{MRC}} \approx \left(\frac{d_1^\tau \gamma_{\text{th}}}{\rho_1}\right)^N \left(\left(\frac{1}{1-\theta}\right)^N \sum_{n=0}^{N} \frac{((1-\theta)\rho_I)^n}{d_I^{n\tau}(N-n)!} + \frac{d_2^{N\tau}}{(\eta\theta)^N \Gamma(N+1)\,\Gamma(N)}\right.$$
$$\left. \times \sum_{i=0}^{N-1} \binom{N-1}{i} (-1)^{N-i-1} \frac{{}_2F_1\left(N, 2N-i-1; 2N-i; 1-\frac{d_1^\tau \rho_I}{d_I^\tau \rho_1}\right)}{2N-i-1}\right). \tag{4.47}$$

Proof. A detailed proof can be found in Appendix V in Reference 40. □

For the special case where the relay is equipped with a single antenna, i.e., $N = 1$, with the help of Eq. (9.121.6) in Reference 44, (4.47) reduces to

$$P_{I\text{out}}^{\text{MRC}} \approx \left(\frac{1}{1-\theta} + \frac{\rho_I}{d_I^\tau} + \frac{d_2^\tau\left(\ln\frac{\rho_1}{d_1^\tau} - \ln\frac{\rho_I}{d_I^\tau}\right)}{\eta\theta}\right)\frac{d_1^\tau \gamma_{\text{th}}}{\rho_1}. \tag{4.48}$$

Theorem 4.9 indicates that a full diversity order of N is still achievable in the presence of CCI for the MRC/MRT scheme. Moreover, from (4.48), we see that the effect of CCI could be either beneficial or detrimental, depending on the relationship between $\rho_I, d_I^\tau, d_2^\tau\ \eta$, and θ, i.e., when $\frac{\rho_I}{d_I^\tau} - \frac{d_2^\tau(\ln\rho_I - \ln d_I^\tau)}{\eta\theta}$ is positive, CCI is detrimental, while when $\frac{\rho_I}{d_I^\tau} - \frac{d_2^\tau(\ln\rho_I - \ln d_I^\tau)}{\eta\theta}$ is negative, CCI becomes beneficial, which suggests that, in wireless-powered relaying systems, CCI could be potentially exploited to improve the performance.

4.4.1.2 Optimal analysis for the parameter θ

We now study the optimal value of θ minimizing the outage probability. Based on the high SNR approximation for $P_{I\text{out}}^{\text{MRC}}$ in (4.47), the optimal θ can be found in the following proposition.

Proposition 4.1. *The optimal $\theta \in (0, 1)$ is a root of the following polynomial:*

$$\sum_{n=0}^{N-1} \mathcal{A}(n)\,(1-\theta)^{n-N-1} - \frac{\mathcal{B}}{\theta^{N+1}} = 0, \tag{4.49}$$

[2] It is worth pointing out that the result in Theorem 4.9 holds for all cases whether the signal power and the CCI power are equal or not.

where $\mathcal{A}(n) = \frac{1}{(N-n-1)!}\frac{\rho_I^n}{d_I^{n\tau}}$ *and*

$$\mathcal{B} = \frac{d_2^{N\tau}}{\eta^N \Gamma^2(N)} \sum_{i=0}^{N-1} \binom{N-1}{i} (-1)^{N-i-1} \frac{{}_2F_1\left(N, 2N-i-1; 2N-i; 1-\frac{\rho_I d_1^\tau}{\rho_1 d_I^\tau}\right)}{2N-i-1}. \quad (4.50)$$

In the special case of $N = 1$, the optimal solution can be expressed in closed-form as follows:

$$\theta_{\text{MRC}}^{\text{opt}} = \frac{\sqrt{\dfrac{d_I^\tau \rho_1 (\ln \rho_I - \ln \rho_1 - \ln d_I^\tau + \ln d_1^\tau)}{\eta (d_1^\tau \rho_I - d_I^\tau \rho_1)}}}{1 + \sqrt{\dfrac{d_I^\tau \rho_1 (\ln \rho_I - \ln \rho_1 - \ln d_I^\tau + \ln d_1^\tau)}{\eta \left(d_1^\tau \rho_I - d_I^\tau \rho_1\right)}}}. \quad (4.51)$$

This simple expression is quite informative, and it can be observed that the optimal θ in (4.51) is a decreasing function of η and ρ_I, and an increasing function of ρ_1, which can be explained as follows:

- As η increases, more transmission power can be collected at the relay, hence the bottleneck of the system performance lies in the SINR of the signal at the input of the information receiver. As a result, we should choose a smaller θ to improve the first-hop performance.
- A large ρ_I provides more energy, while at the same time reduces the SINR of the first-hop transmission. Hence, a smaller θ should be chosen to compensate the loss of the SINR.
- For large ρ_1, in general the first-hop transmission quality is quite good, hence, it is beneficial to have more energy at the relay, i.e., a larger θ is desirable.

4.4.2 ZF/MRT scheme

In order to completely eliminate the CCI, the number of the antennas equipped at the relay should be greater than the number of interferers in the ZF/MRT scheme. Hence, it is assumed that $N > 1$. According to Reference 22, the optimal combining vector $\mathbf{w}_1$ is given by

$$\mathbf{w}_1 = \frac{\mathbf{h}_1^\dagger \mathbf{P}}{\sqrt{\mathbf{h}_1^\dagger \mathbf{P} \mathbf{h}_1}}, \quad (4.52)$$

where $\mathbf{P} = \mathbf{I}_N - \mathbf{h}_I \left(\mathbf{h}_I^\dagger \mathbf{h}_I\right)^{-1} \mathbf{h}_I^\dagger$. Therefore, the end-to-end SINR of the ZF/MRT scheme can be expressed as

$$\gamma_I^{\text{ZF}} = \frac{\gamma_{I1}^{\text{ZF}} \gamma_{I2}^{\text{ZF}}}{\gamma_{I1}^{\text{ZF}} + \gamma_{I2}^{\text{ZF}} + 1}, \quad (4.53)$$

where $\gamma_{I1}^{\text{ZF}} = \left|\mathbf{h}_1^\dagger \mathbf{P} \mathbf{h}_1\right| \frac{(1-\theta)\rho_1}{d_1^\tau}$, and $\gamma_{I2}^{\text{ZF}} = \gamma_{I2}^{\text{MRC}}$.

4.4.2.1 Outage probability

Considering the ZF/MRT scheme, we first present a lower bound for the outage probability in Theorem 4.10 below.

Theorem 4.10. *If $\rho_1 \neq \rho_I$, the outage probability of the ZF/MRT scheme can be lower bounded as*

$$P_{I\text{out}}^{\text{LZF}} = 1 - F_1^{\text{ZF}} F_2^{\text{ZF}}, \tag{4.54}$$

where

$$F_1^{\text{ZF}} = \frac{\Gamma\left(N-1, \frac{d_1^{\tau}\gamma_{\text{th}}}{(1-\theta)\rho_1}\right)}{\Gamma(N-1)} \tag{4.55}$$

and $F_2^{\text{ZF}} = F_2^{\text{MRC}}$.

Proof. A detailed proof can be found in Reference 40. □

To gain further insights, next we look into the high SNR region and present a simple and informative approximation for the outage probability.

Theorem 4.11. *In the high-SNR region, i.e., $\rho_1 \to \infty$, the outage probability of the ZF/MRT scheme can be approximated as*

$$P_{I\text{out}}^{\text{ZF}} \approx \frac{1}{(N-1)!}\left(\frac{d_1^{\tau}\gamma_{\text{th}}}{(1-\theta)\rho_1}\right)^{N-1}. \tag{4.56}$$

Proof. A detailed proof can be found in Reference 40. □

Theorem 4.11 indicates that the achievable diversity order of the ZF/MRT scheme is $N-1$. Compared with the MRC/MRT scheme, the ZF/MRT scheme incurs a diversity loss of one. This is an intuitive and satisfying result since one degree of freedom is used for the elimination of the CCI.

4.4.3 MMSE/MRT scheme

In the case of the MMSE/MRT scheme, $\mathbf{w}_1$ is given by

$$\mathbf{w}_1 = \mathbf{h}_1^{\dagger}\left(\mathbf{h}_1\mathbf{h}_1^{\dagger} + \mathbf{h}_I\mathbf{h}_I^{\dagger} + \frac{d_I^{\tau}}{(1-\theta)\rho_I}\mathbf{I}\right)^{-1}. \tag{4.57}$$

Therefore, the end-to-end SINR can be expressed as

$$\gamma_I^{\text{MMSE}} = \frac{\gamma_{I1}^{\text{MMSE}}\gamma_{I2}^{\text{MMSE}}}{\gamma_{I1}^{\text{MMSE}} + \gamma_{I2}^{\text{MMSE}} + 1}, \tag{4.58}$$

where $\gamma_{I1}^{\text{MMSE}} = \frac{d_I^{\tau}\rho_1}{d_1^{\tau}\rho_I}\mathbf{h}_1^{\dagger}\mathbf{R}^{-1}\mathbf{h}_1$, $\mathbf{R} = \mathbf{h}_I\mathbf{h}_I^{\dagger} + \frac{d_I^{\tau}}{(1-\theta)\rho_I}\mathbf{I}$ and $\gamma_{I2}^{\text{MMSE}} = \gamma_{I2}^{\text{MRC}}$.

4.4.3.1 Outage probability

In order to present an outage lower bound, first notice that the end-to-end SINR expression in (4.58) can be tightly upper bounded by

$$\gamma_I^{\text{MMSE}} \leq \gamma_I^{\text{up}} = \min\left(\gamma_{I1}^{\text{MMSE}}, \gamma_{I2}^{\text{MMSE}}\right). \tag{4.59}$$

Hence, the outage lower bound of the system can be computed by

$$P_{I\text{out}}^{\text{LMMSE}} = \text{Prob}\left(\gamma_I^{\text{up}} < \gamma_{\text{th}}\right). \tag{4.60}$$

Now, starting from (4.44), we are ready to establish the lower outage bound of the MMSE/MRT scheme, which is given in the following theorem.

Theorem 4.12. *If $\rho_1 \neq \rho_I$, the outage probability of the MMSE/MRT scheme can be lower bounded as*

$$P_{I\text{out}}^{\text{LMMSE}} = 1 - F_1^{\text{MMSE}} F_2^{\text{MMSE}}, \tag{4.61}$$

where

$$F_1^{\text{MMSE}} = \frac{\Gamma\left(N, \frac{d_1^\tau \gamma_{\text{th}}}{(1-\theta)\rho_1}\right)}{\Gamma(N)} - \exp\left(-\frac{d_1^\tau \gamma_{\text{th}}}{(1-\theta)\rho_1}\right)\frac{(1-\theta)\rho_I}{d_I^\tau \Gamma(N)} \times \left(\frac{d_1^\tau \gamma_{\text{th}}}{(1-\theta)\rho_1}\right)^N {}_2F_1\left(2, 1; 2; -\frac{d_1^\tau \rho_I}{d_I^\tau \rho_1}\gamma_{\text{th}}\right) \tag{4.62}$$

and $F_2^{\text{MMSE}} = F_2^{\text{MRC}}$.

Proof. A detailed proof can be found in Reference 40. □

To gain further insights, we now look into the high SNR region and present a simple approximation for the outage probability.

Theorem 4.13. *In the high-SNR region, i.e., $\rho_1 \to \infty$, the outage probability of the MMSE/MRT scheme can be approximated as*

$$P_{I\text{out}}^{\text{MMSE}} \approx \left(\frac{d_1^\tau \gamma_{\text{th}}}{\rho_1}\right)^N \left(\left(\frac{1}{N!} + \frac{(1-\theta)\rho_I}{d_I^\tau \Gamma(N)}\right)\frac{1}{(1-\theta)^N} + \frac{\left(\frac{d_2^\tau}{\eta\theta}\right)^N}{\Gamma(N+1)\,\Gamma(N)} \times \sum_{i=0}^{N-1}\binom{N-1}{i}(-1)^{N-i-1} \times \frac{{}_2F_1\left(N, 2N-i-1; 2N-i; 1-\frac{d_1^\tau \rho_I}{d_I^\tau \rho_1}\right)}{2N-i-1}\right). \tag{4.63}$$

Proof. A detailed proof can be found in Reference 40. □

Theorem 4.13 indicates that the MMSE/MRT scheme achieves a diversity order of N, the same as the MRC/MRT scheme.

A close observation of (4.47), (4.56), and (4.63) reveals that the difference among all three schemes only lies in their first terms, which can be expressed as follows:

$$\begin{aligned} a_{\text{MRC}} &= \sum_{n=0}^{N} \frac{((1-\theta)\rho_I)^n}{d_I^{n\tau}(N-n)!}, \\ a_{\text{ZF}} &= \frac{1}{(N-1)!}, \\ a_{\text{MMSE}} &= \frac{1}{N!} + \frac{(1-\theta)\rho_I}{d_I^\tau (N-1)!}. \end{aligned} \tag{4.64}$$

It can be easily observed that a_{MMSE} is strictly smaller than a_{MRC}, since a_{MMSE} only includes the first two terms of a_{MRC}. As such, we conclude that the MMSE/MRT scheme always achieves a strictly better outage performance than the MRC/MRT scheme due to the higher array gain. For the ZF/MRT scheme, although a diversity loss leads to its inferior performance in the high-SNR region, it should be noted that a_{ZF} is generally smaller than a_{MRC}, which means that the ZF/MRC scheme has a larger array gain than the MRC/MRT scheme. Therefore, in the low-SNR region, the ZF/MRT scheme may achieve better outage performance than the MRC/MRT scheme.

4.4.4 Numerical results

Now we show some numerical results to verify the analysis and reveal the effect of various system parameters on the outage probability of the considered SWIPT relay system. Figure 4.6 shows the impact of the number of relay antennas N on the outage probability. We observe that the proposed lower bounds in (4.45), (4.54) and (4.61) and the high SNR approximations in (4.47), (4.56) and (4.63) remain sufficiently accurate for a wide range of SNRs. In addition, the lower bounds become almost exact in the high SNR regime. Furthermore, it can be seen that full diversity order of N can be achieved by the MRC/MRT and MMSE/MRT schemes. On the other hand, the ZF/MRT scheme only attains a diversity order of $N-1$, as predicated by our analysis. Finally, the MMSE/MRT scheme always achieves the lowest outage probability, and the ZF/MRT scheme is superior to the MRC/MRT scheme in the low SNR regime, while the opposite holds in the high SNR regime.

Figure 4.7 examines the effect of CCI on the outage performance of the considered SWIPT relay system. The number of relay antennas $N = 4$. For illustration purpose, the scenario without CCI is also included for comparison. We observe that, for the MRC/MRT scheme, the outage probability improves slightly for smaller ρ_I (i.e., $\rho_I < 0\,\text{dB}$), and then deteriorates rapidly as the CCI becomes stronger. For example, with $\rho_I = 20\,\text{dB}$, the outage probability almost approaches one. On the other hand, the outage probability of the other two schemes do not show a significant degradation. The minimum outage of the MMSE/MRT scheme occurs approximately at $\rho_I = 6\,\text{dB}$. Finally, we see that as ρ_I increases, the outage probabilities of ZF/MRT and MMSE/MRT schemes attain a floor value. These observations clearly manifests the double-role of the CCI, i.e., corrupting the desired signal and supplying extra energy. For the MRC/MRT scheme, when CCI is too strong, the disadvantage of the CCI dominates. However, by employing sophisticated interference mitigation schemes, example, MMSE/MRT, such an undesirable effect could be substantially mitigated, and the CCI could be effectively exploited.

4.5 Conclusions

We reviewed recent advances of linear processing techniques for multi-antenna relaying systems with interference. We showed that with simple linear processing schemes

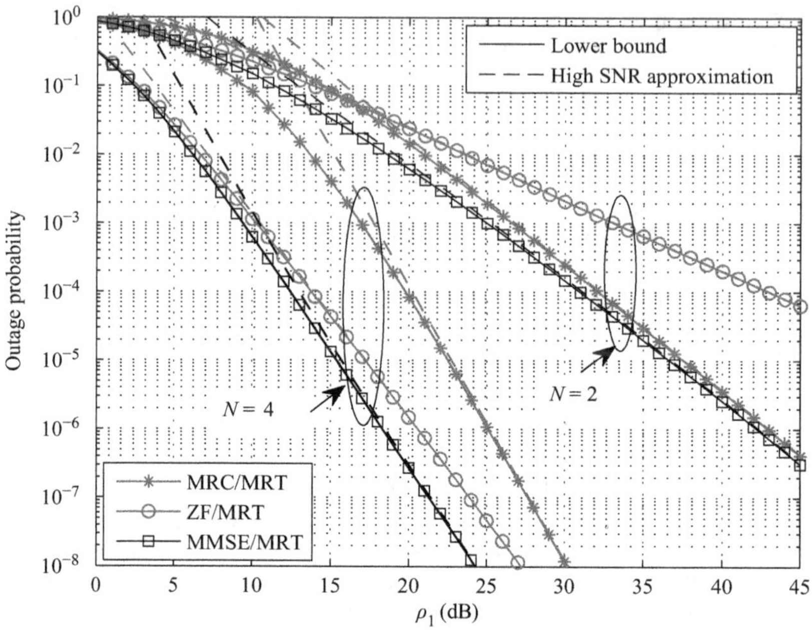

Figure 4.6 Impact of N on the outage performance: $\gamma_{th} = 10$ *dB,* $\eta = 0.2$*,* $\theta = 0.5$*,* $\rho_I = 9.5$ *dB,* $\tau = 2$*, and* $d_1 = d_2 = d_I = 1$

such as MRC/MRT, ZF/MRT, or MMSE/MRT, multiple antennas can be efficiently exploited to suppress or cancel CCI in dual-hop AF relaying systems. The performance of these schemes depend on the SNR operation point of the system. Specifically, the MMSE/MRT scheme achieves the best outage performance, and the ZF/MRT scheme outperforms the MRC/MRT scheme in the low-SNR regime, while becomes inferior to the MRC/MRT scheme in the high-SNR regime. Moreover, increasing the number of antennas at the relay and using advanced linear processing schemes such as ZF/MRT or MMSE/MRT are effective solutions to mitigate the effect of interference.

For SWIPT systems, interference can be regarded as an extra source of energy, and the use of multiple antenna techniques is an effective means of manipulating the CCI. Depending on the adopted linear processing scheme, CCI can be potentially exploited to significantly enhance the system performance. The CCI is always a desirable factor for the MMSE/MRT scheme, and the stronger the CCI, the better the performance. However, in MRC/MRT and ZF/MRT schemes, interference could be detrimental for the performance. For instance, strong CCI degrades the performance of the MRC/MRT scheme, whereas with the ZF/MRT the performance deteriorates due to weak CCI.

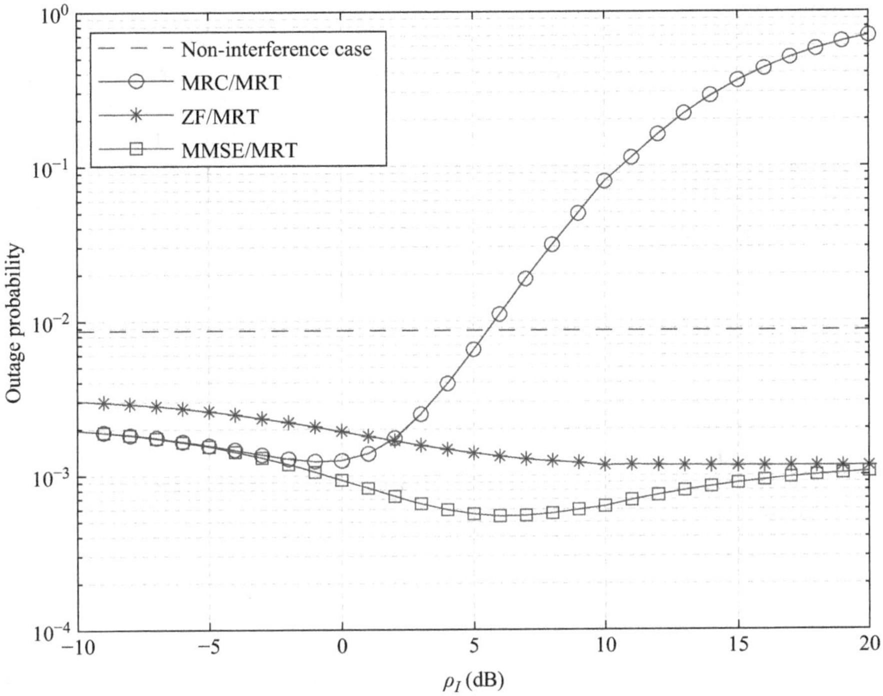

Figure 4.7 Impact of CCI on the system performance: $\gamma_{th} = 0$ dB, $\eta = 0.2$, $\theta = 0.5$, $\rho_1 = 10$ dB, $\tau = 2$, and $d_1 = d_2 = d_I = 1$

4.5.1 Future directions

- In practice, perfect CSI is difficult to obtain and quality of the acquired CSI will have a significant impact on the performance of the linear processing schemes, under investigation. Therefore, the performance of the presented linear combining schemes under imperfect CSI is an interesting research topic of great practical value.
- Increasing the number of antennas was shown to be a promising solution for linear CCI suppression in relaying systems. However, this also increases the implementation complexity. Moreover, energy consumption due to additional hardware complexity is a concern for system design. Therefore, low complexity and energy efficient implementation of linear processing schemes is of practical value.
- This chapter focused on dual-hop relaying systems. However, another promising architecture for cellular cooperative networks is multi-hop relaying. Therefore, extending current results to embrace multi-hop MIMO relaying with linear processing schemes in the presence of CCI is an important research direction. Moreover, performance of linear processing schemes for distributed relay architectures using techniques such as coordinated multi-point (CoMP) is worthwhile

to investigate. Performance analysis for the cases of efficient relaying protocols (e.g., two-way relaying, incremental relaying) could also be considered.

- Wireless relaying has found useful applications in several emerging paradigms such as physical layer security. The role of interference and linear processing need to be well understood for efficient design of such secure communication systems.
- Interference exploitation is a relatively new topic, and there is plenty of room for further investigation. For example, how to effectively exploit CCI in the presence channel estimation errors is an open research problem.

References

[1] I. Akyildiz, D. M. Gutierrez-Estevez, and E. C. Reyes, "The evolution of 4G systems: LTE-Advanced," *Phys. Commun.*, vol. 3, no. 4, pp. 217–244, 2010.

[2] W. Guo and T. O'Farrell, "Relay deployment in cellular networks: Planning and optimization," *IEEE J. Select. Areas Commun.*, vol. 30, no. 8, pp. 1597–1606, Sept. 2012.

[3] R. Pabst *et al.*, "Relay-based deployment concepts for wireless and mobile broadband radio," *IEEE Commun. Mag.*, vol. 42, pp. 80–89, Sept. 2004.

[4] J. Cho and Z. J. Haas, "On the throughput enhancement of down-stream channel in cellular radio networks through multihop relaying," *IEEE J. Select. Areas Commun.*, vol. 22, no. 7, pp. 1206–1219, Sept. 2004.

[5] M. Dohler and Y. Li, *Cooperative Communications: Hardware, Channel & PHY.* John Wiley & Sons, Chichester, UK, 2010.

[6] M. O. Hasna and M.-S. Alouini, "End-to-end performance of transmission systems with relays over Rayleigh-fading channels," *IEEE Trans. Wireless Commun.*, vol. 2, no. 6, pp. 1126–1131, Nov. 2003.

[7] J. G. Andrews, S. Buzzi, W. Choi, *et al.*, "What will 5G be?," *IEEE J. Select. Areas Commun.*, vol. 32, pp. 1065–1082, June 2014.

[8] I. Ku, C.-X. Wang, and J. Thompson, "Spectral, energy and economic efficiency of relay-aided cellular networks," *IET Commun.*, vol. 7, no. 14, pp. 1476–1486, 2013.

[9] A. Abu-Dayya and N. Beaulieu, "Outage probability of cellular mobile radio systems with multiple Nakagami interferers," *IEEE Trans. Veh. Technol.*, vol. 41, no. 4, pp. 757–768, Nov. 1991.

[10] J. G. Andrews, "Interference cancellation for cellular systems: A contemporary overview," *IEEE Wireless Commun. Mag.*, vol. 12, no. 2, pp. 19–29, Apr. 2005.

[11] I. Krikidis, J. Thompson, S. McLaughlin, and N. Goertz, "Max–min relay selection for legacy amplify-and-forward systems with interference," *IEEE Trans. Wireless Commun.*, vol. 8, no. 6, pp. 3016–3027, June 2009.

[12] C. Zhong, S. Jin, and K. K. Wong, "Dual-hop system with noisy relay and interference-limited destination," *IEEE Trans. Commun.*, vol. 58, no. 6, pp. 764–768, Mar. 2010.

[13] H. A. Suraweera, H. K. Garg, and A. Nallanathan, "Performance analysis of two hop amplify-and-forward systems with interference at the relay," *IEEE Commun. Lett.*, vol. 14, no. 8, pp. 692–694, Aug. 2010.

[14] I. Trigui, S. Affes, and A. Stephenne, "On the ergodic capacity of amplify-and-forward relay channels with interference in Nakagami-*m* fading," *IEEE Trans. Commun.*, vol. 61, no. 8, pp. 3136–3145, Aug. 2013.

[15] F. Al-Qahtani, T. Q. Duong, C. Zhong, K. Qaraqe, and H. Alnuweiri, "Performance analysis of dual-hop AF systems with interference in Nakagami-*m* fading channels," *IEEE Signal Process. Lett.*, vol. 18, no. 8, pp. 454–457, Aug. 2011.

[16] H. A. Suraweera, D. S. Michalopoulos, and C. Yuen, "Performance analysis of fixed gain relay systems with a single interferer in Nakagami-*m* fading channels," *IEEE Trans. Veh. Technol.*, vol. 61, no. 3, pp. 1457–1463, Mar. 2012.

[17] S. Ikki and S. Aïssa, "Performance analysis of dual-hop relaying systems in the presence of co-channel interference," in *Proc. IEEE Global Commun. Conf. (GLOBECOM 2010)*, Miami, FL, Dec. 2010, pp. 1–5.

[18] A. Salhab, F. Al-Qahtani, S. Zummo, and H. Alnuweiri, "Exact outage probability of opportunistic DF relay systems with interference at both the relay and the destination over Nakagami-*m* fading channels," *IEEE Trans. Veh. Technol.*, vol. 62, no. 2, pp. 920–927, Feb. 2013.

[19] S. Yang and L. Hanzo, "Fifty years of MIMO detection: The road to large-scale MIMOs," *IEEE Commun. Surveys Tutorials*, vol. 17, no. 4, pp. 1941–1988, Fourth quarter, 2015.

[20] H. Ding, C. He, and L. Jiang, "Performance analysis of fixed gain MIMO relay systems in the presence of co-channel interference," *IEEE Commun. Lett.*, vol. 16, no. 7, pp. 1133–1136, July 2012.

[21] C. Zhong, H. A. Suraweera, A. Huang, Z. Zhang, and C. Yuen, "Outage probability of dual-hop multiple antenna AF relaying systems with interference," *IEEE Trans. Commun.*, vol. 61, no. 1, pp. 108–119, Jan. 2013.

[22] G. Zhu, C. Zhong, H. A. Suraweera, Z. Zhang, and C. Yuen, "Outage probability of dual-hop multiple antenna AF systems with linear processing in the presence of co-channel interference," *IEEE Trans. Wireless Commun.*, vol. 13, no. 4, pp. 2308–2321, Apr. 2014.

[23] G. Zhu, C. Zhong, H. A. Suraweera, Z. Zhang, C. Yuen, and R. Yin, "Ergodic capacity comparison of different relay precoding schemes in dual-hop AF systems with co-channel interference," *IEEE Trans. Commun.*, vol. 62, no. 7, pp. 2314–2328, July 2014.

[24] K. S. Karthik and B. Ramamurthi, "A two-hop AF relaying scheme with interference suppression at the relay," *IEEE Trans. Veh. Technol.*, vol. 63, no. 7, pp. 3469–3474, Sept. 2014.

[25] N. Suraweera and N. C. Beaulieu, "Optimum combining with joint relay and antenna selection for multiple-antenna relays in the presence of co-channel interference," *IEEE Commun. Lett.*, vol. 18, no. 8, pp. 1459–1462, Aug. 2014.

[26] J. Guo, C. Pei, and H. Yang, "Performance analysis of TAS/MRC in MIMO relay systems with outdated CSI and co-channel interference," *IEEE Trans. Wireless Commun.*, vol. 13, no. 9, pp. 4848–4856, Sept. 2014.

[27] M. Joham, W. Utschick, and J. A. Nossek, "Linear transmit processing in MIMO communications systems," *IEEE Trans. Signal Process.*, vol. 53, no. 8, pp. 2700–2712, Aug. 2005.

[28] M. Vu and A. Paulraj, "MIMO wireless linear precoding," *IEEE Signal Process. Mag.*, vol. 24, no. 5, pp. 86–105, Sept. 2007.

[29] W. Zhang, X. Ma, B. Gestner, and D. V. Anderson, "Designing low-complexity equalizers for wireless systems," *IEEE Commun. Mag.*, vol. 47, pp. 56–64, Jan. 2009.

[30] A. H. Mehana and A. Nosratinia, "Diversity of MIMO linear precoding," *IEEE Trans. Inf. Theory*, vol. 60, no. 2, pp. 1019–1038, Feb. 2014.

[31] G. Zheng, I. Krikidis, C. Masouros, S. Timotheou, D. Toumpakaris, and Z. Ding, "Rethinking the role of interference in wireless networks," *IEEE Commun. Mag.*, vol. 52, no. 11, pp. 152–158, Nov. 2014.

[32] C. Masouros, "Correlation rotation linear precoding for MIMO broadcast channel," *IEEE Trans. Signal Process.*, vol. 59, no. 1, pp. 252–262, Jan. 2011.

[33] C. Masouros, T. Ratnarajah, M. Sellathurai, C. Papadias, and A. Shukla, "Known interference in wireless communications: A limiting factor or a potential source of green signal power?," *IEEE Commun. Mag.*, vol. 51, no. 10, pp. 162–171, Oct. 2013.

[34] C. Masouros and G. Zheng, "Exploiting known interference as green signal power for downlink beamforming optimization," *IEEE Trans. Signal Process.*, vol. 63, no. 14, pp. 3628–3640, July 2015.

[35] L. R. Varshney, "Transporting information and energy simultaneously," in *Proc. IEEE Intl. Symp. Inf. Theory (ISIT 2008)*, Toronto, Canada, July 2008, pp. 1612–1616.

[36] P. Grover and A. Sahai, "Shannon meets Tesla: Wireless information and power transfer," in *Proc. IEEE Intl. Symp. Inf. Theory (ISIT 2010)*, Austin, TX, June 2010, pp. 2363–2367.

[37] R. Zhang and C. K. Ho, "MIMO broadcasting for simultaneous wireless information and power transfer," *IEEE Trans. Wireless Commun.*, vol. 12, no. 5, pp. 1989–2001, May 2013.

[38] A. A. Nasir, X. Zhou, S. Durrani, and R. A. Kennedy, "Relaying protocols for wireless energy harvesting and information processing," *IEEE Trans. Wireless Commun.*, vol. 12, no. 7, pp. 3622–3636, July 2013.

[39] I. Krikidis, S. Sasaki, S. Timotheou, and Z. Ding, "A low complexity antenna switching for joint wireless information and energy transfer in MIMO relay channels," *IEEE Trans. Commun.*, vol. 62, no. 5, pp. 1577–1587, May 2014.

[40] G. Zhu, C. Zhong, H. A. Suraweera, G. K. Karagiannidis, Z. Zhang, and T. A. Tsiftsis, "Wireless information and power transfer in relay systems with multiple antennas and interference," *IEEE Trans. Commun.*, vol. 63, no. 4, pp. 1400–1418, Apr. 2015.

[41] Z. Ding, C. Zhong, D. W. K. Ng, *et al.*, "Application of smart antenna technologies in simultaneous wireless information and power transfer," *IEEE Commun. Mag.*, vol. 53, no. 4, pp. 86–93, Apr. 2015.

[42] E. G. Larsson, F. Tufvesson, O. Edfors, and T. L. Marzetta, "Massive MIMO for next generation wireless systems," *IEEE Commun. Mag.*, vol. 52, no. 2, pp. 186–195, Feb. 2014.

[43] H. Gao, P. J. Smith, and M. V. Clark, "Theoretical reliability of MMSE linear diversity combining in Rayleigh-fading additive interference channels," *IEEE Trans. Commun.*, vol. 46, no. 5, pp. 666–672, May 1998.

[44] I. S. Gradshteyn and I. M. Ryzhik, *Tables of Integrals, Serious and Products*, sixth ed., Academic Press, San Diego, CA, 2007.

[45] Z. Wang and G. B. Giannakis, "A simple and general parameterization quantifying performance in fading channels," *IEEE Trans. Commun.*, vol. 51, no. 8, pp. 1389–1398, Aug. 2003.

[46] D. Katselis, "On estimating the number of co-channel interferers in MIMO cellular systems," *IEEE Signal Process. Lett.*, vol. 18, no. 6, pp. 379–382, June 2011.

[47] R. Narasimhan and S. Cheng, "Channel estimation and co-channel interference rejection for LTE-Advanced MIMO uplink," in *Proc. IEEE Wireless Commun. and Networking Conf. (WCNC 2012)*, Shanghai, China, Apr. 2012, pp. 1–4.

[48] L. Liu, R. Zhang, and K. Chua, "Wireless information and power transfer: A dynamic power splitting approach," *IEEE Trans. Commun.*, vol. 61, no. 9, pp. 3990–4001, Sept. 2013.

Chapter 5
Relaying in full-duplex radio communication systems

Taneli Riihonen and Xiaodong Wang

5.1 Introduction

As explained in many chapters of this book, relaying in principle means indirect data transmission from a source node to a destination node via intermediate relay nodes (RNs), e.g., not so unlike repeaters, gap-fillers, or signal boosters do in old-fashioned wireless systems. On the other hand, the field of information theory has addressed many classical questions related to the performance of idealistic relay links. Relay transceivers in these original research areas conventionally operate in a *full-duplex* (FD) mode, while later studies on cooperative communication deemed it impossible to begin with and focused almost invariably on *half-duplex* (HD) relaying.

However, modern relay transceivers such as those assumed in the context of this chapter are much more advanced than simple analog amplifiers in terms of their processing capabilities so that, e.g., they can be made robust to interference. Time has thus become right for us to return to the origins of relaying by promoting FD operation as a practical and highly efficient design choice whose technical difficulties can be now overcome. With this background, this chapter provides an overview of essential aspects to be considered when introducing FD operation into relaying systems. Two particular contributions are central for this development, namely introducing transmit power control for FD relaying and comparing the performance of FD relay links to that of their HD counterparts.

The main focus herein is on infrastructure relaying although the considered systems have their relatives with cable-connected distributed antennas and mobile relays instead. Thus, the considered relay transceivers are fixed nodes and belong to infrastructure deployed by a network operator. This scope choice differs from earlier mainstream literature which is inclined toward ad hoc mobile relaying by user terminals. When discussing cellular systems, the link elements may be referred to as a base station (BS), a RN, and a user equipment (UE). The BS and RNs form a "backhaul link" while UEs are always connected to a "service link". There is typically only one relay per each group of destinations (except for special hand-over situations) while backhaul links may involve multihop relaying. Furthermore, the study is formulated in an implicit context to concern a single subcarrier within orthogonal frequency-division multiplexing (OFDM) transmission.

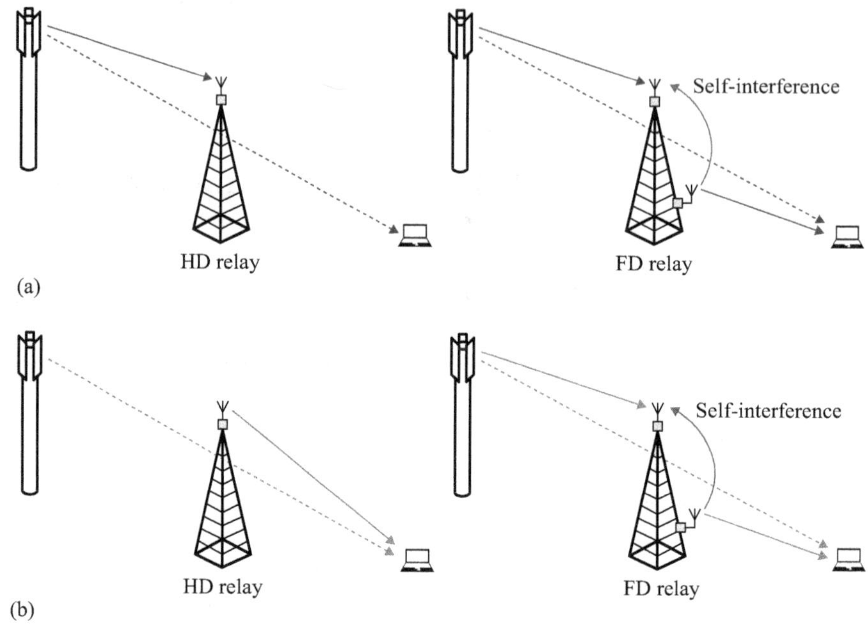

Figure 5.1 Comparison of HD and FD relays when two orthogonal (in time or frequency) channels are allocated for two-hop communication

5.1.1 Duplex modes for relay systems

The choice between FD and HD operation modes is a fundamental key factor determining the spectral efficiency of relay links [25]. They can be formally defined as follows (cf. Figure 5.1).

FD designates or pertains to a mode of operation by which information is transmitted to and from a relay in two directions simultaneously on the same physical channel.

HD designates or pertains to a mode of operation by which information is transmitted to and from a relay in two directions, but not simultaneously on the same physical channel.

In particular, HD operation needs to allocate orthogonal channels, which are separated in time or frequency, before and after each relay while FD relaying exploits frequency reuse. However, FD operation induces processing delay such that relayed signals are subject to an offset that is pronounced in OFDM with frequency-domain processing. Most importantly, FD operation is subject to severe loopback self-interference which is discussed in the following section. In fact, communication theory has so far considered mostly applications, e.g., user cooperation, in which the self-interference problem can be reasonably considered to be insurmountable.

In information theory, FD relays provide an idealistic benchmark when deriving capacity results for generic relay channels [3, 6, 10, 33, 34]. Studies in this context assume that self-interference can be eliminated perfectly and, consequently, the potential existence of residual distortion is neglected. On the other hand, practical systems have exploited the simplest form of FD relays, i.e., analog repeaters, since the early days of wireless communication, e.g., in digital television broadcasting. In this more down-to-earth context, the existence of self-interference is widely acknowledged although the field typically adheres to the FD mode from the start without rationalizing whether it is the proper choice. Yet HD relaying has received the main attention since the theory of cooperative communications was developed in the pioneering works, e.g., [8, 14, 15, 17, 30, 31].

Instead of considering actual FD operation, earlier research has focused on finding detours to overcome the limitations of the HD mode. The recent overview article [4] summarizes comprehensively key techniques for "combating the HD constraint", not so unlike the introduction in Reference 25. Especially, it is straightforward to achieve diversity gain with the HD mode but the allocation of two orthogonal channels imposes severe degradation in the multiplexing gain [14]. *Non-orthogonal relaying* refers to systems where the BS may transmit new data during RN transmission and the UE applies extra techniques for decoding the superimposed transmissions [2, 17]. This can be also used to gain an additional diversity path without essentially reducing the effect of the HD constraint [15]. *Successive relaying* systems are based on the idea that two alternating HD RNs can emulate the FD mode [5, 19]. This requires the cancellation of inter-relay interference instead of self-interference in FD relaying; the former signal is unknown unlike the latter which makes the task more challenging. *Two-way HD relaying* schemes [18] relax the orthogonality of downlink (DL) and uplink (UL) transmissions after proper signal design and receiver structures facilitate interference mitigation. In this case, self-interference echoes from the terminals back to themselves via the RN. Other related concepts are discussed in References 7 and 16.

Common for the above schemes is to relax the orthogonality of HD transmissions by providing means of suppressing the resulting interference or tolerating its presence. This is both in contrast with and, also, similar to FD relaying where self-interference emerges inherently without any special design that renders spectrum reuse. Eventually, relaying should still be contrasted with the conventional scenario of not using relays at all but directly transmitting from the source to the destination. This reference case is referred to as a direct transmission (DT) mode when considering degenerate FD relaying where relays are switched off and destinations receive only direct signals from sources.

Our research field has only recently begun to promote FD relaying under self-interference as a progressive design choice for future cooperative communication systems; some original works can be found in References 13, 22, 25, 26 and 32. Within a short period of time, FD radio communication systems have already been associated with many other timely schemes such as beamforming optimization in multiantenna relays [26, 32, 35], relay selection [12], distributed space–time codes [11] as well as wireless information and power transfer with energy harvesting [36].

5.1.2 Loopback self-interference

FD relaying achieves twice the end-to-end symbol rate of HD relaying per allocated channel resources. This causes inevitably a feedback loop (cf. Figure 5.1) where signals transmitted by relays leak back to their receive side. The phenomenon can be referred to, in short, as self-, loop(back) or echo interference, and it calls for gain control [22] to avoid saturation or oscillation in conventional repeaters and for efficient mitigation [26] in modern FD radios.

A FD relay needs separate antenna elements for reception and transmission (although the number of transceiver chains can be still conserved!) in contrast to HD relays, which provides passive physical isolation and facilitates active signal processing for additional mitigation. Thus, mobile terminals are almost invariably assumed to act as relays in the HD mode due to their small size. Using two arrays in the relay is useful not only for the FD mode due to interference mitigation but the HD mode benefits from it as well because antenna configurations, e.g., location and beam pattern, can be optimized separately in two directions while the placement of a single array is a compromise over the quality of the two hops. For example, both FD and HD can avoid building wall penetration loss which is inherent for the DT mode, and the self-interference is also suppressed by the very same effect in the FD mode. At the same time, beamforming using both antennas in HD relaying is inefficient, because the branches are highly imbalanced. Two-array deployment is especially advantageous in scenarios where the coverage of a macro-BS is extended to a building interior, to a metro tunnel, or to a shadowed area between buildings. A highly directive backhaul antenna can then be placed on the roof of a building and pointed toward the BS possibly with a line-of-sight (LOS) connection, and service antennas are placed on the street-level or indoors with wider beam pattern toward the covered area. Another reasonable scenario could be RN deployment on a (lamp) pole where sufficient isolation is ensured by vertical distance or an isolating plate between the antennas and by exploiting their directivity.

Assuming antenna partitioning which divides transmit and receive elements into separate arrays, three basic types of countermeasures against loopback interference can be broadly classified as follows: (1) the features of propagation paths between the antennas; (2) the directivity characteristics of the antennas; and (3) signal processing techniques for interference mitigation. The original list is rephrased above to clarify how these concepts are dealt with in References 23 and 26. Namely, Countermeasures (1) and (2) represent together *passive physical isolation* while Countermeasure (3) covers both *time-domain cancellation* and *spatial-domain suppression* although applying the latter is not possible for single-input single-output (SISO) relays at hand due to the lack of capability for multiantenna beamforming.

The self-interference situation of FD relays is extremely challenging. Since mitigation cannot be perfect in practice, the relays are likely subject to significant residual distortion, effect of which may be tolerable but should not be neglected. In particular, a relay's input channel may remain the bottleneck of the system which can still be compensated by decreasing transmit power at the cost of degrading the output channel. Hence, firstly, this chapter presents transmit power control schemes for FD relay links taking into account the effect of residual loopback interference that remains after imperfect mitigation. Throughout the analysis, the amplify-and-forward (AF) and decode-and-forward (DF) protocols are analyzed side-by-side in the same framework; the latter avoids an infinite feedback loop through the relay that is inherent for the former. With the proposed schemes, the end-to-end signal-to-interference-and-noise ratio (SINR) can be maximized while, at the same time, transmit power is decreased. Related target-SINR approach allows to relieve the need for the feedback of channel state information (CSI) on the hop after the relay. In addition to SINR illustrations, the system performance is evaluated by deriving closed-form outage probability expressions for all the transmit power control schemes which covers both DL and UL relaying.

Yet successful design of FD relaying should be always contrasted with the option to revert back to conventional HD relaying or even to plain DT. This requires analytical comparison between the choices and poses a fundamental rate–interference trade-off: The HD mode avoids inherently self-interference at the cost of halving end-to-end symbol rate while the FD mode offers full symbol rate but, in practice, suffers from self-interference even after mitigation and power control. Hence, secondly, this chapter introduces the concept of hybrid relaying to switch opportunistically between FD relaying, HD relaying, and the DT mode, which serves also as a thorough comparison between the modes. In particular, the combination of opportunistic mode selection and transmit power adaptation is developed for maximizing instantaneous and average link capacity after noting that the trade-off favors alternately the modes during operation. The analysis covers both AF and DF relaying as well as reflects the difference of DL and UL systems. In summary, the results show that the proposed schemes offer significant rate gain over a system design that is confined to any mode without rationalization.

5.1.3 Organization of the chapter

The remainder of this chapter[1] is organized as follows. The considered system and signal models are first specified in Section 5.2. After a brief introduction to the general purpose of power control and conventional methods, Section 5.3 features the main results of the first part of this study, namely the design and analysis of transmit power adaptation schemes for FD relays, while associated numerical results explain their principles and performance. In the second part of this study in Section 5.4, the main analytical results are two-fold: Section 5.4.1 focuses on short-term capacity analysis assuming static channels and instantaneous CSI while Section 5.4.2 proceeds to the case of fading channels and long-term capacity optimization using statistical CSI in infrastructure-based relaying. Finally, Section 5.5 concludes with a summary of observations.

[1] This chapter is partially based on the results originally presented in References 21, 22, 24, 25, 28, and 29.

5.2 System model

This section specifies a signal model for a two-hop OFDM link and derives expressions for the effective SINR achieved by end-to-end transmission via a relay. Both AF and DF protocols are considered like everywhere in this chapter but the scope of Section 5.3 is mainly on the FD mode. The HD counterparts are added in by Section 5.4 which works on the same setup to analyze trade-offs between relaying modes with or without power control.

The study of this chapter considers generic two-hop systems in which a source (S) communicates to a scheduled destination (D) via a relay (R) while focusing on coverage extension scenarios in which direct source-to-destination (SD) connection is weak and relays are deployed to help other nodes without having own data to transmit. The overall system setup is illustrated in Figure 5.2; it comprises four slowly

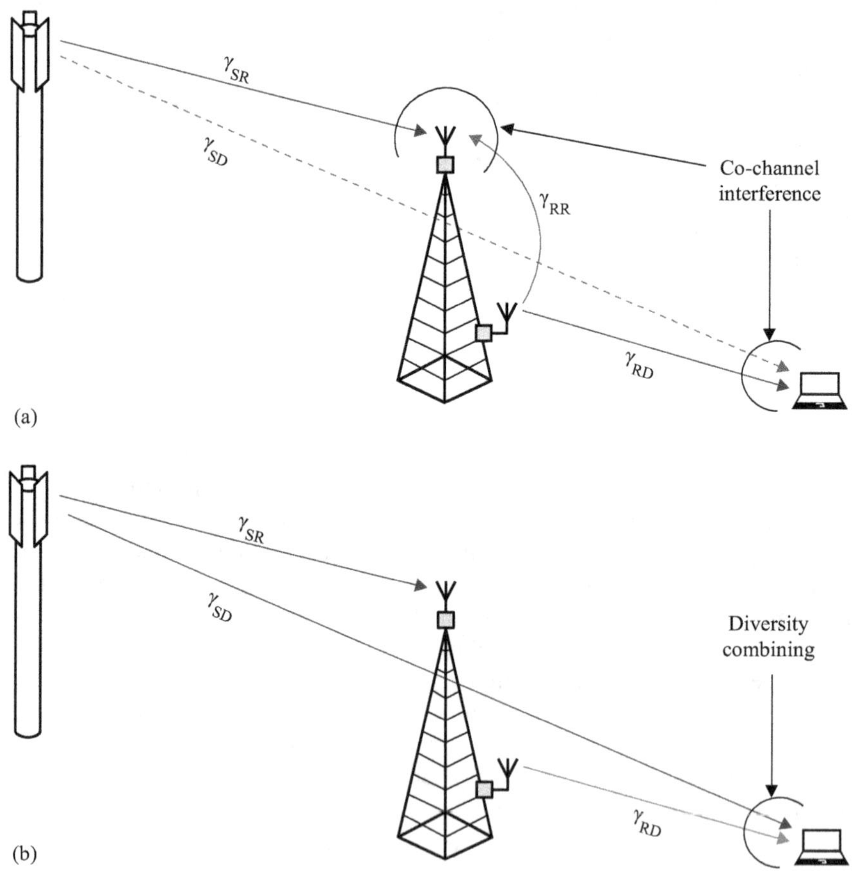

Figure 5.2 Two operation modes of a two-hop relay link in which γ_{SR}, γ_{SD}, γ_{RR}, and γ_{RD} denote the instantaneous SNRs of the narrowband channels. (a) FD mode and (b) HD mode

time-varying multipath channels which model the source-to-relay (SR), loopback relay-to-relay (RR), relay-to-destination (RD) and SD links. The system is parametrized with corresponding channel signal-to-noise ratios (SNRs), which are defined as $\gamma_{SR} \triangleq |h_{\text{SR}}|^2/\sigma_{\text{R}}^2$, $\gamma_{\text{RD}} \triangleq |h_{\text{RD}}|^2/\sigma_{\text{D}}^2$ and $\gamma_{\text{SD}} \triangleq |h_{\text{SD}}|^2/\sigma_{\text{D}}^2$. In plain words, these variables represent SNR levels that single-hop transmission could achieve over each channel with maximum power usage. Likewise, γ_{RR} represents the level of residual distortion due to loopback interference.

The channels are modeled to be frequency-flat and quasi-static, i.e., they may represent equivalent subchannels when the signals are transmitted on narrowband subcarriers in a full OFDM framework. This means that the channels are assumed to remain approximately stationary during a relatively long observation period and their multipath delay spread is assumed to be shorter than the cyclic prefix of OFDM symbols. By assuming also decent frequency and time synchronization, all subcarriers become orthogonal and subject to flat fading although frequency selectivity still causes different channel coefficients for each subcarrier since wideband transmission is assumed implicitly. In summary, there is no significant inter-carrier interference (ICI) at all and no other inter-symbol interference than that due to FD operation.

The novelty of this study lies in taking explicitly account of residual loopback self-interference that is inevitable in nearly all practical FD relays. As per Figure 5.2, the system thus chooses upfront or switches opportunistically between FD relaying, with or without power control, under co-channel interference and HD relaying, with or without diversity combining, under reduced spectral efficiency.

5.2.1 End-to-end signal models

Given the foregoing assumptions, the following signal models correspond to a single arbitrary OFDM subcarrier in frequency-domain after multiuser scheduling is completed. Yet subcarrier and user indices are omitted, and the system is studied at generic time instant i, i.e., when the ith OFDM symbol is being forwarded by the relay. The signal model is illustrated in Figure 5.3.

The source (S) transmits signal $x_{\text{S}}[i]$ to the relay (R) with constant normalized transmit power $p_{\text{S}} \triangleq \mathcal{E}\{|x_{\text{S}}[i]|^2\} = 1$ which is also overheard by the destination (D). The link operates in a single-frequency FD mode, i.e., the relay concurrently receives signal $y_{\text{R}}[i]$ and transmits signal $x_{\text{R}}[i]$ for which $p_{\text{R}} \triangleq \mathcal{E}\{|x_{\text{R}}[i]|^2\} \leq 1$. Thus, the relay input signal is a combination of the desired source signal, self-interference, and noise:

$$y_{\text{R}}[i] = h_{\text{SR}}x_{\text{S}}[i] + h_{\text{RR}}x_{\text{R}}[i] + n_{\text{R}}[i], \tag{5.1}$$

where h_{SR} is the flat-fading SR channel, h_{RR} is the loopback channel between the transmit antenna and the receive antenna of the relay, and the power of the thermal noise term is $\mathcal{E}\{|n_{\text{R}}[i]|^2\} \triangleq \sigma_{\text{R}}^2$.

The desired signal component in (5.1) is $h_{\text{SR}}x_{\text{S}}[i]$, and the relay may apply mitigation [26] to partly eliminate the adverse component $h_{\text{RR}}x_{\text{R}}[i]$. However, in practice, total interference elimination cannot be achieved due to non-ideal side information and signal processing, e.g., the dynamic range of the receiver front-end may be insufficient for both the useful and the interference signal causing quantization noise.

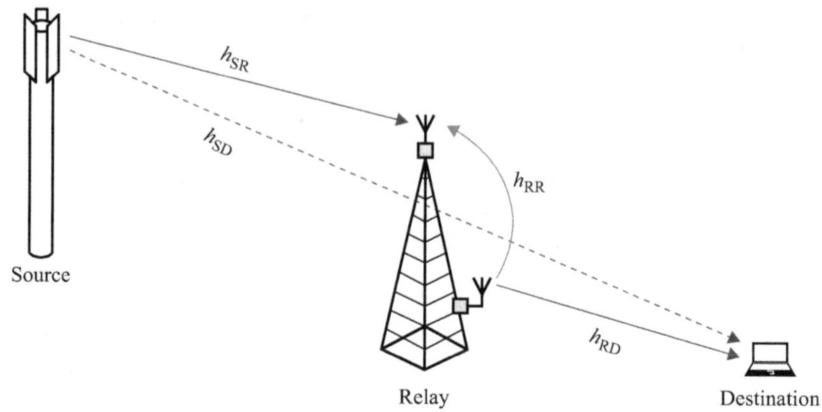

Figure 5.3 FD relay link in the presence of self-interference where narrowband wireless channels are denoted by h_{SR}, h_{SD}, h_{RR}, and h_{RD}

After mitigation, the residual self-interference distortion and noise can be denoted with an aggregate term $\hat{n}_R[i]$ translating effectively (5.1) to

$$\hat{y}_R[i] = h_{SR}x_S[i] + \hat{n}_R[i], \tag{5.2}$$

where $\mathcal{E}\{|\hat{n}_R[i]|^2\} \triangleq (p_R\gamma_{RR} + 1)\sigma_R^2$ and γ_{RR} represents the normalized gain of residual interference due to imperfect mitigation. As each combination of mitigation techniques is characterized by different implementation-specific residual interference levels, parametrization by γ_{RR} makes the analysis more generic. In fact, the model covers also the extreme case where the relay uses a single antenna for both transmission and reception which yields a large value for γ_{RR}. Likewise, $\gamma_{RR} \triangleq |h_{RR}|^2/\sigma_R^2$ when only passive physical antenna isolation is used without active mitigation. A more versatile model is proposed and analyzed in Reference 9.

As described shortly, $\hat{y}_R[i]$ is translated to $x_R[i]$ in an AF or DF manner:

$$x_R[i] \triangleq \hat{f}\left(\hat{y}_R\left[i - \tau_R\right], \hat{y}_R\left[i - (\tau_R + 1)\right], \hat{y}_R\left[i - (\tau_R + 2)\right], \ldots\right), \tag{5.3}$$

in which delay $\tau_R \geq 1$ is required to allow frequency-domain subcarrier-wise processing. Especially, each OFDM symbol must be received in its entirety before demodulation using the discrete Fourier transform.

Finally, the received signal in the destination is given by

$$y_D[i] = h_{RD}x_R[i] + h_{SD}x_S[i] + n_D[i], \tag{5.4}$$

where respective h_{RD} and h_{SD} denote flat-fading RD and SD channels, and $n_D[i]$ is thermal noise with $\mathcal{E}\{|n_D[i]|^2\} \triangleq \sigma_D^2$.

5.2.1.1 FD AF relaying

When adopting the linear AF protocol, the relay simply increases (by factor $\beta > 0$) the amplitude of the post-processing input signal obtained after self-interference mitigation. This aims at boosting the power of the desired signal sufficiently high to facilitate decoding in the destination even after the attenuation of the second-hop channel. Unfortunately, also the residual self-interference and noise components are amplified and forwarded to the destination at the same time.

The output signal of the AF relay can be expressed as

$$x_R[i] = \beta\, \hat{y}_R\,[i - \tau_R] = \beta\, h_{SR} x_S\,[i - \tau_R] + \beta\, \hat{n}_R\,[i - \tau_R]\,, \tag{5.5}$$

where the last form is obtained by substituting (5.2). In particular, the amplification factor β (or equivalently the relay gain β^2) shall be the key parameter for all the concepts discussed in this chapter. In (5.5), the amplitude of the desired signal is linearly proportional to β, while residual self-interference is not only amplified by β but also its input level, namely $p_R(\beta)\,\gamma_{RR}\,\sigma_R^2$, already increases monotonically as a function of β.

The case of using only passive antenna isolation lends itself to illustrating the problem of signal circulation through the relay and the feedback loop. Then $\hat{n}_R[i] = h_{RR}x_R[i] + n_R[i]$ in (5.2) and recursive substitution with (5.5) reveal that the relay transmits an infinite sum of echoes:

$$x_R[i] = \beta \sum_{j=1}^{\infty} (h_{RR}\beta)^{j-1}(h_{SR}x_S[i - j\tau_R] + n_R[i - j\tau_R]). \tag{5.6}$$

The expected value of $|x_R[i]|^2$ may diverge, i.e., the system is unstable, unless the gain setting satisfies $|h_{RR}\beta| < 1$. With active mitigation, $\hat{n}_R[i]$ is caused by $h_{RR}x_R[i]$ through a complex, possibly unknown nonlinear process but, in principle, residual self-interference power is subject to an analogous feedback loop.

As an alternative for parametrization with the amplification factor β, one may normalize protocols to guarantee constant transmit power p_R for the relay. Firstly, the relay input power can be calculated from (5.2) as

$$\mathcal{E}\left\{\left|\hat{y}_R[i]\right|^2\right\} = |h_{SR}|^2 + (p_R\gamma_{RR} + 1)\sigma_R^2, \tag{5.7}$$

assuming that $x_S[i]$ and $\hat{n}_R[i]$ (or equivalently $x_S[i - \tau_R]$) are uncorrelated. With any given value of p_R, (5.5) and the definition of p_R then imply $\mathcal{E}\{|x_R[i]|^2\} = \beta^2\mathcal{E}\{|\hat{y}_R[i - \tau_R]|^2\} = p_R$, and the amplification factor becomes

$$\beta(p_R) = \sqrt{\frac{p_R}{|h_{SR}|^2 + (p_R\gamma_{RR} + 1)\sigma_R^2}}. \tag{5.8}$$

The resulting gain level is always bounded, e.g., $\beta^2 < 1/(\gamma_{RR}\sigma_R^2)$, which guarantees the stability of the relay and prevents oscillation.

5.2.1.2 FD DF relaying

When adopting the nonlinear DF protocol, the relay regenerates the original signal transmitted by the source. Assuming proper rate adaptation, modulation and coding schemes are selected for the end-to-end link according to the weakest hop to guarantee reliable decoding and re-encoding process $\hat{f}(\cdot)$ from $\hat{y}_R[i]$ to $x_R[i]$. Thus, the destination receives a *clean* amplified signal from the relay unlike with AF relaying. However, residual interference lowers the achievable transmission rate in the first hop which indirectly affects the second hop as well.

Throughout the chapter, transmit power control is formulated in terms of the amplification factor β which is inherent in the context of the AF protocol. The same framework can be utilized also with the DF protocol by recognizing that $h_{SR}x_S[i]$ in (5.2) represents the clean signal before amplification in the relay. Incorporating the processing delay $\tau_R > 0$, the effective output signal of the DF relay can be thus expressed as

$$x_R[i] = \beta\, h_{SR}x_S[i - \tau_R]. \tag{5.9}$$

The amplification factor can be determined by setting relay transmit power to p_R since (5.9) yields $\mathcal{E}\{|x_R[i]|^2\} = \beta^2|h_{SR}|^2 = p_R$ and, consequently,

$$\beta(p_R) = \sqrt{\frac{p_R}{|h_{SR}|^2}}. \tag{5.10}$$

This can be contrasted with (5.8) which is lower due to noise amplification.

5.2.1.3 DT without relaying

As a degenerate case of the FD mode, the system may shut down the relay and revert to a DT mode, in which the source transmits $x_S[i]$ over channel h_{SD} and the destination decodes noisy signal:

$$y_D[i] = h_{SD}x_S[i] + n_D[i]. \tag{5.11}$$

Thus, the link avoids both the co-channel interference of the FD mode and the extra resource usage of the HD mode. In "selective"/"incremental" relaying [14], the DT mode is used with relatively strong direct connection.

5.2.1.4 HD relaying with diversity combining

The end-to-end transmission of one symbol can be spread over two orthogonal time slots to eliminate self-interference. This is common in earlier literature, e.g., with "fixed" relaying [14]. Analytically, such separation could as well be achieved in frequency-domain but practical implementation may impose other constraints for time and frequency duplexing.

In the HD mode, the source transmits $x_S[2i]$ only during the even time slots while the relay receives $y_R[2i]$ during the even time slots and transmits $x_R[2i + 1]$ during the uneven time slots. Thereby, the noisy received signals in the relay and in the destination are expressed as

$$y_R[2i] = h_{SR}x_S[2i] + n_R[2i], \tag{5.12a}$$

$$y_D[2i] = h_{SD}x_S[2i] + n_D[2i], \tag{5.12b}$$

$$y_D[2i + 1] = h_{RD}x_R[2i + 1] + n_D[2i + 1]. \tag{5.12c}$$

As for relaying protocols, processing steps to obtain x_R from y_R are equivalent to those described above, for the FD mode except processing delay τ_R is always uneven and self-interference is inherently avoided.

5.2.2 Signal-to-interference-and-noise ratios

The end-to-end SINRs are used for measuring and optimizing the performance observed by the destination. At time instant i, the receiver aims at decoding $y_{\mathrm{D}}[i]$ given in (5.4); with both protocols, the desired signal is $x_{\mathrm{S}}[i-\tau_{\mathrm{R}}]$ included within $x_{\mathrm{R}}[i]$ and the other components act as interference and noise. Useful auxiliary variables to be kept in mind are the channel SNRs which are given by $\gamma_{\mathrm{SR}} \triangleq |h_{\mathrm{SR}}|^2/\sigma_{\mathrm{R}}^2$, $\gamma_{\mathrm{RD}} \triangleq |h_{\mathrm{RD}}|^2/\sigma_{\mathrm{D}}^2$, and $\gamma_{\mathrm{SD}} \triangleq |h_{\mathrm{SD}}|^2/\sigma_{\mathrm{D}}^2$.

The weak direct-link signal $h_{\mathrm{SD}}x_{\mathrm{S}}[i]$ in (5.4) is treated as co-channel interference embedded in the destination receiver noise which is justified when the relay is employed for coverage extension. In hotspot scenarios, where the direct and relayed transmissions are both strong, equalization or multiantenna techniques could facilitate signal separation and two-branch diversity combining although the next section shows that simple switching between DT and relaying is already sufficient.

5.2.2.1 FD AF relaying

The end-to-end SINR of the AF protocol is derived below with two intermediate steps expressing relay output power and destination input power.

By assuming that all successive signal and noise samples are mutually independent and that the channels vary slowly, the transmit power of the AF relay can be calculated from (5.5) to (5.7) as

$$\mathcal{E}\{|x_{\mathrm{R}}[i]|^2\} = \beta^2 \sum_{j=1}^{\infty} \left(|h_{\mathrm{RR}}|^2 \beta^2\right)^{j-1} \left(|h_{\mathrm{SR}}|^2 + \sigma_{\mathrm{R}}^2\right) = \frac{\beta^2 \left(|h_{\mathrm{SR}}|^2 + \sigma_{\mathrm{R}}^2\right)}{1 - \sigma_{\mathrm{R}}^2 \beta^2 \gamma_{\mathrm{RR}}} \tag{5.13}$$

for the respective cases of passive physical isolation and active mitigation. Especially, this shows that the relay gain must be limited according to

$$|h_{\mathrm{RR}}|^2 \beta^2 < 1 \tag{5.14a}$$

or, with active mitigation,

$$\sigma_{\mathrm{R}}^2 \beta^2 < \frac{1}{\gamma_{\mathrm{RR}}} \tag{5.14b}$$

to prevent oscillation and guarantee finite relay transmit power. All the gain control methods considered herein ensure this condition by design.

Again by requiring signal and noise independence, the received signal power in the destination is calculated from (5.4) as

$$\mathcal{E}\left\{|y_{\mathrm{D}}[i]|^2\right\} = |h_{\mathrm{RD}}|^2 \mathcal{E}\left\{|x_{\mathrm{R}}[i]|^2\right\} + |h_{\mathrm{SD}}|^2 + \sigma_{\mathrm{D}}^2. \tag{5.15}$$

This expression with the substitution of (5.13) can be further reorganized as a sum of desired signal power, residual loopback interference power, SD interference power, and noise power, respectively:

$$\begin{aligned}\mathcal{E}\left\{|y_{\mathrm{D}}[i]|^2\right\} &= |h_{\mathrm{SR}}|^2\beta^2|h_{\mathrm{RD}}|^2 \\ &\quad + \left(|h_{\mathrm{SR}}|^2+\sigma_{\mathrm{R}}^2\right)\beta^2|h_{\mathrm{RD}}|^2\frac{\sigma_{\mathrm{R}}^2\beta^2\,\gamma_{\mathrm{RR}}}{1-\sigma_{\mathrm{R}}^2\beta^2\,\gamma_{\mathrm{RR}}} \\ &\quad + |h_{\mathrm{SD}}|^2+\beta^2|h_{\mathrm{RD}}|^2\sigma_{\mathrm{R}}^2+\sigma_{\mathrm{D}}^2. \end{aligned} \tag{5.16}$$

Finally, the SINR of AF relaying can be stated based on (5.16) as

$$\gamma = \frac{\sigma_{\mathrm{R}}^2\beta^2\,\gamma_{\mathrm{SR}}\gamma_{\mathrm{RD}}}{(\gamma_{\mathrm{SR}}+1)\sigma_{\mathrm{R}}^2\beta^2\,\gamma_{\mathrm{RD}}\dfrac{\sigma_{\mathrm{R}}^2\beta^2\,\gamma_{\mathrm{RR}}}{1-\sigma_{\mathrm{R}}^2\beta^2\,\gamma_{\mathrm{RR}}}+\sigma_{\mathrm{R}}^2\beta^2\,\gamma_{\mathrm{RD}}+\gamma_{\mathrm{SD}}+1}. \tag{5.17}$$

5.2.2.2 FD DF relaying

The end-to-end SINR of the DF protocol is an "equivalent" variable which corresponds to the SINR of the AF protocol resulting in the same transmission rate, i.e.,

$$\log_2(1+\gamma) \triangleq \min\{\log_2(1+\gamma_{\mathrm{R}}), \log_2(1+\gamma_{\mathrm{D}})\} \tag{5.18}$$

for which

$$\gamma_{\mathrm{R}} \triangleq \frac{\gamma_{\mathrm{SR}}}{\sigma_{\mathrm{R}}^2\beta^2\,\gamma_{\mathrm{SR}}\gamma_{\mathrm{RR}}+1}, \tag{5.19a}$$

$$\gamma_{\mathrm{D}} \triangleq \frac{\sigma_{\mathrm{R}}^2\beta^2\,\gamma_{\mathrm{SR}}\gamma_{\mathrm{RD}}}{\gamma_{\mathrm{SD}}+1} \tag{5.19b}$$

are given by (5.7) and (5.15), respectively. Especially, one should not mistake the equivalent SINR of the DF protocol for a physical measurable quantity since it is only adopted for handling both protocols in the same framework. Finally,

$$\gamma = \min\{\gamma_{\mathrm{R}}, \gamma_{\mathrm{D}}\} = \min\left\{\frac{\gamma_{\mathrm{SR}}}{\sigma_{\mathrm{R}}^2\beta^2\,\gamma_{\mathrm{SR}}\gamma_{\mathrm{RR}}+1}, \frac{\sigma_{\mathrm{R}}^2\beta^2\,\gamma_{\mathrm{SR}}\gamma_{\mathrm{RD}}}{\gamma_{\mathrm{SD}}+1}\right\}. \tag{5.20}$$

5.3 Transmit power control in FD relaying

Presumably most FD relays will need to tolerate residual self-interference adapting to its presence as best they can. To that end, inherent imbalance in FD relay links offers an opportunity to minimize the effect of residual distortion. In particular, loopback interference distorts the channel before a relay, but not the one after it, and the trade-off between their qualities can be controlled with relay transmit power. This section shows that the list of countermeasures given in Section 5.1.2 can be complemented with another one working against residual distortion after the other three: (4) the optimization of relay gain. The completed list of Countermeasures (1)–(4) was first compiled in Reference 20. In particular, Countermeasure (4) is motivated by observing that useful signal power at a destination receiver is linearly proportional to relay gain while residual self-interference level at the relay itself may not be. Gain setting that is considerably below its maximum allowed value may thus render improved effective SINR for the end-to-end link although, obviously, backing off too much will not come out well either. This poses a challenge to choose optimal relay gain, or equivalently transmit power, and maximize the end-to-end performance by balancing the wireless links before and after the relay.

Instead of end-to-end performance optimization, relay gain control is conventionally applied only to satisfy certain constraints pertaining to transceiver electronics and relay installation. In particular, variable gain adaptation may be necessary for maintaining relay transmit power at a constant level below limitations set by regulations or the capability of high-power amplifiers. Likewise, the gain of the feedback loop sets a maximum level for the gain of an AF relay to guarantee that loopback echoes fade away in a reasonable time or, otherwise, the signal of interest will be distorted due to the oscillation or saturation of transceivers. A conventional approach may thus adopt constant amplification, but variable transmit power, by setting relay gain to leave a fixed safety margin w.r.t. an estimate of the maximum possible level.

This section formalizes the gain control challenge as optimization problems to maximize the effective end-to-end SINRs. In particular, new methods are developed for choosing the amplification factor β. The system framework and analysis accommodate both common relaying protocols: AF as well as DF. To that end, SINR maximization is formalized as an optimization task:

$$\beta^* = \arg\max_{\beta} \gamma, \tag{5.21}$$

for which γ is given by (5.17) and (5.20) with AF and DF relaying, respectively. Ideal CSI is required for this purpose but the assumption can be relaxed with suboptimal target-SINR schemes. The analysis is formulated in terms of β which can be translated to

$$p_{\mathrm{R}}(\beta^2) = \begin{cases} \dfrac{\sigma_{\mathrm{R}}^2\beta^2(\gamma_{\mathrm{SR}}+1)}{1-\sigma_{\mathrm{R}}^2\beta^2\,\gamma_{\mathrm{RR}}}, & \text{with AF,} \\ \sigma_{\mathrm{R}}^2\beta^2\gamma_{\mathrm{SR}}, & \text{with DF.} \end{cases} \tag{5.22}$$

Especially, the values of β^* solved shortly yield (5.56) in the next section.

As solutions to the optimization problems, explicit expressions are provided for the optimal relay amplification factors which can be translated to optimal transmit power levels. These are contrasted with conventional methods to prove that transmit power constraints are still satisfied and relay oscillation is avoided by design. The optimal solutions are modified to target-SINR approaches which reduce the need for CSI feedback at the cost of rendering suboptimal, but still improved, performance. Finally, the principles of gain control are illustrated with numerical SINR results, and unified outage probability analysis is conducted for the considered gain control schemes by deriving closed-form performance expressions in both DL and UL relaying. This also lays foundations for the following section which compares HD and FD relaying, with and without power control, in the same setup.

In an OFDM context, the study concerns a single-subcarrier relaying model with frequency-domain processing due to which loopback interference manifests itself as residual distortion after mitigation; the effect of ICI due to excessive multipath delay spread from FD operation is analyzed in Reference 20 in the case of an explicit wideband OFDM model. The results can be also interpreted to concern an end-to-end orthogonal spatial stream in an MIMO context, while an extension with power allocation between streams can be found in Reference 27.

5.3.1 AF relaying

As summarized to begin with, conventional methods render suboptimal SINR and transmit power limits may be exceeded due to fading channels.

The gain of AF repeaters is often set, after measuring the loopback gain $|h_{RR}|^2$, by choosing constant amplification which satisfies the stability conditions in (5.14) with a *fixed* gain margin $\Delta_{GM} > 1$. This yields

$$\beta = \frac{1}{\sqrt{\Delta_{GM}}|h_{RR}|} \tag{5.23a}$$

or, with active mitigation,

$$\beta = \frac{1}{\sigma_R} \cdot \frac{1}{\sqrt{\Delta_{GM}\gamma_{RR}}}. \tag{5.23b}$$

However, condition $p_R \leq 1$ for (5.22) may be violated, e.g., if γ_{SR} fluctuates.

Another conventional method let the relay gain vary in such a way that maximum transmission power is used constantly. In particular, the substitution of $p_R \triangleq \mathcal{E}\{|x_R[i]|^2\} = 1$ in (5.8) yields the maximum amplification factor as

$$\beta = \frac{1}{\sigma_R} \cdot \frac{1}{\sqrt{\gamma_{SR} + \gamma_{RR} + 1}}, \tag{5.24}$$

which also guarantees that (5.14) holds. Thus, this method can be considered to exploit the other conventional method with an *adaptive* gain margin $\Delta_{GM} = 1 + (\gamma_{SR} + 1)/\gamma_{RR}$. By substituting (5.24) into (5.17), the end-to-end SINR with maximum transmit power usage becomes

$$\gamma = \frac{\gamma_{SR}\gamma_{RD}}{\gamma_{SR}(\gamma_{SD} + 1) + \gamma_{RD}(\gamma_{RR} + 1) + (\gamma_{SD} + 1)(\gamma_{RR} + 1)}. \tag{5.25}$$

It should be noted that implementation is actually simpler than it looks, since the relay can just adaptively drive its transmit power to a constant level without using explicit CSI, and this method also maximizes the end-to-end SINR if there is zero or only negligible residual self-interference.

Neither simple gain control method offers optimal end-to-end performance in the presence of significant residual self-interference though. In particular, (5.16) shows that the desired signal power is linear as a function of the relay gain β^2 but the interference power is highly nonlinear. Increasing gain can thus raise the self-interference level faster than the desired signal level leading eventually to reduced effective SINR.

It turns out that (5.17) has a single maximum point at $0 \leq \beta^2 < 1/(\sigma_R^2 \gamma_{RR})$. By finding the correct root of the derivative of γ in terms of β^2, the optimal amplification factor can be consequently solved for (5.21) as follows.

Optimal amplification factor in AF relaying

$$\beta^* = \frac{1}{\sigma_R} \cdot \sqrt{\frac{\gamma_{SD} + 1}{\gamma_{RR}(\gamma_{SD} + 1) + \sqrt{(\gamma_{SR} + 1)\,\gamma_{RD}\gamma_{RR}(\gamma_{SD} + 1)}}}. \tag{5.26}$$

The solution obviously satisfies (5.14) too. Actually, this represents an enhanced version of the conventional method with an *optimal adaptive* gain margin

$$\Delta_{GM}^* = 1 + \sqrt{\frac{(\gamma_{SR} + 1)\gamma_{RD}}{\gamma_{RR}(\gamma_{SD} + 1)}}. \tag{5.27}$$

By substituting (5.26) into (5.17), the maximum end-to-end SINR becomes

$$\gamma^* = \frac{\gamma_{SR}\gamma_{RD}}{\gamma_{RD} + \gamma_{RR}(\gamma_{SD} + 1) + 2\sqrt{(\gamma_{SR} + 1)\,\gamma_{RD}\gamma_{RR}(\gamma_{SD} + 1)}}. \tag{5.28}$$

In practice, $\beta = \min\{\beta^*, (1/\sigma_R) \cdot (1/\sqrt{\gamma_{SR} + \gamma_{RR} + 1})\}$ is employed to satisfy also a transmit power constraint $p_R \leq 1$. By solving $(\sigma_R \beta^*)^2 \geq (\gamma_{SR} + \gamma_{RR} + 1)^{-1}$, the essential condition for residual loopback interference is shown to be

$$\gamma_{RR} \leq \frac{(\gamma_{SR} + 1)(\gamma_{SD} + 1)}{\gamma_{RD}} \tag{5.29}$$

for determining whether the transmit power limitation restricts the relay to use maximum gain instead of the optimal amplification factor.

An inevitable drawback of the above method is that it exploits the knowledge of γ_{RD} and γ_{SD} which can be estimated only at the destination; a feedback channel is thus required for conveying CSI to the relay. Instead, one may consider restricted gain control that results in consuming a constant fraction ($p_R \leq 1$) of the maximum allowed power although, e.g., γ_{RD} varies in DL relaying. Expressions for β and the end-to-end SINR γ are given by (5.8) and (5.17) with (5.8), respectively. To avoid using CSI from the destination, target end-to-end SINR $\hat{\gamma} < \gamma_{SR}$ can then be chosen aiming at designing β to result in optimal performance only when $\gamma \approx \hat{\gamma}$.

With the target-SINR scheme, p_R is selected in such a way that the end-to-end SINR is maximized whenever it coincides with $\hat{\gamma}$ because γ_{RD} and γ_{SD} meet their respective target values. Thus, by eliminating $\gamma_{RD}/(\gamma_{SD} + 1)$ from the equation pair $\{\hat{\gamma} = \gamma,\ \hat{\gamma} = \gamma^*\}$, the desired normalized transmit power can be shown to admit the form:

$$p_R = \frac{\sqrt{\gamma_{SR}(\gamma_{SR} + 1)\hat{\gamma}(\hat{\gamma} + 1)}}{\hat{\gamma}\,\gamma_{RR}} - \frac{\gamma_{SR} + 1}{\gamma_{RR}} \tag{5.30}$$

when the target SINR satisfies

$$\hat{\gamma} > \frac{\gamma_{\mathrm{SR}}(\gamma_{\mathrm{SR}} + 1)}{\gamma_{\mathrm{SR}} + 2\gamma_{\mathrm{SR}}\gamma_{\mathrm{RR}} + (\gamma_{\mathrm{RR}} + 1)^2}, \tag{5.31}$$

and otherwise $p_{\mathrm{R}} = 1$ due to the transmit power constraint.

5.3.2 DF relaying

With the DF protocol, the nonlinear de/encoding process effectively cuts the feedback loop through the relay. Thus, self-interference echoes do not remain circulating endlessly as with AF relaying, and the gain-margin method is irrelevant because relay stability is guaranteed by default.

The gain of DF relays is conventionally set to yield maximum transmit power usage constantly. In such cases, the substitution of $p_{\mathrm{R}} = 1$ in (5.10) yields the maximum amplification factor and the corresponding SINR as

$$\beta = \frac{1}{\sigma_{\mathrm{R}}} \cdot \frac{1}{\sqrt{\gamma_{\mathrm{SR}}}}, \tag{5.32}$$

$$\gamma = \min\left\{\frac{\gamma_{\mathrm{SR}}}{\gamma_{\mathrm{RR}} + 1}, \frac{\gamma_{\mathrm{RD}}}{\gamma_{\mathrm{SD}} + 1}\right\}, \tag{5.33}$$

respectively, which is actually equivalent to using (5.23) with $\Delta_{\mathrm{GM}} = 1$.

Transmit power optimization is now motivated by the fact that, in (5.20), γ_{R} decreases and γ_{D} increases in terms of β^2. Thus, their minimum is maximized when $\gamma_{\mathrm{R}} = \gamma_{\mathrm{D}}$ which renders the solution for (5.21) as follows:

Optimal effective amplification factor in DF relaying

$$\beta^* = \frac{1}{\sigma_{\mathrm{R}}} \cdot \sqrt{\frac{\sqrt{\gamma_{\mathrm{RD}}^2 + 4\gamma_{\mathrm{SR}}\gamma_{\mathrm{RD}}\gamma_{\mathrm{RR}}(\gamma_{\mathrm{SD}} + 1)} - \gamma_{\mathrm{RD}}}{2\gamma_{\mathrm{SR}}\gamma_{\mathrm{RD}}\gamma_{\mathrm{RR}}}}. \tag{5.34}$$

A substitution in (5.20) yields the maximum end-to-end SINR as

$$\gamma^* = \frac{\gamma_{\mathrm{RD}}}{\gamma_{\mathrm{RR}}(\gamma_{\mathrm{SD}} + 1)}\left(\sqrt{\frac{\gamma_{\mathrm{SR}}\gamma_{\mathrm{RR}}(\gamma_{\mathrm{SD}} + 1)}{\gamma_{\mathrm{RD}}} + \frac{1}{4}} - \frac{1}{2}\right). \tag{5.35}$$

However, the transmit power constraint $p_{\mathrm{R}} \leq 1$ clips the SINR at (5.33) if

$$\gamma_{\mathrm{RR}} \leq \frac{\gamma_{\mathrm{SR}}(\gamma_{\mathrm{SD}} + 1)}{\gamma_{\mathrm{RD}}} - 1, \tag{5.36}$$

which is solved from $(\sigma_{\mathrm{R}}\beta^*)^2 \geq \gamma_{\mathrm{SR}}^{-1}$. This means in practice that the relay uses $\beta = \min\{\beta^*, (1/\sigma_{\mathrm{R}}) \cdot (1/\sqrt{\gamma_{\mathrm{SR}}})\}$.

Expressions for β and γ are given by (5.10) and (5.20) with (5.10), respectively, when a constant fraction ($p_{\mathrm{R}} \leq 1$) of the maximum allowed power is used. Especially, optimal SINR can be approached without using CSI from the destination in the vicinity of a target value $\hat{\gamma}$ by choosing

$$p_{\mathrm{R}} = \frac{\gamma_{\mathrm{SR}} - \hat{\gamma}}{\hat{\gamma}\,\gamma_{\mathrm{RR}}} \tag{5.37}$$

when

$$\hat{\gamma} > \frac{\gamma_{\mathrm{SR}}}{\gamma_{\mathrm{RR}} + 1}, \tag{5.38}$$

and otherwise $p_{\mathrm{R}} = 1$ due to the transmit power constraint; this result is solved from the equation pair $\{\hat{\gamma} = \gamma,\ \hat{\gamma} = \gamma^*\}$ as with AF relaying.

5.3.3 Performance analysis

The performance of FD relaying and the transmit power control schemes are analyzed next with both short-term and long-term perspectives. Firstly, instantaneous end-to-end SINR lends itself to illustrating the principles of transmit power control. Secondly, outage probability statistics characterize relaying over fading channels. While the short-term analysis is not specific to any system scenario, the long-term analysis concentrates on cellular relaying where DL and UL directions turn out to be rather different in nature. Further outage analysis and more numerical results can be found in References 22 and 28.

5.3.3.1 SINRs vs. relay gain

The analytical SINR expressions are illustrated in Figure 5.4 for a fixed example scenario where $\gamma_{\mathrm{SR}} = 15$ dB and $\gamma_{\mathrm{RD}}/(\gamma_{\mathrm{SD}} + 1) = 20$ dB. Only the achievable SINR area bounded by the cases of no/low residual self-interference and maximum power usage is shown. In particular, the figures display the performance of the different gain control methods in terms of both the residual self-interference level γ_{RR} and the relay gain β^2. The DF protocol reflects similar general behavior as the AF protocol but there is much sharper transition from the second hop being the bottleneck (low β^2) to the case where self-interference limits the overall performance (high β^2).

For each fixed value of γ_{RR}, the end-to-end SINR has a distinct global maximum (in terms of the relay gain β^2) which is reached with optimal gain control. For reasonably large γ_{RR}, the optimal β^* provides both significant SINR improvement and transmit power reduction compared to maximum power usage with $p_{\mathrm{R}} = 1$ in β. Inequalities (5.29) and (5.36) show that maximum transmit power is optimal due to the transmit power constraint when $\gamma_{\mathrm{RR}} < -4.9$ dB with the AF protocol while the DF protocol benefits from transmit power reduction with any γ_{RR} since $\gamma_{\mathrm{SR}} < 20$ dB.

The SINR loss of using the target-SINR approach instead of the optimal gain can be also inferred from Figure 5.4. The free design parameter $\hat{\gamma}$ defines the SINR operation area, where the performance of the target-SINR approach is nearly optimal, and the loss is rather small in a large gain range around it, but the DF protocol is more sensitive. In particular, the target-SINR approach with $\hat{\gamma} = 12$ is optimal only when $\gamma_{\mathrm{RR}} \approx 3.0$ dB (resp. $\gamma_{\mathrm{RR}} \approx 11.3$ dB), but performance is always improved over setting

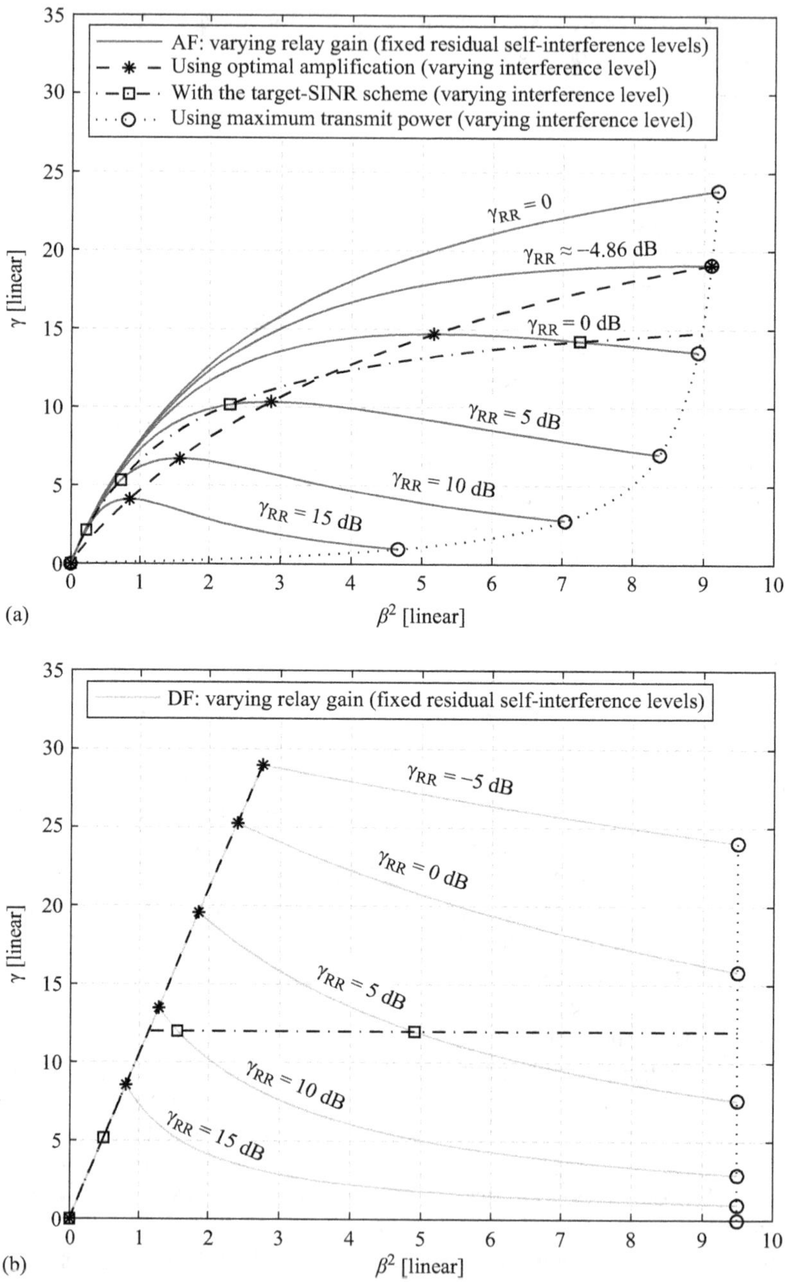

Figure 5.4 Effective end-to-end SINR with the different transmit power control methods when channel SNRs are given by $\gamma_{SR} = 15$ dB, $\gamma_{RD} = 23$ dB, and $\gamma_{SD} = 0$ dB. The target SINR is set at $\hat{\gamma} = 12$, and a constant value, i.e., $\sigma_R^2 = \frac{1}{300}$ to be specific, is chosen for the thermal noise level in order to scale the relay gain level within the single-digit range for illustration purposes. (a) AF relaying and (b) DF relaying

$p_{\mathrm{R}} = 1$ in β except when $\gamma_{\mathrm{RR}} \leq 0.9$ dB (resp. $\gamma_{\mathrm{RR}} \leq 2.1$ dB) with the AF (resp. DF) protocol and the transmit power constraints become activated.

5.3.3.2 Outage probability analysis

The following analysis concentrates on the long-term performance of a cellular RN deployed for extending BS coverage area; this is especially reflected in the choice of fading distributions for the wireless channels. In particular, the performance of the different gain control schemes is evaluated in terms of outage probability $P_{\mathrm{out}}(\gamma_{\mathrm{th}})$ which is directly given by the cumulative distribution function (CDF) of SINR:

$$P_{\mathrm{out}}(\gamma_{\mathrm{th}}) \triangleq F_{\gamma}(\gamma_{\mathrm{th}}) \triangleq \mathrm{Prob}(\gamma < \gamma_{\mathrm{th}}). \tag{5.39}$$

In other words, the analysis quantifies the probability, Prob(·), that the end-to-end SINR falls below a target threshold γ_{th} due to fading and self-interference. Since this characterizes any arbitrary subcarrier, joint performance over the whole OFDM bandwidth is also explained implicitly.

The RN is likely a fixed transceiver, like the BS of the cellular system, in order to facilitate FD operation. Thus, it is reasonable to assume that the backhaul (BS-to-RN or RN-to-BS) channel is static, which is the closest approximation for Rician fading with a strong LOS component that could be expected in practice. The destination is assumed to be a mobile UE without LOS to the RN. Thus, the service (RN-to-UE or UE-to-RN) channel admits Rayleigh-fading statistics; when denoting the average service-link SNRs by $\bar{\gamma}_{\mathrm{RD}} \triangleq \mathcal{E}\{\gamma_{\mathrm{RD}}\}$ and $\bar{\gamma}_{\mathrm{SR}} \triangleq \mathcal{E}\{\gamma_{\mathrm{SR}}\}$, the respective DL and UL CDFs are given by

$$F_{\gamma_{\mathrm{RD}}}(x) = 1 - \exp\left(-\frac{x}{\bar{\gamma}_{\mathrm{RD}}}\right), \tag{5.40}$$

$$F_{\gamma_{\mathrm{SR}}}(x) = 1 - \exp\left(-\frac{x}{\bar{\gamma}_{\mathrm{SR}}}\right) \tag{5.41}$$

for $x > 0$.

The focus on coverage extension implies $\gamma_{\mathrm{SD}} \ll 0$ dB and $\sigma_{\bar{\mathrm{D}}}^2 \approx (\gamma_{\mathrm{SD}} + 1)\sigma_{\mathrm{D}}^2$. However, the case of a strong direct link can be still approximately covered by setting $\gamma_{\mathrm{SD}} = 0$ and replacing γ_{RD} with $\frac{\gamma_{\mathrm{RD}}}{\mathcal{E}\{\gamma_{\mathrm{SD}}\}+1}$ while Reference 28 provides exact complementary analysis where γ_{SD} is fading.

With each power control scheme, the end-to-end SINR is a specific function of the fading channel's SNR, i.e., $\gamma = \gamma_{\mathrm{DL}}(\gamma_{\mathrm{RD}})$ or $\gamma = \gamma_{\mathrm{UL}}(\gamma_{\mathrm{SR}})$, as shown in the previous sections. After determining the respective inverse functions, namely $\gamma_{\mathrm{RD}} = \gamma_{\mathrm{DL}}^{-1}(\gamma)$ and $\gamma_{\mathrm{SR}} = \gamma_{\mathrm{UL}}^{-1}(\gamma)$, for $0 \leq \gamma < \gamma_{\max}$ where the boundary value $\gamma_{\max}$ denotes the SINR limit that is approached when $\gamma_{\mathrm{RD}} \to \infty$ or $\gamma_{\mathrm{SR}} \to \infty$, outage probabilities are directly given by transforms of random variables. The final outage probability expressions and the inverse functions can be summarized for the different transmit power control methods and system variations (DL vs. UL and AF vs. DF) as follows.

Outage probability in FD relaying

$$P_{\text{out}}^{\text{DL}}(\gamma_{\text{th}}) = F_{\gamma_{\text{RD}}}\left(\gamma_{\text{DL}}^{-1}(\gamma_{\text{th}})\right) = 1 - \exp\left(-\frac{\gamma_{\text{DL}}^{-1}(\gamma_{\text{th}})}{\bar{\gamma}_{\text{RD}}}\right), \tag{5.42a}$$

$$P_{\text{out}}^{\text{DL}}(\gamma_{\text{th}}) = F_{\gamma_{\text{SR}}}\left(\gamma_{\text{UL}}^{-1}(\gamma_{\text{th}})\right) = 1 - \exp\left(-\frac{\gamma_{\text{UL}}^{-1}(\gamma_{\text{th}})}{\bar{\gamma}_{\text{SR}}}\right) \tag{5.42b}$$

for $\gamma_{\text{th}} < \gamma_{\max}$ and $P_{\text{out}}(\gamma_{\text{th}}) = 1$ otherwise.

In DL AF relaying, the inverse of (5.17) including (5.8) becomes

$$\gamma_{\text{DL}}^{-1}(x) = \frac{\gamma_{\text{SR}} + p_{\text{R}}\gamma_{\text{RR}} + 1}{\gamma_{\text{SR}} - (p_{\text{R}}\gamma_{\text{RR}} + 1)x} \cdot \frac{x}{p_{\text{R}}} \tag{5.43a}$$

with $\gamma_{\max} = \gamma_{\text{SR}}/(p_{\text{R}}\gamma_{\text{RR}} + 1)$, while the inverse of (5.28) is given by

$$\gamma_{\text{DL}}^{-1}(x) = \frac{\left(\sqrt{(\gamma_{\text{SR}} + 1)\gamma_{\text{RR}}\,x} + \sqrt{\gamma_{\text{SR}}\gamma_{\text{RR}}x(x+1)}\right)^2}{(\gamma_{\text{SR}} - x)^2} \tag{5.43b}$$

with $\gamma_{\max} = \gamma_{\text{SR}}$.

In DL DF relaying, the inverse of (5.20) including (5.10) becomes

$$\gamma_{\text{DL}}^{-1}(x) = \frac{x}{p_{\text{R}}} \tag{5.44a}$$

with $\gamma_{\max} = \gamma_{\text{SR}}/(p_{\text{R}}\gamma_{\text{RR}} + 1)$, while the inverse of (5.35) is given by

$$\gamma_{\text{DL}}^{-1}(x) = \frac{\gamma_{\text{RR}}\,x^2}{\gamma_{\text{SR}} - x} \tag{5.44b}$$

with $\gamma_{\max} = \gamma_{\text{SR}}$.

In UL AF relaying, the inverse of (5.17) including (5.8) becomes

$$\gamma_{\text{UL}}^{-1}(x) = \frac{p_{\text{R}}\gamma_{\text{RD}} + 1}{p_{\text{R}}\gamma_{\text{RD}} - x} \cdot (p_{\text{R}}\gamma_{\text{RR}} + 1)x \tag{5.45a}$$

with $\gamma_{\max} = p_{\text{R}}\gamma_{\text{RD}}$, while the inverse of (5.28) is given by

$$\gamma_{\text{UL}}^{-1}(x) = \frac{\left(\sqrt{\gamma_{\text{RR}}}\,x + \sqrt{(\gamma_{\text{RD}} + \gamma_{\text{RR}}x)(x+1)}\right)^2}{\gamma_{\text{RD}}} - 1 \tag{5.45b}$$

with $\gamma_{\max} \to \infty$.

In UL DF relaying, the inverse of (5.20) including (5.10) becomes

$$\gamma_{\text{UL}}^{-1}(x) = (p_{\text{R}}\gamma_{\text{RR}} + 1)x \tag{5.46a}$$

with $\gamma_{\max} = p_{\mathrm{R}}\gamma_{\mathrm{RD}}$, while the inverse of (5.35) is given by

$$\gamma_{\mathrm{UL}}^{-1}(x) = \frac{(\gamma_{\mathrm{RD}} + \gamma_{\mathrm{RR}}x)x}{\gamma_{\mathrm{RD}}} \tag{5.46b}$$

with $\gamma_{\max} \to \infty$.

In addition, power constraints may be invoked such that a)-expressions replace b)-expressions if the inequality (5.29), or (5.36), is true when setting $\gamma_{\mathrm{RD}} = \gamma_{\mathrm{DL}}^{-1}(x)$ (resp. $\gamma_{\mathrm{SR}} = \gamma_{\mathrm{UL}}^{-1}(x)$) in the DL (resp. UL) direction. One may also note from the above that the analysis of the target-SINR approach is omitted in the UL direction. This is reasonable because the cost of acquiring CSI for the static backhaul channel is relatively low.

The outage probability expressions are illustrated by Figure 5.5 in an example setup where the static backhaul SNR is 20 dB and the average SNR of the fading service link is 25 dB. In general, it is noted that the bottleneck of the end-to-end link is the first hop due to residual self-interference and the relay should often back off from its maximum transmit power.

The DL relaying direction is shown in Figure 5.5(a). With both protocols, the maximum achievable SINR is only 4.9 dB (resp. 13.8 dB) when $\gamma_{\mathrm{RR}} = 15$ dB (resp. $\gamma_{\mathrm{RR}} = 5$ dB) while optimal transmit power control minimizes the effect of residual self-interference and, in theory, the maximum SINR increases to 20 dB irrespective of γ_{RR}. Furthermore, at the median level (γ_{th} with $P_{\mathrm{out}}(\gamma_{\mathrm{th}}) = 0.5$), the gain of transmit power optimization over maximum power usage is approximately 5.8 dB (resp. 1.3 dB) when $\gamma_{\mathrm{RR}} = 15$ dB (resp. $\gamma_{\mathrm{RR}} = 5$ dB) with AF relaying and the DF protocol gains roughly 3 dB more.

The target-SINR approach is nearly optimal when the threshold SINR is close to its target as intended. When $\gamma_{\mathrm{RR}} = 15$ dB (resp. $\gamma_{\mathrm{RR}} = 5$ dB), $\hat{\gamma}$ is set to 10 dB (resp. 15 dB) resulting in the transmit power fraction $p_{\mathrm{R}} = 0.14$ (resp. $p_{\mathrm{R}} = 0.34$) in AF relaying, i.e., power savings are also significant although the outage probability remains often slightly suboptimal. The corresponding power fraction is 0.28 (resp. 0.68) in DF relaying where higher, roughly doubled, transmit power can be used since the nonlinear de/encoding process prevents infinite interference circulation through the relay. However, the target-SINR approach saturates the outage probability of the DF protocol for $\gamma_{\mathrm{th}} > \hat{\gamma}$ by rendering too high transmit power when the second hop observes a better-than-expected SNR level.

The UL relaying direction is shown in Figure 5.5(b); general observations from therein are in line with the DL case. Yet the two directions are rather different because self-interference distorts a fading channel in the UL case. In particular, fading amplifies the effect of self-interference and transmit power optimization such that both ends of the SINR range are achieved with higher probability than in DL relaying. The theoretical maximum of SINR is 20 dB even without transmit power optimization although it is rarely met in practice due to self-interference. As indicated by (5.29) and (5.36), the maximum transmit power is optimal, i.e., the optimal gain factor is clipped due to power constraints, when γ_{SR} and the end-to-end SINR fluctuate to a large level and, consequently, the UL outage probabilities coincide for large γ_{th}; the converse case holds true in the DL case where γ_{RD} is fading.

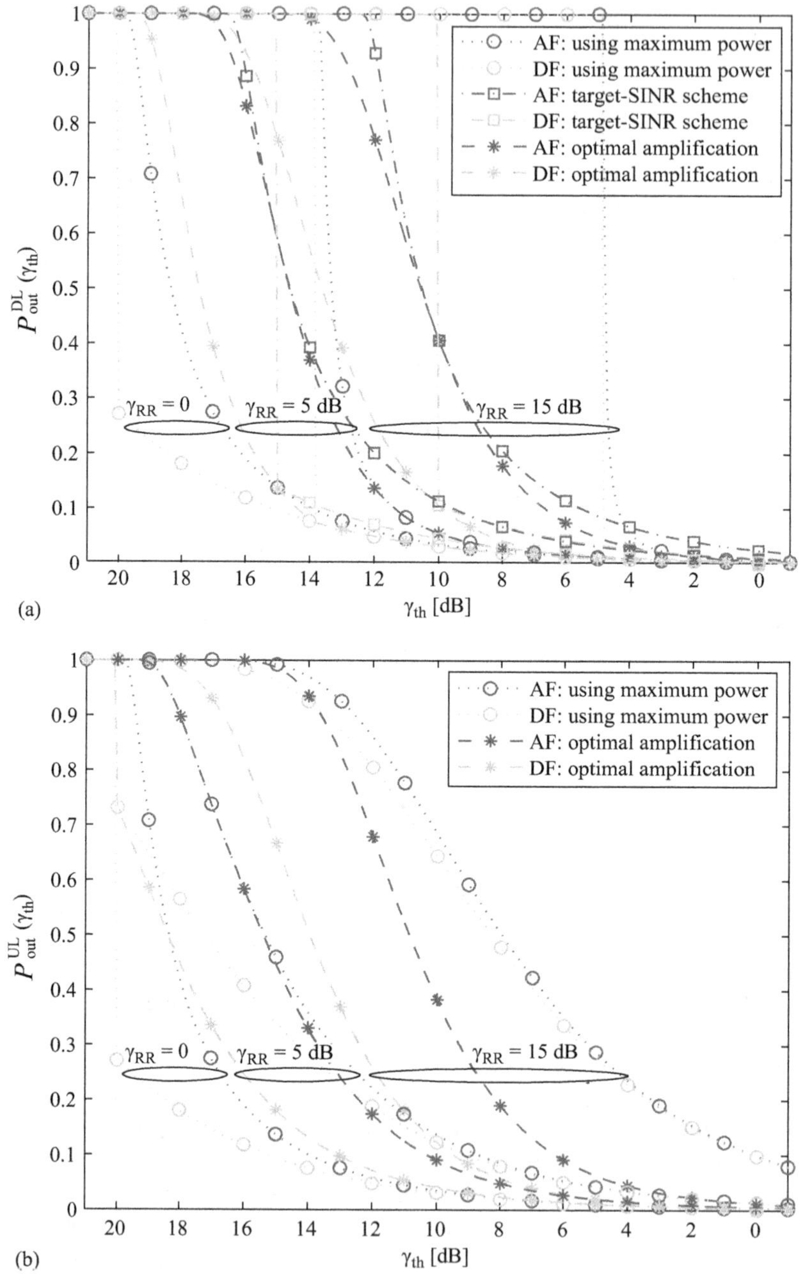

Figure 5.5 Effective end-to-end outage probability in terms of the threshold SNR with the different transmit power control methods when the transmission directions have the same (average) backhaul- and service-link SNRs and $\gamma_{SD} \approx 0$. In the DL direction, the target SINR is set at $\hat{\gamma} = 10$ dB (resp. $\hat{\gamma} = 15$ dB) when $\gamma_{RR} = 15$ dB (resp. $\gamma_{RR} = 5$ dB). (a) DL relaying when $\gamma_{SR} = 20$ dB and $\bar{\gamma}_{RD} = 25$ dB and (b) UL relaying when $\bar{\gamma}_{SR} = 25$ dB and $\gamma_{RD} = 20$ dB

5.4 FD vs. HD relaying

The presence of (residual) self-interference is undisputed with FD operation. Scanning through the related literature, the prior art turns out to be polarized such that its effect is considered either ultimately destructive or virtually negligible. In practice, the truth may lie in between the two extremes, especially with infrastructure RNs and efficient mitigation solutions. This section charts what happens between the extremes when making a calculated choice between the modes. In terms of both instantaneous and average link capacity, this study provides explicit conditions under which one relaying mode can be preferred over the other and evaluates the benefit of opportunistic switching between modes which is referred to as "hybrid FD/HD relaying". All essential aspects of hybrid FD/HD relaying are elaborated within a unified framework, and the treatment is not limited to a specific example setup but covers all typical scenarios encountered in infrastructure relaying: The study illustrates the differences of using instantaneous or statistical CSI, employing AF or DF protocol, communicating in DL or UL direction as well as variations of transmit power normalization and adaptation.

More specifically, this section contributes to the design and analysis of advanced relaying technologies as follows. The study rationalizes the trade-off between FD and HD modes by explicitly modeling the residual self-interference remaining in practice after mitigation. Still the FD mode is shown to be a feasible choice for infrastructure-based relay links in contrast to cooperative communication systems, and it can also achieve higher capacity than the HD mode. The analysis indicates that in many cases neither mode is the obvious choice and premature mode selection in early design stage leads to inefficient spectrum utilization. Thus, new hybrid FD/HD relaying schemes are introduced to facilitate switching optimally between the modes based on either instantaneous or statistical CSI.

Optimal short-term and long-term transmit power controls are adopted to minimize the effect of residual self-interference. Opportunistic mode switching is optimized also when combined with these techniques. It is observed that the employment of transmit power adaptation changes drastically the trade-off between the FD mode and the HD mode, making the choice of the former even more attractive. Throughout the section, the performance of the system with opportunistic mode switching and transmit power adaptation is evaluated by deriving new closed-form expressions for the instantaneous and average link capacity. Consequently, the study demonstrates that FD and hybrid FD/HD modes offer great potential for increasing spectral efficiency w.r.t. the conventional HD mode and plain DT.

The presented results are useful in two ways. Firstly, they show how to improve the efficiency of spectrum utilization during relay operation by introducing new opportunistic protocols. Secondly, they apply as system design guidelines by determining the suitable use cases for the DT, FD, HD, and hybrid FD/HD modes, and by setting targets for the performance of interference mitigation when the FD mode is preselected.

While the short-term performance is not tied to fading statistics, the analysis of average link capacity concerns an infrastructure RN deployed for extending the coverage area of a cellular BS. Thus, the backhaul link is fixed while the service link to/from a UE is subject to fast fading. Instead of capacity, the results can be reproduced also using outage probability [28].

When it comes to the rest of the chapter, system models are first formulated based on the signal models of the previous sections for two-hop relay links that, as a novel factor, account for the residual self-interference remaining after any combination of mitigation schemes, e.g., as discussed in References 23 and 26. Section 5.4.1 is devoted to hybrid FD/HD and transmit power adaptation schemes which optimize the instantaneous link capacity based on instantaneous CSI. The focus of Section 5.4.2 is on average link capacity when exploiting statistical CSI. In particular, the comparison of the proposed schemes explains the differences of DT, FD, HD, and hybrid FD/HD, AF and DF, DL and UL directions as well as short-term and long-term performance optimization. Finally, Section 5.5 summarizes the observations: Hybrid FD/HD relaying is an attractive concept for infrastructure relays whenever the self-interference can be first sufficiently mitigated.

5.4.1 *Analysis of short-term performance*

This section studies link capacity with hybrid FD/HD relaying and transmit power adaptation in the case of static channels and, thus, exploiting instantaneous CSI. Alternatively, the study can be interpreted to explain instantaneous snapshot performance within channel coherence time in a slow-fading environment. This is referred to as short-term link performance analysis where γ_{SR}, γ_{RR}, γ_{RD}, and γ_{SD} are regarded as constants representing the instantaneous channel states. The key optimization parameters are transmit powers in the source and in the relay, i.e., $p_{\mathrm{S}} \triangleq \mathcal{E}\{|x_{\mathrm{S}}[i]|^2\}$ and $p_{\mathrm{R}} \triangleq \mathcal{E}\{|x_{\mathrm{R}}[i]|^2\}$, respectively.

To begin with, Section 5.4.1.1 presents the instantaneous link capacities of the three conventional fixed modes: DT, FD, and HD. The system is directly characterized by their performance when any mode is chosen upfront during design stage. The new concept of hybrid switching between the modes is developed later in this section by formulating corresponding cross-over boundaries.

5.4.1.1 Instantaneous link capacity

The instantaneous link capacities achieved with the different conventional fixed operation modes can be expressed as

$$C_{\mathrm{FD}} = \log_2(1 + \gamma_{\mathrm{FD}}), \tag{5.47a}$$

$$C_{\mathrm{HD+MRC}} = \log_2(1 + \hat{\gamma}_{\mathrm{HD+MRC}}) \triangleq \frac{1}{2}\log_2(1 + \gamma_{\mathrm{HD+MRC}}), \tag{5.47b}$$

$$C_{\mathrm{DT}} = \log_2(1 + \gamma_{\mathrm{SD}}), \tag{5.47c}$$

for which the end-to-end SINRs, namely γ_{FD}, $\gamma_{\mathrm{HD+MRC}}$, and γ_{SD}, are specified shortly; without maximum ratio combining (MRC), the link capacity of HD relaying

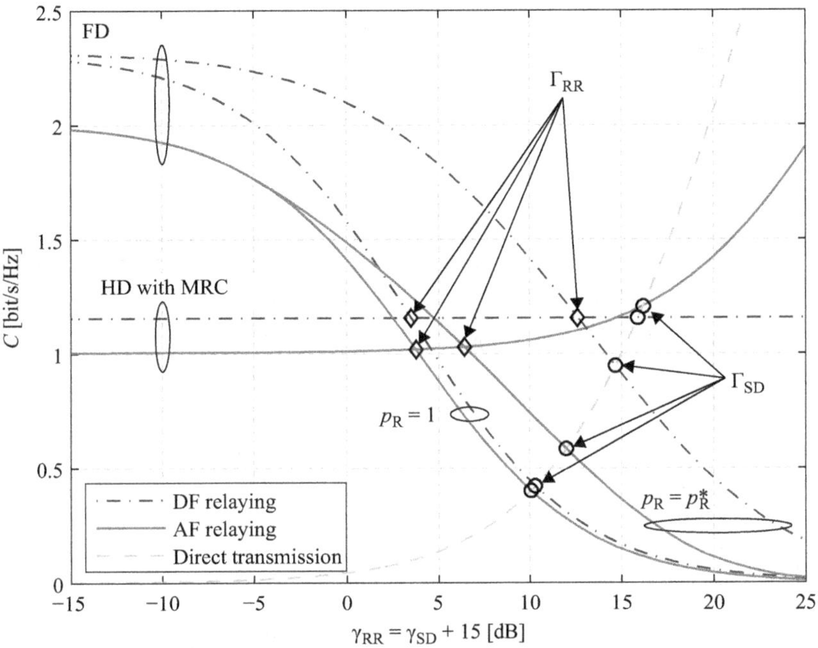

Figure 5.6 Instantaneous link capacities when $\gamma_{SR} = 6$ dB and $\gamma_{RD} = 12$ dB. The FD mode is preferred over the HD mode if $\gamma_{RR} < \Gamma_{RR}$, and DT is preferred over relaying when $\gamma_{SD} > \Gamma_{SD}$. Transmit power adaptation ($p_R = p_R^*$) improves capacity w.r.t. maximum power usage ($p_R = 1$)

is denoted by

$$C_{HD} = \log_2(1 + \hat{\gamma}_{HD}) \triangleq \frac{1}{2}\log_2(1 + \gamma_{HD}). \tag{5.48}$$

Capacities are illustrated in Figure 5.6.

It is worth emphasizing that (5.47b) exploits definition

$$\hat{\gamma}_{HD+MRC} \triangleq \sqrt{1 + \gamma_{HD+MRC}} - 1, \tag{5.49}$$

which represents an equivalent SNR level after compensating for the pre-log factor 1/2 of the capacity expression. This notation is adopted to allow capacity comparison indirectly in terms of only end-to-end SINRs by making the pre-log factors the same in all modes. Likewise, the S(I)NR of the DF protocol is not a real measurable quantity, but a "capacity-equalizing" variable adopted for handling both protocols in parallel as before.

In the FD mode, respective received signal powers in the relay and in the destination are given by $\mathcal{E}\{|\hat{y}_R[i]|^2\}$, as shown in (5.7), and

$$\mathcal{E}\left\{|y_D[i]|^2\right\} = p_R|h_{RD}|^2 + p_S|h_{SD}|^2 + \sigma_D^2, \tag{5.50}$$

where terms $p_S|h_{SR}|^2$ and $p_R|h_{RD}|^2$ are useful signal power and co-channel interference produces $p_R\gamma_{RR}\sigma_R^2$ and $p_S|h_{SD}|^2$. The SINRs of the two hops can be thus compactly defined as

$$\gamma_R \triangleq \frac{p_S\gamma_{SR}}{p_R\gamma_{RR}+1}, \tag{5.51a}$$

$$\gamma_D \triangleq \frac{p_R\gamma_{RD}}{p_S\gamma_{SD}+1}. \tag{5.51b}$$

Finally, following the derivations of Section 5.2.2, the above parameters allow to state the instantaneous end-to-end SINR of the FD mode as

$$\gamma_{FD} = \begin{cases} \dfrac{\gamma_R\gamma_D}{\gamma_R+\gamma_D+1}, & \text{with AF,} \\ \min\{\gamma_R,\gamma_D\}, & \text{with DF.} \end{cases} \tag{5.52}$$

These expressions update the earlier models presented for the HD mode, e.g., in Reference 14, by explicitly including the effects of the residual loopback self-interference and overheard direct-link transmission.

In the HD mode, the destination receives two independent copies of each transmitted symbol due to the adopted repetition-based protocols. The de facto standard way to benefit from such diversity is to apply MRC. Following the derivations of Reference 14, the end-to-end SNR of the HD mode with MRC can be expressed as

$$\gamma_{HD+MRC} = \begin{cases} p_S\gamma_{SD} + \dfrac{p_S\gamma_{SR}p_R\gamma_{RD}}{p_S\gamma_{SR}+p_R\gamma_{RD}+1}, & \text{with AF,} \\ \min\{p_S\gamma_{SR}, p_S\gamma_{SD}+p_R\gamma_{RD}\}, & \text{with DF.} \end{cases} \tag{5.53}$$

Without MRC, the end-to-end SNR is γ_{HD} obtained from above expressions simply by setting $\gamma_{SD}=0$ and, consequently, $\hat{\gamma}_{HD} \triangleq \sqrt{1+\gamma_{HD}}-1$.

If the direct link is strong in FD relaying, the destination may conversely regard respective $p_S|h_{SD}|^2$ and $p_R|h_{RD}|^2$ as useful and interference power, i.e., $\gamma_D \triangleq \frac{p_S\gamma_{SD}}{p_R\gamma_{RD}+1}$ in contrast to (5.51). Obviously, the system should shut down the relay completely in this case. In the DT mode, the end-to-end SNR is thus simply given by the SD link SNR:

$$\gamma_{DT} = p_S\gamma_{SD}. \tag{5.54}$$

Effectively $p_R=0$ and $\gamma_{FD}=\gamma_D=p_S\gamma_{SD}$ replace (5.51) and (5.52), respectively, when the DT mode is understood as a degenerate case of the FD mode.

5.4.1.2 Transmit power adaptation for FD relaying

Transmit power adaptation is proposed in the previous section as a technique to mitigate the effect of residual self-interference, and now it shall be integrated with the concept of hybrid mode switching.

Transmission power levels are subject to individual constraints throughout the chapter: $p_S \leq 1$ and $p_R \leq 1$. This is practical for infrastructure-based deployments where the total energy consumption of a system is less critical than each transmitter's power limitations set by regulations and the physical capabilities of radio circuitry. In this chapter, the performance of infrastructure relay links will not be optimized under a joint sum power constraint ($p_S + p_R \leq 2$), which may be relevant for cooperative communications instead, but this complementary analysis can be found in Reference 29.

Without transmit power optimization in the FD mode, the relay simply uses the maximum allowed power, i.e., $p_R = 1$. The remaining chapter concentrates on power adaptation in the relay implying that the source uses always the maximum transmit power, i.e., $p_S = 1$. This is justified in coverage extension scenarios: $\gamma_R \triangleq (p_S\gamma_{SR})/(p_R\gamma_{RR} + 1)$ is then maximized and, because typically $\gamma_{SD} \ll \gamma_{RR}$, the effect on $\gamma_D \triangleq (p_R\gamma_{RD})/(p_S\gamma_{SD} + 1) \simeq p_R\gamma_{RD}$ is insignificant; the best performance under constraint $p_S \leq 1$ is thus achieved when $p_S \simeq 1$.

Transmit power adaptation in FD relaying is motivated by the observation that the end-to-end performance is limited by the weakest hop: If the limiting factor is the first hop due to excessive self-interference, the link capacity can be, in fact, *improved* by *decreasing* the relay transmit power. This reasoning can be formulated as an optimization problem

$$p_R^* = \arg\max_{p_R} C_{FD} \text{ subject to } p_R \leq 1, \tag{5.55}$$

for which C_{FD} is given by (5.47a) with (5.51) and (5.52). This power control problem is already solved in a previous section for both protocols, and the result can be restated as follows.

Optimal relay transmit power

The optimal gain factors β^* obtained in Section 5.3 imply that

$$p_R^* = \begin{cases} \min\left\{1, \dfrac{1}{\gamma_{RR}}\sqrt{\dfrac{(\gamma_{SR}+1)\gamma_{RR}(\gamma_{SD}+1)}{\gamma_{RD}}}\right\}, & \text{with AF,} \\ \min\left\{1, \dfrac{1}{\gamma_{RR}}\left(\sqrt{\dfrac{\gamma_{SR}\gamma_{RR}(\gamma_{SD}+1)}{\gamma_{RD}} + \dfrac{1}{4}} - \dfrac{1}{2}\right)\right\}, & \text{with DF.} \end{cases} \tag{5.56}$$

As a consequence, the transmit power constraint is invoked ($p_R^* = 1$) when

$$\gamma_{RR} \leq \begin{cases} \dfrac{(\gamma_{SR}+1)(\gamma_{SD}+1)}{\gamma_{RD}}, & \text{with AF,} \\ \dfrac{\gamma_{SR}(\gamma_{SD}+1)}{\gamma_{RD}} - 1, & \text{with DF.} \end{cases} \tag{5.57}$$

In particular, these inequalities show that the maximum transmit power is optimal only in the case of weak interference while otherwise link performance can be improved by backing off from the maximum. Furthermore, the limitation appears to be the most critical for the AF protocol.

The substitution of (5.51) and (5.56) in (5.52) yields end-to-end SINRs with optimal transmit power adaptation as

$$\gamma_{\mathrm{FD}}^{*}=\begin{cases}\dfrac{\gamma_{\mathrm{SR}}\gamma_{\mathrm{RD}}}{\gamma_{\mathrm{RD}}+\gamma_{\mathrm{RR}}(\gamma_{\mathrm{SD}}+1)+2\sqrt{(\gamma_{\mathrm{SR}}+1)\gamma_{\mathrm{RD}}\gamma_{\mathrm{RR}}(\gamma_{\mathrm{SD}}+1)}}, & \text{with AF,}\\[2ex] \dfrac{\gamma_{\mathrm{RD}}}{\gamma_{\mathrm{RR}}(\gamma_{\mathrm{SD}}+1)}\left(\sqrt{\dfrac{\gamma_{\mathrm{SR}}\gamma_{\mathrm{RR}}(\gamma_{\mathrm{SD}}+1)}{\gamma_{\mathrm{RD}}}+\dfrac{1}{4}}-\dfrac{1}{2}\right), & \text{with DF.}\end{cases} \tag{5.58}$$

This expression disregards (for clarity and compactness) the saturation of transmit power such that $p_{\mathrm{R}}^{*}\leq 1$. Yet the following analysis still considers constrained power adaptation for which the SINRs coincide with those of the unconstrained case given above whenever the conditions in (5.57) are *not* satisfied; the maximum transmit power is used otherwise and the SINRs are given by (5.52) with the substitution of (5.51) where $p_{\mathrm{S}}=p_{\mathrm{R}}=1$.

5.4.1.3 Transmit power in DT and HD relaying

Under independent power constraints, the performance of the HD mode is optimized simply by using maximum power in both transmitters. Likewise, the DT mode simply adopts maximum transmit power for obtaining the highest capacity. Thus, the analysis shall focus on power adaptation only in the FD mode while $p_{\mathrm{S}}=p_{\mathrm{R}}\in\{1,2\}$ is set for the other modes.

The comparison of the FD and DT modes to the HD mode is not totally straightforward because there are two different ways to normalize transmit power in the HD mode; both of them are common in related literature.

Energy normalization: Since the source and the relay transmit only half of the time when compared to the FD mode, the HD mode should use double power ($p_{\mathrm{S}}=p_{\mathrm{R}}=2$) to equalize the total average transmitted energy in all modes. Further analysis for this normalization is available in Reference 24 but it is used as a reference case herein too.

Power normalization: The present study mainly adopts the other normalization scheme, i.e., the same constraints $p_{\mathrm{S}}\leq 1$ and $p_{\mathrm{R}}\leq 1$ are set for all modes. This is because hybrid FD/HD relaying uses the same radio equipment for both modes while doubled instantaneous power in the HD mode would imply that the emission regulations or the radio front-ends were different. Furthermore, total energy consumption is a less critical factor for cellular relays connected to a power grid than for battery-operated relays in user cooperation.

The *energy normalization* scheme is relevant in conjunction with the joint transmit power constraint, i.e., $p_{\mathrm{S}}+p_{\mathrm{R}}\leq 4$, since both concepts share energy budget between two *separate* nodes. However, $p_{\mathrm{S}}=p_{\mathrm{R}}=2$ becomes suboptimal then and also the HD mode needs transmit power adaptation.

5.4.1.4 Hybrid relaying modes

The need for proper mode selection is motivated by Figure 5.6 which shows the instantaneous link capacities given by (5.47a)–(5.47c) in terms of γ_{RR} and γ_{SD}. Especially, this reveals two relevant break-even levels for the SNRs. Sufficient and necessary condition for the FD mode to outperform the HD mode is that the residual self-interference gain γ_{RR} is below a break-even self-interference level Γ_{RR}. Similarly, if the direct-link SNR γ_{SD} is above a break-even level Γ_{SD}, the DT mode is superior to either relaying mode. These break-even levels can be formally defined such that

$$C_{FD} \geq C_{HD} \quad \text{if and only if } \gamma_{RR} \leq \Gamma_{RR}, \tag{5.59a}$$

$$C_{HD+MRC} \geq C_{DT} \text{ or } C_{FD} \geq C_{DT} \quad \text{if and only if } \gamma_{SD} \leq \Gamma_{SD}. \tag{5.59b}$$

For end-to-end performance optimization, opportunistic hybrid relaying switches to the best mode based on instantaneous CSI, i.e., the values of γ_{SR}, γ_{RD}, γ_{RR}, and γ_{SD}. The resulting instantaneous link capacity becomes

$$C_{\text{hybrid}} \triangleq \max\{C_{FD}, C_{HD+MRC}, C_{DT}\}, \tag{5.60}$$

while link capacity is fixed to either C_{FD}, C_{HD+MRC}, or C_{DT} in conventional reference cases. With (5.60), Figure 5.6 shows readily the advantage of hybrid relaying w.r.t. conventional modes: Each mode achieves in turn much higher link capacity than the others in different channel states.

In what follows, the explicit implementation of the hybrid relaying mode is formulated by deriving the switching boundaries between the modes. In particular, expressions are given for break-even levels Γ_{SD} and Γ_{RR}.

The direct-link SNR γ_{SD} is the key parameter for switching between DT and relaying modes. Its effect can be quantified with closed-form expressions for Γ_{SD} in terms of γ_{SR}, γ_{RD}, and γ_{RR} as summarized in the following property.

Property: Switching between DT and relaying

The DT mode is preferred over FD relaying if

$$\gamma_{SD} \geq \Gamma_{SD} = \begin{cases} \dfrac{\gamma_R + \gamma_{RD} + 1}{\gamma_R + 1}\left(\sqrt{\dfrac{\gamma_R(\gamma_R + 1)\gamma_{RD}}{(\gamma_R + \gamma_{RD} + 1)^2} + \dfrac{1}{4}} - \dfrac{1}{2}\right), & \text{with AF,} \\ \min\left\{\gamma_R, \sqrt{\gamma_{RD} + \dfrac{1}{4}} - \dfrac{1}{2}\right\}, & \text{with DF,} \end{cases} \tag{5.61a}$$

and the DT mode is preferred over HD relaying (which uses MRC) if

$$\gamma_{SD} \geq \Gamma_{SD} = \begin{cases} \sqrt{\dfrac{\gamma_{SR}\gamma_{RD}}{\gamma_{SR} + \gamma_{RD} + 1} + \dfrac{1}{4}} - \dfrac{1}{2}, & \text{with AF,} \\ \min\left\{\sqrt{\gamma_{SR} + 1} - 1, \sqrt{\gamma_{RD} + \dfrac{1}{4}} - \dfrac{1}{2}\right\}, & \text{with DF,} \end{cases} \tag{5.61b}$$

without transmit power adaptation. As before, $\gamma_R \triangleq \gamma_{SR}/(\gamma_{RR} + 1)$.

The break-even levels of (5.61a) and (5.61b) are obtained by solving γ_{SD} in terms of the other parameters of (5.51)–(5.54) from the respective inequalities $C_{DT} \geq C_{FD}$ and $C_{DT} \geq C_{HD+MRC}$ for which C_{FD}, C_{HD+MRC}, and C_{DT} are given in (5.47a)–(5.47c) and $p_S = p_R = 1$ in the absence of power adaptation.

With optimal transmit power adaptation, similar break-even level Γ_{SD} for both AF and DF can be (in theory) solved explicitly from $\gamma_{SD} \geq \gamma^*_{FD}$ with the substitution of (5.58). However, the expressions are in practice too tedious to be presented here, because they correspond to the roots of fourth-order polynomials. A lookup table of numerical values can be used for mode switching in this case. Moreover, power adaptation reduces only slightly the range of direct-link SNRs that guarantee the superiority of the DT mode over the FD mode. Thus, nothing essential is left unseen.

The switching boundaries as per (5.61a) and (5.61b) are illustrated in Figure 5.7 for the case in which the residual self-interference is negligible to emphasize the effect of the direct link. Thereby, the FD mode dominates the choice between relaying and mere DT. In fact, adding loopback interference into consideration just further extends the range of γ_{SR} and γ_{RD} for which the DT mode is preferred. Furthermore, MRC does not offer significant gain for the HD mode which is also demonstrated by Figure 5.6 in which the link capacity of the AF protocol increases only slightly before the DT mode becomes preferable. In fact, MRC is not beneficial at all for the DF

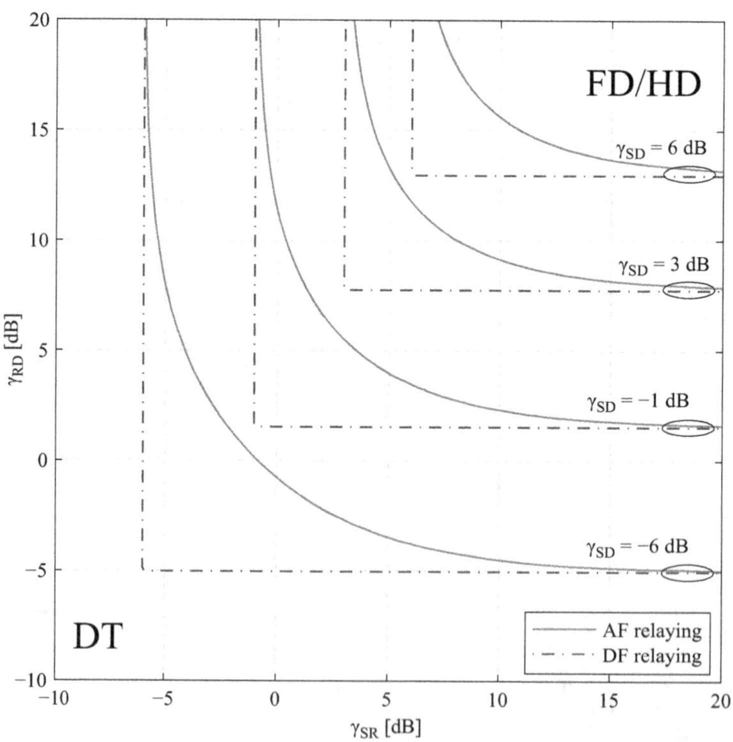

Figure 5.7 Switching boundaries γ_{SD} between DT and either of the relaying modes when $\gamma_{RR} = 0$. FD or HD relaying is preferred over DT inside the boundaries

protocol in Figure 5.6 because $\gamma_{SR} < \gamma_{RD}$. It should be especially noted that neither relaying mode is preferred in a large range of practical SR and RD channel states with a relatively weak direct link, e.g., even $\gamma_{SD} < 3$ dB. Thus, it is beneficial to retain the possibility to revert to a plain DT mode instantaneously, because it is probable that the direct link is momentarily usable although it remains blocked on average in coverage extension scenarios.

As demonstrated in Figure 5.6, the residual loopback channel gain γ_{RR} is the key parameter for switching between relaying modes. Explicit closed-form expressions are derived for break-even levels Γ_{RR} in terms of γ_{SR}, γ_{RD}, and γ_{SD} such that the switching conditions are summarized in the following properties.

Property: Switching between FD and HD relaying without transmit power adaptation

The HD mode is preferred over the FD mode without transmit power adaptation if $\gamma_D < \hat{\gamma}_{HD+MRC} \triangleq \sqrt{1 + \gamma_{HD+MRC}} - 1$ for which γ_{HD+MRC} is given in (5.53); otherwise, the FD mode is preferred over the HD mode if

$$\gamma_{RR} < \Gamma_{RR} = \begin{cases} \dfrac{\gamma_{SR}(\gamma_D - \hat{\gamma}_{HD+MRC})}{\hat{\gamma}_{HD+MRC}(\gamma_D + 1)} - 1, & \text{with AF,} \\ \dfrac{\gamma_{SR}}{\hat{\gamma}_{HD+MRC}} - 1, & \text{with DF.} \end{cases} \tag{5.62}$$

As before, $\gamma_D \triangleq \gamma_{RD}/(\gamma_{SD} + 1)$.

After setting $p_R = 1$ for the FD mode as power adaptation is not used, the break-even levels in (5.62) are obtained by solving γ_{RR} in terms of the other parameters of (5.51) and (5.52) from the inequality $C_{FD} \geq C_{HD+MRC}$ for which C_{FD} and C_{HD+MRC} are given in (5.47a) and (5.47b), respectively.

Transmit power adaptation is beneficial for the FD mode in the presence of strong self-interference as discussed in earlier in this chapter. Consequently, lowering transmit power from its maximum guarantees the superiority of the FD mode in an extended region of channel states as observed next.

Property: Switching between FD and HD relaying with transmit power adaptation

With optimal *unconstrained* transmit power settings, the FD mode is preferred over the HD mode if

$$\gamma_{RR} < \Gamma_{RR} = \begin{cases} \dfrac{\gamma_{RD}}{\hat{\gamma}_{HD+MRC}(\gamma_{SD} + 1)}\Big[\gamma_{SR} + (2\gamma_{SR} + 1)\hat{\gamma}_{HD+MRC} \\ \quad -2\sqrt{\gamma_{SR}(\gamma_{SR} + 1)\hat{\gamma}_{HD+MRC}(\hat{\gamma}_{HD+MRC} + 1)}\Big], & \text{with AF,} \\ \dfrac{\gamma_{RD}}{\hat{\gamma}_{HD+MRC}(\gamma_{SD} + 1)}\left(\dfrac{\gamma_{SR}}{\hat{\gamma}_{HD+MRC}} - 1\right), & \text{with DF,} \end{cases} \tag{5.63}$$

for which $\hat{\gamma}_{HD+MRC} \triangleq \sqrt{1 + \gamma_{HD+MRC}} - 1$ and γ_{HD+MRC} is given in (5.53).

The conditions in (5.57) are not invoked with unconstrained power adaptation and, thus, γ_{FD} is given by (5.58). The break-even levels in (5.63) are then obtained by solving γ_{RR} from the inequality $C_{FD} \geq C_{HD+MRC}$ for which C_{FD} and C_{HD+MRC} are given in (5.47a) and (5.47b), respectively.

Naturally, one will always in practice apply *constrained* power adaptation for which the foregoing unconstrained case represents a subsolution. However, it is remarkable that the activation of transmit power limitations does not affect at all the break-even self-interference level Γ_{RR} given in (5.63) for the DF protocol. On the contrary in the case of AF relaying, the break-even self-interference level Γ_{RR} is given by (5.63) only if

$$\sqrt{\frac{(\gamma_{SR}+1)[(\gamma_{SR}+1)(\gamma_{SD}+1)+\gamma_{RD}]}{\gamma_{RD}+\gamma_{SD}+1}} \geq \frac{(\gamma_{SR}+1)(\gamma_{SD}+1)}{\gamma_{RD}}, \tag{5.64}$$

while otherwise it is as with maximum power usage ($p_R^* = 1$) shown in (5.62). Unfortunately, this inequality does not admit a compact solution in terms of any variable which hides explicit limiting conditions.

The switching boundaries of (5.62) are illustrated in Figure 5.8. Without transmit power adaptation, the choice between FD and HD relaying is mostly defined by the

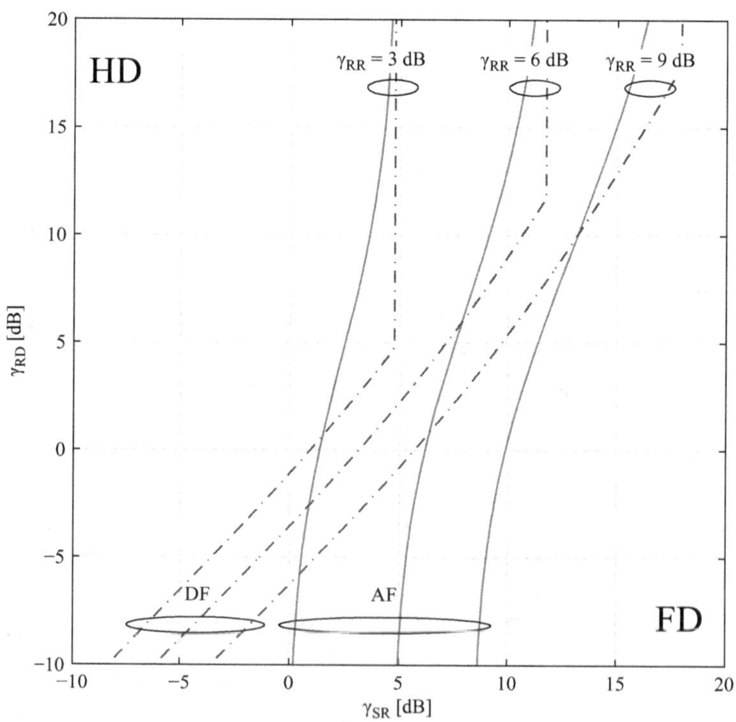

Figure 5.8 Switching boundaries γ_{RR} between the two relaying modes when $\gamma_{SD} = 0$ and maximum transmit power is used ($p_R = 1$). The FD mode is preferred over the HD mode on the right-hand side of the boundaries, and vice versa on the left-hand side

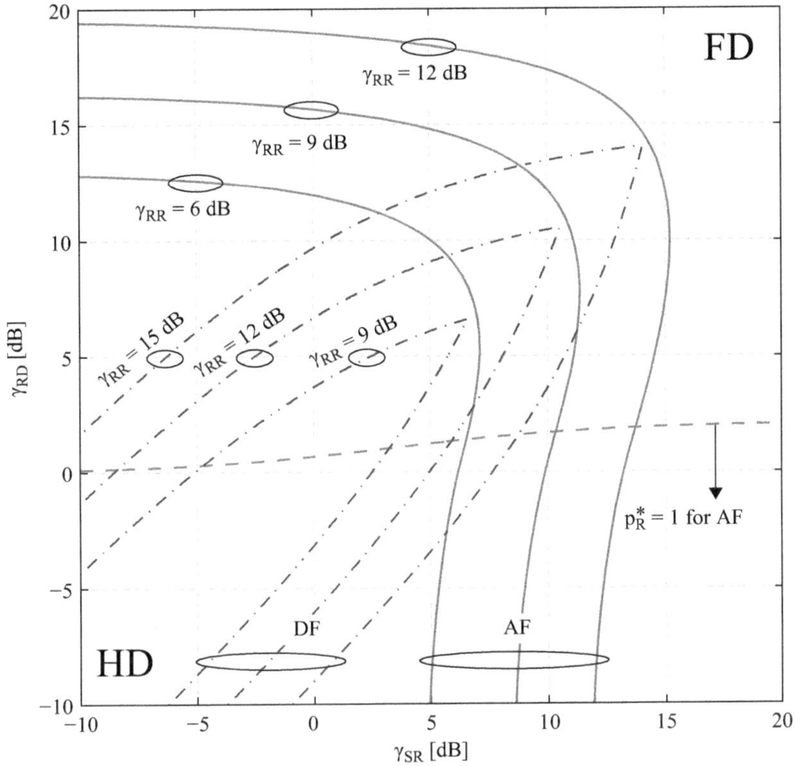

Figure 5.9 Switching boundaries γ_{RR} between the two relaying modes when $\gamma_{SD} = 0$ and the relay transmit power is optimized ($p_R = p_R^*$). The FD mode is preferred over the HD mode outside the boundaries

first-hop SNR such that the HD mode is preferred for small γ_{SR}. Figure 5.9 illustrates the boundaries for optimal transmit power adaptation according to (5.63); now the FD mode is preferred in an extended region of channel states and, in particular, with small γ_{SR} and large γ_{RD}. The right-hand side of the condition in (5.64) corresponds to the power limitation given in (5.57) indicating that the limitation is effective only with small γ_{RD} as also demonstrated in Figure 5.9. Both with and without power adaptation, one should notice that the FD mode is preferable with all γ_{RD} if γ_{SR} is large enough. According to (5.64), the power constraint limits Γ_{RR} with AF relaying below the dashed curve.

For comparison with Figures 5.8 and 5.9, Figure 5.10 illustrates the switching boundaries when using *energy normalization* instead of *power normalization* assumed above. This clearly favors the HD mode but the FD mode is still useful in a large region of practical channel states and the gain of transmit power adaptation is consistent for both normalization schemes.

One should also recall that the foregoing analysis assumes that the relay is implemented with a single transmitter–receiver pair per two antenna elements. An extra

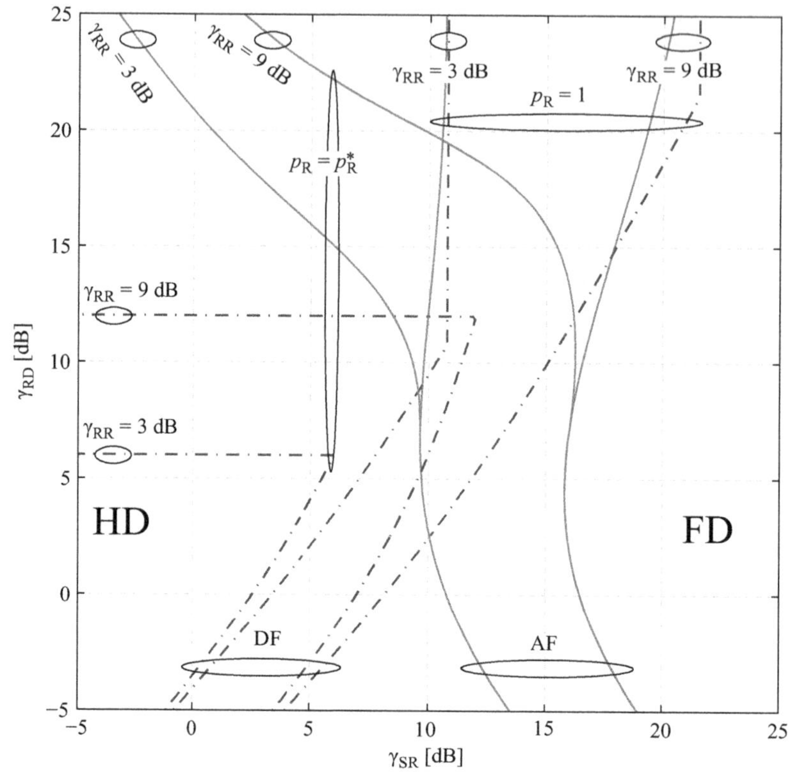

Figure 5.10 Switching boundaries between the two relaying modes when $\gamma_{SD} = 0$ and energy normalization ($p_S = p_R = 2$) is assumed for the HD mode. The FD mode is preferred over the HD mode outside and on the right-hand side of the boundaries, and vice versa on the left-hand side

transceiver would allow the HD mode to exploit receive and transmit beamforming and achieve higher γ_{SR} and γ_{RD} than the FD mode. Figure 5.10 illustrates also this case as two-antenna beamforming brings ideally (with balanced branches) up to three-decibel SNR gain for both hops which is equivalent to doubled transmit powers assumed with energy normalization. However, in practice, the potential beamforming gain becomes insignificant and, thus, not worth the cost of another transceiver, because the branches are highly imbalanced in typical target scenarios.

5.4.2 Analysis of long-term performance

Proceeding to relaying over fading channels, objective shifts next to optimizing long-term link capacity using statistical CSI. In particular, DL and UL directions are now studied separately as they are of different nature due to asymmetry caused by setting scope on fixed relays. The key system parameters herein are the *average* channel SNRs: $\bar{\gamma}_{SR} \triangleq \mathcal{E}\{\gamma_{SR}\}$, $\bar{\gamma}_{RR} \triangleq \mathcal{E}\{\gamma_{RR}\}$, $\bar{\gamma}_{RD} \triangleq \mathcal{E}\{\gamma_{RD}\}$, and $\bar{\gamma}_{SD} \triangleq \mathcal{E}\{\gamma_{SD}\}$.

The BS and the RN are assumed to be part of fixed infrastructure, and LOS connection is often achieved by planning RN antenna locations properly. Thereby, it is reasonable to model the backhaul channels and the residual self-interference gain as static, i.e., $\gamma_{\mathrm{SR}} = \bar{\gamma}_{\mathrm{SR}}$ (in DL), $\gamma_{\mathrm{RD}} = \bar{\gamma}_{\mathrm{RD}}$ (in UL), and $\gamma_{\mathrm{RR}} = \bar{\gamma}_{\mathrm{RR}}$. The UE is assumed to be a mobile terminal without LOS to the RN which justifies to conventionally model the service channels by Rayleigh fading, i.e., γ_{RD} (in DL) and γ_{SR} (in UL) become exponential random variables with probability density functions $f_{\gamma_{\mathrm{RD}}}(x) = (1/\bar{\gamma}_{\mathrm{RD}})\exp(-x/\bar{\gamma}_{\mathrm{RD}})$ and $f_{\gamma_{\mathrm{SR}}}(x) = (1/\bar{\gamma}_{\mathrm{SR}})\exp(-x/\bar{\gamma}_{\mathrm{SR}})$. Finally, scope on coverage extension implies $\bar{\gamma}_{\mathrm{SD}} \ll \bar{\gamma}_{\mathrm{RD}}$ rendering the contribution of BS-to-UE and UE-to-BS channels insignificant for average link capacity, although they are intermittently useful for diversity. Thus, the following long-term performance analysis sets $\gamma_{\mathrm{SD}} \approx 0$, i.e., interference from the direct link is effectively embedded in thermal noise in FD relaying, and MRC can be omitted in HD relaying.

5.4.2.1 Average link capacity

Average link capacity shall be denoted by $\bar{C}_{\mathrm{FD}}$ (resp. $\bar{C}_{\mathrm{HD}}$) if the fixed FD (resp. HD) mode is preselected during system design. Additional superscripts help to distinguish between different variations, e.g., $\bar{C}_{\mathrm{FD}}^{\mathrm{DL-AF}}$ means naturally the average DL capacity of FD AF relaying.

Omitting subscripts FD and HD as well as superscripts AF and DF since the following expressions look identical for these variations, DL and UL capacities are calculated by averaging over fading distributions as

$$\bar{C}_{\mathrm{DL}} \triangleq \mathcal{E}\{C(\bar{\gamma}_{\mathrm{SR}}, \gamma_{\mathrm{RD}})\} = \int_0^\infty C(\bar{\gamma}_{\mathrm{SR}}, x) f_{\gamma_{\mathrm{RD}}}(x)\,\mathrm{d}x, \tag{5.65a}$$

$$\bar{C}_{\mathrm{UL}} \triangleq \mathcal{E}\{C(\gamma_{\mathrm{SR}}, \bar{\gamma}_{\mathrm{RD}})\} = \int_0^\infty C(x, \bar{\gamma}_{\mathrm{RD}}) f_{\gamma_{\mathrm{SR}}}(x)\,\mathrm{d}x, \tag{5.65b}$$

respectively. Above $C(\gamma_{\mathrm{SR}}, \gamma_{\mathrm{RD}})$ is substituted by $C_{\mathrm{FD}}(\gamma_{\mathrm{SR}}, \gamma_{\mathrm{RD}})$ from (5.47a) with the FD mode and by $C_{\mathrm{HD}}(\gamma_{\mathrm{SR}}, \gamma_{\mathrm{RD}})$ from (5.47b) with the HD mode; the instantaneous link SNRs are defined in (5.51)–(5.53) for AF and DF relaying.

With both relaying protocols, the average link capacity admits closed-form expressions which can be neatly presented in terms of auxiliary integrals $\mathcal{I}_{\mathrm{AF}}$ and $\mathcal{I}_{\mathrm{DF}}$ (or their variations) derived below; all of them are based on the exponential integral $E_1(z) = \int_z^\infty (1/t)\exp(-t)\,\mathrm{d}t$ [1, Eq. 5.1.1]. The final capacity expressions are collected in Table 5.1 for concise reference.

The integral needed for the AF protocol is calculated with integration by parts as follows:

$$\begin{aligned}\mathcal{I}_{\mathrm{AF}}^{\mathrm{tail}}(a,b,c) &\triangleq \int_c^\infty \log_2\left(1+\frac{ax}{a+1+x}\right)\frac{\exp\left(-\frac{x}{b}\right)}{b}\,\mathrm{d}x \\ &= e^{\frac{1}{b}}\frac{E_1\left(\frac{c+1}{b}\right) - e^{\frac{a}{b}}E_1\left(\frac{a+c+1}{b}\right)}{\log_e(2)} - e^{-\frac{c}{b}}\log_2\left(\frac{a+c+1}{(a+1)(c+1)}\right),\end{aligned} \tag{5.66a}$$

Table 5.1 *Closed-form expressions for average DL and UL capacity with AF and DF relaying. With different sub/superscripts, constants χ represent the switching points between the FD and HD modes and integrals $\mathcal{I}$ are given by (5.66) and (5.67)*

Static mode selection

$$\bar{C}_{\text{FD}}^{\text{DL-AF}} = \mathcal{I}_{\text{AF}}\left(\frac{\bar{\gamma}_{\text{SR}}}{\bar{p}_{\text{R}}\bar{\gamma}_{\text{RR}}+1}, \bar{p}_{\text{R}}\bar{\gamma}_{\text{RD}}\right) \qquad \bar{C}_{\text{HD}}^{\text{DL-AF}} = \frac{1}{2}\mathcal{I}_{\text{AF}}(\bar{\gamma}_{\text{SR}}, \bar{\gamma}_{\text{RD}}) \qquad \bar{C}_{\text{FD}}^{\text{UL-AF}} = \mathcal{I}_{\text{AF}}\left(\bar{p}_{\text{R}}\bar{\gamma}_{\text{RD}}, \frac{\bar{\gamma}_{\text{SR}}}{\bar{p}_{\text{R}}\bar{\gamma}_{\text{RR}}+1}\right) \qquad \bar{C}_{\text{HD}}^{\text{UL-AF}} = \frac{1}{2}\mathcal{I}_{\text{AF}}(\bar{\gamma}_{\text{RD}}, \bar{\gamma}_{\text{SR}})$$

$$\bar{C}_{\text{FD}}^{\text{DL-DF}} = \mathcal{I}_{\text{DF}}\left(\frac{\bar{\gamma}_{\text{SR}}}{\bar{p}_{\text{R}}\bar{\gamma}_{\text{RR}}+1}, \bar{p}_{\text{R}}\bar{\gamma}_{\text{RD}}\right) \qquad \bar{C}_{\text{HD}}^{\text{DL-DF}} = \frac{1}{2}\mathcal{I}_{\text{DF}}(\bar{\gamma}_{\text{SR}}, \bar{\gamma}_{\text{RD}}) \qquad \bar{C}_{\text{FD}}^{\text{UL-DF}} = \mathcal{I}_{\text{DF}}\left(\bar{p}_{\text{R}}\bar{\gamma}_{\text{RD}}, \frac{\bar{\gamma}_{\text{SR}}}{\bar{p}_{\text{R}}\bar{\gamma}_{\text{RR}}+1}\right) \qquad \bar{C}_{\text{HD}}^{\text{UL-DF}} = \frac{1}{2}\mathcal{I}_{\text{DF}}(\bar{\gamma}_{\text{RD}}, \bar{\gamma}_{\text{SR}})$$

Hybrid FD/HD relaying

$$\text{FD if } \gamma_{\text{RD}} < \chi_{\text{hybrid}}^{\text{DL-AF}} \triangleq \frac{(\bar{\gamma}_{\text{SR}}+1)^2 - \bar{\gamma}_{\text{RR}}^2}{\bar{\gamma}_{\text{RR}}^2 - (\bar{\gamma}_{\text{SR}}+1)}$$

$$\bar{C}_{\text{hybrid}}^{\text{DL-AF}} = \begin{cases} \bar{C}_{\text{HD}}^{\text{DL-AF}}, & \bar{\gamma}_{\text{SR}} < \bar{\gamma}_{\text{RR}} - 1 \\ \bar{C}_{\text{FD}}^{\text{DL-AF}}, & \bar{\gamma}_{\text{SR}} > \bar{\gamma}_{\text{RR}}^2 - 1 \\ \mathcal{I}_{\text{AF}}^{\text{head}}\left(\frac{\bar{\gamma}_{\text{SR}}}{\bar{\gamma}_{\text{RR}}+1}, \bar{\gamma}_{\text{RD}}, \chi_{\text{hybrid}}^{\text{DL-AF}}\right) + \frac{1}{2}\mathcal{I}_{\text{AF}}^{\text{tail}}(\bar{\gamma}_{\text{SR}}, \bar{\gamma}_{\text{RD}}, \chi_{\text{hybrid}}^{\text{DL-AF}}), & \text{otherwise} \end{cases}$$

$$\text{FD if } \gamma_{\text{SR}} > \chi_{\text{hybrid}}^{\text{UL-AF}} \triangleq \sqrt{\bar{\gamma}_{\text{RR}}^2(\bar{\gamma}_{\text{RD}}+1) + \frac{1}{4}\bar{\gamma}_{\text{RD}}^2} - \frac{1}{2}\bar{\gamma}_{\text{RD}} - 1$$

$$\bar{C}_{\text{hybrid}}^{\text{UL-AF}} = \frac{1}{2}\mathcal{I}_{\text{AF}}^{\text{head}}(\bar{\gamma}_{\text{RD}}, \bar{\gamma}_{\text{SR}}, \chi_{\text{hybrid}}^{\text{UL-AF}}) + \mathcal{I}_{\text{AF}}^{\text{tail}}\left(\bar{\gamma}_{\text{RD}}, \frac{\bar{\gamma}_{\text{SR}}}{\bar{\gamma}_{\text{RR}}+1}, \chi_{\text{hybrid}}^{\text{UL-AF}}\right)$$

$$\text{FD if } \gamma_{\text{RD}} < \chi_{\text{hybrid}}^{\text{DL-DF}} \triangleq \frac{2\bar{\gamma}_{\text{SR}}(\bar{\gamma}_{\text{RR}}+1) + \bar{\gamma}_{\text{SR}}^2}{(\bar{\gamma}_{\text{RR}}+1)^2}$$

$$\bar{C}_{\text{hybrid}}^{\text{DL-DF}} = \begin{cases} \bar{C}_{\text{FD}}^{\text{DL-DF}}, & \bar{\gamma}_{\text{SR}} > \bar{\gamma}_{\text{RR}}^2 - 1 \\ \mathcal{I}_{\text{DF}}^{\text{head}}\left(\frac{\bar{\gamma}_{\text{SR}}}{\bar{\gamma}_{\text{RR}}+1}, \bar{\gamma}_{\text{RD}}, \chi_{\text{hybrid}}^{\text{DL-DF}}\right) + \frac{1}{2}\mathcal{I}_{\text{DF}}^{\text{tail}}(\bar{\gamma}_{\text{SR}}, \bar{\gamma}_{\text{RD}}, \chi_{\text{hybrid}}^{\text{DL-DF}}), & \text{otherwise} \end{cases}$$

$$\text{FD if } \gamma_{\text{SR}} > \chi_{\text{hybrid}}^{\text{UL-DF}} \triangleq \begin{cases} \bar{\gamma}_{\text{RR}}^2 - 1, & \bar{\gamma}_{\text{RD}} > \bar{\gamma}_{\text{RR}}^2 - 1 \\ (\bar{\gamma}_{\text{RR}}+1)(\sqrt{\bar{\gamma}_{\text{RD}}+1} - 1), & \text{otherwise} \end{cases}$$

$$\bar{C}_{\text{hybrid}}^{\text{UL-DF}} = \frac{1}{2}\mathcal{I}_{\text{DF}}^{\text{head}}(\bar{\gamma}_{\text{RD}}, \bar{\gamma}_{\text{SR}}, \chi_{\text{hybrid}}^{\text{UL-DF}}) + \mathcal{I}_{\text{DF}}^{\text{tail}}\left(\bar{\gamma}_{\text{RD}}, \frac{\bar{\gamma}_{\text{SR}}}{\bar{\gamma}_{\text{RR}}+1}, \chi_{\text{hybrid}}^{\text{UL-DF}}\right)$$

in which constants a, b, and c are real-valued and non-negative. Based on the above expression, the other two integrals are given by

$$\mathcal{I}_{\mathrm{AF}}(a,b) = \lim_{c\to 0} \mathcal{I}_{\mathrm{AF}}^{\mathrm{tail}}(a,b,c) = e^{\frac{1}{b}} \frac{E_1\left(\frac{1}{b}\right) - e^{\frac{a}{b}} E_1\left(\frac{a+1}{b}\right)}{\log_e(2)}, \tag{5.66b}$$

$$\mathcal{I}_{\mathrm{AF}}^{\mathrm{head}}(a,b,c) = \mathcal{I}_{\mathrm{AF}}(a,b) - \mathcal{I}_{\mathrm{AF}}(a,b,c). \tag{5.66c}$$

Similarly, integration by parts yields the integral needed for the DF protocol as follows:

$$\mathcal{I}_{\mathrm{DF}}^{\mathrm{head}}(a,b,c) \triangleq \int_0^c \log_2(1+\min\{a,x\}) \frac{\exp\left(-\frac{x}{b}\right)}{b} \mathrm{d}x \tag{5.67a}$$

$$= \begin{cases} e^{\frac{1}{b}} \dfrac{E_1\left(\frac{1}{b}\right) - E_1\left(\frac{c+1}{b}\right)}{\log_e(2)} - e^{-\frac{c}{b}} \log_2(c+1), & c \le a, \\ e^{\frac{1}{b}} \dfrac{E_1\left(\frac{1}{b}\right) - E_1\left(\frac{a+1}{b}\right)}{\log_e(2)} - e^{-\frac{c}{b}} \log_2(a+1), & c > a. \end{cases}$$

And the other integrals are given by

$$\mathcal{I}_{\mathrm{DF}}(a,b) = \lim_{c\to\infty} \mathcal{I}_{\mathrm{DF}}^{\mathrm{head}}(a,b,c) = e^{\frac{1}{b}} \frac{E_1\left(\frac{1}{b}\right) - E_1\left(\frac{a+1}{b}\right)}{\log_e(2)}, \tag{5.67b}$$

$$\mathcal{I}_{\mathrm{DF}}^{\mathrm{tail}}(a,b,c) = \mathcal{I}_{\mathrm{DF}}(a,b) - \mathcal{I}_{\mathrm{DF}}^{\mathrm{head}}(a,b,c). \tag{5.67c}$$

5.4.2.2 Transmit power adaptation

When using statistical CSI, the relay can adapt only its long-term transmit power, i.e., $p_{\mathrm{R}} = \bar{p}_{\mathrm{R}}$. To obtain average link capacity without transmit power adaptation, $\bar{p}_{\mathrm{R}} = 1$ is substituted in the expressions of Table 5.1. On the other hand, long-term power optimization is formulated as

$$\bar{p}_{\mathrm{R}}^* \triangleq \arg\max_{\bar{p}_{\mathrm{R}}} \bar{C}_{\mathrm{FD}} \text{ subject to } \bar{p}_{\mathrm{R}} \le 1. \tag{5.68}$$

The concave objective function clearly has a global maximum, but it can be determined only numerically. In particular, the derivative $\frac{\mathrm{d}\bar{C}_{\mathrm{FD}}}{\mathrm{d}\bar{p}_{\mathrm{R}}}$ admits a closed-form expression and has exactly one root for $\bar{p}_{\mathrm{R}} \ge 0$ which can be computed with standard numerical tools to obtain $\bar{p}_{\mathrm{R}}^*$.

5.4.2.3 Hybrid relaying modes

With hybrid FD/HD relaying based on *statistical CSI*, the average link capacity becomes similar to its instantaneous counterpart from (5.60):

$$\bar{C}_{\mathrm{hybrid}} = \max\{\bar{C}_{\mathrm{FD}}, \bar{C}_{\mathrm{HD}}\} \tag{5.69}$$

for which $\bar{C}_{\mathrm{FD}}$ and $\bar{C}_{\mathrm{HD}}$ are the average link capacities derived on the previous pages and expressed in Table 5.1. The corresponding break-even residual self-interference level $\bar{\Gamma}_{\mathrm{RR}}$ can be formally defined such that

$$\bar{C}_{\mathrm{FD}} \geq \bar{C}_{\mathrm{HD}} \text{ if and only if } \bar{\gamma}_{\mathrm{RR}} \leq \bar{\gamma}_{\mathrm{RR}}. \tag{5.70}$$

Unlike hybrid FD/HD relaying based on instantaneous CSI, which is considered shortly, the present case does not admit closed-form expressions for switching conditions. Instead, a numerically evaluated lookup table should be precalculated offline after which mode switching becomes a straightforward task based on the average channel SNRs.

To calculate the average link capacity for the reference case of hybrid FD/HD relaying based on *instantaneous CSI* by averaging C_{hybrid} over the fading distributions, one should first recall how the relaying mode is chosen for each channel state as explained in Section 5.4.1.4. In DL relaying, the determining factor is γ_{RD} such that the FD mode is used whenever $\gamma_{\mathrm{RD}} < \chi^{\mathrm{DL}}_{\mathrm{hybrid}}$; the switching boundary $\chi^{\mathrm{DL}}_{\mathrm{hybrid}}$ is solved from (5.62) as given in Table 5.1 for both protocols ($\chi^{\mathrm{DL-AF}}_{\mathrm{hybrid}}$ vs. $\chi^{\mathrm{DL-DF}}_{\mathrm{hybrid}}$). Consequently, the average DL capacity can be divided into two integrals as follows:

$$\begin{aligned}\bar{C}^{\mathrm{DL}}_{\mathrm{hybrid}} &\triangleq \mathcal{E}\{\max\{C_{\mathrm{FD}}(\bar{\gamma}_{\mathrm{SR}},\gamma_{\mathrm{RD}}), C_{\mathrm{HD}}(\bar{\gamma}_{\mathrm{SR}},\gamma_{\mathrm{RD}})\}\} \\ &= \int_0^{\chi^{\mathrm{DL}}_{\mathrm{hybrid}}} C_{\mathrm{FD}}(\bar{\gamma}_{\mathrm{SR}},x) f_{\gamma_{\mathrm{RD}}}(x)\,\mathrm{d}x + \int_{\chi^{\mathrm{DL}}_{\mathrm{hybrid}}}^{\infty} C_{\mathrm{HD}}(\bar{\gamma}_{\mathrm{SR}},x) f_{\gamma_{\mathrm{RD}}}(x)\,\mathrm{d}x. \end{aligned} \tag{5.71}$$

On the contrary, γ_{SR} is the determining factor in UL relaying such that the FD mode is used if $\gamma_{\mathrm{SR}} > \chi^{\mathrm{UL}}_{\mathrm{hybrid}}$ for which the switching boundary $\chi^{\mathrm{UL}}_{\mathrm{hybrid}}$ is solved from (5.62) as shown in Table 5.1 for both protocols ($\chi^{\mathrm{UL-AF}}_{\mathrm{hybrid}}$ vs. $\chi^{\mathrm{UL-DF}}_{\mathrm{hybrid}}$). The average UL capacity $\bar{C}^{\mathrm{UL}}_{\mathrm{hybrid}}$ is also calculated similar to (5.71) by integrating over density function $f_{\gamma_{\mathrm{SR}}}(x)$ and switching mode at $x = \chi^{\mathrm{UL}}_{\mathrm{hybrid}}$. As with (5.65), the capacity expressions can be compactly solved in terms of the auxiliary integrals derived on the previous page. The first (resp. second) integral in (5.71) is translated to $\mathcal{I}^{\mathrm{head}}_{\mathrm{AF}}$ or $\mathcal{I}^{\mathrm{head}}_{\mathrm{DF}}$ (resp. $\mathcal{I}^{\mathrm{tail}}_{\mathrm{AF}}$ or $\mathcal{I}^{\mathrm{tail}}_{\mathrm{DF}}$) depending on the considered protocol (AF or DF).

The final expressions of the above calculations are again collected in Table 5.1 and illustrated with numerical results in Figures 5.11, 5.12(a), and 5.12(b). The switching boundaries χ are also reflected in what follows.

In UL direction, the RN switches frequently between the modes with all $\bar{\gamma}_{\mathrm{SR}}$ and $\bar{\gamma}_{\mathrm{RD}}$, while DL relaying is rather different in this perspective. The following properties show that the RN uses constantly (for all γ_{RD} and $\bar{\gamma}_{\mathrm{RD}}$) the HD mode if $\bar{\gamma}_{\mathrm{SR}}$ is below a certain level with the AF protocol and the FD mode if $\bar{\gamma}_{\mathrm{SR}}$ is above a certain level with both protocols. This is also visualized by Figure 5.8 as a corroboration for the proofs.

Low BS-to-RN backhaul SNR gives an advantage for the HD mode.

Property: Preferred mode with weak backhaul connection

If $\gamma_{\mathrm{SR}} < \gamma_{\mathrm{RR}} - 1$ then $C^{\mathrm{AF}}_{\mathrm{FD}} < C^{\mathrm{AF}}_{\mathrm{HD}}$ for all γ_{RD} and $C^{\mathrm{DF}}_{\mathrm{FD}} < C^{\mathrm{DF}}_{\mathrm{HD}}$ whenever $\gamma_{\mathrm{RD}} > \chi^{\mathrm{DL-DF}}_{\mathrm{hybrid}}$. Consequently, $\bar{C}^{\mathrm{DL-AF}}_{\mathrm{FD}} < \bar{C}^{\mathrm{DL-AF}}_{HD}$ for all $\bar{\gamma}_{\mathrm{RD}}$ if $\bar{\gamma}_{\mathrm{SR}} < \bar{\gamma}_{\mathrm{RR}} - 1$.

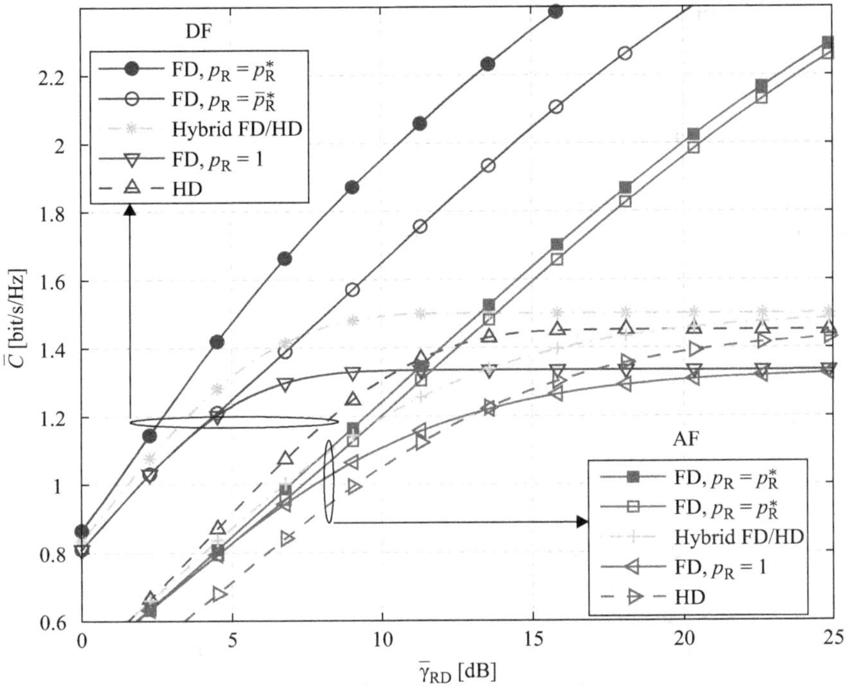

Figure 5.11 Average UL capacities with and without transmit power optimization when $\bar{\gamma}_{SR} = 10$ dB and $\bar{\gamma}_{RR} = 6$ dB. The curves labeled as "hybrid FD/HD" represent instantaneous mode switching under maximum transmit power usage ($p_R = 1$), while the respective curves with transmit power optimization ($p_R = p_R^*$) overlap with the topmost curves labeled as "FD, $p_R = p_R^*$"

The above property can be proven by substituting (5.51)–(5.53) in (5.47a) and (5.47b) with $p_S = p_R = 1$ and $\gamma_{SD} = 0$, after which C_{FD}^{AF}/C_{HD}^{AF} is formulated and discovered to be a monotonically decreasing function in terms of γ_{RD}. Then $C_{FD}^{AF}/C_{HD}^{AF} \leq \lim_{\gamma_{RD}\to 0} C_{FD}^{AF}/C_{HD}^{AF} = 2(\gamma_{SR} + 1)/(\gamma_{SR} + \gamma_{RR} + 1) < 1$ as a limiting case; this condition can be transformed to $\gamma_{SR} < \gamma_{RR} - 1$ concluding the proof for the instantaneous case. The average case follows from the fact that the instantaneous case holds for all γ_{RD}, i.e., for all fading states of the RN-to-UE service channel in DL direction.

Conversely, high BS-to-RN backhaul SNR gives an advantage for the FD mode.

Property: Preferred mode with strong backhaul connection

If $\gamma_{SR} > \gamma_{RR}^2 - 1$ then $C_{FD}^{AF} > C_{HD}^{AF}$ and $C_{FD}^{DF} > C_{HD}^{DF}$ for all γ_{RD}. Consequently, $\bar{C}_{FD}^{DL-AF} > \bar{C}_{HD}^{DL-AF}$ and $\bar{C}_{FD}^{DL-DF} > \bar{C}_{HD}^{DL-DF}$ for all $\bar{\gamma}_{RD}$ if $\bar{\gamma}_{SR} > \bar{\gamma}_{RR}^2 - 1$.

With (5.47)–(5.53), $p_S = p_R = 1$ and $\gamma_{SD} = 0$, one can show that C_{FD}^{DF}/C_{HD}^{DF} is a monotonically decreasing function in terms of γ_{RD} in addition to C_{FD}^{AF}/C_{HD}^{AF}

(which is already dealt with above). With both protocols, the condition from a limiting case, namely $C_{\mathrm{FD}}/C_{\mathrm{HD}} \geq \lim_{\gamma_{\mathrm{RD}}\to\infty} C_{\mathrm{FD}}/C_{\mathrm{HD}} > 1$ with $\lim_{\gamma_{\mathrm{RD}}\to\infty} C_{\mathrm{FD}} = \log_2(1+\gamma_{\mathrm{SR}}/(\gamma_{\mathrm{RR}}+1))$ and $\lim_{\gamma_{\mathrm{RD}}\to\infty} C_{\mathrm{HD}} = \frac{1}{2}\log_2(1+\gamma_{\mathrm{SR}})$, can be transformed to $\gamma_{\mathrm{SR}} > \gamma_{\mathrm{RR}}^2 - 1$. The instantaneous case holds for any γ_{RD} which proves the average case by rendering fading in the DL RN-to-UE service channel ineffective whenever $\bar{\gamma}_{\mathrm{SR}} > \bar{\gamma}_{\mathrm{RR}}^2 - 1$.

Furthermore, the fact that always $\gamma_{\mathrm{SR}} \geq 0$ by definition (since it is a power ratio) proves the following consequence of the latter property which holds true in both DL and UL relaying.

Property: Preferred mode with weak residual self-interference

If $\gamma_{\mathrm{RR}} < 1$ then $C_{\mathrm{FD}}^{\mathrm{AF}} > C_{\mathrm{HD}}^{\mathrm{AF}}$ and $C_{\mathrm{FD}}^{\mathrm{DF}} > C_{\mathrm{HD}}^{\mathrm{DF}}$ for all γ_{SR} and γ_{RD}. Consequently, $\bar{C}_{\mathrm{FD}}^{\mathrm{AF}} > \bar{C}_{\mathrm{HD}}^{\mathrm{AF}}$ and $\bar{C}_{\mathrm{FD}}^{\mathrm{DF}} > \bar{C}_{\mathrm{HD}}^{\mathrm{DF}}$ for all $\bar{\gamma}_{\mathrm{SR}}$ and $\bar{\gamma}_{\mathrm{RD}}$ if $\bar{\gamma}_{\mathrm{RR}} < 1$.

In plain words, the FD mode is always superior to the HD mode if the power of the residual interference can be suppressed below the noise level ($\bar{\gamma}_{\mathrm{RR}} < 0$ dB). This condition could be adopted as an ultimate target for self-interference mitigation although it is difficult to achieve in reality.

Average UL capacity is illustrated in Figure 5.11 after implementing the analytical expressions derived in (5.65) for conventional relaying modes and in (5.71) for hybrid FD/HD relaying as they are summarized in Table 5.1. This again verifies that hybrid mode switching gives significant performance improvement w.r.t. adhering to either conventional mode especially when transmit power adaptation is not used. With hybrid FD/HD relaying, exploiting instantaneous CSI gives the largest performance improvement w.r.t. exploiting only statistical CSI when the UE-to-RN service channel is approximately balanced with the RN-to-BS backhaul channel.

Combining hybrid FD/HD mode switching with either instantaneous or long-term transmit power adaptation gives the largest performance improvement when the backhaul channel is relatively of better quality. This is because the RN may then set transmit power lower to significantly reduce self-interference which makes FD the most frequently chosen mode. In this scenario, instantaneous transmit power adaptation actually renders approximately equal performance for hybrid FD/HD mode switching and static FD relaying due to the very same reason. Furthermore, the DF protocol benefits more from instantaneous transmit power adaptation than the AF protocol. Similar observations can be made also in DL relaying although absolute performance is different in that case.

Finally, Figure 5.12(a) illustrates the benefit of hybrid FD/HD relaying with statistical CSI w.r.t. conventional FD and HD relaying while similar comparison is shown in Figure 5.12(b) for the case of optimal long-term power adaptation; these numerical results are again computed using the expressions of Table 5.1. In particular, switching to proper relaying mode brings up to 50% improvement in link capacity depending on the average channel SNRs. The switching boundaries are smoothed in comparison to those shown in Figures 5.8 and 5.9 for the respective instantaneous case.

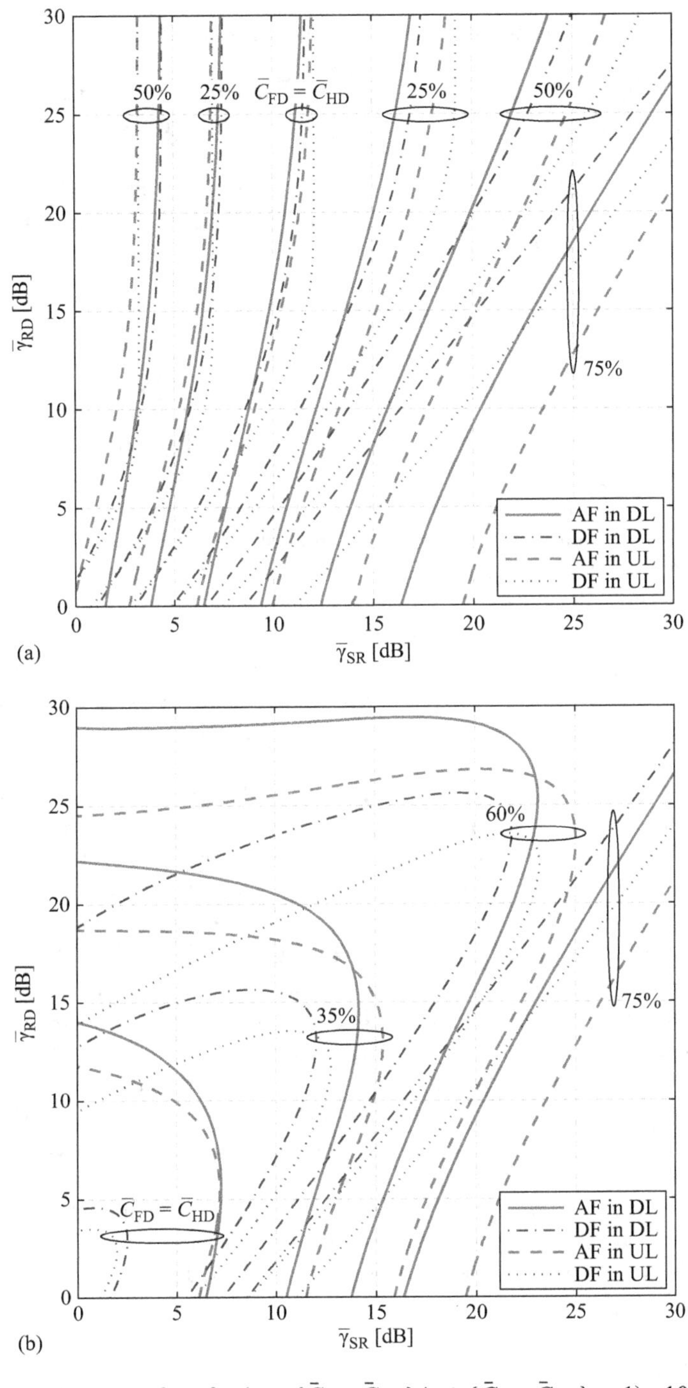

Figure 5.12 Contour plots for $(max\{\bar{C}_{FD}, \bar{C}_{HD}\}/min\{\bar{C}_{FD}, \bar{C}_{HD}\} - 1) \cdot 100\%$ *when* $\bar{\gamma}_{RR} = 6\,dB$*. The FD mode is preferred over the HD mode outside and on the right-hand side of the switching boundaries labeled as "*$\bar{C}_{FD} = \bar{C}_{HD}$*". (a) When using maximum transmit power in the relay (*$\bar{p}_R = 1$*) and (b) when using optimized transmit power in the relay (*$\bar{p}_R = \bar{p}_R^*$*)*

The benefit of long-term transmit power adaptation can be seen by comparing Figure 5.12(a) to 5.12(b) in which the relay transmit power is chosen according to (5.68). Under transmit power adaptation, it is reasonable to choose the HD mode only when both the backhaul and service channels have very low SNR, and otherwise switching to the FD mode improves significantly link capacity. These result plots also reflect the differences of AF and DF protocols as well as DL and UL transmission directions.

5.5 Conclusions

The first part of this chapter studied the effect of residual loopback self-interference in single-frequency FD relay links. Both AF and DF relaying were considered within a unified analytical framework. The study was motivated by the reasonable presumption that, even with efficient self-interference mitigation, performance loss due to residual distortion cannot be completely avoided. Transmit power adaptation discussed herein offers a win-win solution for still alleviating further the remaining problem: With smart relay gain selection, less transmit power is consumed while the end-to-end link quality is also improved. Especially, the chapter presented two reference gain control schemes and two enhanced ones that improve effective signal-to-interference-and-noise ratios. Finally, the performance of the schemes was evaluated also in terms of outage probability.

The second part of this chapter studied the fundamental rate–interference trade-off arising from the choice of using either a FD mode or a HD mode in a relay link. For this purpose, it is essential to model explicitly the effect of residual self-interference which remains inevitably in the FD mode after mitigation and transmit power control. The analysis covered both common relaying protocols, namely AF and DF processing, in both DL and UL directions. Contrary to earlier literature which typically adheres to either mode at an early design stage, the study suggested opportunistic switching between FD and HD modes as a new scheme to optimize relay links' short-term and long-term capacity. The performance was further improved by transmit power adaptation in the spirit of the first part; especially, its implementation was also reflected in the design of opportunistic mode switching. The analysis was conducted by deriving closed-form expressions for link capacity as well as for break-even interference levels and switching boundaries which determine the choice of one mode over the other based on CSI. The respective numerical results illustrated that FD and hybrid FD/HD relaying are indeed attractive techniques for infrastructure-based relay links in which the self-interference can be mitigated efficiently.

References

[1] M. Abramowitz and I. A. Stegun, editors. *Handbook of Mathematical Functions with Formulas, Graphs, and Mathematical Tables*. Dover Publications, New York, United States, 1972.

[2] K. Azarian, H. El Gamal, and P. Schniter. On the achievable diversity-multiplexing tradeoff in half-duplex cooperative channels. *IEEE Transactions on Information Theory*, 51(12):4152–4172, December 2005.

[3] T. M. Cover and A. A. El Gamal. Capacity theorems for the relay channel. *IEEE Transactions on Information Theory*, 25(5):572–584, September 1979.

[4] Z. Ding, I. Krikidis, B. Rong, J. S. Thompson, C. Wang, and S. Yang. On combating the half-duplex constraint in modern cooperative networks: Protocols and techniques. *IEEE Wireless Communications*, 19(6):20–27, December 2012.

[5] Y. Fan, C. Wang, J. S. Thompson, and H. V. Poor. Recovering multiplexing loss through successive relaying using repetition coding. *IEEE Transactions on Wireless Communications*, 6(12):4484–4493, December 2007.

[6] A. Høst-Madsen and J. Zhang. Capacity bounds and power allocation for wireless relay channels. *IEEE Transactions on Information Theory*, 51(6):2020–2040, June 2005.

[7] A. Ikhlef, J. Kim, and R. Schober. Mimicking full-duplex relaying using half-duplex relays with buffers. *IEEE Transactions on Vehicular Technology*, 61(7):3025–3037, September 2012.

[8] M. Janani, A. Hedayat, T. E. Hunter, and A. Nosratinia. Coded cooperation in wireless communications: Space–time transmission and iterative decoding. *IEEE Transactions on Signal Processing*, 52(2):362–371, February 2004.

[9] L. Jiménez Rodríguez, N. H. Tran, and T. Le-Ngoc. Optimal power allocation and capacity of full-duplex AF relaying under residual self-interference. *IEEE Wireless Communications Letters*, 3(2):233–236, April 2014.

[10] G. Kramer, M. Gastpar, and P. Gupta. Cooperative strategies and capacity theorems for relay networks. *IEEE Transactions on Information Theory*, 51(9):3037–3063, September 2005.

[11] I. Krikidis and H. A. Suraweera. Full-duplex cooperative diversity with Alamouti space–time code. *IEEE Wireless Communications Letters*, 2(5):519–522, October 2013.

[12] I. Krikidis, H. A. Suraweera, P. J. Smith, and C. Yuen. Full-duplex relay selection for amplify-and-forward cooperative networks. *IEEE Transactions on Wireless Communications*, 11(12):4381–4393, December 2012.

[13] I. Krikidis, H. A. Suraweera, S. Yang, and K. Berberidis. Full-duplex relaying over block fading channel: A diversity perspective. *IEEE Transactions on Wireless Communications*, 11(12):4524–4535, December 2012.

[14] J. N. Laneman, D. N. C. Tse, and G. W. Wornell. Cooperative diversity in wireless networks: Efficient protocols and outage behavior. *IEEE Transactions on Information Theory*, 50(12):3062–3080, December 2004.

[15] J. N. Laneman and G. W. Wornell. Distributed space–time-coded protocols for exploiting cooperative diversity in wireless networks. *IEEE Transactions on Information Theory*, 49(10):2415–2425, October 2003.

[16] D. S. Michalopoulos and G. K. Karagiannidis. Bypassing orthogonal relaying transmissions via spatial signal separation. *IEEE Transactions on Communications*, 58(10):3028–3038, October 2010.

[17] R. U. Nabar, H. Bölcskei, and F. W. Kneubühler. Fading relay channels: Performance limits and space–time signal design. *IEEE Journal on Selected Areas in Communications*, 22(6):1099–1109, August 2004.

[18] T. J. Oechtering, R. F. Wyrembelski, and H. Boche. Multiantenna bidirectional broadcast channels – Optimal transmit strategies. *IEEE Transactions on Signal Processing*, 57(5):1948–1958, May 2009.

[19] B. Rankov and A. Wittneben. Spectral efficient protocols for half-duplex fading relay channels. *IEEE Journal on Selected Areas in Communications*, 25(2):379–389, February 2007.

[20] T. Riihonen, K. Haneda, S. Werner, and R. Wichman. SINR analysis of full-duplex OFDM repeaters. In *Proc. 20th IEEE International Symposium on Personal, Indoor and Mobile Radio Communications*, September 2009.

[21] T. Riihonen, S. Werner, and R. Wichman. Comparison of full-duplex and half-duplex modes with a fixed amplify-and-forward relay. In *Proc. IEEE Wireless Communications and Networking Conference*, April 2009.

[22] T. Riihonen, S. Werner, and R. Wichman. Optimized gain control for single-frequency relaying with loop interference. *IEEE Transactions on Wireless Communications*, 8(6):2801–2806, June 2009.

[23] T. Riihonen, S. Werner, and R. Wichman. Spatial loop interference suppression in full-duplex MIMO relays. In *Proc. 43rd Annual Asilomar Conference on Signals, Systems, and Computers*, November 2009.

[24] T. Riihonen, S. Werner, and R. Wichman. Rate–interference trade-off between duplex modes in decode-and-forward relaying. In *Proc. 21st IEEE International Symposium on Personal, Indoor and Mobile Radio Communications*, September 2010.

[25] T. Riihonen, S. Werner, and R. Wichman. Hybrid full-duplex/half-duplex relaying with transmit power adaptation. *IEEE Transactions on Wireless Communications*, 10(9):3074–3085, September 2011.

[26] T. Riihonen, S. Werner, and R. Wichman. Mitigation of loopback self-interference in full-duplex MIMO relays. *IEEE Transactions on Signal Processing*, 59(12):5983–5993, December 2011.

[27] T. Riihonen, S. Werner, and R. Wichman. Transmit power optimization for multiantenna decode-and-forward relays with loopback self-interference from full-duplex operation. In *Proc. 45th Annual Asilomar Conference on Signals, Systems, and Computers*, November 2011.

[28] T. Riihonen, S. Werner, R. Wichman, and J. Hämäläinen. Outage probabilities in infrastructure-based single-frequency relay links. In *Proc. IEEE Wireless Communications and Networking Conference*, April 2009.

[29] T. Riihonen, S. Werner, R. Wichman, and E. Zacarías B. On the feasibility of full-duplex relaying in the presence of loop interference. In *Proc. 10th IEEE Workshop on Signal Processing Advances in Wireless Communications*, June 2009.

[30] A. Sendonaris, E. Erkip, and B. Aazhang. User cooperation diversity – Part I: System description. *IEEE Transactions on Communications*, 51(11):1927–1938, November 2003.

[31] A. Sendonaris, E. Erkip, and B. Aazhang. User cooperation diversity – Part II: Implementation aspects and performance analysis. *IEEE Transactions on Communications*, 51(11):1939–1948, November 2003.

[32] H. A. Suraweera, I. Krikidis, G. Zheng, C. Yuen, and P. J. Smith. Low-complexity end-to-end performance optimization in MIMO full-duplex relay systems. *IEEE Transactions on Wireless Communications*, 13(2):913–927, February 2014.

[33] E. C. Van Der Meulen. Three-terminal communication channels. *Advances in Applied Probability*, 3:120–154, 1971.

[34] B. Wang, J. Zhang, and A. Høst-Madsen. On the capacity of MIMO relay channels. *IEEE Transactions on Information Theory*, 51(1):29–43, January 2005.

[35] G. Zheng. Joint beamforming optimization and power control for full-duplex MIMO two-way relay channel. *IEEE Transactions on Signal Processing*, 63(3):555–566, February 2015.

[36] C. Zhong, H. A. Suraweera, G. Zheng, I. Krikidis, and Z. Zhang. Wireless information and power transfer with full duplex relaying. *IEEE Transactions on Communications*, 62(10):3447–3461, October 2014.

Chapter 6
Relay selection in modern communication systems

Constantinos Psomas, Christodoulos Skouroumounis and Ioannis Krikidis

6.1 Introduction

In cooperative networks, multiple relay nodes are employed in order to assist the source transmission through multi-hop relaying or cooperative diversity. To this end, an efficient cooperative technique which is adopted in such networks is relay selection. Specifically, the relay selection technique chooses a single relay or a group of relays to participate in the cooperation process in order to improve the network's performance. The chosen relay nodes are considered to be the best nodes according to some well-defined criteria. This technique has been proven to use the system's resources such as power and bandwidth more efficiently without affecting the system's performance. Its main advantage is the simplicity of its implementation, as it does not require complex physical (PHY) layer transmission techniques, such as distributed space–time codes or explicit synchronization processes, and it is therefore known for the high performance that it provides. Since the seminal work of Bletsas *et al.* [1] where the fundamental benefits from relay selection have been introduced, many papers have appeared on this subject and many relay selection policies have been proposed for different cooperative contexts (see, e.g., References 2–4 and the references therein). This chapter presents relay selection policies under different communication scenarios and applications. Specifically, the chapter focuses on the benefits and efficiency of relay selection in the context of full-duplex (FD) enabled relays (Section 6.2), relays with buffers (Section 6.3) and relays with wireless power transfer (WPT) capabilities (Section 6.4). These scenarios show that relay selection is an efficient PHY layer tool for the design of modern communication networks.

6.2 FD relay selection

FD relaying refers to the ability of a relay node to receive and transmit data in the same frequency band and in the same time slot. This technique uses the channel bandwidth efficiently, as it requires only one channel use for the end-to-end transmission. However, it suffers from loop interference due to the signal leakage between the relay's output and input [5]. Applying the relay selection concept to FD cooperative systems

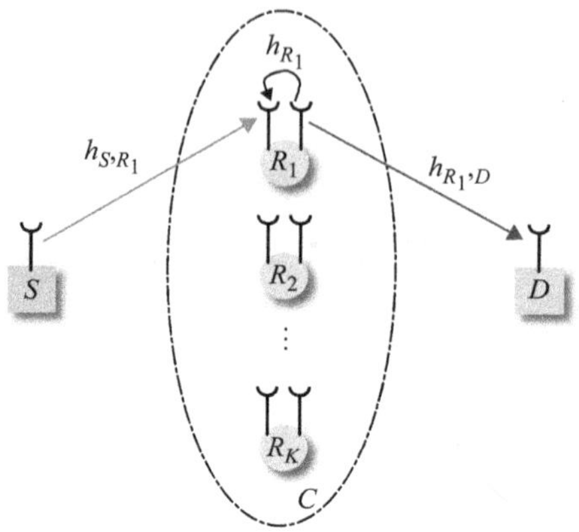

Figure 6.1 System model: FD relaying

provides an efficient approach to combine space diversity benefits with FD spectral efficiency. This section focuses on the relay selection problem in amplify-and-forward (AF) cooperative communication networks with FD capabilities. Several relay selection policies are presented: an optimal relay selection policy under global channel state information (CSI) and numerous suboptimal relay selection policies that incorporate partial CSI such as (a) the source–relay and relay–destination links, (b) the instantaneous loop interference and (c) the source–relay links and the instantaneous loop interference.

6.2.1 System model

A clustered network topology is considered consisting of one source S, one destination D and a cluster $\mathcal{C}$ of K relays R_i, $1 \leq i \leq K$. Figure 6.1 schematically presents the considered system topology. The direct link between S and D is assumed to be strongly attenuated, and communication can be established only via the cooperative relays. This typical assumption corresponds to coverage extension scenarios where the distance between S and D is long, and relays are used in order to maintain connectivity [5, 6] or scenarios where the direct link is in deep shadowing due to surrounding physical obstacles [1, 7]. Each relay employs an AF protocol and is equipped with two antennas (one receive antenna and one transmit antenna) that enable an FD operation at the price of loop interference. Time is considered to be slotted, and in each time slot, only one relay is selected to assist the source transmission. Power allocation issues are not considered, and thus each transmitter (source or relay) is assumed to transmit with a fixed power P. In addition, the information rate is equal to r_0 bits per channel use (BPCU).

All wireless links in the network exhibit non-selective Rayleigh block fading and additive white Gaussian noise (AWGN). In order to simplify the analysis, a partially symmetric clustered structure is assumed where the fading channel coefficients $h_{A,B}$ (for the $A \rightarrow B$ link) remain constant during one slot, but change independently from one slot to another according to a circularly symmetric complex Gaussian distribution with zero mean and variance σ_{SR}^2 and σ_{RD}^2 for $S \rightarrow R_i$ and $R_i \rightarrow D$ links, respectively, where $i = 1, \ldots, K$. Furthermore, in order to reduce the effects of the loop interference on system performance, an imperfect interference cancellation scheme is used at each relay by following the analysis presented in Reference 5. Assume that h_{R_i} denotes the instantaneous residual loop interference channel at the ith relay, which follows a complex Gaussian distribution with zero mean and variance σ_{RR}^2, a common model for the residual loop interference [6].

As for the amplification process at the relay node, a conventional relay gain is assumed that normalizes the instantaneous received power and allows the relay to transmit with a maximum power P [8, Eq. (11)]. In this case, the instantaneous end-to-end capacity of the ith relay is expressed as [5, Eq. (1)]

$$C_{R_i}^{(\mathrm{FD})} = \log_2\left(1 + \frac{\dfrac{\gamma_{S,R_i}}{\gamma_{R_i}+1}\gamma_{R_i,D}}{\dfrac{\gamma_{S,R_i}}{\gamma_{R_i}+1} + \gamma_{R_i,D} + 1}\right), \tag{6.1}$$

where $\gamma_{S,R_i} \triangleq P|h_{S,R_i}|^2$, $\gamma_{R_i,D} \triangleq P|h_{R_i,D}|^2$ and $\gamma_{R_i} \triangleq P|h_{R_i}|^2$ denote the instantaneous signal-to-noise ratios (SNRs) for the $S \rightarrow R_i$, $R_i \rightarrow D$ and $R_i \rightarrow R_i$ (loop interference) links, respectively. Finally, in order to simplify notation, define $\lambda_{AB} \triangleq 1/(P\sigma_{AB}^2)$.

In what follows, the details of different relay selection policies are presented based on the above system model together with the outage probability of each policy. The outage probability can be written as

$$P_* = \mathbb{P}\left\{C_{R_{k*}}^{(\mathrm{FD})} < r_0\right\}, \tag{6.2}$$

where $*$ refers to a relay selection policy. Note that the statistical distributions of $\gamma_{S,R_{k*}}$, $\gamma_{R_{k*},D}$ and $\gamma_{R_{k*}}$ may differ depending on the selection policy. Therefore, any remark concerning the distributions of these random variables is *strictly limited* to the particular selection policy.

6.2.2 Optimal relay selection

The optimal relay selection policy (OS) is based on the capacity expression achieved by the FD transmission. Therefore, based on (6.1), the OS selects the relay k_{OS} where

$$k_{\mathrm{OS}} = \arg\max_i \{\gamma_i\}, \; \gamma_i \triangleq \frac{\dfrac{\gamma_{S,R_i}}{\gamma_{R_i}+1}\gamma_{R_i,D}}{\dfrac{\gamma_{S,R_i}}{\gamma_{R_i}+1} + \gamma_{R_i,D} + 1}. \tag{6.3}$$

Using a simple order statistic result, the outage probability of the OS policy can be written as

$$P_{\mathrm{OS}} = \mathbb{P}\left\{C_{R_{k_{\mathrm{OS}}}}^{(\mathrm{FD})} < r_0\right\} = \left[F_{\gamma_{\mathrm{eq}}}(T)\right]^K, \tag{6.4}$$

where $T \triangleq 2^{r_0} - 1$ and $F_{\gamma_{\mathrm{eq}}}(\cdot)$ is the cumulative distribution function (cdf) of the random variables γ_i and is given by [9]

$$F_{\gamma_{\mathrm{eq}}}(x) = 1 - \lambda_{RD} e^{-\lambda_{RD}x} \int_0^\infty \frac{e^{-\frac{\lambda_{SR}(x+y+1)x}{y} - \lambda_{RD}y}}{1 + \dfrac{\lambda_{SR}(x+y+1)x}{\lambda_{RR}y}} dy. \tag{6.5}$$

As (6.5) does not have a closed-form solution, it can be evaluated numerically using standard software such as Matlab, Maple or Mathematica. To derive an accurate closed-form expression for the outage probability performance, consider the intermediate to high SNR regime, where the outage probability is simplified to

$$P_{\mathrm{OS}} \to \left(\frac{1}{1 + \frac{\lambda_{RR}}{\lambda_{SR} T}} \right)^{K} \quad (\text{for } P \to \infty). \tag{6.6}$$

To derive the above tight lower bound, the inequality where the harmonic mean of two positive numbers can be upper bounded by the minimum of those two numbers is used [10] as well as basic order statistics theory. Given that $\lambda_{RR}/\lambda_{SR}$ remains constant in terms of P, it is obvious from (6.6) that the outage performance of the system asymptotically converges to an error floor and exhibits a zero diversity gain.

6.2.3 Max–min relay selection

The max–min relay selection policy (MM) is the optimal relay selection policy for conventional half-duplex (HD) relaying systems [1]. Here, the performance of this scheme with FD relaying is provided. The max–min relay selection scheme does not consider the loop interference and selects the relay k_{MM} with the best end-to-end link where

$$k_{\mathrm{MM}} = \arg\max_{i} \min \left\{ \gamma_{S,R_i}, \gamma_{R_i,D} \right\}. \tag{6.7}$$

In this case, the outage probability can be written as [9]

$$P_{\mathrm{MM}} = 1 - \int_0^{\infty} \widetilde{F}_{X_2}\left(\frac{(y + T + 1)T}{y} \right) f_{\gamma_{R_{k_{\mathrm{MM}}},D}}(y + T) dy, \tag{6.8}$$

where $\widetilde{F}_X(\cdot) = 1 - F_X(\cdot)$ is the complementary cdf and $X_2 = \frac{\gamma_{S,R_{k_{\mathrm{MM}}}}}{\gamma_{R_{k_{\mathrm{MM}}}}+1}$. The cdf of X_2 is given by

$$\begin{aligned} F_{X_2}(x) = \frac{K}{\lambda_{SR} + \lambda_{RD}} \Bigg[& \sum_{n=0}^{K-1} \frac{(-1)^n \binom{K-1}{n}}{\frac{n}{\lambda_{RD}} + \frac{1}{\lambda_{SR} + \lambda_{RD}}} \left(1 - \frac{e^{-\lambda_{SR} x}}{1 + \frac{\lambda_{SR}}{\lambda_{RR}} x} \right) \\ & - \frac{\lambda_{SR}}{\lambda_{SR} + \lambda_{RD}} \sum_{n=0}^{K-1} \frac{(-1)^n \binom{K-1}{n} \left(1 - \frac{e^{-(n+1)(\lambda_{SR}+\lambda_{RD})x}}{1 + \frac{(n+1)(\lambda_{SR}+\lambda_{RD})x}{\lambda_{RR}}} \right)}{\left(\frac{n}{\lambda_{RD}} + \frac{1}{\lambda_{SR} + \lambda_{RD}} \right)(n+1)} \\ & + \lambda_{SR} \sum_{n=0}^{K-1} \frac{(-1)^n \binom{K-1}{n} \left(1 - \frac{e^{-(n+1)(\lambda_{SR}+\lambda_{RD})x}}{1 + \frac{(n+1)(\lambda_{SR}+\lambda_{RD})x}{\lambda_{RR}}} \right)}{(n+1)} \Bigg]. \end{aligned} \tag{6.9}$$

The probability density function of $\gamma_{R_{k_{\mathrm{MM}}},D}$ is given by

$$f_{\gamma_{R_{k_{\mathrm{MM}}},D}}(x) = K\sum_{n=0}^{K-1}\binom{K-1}{n}\left(\frac{\lambda_{RD}\,(-1)^n\,e^{-\lambda_{RD}x}}{1+\dfrac{n(\lambda_{SR}+\lambda_{RD})}{\lambda_{SR}}}\right.$$

$$\left.+\frac{\lambda_{RD}(-1)^{n+1}e^{-(n+1)(\lambda_{SR}+\lambda_{RD})x}}{1+\dfrac{n(\lambda_{SR}+\lambda_{RD})}{\lambda_{SR}}}+(-1)^n\lambda_{RD}e^{-(n+1)(\lambda_{SR}+\lambda_{RD})x}\right). \quad (6.10)$$

Finally, recalling that $\widetilde{F}_{X_2}(x)=1-F_{X_2}(x)$ and substituting (6.9) and (6.10) into (6.8), the exact outage probability can be evaluated using numerical integration. In order to simplify the analysis and derive a closed-form expression for the error floor, accurate for low loop interference scenarios, the above outage expression can be asymptotically approximated for high SNRs by the first (dominant) term as

$$P_{\mathrm{MM}} \to \frac{\lambda_{SR}}{\lambda_{SR}+\lambda_{RD}}\sum_{n=0}^{K}\binom{K}{n}\frac{(-1)^n\lambda_{RR}}{\lambda_{RR}+nT(\lambda_{SR}+\lambda_{RD})} \quad (\text{for } P\to\infty). \quad (6.11)$$

6.2.4 Loop interference relay selection

The loop interference relay selection policy (LI) selects the relay with the minimum instantaneous loop interference for FD operation and selects the relay k_{LI} where

$$k_{\mathrm{LI}} = \arg\min_i\left\{\gamma_{R_i}\right\}. \quad (6.12)$$

Note that γ_{R_i} models the residual interference after the interference is cancelled at the relay node and therefore cannot be further mitigated but it can be estimated [6]. The selection of the relay is based only on γ_{R_i}, and $\gamma_{S,R_{k_{\mathrm{LI}}}}$ and $\gamma_{R_{k_{\mathrm{LI}}},D}$ are exponentially distributed random variables. Therefore, the exact outage probability can be written as [9]

$$P_{\mathrm{LI}} = \mathbb{P}\left\{C^{(\mathrm{FD})}_{R_{k_{\mathrm{LI}}}} < r_0\right\} 1-\lambda_{RD}Ke^{-\lambda_{RD}T}\int_0^\infty \frac{e^{-\frac{\lambda_{SR}(y+T+1)T}{y}-\lambda_{RD}y}}{K+\dfrac{\lambda_{SR}(y+T+1)T}{\lambda_{RR}y}}dy. \quad (6.13)$$

By using similar simplifications to those used in Sections 6.2.2 and 6.2.3, a tight lower bound for the outage probability at high SNRs can be derived as

$$P_{\mathrm{LI}} \to 1-\frac{\lambda_{RR}K}{\lambda_{RR}K+\lambda_{SR}T} \quad (\text{for } P\to\infty). \quad (6.14)$$

6.2.5 Partial relay selection

For the conventional AF operation [6, Eq. (11)], it is assumed that the relay knows both γ_{S,R_i} and γ_{R_i}. The proposed partial relay selection policy (PS) assumes that the

SNR of the first hop as well as the loop interference is known and selects the relay based on this information. More specifically, the PS policy activates the relay that has the maximum ratio between the signal-to-interference-plus-noise ratio (SINR) of the $S \rightarrow R_i$ link and the loop interference. Hence, the relay k_{PS} is selected where

$$k_{\mathrm{PS}} = \arg\max_i \left\{ \frac{\gamma_{S,R_i}}{\gamma_{R_i}+1} \right\} \approx \arg\max_i \left\{ \frac{\gamma_{S,R_i}}{\gamma_{R_i}} \right\} \quad (\text{for } P \rightarrow \infty). \tag{6.15}$$

In the case of the PS policy, the outage probability $\mathbb{P}\left\{ C_{R_{k_{PS}}}^{(FD)} < r_0 \right\}$ is evaluated by [9]

$$P_{\mathrm{PS}} = 1 - e^{-\lambda_{RD}T} + \lambda_{RD}e^{-\lambda_{RD}T} \int_0^{\infty} \left[1 - \frac{e^{-\frac{\lambda_{SR}(y+1+T)T}{y}}}{1 + \frac{\lambda_{SR}(y+1+T)T}{\lambda_{RR}y}} \right]^K e^{-\lambda_{RD}y}dy. \tag{6.16}$$

By using basic order statistics theory and by following the previous simplifications used in Sections 6.2.2 and 6.2.3, the outage probability for the high SNR regime can be approximated by

$$P_{\mathrm{PS}} \rightarrow \left[\frac{\lambda_{SR}T}{\lambda_{SR}T + \lambda_{RR}} \right]^K \quad (\text{for } P \rightarrow \infty). \tag{6.17}$$

6.2.6 Max–min with loop interference relay selection

The max–min with loop interference relay selection (MMLI) is an improvement of the conventional MM scheme and takes into account the loop interference that affects the relay input. The selection metric is similar to the one used by the MM policy given by (6.7) but updates the first branch with the SINR and thus selects the relay k_{MMLI} where

$$\begin{aligned} k_{\mathrm{MMLI}} &= \arg\max_i \min \left\{ \frac{\gamma_{S,R_i}}{\gamma_{R_i}+1}, \gamma_{R_i,D} \right\} \\ &\approx \arg\max_i \min \left\{ \frac{\gamma_{S,R_i}}{\gamma_{R_i}}, \gamma_{R_i,D} \right\} \quad (\text{for } P \rightarrow \infty). \end{aligned} \tag{6.18}$$

The MMLI policy has a limited practical interest as it requires a full CSI and therefore refers to scenarios where the OS policy can be applied. However, MMLI is used as a reference scheme for comparison purposes. It is worth noting that at high SNRs, the MMLI policy becomes equivalent to the OS scheme and thus converges to the same error floor. This is expected since the selection metric of the MMLI policy is identical with the tight lower bound used for the high SNR regime and therefore it becomes optimal at high SNRs. Therefore, the corresponding outage probability for the high SNR regime can be written as

$$P_{\mathrm{MMLI}} \rightarrow \left(\frac{1}{1 + \frac{\lambda_{RR}}{\lambda_{SR}T}} \right)^K \quad (\text{for } P \rightarrow \infty). \tag{6.19}$$

6.2.7 Optimal relay selection with hybrid relaying

The aforementioned selection schemes suffer from a zero diversity gain due to the nature of the FD transmission and the related loop interference effect. In order to overcome this limitation, an optimal relay selection policy has been proposed based on a dynamic hybrid relaying scheme (HS) that switches between FD and HD relaying. The proposed scheme allows the relay nodes to switch between FD and HD relaying based mainly on the *instantaneous* residual loop interference. The analytical results confirm that for scenarios with a low loop interference, an FD policy is more suitable as it does not suffer from bandwidth-loss effects, while for scenarios with a high interference, an HD policy is more suitable as it eliminates the interference. Given that each relay defines its own (the most efficient) relaying policy, the HS scheme selects the relay that achieves the best performance in terms of outage probability. If the instantaneous capacity of HD relaying is defined as [1, Eq. (26)]

$$C_{R_i}^{(\mathrm{HD})} = \frac{1}{2}\log_2\left(1 + \frac{\gamma_{S,R_i}\gamma_{R_i,D}}{\gamma_{S,R_i} + \gamma_{R_i,D} + 1}\right), \tag{6.20}$$

the HS policy activates the relay k_{HS} which satisfies the following condition:

$$\begin{aligned} k_{\mathrm{HS}} &= \arg\max_i \max\left\{C_{R_i}^{(\mathrm{HD})}, C_{R_i}^{(\mathrm{FD})}\right\} \\ &= \arg\max_i \max\left\{\sqrt{1 + \frac{\gamma_{S,R_i}\gamma_{R_i,D}}{\gamma_{S,R_i} + \gamma_{R_i,D} + 1}}, 1 + \gamma_i\right\}. \end{aligned} \tag{6.21}$$

The above selection policy jointly selects the duplexing mode as well as the relay node that forwards the transmitted signal. Due to this joint optimization, the HS scheme achieves the best performance in comparison to the other selection schemes without suffering from zero diversity limitations. The outage performance of the HS scheme is given by [9]

$$P_{\mathrm{HS}} = \mathbb{P}\left\{C_{R_{k_{HS}}}^{(\mathrm{HD})} < r_0\right\} = [F_{\mathrm{HS}}(T+1)]^K, \tag{6.22}$$

where

$$\begin{aligned} F_{\mathrm{HS}}(x) = 1 &- \lambda_{RD}\lambda_{RR}\int_{x^2-1}^{\infty} e^{-\frac{\lambda_{SR}(x-1)(y+1)}{1+y-x} - \lambda_{RD}y}\left(\frac{1 - e^{-\left(\frac{\lambda_{SR}(x-1)(y+1)}{1+y-x}\right)\frac{xy}{1+y-x^2}}}{\lambda_{RR} + \frac{\lambda_{SR}(x-1)(y+1)}{1+y-x}}\right)dy \\ &- \lambda_{RD}\lambda_{RR}\int_{x-1}^{x^2-1}\frac{e^{-\frac{\lambda_{SR}(x-1)(y+1)}{1+y-x} - \lambda_{RD}y}}{\lambda_{RR} + \frac{\lambda_{SR}(x-1)(y+1)}{1+y-x}}dy \\ &- \lambda_{RD}\int_{x^2-1}^{\infty} e^{-\frac{x\lambda_{RR}y + \lambda_{SR}(x^2-1)(y+1)}{1+y-x^2} - \lambda_{RD}y}dy. \end{aligned} \tag{6.23}$$

The above expression requires the computation of some complicated integrals and thus is a cumbersome numerical problem. However, for high SNRs (e.g., $P \to \infty$), the above expression can be approximated as

$$P_{\text{HS}} \approx \Bigg[\lambda_{SR}((T+1)^2-1)+\lambda_{RD}((T+1)^2-1) \times \left(1-\frac{\lambda_{RR}}{\lambda_{RR}+\lambda_{SR}T}\right)+\lambda_{RD}T\frac{\lambda_{RR}}{\lambda_{RR}+\lambda_{SR}T}\Bigg]^K \rightarrow O(1/P^K). \tag{6.24}$$

The above expression shows that the HS scheme offers a diversity gain equal to the number of the available relays K with a corresponding coding gain $G_c^{(\text{HS})}$ equal to

$$\begin{aligned} G_c^{(\text{HS})} &= \left[(2^{4R_0}-1)\left(\frac{1}{\sigma_{SR}^2}+\frac{1}{\sigma_{RD}^2}\right)-(2^{4R_0}-1)(2^{4R_0}+2)\frac{1}{1+(2^{4R_0}-1)\frac{\sigma_{RR}^2}{\sigma_{SR}^2}}\frac{1}{\sigma_{RD}^2}\right]^K \\ &\leq G_c^{(\text{HD})} = \left[(2^{4R_0}-1)\left(\frac{1}{\sigma_{SR}^2}+\frac{1}{\sigma_{RD}^2}\right)\right]^K, \end{aligned} \tag{6.25}$$

where $G_c^{(\text{HD})}$ is the coding gain for the conventional HD relaying with max–min selection [1].

6.2.8 Numerical results

In this section, numerical examples are given for the outage probability performance of the relay selection schemes presented above. The simulation system uses the parameters $r_0 = 2$ BPCU and $\sigma_{SR}^2 = \sigma_{RD}^2 = 1$, and the considered relay selection policies are: OS, MM, PS, MMLI and HS. The conventional max–min relay selection for HD relaying is also used as a reference selection scheme [1].

Figure 6.2 plots the outage probability as a function of the transmitted power P for the different relay selection schemes with $K = 4$ relays and $\sigma_{RR}^2 = 0.08$. The exact outage probability curves are given by Monte Carlo simulations and perfectly match with the expressions that are obtained by numerically evaluating (6.4), (6.8), (6.13), (6.16) and (6.22). The numerical evaluation of the theoretical expressions corresponds to the solid lines. The first important observation is that all the FD-based relay selection policies converge to an error floor and thus provide a zero diversity gain. In addition, the OS and MMLI policies are almost equivalent and provide the best outage performance, among the FD-based schemes, for all SNRs and converge to the lowest error floor. As for the PS relay selection scheme, it provides the second best outage performance and converges to the same error floor as the OS scheme. This behaviour is justified by (6.6) which shows that the capacity of the system is dominated by the ratio $\gamma_{S,R}/(\gamma_{R,R}+1)$ at high SNRs, and therefore the PS scheme coincides with the OS scheme at the high SNR regime. The LI scheme decides the relay selection only based on the loop interference and given that the considered scenario assumes a low loop interference, the achieved outage probability converges to the worst error floor. The MM relay selection does not take into account the loop interference and thus provides a performance similar to the LI scheme (due to the low loop interference). On the other hand, the HS scheme overcomes the problem of zero diversity that characterizes the FD-based schemes and provides the best outage

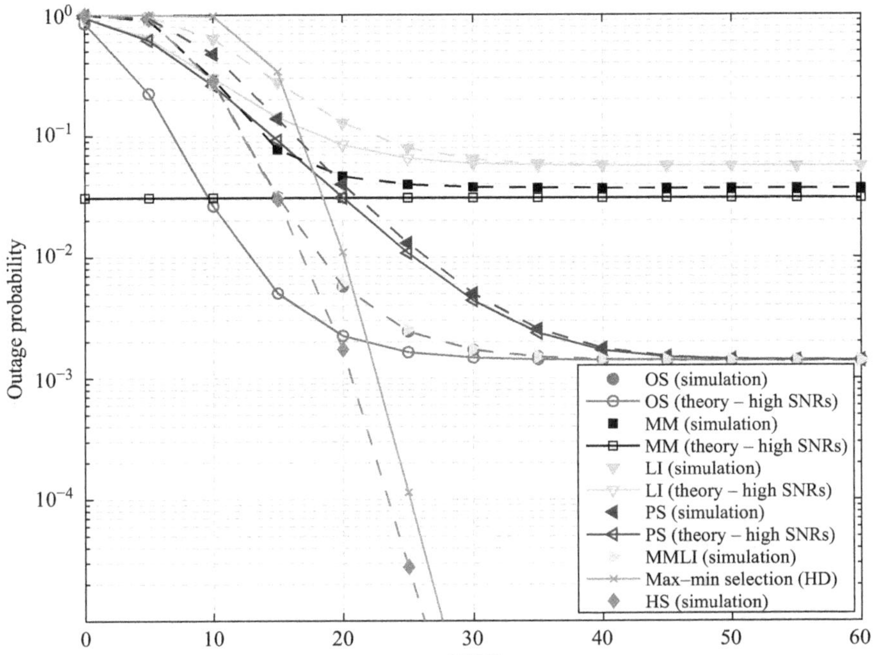

Figure 6.2 Outage probability versus the transmitted power P; $K = 4$ relays, $r_0 = 2$ BPCU, $\sigma_{SR}^2 = \sigma_{RD}^2 = 1$ and $\sigma_{RR}^2 = 0.08$; the dashed lines correspond to the numerical evaluation of (6.4), (6.8), (6.13), (6.16) and (6.22)

performance with a diversity gain equal to $K = 4$. The HS scheme outperforms the conventional max–min selection with HD relaying with a gain of about 1.5 dB at high SNRs. As for the closed-form approximations given in (6.6), (6.11), (6.14) and (6.17), it is clear that they provide efficient lower bounds that become very close at high SNRs and accurately locate the error floors. Figure 6.3 plots the outage probability versus P for $K = 3$ relays and $\sigma_{RR}^2 = 0.01$. The main performance trends follow those in Figure 6.2. It is worth noting that for this scenario, the MM scheme outperforms the LI scheme, and the proposed MM approximation becomes more efficient due to the lower loop interference, which is in line with the discussion in Section 6.2.3.

The performance advantage of the HS scheme is further illustrated in Figure 6.4, where the outage probability is plotted as a function of P with $K = 1, 2, 3$ relays, $\sigma_{RR}^2 = 0.5$, $r_0 = 3$ BPCU and $\sigma_{SR}^2 = \sigma_{RD}^2 = 1$. The performance of the conventional max–min HD relaying selection scheme is depicted for comparison. As can be seen, the HS scheme provides a diversity gain that is equal to the number of the available relays K and a significant coding gain in comparison to the max–min HD selection. Therefore, HS is an efficient technique for relay systems with FD and HD capabilities as it combines the spectral efficiency of FD with the interference efficiency of HD in an instantaneous way (i.e., in each time slot, the HD and OS capacity expressions are

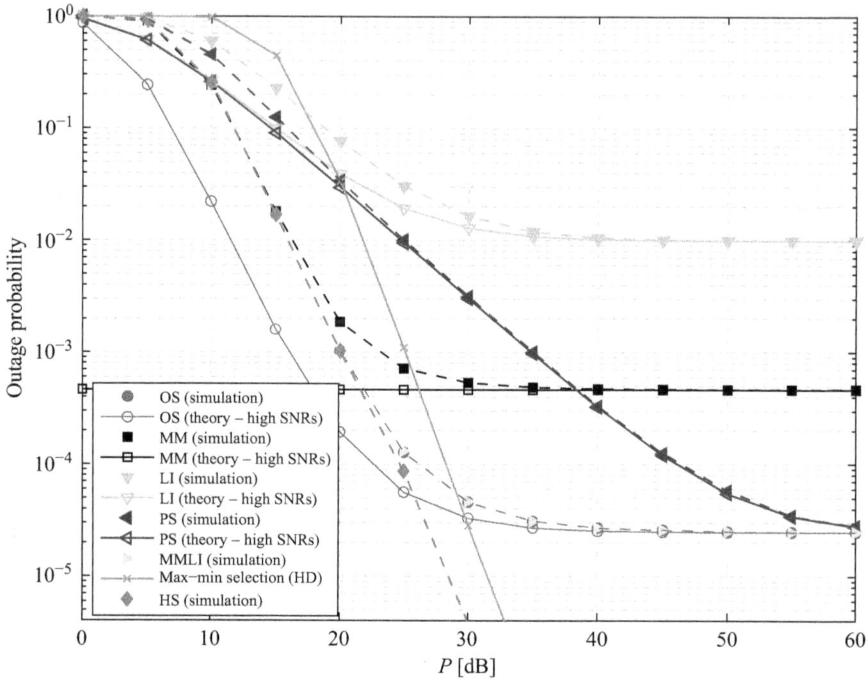

Figure 6.3 Outage probability versus the transmitted power P; $K = 3$ relays, $r_0 = 2$ BPCU, $\sigma^2_{SR} = \sigma^2_{RD} = 1$ and $\sigma^2_{RR} = 0.01$; the dashed lines correspond to the numerical evaluation of (6.4), (6.8), (6.13), (6.16) and (6.22)

generated for each relay and the "best" relay is selected based on (6.21)). In addition, Figure 6.4 depicts the results for the HS scheme using the analysis given in the Appendices of Reference 9.

Finally, Figure 6.5 plots the coding gain gap ($G_c^{(\mathrm{HD})}$ [dB]-$G_c^{(\mathrm{HS})}$ [dB]) based on (6.25), between the HS scheme and the conventional max–min HD selection scheme for different numbers of relays and loop interference values. As can be seen, the HS scheme significantly outperforms the HD scheme and the associated coding gain increases as the number of relays increases. Another important observation is that as the variance of the loop interference increases, the coding gain gap decreases. The main reason for this behaviour is that as the loop interference increases, the HS scheme employs the HD mode more often in order to mitigate the interference which results in the coding gain gap being decreased.

6.3 Buffer-aided relay selection

In this section, relay selection schemes with *buffers of finite size* at the relay nodes are presented. The use of data buffers at the relay nodes helps overcome the limitation of a two-slot transmission since it allows the selection of a different relay for reception

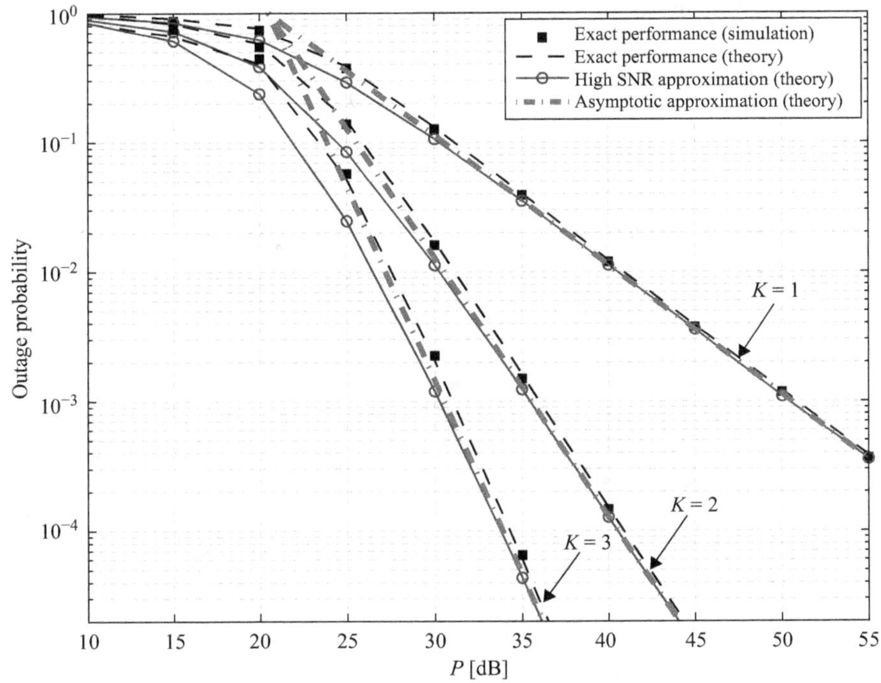

Figure 6.4 Outage probability versus the transmitted power P for the HS scheme; $K = 1,2,3$ relays, $r_0 = 3$ BPCU, $\sigma_{SR}^2 = \sigma_{RD}^2 = 1$ and $\sigma_{RR}^2 = 0.5$

and transmission in order to extract further diversity gains. The benefits of buffers in terms of the throughput has been reported in the literature [3, 11, 12]. Two main selection policies used in the literature are reviewed, namely the max–max and max-link, with particular focus on the max-link policy which fully exploits the available *diversity degrees* offered by the wireless channels through intelligent and dynamic switching between source and relay transmission.

6.3.1 System model

Assume a simple cooperative network consisting of one source S, one destination D and a cluster $\mathcal{C}$ with K decode-and-forward (DF) relays $R_k \in \mathcal{C}$, $1 \leq k \leq K$. All nodes are characterized by the HD constraint and therefore cannot transmit and receive simultaneously. A direct link between the source and the destination does not exist, and communication can be established only via relays (see Section 6.2.1). Figure 6.6 schematically presents the system model. Each relay R_k holds a data buffer Q_k of finite size L (number of data elements). At the beginning of the transmission, each buffer is empty, and the function $0 \leq \Psi(Q_k) \leq L$ gives the number of data elements stored in buffer Q_k. The buffer size $\Psi(Q_k)$ is increased by one when a source packet (data) is correctly decoded at the kth relay and is decreased by one in case

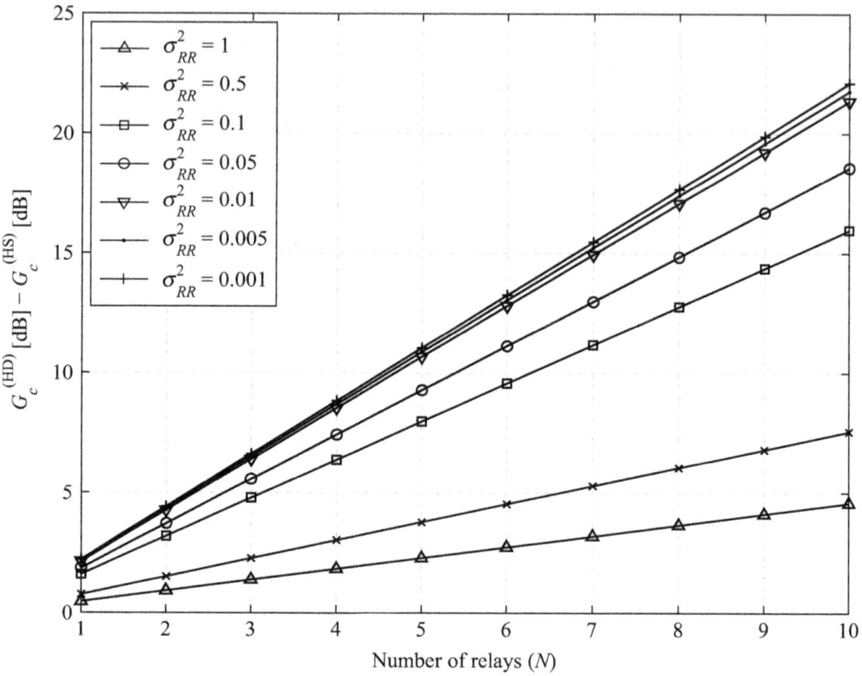

Figure 6.5 Coding gain gap between the HS scheme and the HD relaying scheme with max–min relay selection versus the number of the relays K; $r_0 = 2$ BPCU and $\sigma^2_{SR} = \sigma^2_{RD} = 1$

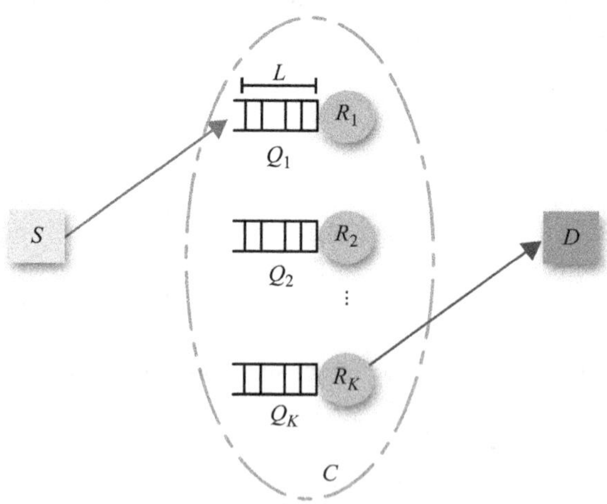

Figure 6.6 System model: buffer-aided relaying

of a successful kth relay transmission. Time is considered to be slotted, and each transmitter (either the source S or a relay R_k) transmits with a fixed power P. The source node is assumed to be saturated (infinite data backlog at the source), and the information rate is equal to r_0 BPCU. The retransmission process is based on an Acknowledgement/Negative-Acknowledgement (ACK/NACK) mechanism, in which short-length error-free packets are broadcasted by the receivers (either a relay R_k or the destination D) over a separate narrow-band channel in order to inform the network of that packet's reception status.

Similarly to the system model in Section 6.2, in order to simplify the analysis, assume that the clustered relay configuration ensures independent and identically distributed (i.i.d.) channel links which means equivalent average SNRs for the links $S \rightarrow R_k$ and $R_k \rightarrow D$ [13]. Furthermore, all wireless links exhibit fading and AWGN. The fading is assumed to be stationary, with frequency non-selective Rayleigh block fading. This means that the fading coefficients $h_{i,j}$ (for the $i \rightarrow j$ link) remain constant during one slot but change independently from one slot to another according to a circularly symmetric complex Gaussian distribution with zero mean and unit variance. The variance of the AWGN is assumed to be normalized with zero mean and unit variance, and therefore the SNR for each link is equal to P. Each link $i \rightarrow j$ is characterized by the success probability $p \triangleq \mathbb{P}\{(1/2)\log_2(1 + P|h_{i,j}|^2) \geq r_0\}$ which denotes the probability that the link $i \rightarrow j$ is not in outage ($\overline{p} = 1 - p$ denotes the outage probability); the factor $1/2$ captures the fact that communication is performed in two time slots [14, 15]. Finally, assume that the destination node has a perfect channel and buffer state information and selects the relays for transmission and reception through an error-free feedback channel [14]. This assumption can be ensured by an appropriate signalling that provides global CSI at the destination node. On the other hand, the destination can trivially know the status of the relay buffers by monitoring the ACK/NACK signalling and the identity of the transmitting/receiving relay.

6.3.2 Max–min relay selection

Recall from Section 6.2, that the max–min relay selection policy [1] is considered the optimal selection scheme for conventional relay setups without buffers and is used as the baseline relay selection strategy. The selected relay is the one that provides the "best" (strongest) end-to-end path between the source and the destination. The max–min selection policy ensures a full diversity, equal to the number of the relays and refers to a standard cooperative protocol where communication is performed in two time slots: (a) in the first time slot, the source transmits and (b) in the second time slot, the selected relay forwards the data received by the source towards the destination. The max–min relay selection policy can be written as

$$R^* = \arg \max_{R_k \in C} \min \left\{ |h_{S,R_k}|^2, |h_{R_k,D}|^2 \right\}, \tag{6.26}$$

where R^* denotes the selected relay. Although the max–min selection scheme achieves a full diversity, it does not take into account the fact that relays are equipped with buffers and their potential benefits. The selected relay forwards the received data (in case of a successful decoding) immediately in the next time slot to the destination, and therefore the ability of the relays to store at least a limited number of data packets is not elaborated.

6.3.3 Max–max relay selection (bound)

The max–max relay selection is the first policy that considered the buffering capability at the relay nodes [14]. The max–max policy selects the end-to-end path that holds the strongest source–relay and relay–destination links. More specifically, it splits the relay selection decision in two parts and selects the relay with the best source–relay link for reception and the relay with the best relay–destination link for transmission. The max–max selection policy respects the conventional two-slot cooperative transmission where the first slot is dedicated for the source transmission and the second slot for the relaying transmission, but the relay node may not be the same for both phases of the protocol. The max–max relay selection policy can be written as

$$R_r^* = \arg\max_{R_k \in C} \left\{ |h_{S,R_k}|^2 \right\}, \quad R_t^* = \arg\max_{R_k \in C} \left\{ |h_{R_k,D}|^2 \right\}, \tag{6.27}$$

where R_r^* and R_t^* denote the relay selected for the first phase and the second phase of the cooperative protocol, respectively. It has been proven that the max–max relay selection policy ensures full diversity equal to the number of the relays and it provides a significant coding gain in comparison to the conventional max–min selection scheme. However, it assumes that no buffer can be empty or full at any time, and thus all relays can always receive or transmit data [14, Sec. III. C]. Therefore, the max–max relay selection considered provides the optimal performance that can be achieved by the max–max relay selection scheme and yields the lowest outage bound.

6.3.4 Max-link relay selection

The previous relay selection schemes are associated with a two-slot cooperative protocol where the schedule for the source and relay transmission is fixed a priori. Here, this limitation is relaxed, and each slot is allowed to be allocated dynamically to the source or to a relay transmission according to the instantaneous quality of the links and the status of the relays' buffers. Specifically, the max-link relay selection scheme exploits fully the flexibility offered by the buffers at the relay nodes and at each slot selects the strongest link for transmission (source or relay transmission) among the available links [3]. A source–relay link is considered to be available when the buffer of the corresponding relay node is not full, while a relay–destination link is considered to be available when the buffer is not empty. The max-link relay selection policy can be analytically expressed as follows:

$$R^* = \arg\max_{R_k \in C} \left\{ \bigcup_{R_k \in C : \Psi(Q_k) \neq L} \{|h_{S,R_k}|^2\} \bigcup_{R_k \in C : \Psi(Q_k) \neq 0} \{|h_{R_k,D}|^2\} \right\}, \tag{6.28}$$

where R^* denotes the selected relay, for either transmission or reception.

6.3.4.1 Outage probability

For the max-link relay selection policy, the outage probability is defined as the probability that the selected link is in outage, i.e.,

$$P_{\text{out}} \triangleq \begin{cases} \mathbb{P}\left(\frac{1}{2}\log_2\left(1+P|h_{S,R^*}|^2\right) < r_0\right) & \text{for relay reception,} \\ \mathbb{P}\left(\frac{1}{2}\log_2\left(1+P|h_{R^*,D}|^2\right) < r_0\right) & \text{for relay transmission.} \end{cases} \tag{6.29}$$

The analysis of the max-link relay selection policy is done by modelling the possible states of the buffers and the transitions between the states as a Markov chain (MC). A state of the MC represents the number of elements at each buffer, and thus $s_l \triangleq (\Psi(Q_1)\Psi(Q_2)\ldots\Psi(Q_K))$ denotes the lth state of the MC with $l \in \mathbb{N}_+$, $1 \leq l \leq (L+1)^K$. The states are predefined in a random way as all the possible combinations ($(L+1)^K$ combinations) of the buffer sizes and are considered as a data input for the investigated algorithm given below.

The state-transition matrix represents the MC of the buffers' states and models the connectivity between them. Let $\mathbf{A}$ denote the $(L+1)^K \times (L+1)^K$ state-transition matrix of the MC, in which the entry $\mathbf{A}_{i,j} = \mathbb{P}(s_j \to s_i) = \mathbb{P}(X_{t+1} = s_i | X_t = s_j)$ is the transition probability to move from state s_i at time t to state s_j at time $(t+1)$. The transition probabilities depend on the status of the relay buffers (number of full elements at each buffer) and the related number of the available links that participate in the relay selection decision. Specifically, when the kth relay node has a full buffer ($\Psi(Q_k) = L$), it cannot receive data and therefore it offers the link $R_k \to D$. Similarly, when the kth relay node has an empty buffer ($\Psi(Q_k) = 0$), it cannot transmit data and therefore it offers only the link $S \to R_k$. Otherwise, the relay node can be used for either transmission or reception and thus it offers two links for selection ($S \to R_k, R_k \to D$). Consequently, for the s_l state of the buffers, the total number of the available links that participate in the max-link selection process is equal to

$$D_l = \sum_{i=1}^{K} \Phi(Q_i), \tag{6.30}$$

where

$$\Phi(Q_i) = \begin{cases} 2 & \text{if } 0 < \Psi(Q_i) < L, \\ 1 & \text{elsewhere.} \end{cases} \tag{6.31}$$

In order to construct the state-transition matrix $\mathbf{A}$, the connectivity between the different states of the buffers needs to be identified. For each time slot, the status of a buffer can be modified as follows: (a) the number of elements can be decreased by one, if a relay node is selected for transmission and the transmission is successful, (b) the number of elements can be increased by one, if the source node is selected for transmission and the transmission is successful and (c) the buffer status remains unchanged in case of outage. In order to formulate the above buffer state connectivity, define the associated set U_l for the state s_l by

$$U_l = \left\{ \bigcup_{1 \le i \le (L+1)^K} s_i : \mathbf{s}_i - \mathbf{s}_l \in \mathcal{Q} \right\}, \tag{6.32}$$

where $\mathbf{s}_l$ denotes a $1 \times K$ vector with entries the number of elements stored at each buffer for the s_l state and $\mathcal{Q} \triangleq \{\bigcup_{1 \le j \le K} \pm \mathbf{I}_{j,\bullet}\}$, where $\mathbf{I}_{j,\bullet}$ denotes the jth row of the identity matrix. The set U_l contains all the buffer states that are connected to the state s_l based on the previous connectivity rule. Given the assumption of i.i.d. symmetric channel links, for the s_l buffer state, the probability to select a specific link is equal to $1/D_l$ and the probability that the selected link is not in outage can be calculated by using order statistics (the maximum among D_l i.i.d. exponential random variables); therefore the probability to leave from the state s_l is equal to

$$p_{D_l} \triangleq \frac{1}{D_l} \left[1 - \left(1 - \exp\left(-\frac{2^{2r_0} - 1}{P} \right) \right)^{D_l} \right]. \tag{6.33}$$

On the other hand the probability to have an outage event and therefore no change in the buffer status is equal to

$$\bar{p}_{D_l} \triangleq 1 - \sum_{i=1}^{D_l} p_{D_l} = \left(1 - \exp\left(-\frac{2^{2r_0} - 1}{P} \right) \right)^{D_l}. \tag{6.34}$$

By using the previous notation, the entries of the state-transition matrix are given as

$$\mathbf{A}_{i,j} = \begin{cases} \bar{p}_{D_j} & \text{if } s_i \notin U_j \\ p_{D_j} & \text{if } s_i \in U_j \\ 0 & \text{elsewhere} \end{cases} \quad \text{for } i,j \in \{1, \ldots, (L+1)^K\}. \tag{6.35}$$

In what follows, the stationary distribution of the MC, denoted by $\boldsymbol{\pi}$, is given. The state-transition matrix $\mathbf{A}$ of the MC that models the buffer states is a *column stochastic*, *irreducible*, *aperiodic* and *reversible* matrix.

Lemma 6.1. ([3]) *The stationary distribution of the column stochastic matrix* $\mathbf{A}$ *of the MC that models the buffer states is given by*

$$\boldsymbol{\pi} = (\mathbf{A} - \mathbf{I} + \mathbf{B})^{-1}\mathbf{b}, \tag{6.36}$$

where $\boldsymbol{\pi}$ *is the stationary distribution,* $\mathbf{b} = (1\ 1\ \ldots\ 1)^T$ *and* $\mathbf{B}_{i,j} = 1,\ \forall i,j$.

Let Θ_i denote the set of states i for which the steady-state probability is the same. Let $\Xi(j, \Theta_i)$ denote the states that state j has to pass through to reach a state $i \in \Theta_i$, with the state in Θ_i included; i.e., if from state i to a state that belongs in Θ_j, say l, go through a state k, then $\Xi(i, \Theta_j) = \{k, l\}$.

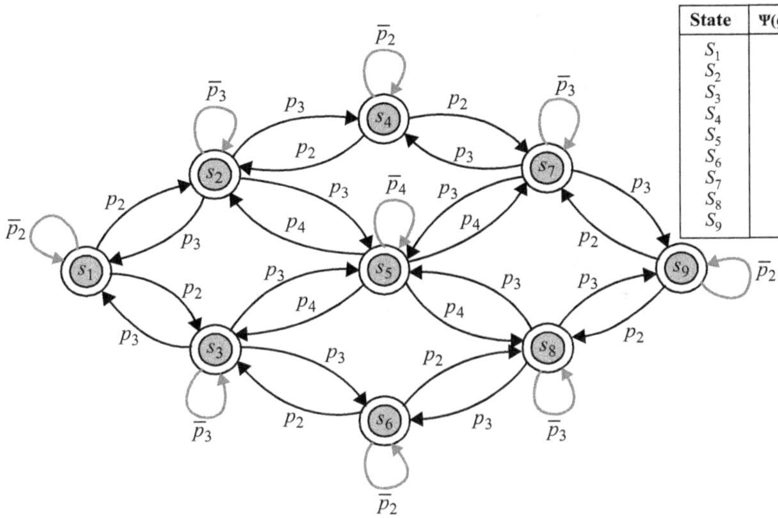

State	$\Psi(Q_1)\Psi(Q_2)$
S_1	00
S_2	01
S_3	10
S_4	02
S_5	11
S_6	20
S_7	12
S_8	21
S_9	22

Figure 6.7 State diagram of the MC representing the states of the buffers and the transitions between them for a case with $K=2$ relays and $L=2$

Lemma 6.2. ([3]) *The steady state for each state $i \in \Theta_i$ is given by*

$$\pi_i = \left(\sum_{j=1}^{(L+1)^K} \frac{\prod_{k \in \Xi(i,\Theta_j)} \mathbf{A}_{i,k}}{\prod_{m \in \Xi(j,\Theta_i)} \mathbf{A}_{j,m}} \right)^{-1}. \tag{6.37}$$

By using the steady state of the MC and the fact that an outage event occurs when there is no change in the buffer status, the outage probability of the system can be expressed as

$$P_{\text{out}} = \sum_{i=1}^{(L+1)^K} \pi_i \bar{p}_{D_i} = \text{diag}(\mathbf{A})\boldsymbol{\pi}. \tag{6.38}$$

The above expression shows that the construction of the state matrix $\mathbf{A}$ and the computation of the related steady state $\boldsymbol{\pi}$ consists of a simple theoretical framework for the computation of the outage probability for the max-link relay selection scheme with finite buffers. For the case of large L, a simplified expression of the above outage probability is proposed in Section 6.3.4.2.

6.3.4.2 Illustrative examples

In order to visualize the above approach for the computation of the achieved outage probability, the analysis is applied to special, illustrative cases. Firstly, a simple case is considered where there are two relays ($K=2$) with maximum buffer size equal to two ($L=2$). Finally, the case where the buffer size is infinitely large ($L \to \infty$) is considered.

Simple example: $K=2$ relays and $L=2$

In this simple case, the state MC that captures the evolution of the buffers is presented in Figure 6.7, and the corresponding state-transition matrix $\mathbf{A}$ becomes equal to

$$\mathbf{A} = \begin{pmatrix} \overline{p_2} & p_3 & p_3 & 0 & 0 & 0 & 0 & 0 & 0 \\ p_2 & \overline{p}_3 & 0 & p_2 & p_4 & 0 & 0 & 0 & 0 \\ p_2 & 0 & \overline{p}_3 & 0 & p_4 & p_2 & 0 & 0 & 0 \\ 0 & p_3 & 0 & \overline{p}_2 & 0 & 0 & p_3 & 0 & 0 \\ 0 & p_3 & p_3 & 0 & \overline{p}_4 & 0 & p_3 & p_3 & 0 \\ 0 & 0 & p_3 & 0 & 0 & \overline{p}_2 & 0 & p_3 & 0 \\ 0 & 0 & 0 & p_2 & p_4 & 0 & \overline{p}_3 & 0 & p_2 \\ 0 & 0 & 0 & 0 & p_4 & p_2 & 0 & \overline{p}_3 & p_2 \\ 0 & 0 & 0 & 0 & 0 & 0 & p_3 & p_3 & \overline{p}_2 \end{pmatrix}. \tag{6.39}$$

By using the method described in the previous section, the steady state of the system for different values of SNR is shown in Table 6.1.

Buffers of infinite size ($L \rightarrow \infty$)

The total number of buffer states is equal to $(L+1)^K$. However, from these $(L+1)^K$ states, $(L-1)^K$ states are neither full nor empty. In Reference 3, it is shown that for large L the $(L-1)^K$ states dominate the system and for $L \rightarrow \infty$ the probability of being in one of these states approaches 1. Hence, as the buffer size L increases the MC approaches a symmetric matrix, and therefore the steady state converges to a uniform

Table 6.1 State probabilities for different SNRs (P); $K=2$ relays, $L=2$, $r_0=1$ BPCU

Buffer state	**π ($P=0$ dB)**	**π ($P=8$ dB)**	**π ($P=20$ dB)**	**π ($P=30$ dB)**
s_1	0.1093	0.0894	0.0834	0.0783
s_2	0.1120	0.1215	0.1250	0.1326
s_3	0.1120	0.1215	0.1250	0.1326
s_4	0.1093	0.0894	0.0834	0.0783
s_5	0.1148	0.1564	0.1666	0.1565
s_6	0.1093	0.0894	0.0834	0.0783
s_7	0.1120	0.1215	0.1250	0.1326
s_8	0.1120	0.1215	0.1250	0.1326
s_9	0.1093	0.0894	0.0834	0.0783

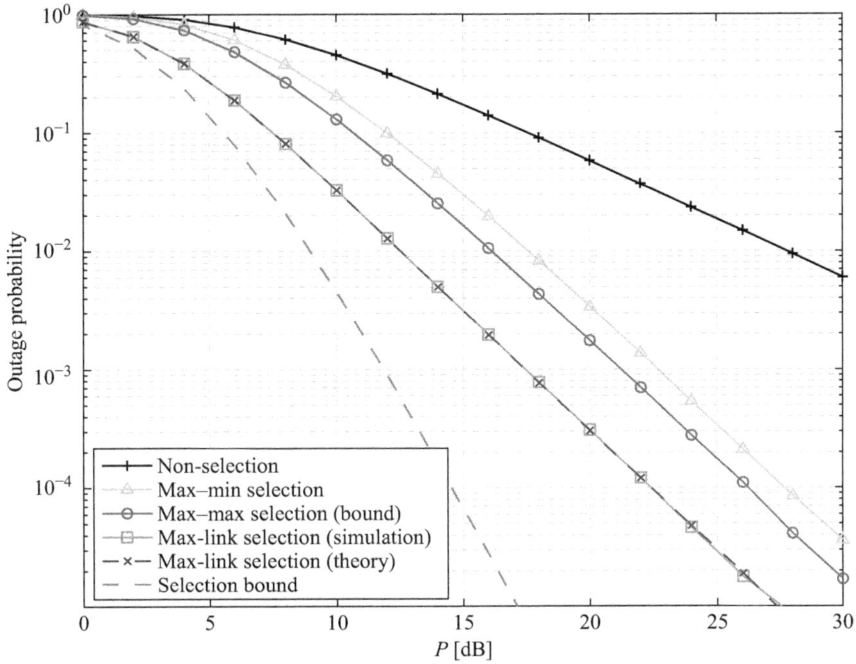

Figure 6.8 Outage probability performance versus P for a simulation setting with $K = 2$ relays, $L = 2$ and $r_0 = 1$ BPCU

distribution; as a result, the steady-state probabilities become equal to $1/(L+1)^K$. Among these states, the total number of states that have i, $0 \leq i \leq K$, full or empty buffers is equal to

$$\nu_i \triangleq \underbrace{2^i \binom{K}{i}}_{\text{full/empty buffers}} \cdot \underbrace{(L-1)^{K-i}}_{\text{no full/empty buffers}} . \tag{6.40}$$

Accordingly, a state with i buffers that are either full or empty achieves an outage probability equal to $\overline{p}_{2K-i}$, as i relays provide only one link in the link-selection process. Therefore, the total outage probability is approximated as

$$P_{\text{out}} = \sum_{i=1}^{(L+1)^K} \pi_i \nu_i = \left(\frac{L-1}{L+1}\right)^K \overline{p}_{2K} + \sum_{i=1}^{K} \frac{2^i \binom{K}{i} (L-1)^{K-i}}{(L+1)^K} \overline{p}_{2K-i}. \tag{6.41}$$

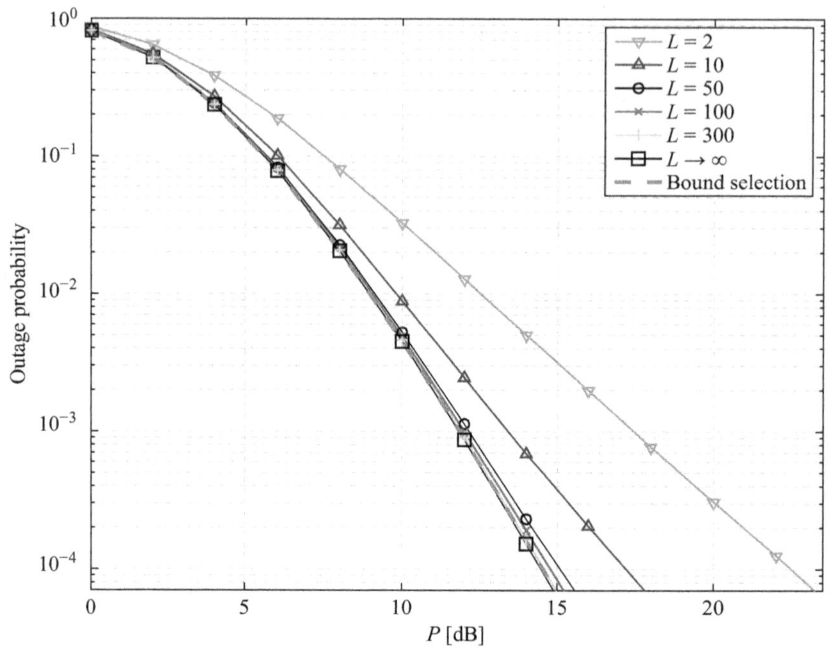

Figure 6.9 Outage probability performance versus P for max-link relay selection with K = 2 relays, L = 2,10,50,100,300, ∞ and $r_0 = 1$ BPCU

For the extreme case with infinite buffer sizes $L \to \infty$, the outage probability is simplified to

$$P_{\text{out}}^{\infty} = \lim_{L\to\infty}\left[\underbrace{\left(\frac{L-1}{L+1}\right)^{K}}_{\to 1}\overline{p}_{2K} + \sum_{i=1}^{K}\underbrace{\frac{2^{i}\binom{K}{i}(L-1)^{K-i}}{(L+1)^{K}}}_{\to 0}\overline{p}_{2K-i}\right] = \overline{p}_{2K}. \tag{6.42}$$

The above expression shows that for infinite buffer sizes, the max-link relay selection policy always selects between $2K$ available links, i.e., all the links become available. Therefore, it ensures a full diversity equal to two times the number of relays.

6.3.5 Numerical results

The simulation system follows the system model described in Section 6.3.1, and the relay selection policies considered are: the non-selection policy (where a fixed relay is selected for relaying), the max–min relay selection, the max–max relay selection (lowest bound) and the investigated max-link relay selection; the selection bound which refers to an ideal max-link selection scheme where all the channel links are always available for selection is used as a reference scheme. Figure 6.8 plots the outage probability versus the transmitted power P (SNR) with $K = 2$ relays, $L = 2$ and $r_0 = 1$ BPCU. As it can be seen, both the max–min and the max–max selection policies achieve a diversity gain equal to the number of relays which, in this case is 2 while

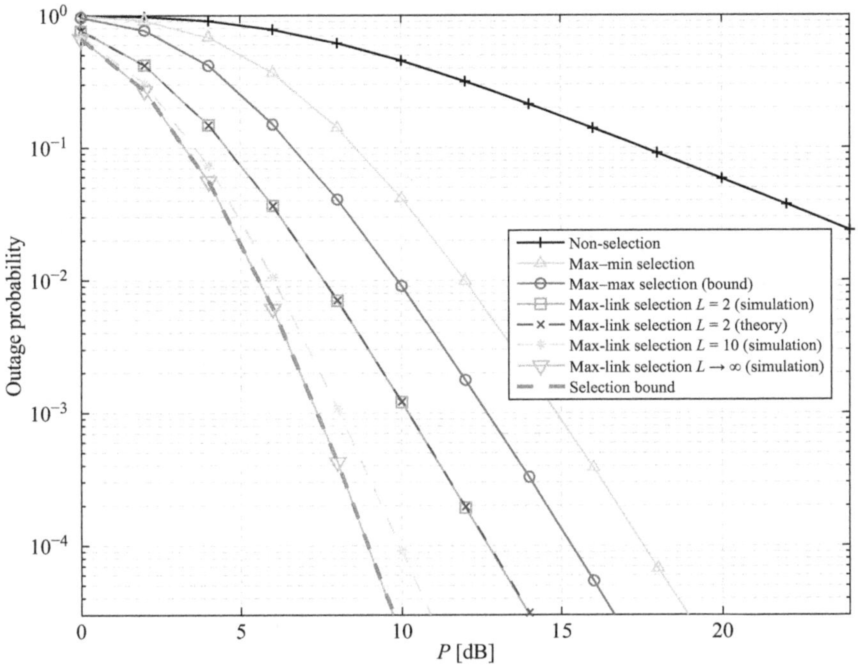

Figure 6.10 Outage probability performance versus P for max-link relay selection with $K=4$ relays, $L=2,10,\infty$ and $r_0=1$ BPCU

the max–max (bound) scheme offers an additional coding gain equal to 1.5 dB in comparison to the max–min scheme at high SNRs; this observation is in line with the analysis presented in Reference 14. On the other hand, the proposed max-link selection policy also achieves a diversity gain equal to $K=2$ and outperforms the max–max (bound) policy with a gain of about 4 dB. It is worth noting that this gain concerns the optimal max–max policy (lowest bound) and therefore the gain for a practical version of the max–max should be higher. This result shows that the max-link selection strategy uses more efficiently the existence of the relay buffers and significantly outperforms the existing selection schemes. However, it is worth noting that for the assumed buffer size $L=2$, the max-link does not provide a further diversity gain in comparison to the max–min and max–max schemes due to the small buffer size; on the other hand, the selection bound provides a diversity gain equal to $2K$.

Figure 6.9 focuses on the max-link selection scheme and shows the impact of the buffer size on the achieved diversity gain. More specifically, Figure 6.9 plots the outage probability performance versus P for different buffer sizes ($L=2, 10, 50, 100, 300, \infty$) with $K=2$ relays and $r_0=1$ BPCU. As can be seen, as the buffer size increases the outage probability performance is improved and approaches the selection bound that offers a diversity gain equal to 4. For a small increase of the buffer size (e.g., $L=10$ packets), the max-link relay selection achieves a diversity gain

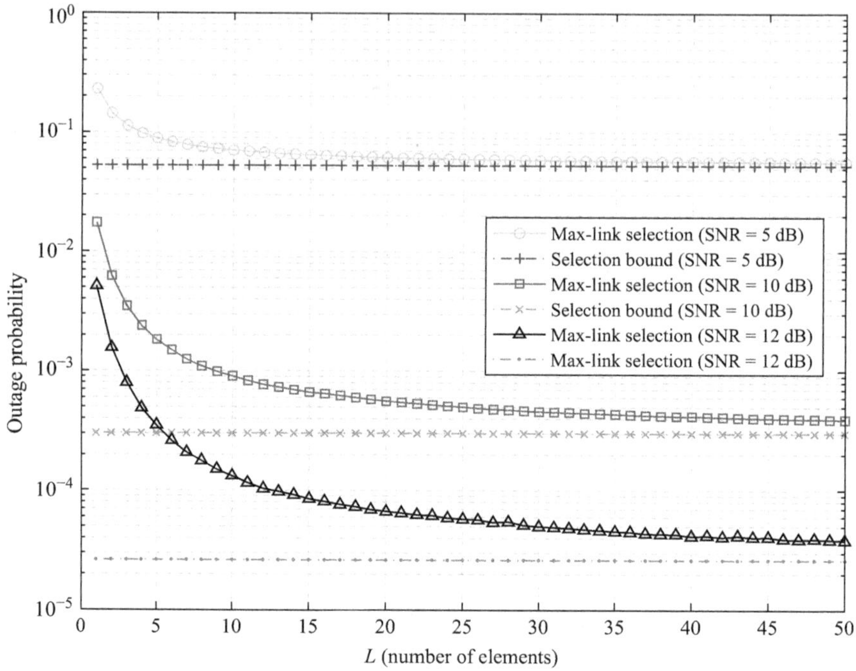

Figure 6.11 Outage probability performance versus the size of the relays' buffers (L) for max-link relay selection with K = 3 relays, P = 5,10,12 dB (SNR) and $r_0 = 1$ BPCU

higher than 2 (the slope of the curve becomes steeper), and therefore it significantly outperforms its previous version with $L = 2$ (i.e., 5 dB gain for an outage probability equal to 10^{-3}). Furthermore, as the size of the buffers increases the diversity order is continuously improved, and for $L = 100$ it approximates the selection bound and achieves a full diversity equal to 4; for the ideal case of $L \to \infty$, the max-link relay selection policy matches the selection bound, an observation that follows the discussion presented in Section 6.3.4.2. These remarks reveal the benefits of the max-link selection and show that a small buffer size (low complexity) can double the diversity gain in comparison to existing relay selection schemes. Figure 6.10 plots the outage probability performance versus P with $K = 4$ relays, $L = 2, 10, \infty$ and $r_0 = 1$ BPCU for all the considered relay selection schemes. As can be seen, the curves follow the previous remarks and validate the observation that the max-link outperforms all the existing relay selection policies and that as the size of the buffers increases, it approaches the selection bound and a diversity order equal to 8.

Figure 6.11 illustrates the outage probability performance of the max-link relay selection scheme versus the size of the relays' buffers L with $K = 3$ relays, $r_0 = 1$ BPCU and $P = 5, 10, 12$ dB; the range of the buffer size L is [1 50] which corresponds to practical implementations with low complexity. As can be seen, as the buffer size is

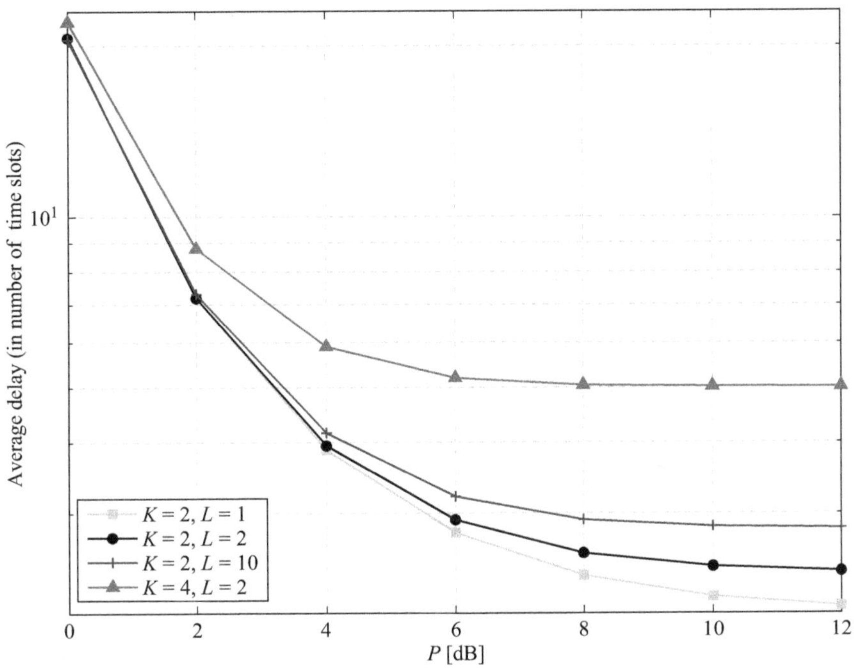

Figure 6.12 Average delay in number of time slots versus P for different system configurations; $K = 2,4$ relays, $L = 1,2,10$ and $r_0 = 1$ BPCU

increased, the achieved performance approximates the selection bound. In addition, a relatively small size buffer is sufficient to achieve the optimal outage performance (e.g., $L = 40$ for $P = 10$ dB); however as the SNR increases and the optimal outage probability is decreased, the required value of L is increased (e.g., $L = 15$ for $P = 5$ dB and $L = 40$ for $P = 10$ dB). Finally, Figure 6.12 plots the average time delay in number of time slots versus P for different system configurations in order to give an idea regarding this important metric. It can be seen that as the number of relays increases, the probability to select a specific relay for transmission decreases and thus the average delay increases. In addition, as the buffer size increases, the probability to have a full relay and thus to select a specific relay decreases, which results in an increase of the average delay. By tolerating a higher delay, the diversity performance improves as the number of relays K or buffer size L increases, as shown in Figures 6.9 and 6.10.

6.4 Wireless powered relay selection with wireless battery charging

This section focuses on wireless powered communication cooperative networks with respect to the relay selection problem [16]. The relays in the network are equipped with two-state storage devices (e.g., batteries) which are charged by the source

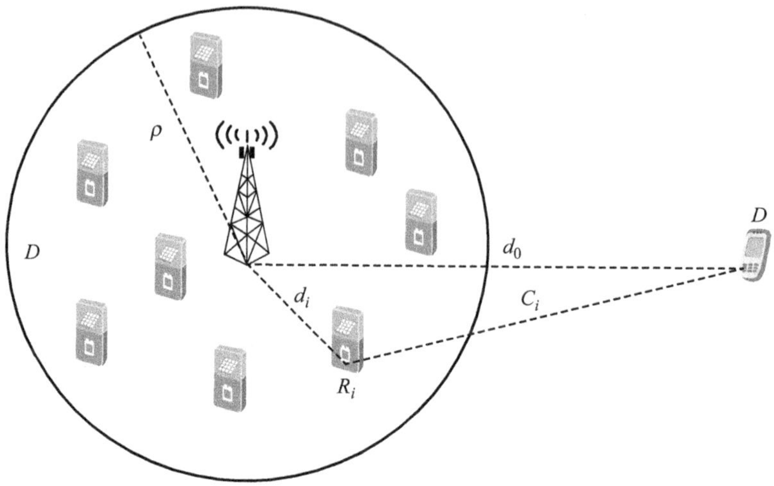

Figure 6.13 System model: wireless powered relaying

transmission; relay nodes become active and can assist source transmission only when they are fully charged. The proposed relay selection policies have different CSI requirements and correspond to different complexities. Analytical results for the outage probability performance of the proposed schemes as well as simplified asymptotic expressions for the high SNR regime are given below. In contrast to the conventional case [17], where a diversity gain equal to one is ensured, the battery model results in a zeroth diversity gain at high SNRs.

6.4.1 System model

Consider a disc $\mathcal{D}$ of radius ρ, where an access point (AP) located at the origin communicates with a destination D via the help of relay nodes R_i. Figure 6.13 schematically presents the system model. The relay nodes are located inside the disc, and their locations form a homogeneous Poisson point process Φ of density λ; K denotes the number of relays. All nodes are equipped with a single antenna, and the AP transmits with a fixed power P and a spectral efficiency r_0 BPCU. The distance between the AP and the destination is denoted by d_0, and the AP does not have a direct link with the destination, e.g., due to obstacles or severe shadowing and path-loss effects. We assume that d_i denotes the Euclidean distance between the AP and R_i, while c_i denotes the distance between R_i and D.

Each relay node employs a DF policy and is equipped with a single battery which can store energy for future use. At the beginning of the broadcast phase, the battery can be either fully charged or empty (two states) [18, Sec. II.A]; a relay with a charged battery is active and can participate in the relaying operation, while a relay with an empty battery is in harvesting mode and uses the AP's signal for WPT purposes.

Assume that the battery can be used for either charging or transceiver operations (a single battery cannot be charged and discharged simultaneously [19, 20]).

The WPT process is based on the AP's transmissions, and WPT from relaying signals is negligible. An empty battery can be fully charged when the input power is larger than a predefined threshold (sensitivity), which depends on the characteristics of the rectenna's circuit. Without loss of generality, it is assumed that this threshold is equal to the size/capacity of the battery and is denoted as P_r, where $P_r = \Psi P$ with $\Psi < 1$. A relay node that is charged and selected for transmission attempts to decode the AP's signal. If the decoding is successful, it transmits with a fixed power P_r, i.e., transmission dominates the total energy consumption at the relay nodes. The battery does not have intermediate energy states/levels and thus is fully discharged at the end of the second/relaying time slot, when it is connected to the transceiver's circuit.

The charging/discharging behaviour of the battery can be represented by a finite-state MC with two states $\{s_0, s_1\}$, where the state s_0 indicates that the battery is empty and the state s_1 that the battery is charged. The state-transition probability matrix $\boldsymbol{P}$ can be written as

$$\boldsymbol{P} = \begin{bmatrix} 1-\pi_0 & \pi_0 \\ \pi_1 & 1-\pi_1 \end{bmatrix}, \tag{6.43}$$

where π_0 denotes the probability that the input power is greater than the sensitivity threshold of the WPT device, and π_1 is the probability that a charged relay node is selected for relaying. Both probabilities are defined in the following section for the proposed relay selection schemes. If $\boldsymbol{\eta} = [\eta_0 \ \eta_1]$ denotes the stationary steady-state probability vector of the MC, by using basic queueing theory,

$$\boldsymbol{\eta P} = \boldsymbol{\eta}. \tag{6.44}$$

The solution of the above system of linear equations gives the steady-state distribution of the battery, which is equal to

$$\eta_0 = \frac{\pi_1}{\pi_0 + \pi_1}, \quad \eta_1 = \frac{\pi_0}{\pi_0 + \pi_1}. \tag{6.45}$$

It is assumed that the wireless links suffer from both small-scale block fading and large-scale path-loss effects. The fading is Rayleigh distributed so the power of the channel fading is an exponential random variable with unit variance. We denote by h_i and g_i, the channel coefficients for the links between AP and R_i and R_i and D, respectively. The path-loss model assumes that the received power is proportional to $1/(1+d^\alpha)$ where d is the Euclidean distance between the transmitter and the receiver, $\alpha > 2$ denotes the path-loss exponent, and we define $\delta \triangleq 2/\alpha$. The instantaneous fading channels are known only at the receivers, except if otherwise defined. In addition, all wireless links exhibit AWGN with variance σ^2; n_i denotes the AWGN at the ith node. To simplify the notation, define $\epsilon \triangleq 2^{2r_0} - 1$ and $\Xi \triangleq \epsilon\sigma^2/P$.

In what follows, several relay selection schemes that correspond to different complexities are presented based on the above system model.

6.4.2 Random relay selection

The random relay selection (RRS) scheme does not require any feedback regarding the battery status or the location of the relay nodes and selects a relay in a random way. It corresponds to a low implementation complexity and is appropriate for networks with strict power/bandwidth constraints. Without loss of generality, consider that the ith relay is selected to assist the source. If the ith relay is fully charged, it attempts to decode the source signal and acts as a relay in case of successful detection. If the ith relay has an empty battery, it switches to harvesting mode and uses the received signal for WPT purposes; in this case, the relay remains inactive during the cooperative slot and an outage event occurs. On the other hand, the non-selected relays with empty batteries switch to harvesting mode and use the source signal for potential charging. During the first time slot, the received signal at the ith relay can be written as

$$y_i = \sqrt{P}\frac{h_i}{\sqrt{1+d_i^{\alpha}}}s + n_i, \tag{6.46}$$

where s denotes the source signal with normalized power. If the ith relay is active, the associated received SNR is equal to

$$\mathsf{SNR}_i = \frac{P|h_i|^2}{(1+d_i^{\alpha})\sigma^2}. \tag{6.47}$$

If the ith relay is inactive (empty battery), the input power at the WPT device is equal to

$$P_h = \zeta\frac{P|h_i|^2}{(1+d_i^{\alpha})}, \tag{6.48}$$

where ζ denotes the WPT conversion efficiency and (6.48) assumes that energy harvesting from AWGN is negligible (very small compared to the desired signal [21]). It is assumed, without loss of generality, that $\zeta = 1$. In the second time slot, the received signal at the destination can be written as

$$y_D = \sqrt{P_r}\frac{g_i}{\sqrt{1+c_i^{\alpha}}}s + n_D, \tag{6.49}$$

and the associated SNR is given by

$$\mathsf{SNR}_D = \frac{P_r|g_i|^2}{(1+c_i^{\alpha})\sigma^2}. \tag{6.50}$$

6.4.2.1 Outage probability

In the RRS scheme, an outage event occurs when (a) there is not any available relay in $\mathcal{D}$, i.e., $K=0$, (b) the disc contains at least one relay ($K \geq 1$) but the selected relay has an empty battery, (c) $K \geq 1$, the selected relay is fully charged but cannot decode the source signal, and (d) $K \geq 1$, the selected relay is fully charged and decodes the source signal but the destination cannot support the targeted spectral efficiency. By analysing the probability of these events, the outage probability achieved by the RRS scheme is given by [16]

$$P_{\mathrm{RRS}} = \exp(-\lambda\pi\rho^2)+(1-\exp(-\lambda\pi\rho^2))\Bigg[1-\eta_1^{\mathrm{RRS}}\frac{\delta}{\pi\rho^4}\frac{\gamma(\delta,\Xi\rho^\alpha)}{\Xi^\delta}$$

$$\times\exp\left(-\Xi\left(1+\frac{1}{\Psi}\right)\right)\int_0^{2\pi}$$

$$\int_0^\rho \exp\left(-\frac{\Xi}{\Psi}\left(x^2+d_0^2-2xd_0\cos(\theta)\right)^{\frac{1}{\delta}}\right)xdxd\theta\Bigg], \tag{6.51}$$

where η_1^{RRS} is the steady-state probability that a relay node is charged at the beginning of the broadcast phase and is given by

$$\eta_1^{\mathrm{RRS}} = \frac{\delta\exp(-\Psi)\frac{\gamma(\delta,\Psi\rho^\alpha)}{\Psi^\delta}}{\delta\exp(-\Psi)\frac{\gamma(\delta,\Psi\rho^\alpha)}{\Psi^\delta}+\frac{1}{\lambda\pi}}. \tag{6.52}$$

Remark 6.1. *For the special case with $P\to\infty$, $P_r\to\infty$, $\Psi=P_r/P$ (constant ratio), $\rho << d_0$, and $K\geq 1$, the outage probability of the RRS scheme is given by the expression*

$$P_{\mathrm{RRS}}^{\infty}\to 1-\eta_1^{\mathrm{RRS}}. \tag{6.53}$$

6.4.3 Relay selection based on the closest distance

The relay selection based on the closest distance (RCS) requires an a priori knowledge of the location of the relay nodes. It is assumed that the AP monitors the location of the relays through a low-rate feedback channel or a global positioning system mechanism and selects the relay node which is closest to the AP. The RCS scheme does not take into account battery status and/or instantaneous fading and also corresponds to a low implementation complexity, specifically for scenarios with low mobility. The relay node R^* selected by the RCS policy can be expressed as

$$R^* = \arg_{R_i\in\Phi}\min_{i=1,\ldots,K} d_i. \tag{6.54}$$

The mathematical description of the RCS scheme follows the equations in Section 6.4.2, by simply replacing the random relay node R_i with R^*.

6.4.3.1 Outage probability

The scenarios for an outage event follow the discussion in Section 6.4.2. The outage probability achieved by the RCS scheme is given by [16]

$$P_{\mathrm{RCS}} = \exp(-\lambda\pi\rho^2)+\left(1-\exp(-\lambda\pi\rho^2)\right)\Bigg[1-\eta_1^{\mathrm{RCS}}\frac{2\pi\lambda^2\exp\left(-\Xi(1+\frac{1}{\Psi})\right)}{[1-\exp(-\pi\lambda\rho^2)]^2}$$

$$\times\int_0^{2\pi}\int_0^\rho\int_0^\rho \exp\Bigg(-\frac{\Xi}{\Psi}(r^2+d_0^2-2rd_0\cos(\theta))^{\frac{1}{\delta}}$$

$$-\lambda\pi r^2-\Xi x^\alpha-\lambda\pi x^2\Bigg)rxdrdxd\theta\Bigg], \tag{6.55}$$

where η_1^{RRS} is the steady-state probability that a relay node is charged at the beginning of the broadcast phase and is given by $\eta_1^{\mathrm{RCS}} = \eta_1^{\mathrm{RRS}}$.

Remark 6.2. *For the special case with $P \to \infty$, $P_r \to \infty$, $\Psi = P_r/P$ (constant ratio), $\rho << d_0$, $K \geq 1$ and $\alpha = 2$, the outage probability of the RCS scheme is given by*

$$P_{\mathrm{RCS}}^{\infty} \to 1 - \eta_1^{\mathrm{RCS}}. \tag{6.56}$$

6.4.4 Distributed beamforming

The distributed beamforming (DB) scheme selects all the relay nodes that are charged at the beginning of the broadcast phase. More specifically, all the relay nodes with fully charged batteries become active and attempt to decode the source signal. The relays that are able to successfully decode the source signal create a virtual multiple antenna array and coherently transmit the signal to the destination. The DB scheme requires a perfect time synchronization between the relay nodes as well as a CSI at the relays; a feedback channel ensures the knowledge of the g_i channel coefficient at the ith relay. The broadcast phase of the DB protocol follows the description of the RRS and RCS schemes and is given by the expressions in (6.46)–(6.48). The received signal at the destination, during the second phase of the protocol, can be written as

$$y_D = \sqrt{P_r}\sum_{i\in\mathcal{C}} \frac{w_i g_i}{\sqrt{1+c_i^{\alpha}}} s + n_D, \tag{6.57}$$

where $w_i = g_i^* / \sqrt{\sum_{i\in\mathcal{C}} |g_i|^2}$ is the precoding coefficient at the ith relay that ensures coherent combination of the relaying signals at the destination, and $\mathcal{C}$ denotes the set of the charged relay nodes (index), which successfully decoded the source message and participate in the relaying transmission. The associated SNR at the destination is given by

$$\mathsf{SNR}_D = P_r \sum_{i\in\mathcal{C}} \frac{|g_i|^2}{(1+c_i^{\alpha})\sigma^2}. \tag{6.58}$$

6.4.4.1 Outage probability

Equivalently to the previous protocols, the outage probability performance of the DB scheme depends on the probability of a battery to be charged at the beginning of the broadcast phase. Based on the description of the DB scheme, an outage event occurs when (a) the relay set $\mathcal{C}$ is empty or (b) when the coherent relaying transmission is in outage, i.e., the destination is not able to decode the relaying signal. Therefore, the outage probability achieved by the DB scheme when $\rho \ll d_0$ is given by [16]

$$P_{\mathrm{DB}} = \sum_{k=0}^{\infty} \frac{\gamma\left(k, \frac{\Xi(1+d_0^2)\sigma^2}{\Psi}\right)}{\Gamma(k)} \exp\left(-\lambda'\pi\rho^2\right) \frac{(\lambda'\pi\rho^2)^k}{k!}, \tag{6.59}$$

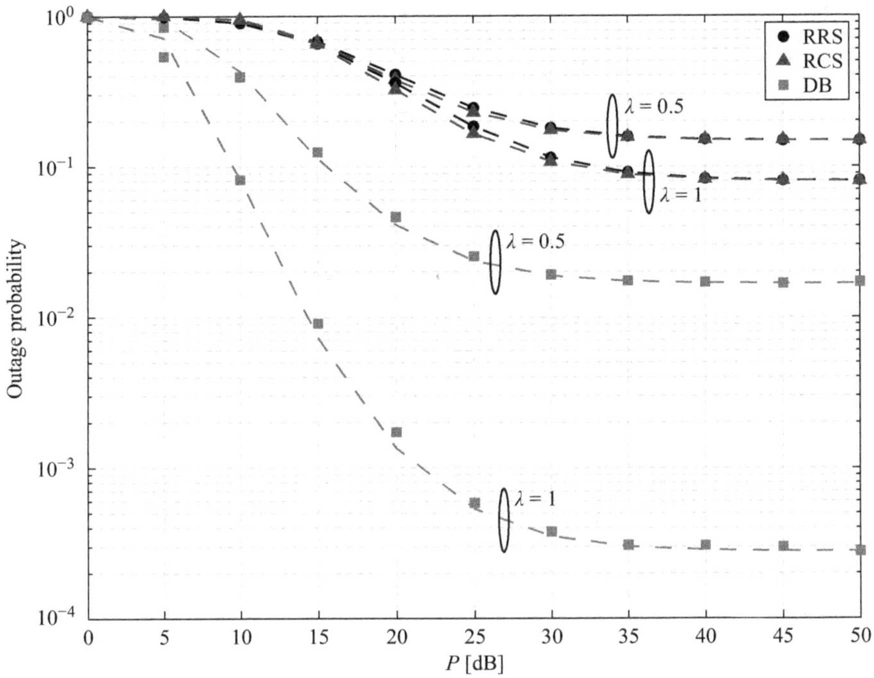

Figure 6.14 Outage probability versus P; $\rho = 3$ m, $d_0 = 2\rho$, $\Psi = 0.1$, $\alpha = 3$, $\sigma^2 = 1$, $r_0 = 0.01$ BPCU and $\lambda = \{1, 0.5\}$; the dashed lines represent the theoretical results

where $\lambda' = \lambda \eta_1^{\mathrm{DB}} \frac{\delta}{\rho^2} \exp(-\Xi) \frac{\gamma(\delta, \Xi\rho^\alpha)}{\Xi^\delta}$, and the steady-state probability η_1^{DB} is given by

$$\eta_1^{\mathrm{DB}} = \frac{\delta \exp(-\Psi) \frac{\gamma(\delta, \Psi\rho^\alpha)}{\Psi^\delta}}{\delta \exp(-\Psi) \frac{\gamma(\delta, \Psi\rho^\alpha)}{\Psi^\delta} + \rho^2}. \tag{6.60}$$

For the high SNR regime, the following simplified expression is given.

Remark 6.3. *For the special case with $P \to \infty$, $P_r \to \infty$, $\Psi = P_r/P$ (constant ratio), the outage probability of the DB scheme is given by*

$$P_{\mathrm{DB}}^{\infty} \to \exp(-\lambda \eta_1^{\mathrm{DB}} \pi \rho^2). \tag{6.61}$$

6.4.5 Numerical results

Figure 6.14 plots the outage probability performance of the proposed relay selection schemes versus the transmitted power P. The first main observation is that the RRS and the RCS schemes converge to the same outage floor at high SNRs, as it has been reported in Remarks 6.1 and 6.2, respectively. However, the RCS scheme slightly

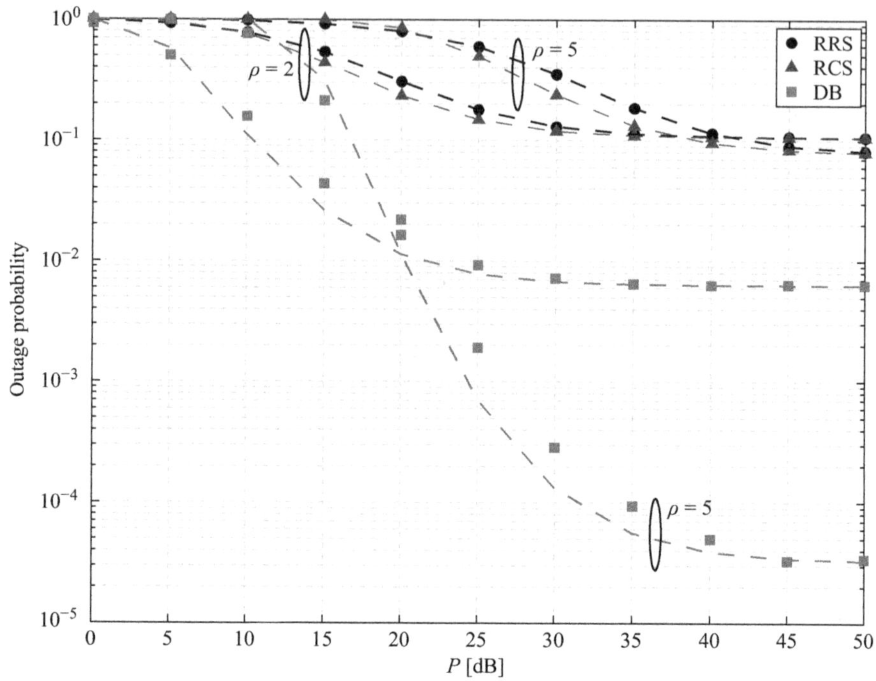

Figure 6.15 Outage probability versus ρ. Simulation parameters: $d_0 = \rho$, $\Psi = 0.1$, $\alpha = 3$, $\sigma^2 = 1$, $r_0 = 0.1$ BPCU and $\lambda = 1$; the dashed lines represent the theoretical results

outperforms the RRS scheme at the moderate SNRs and thus converges to the outage floor much faster. On the other hand, the DB scheme significantly outperforms RRS/RCS schemes; the transmission of the source signal by multiple relays through beamforming boosts the SNR at the destination and improves the outage probability performance. In addition, it can be seen that as the density λ increases, more relays participate in the relaying operation, and therefore the gap between DB and single-relay selection schemes increases.

Figure 6.15 plots the outage probability versus the radius of the disc ρ. The first main observation is that as ρ increases, the performance of the proposed relay selection schemes is improved at high SNRs, i.e., they converge to lower outage floors. This observation was expected, since as the area of the disc increases, the number of the relay nodes increases ($\mathbb{E}[K] = \exp(-\lambda\pi\rho^2)$), and thus (a) more relays participate in the relaying operation (DB scheme), or (b) the probability to select an uncharged relay node is decreased (RRS and RCS schemes). However, for the low and moderate SNRs, it can be seen that as ρ increases, the outage probability performance decreases. For this SNR regime, a larger ρ corresponds to a higher distance between source–relay (and/or relay–destination), and thus the outage probability is dominated by the path-loss degradation effects.

6.5 Conclusion

This chapter focused on relay selection policies in the context of three different communication scenarios: relays with FD radio, buffers and WPT capabilities. Relay selection is an efficient PHY layer tool with a low-complexity implementation due to the fact that it chooses the best relay(s) to retransmit according to some predefined criteria and thus avoids the simultaneous retransmission by all available relays. It is thus well known for the high performance it provides and its efficient use of system resources. The scenarios described above refer to technologies which will have potential applications in modern communication networks. These networks are required to achieve high spectral efficiencies but also have low power consumption; FD radio and WPT can assist towards this goal. Furthermore, modern networks are characterized by their high density and the use of high-frequency bands, i.e., millimetre waves. In such cases, line-of-sight communications are vital, and thus relays will play an important role. The results and observations in this chapter clearly show the major impact that the relay selection technique will have in the design of modern communication networks. Its simplicity and the gains that can be achieved by its implementation with other next-generation technologies make it an attractive application for future communication systems such as 5G.

References

[1] A. Bletsas, A. Khisti, D. Reed, and A. Lippman, "A simple cooperative diversity method based on network path selection," *IEEE J. Select. Areas Commun.*, vol. 24, pp. 659–672, March 2006.

[2] A. S. Ibrahim, A. K. Sadek, W. Su, and K. J. Ray Liu, "Cooperative communications with relay-selection: when to cooperate and whom to cooperate with?," *IEEE Trans. Wireless Commun.*, vol. 7, pp. 2814–2827, July 2008.

[3] I. Krikidis, T. Charalambous, and J. Thompson, "Buffer-aided relay selection for cooperative diversity systems without delay constraints," *IEEE Trans. Wireless Commun.*, vol. 11, no. 5, pp. 1957–1967, May 2012.

[4] D. B. da Costa and S. Aissa, "Performance analysis of relay selection techniques with clustered fixed-gain relays," *IEEE Signal Process. Lett.*, vol. 17, pp. 201–204, February 2010.

[5] T. Riihonen, S. Werner, R. Wichman, and E. B. Zacarias, "On the feasibility of full-duplex relaying in the presence of loop interference," in *Proc. IEEE Workshop Signal Process. Adv. Wireless Commun.*, Perugia, Italy, June 2009, pp. 275–279.

[6] T. Riihonen, S. Werner and R. Wichman, "Mitigation of loopback self-interference in full-duplex MIMO relays," *IEEE Trans. Signal Process.*, vol. 59, pp. 5983–5993, December 2011.

[7] I. Krikidis, J. S. Thompson, S. McLaughlin, and N. Goertz, "Max-min relay selection for legacy amplify-and-forward systems with interference," *IEEE Trans. Wireless Commun.*, vol. 8, pp. 3016–3027, June 2009.

[8] T. Riihonen, S. Werner, and R. Wichman, "Optimized gain control for single-frequency relaying with loop interference," *IEEE Trans. Wireless Commun.*, vol. 8, pp. 2801–2806, June 2009.

[9] I. Krikidis, H. Suraweera, P. Smith, and C. Yuen, "Full-duplex relay selection for amplify-and-forward cooperative networks," *IEEE Trans. Wireless Commun.*, vol. 11, no. 12, pp. 4381–4393, October 2012.

[10] P. A. Anghel and M. Kaveh, "Exact symbol error probability of a cooperative network in a Rayleigh-fading environment," *IEEE Trans. Wireless Commun.*, vol. 3, pp. 1416–1421, September 2004.

[11] N. Zlatanov, R. Schober, and P. Popovski, "Throughput and diversity gain of buffer-aided relaying," in *Proc. IEEE Global Commun. Conf.*, Houston, TX, December 2011, pp. 1–6.

[12] L. Ding, M. Tao, F. Yang, and W, Zhang, "Joint scheduling and relay selection in one- and two-way relay networks with buffering," in *Proc. IEEE Int. Conf. Commun.*, Dresden, Germany, June 2009, pp. 1–5.

[13] K.-S. Hwang, Y. -C. Ko, and M. -S. Alouini, 'Performance analysis of two-way amplify-and-forward relaying with adaptive modulation over multiple relay network," *IEEE Trans. Wireless Commun.*, vol. 59, pp. 402–406, February 2011.

[14] A. Ikhlef, D. S. Michalopoulos, and R. Schober, "Buffers improve the performance of relay selection," in *Proc. IEEE Global Commun. Conf.*, Houston, TX, December 2011, pp. 1–6.

[15] J. N. Laneman, D. N. C. Tse, and G. W. Wornell, "Cooperative diversity in wireless networks: efficient protocols and outage behavior," *IEEE Trans. Inf. Theory*, vol. 50, pp. 3062–3080, December 2004.

[16] I. Krikidis, "Relay selection in wireless powered cooperative networks with energy storage," *IEEE J. Select. Areas Commun.*, vol. 33, pp. 2596–2610, December 2015.

[17] Z. Ding, I. Krikidis, B. Sharif, and H. V. Poor, "Wireless information and power transfer in cooperative networks with spatially random relays," *IEEE Trans. Wireless Commun.*, vol. 13, pp. 4440–4453, August 2014.

[18] S. Lee, R. Zhang, and K. Huang, "Opportunistic wireless energy harvesting in cognitive radio networks," *IEEE Trans. Wireless Commun.*, vol. 12, pp. 4788–4799, September 2013.

[19] I. Krikidis, S. Timotheou, and S. Sasaki, "RF energy transfer for cooperative networks: data relaying or energy harvesting?", *IEEE Commun. Lett.*, vol. 16, pp. 1772–1775, November 2012.

[20] S. Luo, R. Zhang, and Y. J. Lim, "Optimal save-then-transmit protocol for energy harvesting wireless transmitters," *IEEE Trans. Wireless Commun.*, vol. 12, pp. 1196–1207, March 2013.

[21] R. Zhang and C. K. Ho, "MIMO broadcast for simultaneous wireless information and power transfer," *IEEE Trans. Wireless Commun.*, vol. 12, pp. 1989–2001, May 2013.

Chapter 7
Relaying in green communication systems

Pan Cao, Miryam Gonzalez-Perez and John S. Thompson

7.1 Introduction

Energy efficiency of communications has been a long-standing topic of interest both to researchers and industry. Mobile devices, such as smartphones, laptops and tablets, are battery powered and so it makes sense to design communications systems to be as energy efficient as possible in order to promote longer battery lifetimes. In the last 10 years, there has been increasing interest in energy efficient design of the overall wireless networks. The energy costs to operators of powering their networks is becoming increasingly apparent, and many operators have also made pledges to try to reduce their carbon footprints in future. This topic has been studied in detail by several major international research projects as well as international consortia, such as Greentouch [1].

Many previous papers on energy efficiency mainly focus on the energy associated with transmitting radio signals from the antenna. However, when studying mobile networks, it is important to evaluate the overall energy consumption of the network, considering all system components. These evaluations can also take into account the embodied energy associated with manufacturing, installing and maintaining the network over its lifetime as well as the operational energy of actually powering network components. A study in Reference 2 suggests that from 2007 to 2020, the overall carbon footprint of mobile networks may increase threefold. In this context, many researchers have evaluated the potential energy improvements of different communications technologies to mitigate this increase and improve energy efficiency.

When evaluating energy efficiency, it is important to understand that typical networks are not homogeneous, but patterns of use vary significantly, e.g., see Reference 3. It has been observed that the traffic load in different cells of typical networks can vary tremendously, meaning that some cells may be supporting many users or high data rate services (high traffic load), while others may have a very small number of active connections with low bandwidth services (low traffic load). Different techniques are likely to be more applicable for low traffic load conditions than for high load scenarios [4].

Several tutorial papers and books have discussed key concepts for energy saving [4–6] in wireless networks. One important concept for low traffic load conditions relates to energy saving through sleep modes, which switch off unused network components and resources. A second approach which is more applicable at high traffic loads is to improve the energy efficiency of network components when they are switched on. Examples of such approaches include improving the energy efficiency of power amplifiers (PAs) [7], energy efficient packet transmission methods [8] and interference mitigation techniques [9].

Another important approach relates to re-thinking the network topology to improve energy efficiency. The key concept here is to try to reduce the distance between the network and the user, in order to reduce the transmit power required to communicate information to the receiver. If this strategy is successful, it can lead to smaller, lower energy consumption base stations (BSs) and network components reducing the overall energy needed for the network. One method is to use low power picocells and femtocells, in place of existing high power macro-BSs, the effects of which are studied in References 10 and 11. A second method that will be studied in this chapter in more detail is the use of wireless cooperation techniques between BSs, mobile terminals and relays to improve energy consumption [10, 12].

7.2 Motivation

The idea of using relays in wireless communications has been known for a long time. In the last 10–15 years, it has become a very active research topic to evaluate its performance in wireless and cellular networks, e.g., in References 10, 13–15. These methods are often also termed multi-hop communications or wireless cooperation in the literature. From an energy efficiency perspective, relaying is an important approach as locating a relay terminal between the transmitter and receiver results in shorter source–relay and relay–destination distances. This means that the transmit power required on each link is reduced compared to that required for the direct source–destination path. Relay devices are typically simpler, lower power and lower cost devices when compared to wide area macro-BSs. So in some scenarios relays can be used advantageously to reduce the deployment density of macro-BSs, reducing network cost and energy consumption [10]. This means that relay schemes are a potentially important technology for improving the energy efficiency of wireless networks.

This chapter will evaluate different relay schemes from an energy efficiency perspective. It is complementary to the research described in Chapter 8 which focusses mostly on protocol design and energy efficiency evaluations of *amplify-and-forward* (AF) relaying under the assumption of non-existence of direct transmission. This is a simple relaying scheme which uses amplification of received signals to forward messages to the destination terminal from the source terminal. By contrast, the present chapter will provide an energy evaluation of AF with *decode-and-forward* (DF) relaying and with *compress-and-forward* (CF) relaying with the consideration of the cooperation between the relaying transmission and direct transmission. It will also discuss how the CF protocol can be used to support energy efficient multiple antenna communication between a single transmitter and multiple cooperating receivers. Further, methods for modelling the total energy consumption of mobile

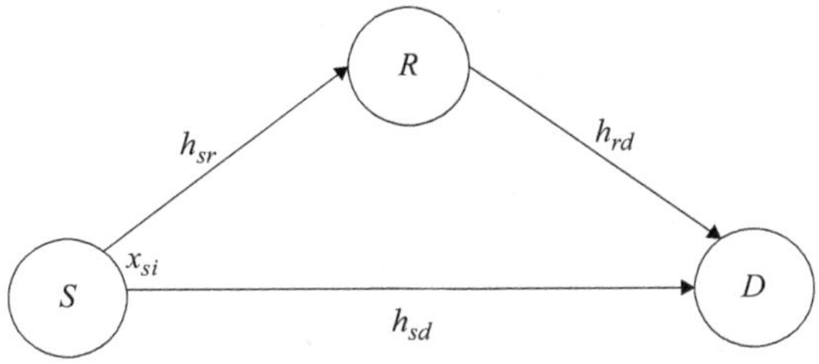

Figure 7.1 Three-node network

and BS devices will be described, as these are important for understanding the energy footprint of mobile networks and how it is affected by implementing schemes such a relay techniques.

The remainder of this chapter is structured as follows. Section 7.3 will describe the system model evaluated in this chapter, including the scenarios under consideration, the communication protocols and metrics for performance evaluation. Section 7.4 presents the performance evaluation and optimization of AF, DF and CF protocols for a three-node scenario. Section 7.5 moves on to address specifically the use of relay protocols to support performance enhancements of closely spaced receivers to form virtual multi-input multi-output (VMIMO) setups. Finally, Section 7.6 presents conclusions to this chapter.

7.3 System model and description

7.3.1 Basic system model

Consider the three-node network shown in Figure 7.1, where the source node (S) transmits to the destination node (D) with the assistance of the relay node (R) [13]. We consider that the source terminal, the destination and relay are equipped with single-antenna transmitter and receiver. Furthermore, we assume that the nodes operate in a half-duplex mode and cannot transmit and receive simultaneously on the same frequency spectrum.

The relay node assists the communication with the destination node by either amplifying-and-forwarding (AF), decoding-and-forwarding (DF) or compressing-and-forwarding (CF) the received signal. In the AF operation mode, the relay node amplifies and re-transmits the signal received, subject to its power constraint. In the DF mode, the signal received from the source terminal is demodulated and decoded before re-transmission. Finally, if a CF operation mode is implemented, a quantized and compressed version of the received signal is forwarded to the destination terminal using a source coding method.

For each of the aforementioned cooperation modes (AF, DF and CF), three different time-division multiple-access (TDMA)-based transmission protocols can be considered. For the transmission protocol Type I, the source node communicates with the relay and destination nodes during the first time slot. In the second time slot,

Table 7.1 TDMA-based protocols. S, R and D stand for the source, relay and destination nodes, respectively. A→ B represents a communication between nodes A and B

Time slot	Type I	Type II	Type III
1	S → R, D	S → R, D	S → R
2	S → D, R → D	R → D	S → D, R → D

both the relay and source nodes communicate with the destination node [13]. During the operation Type II, only the relay node transmits to the destination node over the second time slot [14]. Thus, the third type is identical to Type I apart for the fact that there is no direct transmission from source to destination during the first time slot [16]. The TDMA transmission protocols are summarized in Table 7.1 [13].

Throughout this chapter, we assume a block fading channel model and perfect time synchronization. The subscripts s, r and d are used to classify the transmitted (or received) symbols at the source, relay and destination, respectively. The signal transmitted by the source during the ith time slot is denoted by $x_{s,i}$ for $i = 1, 2$. We assume that $\mathbb{E}[x_{s,i}] = 0$ and $\mathbb{E}[|x_{s,i}|^2] = 1$. The signals received at the relay and destination nodes in the first time slot are given by

$$y_r = f_{sr}\sqrt{P_s}h_{sr}x_{s,1} + z_{r,1}, \tag{7.1}$$

$$y_{d,1} = f_{sd,1}\sqrt{P_s}h_{sd}x_{s,1} + z_{d,1}, \tag{7.2}$$

where $P_s \leq P_s^{\max}$ is the transmit power of the source node over one symbol period and $f_{sr} = 1$ when there is a S → R link and $f_{sr} = 0$ when there is no transmission between the source and relay nodes, the same follows for $f_{sd,1}$ and $f_{sd,2}$, when the S → D direct link is considered in the first time slot and the second slot, respectively. Furthermore, the data channels from the source to the node $j \in \{r, d\}$ are represented by $\sqrt{\alpha_{sj}}\tilde{h}_{sj}$, where $\sqrt{\alpha_{sj}}$ denotes the large-scale fading gain and $\tilde{h}_{sj}$ is the small-scale fading gain modelled as Rayleigh fading. Moreover, $z_{j,i} \sim \mathcal{CN}(0, 1)$ represents the normalized white noise over the ith time slot at the jth node. During the second time slot, the destination node receives a superposition of the relay and source transmission according to

$$y_{d,2} = f_{sd,2}\sqrt{P_s}h_{sd}x_{s,2} + \sqrt{P_r}h_{rd}x_r + z_{d,2}, \tag{7.3}$$

where $P_r \leq P_r^{\max}$ denotes the relay transmit power and x_r is the re-transmitted signal from the relay to the destination node over the second time slot. Note that the source and the destination will be in idle mode if $f_{sd,2} = 0$ and $f_{sd,1} = 0$, respectively.

7.3.2 Metrics for energy evaluation

Wireless relaying is known as an useful technology to provide reliable transmission, high throughput and broad coverage for wireless systems. By utilizing the relay(s),

Table 7.2 Scenario-controlled relaying System(flag1, flag2, flag3)

Transmission types	**Type I**	**Type II**	**Type III**
flag1	1	2	3
Relaying protocols	AF	DF	CF
flag2	1	2	3
Power allocation optimization	Source	Relay	Both
flag3	1	2	3

a long-distance wireless transmission can be broken into two or more short-distance transmission such that the transmit power can be reduced, which in return might degrade the transmission data rate since the half-duplex relaying transmission takes more time than direct transmission. In addition, as the growing demand for energy for information technology comes into conflict with limited energy sources, the trade-off between the data rate and the energy consumption of a relay system becomes a more critical performance metric [4].

The energy efficiency of a relay system mainly depends on its transmission types, AF/DF/CF relaying protocol selection and the transmit strategies design. Thus, for a fixed relay deployment, its energy efficiency can be evaluated from the following three aspects:

1. Transmission types: to evaluate three transmission types defined in Table 7.1 when the relaying protocol is fixed;
2. Relaying protocols: to compare AF/DF/CF three relaying protocols when the transmission type is fixed;
3. Power allocation: to optimize the energy efficient power allocation of the source and the relay and determine the energy efficient locations for the relay or the destination when the transmission type and relaying protocol are fixed.

Based on the above energy evaluation aspects, to be more general, we let System(flag1, flag2, flag3) denote a general relay system, where "flag1", "flag2" and "flag3" are the scenario-controlled indices as defined in Table 7.2. Different relay scenarios can be determined by setting different values to the indices "flag1", "flag2" and "flag3". The relay system defined in Table 7.2 consists of $3 \times 3 \times 3 = 27$ scenarios in total.

Metrics are needed to evaluate the energy efficiency of the different relay system scenarios. For each system scenario, three main metrics in energy efficiency are adopted: energy consumption ratio (ECR), required energy consumption (REC) and energy consumption gain (ECG).

The ECR is used to measure the energy consumed by a system for a data bit communication and is expressed in J/bit as

$$\mathrm{ECR}(\mathrm{flag1}, \mathrm{flag2}, \mathrm{flag3}) = \frac{\mathrm{Energy\ consumed}(\mathrm{flag1}, \mathrm{flag2}, \mathrm{flag3})}{\mathrm{Data\ bits\ communicated}(\mathrm{flag1}, \mathrm{flag2}, \mathrm{flag3})}. \tag{7.4}$$

Another metric directly related to the ECR is its inverse called the energy efficiency ratio (EER) measured in bits/J as

$$\text{EER}(\text{flag1}, \text{flag2}, \text{flag3}) = \frac{1}{\text{ECR}(\text{flag1}, \text{flag2}, \text{flag3})} \tag{7.5}$$

to measure the number of data bits communicated for one joule of energy consumed.

When there exists a transmission requirement, the REC is defined to measure the minimum required energy consumed in joules by a system such that a predefined quality of service (QoS) target is satisfied, i.e.,

$$\text{REC}(\text{flag1}, \text{flag2}, \text{flag3}) = \min_{\text{QoS target is met}} \text{Energy consumed}(\text{flag1}, \text{flag2}, \text{flag3}), \tag{7.6}$$

which is usually used to evaluate the energy saving performance.

The ECG is adopted to directly compare the energy consumed by two systems in a ratio when both are communicating with the same number of data bits or fulfil the same QoS target. For instance, the ECG of the AF relaying protocol under test relative to the reference DF relaying protocol can be expressed as

$$\text{ECG}(\text{flag1}, 1:2, \text{flag3}) = \frac{\text{Energy consumed}(\text{flag1}, 1, \text{flag3})}{\text{Energy consumed}(\text{flag1}, 2, \text{flag3})}, \tag{7.7}$$

where $\text{ECG}(\text{flag1}, 1:2, \text{flag3}) > 1$ indicates an improvement in energy efficiency by the DF relaying protocol compared to the AF relaying protocol. Equivalently, this energy efficiency improvement can be also measured by the energy reduction gain (ERG)

$$\text{ERG}(\text{flag1}, 1:2, \text{flag3}) = 1 - \frac{1}{\text{ECG}(\text{flag1}, 1:2, \text{flag3})}, \tag{7.8}$$

where $\text{ERG}(\text{flag1}, 1:2, \text{flag3}) > 0$ implies an energy reduction by the DF relaying protocol compared to the AF relaying protocol.

As a summary, the energy efficiency improvement can be evaluated by the above metrics as follows:

$$\text{ECR} \downarrow (\text{EER} \uparrow), \ \text{REC} \downarrow, \ \text{ECG} \downarrow (\text{ERG} \uparrow) \Leftrightarrow \text{Energy efficiency} \uparrow. \tag{7.9}$$

In the following, we will show how to evaluate these energy efficiency metrics by formulating optimization problems.

7.3.3 Definition of optimization problems

Without loss of generality, we consider a general two-hop relay system, denoted by System(flag1, flag2, flag3), where the half-duplex cooperative relay transmission combined with the direct transmission can enhance the received signal-to-noise ratio (SNR) by taking two phases as shown in Table 7.1. To simplify the notation, we assume that the full radio spectrum but half of the time resource is used in each phase transmission $S \to R, D$ and $R \to D$.

Let T_c and B_c denote transmission time in seconds and transmission bandwidth in Hz. We assume that the channel state information (CSI) stays constant in the time–frequency chunk $T_c \times B_c$. Then, the amount of transmission bits measured in bits within $T_c \times B_c$ can be computed by

$$\text{Data bits communicated(flag1, flag2, flag3)} = T_c B_c \text{Rate(flag1, flag2, flag3)}, \tag{7.10}$$

where Rate(flag1, flag2, flag3) measured in bits/s/Hz denotes the achievable rate of the relay system System(flag1, flag2, flag3) at the cost of the total energy consumption in joules

$$\text{Energy consumed(flag1, flag2, flag3)} = T_c \text{Power(flag1, flag2, flag3)}, \tag{7.11}$$

where Power(flag1, flag2, flag3) denotes the sum power consumption over the two phases. The power model of Power(flag1, flag2, flag3) will be provided in Section 7.3.4.

Recalling the definitions of energy evaluation metrics, the energy efficient system operation can be determined by minimizing the ECR and REC. Given a typical relay system, i.e., the fixed values of flag1 and flag2 and the value of flag3 $= 3$, the ECR minimization and REC minimization problem can be formulated as follows:

$$\text{(Min-ECR)} \quad \text{minimize}_{\{P_{s,i}\},P_r} \frac{\text{Power(flag1, flag2, 3)}}{B_c \text{Rate(flag1, flag2, 3)}} \tag{7.12a}$$

$$\text{subject to Rate(flag1, flag2, 3)} \geq R^{\min} \tag{7.12b}$$

$$P_{s,i}\text{(flag1, flag2, 3)} \leq P_s^{\max}, \forall i \tag{7.12c}$$

$$P_r\text{(flag1, flag2, 3)} \leq P_r^{\max}, \tag{7.12d}$$

where (7.12b) denotes an achievable rate requirement with the minimum rate target $R^{\min}$, and (7.12c) and (7.12d) denote the transmit power constraints of the source and relay, respectively. Clearly, minimization of the objective (7.12a) is equivalent to maximization of its inverse.

The REC minimization problem is formulated as

$$\text{(Min-REC)} \quad \text{minimize}_{\{P_{s,i}\},P_r} \text{Power(flag1, flag2, 3)} \tag{7.13a}$$

$$\text{subject to Rate(flag1, flag2, 3)} \geq R^{\min} \tag{7.13b}$$

$$P_{s,i}\text{(flag1, flag2, 3)} \leq P_s^{\max}, \forall i \tag{7.13c}$$

$$P_r\text{(flag1, flag2, 3)} \leq P_r^{\max}. \tag{7.13d}$$

Since for the three-node relay system, the achievable rate is monotonically increasing with both the transmit power of the source and the relay, thus the rate constraint (7.13b) will become an equality constraint in the optimization problem, i.e., Rate(flag1, flag2, 3) $= R^{\min}$ in order to reduce the energy consumption. This is not necessary to be always true for the rate constraint (7.12c) in Problem (Min-ECR).

By analysing the activeness of some constraints, Problem (Min-ECR) is equivalent to several simplified problems. The first one is the rate maximization problem

$$\text{(Max-Rate)} \quad \text{maximize}_{\{P_{s,i}\},P_r} \text{Rate(flag1, flag2, 3)} \tag{7.14a}$$

$$\text{subject to } P_{s,i}\text{(flag1, flag2, 3)} \leq P_s^{\max}, \forall i \tag{7.14b}$$

$$P_r\text{(flag1, flag2, 3)} \leq P_r^{\max}, \tag{7.14c}$$

when the rate constraint (7.12b) achieves equality at the optimal operating point, i.e., $\text{Rate}(\text{flag1}, \text{flag2}, 3) = R^{\min}$. The second equivalent problem is Problem (Min-REC) when all its power constraints (7.12c)–(7.12d) become equality constraints, i.e., by the full power transmission. Problem (Min-ECR) is also equivalent to

$$\text{(Min-SoleECR)} \quad \text{minimize}_{\{P_{s,i}\},P_r} \ \frac{\text{Power}(\text{flag1}, \text{flag2}, 3)}{B_c\text{Rate}(\text{flag1}, \text{flag2}, 3)}, \tag{7.15}$$

if all the constraints (7.12b)–(7.12d) do not achieve equality at the optimal operating point and thus they can be removed without influence on the optimal solutions.

For different scenarios, their minimum required power consumption can be computed by optimizing Problem (Min-REC). In order to determine the most energy efficient or energy saving transmission types or relaying protocols, their energy efficiency can be compared by evaluating the ratio of their required power values in (7.13a), using the definitions of the ECG and ERG.

7.3.4 Power consumption model

Even though relay technology was first proposed to extend coverage and improve capacity of wireless systems [17], currently, it is also put forward as an effective approach to improve the energy efficiency [5, 18] of the wireless communication systems. As the power consumption model plays an important role in the analysis of the energy efficiency for relay system, several models have been proposed. A model that evaluates the communication related energy consumption behaviour of a mobile network was presented in Reference 19, where the power consumption was further categorized into three modes: transmission, reception and idle modes. Furthermore, since the BSs have the largest share of the energy consumption of cellular networks, an important trend has become developing models that provide an accurate power consumption for the BSs. The elements which contribute the most with the overall energy consumption of a BS can be divided into two groups: the radio-frequency equipment and the support system. The largest energy consumer in a BS is the PA, which has a share of around 64% of the total energy consumption, meanwhile, the cooling system (17.5%), digital signal processing (10%) and AC/DC conversion elements (7.5%) completed the list [20].

In Reference 21, a non-ideal PA model is considered. Meanwhile, in Reference 22, the authors derived an adaptive transmission power control extension of the raised-fractional root transmitter power consumption model and compared the energy efficiency of a single-hop and two-hop wireless transmission. An upper bound of the energy efficiency for a multi-hop wireless sensor network is presented in Reference 23 based on a power consumption model that incorporates characteristics of a typical low power transceiver. Finally, in Reference 24, a flexible power model that provides realistic power consumption values for a BS over a broad range of configuration and technologies is presented. The model supports different BS architectures, from conventional macro- and picocells to massive multi-input multi-output (MIMO) architectures with hundreds of antennas. This model calculates the power consumption considering the five main components of the BS: the PA, the analog front-end, the digital baseband (BB), the control and network backhaul and the power systems.

In order to quantify the total energy consumption of the relay system in (7.11), both the transmit power and the circuit power consumed at the nodes should be considered. Thus, the total power consumption of the three-node relay system shown in Figure 7.1 is the sum of the transmit power, signal processing and circuit power (including the backhaul power consumption) at the source, the destination and relay nodes, respectively. In this chapter, based on the BS power models introduced in References 11 and 25, the sum power consumption model of a relay system in (7.11) is provided as follows:

$$\begin{aligned}&\text{Power}(\text{flag1},\text{flag2},\text{flag3})\\&\quad=\frac{1}{2}\left(\sum_{i=1}^{2}\left(\frac{1}{\eta_s}P_{s,i}(\text{flag1},\text{flag2},\text{flag3})+P_s^{c,i}(\text{flag1},\text{flag2},\text{flag3})\right)\right)\\&\qquad+\frac{1}{2}\left(\frac{1}{\eta_r}P_r(\text{flag1},\text{flag2},\text{flag3})+\sum_{i=1}^{2}P_r^{c,i}(\text{flag1},\text{flag2},\text{flag3})\right)+\frac{1}{2}\sum_{i=1}^{2}P_d^{c,i}.\end{aligned} \tag{7.16}$$

In (7.16), $P_{s,i}(\text{flag1},\text{flag2},\text{flag3})$ and $P_s^{c,i}(\text{flag1},\text{flag2},\text{flag3})$ in watts denote the sum transmit power of the source(s) with the PA efficiency $\eta_s < 1$ and the sum source circuit power in the phases $i \in \{1,2\}$. When the source(s) do not transmit over the second phase, i.e., $f_{sd,2} = 0$, we have $P_{s,2}(\text{flag1},\text{flag2},\text{flag3}) = 0$ and $P_s^{c,2}(\text{flag1},\text{flag2},\text{flag3}) \ll P_s^{c,1}(\text{flag1},\text{flag2},\text{flag3})$. For the relays power consumption in (7.16), $P_r(\text{flag1},\text{flag2},\text{flag3})$ and $P_r^{c,i}(\text{flag1},\text{flag2},\text{flag3})$ in watts denote the sum transmit power of the relay(s) with the PA efficiency $\eta_r < 1$ and the sum circuit power in the phase $i \in \{1,2\}$, respectively. The transmit power consumption is data traffic dependent, and the circuit power includes both the traffic-dependent signal processing power consumption and the static power consumption. In (7.16), $P_d^{c,i}(\text{flag1},\text{flag2},\text{flag3})$ denotes the signal processing and circuit power consumption of the destination in the phase $i \in \{1,2\}$.

For the power model in (7.16), the signal processing and circuit power of a source (e.g., a BS in downlink) can be expressed as follows:

$$\begin{aligned}P_s^{c,i}(\text{flag1},\text{flag2},\text{flag3}) &= P_s^0(\text{flag1},\text{flag2},\text{flag3})+P_s^{sp,i}(\text{flag1},\text{flag2},\text{flag3})\\&\quad+P_s^{bh,i}(\text{flag1},\text{flag2},\text{flag3}),\end{aligned} \tag{7.17}$$

where P_s^0 denotes the very basic *static* power consumption of a BS to support its idle mode, e.g., the power consumed by the DC–DC power supply, mains supply and active cooling system. This static power P_s^0 is usually different for different types of BSs. And $P_s^{sp,i}$ denotes the power consumption by the BB interface and the signalling of RF transceiver (RF-TRX) of the BS. Since $P_s^{sp,i}$ mainly depends on the bandwidth, the number of antennas and the resolution of the analog-to-digital conversion, it can be approximately expressed as [25]

$$P_s^{sp,i}(\text{flag1}, \text{flag2}, \text{flag3}) = \begin{cases} M \frac{B_c}{10 \text{ MHz}}(P'_{BB} + P'_{RF}) & \text{when } f_{sr,i} + f_{sd,i} > 0 \\ 0 & \text{Otherwise,} \end{cases} \tag{7.18}$$

where M is the number of the source transmit antennas, and P'_{BB} and P'_{RF} are the reference BB and RF related signal processing power consumption terms per 10 MHz bandwidth, respectively. As shown in (7.18), these terms are equal to zero if the source is in idle mode. In (7.17), $P_s^{bh,i}(\text{flag1}, \text{flag2}, \text{flag3})$ measures the power consumption for the backhaul connection, usually including the exchange of the CSI, transmission data and the signalling between coordinated nodes as shown in the following expression:

$$\begin{aligned} P_s^{bh,i}(\text{flag1}, \text{flag2}, \text{flag3}) &= P^{bh,0}\text{Rate}(\text{flag1}, \text{flag2}, \text{flag3}) \\ &\quad + P^{CSI}(\text{flag1}, \text{flag2}, \text{flag3}), \end{aligned} \tag{7.19}$$

where the first term accounts for the power consumption related to transmission data acquisition from the data centre where $P^{bh,0}$ is a constant power in W/bit/Hz, and thus this term increases linearly with the data rate. In (7.19), the second term $P^{CSI}(\text{flag1}, \text{flag2}, \text{flag3})$ depends on the CSI acquisition, e.g., channel training, channel estimation, channel quantization and feedback. In particular, $P^{CSI}(\text{flag1}, \text{flag2}, \text{flag3})$ is an important factor in the imperfect CSI scenario, since there exists a trade-off between robust data rate performance and the power consumption for CSI acquisition. In this chapter, we assume a perfect CSI scenario where the backhaul power consumption can be approximated as a constant term, since $P^{CSI}(\text{flag1}, \text{flag2}, \text{flag3})$ becomes the maximum power consumption for the CSI acquisition and in addition $P^{bh,0}\text{Rate}(\text{flag1}, \text{flag2}, \text{flag3})$ does not influence the optimization of the ECR.

The relay signal processing and circuit power consumption can be similarly modelled by considering it as a picocell or femtocell BS (like the source) or a mobile terminal (like the destination). In particular, $P_r^{c,i}(\text{flag1}, \text{flag2}, \text{flag3})$ highly depends on the relaying protocols according to their processing complexity. In general, we have

$$P_r^{c,i}(\text{flag1}, 1, \text{flag3}) \leq P_r^{c,i}(\text{flag1}, 3, \text{flag3}) \leq P_r^{c,i}(\text{flag1}, 2, \text{flag3}). \tag{7.20}$$

Typically, the circuit power consumption is much larger than differences in power consumed by different relay protocols [12].

Remark 7.1. *The power consumption can be generally classified into two parts: the rate-dependent power and rate-independent power. If the rate-dependent power can be expressed as a linear function of the data rate, it will not influence the solutions of the ECR minimization problem. For the rate-independent power, if the deployment is fixed, this part can be considered as a constant power in transmit/receive strategies design.*

In the next section, the energy efficiency of the system shown in Figure 7.1 will be analysed based on the total power consumption model described in (7.16).

7.4 Comparison of EE of relay protocols for three-node scenario

We consider the energy evaluation of the classical three-node relay system where the source transmits a data symbol only in the first phase and is in idle mode in the second phase, denoted by System(2, flag2, flag3). We describe three typical relaying protocols utilized in the three-node relaying network, investigate their achievable rates, optimize and compare their energy efficiency performance for different scenarios by setting different values to "flag2" and "flag3", respectively.

7.4.1 Relaying protocols

The relaying protocol determines what the relay should do after receiving the signal. Three types of processing are employed at the relay, i.e., to amplify its received signal subject to its power constraint (AF relaying), or to decode, re-encode and re-transmit the message (DF relaying), or to forward a quantized and compressed version of the message employing a source coding method (CF relaying). In the following, we first briefly introduce them and show their different achievable rates.

7.4.1.1 AF relaying protocol

In the AF relaying protocol, the received signal at the relay will be amplified and then forwarded to the destination, which is a linear processing and with lowest complexity. As shown in Table 7.1, in the first phase, a data symbol x_s is transmitted by the source to the destination and the relay by the beamforming vector $w \in \mathbb{C}^{M\times 1}$ satisfying $||w||^2 \leq P_s^{\max}$, which is like the multi-cast transmission.

As shown in (7.1), in the first phase, the information received at the relay and the destination are

$$y_r = h_{sr}^H w x_s + z_{r,1}, \tag{7.21}$$

$$y_{d,1} = h_{sd}^H w x_s + z_{d,1}, \tag{7.22}$$

respectively.

In the second phase, the source is in idle for energy saving purpose, and the relay amplifies and forwards the signal y_r by an amplifier factor a subject to the relay amplifying power budget

$$a \leq \sqrt{\frac{P_r^{\max}}{|h_{sr}^H w|^2 + 1}}. \tag{7.23}$$

Thus, in the second phase, the received signal at the destination is

$$y_{d,2} = h_{sd}^H w x_s + h_{rd} a y_r + z_{d,2}, \tag{7.24}$$

based on which, the data symbol x_s can be detected and yields an achievable rate in the AF relaying protocol [14]

$$\text{Rate}_{AF} = \frac{1}{2}\log_2\left(1 + |h_{sd}^H w|^2 + \frac{|h_{sr}^H w|^2 P_r |h_{rd}|^2}{1 + |h_{sr}^H w|^2 + P_r |h_{rd}|^2}\right), \tag{7.25}$$

where $P_r \triangleq a(|h_{sr}^H w|^2 + 1)$ denotes the relaying power subject to

$$P_r \le P_r^{\max}. \tag{7.26}$$

7.4.1.2 DF relaying protocol

The processing in the first phase of the DF relaying protocol is the same as that in the AF relaying protocol. Instead, in the second phase, x_s will be decoded, re-encoded and re-transmitted at the relay based on y_r. Assuming that x_s is successfully detected as $\widehat{x_s}$, the received signal at the destination in the second phase is

$$y_{d,2} = h_{sd}^H w x_s + h_{rd} P_r \widehat{x_s} + z_{d,2}. \tag{7.27}$$

Based on (7.27), the maximum mutual information with full decoding at the DF relay is shown as [14]

$$\text{Rate}_{DF} = \frac{1}{2}\min\left\{\log_2\left(1 + |h_{sr}^H w|^2\right), \log_2\left(1 + |h_{sd}^H w|^2 + P_r|h_{rd}|^2\right)\right\}, \tag{7.28}$$

where the first term in (7.28) represents the maximum rate at which the relay can reliably decode the source message x_s, and the second term in (7.28) represents the maximum rate at which the destination can reliably decode the source message x_s given repeated transmissions from the source and destination. Requiring both the relay and destination to decode the entire codeword without error results in the minimum of the two mutual information terms in (7.28).

7.4.1.3 CF relaying protocol

In the CF relaying protocol, the relay forwards a quantized and compressed version of the received signal by employing the standard quantization, or some source coding techniques to compress the signal.

According to Proposition 7.1 in Reference 26, in the second phase, the three-node system is equivalent to a system where the destination has two antennas that receive the signals

$$\begin{bmatrix} y_r + z_{cf} \\ y_{d,1} \end{bmatrix}, \tag{7.29}$$

where z_{cf} is the independent and identically distributed circular Gaussian noise (compression noise), which is independent of y_r and $y_{d,1}$. If the relay node employs Wyner–Ziv (WZ) coding [27], one possible source coding technique to compress the signal, the compression noise z_{cf} has the power level:

$$\sigma_{cf}^2 = \frac{\mathbb{E}[|y_r|^2]\mathbb{E}[|y_{d,1}|^2] - |\mathbb{E}[y_{d,1}y_r^*]|^2}{(2^{R_c-1})\mathbb{E}[|y_{d,1}|^2]}, \tag{7.30}$$

where R_c is the capacity of the error-free conference link between the relay and the destination and is given by

$$R_c = \log_2\left(1 + P_r|h_{rd}|^2\right). \tag{7.31}$$

Note that, besides the WZ coding technique, the relay can also employ a standard source coding technique when compressing the signal, at the cost of a slightly lower achievable rate in theory [28].

The achievable rate in the second phase of the CF relaying protocol can be expressed as

$$\text{Rate}_{CF} = \frac{1}{2}\log_2\left(1 + |h_{sd}^H w|^2 + \frac{|h_{sr}^H w|^2}{1+\sigma_{cf}^2}\right), \tag{7.32}$$

where the compression noise has the power

$$\sigma_{cf}^2 = \frac{|h_{sr}^H w|^2 + |h_{sd}^H w|^2 + 1}{P_r |h_{rd}|^2 (|h_{sd}^H w|^2 + 1)}, \tag{7.33}$$

which is derived by both (7.30) and (7.31).

Remark 7.2. *Both the AF achievable rate in* (7.25) *and the CF achievable rate in* (7.32) *depend on the enhanced SNR by both relaying and direct transmission. The only difference is the expression of the SNR of the relaying link because of their different relay processing. The DF achievable rate in* (7.28) *is derived based on the assumption that the received information at the DF relay is full decoded, which restricts the DF rate by the first-hop transmission.*

7.4.2 Optimization and comparison of ECR minimization

As we analysed in Section 7.3.3, Problem (Min-ECR) is a more general problem to evaluate the energy efficiency performance of a system. In this part, the energy efficiency is evaluated and compared by solving the ECR minimization problem of the different relaying protocols.

First, based on Problem (Min-ECR), the ECR minimization problems for the AF, DF and CF relaying protocols are formulated as

$$(\text{Min-ECR}) \begin{cases} \min\limits_{w,P_r} \dfrac{\frac{1}{2}\left(\frac{1}{\eta_s}||w||^2 + P_s^c + \frac{1}{\eta_r}P_r + P_r^c\,\text{Rate}(2,\text{flag2},3)\right)}{\text{Rate}(2,\text{flag2},3)} \\ \text{s.t. Rate}(2,\text{flag2},3) \geq R^{\min} \\ ||w||^2 \leq P_s^{\max},\ P_r \leq P_r^{\max}, \end{cases} \tag{7.34}$$

where Rate(2, flag2, 3) refers to the AF rate in (7.25) when flag2 $= 1$, the DF rate in (7.28) when flag2 $= 2$ and the CF rate in (7.33) when flag2 $= 3$, respectively, and $R^{\min}$ denotes the minimum achievable rate target. Based on the power model in (7.16), P_s^c denotes the sum circuit power consumption of the source in two phases, and $P_r^c(2, \text{flag2}, 3)$ denotes the sum circuit power consumption of the relay in two phases, which might be different for the three types of relaying protocols. In this problem, as flag3 $= 3$, both the source precoder and relay transmit power, i.e., $\{w, P_r\}$, are jointly optimized. The case of flag3 $= 1$ or flag3 $= 2$ when the sole w or P_r is optimized provides a upper bound of the ECR performance to Problem (7.34).

Observe that the ECR minimization problem (7.34) is a fractional program, which is non-convex in general. In order to make it tractable, fractional programming will be employed to solve this problem. First, a general fractional programming framework is provided.

7.4.2.1 Fractional programming

Consider a fractional program

$$(\mathrm{P_G})\quad \max_{x \in \mathcal{S}_x} \frac{f_1(x)}{f_2(x)}, \tag{7.35}$$

where $f_1(x) \geq 0$ is strictly concave, differentiable and monotonically increasing in x, $f_2(x) > 0$ is affine, and $S_x \subseteq \mathbb{R}_+^n$ is a convex set. Problem $(\mathrm{P_G})$ is strictly pseudo-concave since $f_1(x)$ is a strictly concave and differentiable function.

Proposition 7.1. *[29] Any stationary point of pseudo-concave fractional program is its global maximum. If in addition it is strictly pseudo-concave, implying that the global maximum, if exists, is unique.*

Corollary 7.1. *[30] Assume $x^\star$ is a stationary/optimal solution to Problem $(\mathrm{P_G})$. Then, $\frac{f_1(x)}{f_2(x)}$ is monotonically increasing in $f_2(x) \leq f_2(x^\star)$ and monotonically decreasing in $f_2(x) \geq f_2(x^\star)$.*

Proposition 7.2. *[30] If the feasible set $\mathcal{S}_x$ in Problem $(\mathrm{P_G})$ contains the box constraints of $f_1(x)$ and $f_2(x)$ in such a form*

$$\mathcal{S}_x = \{x \in \mathcal{S}_x^0 : f_1(x) \geq \underline{f_1}, \underline{f_2} \leq f_2(x) \leq \overline{f_2}\}, \tag{7.36}$$

where $\mathcal{S}_x^0$ is a convex set, $\underline{f_1}$, $\underline{f_2}$ and $\overline{f_1}$ denote the feasible lower/upper bounds of the box constraints. The bounds are feasible if $\underline{f_1} \leq \max_{x \in \mathcal{S}_x^0, f_2(x) = \overline{f_2}} f_1(x)$ is fulfilled.

Let $x^\star$ be an optimal solution to the following program without considering the box constraints

$$(\mathrm{P_G^0})\quad \max_{x \in \mathcal{S}_x^0} \frac{f_1(x)}{f_2(x)}, \tag{7.37}$$

then we can simplify Problem $(\mathrm{P_G})$ to one of the following equivalent problems.

- ***Case 1:** If $\underline{f_2} \leq f_2(x^\star) \leq \overline{f_2}$ and $f_1(x^\star) \geq \underline{f_1}$, then Problem $(\mathrm{P_G})$ is equivalent to Problem $(\mathrm{P_G^0})$ without box constraints, i.e.,*

$$\max_{x \in \mathcal{S}_x} \frac{f_1(x)}{f_2(x)} = \max_{x \in \mathcal{S}_x^0} \frac{f_1(x)}{f_2(x)} = \frac{f_1(x^\star)}{f_2(x^\star)}. \tag{7.38}$$

- ***Case 2:** If $\underline{f_2} \leq f_2(x^\star) \leq \overline{f_2}$ and $\underline{f_1} > f_1(x^\star)$, then Problem $(\mathrm{P_G})$ can be solved by minimizing $f_2(x)$, i.e.,*

$$\max_{x \in \mathcal{S}_x} \frac{f_1(x)}{f_2(x)} = \frac{\underline{f_1}}{\max_{x \in \mathcal{S}_x^0, f_1(x) = \underline{f_1}} f_2(x)}. \tag{7.39}$$

- ***Case 3:** If $\overline{f_2} < f_2(x^\star)$, then Problem $(\mathrm{P_G})$ can be solved by maximizing $f_1(x)$, i.e.,*

$$\max_{x \in \mathcal{S}_x} \frac{f_1(x)}{f_2(x)} = \frac{\max_{x \in \mathcal{S}_x^0, f_2(x) = \overline{f_2}} \frac{f_1(x)}{f_2(x)}}{\overline{f_2}}. \tag{7.40}$$

Note that Proposition 7.2 is helpful for us not only to determine the feasible bounds of $f_1(x)$ and $f_2(x)$ but also to gain more insight into fractional programs. More precisely, the fractional program with box constraints can be equivalently reduced to a fractional program without box constraints or a problem without the fractional structure.

Various approaches in Reference 29 can solve Problem ($\mathrm{P}_{\mathrm{G}}^{0}$), e.g., the parametric convex program. Then, Problem ($\mathrm{P}_{\mathrm{G}}^{0}$) is equivalent to

$$F(\zeta) = \max_{x \in \mathscr{S}_x^0} f_1(x) - \zeta f_2(x). \tag{7.41}$$

Given ζ, (7.41) is a strictly convex problem w.r.t. x and has an optimal solution $x^\star(\zeta)$. Furthermore, $F(\zeta)$ is continuous, convex, strictly decreasing in ζ [31]. In particular, $F(\zeta) = 0$ implies that $\zeta = \zeta^\star$ where $\zeta^\star = \max_{x \in \mathscr{S}_{x^0}} \frac{f_1(x)}{f_2(x)}$. Thus, solving Problem ($\mathrm{P}_{\mathrm{G}}^{0}$) is equivalent to finding the root of the equation

$$F(\zeta^\star) = f_1(x^\star(\zeta^\star)) - \zeta^\star f_2(x^\star(\zeta^\star)) = 0, \tag{7.42}$$

which can be solved by the Dinkelbach method [31] in Algorithm 1 as follows.

Algorithm 1 is in the spirit of the generalized Newton method and has a super-linear convergence behaviour and yields a unique optimal solution for a strictly pseudo-concave fractional program. Since it is equivalent to maximize a fraction as to minimize its inverse, this above parametric approach will be utilized to solve the ECR minimization problem for the three relaying protocols.

Algorithm 1 The Dinkelbach method to solve (7.42)

Input: ζ_0 satisfying $F(\zeta_0) \geq 0$, tolerance ϵ, $n = 0$. ;
Output: Optimal $\zeta^\star$ and $x^\star$. ;
while $|F(\zeta_n)| > \epsilon$ **do**
  Given $\zeta = \zeta_n$, solve optimal $x_n^\star(\zeta)$ to (7.42);
  $\zeta_{n+1} = \dfrac{f_1(x_n^\star(\zeta))}{f_2(x_n^\star(\zeta))}$;
  $n++$.
Consider $\zeta^\star = \zeta_n$ and $x^\star = x_n^\star(\zeta)$.

7.4.2.2 AF-ECR minimization problem and solutions

In the AF relaying protocol when flag2 $= 1$, Problem (7.34) specifies

$$\begin{cases} \min\limits_{\vec{w}, P_s, P_r} \dfrac{\frac{1}{\eta_s} P_s + \frac{1}{\eta_r} P_r + P_{AF}^c}{\log_2\left(1 + P_s |h_{sd}^H \vec{w}|^2 + \dfrac{P_s |h_{sr}^H \vec{w}|^2 P_r |h_{rd}|^2}{1 + P_s |h_{sr}^H \vec{w}|^2 + P_r |h_{rd}|^2}\right)} \\ \text{s.t. } \dfrac{1}{2} \log_2\left(1 + P_s |h_{sd}^H \vec{w}|^2 + \dfrac{P_s |h_{sr}^H \vec{w}|^2 P_r |h_{rd}|^2}{1 + P_s |h_{sr}^H \vec{w}|^2 + P_r |h_{rd}|^2}\right) \geq R^{\min} \\ \|\vec{w}\| = 1, \;\; P_s \leq P_s^{\max}, \;\; P_r \leq P_r^{\max}, \end{cases} \tag{7.43}$$

where $P_{AF}^c \triangleq P_s^c + P_r^c(2,1,3)$ denotes the sum circuit power consumption of the source and the AF relay in the two phases, and $w \triangleq \sqrt{P_s}\vec{w}$ where P_s and $\vec{w} \triangleq \frac{w}{||w||}$ denote the source transmit power and beamforming pattern. Since it is difficult to optimize w and P_r directly in Problem (7.34) for the AF relaying protocol, our objective is to jointly optimize $\vec{w}$, P_s and P_r in Problem (7.43).

With the fixed P_s and P_r, the beamforming pattern $\vec{w}$ is optimized in closed-form as follows:

Proposition 7.3. *The optimal source beamforming pattern of Problem* (7.43) *in the AF relaying protocol is*

$$\vec{w}^\star = u_{\max}\left(h_{sd}h_{sd}^H + B^{-\frac{1}{2}}h_{sr}h_{sr}^H B^{-\frac{1}{2}}\right), \tag{7.44}$$

where $(\lambda_{\max}(X), u_{\max}(X))$ *denotes the pair of the largest eigenvalue and its corresponding eigenvector of a matrix* X, *and*

$$B \triangleq \frac{P_s}{P_r|h_{rd}|^2}h_{sr}h_{sr}^H + \left(\frac{1}{P_r|h_{rd}|^2} + 1\right)I \tag{7.45}$$

is a full-rank Hermitian matrix.

This result can be proved by solving the rate maximization problem, since the sum power consumption is independent of the source beamforming pattern. In addition, the achievable rate is monotonically increasing with both $|h_{sd}^H w|^2$ and $|h_{sr}^H w|^2$, and it can be maximized by the optimized beamforming pattern in (7.44).

With the fixed $\vec{w}$ in (7.44), Problem (7.43) becomes a power allocation problem, and the ECR is pseudo-convex with respect to P_s and P_r. Therefore, P_s and P_r can be optimized individually by the fractional programming and convex optimization.

Therefore, Problem (7.43) can be solved suboptimally but efficiently by an alternating optimization algorithm as described in Algorithm 2.

Algorithm 2 The alternating optimization algorithm to solve (7.43)

Data: Initial $P_s^{(0)} \le P_s^{\max}$ and $P_r^{(0)} \le P_r^{\max}$, tolerance ϵ, $i = 0$.
Result: Suboptimal solution $\{P_s, P_r, \vec{w}\}$.
repeat:
  $i \leftarrow i + 1$; ;
  Step. 1: Given $P_s^{(i-1)}, P_r^{(i-1)}$, solve the optimal $\vec{w}^{(i)}$ to Problem (7.43); ;
  Step. 2: Given $P_s^{(i-1)}, \vec{w}^{(i)}$, solve the optimal $P_r^{(i)}$ to Problem (7.43); ;
  Step. 3: Given $P_r^{(i)}, \vec{w}^{(i)}$, solve the optimal $P_s^{(i)}$ of Problem (7.43), where the value of the objective function is $\mathrm{ECR}^{(i)}$. ;
until $|\mathrm{ECR}^{(i)} - \mathrm{ECR}^{(i-1)}| \le \epsilon$;

As shown in Algorithm 2, the variables $\vec{w}$, P_r and P_s are optimized alternatively in *Step.1–Step.3* in each iteration. The convergence of Algorithm 2 is guaranteed, since the objective function is monotonically decreasing and in addition the objective function is lower bounded. Note that the global optimality of the solutions obtained by Algorithm 2 is not guaranteed.

7.4.2.3 DF-ECR minimization problem and solutions

In the DF relaying protocol when flag2 = 2, Problem (7.34) becomes

$$\begin{cases} \min\limits_{W\succeq 0, P_r} \dfrac{\frac{1}{\eta_s}\mathrm{Tr}(W) + \frac{1}{\eta_r}P_r + P_{DF}^c}{\log_2\left(1+\min\{\mathrm{Tr}(h_{sr}h_{sr}^H\mathrm{Tr}(W), \mathrm{Tr}(h_{sd}h_{sd}^H\mathrm{Tr}(W) + P_r|h_{rd}|^2\}\right)} \\ \text{s.t. } \frac{1}{2}\log_2\left(1+\min\{\mathrm{Tr}(h_{sr}h_{sr}^H\mathrm{Tr}(W), \mathrm{Tr}(h_{sd}h_{sd}^H\mathrm{Tr}(W) + P_r|h_{rd}|^2\}\right) \geq R^{\min} \\ \mathrm{Tr}(W) \leq P_s^{\max}, \ \ \mathrm{rank}(W) = 1, \ \ P_r \leq P_r^{\max}, \end{cases} \tag{7.46}$$

where $P_{DF}^c \triangleq P_s^c + P_r^c(2,2,3)$ denotes the sum circuit power consumption of the source and the DF relay in the two phases, and $W \triangleq ww^H$ as a semidefinite matrix with a rank-one constraint is used to express the quadric terms of w by the linear terms of W, which is also called semidefinite programming (SDP) [32]. However, it is still hard to solve this problem because of the rank-one constraint. Now, a two-step "relaxation-and-restriction" algorithm is developed to solve Problem (7.46) optimally and efficiently.

First, by relaxing the constraint rank$(W) = 1$, Problem (7.46) becomes a jointly pseudo-convex problem with respect to W and P_r, which can be optimally solved by employing the fractional programming and the SDP solver in CVX. Second, $W^\star$ solved in the first step might be not rank-one, but a rank-one solution $w^\star$ can be extracted from $W^\star$ by the matrix rank-one decomposition technique. Note that the obtained $w^\star$ is optimal to Problem (7.46), since the matrix rank-one decomposition is tight/optimal when the number of constraints of W is no more than four [33].

7.4.2.4 CF-ECR minimization problem and solutions

For the CF relaying protocol when flag2 = 2, Problem (7.34) becomes

$$\begin{cases} \min\limits_{w, P_r} \dfrac{\frac{1}{\eta_s}||w||^2 + \frac{1}{\eta_r}P_r + P_{CF}^c}{\log_2\left(1 + w^H(h_{sd}h_{sd}^H + \frac{1}{1+\sigma_{cf}^2}h_{sr}h_{sr}^H)w\right)} \\ \text{s.t. } \frac{1}{2}\log_2\left(1 + w^H(h_{sd}h_{sd}^H + \frac{1}{1+\sigma_{cf}^2}h_{sr}h_{sr}^H)w\right) \geq R^{\min} \\ w^H\left((\sigma_{cf}^2P_r|h_{rd}|^2 - 1)h_{sd}h_{sd}^H - h_{sr}h_{sr}^H\right)w = 1 - \sigma_{cf}^2P_r|h_{rd}|^2 \\ ||w||^2 \leq P_s^{\max}, \ \ P_r \leq P_r^{\max}, \end{cases} \tag{7.47}$$

where $P_{CF}^c \triangleq P_s^c + P_r^c(2,3,3)$ denotes the sum circuit power consumption of the source and the CF relay in the two phases. This problem has a very complicated structure, since the relaying SNR will have a fourth-order term for w after plugging the expression of σ_{cf}^2 into the rate. In order to make the problem tractable, we consider σ_{cf}^2 is an auxiliary variable and also employ the SDP by defining $W \triangleq ww^H$. Then, Problem (7.47) is equivalent to

$$\begin{cases} \min\limits_{W \succeq 0, \sigma_{cf}^2, P_r} \dfrac{\frac{1}{\eta_s}\mathrm{Tr}(W) + \frac{1}{\eta_r}P_r + P_{AF}^c}{\log_2\left(1 + \mathrm{Tr}((h_{sd}h_{sd}^H + \frac{1}{1+\sigma_{cf}^2}h_{sr}h_{sr}^H)W)\right)} \\ \text{s.t. } \dfrac{1}{2}\log_2\left(1 + \mathrm{Tr}((h_{sd}h_{sd}^H + \frac{1}{1+\sigma_{cf}^2}h_{sr}h_{sr}^H)W)\right) \geq R^{\min} \\ \mathrm{Tr}\left(\left((\sigma_{cf}^2 P_r|h_{rd}|^2 - 1)h_{sd}h_{sd}^H - h_{sr}h_{sr}^H\right)W\right) \geq 1 - \sigma_{cf}^2 P_r|h_{rd}|^2 \\ \mathrm{Tr}(W) \leq P_s^{\max}, \ \ \mathrm{rank}(W) = 1, \ \ P_r \leq P_r^{\max}. \end{cases} \tag{7.48}$$

By fixing σ_{cf}^2 and P_r, Problem (7.48) is pseudo-convex with respect to W after relaxing the rank-one constraint. Then, with the application of the two-step "relaxation-and-restriction" algorithm used in the optimization of (7.46), an optimal w can be computed from the optimized W. With the fixed P_r and w, the optimal σ_{cf}^2 can be easily obtained by (7.33), since the achievable rate is monotonically decreasing with the value of σ_{cf}^2 and thus the second constraint in Problem (7.48) will become an equality constraint as (7.33). The optimizing P_r by fixing W and σ_{cf}^2 yields the solution

$$P_r = \frac{|h_{sr}^H w|^2 + |h_{sd}^H w|^2 + 1}{\sigma_{cf}^2|h_{rd}|^2(|h_{sd}^H w|^2 + 1)}. \tag{7.49}$$

Since each individual term w, σ_{cf}^2 or P_r can be optimally solved by fixing other variables, the alternating optimization algorithm can be also applied to solve Problem (7.48) by optimizing w, σ_{cf}^2 or P_r alternatively in each iteration as shown in Algorithm 2. Thus, a converged solution can be always obtained.

Remark 7.3. *The ECR minimization problem in the DF relaying protocol is jointly and optimally solved. The ECR minimization problems for the AF and CF relaying protocols are suboptimally but efficiently solved by the alternating optimization algorithm with the guaranteed convergence. Note that the ECR minimization problems for all the three relaying protocols with* flag3 = 1 *or* flag3 = 2 *can be optimally solved.*

7.4.3 Study of special cases

In the AF/DF/CF relaying protocol, the optimal beamforming pattern $\overrightarrow{w}$ should be a combination or trade-off of the two directions h_{sr} and h_{sd}, since each achievable rate is increasing with both the terms $|h_{sd}^H w|^2$ and $|h_{sd}^H w|^2$. Based on their rate expression, a characterization of $\overrightarrow{w}$ is provided as follows.

Proposition 7.4. *The optimal beamforming pattern $\overrightarrow{w}$ in Problem* (7.34) *can be characterized by*

$$\overrightarrow{w} = u_{\max}(\sqrt{\rho} h_{sd} h_{sd}^H + \sqrt{1-\rho} h_{sr} h_{sr}^H), \tag{7.50}$$

where $0 \leq \rho \leq 1$ is a variable to adjust different weights for the beamforming in the two directions h_{sr} and h_{sd}.

In different relaying protocols, the values of the optimal ρ might be different. A direct and easy way to determine ρ is the bisection search. The optimal beamforming pattern will be $\overrightarrow{w} = \overrightarrow{h}_{sd}$ ($\overrightarrow{w} = \overrightarrow{h}_{sr}$) when the SNR gain through the direction transmission is significantly larger (smaller) than the relaying transmission because of $\rho \to 0$ ($\rho \to 1$). This motivates us to analyse the performance with $\rho = 1$ and $\rho = 0$, since both serve as the upper bounds (maybe also tight) of the achievable ECR.

Without loss of generality, we focus on the case of $\rho = 1$ (the case $\rho = 0$ can be studied similarly). In this case, we have $\overrightarrow{w} = \overrightarrow{h}_{sd}$, and then the AF rate in (7.25), the DF rate in (7.28) and the CF rate in (7.33) become

$$\text{Rate}_{AF}^{sd} = \frac{1}{2}\log_2\left(1 + P_s||h_{sd}||^2 + \frac{P_s|h_{sr}^H \overrightarrow{h}_{sd}|^2 P_r|h_{rd}|^2}{1 + |h_{sr}^H \overrightarrow{h}_{sd}|^2 + P_r|h_{rd}|^2}\right), \tag{7.51}$$

$$\text{Rate}_{DF}^{sd} = \frac{1}{2}\min\left\{\log_2\left(1 + P_s|h_{sr}^H \overrightarrow{h}_{sd}|^2\right), \log_2\left(1 + P_s||h_{sd}||^2 + P_r|h_{rd}|^2\right)\right\}, \tag{7.52}$$

$$\text{Rate}_{CF}^{sd} = \frac{1}{2}\log_2\left(1 + P_s||h_{sd}||^2 + \gamma_{CF}^{sd}\right), \tag{7.53}$$

where

$$\gamma_{CF}^{sd} \triangleq \frac{P_s|h_{sr}^H \overrightarrow{h}_{sd}|^2 P_r|h_{rd}|^2(P_s||h_{sd}||^2 + 1)}{(P_s||h_{sd}||^2 + 1)(P_r|h_{rd}|^2 + P_s|h_{sr}^H \overrightarrow{h}_{sd}|^2) + P_s|h_{sr}^H \overrightarrow{h}_{sd}|^2}. \tag{7.54}$$

Based on these rates (7.55)–(7.57) in special case, the energy efficient power allocation problem can be further solved. Instead of the instantaneous performance, it is also interesting to investigate the average performance with respect to the small-scale fading.

Proposition 7.5. *When $\overrightarrow{w} = \overrightarrow{h}_{sd}$, the average rates with respect to the small-scale fading for the AF/DF/CF relaying protocols can be approximately expressed as*

$$\overline{\text{Rate}}_{AF}^{sd} \approx \frac{1}{2}\log_2\left(1+\alpha_{sd}MP_s+\frac{\alpha_{sr}P_s\alpha_{rd}P_r}{1+\alpha_{sr}P_s+\alpha_{rd}P_r}\right), \tag{7.55}$$

$$\overline{\text{Rate}}_{DF}^{sd} \approx \frac{1}{2}\log_2\left(1+\min\left\{\alpha_{sr}P_s,\ \alpha_{sd}MP_s+\alpha_{rd}P_r\right\}\right), \tag{7.56}$$

$$\overline{\text{Rate}}_{CF}^{sd} \approx \frac{1}{2}\log_2\left(1+\alpha_{sd}MP_s+\frac{\alpha_{sr}P_s\alpha_{rd}P_r(\alpha_{sd}MP_s+1)}{(\alpha_{sd}MP_s+1)(\alpha_{rd}P_r+\alpha_{sr}P_s)+\alpha_{sr}P_s}\right). \tag{7.57}$$

The proof is based on the fact that h_{sd}, h_{sr} and h_{rd} are independent in distribution and the widely used approximation of $\mathbb{E}\left\{\log\left(1+\frac{g_1(x)}{g_2(x)}\right)\right\} \approx \log\left(1+\frac{\mathbb{E}\{g_1(x)\}}{\mathbb{E}\{g_2(x)\}}\right)$. The details are omitted here.

Based on the average rate expression, the power allocation can be solved, then the optimized power allocation and beamforming patterns are fixed within each large-scale fading duration. Furthermore, it is interesting to determine the conditions for the relaying protocols outperforming the direct transmission and also the energy efficient locations for the relay or the destination when the location of the source is fixed.

In terms of the average rate performance, a relaying protocol is preferred compared with direct transmission if it achieves a higher average transmission rate. For example, a sufficient condition for the AF relaying protocol is

$$\frac{\alpha_{sr}\alpha_{rd}P_r}{1+\alpha_{sr}P_s+\alpha_{rd}P_r} \geq \alpha_{sd}^2M^2P_s+\alpha_{sd}M \ \Rightarrow\ \overline{\text{Rate}}_{AF}^{sd} \geq \overline{\text{Rate}}_{DT}^{sd}, \tag{7.58}$$

where $\overline{\text{Rate}}_{DT}^{sd} = \log_2(1+\alpha_{sd}MP_s)$ is the average direct transmission data rate. Following a similar approach, sufficient conditions for the DF relaying protocols and CF relaying protocols can be also derived.

For the ECR performance, a relaying protocol might be still preferred even if its rate is lower than the direct transmission, since in the second phase the idle source saves not only its own transmit power consumption but also its circuit power consumption. In general, a relaying protocol outperforms the direction transmission in ECR performance if

$$\frac{\frac{1}{2}\left(\frac{1}{\eta_s}P_s+P_s^c+\frac{1}{\eta_s}P_r+P_r^c(2,\text{flag2},\text{flag3})\right)}{\text{Rate}(2,\text{flag2},\text{flag3})} \leq \frac{\frac{1}{\eta_s}P_s+P_s^{c,dt}}{\text{Rate}_{DT}}. \tag{7.59}$$

Let $\Delta_{Rate} = 2\text{Rate}(2,\text{flag2},\text{flag3}) - \text{Rate}_{DT} > 0$ and $\Delta_{P_s^c} = P_s^{c,dt} - P_s^c > 0$ be the rate gap and the sum circuit power consumption of the source in the two phases of a typical relaying protocol compared with the direct transmission, and then (7.59) can be written as

$$\frac{\frac{1}{\eta_s}P_s + P_s^{c,dt} - \Delta_{P_s^c} + \frac{1}{\eta_s}P_r + P_r^c(2, \text{flag2}, \text{flag3})}{\text{Rate}_{DT} + \Delta_{Rate}} \le \frac{\frac{1}{\eta_s}P_s + P_s^{c,dt}}{\text{Rate}_{DT}} \tag{7.60}$$

$$\Leftrightarrow \frac{\frac{1}{\eta_s}P_r + P_r^c(2, \text{flag2}, \text{flag3}) - \Delta_{P_s^c}}{\Delta_{Rate}} \le \frac{\frac{1}{\eta_s}P_s + P_s^{c,dt}}{\text{Rate}_{DT}}, \tag{7.61}$$

which implies that the increased rate by the two-hop SNR-enhanced transmission should be more energy efficient than direction transmission. Further, how much energy can be saved by the source in the second phase is also an important factor for the energy efficiency comparison.

Remark 7.4. *If the average rate is evaluated for condition* (7.61) *and in addition the transmit power values* $\{P_s, P_r\}$ *are fixed, then we can determine the energy efficient relay placement when the location of the destination is fixed or the energy efficient area for the destination when the location of the relay is fixed.*

7.4.4 Numerical results

In this section, the performance of the proposed ECR minimization algorithms for the three-node AF/DF/CF relay system is evaluated. Let d_{sd}, d_{sr} and d_{rd} denote the distances of the links $S \rightarrow D, S \rightarrow R$ and $R \rightarrow D$, respectively. In the simulation, we assume that $d_{sd} = 100$ m and the system parameters in Table 7.3.

In Figure 7.2, the ECR performance of the three relaying protocols is compared based on the same small-scale fading channels, but the relay is located between the source and the destination with the ratio $d_{sr}/d_{sd} \in [0.1, 0.9]$. The source and the relay transmit power are jointly optimized when the minimum rate requirement is $1.1 \times \text{Rate}_{DT}$. Observe that the instantaneous ECR performance is varying with the relay's location and also the relaying protocols. Based on the ECR performance in Figure 7.2, the CF relaying protocol and the AF relaying protocol might be the best choice when the relay is very close to the source and close to the destination, respectively. Otherwise, the DF relaying achieves the best ECR performance.

Besides evaluating the instantaneous ECR performance, the average ECR performance on the AF/DF/CF relay's placement with the fixed $d_{sd} = 100$ m and the fixed transmit power is simulated in Figures 7.3 and 7.4 when $\vec{w} = \vec{h}_{sd}$ and in Figures 7.5 and 7.6 when $\vec{w} = \vec{h}_{sr}$, respectively. In Figures 7.3–7.6, the dark region

Table 7.3 Three-node relay system parameters

Circuit power	**Reference 25**	***M***	**4**
Noise power	−104 dBm/Hz	$(P_s^{\max}, P_r^{\max})$	(40, 10) W
Path loss	Reference 34	(η_s, η_r)	(1/3, 1/3)

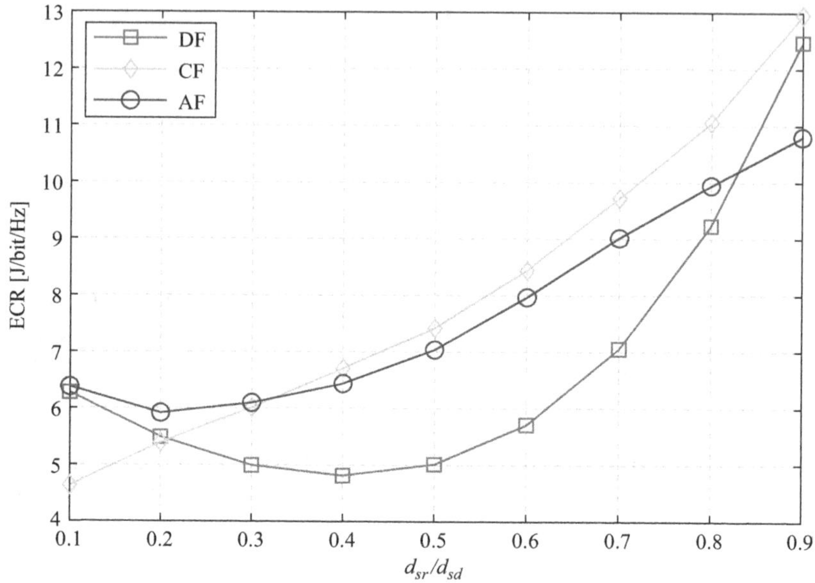

Figure 7.2 An instantaneous ECR performance comparison example with flag3 = 3

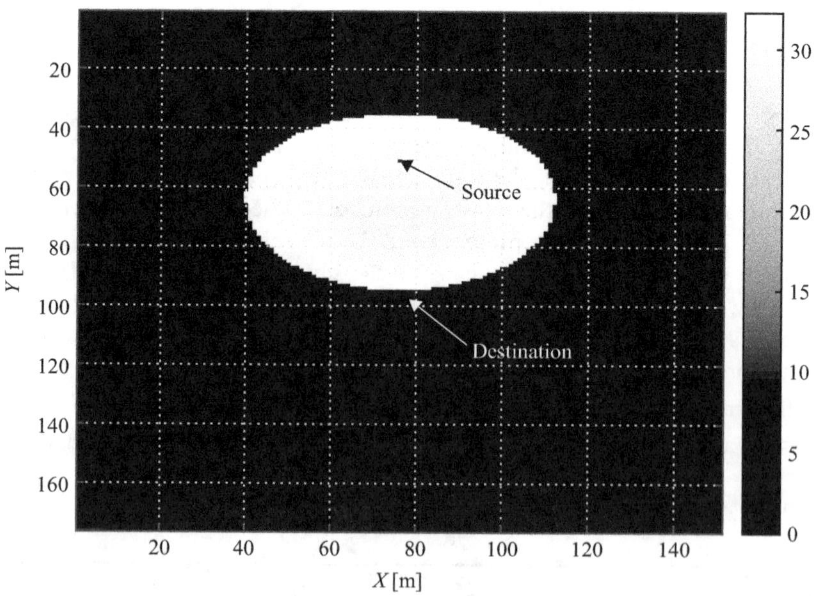

Figure 7.3 Average ECR performance versus AF relay placement when $w||h_{sd}$

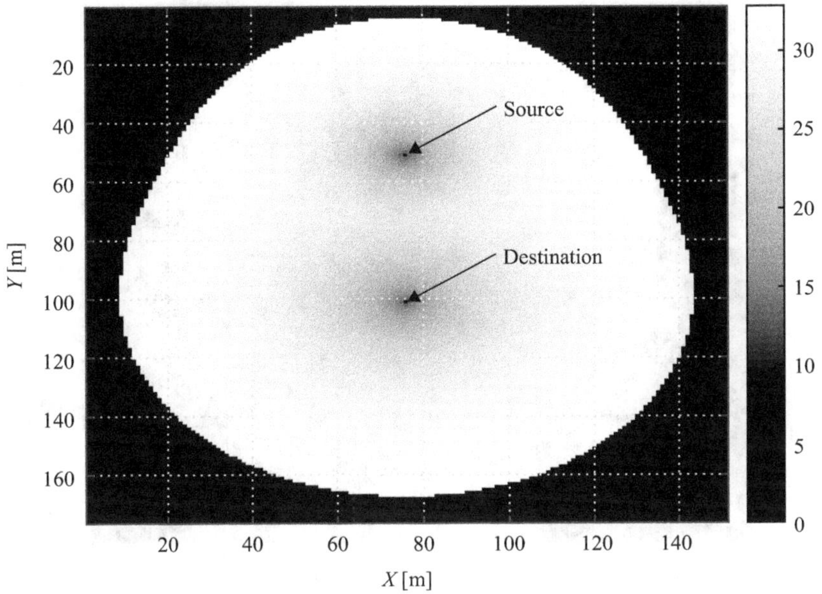

Figure 7.4 Average ECR performance versus DF relay placement when $w||h_{sd}$

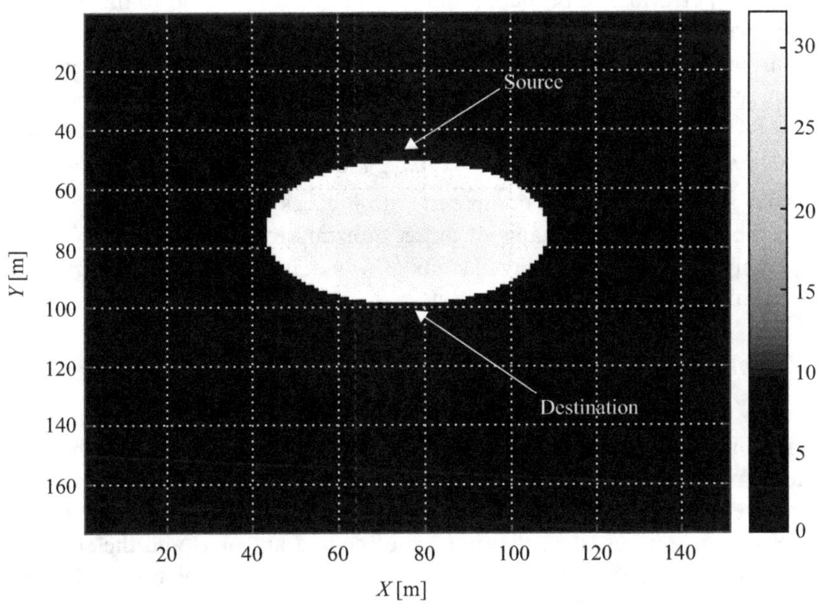

Figure 7.5 Average ECR performance versus AF relay placement when $w||h_{sr}$

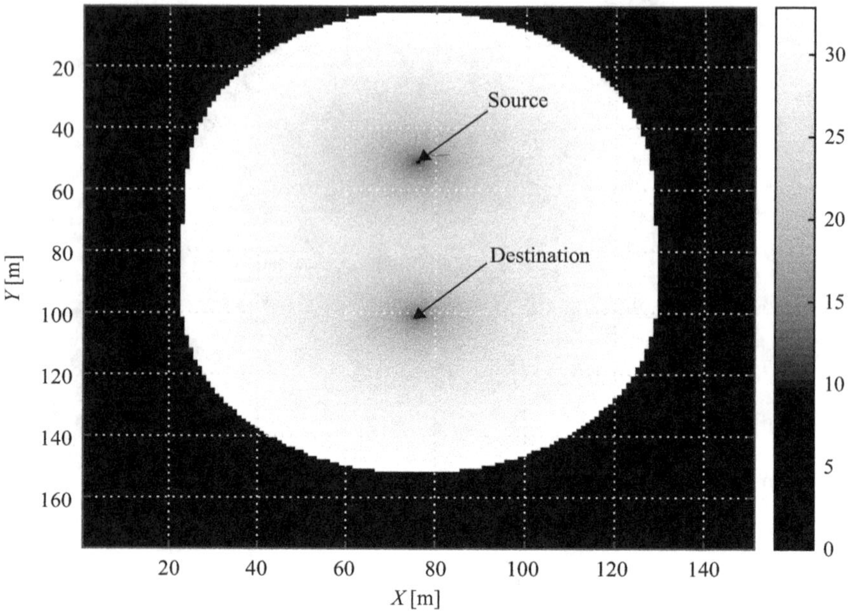

Figure 7.6 Average ECR performance versus DF relay placement when $w||h_{sr}$

denotes the relay placement where a certain rate requirement cannot be met. Since the average ECR performance of the CF relaying protocol is similar to the AF relaying protocol, that case is not shown here. Figure 7.3 implies that the AF relay should be located around the middle point between the source and the destination to achieve the minimum average ECR performance. However, as shown in Figure 7.4, the energy efficient locations of the DF relay are around the source and the destination. With the same system parameters but $\vec{w} = \vec{h}_{sr}$, the feasible relay placement areas in Figures 7.5 and 7.6 are shrunk compared with Figures 7.3 and 7.4, since $\vec{w} = \vec{h}_{sd}$ achieves the largest channel gains of direct transmission. In this case, ECR performance highly depends on the relay placement, which implies that the energy efficient relay placement areas are nearly symmetric.

7.4.5 Conclusion

In this part, general concepts and definitions of the energy evaluation metrics, the corresponding optimization problems and optimization algorithms and techniques are provided. For a classical three-node relay system, AF, CF and DF relaying protocols are compared in terms of the ECR performance. The ECR performance is varying with the system settings, e.g., the number of source antennas and the source/relay transmit power. Implied by the numerical results, the DF relaying protocol in general has the best ECR performance compared with the AF and CF relaying protocols, although it is the most complicated one as the relay needs to decode and re-code the

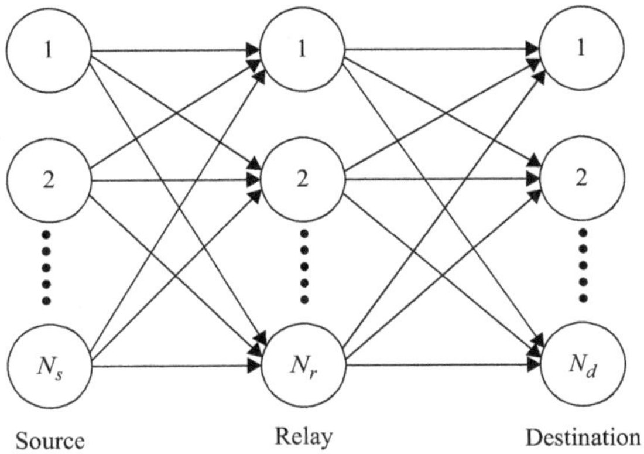

Figure 7.7 Block diagram of a relay network with multiple source, relay and destination nodes

signal and in general the relay signal processing power is not the main part of the overall system power consumption.

7.5 Energy efficiency analysis of relay cooperation

Cooperative systems are considered an effective technique to improve spectral and energy efficiency of wireless networks without requiring an increase on the number of antenna elements per device. A cooperative system, also known as virtual or distributed system, uses distributed antennas on multiple devices to achieve some of the benefits provided by conventional MIMO systems. The basic idea of a cooperative wireless system is to group an arbitrary number of single-antenna devices into virtual antenna arrays to mimic and exploit the gains of multi-antenna systems by means of wireless cooperation [35–37]. In this section, we extend the application of the classic three-node relay network to a system with multiple source, relay and destination terminals, as shown in Figure 7.7.

A VMIMO system lies in the integration of MIMO and relay technologies, and different gains can be obtained according to the system's design. On the one hand, from the MIMO perspective, VMIMO configurations are designed to achieve diversity gains, multiplexing or beamforming gains. On the other hand, from the relay perspective, VMIMO systems can adopt either AF, DF, CF or even a combination of these schemes to complete the transmission in one hop or multiple hops [38]. Even though a cooperative MIMO system shows an increase of the system complexity and a larger signalling overhead required for supporting the wireless cooperation, a cost–effective trade-off regarding capacity, cell edge throughput, coverage, mobility and complexity is obtained. These improvements hinge on the usage of distributed antennas, which increases the system capacity by means of spatial diversity [35].

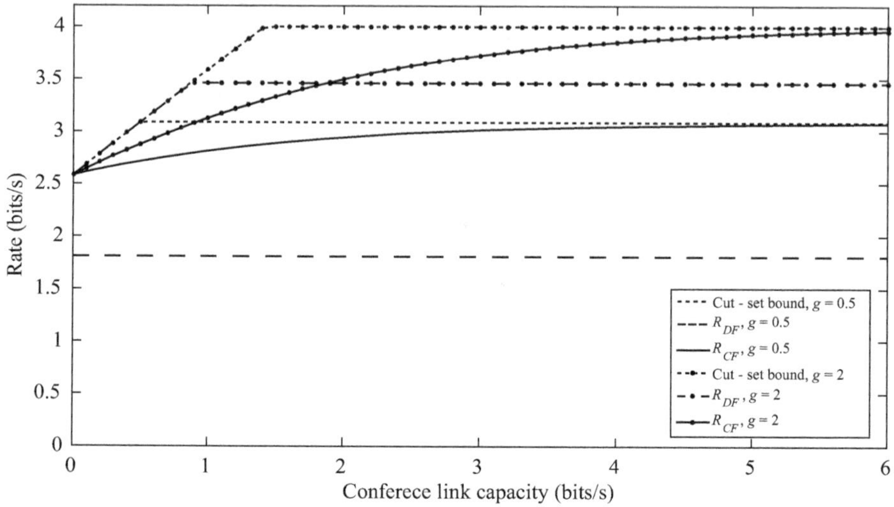

Figure 7.8 Achievable rates when weak (g = 0.5) and strong (g = 2) relay channels are considered

The distributed nature of the VMIMO systems leads to a number of challenges from synchronization to cluster formation. Nevertheless, we can also expect the following gains [36]:

- *Power gain* as each of the relays adds additional transmit power that is combined in the destination terminal.
- *Macrodiversity gain* that allows combating shadowing, when the message is broadcasted to all nearby nodes.
- *Diversity gain* in the presence of fading.

Aiming to improve the EER of a VMIMO, two case studies of current research are resource allocation and interference suppression. In this chapter, we focus on the resource allocation approach where the EER of a VMIMO system can be improved through an effective power allocation between the transmitter and the relay, the bandwidth allocation between the data and cooperation channels, the selection of the protocols used to carry out the cooperation according to the distance between nodes, the optimization of the number of allowed hops and even taken into account the availability of resource blocks.

When the separation between the relay and destination nodes varies, the AF, DF and CF methods exhibit different performance in terms of capacity. If the relay is closer to the destination than the source node, the CF protocol provides better capacity performance compared with DF [39, 40]. Meanwhile, the AF protocol can be seen as a special case of CF, thus, the relay simply scales and forwards the received signal without performing any particular BB processing. If we consider that the power consumed for each of the three protocols is quite small compared with the RF circuitry, the slightly different power consumption between the protocols can be neglected.

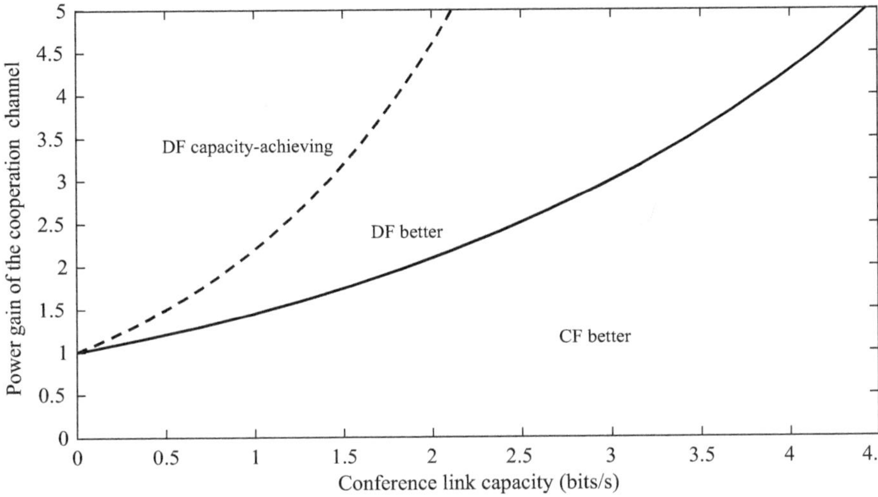

Figure 7.9 Best cooperation protocol based on the capacity and power gain of the conference link

Therefore, for a given transmission power, CF has a higher EER than DF. As can be seen from Figure 7.8, when the relay channel is weak, DF fails to take advantage of the cooperative system, since its performance is limited by the rate at which the relay can correctly decode the received signal. On the other hand, when a reliable channel is considered, DF outperforms CF and meets the cut-set bound for small values of the capacity (C) of the conference link [41].

Moreover, Figure 7.9 shows the regions where each of the protocols performs the best. As can be seen, for large values of $g = ||h_{sr}||^2/||h_{sd}||^2$ relatively to 2^C, the DF not only outperforms the CF protocol but also is capacity-achieving. On the other hand, CF achieves a higher data rate for the region on the right-hand side of solid line [41].

On the other hand, two different scenarios regarding the bandwidth allocation between the data channel and the cooperation channel can be considered [12]. If the separation between the bands used for the cooperation can be spatially reused all over the relay nodes of the network, then, the relay terminal can use a different channel to communicate with the destination node as the bandwidth cost for a particular cooperating pair is negligible [15]. Thus, it is assumed that the data channel and the cooperation channel have equal bandwidths. Meanwhile, when the spatial reuse of the cooperation bands is not considered, the cooperation channel should share the band assigned for the data transmission.

As was mentioned previously, AF can be considered as a special case of the CF protocol, since the relay only scales and forwards the analog signal waveform without any particular digital processing, the bandwidth allocation for both channels (data and cooperation) remain the same. Nevertheless, when the CF protocol is employed and the received signal is quantized and re-encoded, it is possible to optimize the bandwidth requirements for the cooperation channel. Thus, the CF protocol performs

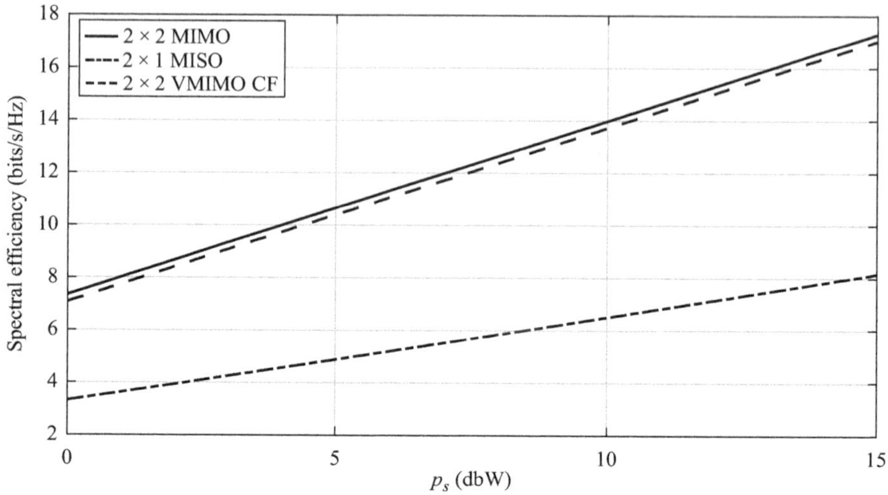

Figure 7.10 Best cooperation protocol based on the capacity and power gain of the conference link

better than or equal to AF in terms of the EER metric. Furthermore, the VMIMO system optimally allocates its transmit power and bandwidth, it outperforms multi-input and single-output (MISO) for a wide range of spectral efficiency (SE).

An efficient way to improve the EER of a cooperative system is to formulate an optimization problem subject to the power allocation ratio between the relay and source [12]. Figure 7.10 shows the SE performance of a 2 × 1 MISO system, 2 × 2 MIMO system and 2 × 2 VMIMO system, where the physical channel propagation parameters are adopted from the 3GPP LTE standard models [12]. The relay implements a CF protocol through an out-of-band short-range cooperation channel, and it is assumed that the maximum transmit power of the relay is smaller than that of the source terminal. From Figure 7.10, it can be noticed that the SE achieved by the cooperative MIMO system is comparable with that of a traditional MIMO system and much larger of that of a MISO system.

Figure 7.11 shows the EER of the VMIMO system previously mentioned, and Figure 7.12 uses the ratios EER_{MIMO}/EER_{VMIMO} and EER_{MISO}/EER_{VMIMO} to compare the EER performance of a VMIMO system with those of the MIMO and MISO systems. As can be seen, when the SE of a VMIMO increases in the low-SE regime, VMIMO demonstrates a much better EER performance than the non-cooperative MISO system and performs close to the ideal MIMO case in terms of EER. Thus, with the help from the relay, VMIMO enables the source terminal to become more energy efficient and allow the user devices to achieve longer battery life [42].

In practical wireless networks, including cellular and/or ad hoc networks, several users are simultaneously served by the BS. Thus, deployment of multiple relays is a natural approach to increase the system performance. Considering that the density of the cell is higher, the terminals are significantly closer to one or more relays

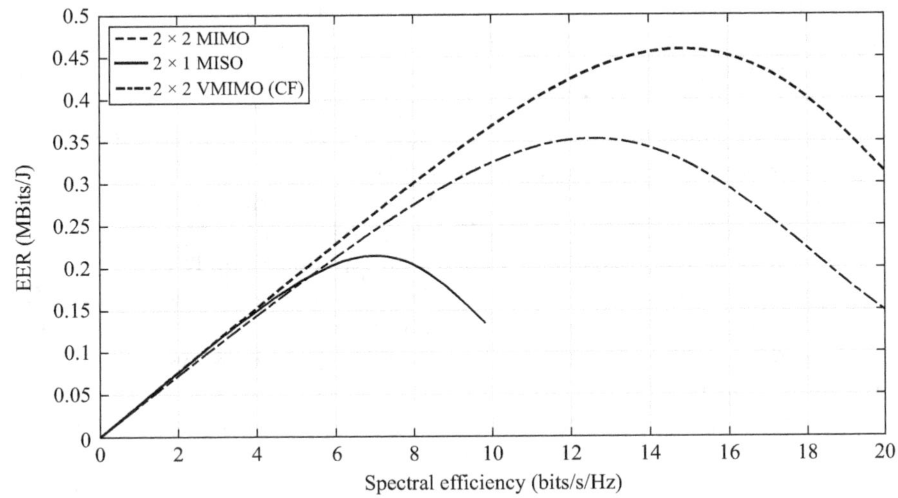

Figure 7.11 *Comparison of the energy efficiency performance of the VMIMO with CF and the MISO and MIMO systems for unit bandwidth*

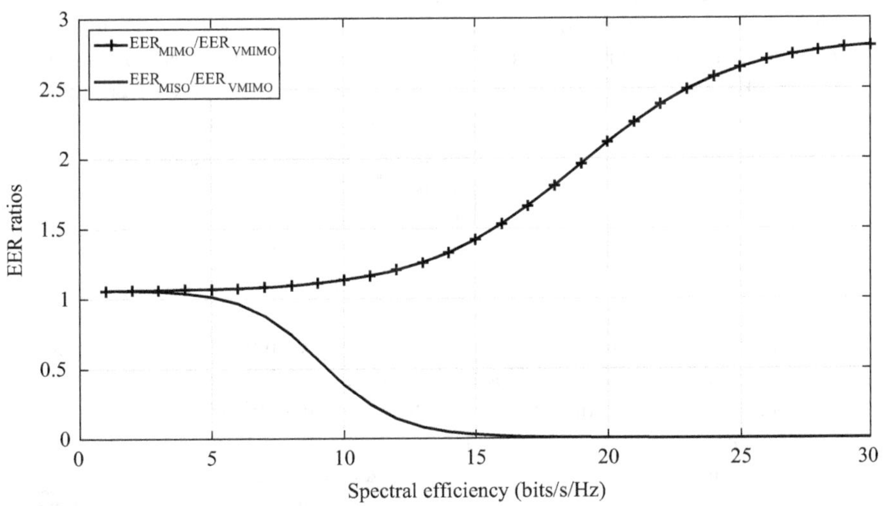

Figure 7.12 *Comparison of the energy efficiency ratios of the VMIMO system with CF and the MISO and MIMO systems*

than the BS. This means that the propagation loss from the BS to the destination is larger than that from a nearby relay to the destination. Thus, higher data rates on the link between the relays and destination terminals can be achieved. This fact not only potentially solves the coverage problem but also increases the data rate at the destination node [36]. Additionally, the data and relay links could use a different spectrum band yielding significant gains from load balancing through the relays [43]. Because of the broadcast nature of the wireless channel, we can model the network as a multiple-user multiple-relay system. In this scenario, the trade-off between decreasing the energy consumption for data transmission and decreasing the overhead energy consumption remain in the CSI acquisition. Thus, the right amount of instantaneous local cooperation should be used to obtain maximum energy savings. A cooperative communication scheme reduces energy consumption compared to non-cooperative schemes and cooperative schemes that use either a single relay; energy savings up to 16% were observed after CSI energy overhead was accounted for [44].

Several approaches aiming to exploit the broadcast nature of the wireless channel have been developed. With the assumption of perfect CSI at the relays, cooperative beamforming algorithms can be used to transmit data from the relays to the destination while saving energy. In this particular approach, the relays linearly weight their transmit signals so that they add up coherently at the destination node. Obtaining and exploiting CSI is an additional challenge for cooperative beamforming. Based on the assumption that a training process to obtain the CSI at the destination and a feedback process from the destination to the relays is performed to enable an energy efficient cooperation, the cost for acquiring CSI is computed in Reference 44. For fading channels, with a power hungry feedback link, there is a trade-off between reducing the power consumption for data transmission by increasing the number of relays and decreasing the overhead power consumed by the CSI by using less relays.

In Reference 45, an analysis of the EER of an adaptive cooperative beamforming system where the transmit power is adapted according to the CSI is presented. Aiming to provide an accurate EER analysis, the algorithm considers the overhead power consumption incurred by the beamforming processing. As shown in Figure 7.13 the cooperative beamforming system can provide higher EER, using either fixed or adaptive modulation, compared to the direct transmission communication system. Moreover, when the distance between the nodes increases, the cooperative system outperforms direct transmission since the power required for transmission becomes the dominant factor compared to the circuit power.

Furthermore, the EER of the system can also be evaluated when the number of available resource blocks is constrained and multiple users are served by the BS. In Reference 46, an analysis of the EER performance when a DF protocol Type II is used for a very large antenna array is presented. Using the protocol II for the communication is the logical choice in a scenario where the source terminal engages in data reception from another terminal in the network over the second time slot thereby rendering it unable to transmit. Moreover, as the source terminal is silent over the second time slot, this protocol is more efficient than Protocols I and III in terms of battery life [13]. Additionally, it is assumed that the user terminals (UTs) are closely located with non-light-sight conditions, where the rich scattering allows spatial separation of the intended signal for each UT [47], then it can be considered that the destination and relays nodes are equally distanced from the transmitter.

As was mentioned in Section 7.3.4, the total energy consumption of a classical three-node relay system is the sum of the power consumed at the source, the destination and relay nodes, plus a constant power term accounting for the signal processing and

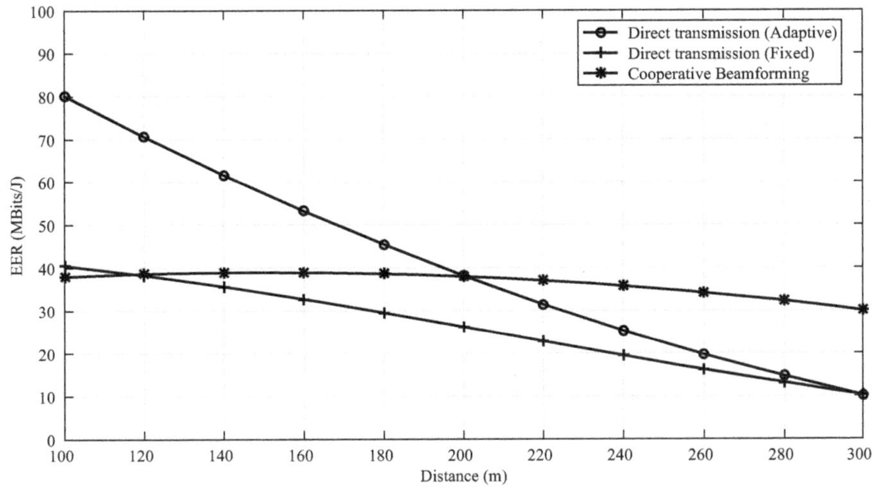

Figure 7.13 Energy efficiency comparison of a CBF and direct transmission communication with 4-QAM

circuit power. In Section 7.4, we mainly optimize and evaluate the energy efficiency of a relay system with respect to the relaying protocol selection and the transmit/receive strategy. However, the system parameters also have an important effect on the energy efficiency, e.g., the modulation schemes and the number of active antennas. It is significant to evaluate the effect of these system parameters on energy efficiency in system design. For this purpose, similar to (7.16), a parameterized BS power consumption model can be also described as [46]

$$P_{\text{BS}} = \frac{1}{\eta_s} P_{\text{s}} + MP_{\text{cs}} + mKP_{\text{cod}} + P_{\text{syn}}, \tag{7.62}$$

where η_s denotes the efficiency of the PA and m is the number of bits transmitted per data symbol. On the other hand, the power required to run the circuit components (including filters, converters and mixers) is denoted as P_{cs}, and it is linearly dependent on M [48]. Further, P_{cod} is the power coding and scales with the number of users terminals K and the modulation efficiency. Meanwhile, P_{syn} represents the power consumed by the single local oscillator. Meanwhile, the total power consumption at the receiver side, i.e., the power consumed by the UTs and the power required to carry out the wireless cooperation, can be written as

$$P_{\text{UT}} = KP_{\text{ck}} + rN_dP_{\text{r}} + mrN_dP_{\text{cod}} + m(K + rN_d)P_{\text{dec}}, \tag{7.63}$$

where P_{ck} is the circuit power consumption per UT and N_d is the number of destination nodes and r the number of relays cooperating with each destination node through one-shot conference links. Considering that the BS and relay nodes are geographically separated and hence, powered by separate power supplies [49], it is a common practice to define $\xi = P_{\text{s}}/P_{\text{r}}$, thus the relay decides its own transmit power with the knowledge

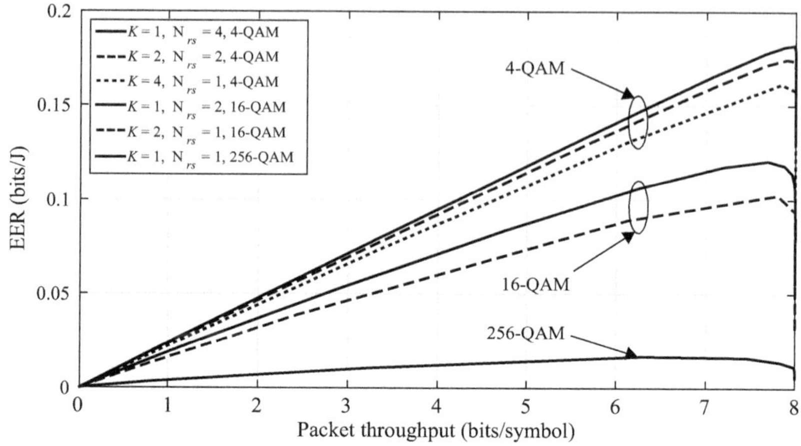

Figure 7.14 Energy efficiency of a VMIMO with $N_t = 10$, $N_d = 1$ and a target rate of 8 bits/symbol

of the value of P_s [12]. The powers for re-coding prior to the re-transmission and decoding are P_{cod} and P_{dec}, respectively, and depend on the number of bits processed. On the other hand, P_0 is a constant accounting for the fixed power consumption required for load-independent power of the backhaul infrastructure, control signalling and site-cooling [48]. Finally, the power consumption of the cooperative system with a very large antenna array can be written as

$$P_{tot} = \frac{1}{2}(P_{BS} + P_{UT} + P_0). \tag{7.64}$$

The study of the EER for a very large antenna system when the number of resource blocks (one slot in both time and frequency domains) per destination node is constrained can be classified into two possible scenarios. First, the target rate for a destination node can be achieved using more that one resource block when the number of UTs served by the BS is reduced. Second, if the number of resource blocks per UT is restricted, the BS can choose either to increase the number of relay nodes serving the destination node or to increase the modulation order per beam.

The results presented in Figure 7.14 assume a bit-interleaved coded-modulation cooperative system that implements a DF protocol Type II with 4-QAM Gray mapping modulation and a Rate-1/2 convolutional code. A multi-antenna BS performs a zero-forcing precoding before transmission over a Rayleigh block-fading channel. From Figure 7.14, can be noticed that for a given target throughput the EER grows until the system reaches the target rate. Thus, once the maximum packet throughput delivered by the system is reached, increasing the transmit power does not affect the output of the system regarding the throughput, but it does in terms of EER, i.e., since the BS is consuming more power at the transmission achieving the same throughput, then the EER of the system drops as can be seen. Figure 7.14 shows that the system without constrained in the use of resource blocks shows the best performance. However, a good and practical trade-off between the EER performance and the required resources is shown for a system than combines MIMO and relays technologies. Thus, a good trade-off between the EER and the required resource blocks is observed when a low modulation order is employed and a wireless cooperation between UTs is allowed, making cooperative system a viable option for the upcoming large wireless systems. For very large antenna VMIMO systems, EER improvements can be expected when the gain in terms of capacity due to the cooperative system compensates for the extra power required at the relays.

7.6 Conclusions

This chapter has discussed energy efficiency issues and comparisons among relay protocols. A widely applicable system model has been introduced along with relevant energy efficiency metrics. In addition, the importance of using models that can compute the overall energy consumption (not just the transmitted radio-frequency power) has been emphasized. We have discussed in detail the energy efficiency optimization of different relay protocols including AF, DF and CF. Mathematical methods including a general fractional programming framework to minimize the energy consumption ratio have been presented and the best energy efficiency regions for deploying relays have also been presented. This chapter has also studied the benefits of using relaying to form VMIMO network configurations. The performance trade-offs for different relay protocols in this scenario have been discussed and compared. The spectral and energy efficiency comparisons of different relay setups with varying numbers of antennas have also been discussed.

Acknowledgements

John Thompson and Pan Cao acknowledge funding of this work by the UK EPSRC SERAN Project (EP/L026147/1). Miryam Gonzalez-Perez acknowledges funding of her PhD studies by Conacyt, Mexico. Simulation data is available at http://www.homepages.ed.ac.uk/jst/.

References

[1] Greentouch Organisation, "Project website." http://www.greentouch.org.

[2] A. Fehske, G. Fettweis, J. Malmodin, and G. Biczok, "The global footprint of mobile communications: The ecological and economic perspective," *IEEE Communications Magazine*, vol. 49, pp. 55–62, August 2011.

[3] P. Frenger, C. Friberg, Y. Jading, M. Olsson, and O. Persson, "Radio network energy performance: Shifting focus from power to precision," *Ericsson Review*, vol. 2, pp. 2–9, February 2014.

[4] C. Han, T. Harrold, S. Armour, *et al.*, "Green radio: Radio techniques to enable energy-efficient wireless networks," *IEEE Communications Magazine*, vol. 49, pp. 46–54, June 2011.

[5] Z. Hasan, H. Boostanimehr, and V. Bhargava, "Green cellular networks: A survey, some research issues and challenges," *IEEE Communications Surveys Tutorials*, vol. 13, pp. 524–540, Fourth 2011.

[6] J. Wu, S. Rangan, and H. Zhang (Eds.), *Green Communications: Theoretical Fundamentals, Algorithms and Applications*. CRC Press, September 2012, Boca Raton, FL.

[7] K. Mimis, K. Morris, S. Bensmida, and J. McGeehan, "Multichannel and wideband power amplifier design methodology for 4G communication systems based on hybrid class-J operation," *IEEE Transactions on Microwave Theory and Techniques*, vol. 60, no. 8, pp. 2562–2570, 2012.

[8] S. Videv, J. Thompson, H. Haas, and P. Grant, "Resource allocation for energy efficient cellular systems," *EURASIP Journal on Wireless Communications and Networking*, vol. 2012, pp. 1–15, December 2012.

[9] C. Masouros, T. Ratnarajah, M. Sellathurai, C. Papadias, and A. Shukla, "Known interference in the cellular downlink: A performance limiting factor or a source of green signal power?," *IEEE Communications Magazine*, vol. 51, pp. 162–171, October 2013.

[10] C. Khirallah, J. Thompson, and H. Rashvand, "Energy and cost impacts of relay and femtocell deployments in long-term-evolution advanced," *IET Communications*, vol. 5, pp. 2617–2628, December 2011.

[11] F. Richter, A. Fehske, and G. Fettweis, "Energy efficiency aspects of base station deployment strategies for cellular networks," in *2009 IEEE 70th Vehicular Technology Conference Fall (VTC 2009-Fall)*, pp. 1–5, September 2009.

[12] J. Jiang, M. Dianati, M. Imran, R. Tafazolli, and S. Zhang, "Energy-efficiency analysis and optimization for virtual-MIMO systems," *IEEE Transactions on Vehicular Technology*, vol. 63, pp. 2272–2283, June 2014.

[13] R. Nabar, H. Bolcskei, and F. Kneubuhler, "Fading relay channels: Performance limits and space–time signal design," *IEEE Journal on Selected Areas in Communications*, vol. 22, pp. 1099–1109, August 2004.

[14] J. Laneman, D. Tse, and G. W. Wornell, "Cooperative diversity in wireless networks: Efficient protocols and outage behavior," *IEEE Transactions on Information Theory*, vol. 50, pp. 3062–3080, December 2004.

[15] C. Ng, N. Jindal, A. Goldsmith, and U. Mitra, "Capacity gain from two-transmitter and two-receiver cooperation," *IEEE Transactions on Information Theory*, vol. 53, pp. 3822–3827, October 2007.

[16] R. Nabar and H. Bolcskei, "Space–time signal design for fading relay channels," in *Global Telecommunications Conference, 2003. GLOBECOM '03. IEEE*, vol. 4, pp. 1952–1956, December 2003.

[17] J. Sydir and R. Taori, "An evolved cellular system architecture incorporating relay stations," *IEEE Communications Magazine*, vol. 47, pp. 115–121, June 2009.

[18] S. Mclaughlin, P. M. Grant, J. Thompson, H. Haas, D. Laurenson, C. Khirallah, Y. Hou, and R. Wang, "Techniques for improving cellular radio base station energy efficiency," *IEEE Wireless Communications*, vol. 18, pp. 10–17, October 2011.

[19] L. Feeney, "An energy consumption model for performance analysis of routing protocols for mobile ad hoc networks," *Mobile Networks and Applications*, vol. 6, no. 3, pp. 239–249, 2001.

[20] T. Chen, H. Kim, and Y. Yang, "Energy efficiency metrics for green wireless communications," in *2010 International Conference on Wireless Communications and Signal Processing (WCSP)*, pp. 1–6, October 2010.

[21] Q. Cui, X. Yang, J. Hamalainen, X. Tao, and P. Zhang, "Optimal energy-efficient relay deployment for the bidirectional relay transmission schemes," *IEEE Transactions on Vehicular Technology*, vol. 63, pp. 2625–2641, July 2014.

[22] G. Dimic, N. Zogovic, and D. Bajic, "Energy efficiency of supportive relay with a novel wireless transmitter power consumption model," in *2012 International Conference on Selected Topics in Mobile and Wireless Networking (iCOST)*, pp. 1–6, July 2012.

[23] Q. Wang, M. Hempstead, and W. Yang, "A realistic power consumption model for wireless sensor network devices," in *2006 3rd Annual IEEE Communications Society on Sensor and Ad Hoc Communications and Networks, 2006. SECON '06*, vol. 1, pp. 286–295, September 2006.

[24] B. Debaillie, C. Desset, and F. Louagie, "A flexible and future-proof power model for cellular base stations," in *2015 IEEE 81st Vehicular Technology Conference (VTC Spring)*, pp. 1–7, May 2015.

[25] H. Holtkamp, G. Auer, V. Giannini, and H. Haas, "A parameterized base station power model," *IEEE Communications Letters*, vol. 17, pp. 2033–2035, November 2013.

[26] A. Host-Madsen, "Capacity bounds for cooperative diversity," *IEEE Transactions on Information Theory*, vol. 52, pp. 1522–1544, April 2006.

[27] A. Wyner and J. Ziv, "The rate-distortion function for source coding with side information at the decoder," *IEEE Transactions on Information Theory*, vol. 22, pp. 1–10, January 1976.

[28] J. Jiang, J. Thompson, P. Grant, and N. Goertz, "Practical compress-and-forward cooperation for the classical relay network," in *EUSIPCO*, pp. 2421–2425, August 2009.

[29] S. Schible, "Fractional programming," *Zeitschrift für Operations Theory and Applications*, vol. 27, no. 1, pp. 347–352, 1983.

[30] P. Cao, Z. Chong, Z. Ho, and E. Jorswieck, "Energy-efficient power allocation for amplify-and-forward MIMO relay channel," in *CAMAD*, pp. 333–337, September 2012.

[31] W. Dinkelbach, "On nonlinear fractional programming," *Management Science*, vol. 13, no. 7, pp. 492–498, 1967.

[32] Z.-Q. Luo, W.-K. Ma, A.-C. So, Y. Ye, and S. Zhang, "Semidefinite relaxation of quadratic optimization problems," *IEEE Signal Processing Magazine*, vol. 27, pp. 20–34, May 2010.

[33] W. Ai, Y. Huang, and S. Zhang, "New results on Hermitian matrix rank-one decomposition," *Mathematical Programming: Series A*, vol. 128, pp. 253–283, June 2011.

[34] M. Riback, J. Medbo, J.-E. Berg, F. Harrysson, and H. Asplund, "Carrier frequency effects on path loss," in *IEEE VTC Spring*, vol. 6, pp. 2717–2721, May 2006.

[35] C.-X. Wang, X. Hong, X. Ge, X. Cheng, G. Zhang, and J. Thompson, "Cooperative MIMO channel models: A survey," *IEEE Communications Magazine*, vol. 48, pp. 80–87, February 2010.

[36] R. Pabst, B. Walke, D. Schultz, *et al.*, "Relay-based deployment concepts for wireless and mobile broadband radio," *IEEE Communications Magazine*, vol. 42, pp. 80–89, September 2004.

[37] A. Goldsmith and S. Wicker, "Design challenges for energy-constrained ad hoc wireless networks," *IEEE Wireless Communications*, vol. 9, pp. 8–27, August 2002.

[38] X. Hong, Y. Jie, C.-X. Wang, J. Shi, and X. Ge, "Energy-spectral efficiency trade-off in virtual MIMO cellular systems," *IEEE Journal on Selected Areas in Communications*, vol. 31, pp. 2128–2140, October 2013.

[39] A. Host-Madsen and J. Zhang, "Capacity bounds and power allocation for wireless relay channels," *IEEE Transactions on Information Theory*, vol. 51, pp. 2020–2040, June 2005.

[40] J. Jiang, J. Thompson, H. Sun, and P. Grant, "Performance assessment of virtual multiple-input multiple-output systems with compress-and-forward cooperation," *IET Communications*, vol. 6, pp. 1456–1465, July 2012.

[41] C. Ng, I. Maric, A. Goldsmith, S. Shamai, and R. Yates, "Iterative and one-shot conferencing in relay channels," in *Information Theory Workshop, 2006. ITW '06 Punta del Este. IEEE*, pp. 193–197, March 2006.

[42] W. Zhang, D. Duan, and L. Yang, "Relay selection from a battery energy efficiency perspective," in *Military Communications Conference, 2009. MILCOM 2009. IEEE*, pp. 1–7, October 2009.

[43] H. Wu, C. Qiao, S. De, and O. Tonguz, "Integrated cellular and ad hoc relaying systems: iCAR," *IEEE Journal on Selected Areas in Communications*, vol. 19, pp. 2105–2115, October 2001.

[44] R. Madan, N. Mehta, A. Molisch, and J. Zhang, "Energy-efficient cooperative relaying over fading channels with simple relay selection," *IEEE Transactions on Wireless Communications*, vol. 7, pp. 3013–3025, August 2008.

[45] G. Lim and L. Cimini, "Energy efficiency of cooperative beamforming in wireless ad-hoc networks," in *2012 IEEE International Conference on Communications (ICC)*, pp. 4039–4043, June 2012.

[46] M. G. Gonzalez-Perez and J. Thompson, "Analysis of the energy efficiency of a virtual MIMO system with a large antenna array," in *EUROCON2015*, vol. 1, pp. 1–6, September 2015.

[47] X. Gao, O. Edfors, F. Rusek, and F. Tufvesson, "Massive MIMO performance evaluation based on measured propagation data," *IEEE Transactions on Wireless Communications*, vol. 14, no. 7, pp. 3899–3911, July 2015.

[48] E. Bjornson, L. Sanguinetti, J. Hoydis, and M. Debbah, "Optimal design of energy-efficient multi-user MIMO systems: Is massive MIMO the answer?," *IEEE Transactions on Wireless Communications*, vol. 14, pp. 3059–3075, June 2015.

[49] Y. Liang, V. Veeravalli, and H. Poor, "Resource allocation for wireless fading relay channels: Max–min solution," *IEEE Transactions on Information Theory*, vol. 53, pp. 3432–3453, October 2007.

Chapter 8
Energy-efficient relaying

Alessio Zappone and Eduard A. Jorswieck

8.1 Introduction

The next generations of wireless cellular communications will deploy flexible heterogeneous architectures including smart relays, femto, and macro base stations [28]. It is expected that 50 billion connected devices will operate by 2020, and that a lot of new services will be implemented including e-health, e-banking, e-learning, and so on. This requires both the data rates to grow by a factor of 1000 and the communication latency to be drastically reduced [8]. These requirements are to be realized with similar or even lower energy consumption as in current wireless communications. Therefore, we need an energy-efficient design of relay-assisted cellular networks.

8.1.1 Energy efficiency

A short introduction to energy efficiency is provided in this section. For a more extensive overview we refer to Reference 32.

In the definition of energy efficiency, we follow the physical meaning of efficiency: the energy efficiency is defined as the system benefit–cost ratio in terms of amount of data reliably transmitted over the energy that is required to do so [13, 32].

As far as wireless communications are concerned, the cost is represented by the amount of energy consumed to operate the whole transceiver chain. Let us consider a single link, and as a first introductory example assume a single-antenna single-carrier system. The transmit power is p and the signal-to-noise ratio (SNR) at the receiver is γ. If the transmission takes T seconds to complete, the consumed energy will be

$$E = T(\mu p + P_{\mathrm{c}}) \text{ [Joule]} \tag{8.1}$$

wherein $\mu = 1/\eta$, with η the efficiency of the transmit power amplifier, while P_{c} includes the power dissipated in all other circuit blocks of the transmitter and receiver to operate the terminals. Typically, the most power-hungry components are mixers, frequency synthesizers, active filters, and digital-to-analog converters (DACs), but in some cases (e.g., for macro-base-stations), significant power consumptions are also due to other components like cooling equipment. The terms included in P_c represent a fixed power cost which is dissipated even if the radiated power p is zero.

On the other hand, a measure of the benefit of a communication system is the system capacity, thereby defining the income in the interval T as

$$TW \log_2(1 + \gamma) \tag{8.2}$$

with W the communication bandwidth and γ the communication signal-to-noise ratio (SNR). We should observe that this metric does not account for the actual communication bit error rate (BER). For this reason, another proposed measure for the system benefit is $TR(1 - e^{-\gamma})$, with R the communication rate and $(1 - e^{-\gamma})$ an approximation of the BER. In both cases, we can define the energy efficiency (EE) of a point-to-point single-antenna single carrier system as

$$\mathrm{EE} = \frac{Tf(\gamma)}{T(\mu p + P_c)} = \frac{f(\gamma)}{\mu p + P_c} \quad \text{[bit/Joule]} \tag{8.3}$$

wherein f can be either $W \log_2(1 + \gamma)$ or $R(1 - e^{-\gamma})$. In both cases, the energy efficiency (8.3) is measured in bit/joule, thereby measuring the amount of data that can be reliably transmitted per Joule of consumed energy. One important property of the energy efficiency function in (8.3), which makes it fundamentally different from common signal-to-interference-plus-noise ratio (SINR)-based utility functions is its unimodality in the transmit power. Otherwise stated, unlike more traditional performance functions, the energy efficiency is not monotonically increasing in the transmit power, but instead admits an optimum transmit power level. Therefore, designing a system for energy efficiency maximization does not necessarily lead to using all of the available power, but instead yields the power level which strikes the optimal balance between rate maximization and power consumption minimization.

The extension of the energy efficiency function in (8.3) to single-user multi-antenna and multi-carrier systems is straightforward [32]. However, for multi-hop and multi-user systems, where multiple links with corresponding energy efficiencies exist, it is not immediately clear how to extend the definition in (8.3). The most well-established approach is always based on the physical interpretation of energy efficiency as the benefit–cost ratio. This leads to the definition of the so-called global energy efficiency (GEE), as the ratio of the system sum-benefit, over the total system power consumption, namely

$$\mathrm{GEE} = \frac{\sum_{\ell=1}^{L} f(\gamma_\ell)}{\sum_{\ell=1}^{L} \mu_\ell p_\ell + P_{c,\ell}} \tag{8.4}$$

wherein L is the number of communication links in the network. In the following, the focus will be on the GEE in (8.4), which will be used for system design. However, we should mention that an alternative approach exists, which is based on the theory of multi-objective optimization. This approach regards the individual energy efficiencies of the different links as the objectives of a multi-objective maximization problem. Different points on the Pareto-boundary of the resulting multi-objective optimization problem can be obtained by defining the system energy efficiency as the arithmetic or geometric mean of the individual energy efficiencies, or as the minimum of the individual energy efficiencies.

In any case, regardless of the particular choice, the energy efficiency is always defined through ratios. Thus, the most suited tool for energy efficiency analysis and optimization is *fractional programming theory*. A comprehensive overview of fractional programming theory for energy efficiency optimization is provided in Reference 32. Here, it suffices to recall that in general a fractional function is neither convex nor concave, even if the numerator and denominator are both affine functions. Nevertheless, fractional programming theory provides a set of methods to optimize general fractional functions, and the maximization can be carried out with polynomial complexity provided the numerator and denominator are concave and convex, respectively, and the enforced constraints are convex.

8.1.2 Recent results

The information-theoretic characterization of relay systems has started more than 30 years ago [6]. Among more recent results in the field of resource allocation for relay-assisted systems, we mention [9, 16], where cooperative communications are proposed to improve coverage and reliability. Further rate results for the multi way relay channel (MWRC) are available in Reference 11, and for the non-regenerative relay channel in Reference 26. In Reference 20, the common-rate capacity of the additive white Gaussian noise (AWGN) MWRC with full message exchange is derived and it is shown that for three and more users this capacity is achieved by decode-and-forward (DF) for SNRs below 0 dB and compute-and-forward (CF) otherwise.

An overview of energy-efficient wireless communication can be found in Reference 15. As for energy-efficient relaying, in Reference 17, relay selection with an energy consumption model to satisfy outage constraints is considered. In Reference 23, relaying and multi-cell multiple-input multiple-output (MIMO) transmission are investigated as approaches for improving resource reuse, by providing a more flexible organization of cellular networks. Furthermore, in Reference 24, the energy efficiency of a multi-carrier relaying channel is optimized. Finally, in Reference 25, an energy-efficient analysis of a single-user, relay-assisted network is provided, while in Reference 14, the spectral-energy efficiency trade-off for relay-based networks is studied.

8.2 Energy-efficient relay protocol design

The first question to ask when designing a relay-assisted system is about the particular protocol to be implemented at the relay. The two first and most well-known relay protocols are surely amplify-and-forward (AF) and DF. In addition, more recently, other schemes have been introduced, among which we mention CF.

When designing the system resources to optimize traditional, non-energy-efficient performance measures like achievable rate and BER, a clear trade-off exists between AF and more complex schemes like DF. On the one hand, AF is simpler, requiring less computational complexity and less feedback overhead, but on the other hand, the performance of AF is typically worse than that of more complex protocols like DF in terms of BER and data rate. However, when the performance metric to optimize is the system energy efficiency, this trade-off becomes less strict, and there are situations in which AF, besides allowing for a simpler system implementation, also grants a higher energy efficiency. The main reason why this happens, is that the larger complexity of DF and of other sophisticated relay protocols, does not only affect the practical implementation of the system, but also directly influences the system energy efficiency. The more complex the scheme to implement, the larger the circuit power consumption, and this directly translates into an energy efficiency reduction. In the rest of this section, we will support this point, by considering the energy-efficient design of a MWRC.

8.2.1 Power control and relay design in three-way relay channels

We consider a three-user single-input single-output (SISO) MWRC system in which the three users communicate with each other via a relay node. A symmetric scenario is considered with a circular (i.e., partial) message exchange. We assume full-duplex transmission and consider a scenario in which no direct user-to-user link is available. For all $k = 1, \ldots, K$, the channel input s_k from user k is such that $\mathbb{E}[|s_k|^2] \le P_{\mathrm{S}}$, whereas the channel input from the relay s_{R} is such that $\mathbb{E}[|s_{\mathrm{R}}|^2] \le P_{\mathrm{R}}$. Let us also denote by σ_{S}^2 and σ_{R}^2 the noise powers at each source and at the relay, respectively.

For the considered system, we consider the AF and DF protocols. Moreover, when AF is used we consider two different implementations of the receiver, namely simultaneous non-unique decoding (SND) and treating interference as noise (IAN). The corresponding achievable sum-rates are obtained as [19]:

$$R_{\Sigma}^{\mathrm{AF-SND}} = \frac{3}{2}\log_2\left(1 + \frac{2p_s p_r}{\sigma_{\mathrm{r}}^2 p_r + 3p_s\sigma_{\mathrm{s}}^2 + \sigma_{\mathrm{s}}^2\sigma_{\mathrm{r}}^2}\right) \tag{8.5}$$

$$R_{\Sigma}^{\mathrm{AF-IAN}} = \log_2\left(1 + \frac{3p_s p_r}{\sigma_{\mathrm{r}}^2 p_r + 3p_s\sigma_{\mathrm{s}}^2 + \sigma_{\mathrm{s}}^2\sigma_{\mathrm{r}}^2}\right) \tag{8.6}$$

$$R_{\Sigma}^{\mathrm{DF}} = \min\left\{\frac{3}{2}\log_2\left(1 + \frac{p_r}{\sigma_{\mathrm{s}}^2}\right), \log_2\left(1 + \frac{3p_s}{\sigma_{\mathrm{r}}^2}\right)\right\} \tag{8.7}$$

and the corresponding system energy efficiencies are written as:

$$\mathrm{EE}^{\mathrm{AF-SND}} = \frac{\frac{3}{2}\log_2\left(1 + \frac{2p_s p_r}{\sigma_{\mathrm{r}}^2 p_r + 3p_s\sigma_{\mathrm{s}}^2 + \sigma_{\mathrm{s}}^2\sigma_{\mathrm{r}}^2}\right)}{\mu_{\mathrm{s}} p_{\mathrm{s}} + \mu_{\mathrm{r}} p_{\mathrm{r}} + P_{\mathrm{c}}^{\mathrm{AF-SND}}} \tag{8.8}$$

$$\mathrm{EE}^{\mathrm{AF-IAN}} = \frac{\log_2\left(1 + \frac{3p_s p_r}{\sigma_{\mathrm{r}}^2 p_r + 3p_s\sigma_{\mathrm{s}}^2 + \sigma_{\mathrm{s}}^2\sigma_{\mathrm{r}}^2}\right)}{\mu_{\mathrm{s}} p_{\mathrm{s}} + \mu_{\mathrm{r}} p_{\mathrm{r}} + P_{\mathrm{c}}^{\mathrm{AF-IAN}}} \tag{8.9}$$

$$\mathrm{EE}^{\mathrm{DF}} = \frac{\min\left\{\frac{3}{2}\log_2\left(1 + \frac{p_r}{\sigma_{\mathrm{s}}^2}\right), \log_2\left(1 + \frac{3p_s}{\sigma_{\mathrm{r}}^2}\right)\right\}}{\mu_{\mathrm{s}} p_{\mathrm{s}} + \mu_{\mathrm{r}} p_{\mathrm{r}} + P_{\mathrm{c}}^{\mathrm{DF}}} \tag{8.10}$$

wherein μ_s and μ_r represent the drain efficiencies of the source and relay amplifiers respectively, whereas P_c^{AF-SND}, P_c^{AF-IAN}, and P_c^{DF} are the circuit power dissipated to implement the system with AF and DF relaying, respectively.

The energy-efficient power control problem can be cast as the maximization of EE^{AF-SND}, EE^{AF-IAN}, and EE^{DF} with respect to $p_s \in [0; P_{s,\max}]$ and $p_r \in [0; P_{\max,r}]$. When DF is used, we have to solve the fractional program

$$\max_{(p_s,p_r)} \frac{\min\left\{\frac{3}{2}\log_2\left(1+\frac{p_r}{\sigma_s^2}\right), \log_2\left(1+\frac{3p_s}{\sigma_r^2}\right)\right\}}{\mu_s p_s + \mu_r p_r + P_c^{DF}} \tag{8.11a}$$

$$\text{s.t. } p_s \in [0; P_{\max,s}],\ p_r \in [0; P_{\max,r}] \tag{8.11b}$$

As shown in Reference 19, (8.11) can be globally solved by directly employing fractional programming theory, because the numerator and denominator of (8.11a) are concave and affine, respectively, and the constraints are also affine.

Instead, when AF is used, we have to solve the problem

$$\max_{(p_s,p_r)} \frac{\log_2\left(1+\dfrac{zp_sp_r}{\sigma_r^2 p_r + 3p_s\sigma_s^2 + \sigma_s^2\sigma_r^2}\right)}{\mu_s p_s + \mu_r p_r + P_c} \tag{8.12a}$$

$$\text{s.t. } p_s \in [0; P_{\max,s}],\ p_r \in [0; P_{\max,r}] \tag{8.12b}$$

wherein $z = 2, P_c = P_c^{AF-SND}$ if SND is used, while $z = 3, P_c = P_c^{AF-IAN}$ if IAN is used. Problem (8.12) is more challenging than (8.11), because the numerator of (8.12a) is not jointly concave in (p_s, p_r), which prevents the direct use of fractional programming. In Reference 19, it is shown that the global solution of (8.12) can be obtained at the price of an exponential computational complexity, by combining fractional programming theory with monotonic optimization theory. However, in Reference 19, it is also shown that virtually optimal performance can be obtained with limited complexity by resorting to the alternating optimization algorithm, iteratively solving (8.12) with respect to p_s for fixed p_r and with respect to p_r for fixed p_r, and solving each sub-problem by means of fractional programming.

In order to fairly compare the resulting optimum energy efficiencies with DF and AF, it is necessary to properly model the circuit power consumption terms P_c^{AF-SND}, P_c^{AF-IAN}, and P_c^{DF}, accounting for the different hardware power consumptions of the different schemes. For each scheme, the power consumption is given by

$$P_c = 3P_{c,s} + P_{c,r} \tag{8.13}$$

wherein $P_{c,s}$ and $P_{c,r}$ are the hardware power required to operate each source and the relay, respectively.

Let us analyze the term $P_{c,s}$ first. A power consumption term which is common to both AF and DF is the power dissipated due to analog operations in the receiver front-end, P_{rx}, in the transmitter front-end, P_{tx}, in the digital-to-analog converters

(DACs), P_{DAC}, and in the analog-to-digital converters (ADCs), P_{ADC}. Let us denote the sum of all these powers by

$$P_{\mathrm{c,analog}} = P_{rx} + P_{\mathrm{ADC}} + P_{\mathrm{DAC}} + P_{tx} \tag{8.14}$$

As for the digital part of the system, the power consumption is related to the power dissipated for digital signal processing operations, P_{DSP}, which can be modeled as a multiple of the power P_{dec} required by a single-user decoder.[1] Then, we have

$$P_{\mathrm{c,S}}^{\mathrm{AF-IAN}} = P_{\mathrm{c,analog}} + P_{\mathrm{dec}} \tag{8.15}$$

$$P_{\mathrm{c,S}}^{\mathrm{AF-SND}} = P_{\mathrm{c,analog}} + 2P_{\mathrm{dec}} \tag{8.16}$$

$$P_{\mathrm{c,S}}^{\mathrm{DF}} = P_{\mathrm{c,analog}} + 2P_{\mathrm{dec}} \tag{8.17}$$

where we have accounted for the fact that SND reception requires to decode two messages, and the same is true when the relay uses decode-and-forward (DF).

Next, let us focus on the term $P_{\mathrm{c,r}}$. When DF is used, all messages need to be decoded at the relay, whereas no digital operation is required for AF operation, which directly amplifies the analog signal. So, we have

$$P_{\mathrm{c,r}}^{\mathrm{AF-SND}} = P_{\mathrm{c,r}}^{\mathrm{AF-IAN}} = P_{rx} + P_{tx} \tag{8.18}$$

$$P_{\mathrm{c,r}}^{\mathrm{DF}} = P_{\mathrm{c,analog}} + 3P_{\mathrm{dec}} \tag{8.19}$$

Using this model to evaluate the circuit power consumption terms in (8.11) and (8.12), and solving the optimization problems, it is possible to obtain a fair comparison of the energy efficiency with the different schemes. The results are shown in Figure 8.1. We can see that in the saturation region of the energy efficiency, which is the region of interest, DF is outperformed by AF. The reasons for this behavior are essentially two:

- As already anticipated, DF consumes a much larger circuit power than AF, which decreases the energy efficiency.
- While DF is able to grant a higher rate than AF, this effect is more significant for large transmit powers. However, this regime is not reached when maximizing the energy efficiency, because the energy efficiency is not maximized by using all of the available power. Otherwise stated, if $P_{\mathrm{s,max}}$ and $P_{\mathrm{r,max}}$ are large, the power levels $(p_{\mathrm{s}}^*, p_{\mathrm{r}}^*)$ which solve (8.11) and (8.12) will be such that $p_{\mathrm{s}} < P_{\mathrm{s,max}}$ and $p_{\mathrm{r}} < P_{\mathrm{r,max}}$.

Instead, we see that for lower values of the SNR, the DF scheme slightly outperforms both AF schemes. This is explained noticing that in this region, the optimal solution of (8.11) and (8.12) is to transmit at maximum power, and therefore energy efficiency maximization is equivalent to achievable sum-rate maximization.

These considerations motivate us to consider AF relaying as far as the energy-efficient design of relay-assisted systems is concerned. The rest of this chapter will present several energy-efficient resource allocation schemes for AF relay-assisted

[1] For realistic numbers of P_{rx}, P_{ADC}, P_{DAC}, P_{tx}, and P_{dec} we refer to Reference 19 and references therein.

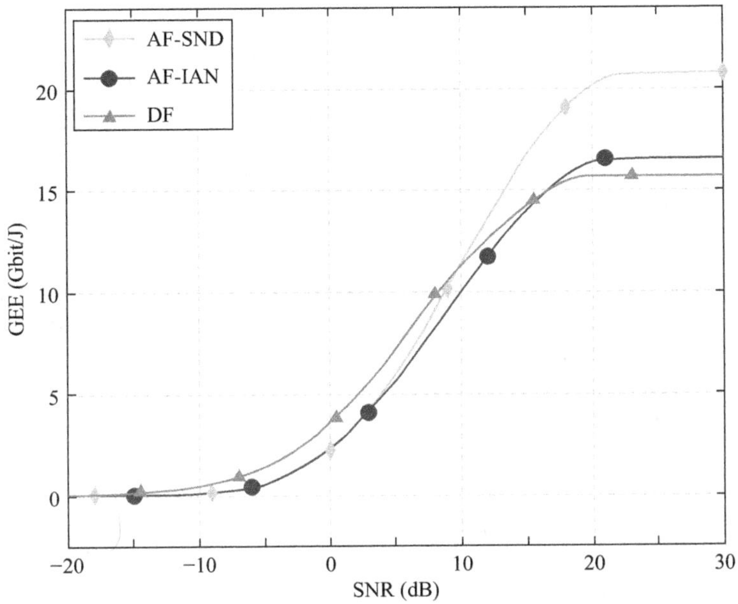

Figure 8.1 *GEE in the three-user MWRC of (1) AF-SND; (2) AF-IAN; and (3) DF as a function of SNR* $= P_{max,s}/\sigma_s^2 = P_{max,r}/\sigma_r^2$

systems. The focus will be on MIMO systems, but most results can be extended to multi-carrier settings. Three main scenarios will be considered. In Section 8.3, the problem of power allocation will be considered, assuming that other physical layer resources are kept fixed and that single-stream transmission is used. This scenario represents the basis to tackle the more involved problems presented in Section 8.4, where transmit powers allocation is performed jointly with receiver and relay design, and in Section 8.5 where the assumption of single-stream transmission is relaxed and a multi-stream MIMO system is considered. Both centralized and distributed algorithms will be presented and compared in terms of performance, complexity, and ease of implementation.

8.3 Energy-efficient power allocation with AF relaying

Let us consider a relay-assisted network with K transmitters equipped with N_T antennas each, which communicate with M receivers, equipped with N_R antennas each, through one AF relay, equipped with N_A antennas.

Let us denote by p_k, q_k, s_k, and H_k the kth user's transmit power, beamforming vector, unit-norm information symbol, and matrix channel to the relay, respectively. Then, the signal y_r received at the relay is expressed as

$$y_r = \sqrt{p_k} H_k q_k s_k + \sum_{\ell \neq k} \sqrt{p_\ell} H_\ell q_\ell s_\ell + w_r \tag{8.20}$$

with w_r the thermal noise at the relay, modeled as a zero-mean complex Gaussian vector with covariance matrix $\sigma_r^2 I_{N_A}$. In order to avoid amplifier saturation, the signal y_r must be normalized by its power P_t, before it can be amplified. Assuming that the information symbols from the different users are statistically independent, and also independent from the thermal noise, we can compute P_t as

$$P_t = \sum_{k=1}^{K} p_k \left\| H_k q_k \right\|^2 + \sigma_r^2 N_A \tag{8.21}$$

After normalization, the signal is amplified by the AF matrix A, subject to the power constraint $\mathrm{tr}(AA^H) \leq P_r^{\max}$, with $P_r^{\max}$ the available relay power, and finally forwarded to the receivers. Denoting by G_m the vector channel from the relay to receiver m, the signal y_m at receiver m is expressed as

$$y_m = \sqrt{\frac{1}{P_t}} G_m A y_r + w_m \tag{8.22}$$

with w_m the thermal noise at receiver m, modeled as a complex Gaussian vector with zero mean and covariance matrix $\sigma_m^2 I_{N_R}$. Plugging (8.20) and (8.21) into (8.22) and elaborating, the signal to interference plus noise ratio (SINR) γ_k enjoyed by the kth transmitter at its intended receiver $m = a(k)$, after linear reception by the filter c_k, takes the general form:

$$\gamma_k = \frac{\alpha_k p_k}{1 + \phi_k p_k + \sum_{\ell \neq k} \omega_{\ell,k} p_\ell} \tag{8.23}$$

wherein we have defined the non-negative coefficients

$$\alpha_k = \frac{\left| c_k^H G_{a(k)} A H_k q_k \right|^2}{\sigma_r^2 \left(\sigma_{a(k)}^2 N_A \|c_k\|^2 + \left\| c_k^H G_{a(k)} A \right\|^2 \right)} \tag{8.24}$$

$$\phi_k = \frac{\left\| H_k^H q_k \right\|^2 \|c_k\|^2 \sigma_{a(k)}^2}{\sigma_r^2 \left(\sigma_{a(k)}^2 N_A \|c_k\|^2 + \left\| c_k^H G_{a(k)} A \right\|^2 \right)} \tag{8.25}$$

$$\omega_{\ell,k} = \frac{\left| c_k^H G_{a(k)} A H_\ell q_\ell \right|^2 + \left\| H_\ell q_\ell \right\|^2 \sigma_{a(k)}^2 \|c_k\|^2}{\sigma_r^2 \left(\sigma_{a(k)}^2 N_A \|c_k\|^2 + \left\| c_k^H G_{a(k)} A \right\|^2 \right)} \tag{8.26}$$

Based on (8.23), the energy efficiency of user k and the system global energy efficiency (GEE) are written as

$$\mathrm{EE}_k = \frac{B \log_2 (1 + \gamma_k)}{\mu_k p_k + P_{o,k}} \tag{8.27}$$

$$\mathrm{GEE} = \frac{B \sum_{k=1}^{K} \log_2 (1 + \gamma_k)}{\sum_{k=1}^{K} \mu_k p_k + P_{c,k}} \tag{8.28}$$

The rest of this section is devoted to the description of both centralized and distributed energy-efficient power allocation algorithms. In the centralized setting, the problem will be formulated as the maximization of the GEE with respect to all of the system transmit powers. Instead, the distributed setup will consider the competitive problem in which the different users aim at maximizing their own energy efficiencies by tuning their own transmit powers. Both scenarios will be tackled considering maximum power and minimum rate constraints.

8.3.1 Centralized power allocation

The centralized energy-efficient power control problem is formulated as the optimization program

$$\max_{\{p_k\}_{k=1}^K} \frac{B\sum_{k=1}^K \log_2(1+\gamma_k)}{\sum_{k=1}^K \mu_k p_k + P_{c,k}} \tag{8.29a}$$

$$\text{s.t. } 0 \le p_k \le P_{\max,k},\ \forall\, k = 1,\ldots,K \tag{8.29b}$$

$$\log_2(1+\gamma_k) \ge R_{\min,k},\ \forall\, k = 1,\ldots,K \tag{8.29c}$$

with $P_{\max,k}$ and $R_{\min,k}$ the kth user's maximum feasible power and minimum acceptable rate, respectively. We should observe that considering at the same time a maximum power constraint and a minimum rate requirement might make Problem (8.29) unfeasible. A necessary and sufficient condition such that (8.29) is feasible is provided in the next proposition proved in Reference 34.

Proposition 8.1. *Let* $\mathbf{F} \in \mathbb{C}^{K\times K}$ *be a matrix whose* (k,j)*th element is defined as*

$$[\mathbf{F}]_{k,j} \triangleq \begin{cases} 0 & j = k \\ \dfrac{\omega_{k,j}(2^{R_{\min,k}}-1)}{\alpha_k - \phi_k(2^{R_{\min,k}}-1)} & j \ne k \end{cases} \tag{8.30}$$

and denote by $\rho_{\mathbf{F}}$ *its spectral radius. The feasible set of* (8.29) *is not empty if and only if:*

$$\frac{\alpha_k}{\phi_k} \ge 2^{R_{\min,k}} - 1,\ \forall\, k = 1,\ldots,K \tag{8.31}$$

$$\rho_{\mathbf{F}} < 1 \quad \text{and} \quad (I-F)^{-1}\underline{\mathbf{s}} \le \mathbf{P}_{\max} \tag{8.32}$$

where $\mathbf{P}_{\max} = [P_{\max,1}, P_{\max,2},\ldots,P_{\max,K}]^T \in \mathbb{R}_+^{K\times 1}$ *and* $\mathbf{s} \in \mathbb{R}_+^K$ *has elements given by* $[\underline{\mathbf{s}}]_k \triangleq \sigma_k^2(2^{R_{\min,k}}-1)(\alpha_k - \phi_k\underline{\gamma}_k)^{-1}$.

After providing a way to easily test the feasibility of (8.29), we turn our attention to determining its solution. Problem (8.29) is a fractional problem, in which the numerator of (8.29a) is not jointly concave in all transmit powers, due to the interference coefficients $\omega_{\ell,k}$. This prevents the direct use of fractional programming theory to find the global solution of (8.29a) with limited complexity. Therefore, in the following, we trade-off complexity and global optimality, aiming at developing low-complexity algorithms to find possibly sub-optimal solutions of (8.29a), which nevertheless enjoy strong optimality claims. To this end, we resort to the framework of *sequential fractional programming*, which allows one to determine a solution of the Karush Kuhn Tucker (KKT) first-order optimality conditions of (8.29) by solving a sequence of easier fractional problems. In order to elaborate further, we need to recall the following result from Reference 18.

Proposition 8.2. *Assume $\mathcal{M}$ is a maximization problem with objective $f_0(x)$ and constraint functions $f_i(x) \geq 0$, for $i = 1, \ldots, I$. Consider a sequence of approximate Problems $\{\mathcal{M}^{(j)}\}_j$ with objectives $\{\tilde{f}_0^{(j)}(x)\}_j$ and constraints $\{\tilde{f}_i^{(j)}(x)\}_{j,i}$. Let $\tilde{x}^{(j)}$ be the optimal solution of $\mathcal{M}^{(j)}$ and assume that $\forall j, i\ \tilde{f}_i^{(j)}(x)$ is such that*

1. $\tilde{f}_i^{(j)}(x) \leq f_i(x)$, *for all* x;
2. $\tilde{f}_i^{(j)}\left(\tilde{x}^{(j-1)}\right) = f_i\left(\tilde{x}^{(j-1)}\right)$;
3. $\nabla\tilde{f}_i^{(j)}\left(\tilde{x}^{(j-1)}\right) = \nabla f_i\left(\tilde{x}^{(j-1)}\right)$.

Then, the sequence $\{\tilde{x}^{(j)}\}_j$ converges to a point $\tilde{x}$, which fulfills the KKT optimality conditions of the original problem $\mathcal{M}$.

Thus, leveraging Proposition 8.2, we can fulfill the KKT conditions of a difficult problem, by solving a sequence of auxiliary problems. Clearly, the critical issue in this approach is to find suitable approximations $\{f_i\}_i$ which fulfill the listed requirements, while at the same time resulting in simpler optimization problems. As far as GEE maximization is concerned, this can be accomplished by leveraging the following lower-bound of the logarithmic function [21]. Specifically, $\forall \gamma,\ \tilde{\gamma} \geq 0$ we have

$$\log_2(1+\gamma) \geq a\log_2\gamma + b \tag{8.33}$$

with

$$a = \frac{\tilde{\gamma}}{1+\tilde{\gamma}} \quad \text{and} \quad b = \log_2(1+\tilde{\gamma}) - \frac{\tilde{\gamma}}{1+\tilde{\gamma}}\log_2\tilde{\gamma}. \tag{8.34}$$

The right-hand side (RHS) and left-hand side (LHS) of (8.33) are equal at $\gamma = \tilde{\gamma}$, and the same holds for their derivatives with respect to γ evaluated at $\gamma = \tilde{\gamma}$. Therefore, we may lower bound the GEE as follows:

$$\begin{aligned} \text{GEE} \geq \frac{\sum_{k=1}^{K}\left[a_k\log_2(\gamma_k) + b_k\right]}{p_c + \sum_{k=1}^{K}\mu_k p_k} = \frac{\sum_{k=1}^{K}\left[b_k + a_k\log_2(\alpha_k p_k)\right]}{p_c + \sum_{k=1}^{K}\mu_k p_k} \\ - \frac{\sum_{k=1}^{K}\left[a_k\log_2\left(1+\phi_k p_k + \sum_{\ell\neq k}\omega_{\ell,k}p_\ell\right)\right]}{p_c + \sum_{k=1}^{K}\mu_k p_k} = \widetilde{\text{GEE}} \end{aligned} \tag{8.35}$$

Next, applying the variable change $p_k = 2^{q_k}$, one gets

$$\widetilde{\text{GEE}} = \frac{\sum_{k=1}^{K}\left[b_k + a_k \log_2(\alpha_k) + a_k q_k\right]}{p_c + \sum_{k=1}^{K} \mu_k 2^{q_k}} - \frac{\sum_{k=1}^{K}\left[a_k \log_2\left(1 + \phi_k 2^{q_k} + \sum_{\ell \neq k} \omega_{\ell,k} 2^{q_\ell}\right)\right]}{p_c + \sum_{k=1}^{K} \mu_k 2^{q_k}} \tag{8.36}$$

Algorithm 1 GEE maximization.

1: Test feasibility by Proposition 8.1.
2: **if** Feasible **then**
3: Set $j = 0$ and choose any feasible $p^{(0)}$;
4: **repeat**
5: $j = j + 1$;
6: Given $p_0^{(j-1)}$, for all $k = 1, \ldots, K$ compute $a_k^{(j)}$ and $b_k^{(j)}$ as in (8.34);
7: Set $p_k^{(j)} = 2^{q_k^{(j)}}$, with $\{q_k^{(j)}\}_k$ the solution of (8.37) with parameters $a_k^{(j)}$ and $b_k^{(j)}$;
8: **until** convergence
9: **end if**

Using the above results, the solution to (8.29) can be lower bounded by solving the approximate problem

$$\max_{\{q_k\}_{k=1}^{K}} \widetilde{\text{GEE}}\left(\{q_k\}_{k=1}^{K}\right) \tag{8.37a}$$

$$\text{s.t. } 2^{q_k} \leq P_{\max,k}, \quad \forall\, k = 1, \ldots, K \tag{8.37b}$$

$$\log_2\left(1 + \frac{\alpha_k 2^{q_k}}{1 + \phi_k 2^{q_k} + \sum_{\ell \neq k} \omega_{\ell,k} 2^{q_\ell}}\right) \geq R_{\min,k}, \quad \forall\, k = 1, \ldots, K \tag{8.37c}$$

Observe now that for any given $\{a_k\}_k$ and $\{b_k\}_k$, the numerator and the denominator of (8.36) are both differentiable, and, respectively, concave[2] and convex in $\{q_k\}_k$. Moreover, (8.37b) is a convex constraint, while (8.37c) can be reformulated in convex form by first rewriting it as

$$2^{q_k}\left(\alpha_k - \underline{\gamma}_k \phi_k\right) \geq \underline{\gamma}_k \left(1 + \sum_{\ell \neq k} \omega_{\ell,k} 2^{q_\ell}\right) \tag{8.38}$$

[2] Recall that the log-sum-exp function is convex [4].

and then applying the logarithm function to both sides[3]

$$q_k - \log_2\left(1 + \sum_{\ell \neq k} \omega_{\ell,k} 2^{q_\ell}\right) + \log_2\left(\frac{\alpha_k - \underline{\gamma}_k \phi_k}{\underline{\gamma}_k}\right) \geq 0 \tag{8.39}$$

As a consequence, (8.37) is a fractional problem in which the objective has a concave numerator and convex denominator, and the constraints are convex. Therefore, (8.37) can be globally and efficiently solved by means of standard fractional programming methods. This allows one to formulate the power allocation algorithm in Algorithm 1, which works by solving a sequence of fractional problems of the form of (8.37).

Proposition 8.3. *Algorithm 1 monotonically increases the GEE value and converges to a point fulfilling the KKT conditions of the original problem* (8.29).

8.3.2 Distributed power allocation

As opposed to the centralized setting in which all network nodes cooperate to maximize a system-wide performance functions as the GEE, in a competitive scenario the network nodes tune their transmit powers in order to maximize their individual energy efficiencies. This problem can be mathematically formulated as the K coupled problems:

$$\arg\max_{p_k} \quad \mathrm{EE}_k(p_k, \mathbf{p}_{-k}) \quad \forall k \tag{8.40a}$$

$$\text{s.t. } 0 \leq p_k \leq P_{\max,k}, \quad \forall\, k = 1, \ldots, K \tag{8.40b}$$

$$\log_2(1 + \gamma_k) \geq R_{\min,k}, \quad \forall k = 1, \ldots, K \tag{8.40c}$$

where $\mathbf{p}_{-k} = [p_1, \ldots, p_{k-1}, p_{k+1}, \ldots, p_K]^T$ is the interference vector containing all powers except user ks.

A mathematical tool which allows one to analyze coupled problems like (8.40) is game theory. More in general, game theory provides a mathematical framework to analyze the interactions among autonomous decision-makers with contrasting interests. Applying a game-theoretic formulation, the competitive power allocation problem can be modeled as the non-cooperative game in normal form

$$\mathcal{G} \triangleq \left\{\mathcal{K}, \{\mathcal{P}_k(\mathbf{p}_{-k})\}_k, \{u_k(p_k, \mathbf{p}_{-k})\}_k\right\} \tag{8.41}$$

wherein $\mathcal{K} = [1, 2, \ldots, K]$ is the set of players, $\mathcal{P}_k(\mathbf{p}_{-k}) = [0; P_{\max,k}] \cup \{p_k : \log_2(1 + \gamma_k(p_k, \mathbf{p}_{-k})) \geq R_{\min,k}\}$ is player k's strategy set, $u_k(\mathbf{p}) = \mathrm{EE}_k(p_k, \mathbf{p}_{-k})$ is player k's utility function. The K coupled problems in (8.40) define the best response dynamics (BRD) of the game $\mathcal{G}$, while the solution of the kth problem in (8.40) is the kth player's best response to the other players' choices, which we denote by $\mathcal{B}_k(\mathbf{p}_{-k})$. Any fixed point of the BRD is a Nash equilibrium of the game.

The Nash equilibrium plays a central role in game theory, as it represents a predictable outcome of a game, and is therefore the most widely used solution concept in the context of non-cooperative games. Unfortunately, in general a non-cooperative

[3] Note that this is always possible if the problem is feasible.

game is not guaranteed to admit any equilibrium, and even if one exists, it might not be unique, or the convergence of the BRD is not guaranteed. Moreover, in the BRD (8.40), a coupling exists not only in the utility functions, as is the case in regular non-cooperative games, but also in the strategy sets, due to the presence of the minimum rate constraints. Similar non-cooperative games are termed generalized non-cooperative games, and their analysis is more involved than for regular games [1, 7]. Nevertheless, for the case at hand it is possible to show that under mild conditions, a unique generalized Nash equilibrium exists, and the BRD converges to the equilibrium. The detailed analysis is reported in Reference 34. In the following, we recall the main results from Reference 34.

The best response $\mathscr{B}_k(\mathbf{p}_{-k})$ can be obtained in closed-form as shown next.

Lemma 8.1. *If*

$$P_{\max,k} \geq \gamma_{\min,k} \frac{1 + \sum_{\ell \neq k} \omega_{\ell,k} P_{\max,\ell}}{\alpha_k - \phi_k \gamma_{\min,k}}, \ \forall k \tag{8.42}$$

then $\mathscr{B}_k(\mathbf{p}_{-k})$ *takes the form*

$$\mathscr{B}_k(\mathbf{p}_{-k}) = \min\{P_{\max,k}, \max\{p_k^\star, P_{\min,k}\}\} \tag{8.43}$$

wherein

$$P_{\min,k}(\mathbf{p}_{-k}) \triangleq \gamma_{\min,k} \frac{1 + \sum_{\ell \neq k} \omega_{\ell,k} p_\ell}{\alpha_k - \phi_k \gamma_{\min,k}} \tag{8.44}$$

and $p_k^\star$ *is the unique stationary point of* EE_k*, found as the unique, non-negative solution of the equation:*

$$(\mu_k p_k + P_{c,k}) f'(\gamma_k) \frac{\partial \gamma_k}{\partial p_k} = \mu_k f(\gamma_k) \tag{8.45}$$

Next, assuming (8.42) holds, it is possible to show that the power control game has a unique generalized Nash equilibrium, and that the BRD (8.40) always converges [34]. The proof exploits the quasi-concavity of EE_k in p_k, and leverages the framework of standard functions [29].

8.3.2.1 Distributed implementation

After ensuring that the BRD of the game converges to the unique equilibrium of the game, it is natural to develop a power allocation algorithm by iteratively computing the best response (8.43) until the equilibrium is reached. The formal algorithm is stated in Algorithm 2.

Algorithm 2 Iterative algorithm to solve (8.40).

For all $k = 1, \ldots, K$, initialize $p_k \in \mathbb{R}_+$ in the feasible set;
repeat
    for $k = 1$ to K **do**
        Update p_k according to (8.43);
    end for
until convergence

At a first look, it would seem that Algorithm 2 cannot be implemented in a distributed fashion, because each player k needs to know the other players' channels and power levels to compute its best response. However, this is not true, and Algorithm 2 can indeed be implemented in a self-organizing way. To see this, observe that in order to compute its best response, player k needs to know the parameters α_k, ϕ_k, and the aggregate parameter

$$\Omega_k = 1 + \sum_{j \neq k} p_j \omega_{k,j} \tag{8.46}$$

The parameters α_k and ϕ_k depend only on player k's own channels and filters, as well as on global system parameters, and therefore can be easily computed locally. As for Ω_k, it can be computed from the SINR expression as we have

$$\Omega_k = \frac{\alpha_k p_k}{\gamma_k} - \phi_k p_k \tag{8.47}$$

Then, Ω_k can also be computed locally by having the receiver associated to transmitter k measure the SINR γ_k and feed it back to transmitter k. We remark that this only requires the existence of a feedback channel between transmitter k and the corresponding receiver, which is typically available in wireless cellular networks. Therefore, besides being guaranteed to converge to the unique generalized Nash equilibrium, Algorithm 2 can also be implemented in a fully decentralized fashion.

8.4 Energy-efficient resource allocation with multiple-antenna AF relaying

Section 8.3 has dealt with the power allocation problem at the transmitters, assuming all other physical layer resources are fixed. This section extends the framework of Section 8.3 to the case in which the transmit powers are to be jointly allocated together with the relay AF matrix A and with the users' receive filters $\{c_k\}_{k=1}^K$. A similar system model as in Section 8.3 will be considered, but the focus will be on interference channels, i.e., $K = M$. This allows one to simplify the notation, denoting a transmitter and its intended receiver by the same index. So, for all k, we have $a(k) = k$.

The approach will be to decouple relay design from the design of the receive filters and transmit powers, which will be accomplished by means of a universal relay interference neutralization scheme. Next, the optimal receive filters will be derived and plugged into the energy efficiency expression to optimize. Finally, based on the resulting energy efficiency expression, both centralized and distributed power allocation will be performed.

8.4.1 Relay interference neutralization

A natural approach for relay design would be to allocate the AF matrix A so as to maximize a system-wide performance function like the GEE in (8.27). However, this approach comes with a high computational complexity, because maximizing the GEE with respect to A is a non-convex problem. Moreover, the optimal A would be coupled with the users' transmit powers and receive filters, which would require to solve the non-convex GEE maximization problem with respect to A every time the transmit powers are updated during the resource allocation phase.

Instead, we look for a simpler relay design scheme, which is affordable for layer-1 devices such as AF relays. To this end, we introduce a relay precoding scheme which is able to neutralize multi-user interference at all receivers, for any choice of the users' transmit powers and receive filters. This approach has several advantages:

- it significantly reduces the complexity of the overall resource allocation algorithm because it completely decouples relay design from the design of the other network resources;
- as it will be shown, relay interference neutralization only requires standard algebraic operations and therefore has a very limited computational burden;
- neutralizing multi-user interference also allows for simpler data decoding at the receivers, as single-user receivers can be used instead of multi-user detection structures.

In order to obtain a universal interference neutralization scheme that does not depend on the other network resources, we proceed as follows. For any choice of the receive filters, beamforming vectors, and transmit powers, interference neutralization at the relay can be achieved by enforcing the conditions

$$G_k A H_\ell = 0_{N_R, N_T}, \quad \forall\, \ell \neq k \tag{8.48}$$

Additionally, in order to guarantee a desired effective channel for each user k, the following conditions should also be enforced

$$G_k A H_k = a\widetilde{A}_k, \quad \forall\, k \tag{8.49}$$

with the scalar a to be set so as to meet the relay power constraint, and with the $N_R \times N_T$ matrices $\widetilde{A}_k$ to be adjusted according to some quality of service (QoS) policy among the users. Conditions (8.49) can be regarded as a form of channel equalization among the users. If $\widetilde{A}_1 = \widetilde{A}_2 = \cdots = \widetilde{A}_K$, each transmitter–receiver pair has the same equivalent channel $G_k A H_k$. Instead, if some users belong to a different priority class, the corresponding $\widetilde{A}_k$ can be tuned to grant these users a better equivalent channel. We allocate the matrices $\{\widetilde{A}_k\}_{k=1}^K$ according to a channel-matching criterion [22]. Specifically, denoted by $U_{H,k}\Lambda_{H,k}^{1/2}V_{H,k}^H$ and $U_{G,k}\Lambda_{G,k}^{1/2}V_{G,k}^H$ the singular value decomposition (SVDs) of H_k and G_k, respectively, for all $k = 1, \ldots, K$ we impose $G_k A H_k = aU_{G,k}\Lambda_{G,k}^{1/2}\Lambda_{H,k}^{1/2}V_{H,k}^H = a\widetilde{A}_k$.

Applying the vectorization operator, and defining $a_k = \mathrm{vec}(\widetilde{A}_k)$ for all $k = 1, \ldots, K$, (8.48) and (8.49) can be equivalently rewritten as

$$\mathrm{vec}\,(G_k A H_k) = a a_k \tag{8.50}$$

$$\mathrm{vec}\,(G_k A H_\ell) = 0_{N_R N_T} \tag{8.51}$$

which, exploiting the identity $\mathrm{vec}(XYZ) = (Z^T \otimes X)\mathrm{vec}(Y)$, become

$$\left(H_k^T \otimes G_k\right)\mathrm{vec}(A) = a a_k \tag{8.52}$$

$$\left(H_\ell^T \otimes G_k\right)\mathrm{vec}(A) = 0_{N_R N_T} \tag{8.53}$$

Stacking together the equations for each user k, we obtain the linear system

$$C\text{vec}(A) = av \tag{8.54}$$

with $v^T = [a_1^T, 0^T_{(K-1)N_R N_T}, \ldots, 0^T_{(K-1)N_R N_T}, a_K^T]$ and

$$C = \begin{bmatrix} \widetilde{H}_1 \\ \vdots \\ \widetilde{H}_K \end{bmatrix}, \quad \widetilde{H}_k = \begin{bmatrix} H_k^T \otimes G_1 \\ \vdots \\ H_k^T \otimes G_K \end{bmatrix} \tag{8.55}$$

Thus, the interference neutralization problem can be equivalently reformulated as the system of linear equations (8.54), which admits at least one solution if the $K^2 N_R N_T \times N_A^2$ matrix C has rank $r = K^2 N_R N_T$. Under such an assumption, from (8.54) we obtain

$$\text{vec}(A) = aC^+ v \tag{8.56}$$

where $a \leq \frac{\sqrt{P_R}}{\|C^+ v\|}$ ensures that the relay power constraint is met, because $\text{tr}(AA^H) = \|\text{vec}(A)\|^2$.

It should be stressed that the considered approach is different from interference alignment techniques which design the beamforming vectors and AF matrix to constrain the interference at each receiver in a suitable vector space, in order to eliminate it by zero forcing reception. Here, we neither assume nor impose any particular configuration of the interference at each receiver. The beamforming vectors and receive filters can be allocated for EE maximization, or according to any other criterion, and the resulting interference is nulled out by the proposed relay processing.

8.4.1.1 Required number of relay antennas

An important issue to investigate is the number of relay antennas which are necessary in order to achieve perfect interference neutralization. From (8.56), we deduce that in order to solve the system the matrix C needs to have rank $K^2 N_R N_T$. However, in order for this to happen with probability one, it is not enough to have $N_A^2 \geq K^2 N_R N_T$, even if the entries of the coefficient matrix C are obtained from independent channel realizations. To see this, note that, for all $k = 1, \ldots, K$, the $K N_R N_T \times N_A^2$ matrix $\widetilde{H}_k$ can be written as

$$\widetilde{H}_k = H_k^T \otimes \widetilde{G} \tag{8.57}$$

wherein $\widetilde{G}$ is the $K N_R \times N_A$ matrix obtained by stacking together the channel matrices $\{G_k\}_{k=1}^K$. Now, since the rank of a Kronecker product is equal to the product of the ranks of the factors, we obtain

$$\text{rank}(\widetilde{H}_k) = \text{rank}(H_k)\text{rank}(\widetilde{G}) = \min(N_T, N_A)\min(N_A, K N_R) \tag{8.58}$$

where the last equality holds with probability one, since the entries of H_k and $\widetilde{G}$ are realizations of independent random variables. Consequently, in order to guarantee that the rank of C is $r = K^2 N_R N_T$ with probability 1, it is necessary to have both $N_A \geq K N_R$ and $N_A \geq N_T$ so that each block $\widetilde{H}_k$ in C provides $K N_R N_T$ independent equations. Then, the condition for the achievability of interference neutralization can be expressed in compact form as

$$N_A \geq \max\left\{K\sqrt{N_R N_T}, KN_R, N_T\right\} \tag{8.59}$$

If we consider the special case $N_T = N_R = 1$, we obtain $N_A \geq K$, which is known to hold in the particular case of single-antenna terminals [12].

It is seen that the number of relay antennas that are required in order to achieve IN grows linearly in the system parameters. Two comments should be made regarding this point.

- Having light infrastructure nodes with tens of antennas is not only technically feasible, but also appears as a likely scenario in future wireless networks. Two popular trends in wireless communications go in this direction, namely massive MIMO systems, where a large number of antennas are used, and mmWave communications, which allows one to reduce antenna sizes, thereby making the deployment of a large number of antennas feasible.
- Even if not enough antennas can be deployed to exactly solve (8.54), still, computing (8.56) is always possible and yields the least squares solution of (8.54), i.e., the element in the column span of C with minimum Euclidean distance from v. In light of this consideration, it is expected that allocating the relay matrix according to (8.56) performs well even when perfect interference neutralization cannot be achieved. Some numerical results will be provided to corroborate this point.

8.4.2 Receive filters allocation

Let us denote by A_{IN} the interference-neutralizing AF matrix. Then, for all $k = 1, \ldots, K$, (8.23) becomes

$$\gamma_k = \frac{p_k \left|c_k^H G_k A_{IN} H_k q_k\right|^2}{\sigma_k^2 P_t \|c_k\|^2 + \sigma_r^2 c_k^H G_k A_{IN} A_{IN}^H G_k^H c_k} \tag{8.60}$$

Inspecting (8.60), leads to two significant observations:

1. Even if the relay precoding has suppressed the spatial multi-user interference, the users' SINRs are still coupled in the transmit powers, due to the relay normalization term P_t. This makes the power control problem non-trivial, despite the use of interference neutralization.
2. As a result of the relay processing, the overall noise term is colored. As a consequence, for any link k, the optimal receive structure is known to be a whitening filter followed by matched-filtering [27]. For the case at hand, such a receive structure is equivalent to a linear minimum mean square error (LMMSE) receiver.

Thus, for any k, the optimal receive filter c_k is given by

$$c_k = \sqrt{\frac{p_k}{P_t}} Z_k^{-1} G_k A_{IN} H_k q_k \tag{8.61}$$

wherein $Z_k = \frac{\sigma_r^2}{P_t} G_k A_{IN} A_{IN}^H G_k^H + \sigma_k^2 I_{N_R}$, being the colored noise covariance matrix of user k. Plugging (8.61) into (8.60) leads to

$$\gamma_k^{\text{LMMSE}} = \frac{p_k}{P_t} q_k^H H_k^H A_{IN}^H G_k^H Z_k^{-1} G_k A_{IN} H_k q_k \tag{8.62}$$

To elaborate further, let us plug the expression of P_t into (8.62), which yields

$$\gamma_k^{\text{LMMSE}} = p_k v_k^H \left(\sigma_r^2 G_k A_{IN} A_{IN}^H G_k^H + P_t \sigma_k^2 I_{N_R}\right)^{-1} v_k \tag{8.63}$$

$$= p_k v_k^H \left(R_k + \phi_k p_k I_{N_R}\right)^{-1} v_k \tag{8.64}$$

where we have defined

$$v_k = G_k A_{IN} H_k q_k \tag{8.65}$$

$$\phi_k = \sigma_k^2 \left\| H_k q_k \right\|^2 \tag{8.66}$$

$$R_k = \sigma_r^2 G_k A_{IN} A_{IN}^H G_k^H + \sigma_k^2 \left(\sigma_r^2 N_A + \sum_{\ell \neq k} p_\ell \left\| H_\ell q_\ell \right\|^2\right) I_{N_R} \tag{8.67}$$

The SINR in (8.63) can be further simplified by observing that the eigenvalues of R_k are equal to

$$\lambda(R_k) = \{\lambda_{i,k}\}_{i=1}^{N_R} + \sigma_k^2 \sigma_r^2 N_A + \sum_{\ell \neq k} p_\ell \omega_{\ell,k} \tag{8.68}$$

$$\omega_{\ell,k} = \sigma_k^2 \left\| H_\ell q_\ell \right\|^2 \tag{8.69}$$

and $\{\lambda_{i,k}\}_{i=1}^{N_R}$ the eigenvalues of $\sigma_r^2 G_k A_{IN} A_{IN}^H G_k^H$. Finally, denoting by $\{u_{i,k}\}_{i=1}^{N_D}$ the eigenvectors of $G_k A_{IN} A_{IN}^H G_k^H$, and defining

$$\alpha_{i,k} = \left| v_k^H u_{i,k} \right|^2 \tag{8.70}$$

$$\tau_{i,k} = \lambda_{i,k} + \sigma_k^2 \sigma_r^2 N_A \tag{8.71}$$

we can finally express (8.63) as

$$\gamma_k^{\text{LMMSE}} = \sum_{i=1}^{N_R} \frac{p_k \alpha_{i,k}}{\tau_{i,k} + \phi_k p_k + \sum_{\ell \neq k} \omega_{\ell,k} p_\ell} \tag{8.72}$$

8.4.3 Centralized power allocation

The final ingredient for the development of a joint resource allocation algorithm is to allocate the transmit powers, based on the SINR expression (8.72). From a centralized point of view, the problem is cast as

$$\max_{\{p_k\}_{k=1}^K} \frac{\sum_{k=1}^K \log_2\left(1 + \gamma_k^{\text{LMMSE}}\right)}{\sum_{k=1}^K \mu_k p_k + P_{c,k}} \tag{8.73a}$$

$$\text{s.t. } 0 \leq p_k \leq P_{\max,k}, \quad \forall\, k = 1, \ldots, K \tag{8.73b}$$

$$\log_2\left(1 + \gamma_k^{\text{LMMSE}}\right) \geq R_{\min,k} \tag{8.73c}$$

Problem (8.73) has a similar structure as Problem (8.29), with the additional challenge of the more involved SINR expression (8.72). It easy to recognize that (8.72) is more general than the SINR in (8.23), which is obtained as the special case in which $N_{\mathrm{R}} = 1$. Indeed, in this case, LMMSE reception becomes equivalent to matched filtering.

In order to tackle (8.73), the framework of sequential fractional programming will be used again. However, in order to find suitable approximations of the GEE, the lower-bound (8.33) alone is not enough, due to the inner sum over N_{R}. Instead, an additional bound is required, based on the arithmetic–geometric mean inequality. Specifically, for any non-negative $\{\beta_i \geq 0\}_{i=1}^{n}$, with $\sum_{i=1}^{n} \beta_i = 1$, and differentiable functions $g_i : p \in \mathbb{R}^K \to \mathbb{R}_+$, $i = 1, \ldots, N_{\mathrm{R}}$, it holds that [5, 33]:

$$\sum_{i=1}^{N_{\mathrm{R}}} g_i(p) \geq \prod_{i=1}^{N_{\mathrm{R}}} (g_i(p)/\beta_i)^{\beta_i} \tag{8.74}$$

Exploiting the bounds in (8.74) and (8.33), a suitable approximation of the objective of Problem (8.73) can be obtained as follows. First of all, let us define the ith summand in (8.72) as the function $\gamma_{i,k}$, which allows us to express (8.72) as

$$\gamma_k^{\mathrm{LMMSE}}(p) = \sum_{i=1}^{N_{\mathrm{R}}} \frac{p_k \alpha_{i,k}}{\tau_{i,k} + \phi_k p_k + \sum_{\ell \neq k} \omega_{\ell,k} p_\ell} = \sum_{i=1}^{N_{\mathrm{R}}} \gamma_{i,k}(p) \tag{8.75}$$

Next, for any p_0 with non-negative components, and for all $i = 1, \ldots, N_{\mathrm{R}}$, $k = 1, \ldots, K$, we give the following definitions after (8.74) and (8.33):

$$\beta_{i,k} = \frac{\gamma_{i,k}(p_0)}{\gamma_k^{\mathrm{LMMSE}}(p_0)}, \quad a_k = \frac{\gamma_k^{\mathrm{LMMSE}}(p_0)}{1 + \gamma_k^{\mathrm{LMMSE}}(p_0)} \tag{8.76}$$

$$b_k = \log\left(1 + \gamma_k^{\mathrm{LMMSE}}(p_0)\right) - \frac{\gamma_k^{\mathrm{LMMSE}}(p_0)}{1 + \gamma_k^{\mathrm{LMMSE}}(p_0)} \log\left(\gamma_k^{\mathrm{LMMSE}}(p_0)\right)$$

Finally, for all $p \succeq 0$, we approximate the objective function of (8.73) by applying the two bounds (8.74) and (8.33) one after the other. This yields:

$$\mathrm{GEE}(p) \geq \frac{\sum_{k=1}^{K} a_k \log_2\left(\prod_{i=1}^{N_{\mathrm{R}}} \left(\frac{\gamma_{i,k}(p)}{\beta_{i,k}}\right)^{\beta_{i,k}}\right) + b_k}{\sum_{k=1}^{K} \mu_k p_k + P_{c,k}} = \widetilde{\mathrm{GEE}}(p) \tag{8.77}$$

In Reference 33, it is shown that the bound in (8.77) fulfills the assumptions of Proposition 8.2. Moreover, by the variable change $p_k = 2^{q_k}$ for all $k = 1, \ldots, K$, the numerator of (8.77) can be shown to be concave, following similar steps as in Section 8.3.

As for the QoS constraint, following the steps taken for Problem (8.29) does not allow one to reformulate the constraint in convex form. However, we can exploit again the bound (8.74) to obtain a concave lower-bound of the constraint function. Specifically, we can first express the QoS constraint as

$$\log_2\left(\sum_{i=1}^{N_R}\gamma_{i,k}\right) \geq \log_2\left(2^{R_{\min,k}}-1\right) \tag{8.78}$$

and then apply (8.74) to lower-bound the LHS of (8.78) as

$$\log_2\left(\sum_{i=1}^{N_R}\gamma_{i,k}\right) \geq \log_2\left(\prod_{i=1}^{N_R}\left(\frac{\gamma_{i,k}(p)}{\beta_{i,k}}\right)^{\beta_{i,k}}\right) \tag{8.79}$$

Finally, upon applying the variable change $p_k = 2^{q_k}$, we obtain that, similarly to the numerator of (8.77), the RHS of the constraint function in (8.79) is concave. Thus, the approximate fractional problem to solve in each step of the sequential algorithm is formulated as

$$\max_{\{q_k\}_{k=1}^K} \widetilde{\mathrm{GEE}}\left(\{q_k\}_{k=1}^K\right) \tag{8.80a}$$

$$\text{s.t. } 2^{q_k} \leq P_{\max,k}, \quad \forall\, k = 1, \ldots, K \tag{8.80b}$$

$$\log_2\left(\prod_{i=1}^{N_R}\left(\frac{\gamma_{i,k}(p)}{\beta_{i,k}}\right)^{\beta_{i,k}}\right) \geq \log_2\left(2^{R_{\min,k}}-1\right) \tag{8.80c}$$

Finally, we can formulate a centralized resource allocation algorithm as in Algorithm 3.

Algorithm 3 Cooperative, energy-efficient resource allocation with relay IN.

Set vec(A) as in (8.56) and obtain the corresponding A_{IN};
Set $j = 0$ and choose any feasible $p_0^{(j)}$;
repeat
  $j = j + 1$;
  Given $p_0^{(j-1)}$, for all $k = 1, \ldots, K$, $i = 1, \ldots, N_R$ compute $a_k^{(j)}, b_k^{(j)}, \beta_{i,k}^{(j)}$ according to (8.76);
  Set $p_k^{(j)} = 2^{q_k^{(j)}}$, with $\{q_k^{(j)}\}_k$ the solution of (8.80) with parameters $a_k^{(j)}, b_k^{(j)}, \beta_{i,k}^{(j)}$;
until convergence
For all $k = 1, \ldots, K$, set c_k to the LMMSE receiver;

Algorithm 3 enjoys similar properties as Algorithm 1 and is designed to be implemented in a centralized fashion. Based on global channel state information, a computational center runs Algorithm 3 and then enforces the resulting resource allocation on the users. In the considered system, this role can be carried out by the relay, which can estimate the first-hop channels by training sequences and the second-hop channels either exploiting channel reciprocity in case of time division duplex (TDD) operation, or by having the receivers estimate and feedback the channels.

8.4.4 Distributed power allocation

Also in this scenario, a game-theoretic approach can be used to develop a distributed power control algorithm. The problem is mathematically formulated as the K coupled problems:

$$\arg\max_{p_k} \quad \frac{\log_2\left(1+\gamma_k^{\text{LMMSE}}\right)}{\mu_k p_k + P_{c,k}}, \quad \forall k \tag{8.81a}$$

$$\text{s.t. } 0 \leq p_k \leq P_{\max,k}, \quad \forall\, k = 1, \ldots, K \tag{8.81b}$$

wherein with respect to (8.40), we have not considered QoS constraints. Due to the SINR expression (8.72), accounting for QoS constraints in (8.81) leads to a very involved mathematical analysis. A theoretical characterization of the corresponding generalized Nash equilibrium problem is still an open problem.

The K coupled problems in (8.81) define the BRD of the non-cooperative game in normal form

$$\mathcal{G} = \left\{\mathcal{K}, \{\mathcal{S}_k\}_{k=1}^{K}, \{u_k\}_{k=1}^{K}\right\} \tag{8.82}$$

wherein $\mathcal{K} = \{1, 2, \ldots, K\}$ is the set of players, $\mathcal{S}_k = [0; P_{\max,k}]$ is the kth player's strategy set, and $u_k = \text{EE}_k$ is the kth player's utility function.

Despite the more involved SINR expression, it is possible to show that a unique Nash equilibrium exists, and that the BRD is always guaranteed to converge. The main observation behind the existence result is that although more involved than (8.23), (8.72) is still a concave function of the useful power p_k. Exploiting this observation, it is possible to show that a Nash equilibrium exists and that the best response of player k is:

$$\mathcal{B}_k(\mathbf{p}_{-k}) = \min\left\{P_{\max,k}, p_k^{\star}\right\} \tag{8.83}$$

with $p_k^{\star}$ the unique stationary point of EE_k, which is found as the unique positive solution of the equation

$$(\mu_k p_k + P_{c,k}) f'\left(\gamma_k^{\text{LMMSE}}\right) \frac{\partial \gamma_k^{\text{LMMSE}}}{\partial p_k} = \mu_k f\left(\gamma_k^{\text{LMMSE}}\right) \tag{8.84}$$

Instead, the uniqueness of the equilibrium and the convergence of the BRD follow upon showing that the best response of the game is a standard function [33].

Finally, having ensured that the BRD of the non-cooperative game converges to a Nash equilibrium, it is possible to allocate the transmit powers by implementing the BRD of the game, as done in Algorithm 2. Merging this power control routine with the allocation of the relay matrix and of the receive filters leads to the following distributed resource allocation algorithm.

Algorithm 4 enjoys similar properties as Algorithm 2 and it can be implemented in a distributed fashion by a similar approach as in Section 8.3. In particular, in order to compute its best response, each transmitter needs to compute the coefficient

$$\Omega_k = \sigma_k^2 \left(\sigma_r^2 N_A + \sum_{\ell \neq k} p_\ell \left\| H_\ell q_\ell \right\|^2 \right) \tag{8.85}$$

which depends on the other transmitters' channels and power levels. This can be accomplished locally, by having each receiver k measure the SINR γ_k and feed it back to the intended transmitter. At this point, each transmitter k can solve for Ω_k in the expression of γ_k^{LMMSE} and then compute (8.83).

Algorithm 4 Competitive resource allocation with relay interference neutralization.

For all $k = 1, \ldots, K$, initialize $p_k \in [0; P_{\max,k}]$.
Set vec(A) as in (8.56) and obtain the corresponding A_{IN}.
repeat
  for Each player k **do**
    Update p_k according to (8.83).
  end for
until convergence
For all $k = 1, \ldots, K$, set c_k to the LMMSE receiver

8.5 Multi-stream MIMO AF system

So far, we have considered MIMO AF systems in which a single data stream was transmitted by the sources. If instead a multi-stream transmission is allowed, the resource allocation problem becomes much more challenging. In this section, we take a step in this direction, considering a single-user system, with one source, one half-duplex AF relay, and one destination, which are equipped with N_T, N_A, and N_R antennas, respectively. Let s be the source's unit-norm symbol vector, and $x = Q^{1/2}s$, with $Q = E\left[xx^H\right]$ being the source transmit covariance matrix. Let us also denote by H and G the source–relay and relay–destination channels. Then, the signals y_r and y received at the relay and destination, respectively, can be written as

$$y_r = HQ^{1/2}s + n_r \tag{8.86}$$

$$y = GAHQ^{1/2}s + GAn_R + n \tag{8.87}$$

with n_r and n being the thermal noise at relay and destination, modeled as zero-mean complex circular Gaussian vectors with covariance matrices $\sigma_r^2 I_{N_A}$ and $\sigma^2 I_{N_R}$, respectively. For future reference, denote the SVDs of the channels by $H = U_H \Lambda_H^{1/2} V_H^H$, $G = U_G \Lambda_G^{1/2} V_G^H$, while the eigenvalue decomposition of Q and SVD of A are given by $Q = U_Q \Lambda_Q U_Q^H$ and $A = U_A \Lambda_A^{1/2} V_A^H$.

The resource allocation problem subject to maximum power and minimum rate constraints can be formulated as the maximization program

$$\begin{aligned}
&\max_{Q\succeq 0, A} \frac{\log\left|I_{N_D} + W^{-1/2}GAHQH^HA^HG^HW^{-1/2}\right|}{\operatorname{tr}(Q) + \operatorname{tr}\left(A\left(HQH^H + \sigma_r^2 I_{N_A}\right)A^H\right) + P_c} \\
&\text{s.t.} \quad \log\left|I_{N_D} + W^{-1/2}GAHQH^HA^HG^HW^{-1/2}\right| \geq R_{\min} \\
&\quad \operatorname{tr}(Q) \leq P_{\max}, \quad \operatorname{tr}\left(A\left(HQH^H + \sigma_r^2 I_{N_A}\right)A^H\right) \leq P_{\mathrm{r,max}}
\end{aligned} \tag{8.88}$$

wherein $W = \sigma^2 I_{N_\mathrm{R}} + \sigma_\mathrm{r}^2 GAA^HG^H$ is the overall noise covariance matrix at the receiver, which is itself a function of the AF matrix A.

Problem (8.88) is a challenging matrix-valued fractional problem whose objective has a numerator which is not jointly convex in (Q, A). In order to tackle such a problem, we will first determine the optimal eigenvectors of Q and A, and then focus on the resulting power allocation problem.

As far as deriving the optimal eigenvectors of Q and A is concerned, it should be observed that, while the numerator of the objective is well known to be maximized by diagonalizing the channel matrices and arranging the eigenvalues of AA^H and Q in decreasing order, the same allocation of A and Q would actually maximize the denominator, which is instead minimized by arranging the eigenvalues of AA^H in increasing order, when the eigenvalues of Q are arranged in decreasing order. Therefore, concluding that diagonalization is optimal is not obvious and needs to be proved. This is accomplished in the following proposition from Reference 31.

Proposition 8.4. *Consider Problem (8.88). The optimal Q and A are such that $U_Q = V_H$, $U_A = V_G$, and $V_A = U_H$.*

Proposition 8.4 allows to recast the matrix-valued Problem (8.88) into a vector-valued problem. Specifically, denoting by $\lambda_{i,G}$, $\lambda_{i,A}$, $\lambda_{i,H}$, and $\lambda_{i,Q}$, the generic (i,i) entry of the matrices $(\Lambda_G^{1/2}\Lambda_G^{H/2})$, Λ_A, $(\Lambda_H^{1/2}\Lambda_H^{H/2})$, and Λ_Q, respectively, and by λ_Q and λ_A the vectors $\{\lambda_{i,Q}\}_{i=1}^{N_T}$ and $\{\lambda_{i,A}\}_{i=1}^{N_A}$, Problem (8.88) can be expressed as

$$\max_{\lambda_Q,\lambda_A} \frac{\sum_{i=1}^{N_T} \log\left(1 + \frac{\lambda_{i,A}\lambda_{i,Q}\lambda_{i,H}\lambda_{i,G}}{\sigma^2 + \sigma_\mathrm{r}^2\lambda_{i,A}\lambda_{i,G}}\right)}{\sum_{i=1}^{N_T}\lambda_{i,Q} + \sum_{i=1}^{N_A}\lambda_{i,A}(\lambda_{i,H}\lambda_{i,Q} + \sigma_\mathrm{r}^2) + P_\mathrm{c}} \tag{8.89a}$$

$$\text{s.t.} \sum_{i=1}^{N_T} \log\left(1 + \frac{\lambda_{i,A}\lambda_{i,Q}\lambda_{i,H}\lambda_{i,G}}{\sigma^2 + \sigma_\mathrm{r}^2\lambda_{i,A}\lambda_{i,G}}\right) \geq R_{\min} \tag{8.89b}$$

$$\lambda_{i,Q} \geq 0, \quad \forall i = 1, \ldots, N_T, \quad \lambda_{i,A} \geq 0, \quad \forall i = 1, \ldots, N_A \tag{8.89c}$$

$$\sum_{i=1}^{N_T} \lambda_{i,Q} \leq P_{\max}, \quad \sum_{i=1}^{N_A} \lambda_{i,A}\left(\lambda_{i,H}\lambda_{i,Q} + \sigma_\mathrm{r}^2\right) \leq P_{\mathrm{r,max}} \tag{8.89d}$$

Problem (8.89), although being a vector-valued, simpler problem than (8.88), is still challenging, because the numerator of (8.89a) is not jointly concave in λ_A and λ_Q. This prevents the direct use of fractional programming. Instead, in order to tackle (8.89), it is possible to combine fractional programming with the alternating maximization algorithm 3. The idea is to alternatively optimize λ_A for fixed λ_Q, and λ_Q for fixed λ_A. The motivation for this approach lies in the observation that when either λ_A or λ_Q is fixed, the numerator of (8.89a) is concave in the other variable. Therefore, by fixing one of the two vector variables λ_A or λ_Q, it is possible to globally solve (8.89) with respect to the other variable by means of fractional programming, with limited complexity.

Algorithm 5 Alternating maximization for Problem (8.89).

Initialize $\lambda_Q^{(0)}$ to a feasible value. Set $j = 0$;
repeat
  Given $\lambda_Q^{(j)}$, solve Problem (8.89) with respect to λ_A to obtain the optimal $\lambda_A^{(j+1)}$;
  Given $\lambda_A^{(j+1)}$, solve Problem (8.89) with respect to λ_Q to obtain the optimal $\lambda_Q^{(j+1)}$;
  $j = j + 1$;
until convergence

Algorithm 5 is guaranteed to converge in the value of the objective, because (8.89a) is upper-bounded, and after each iteration its value is not decreased. However, convergence to the global solution of (8.89) is not guaranteed, since (8.89a) is not jointly pseudo-concave with respect to both λ_Q and λ_A, and also due to the presence of constraints which are coupled in the two vector variables.

However, the complexity of Algorithm 5 appears very limited, as each sub-problem is a fractional program with concave numerator, convex denominator, and convex constraints, which can therefore be solved by means of standard fractional programming methods with polynomial complexity.

The considered single-user scenario can be extended to the scenario in which the source and the relay only know the statistics of the channels H and G [31], or to a multiple access system with many sources, one relay, and one destination, which models the uplink of a single-cell communication system [30]. Instead, the case of an interference network with multi-stream transmission is still an open problem.

8.6 Numerical results

In this section, we present a numerical analysis of the performance of the proposed algorithms for a K-user AF relay-assisted MIMO interference channel (IC). For all $k = 1, \ldots, K$, the channel matrices have been generated as $H_k = \mathrm{PL}_{k,t}\widetilde{H}_k$ and $G_k = \mathrm{PL}_{k,r}\widetilde{G}_k$, with $\mathrm{PL}_{k,t}$ and $\mathrm{PL}_{k,r}$ the path losses of the kth transmitter–relay link, and of the kth relay–receiver link, respectively, and with $\widetilde{H}_k$ and $\widetilde{G}_k$ realizations of zero-mean Gaussian random matrices whose entries have variance 1. Following Reference 10, the path-losses have been modeled as $\mathrm{PL}_{k,t} = \mathrm{PL}_0 \left(\frac{d_{k,t}}{d_0}\right)^4$

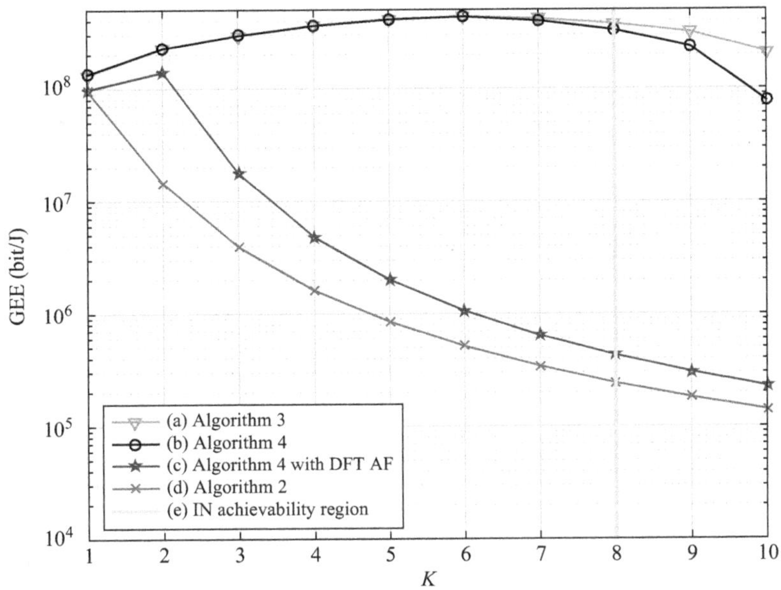

Figure 8.2 $N_T = 1$, $N_A = 16$, $N_R = 2$; *average GEE for: (a) Algorithm 3; (b) Algorithm 4; (c) Algorithm 4 with DFT relay matrix; (d) Algorithm 2; and (e) IN is perfectly achieved for* $K \leq 8$

and $\mathrm{PL}_{k,r} = \mathrm{PL}_0 \left(\frac{d_{k,r}}{d_0}\right)^4$, respectively, where PL_0 is the free-space attenuation at the reference distance $d_0 = 100$ m with a carrier frequency of 1800 MHz, while $d_{k,t}$ and $d_{k,r}$ are the distances between the kth transmitter and the relay and between the relay and the kth receiver, which have been randomly generated in the interval [100; 500] m. The users' maximum transmit powers and circuit powers have been set to $P_{\mathrm{max},k} = 0$ dBW and $P_{c,k} = 10$ dBm for all $k = 1, \ldots, K$, while the available relay power to $P_{\mathrm{r}} = 10$ dBW. The noise powers have been set to $\sigma_{\mathrm{r}}^2 = \sigma_{a(k)}^2 = \sigma^2$, for all $k = 1, \ldots, K$, with $\sigma^2 = F\mathcal{N}_0 B$, wherein $B = 180$ kHz is the communication bandwidth, $F = 3$ dB is the receiver noise figure, and $\mathcal{N}_0 = -174$ dBm/Hz is the receive power spectral density [2]. No minimum rate constraints have been enforced.

Figures 8.2 and 8.3 show the average[4] GEE versus the number of users' pairs K, for the following resource allocation schemes:

- Algorithm 3 which employs relay IN for centralized GEE maximization.
- Algorithm 4 for distributed energy-efficient resource allocation.
- Algorithm 4, but without the use of relay IN. Instead, a discrete Fourier transform (DFT) relay matrix is used, scaled to meet the relay power constraint, as proposed in Reference 22.
- Algorithm 2, for distributed energy-efficient resource allocation.

[4] The presented results have been obtained upon averaging over 10^4 independent system scenarios.

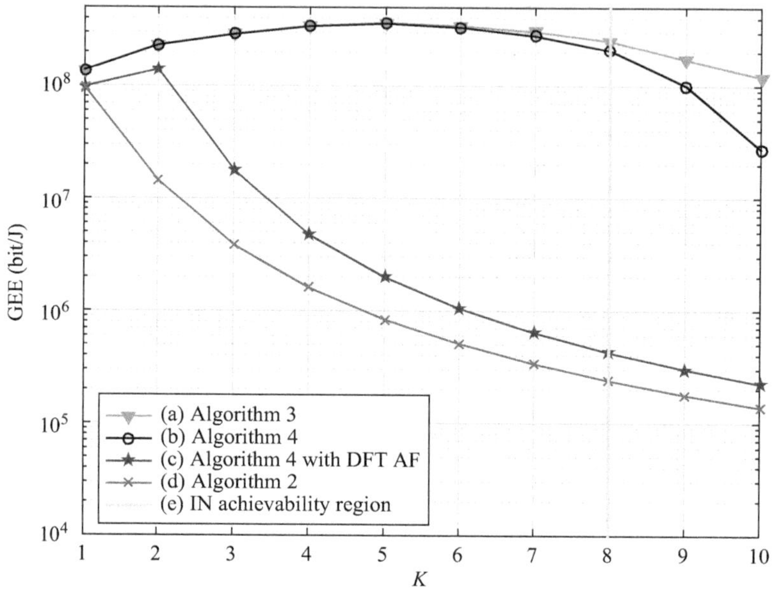

Figure 8.3 $N_T = 2$, $N_A = 16$, $N_R = 2$; *average GEE for: (a) Algorithm 3; (b) Algorithm 4; (c) Algorithm 4 with DFT relay matrix; (d) Algorithm 2; and (e) IN is perfectly achieved for* $K \leq 8$

In Figures 8.2 and 8.3, we set $N_{\mathrm{R}} = 2$ and $N_A = 16$, while the number of transmit antennas is $N_T = 1$ in Figure 8.2, and $N_T = 2$ in Figure 8.3. In both cases, perfect IN is possible up to $K \leq 8$ users' pairs. We can see that the schemes which employ relay IN perform significantly better than other schemes, even when perfect IN is not possible. Remarkably, the results also show that when IN is used, centralized and distributed resource allocation perform virtually the same. Instead, a non-negligible gap emerges when IN is no longer feasible. Moreover, we observe that when $N_T = 2$, a small gap between Algorithms 3 and 4 appears also at the boundary of the IN achievable region, i.e., for $K = 8$ users. This can be explained noticing that, although the number of interferers that can be neutralized is the same, the number of interfering streams has been doubled since $N_T = 2$. As a consequence, more power is required to suppress all interfering streams, which reduces the power that can be given to the intended signals.

Figure 8.4 considers a similar scenario as Figure 8.3, with the difference that the reported metric is the average system sum-rate in bit/s/Hz, rather than the GEE. The behavior of the curves is in line with previous illustrations, with the schemes using IN significantly outperforming other schemes.

Finally, Table 8.1 reports the average number of iterations required for the proposed Algorithms 3 and 4 to converge. It is seen that convergence occurs in a handful of iterations, especially when perfect IN is achieved, thus confirming that the proposed algorithms are particularly suited to being implemented in practical systems.

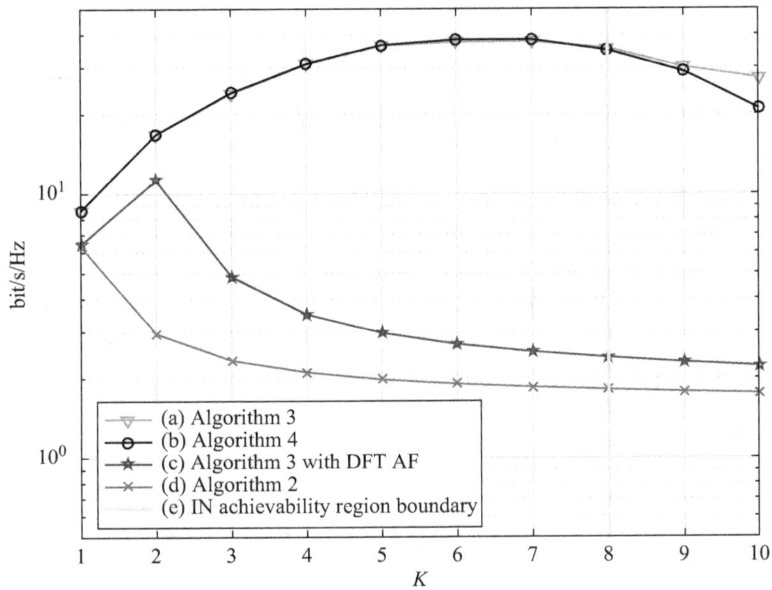

Figure 8.4 $N_T = 2$, $N_A = 16$, $N_R = 2$; *average sum-rate for: (a) Algorithm 3; (b) Algorithm 4; (c) Algorithm 4 with DFT relay matrix; (d) Algorithm 2; and (e) IN is perfectly achieved for* $K \leq 8$

Table 8.1 $N_T = 2$, $N_A = 16$, $N_R = 2$; *average number of iterations needed to converge versus the number of active users K. Convergence is declared when* $\|p^{(n)} - p^{(n-1)}\|^2 \leq 10^{-8}$

	Algorithm 3	**Algorithm 4**
$K = 1$	1	1
$K = 2$	1.43	1.28
$K = 3$	1.90	1.71
$K = 4$	2.05	1.93
$K = 5$	2.29	1.99
$K = 6$	2.73	2.06
$K = 7$	3.46	2.37
$K = 8$	5.98	3.41
$K = 9$	6.15	4.10
$K = 10$	6.88	5.12

We remark that in all considered algorithms, the complexity of each iteration is related to the maximization of a fractional function with concave numerator, convex denominator, subject to affine constraints. As already mentioned, this can be accomplished with polynomial complexity, by means of fractional programming theory [32].

8.7 Conclusions

This chapter has dealt with the problem of energy-efficient design of relay-assisted communication systems. After motivating the need for energy efficiency in wireless communications, and the use of relaying as an essential means to increase coverage and reliability, the notion of bit/Joule energy efficiency of a communication system has been formally defined.

After defining the performance metric to optimize, the first problem to be analyzed has been the selection of the most energy-efficient relaying protocol. It has been observed how the traditional trade-off between AF and DF (or other even more complex schemes) in terms of performance and ease of implementation is less strict as far as energy efficiency maximization is concerned. Indeed, besides being simpler to implement, in many practical systems AF ensures the best energy efficiency, too.

Motivated by this observation, the rest of the chapter has focused on the design of AF wireless networks. A general interference network model has been considered, in which multiple antennas are deployed at the sources, at the relay, and at the receivers. At first, power control algorithms have been developed, for fixed relay precoding matrix and matched-filter reception, assuming a single-stream transmission. Next, the analysis has been extended to the more involved scenario in which also the relay precoding matrix and the receive filters are included in the optimization process. A two-step resource allocation process has been developed. First, the relay allocates its precoding matrix according to an IN neutralization criterion, and then the transmitters and receivers allocate their transmit powers and receive filters. Both centralized and distributed resource allocation algorithms have been studied, considering both maximum power and minimum rate constraints. Finally, the assumption of single-stream transmission has been relaxed, too, and energy efficiency optimization has been carried out in a multi-stream MIMO system.

The last section of the chapter provides a numerical analysis to evaluate and compare the performance and complexity of the different resource allocation methods. The results indicate that the use of IN at the relay is able to grant a significant performance improvement over other relay precoding matrix allocations leading to an energy efficient design of relaying systems.

References

[1] K. J. Arrow and G. Debreu. Existence of an equilibrium for a competitive economy. *Econometrica*, 22(3):265–290, July 1954.

[2] G. Auer, V. Giannini, C. Desset, *et al.* How much energy is needed to run a wireless network? *IEEE Wireless Communications*, 18(5):40–49, 2011.

[3] D. P. Bertsekas. *Nonlinear Programming*. Athena Scientific, Belmont, MA, 1999.

[4] S. P. Boyd and L. Vandenberghe. *Convex Optimization*. Cambridge University Press, Cambridge, 2004.

[5] M. Chiang, C. Wei, D. P. Palomar, D. O'Neill, and D. Julian. Power control by geometric programming. *IEEE Transactions on Wireless Communications*, 6(7):2640–2651, July 2007.

[6] T. Cover and A. E. Gamal. Capacity theorems for the relay channel. *IEEE Transactions on Information Theory*, 25(5):572–584, September 1979.

[7] G. Debreu. A social equilibrium existence theorem. In *National Academy of Sciences*, vol. 38, pp. 886–893, 1952.

[8] G. P. Fettweis. The tactile internet: Applications and challenges. *Vehicular Technology Magazine, IEEE*, 9(1):64–70, March 2014.

[9] F. H. P. Fitzek and M. M. S. Katz. *Cooperation in Wireless Networks: Principles and Applications*. Springer, Berlin, 2006.

[10] A. Goldsmith. *Wireless Communications*. Cambridge University Press, Cambridge, 2005.

[11] D. Gunduz, A. Yener, A. Goldsmith, and H. V. Poor. The multiway relay channel. *IEEE Transactions on Information Theory*, 59(1):51–63, January 2013.

[12] Z. K. M. Ho and E. A. Jorswieck. Instantaneous relaying: Optimal strategies and interference neutralization. *IEEE Transactions on Signal Processing*, 60(12):6655–6668, December 2012.

[13] C. Isheden, Z. Chong, E. A. Jorswieck, and G. P. Fettweis. Framework for link-level energy efficiency optimization with informed transmitter. *IEEE Transactions on Wireless Communications*, 11(8):2946–2957, August 2012.

[14] I. Ku, C.-X. Wang, and J. Thompson. Spectral-energy efficiency tradeoff in relay-aided cellular networks. *IEEE Transactions on Wireless Communications*, 12(10):4970–4982, October 2013.

[15] G. Y. Li, Z.-K. Xu, C. Xiong, *et al*. Energy-efficient wireless communications: tutorial, survey, and open issues. *IEEE Wireless Communications*, 18(6):28–35, December 2011.

[16] K. J. R. Liu, A. K. Sadek, W. Su, and A. Kwasinski. *Cooperative Communications and Networking*. Cambridge University Press, Cambridge, 2009.

[17] R. Madan, N. B. Mehta, A. F. Molisch, and J. Zhang. Energy-efficient cooperative relaying over fading channels with simple relay selection. *IEEE Transactions on Wireless Communications*, 7(8):3013–3025, August 2008.

[18] B. R. Marks and G. P. Wright. A general inner approximation algorithm for non-convex mathematical programs. *Operations Research*, 26(4):681–683, 1978.

[19] B. Matthiesen, A. Zappone, and E. A. Jorswieck. Resource allocation for energy-efficient 3-way relay channels. *IEEE Transactions Wireless Communications*, vol. 14, no. 8, August 2015.

[20] L. Ong, C. M. Kellett, and S. J. Johnson. Capacity theorems for the AWGN multi-way relay channel. In *Information Theory Proceedings (ISIT), 2010 IEEE International Symposium on*, pp. 664–668, June 2010.

[21] J. Papandriopoulos and J. S. Evans. Low-complexity distributed algorithms for spectrum balancing in multi-user DSL networks. In *Proceedings of 2006 IEEE International Conference on Communications*, pp. 3270–3275, Istanbul, Turkey, June 2006.

[22] F. Roemer, M. Haardt, and E. A. Jorswieck. Efficient spatial processing and resource allocation for amplify and forward two-way relaying. In *Cross Layer Designs in WLAN Systems*. Troubador Publishing Ltd., Leicester, June 2011.

[23] P. Rost, G. Fettweis, and J. N. Laneman. Energy- and cost-efficient mobile communication using multi-cell MIMO and relaying. *IEEE Transactions on Wireless Communications*, 11(9):3377–3387, September 2012.

[24] C. Sun, Y. Cen, and C. Yang. Energy efficient OFDM relay systems. *IEEE Transactions on Communications*, 61(5):1797–1809, May 2013.

[25] C. Sun and C. Yang. Energy efficiency analysis of one-way and two-way relay systems. *EURASIP Journal on Wireless Communications and Networking*, vol. 46, pp. 1–18, February 2012.

[26] X. Tang and Y. Hua. Optimal design of non-regenerative MIMO wireless relays. *IEEE Transactions on Wireless Communications*, 6(4):1398–1407, April 2007.

[27] S. Verdú. *Multiuser Detection*. Cambridge University Press, Cambridge, 1998.

[28] C.-X. Wang, F. Haider, X. Gao, *et al.* Cellular architecture and key technologies for 5G wireless communication networks. *IEEE Communications Magazine*, 52(2):122–130, February 2014.

[29] R. D. Yates. A framework for uplink power control in cellular radio systems. *IEEE Journal on Selected Areas in Communications*, 13(7):1341–1347, September 1995.

[30] A. Zappone, P. Cao, and E. A. Jorswieck. Energy efficiency optimization in multiuser relay-assisted MIMO systems. In *Asilomar Conference on Signals, Systems, and Computers*, Pacific Grove, CA, November 2013.

[31] A. Zappone, P. Cao, and E. A. Jorswieck. Energy efficiency optimization in relay-assisted MIMO systems with perfect and statistical CSI. *IEEE Transactions on Signal Processing*, 62(2):443–457, January 2014.

[32] A. Zappone and E. A. Jorswieck. Energy efficiency in wireless networks via fractional programming theory. *Foundations and Trends® in Communications and Information Theory*, 11(3–4):185–396, 2015.

[33] A. Zappone, E. A. Jorswieck, and S. Buzzi. Energy efficiency and interference neutralization in two-hop MIMO interference channels. *IEEE Transactions on Signal Processing*, 62(24):6481–6495, December 2014.

[34] A. Zappone, L. Sanguinetti, G. Bacci, E. A. Jorswieck, and M. Debbah. Energy-efficient power control: A look at 5G wireless technologies. *IEEE Transactions on Signal Processing*, 64(7), April 2016.

Chapter 9
Cognitive relaying for information and energy cooperation

Gan Zheng

9.1 Introduction

The rapidly growing wireless networks come with great challenges such as fading degradation, spectrum scarcity and energy constraints. This chapter introduces advanced relaying techniques in cognitive radio networks in which the primary and the secondary systems actively seek for cooperation opportunities, that can deal with these challenges in an effective way. We focus on cooperation at both information level and energy level, such that the primary users' (PUs) performance can be boosted, the secondary users (SUs) have much higher chance for spectrum access, and energy flow is optimized to benefit both systems.

9.1.1 Wireless challenges and motivation

To achieve highly efficient and reliable communications over the wireless medium has been a major challenge due to channel fading characteristics and the spectrum scarcity, as well as the high-energy consumption that receive growing concerns globally.

It is predicted that by 2017, the number of mobile devices will exceed 10 billion, and more than 5 million base stations (BSs) will consume 5.5 GW of power worldwide, resulting in more than 30 Megatons of carbon emission every year. It goes without saying that this would mean huge spectrum and energy costs on mobile operators and therefore efficient use of available spectrum and energy resources is of paramount importance.

On the other hand, recent years have witnessed a reform in spectrum allocation around the world, moving from exclusive-use licensing model to the more relaxed trading and liberalization model. The aim of such move is to tackle the inefficiency problem in static spectrum allocation, giving birth to cognitive radio technologies [1], where secondary radio terminals optimize their parameters on the fly based on instantaneous observations of the radio environment and the primary terminals, and creating scenarios where a hierarchy of primary and secondary users are sharing the same band.

Cognitive radio was originally proposed to improve spectral usage while energy issue has been largely overlooked. This is partly because the primary and secondary systems used to compete with each other which makes common goals such as energy saving difficult. Although competition is the nature of cognitive radio networks, the introduction of cooperation can actually benefit both the primary and the secondary systems to improve efficiency. This is particularly useful when the primary transmitter (PT) is a spectrum-hungry terminal, such as a macrocell BS. In this regard, advanced cognitive relaying is a key enabling technique that can facilitate the cooperation between the two systems.

Competition and cooperation in a cognitive network scenario is shown in Figure 9.1 where a secondary link coexists with a a primary link providing a large-area wireless coverage. In the conventional set-up, the PT and secondary transmitters (STs) compete in the same radio resources and they interfere with each other, as seen in Figure 9.1(a). A vast amount of efforts in the literature dealt with this situation by suppressing the interference between the PU and the SUs. In contrast, Figure 9.1(b) shows the new *paradigm of cooperation* between the primary and secondary systems that can turn unwanted interference to useful signals.

In particular, as illustrated in Figure 9.1(b), the STs help relay the traffic from the PT, and in return can gain access to the primary spectrum. As a result, the interfering links now convey useful signals, leading to a win–win situation that gives incentives to both the PU and SU systems to cooperate. By letting the SUs access the same spectrum, if done wisely, the PU can achieve much better rate performance with the same energy consumption. In addition, the interfering links can also be useful vehicles for energy cooperation enabled by the recently proposed radio frequency (RF) energy harvesting (EH) or wireless energy transfer techniques [2]. This motivates the cooperation between the primary and secondary systems at both information and energy levels, i.e., the PT will transmit both information and energy to the ST, in exchange for the ST to relay the primary information. Compared to the information-only cooperation, this scheme creates even stronger incentives for both systems to cooperate and substantially improves the system's overall spectrum efficiency. This is especially useful, when the ST is a low-power relay node rather than a powerful BS. Even the ST has good channel quality but is energy constrained, so it cannot help to serve the PU. Obviously, such a paradigm is a result of cognitive cooperation by recycling the interference links between the primary and secondary systems. This chapter will focus on exploring the interference links to achieve high spectral efficiency and high energy efficiency cognitive radio systems.

9.1.2 Literature review

9.1.2.1 Information cooperation

Early works about information cooperation in cognitive radio networks focus on information-theoretical study. The overlay scenarios in which the ST serves both PUs and SUs using advanced precoding such as dirty-paper coding (DPC) [3] provide an information-theoretic performance upper bound. Using multiple antennas at the ST, the optimal beamforming is studied using DPC [4] and linear precoding [5]. However, despite capacity-achieving, DPC loses its meaning in a practical context as it would require the ST to possess non-causal information about the PT. For this reason, it

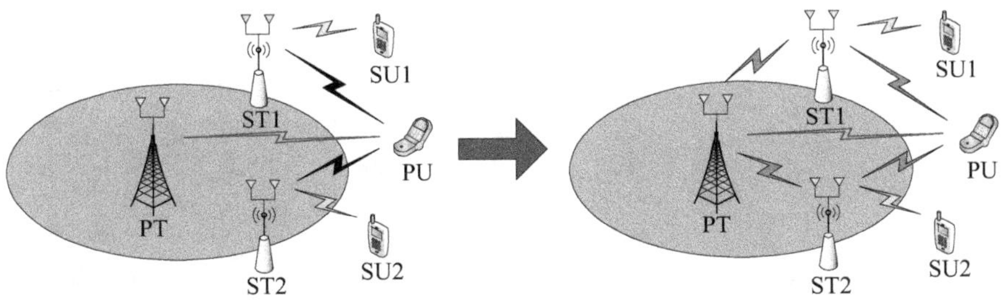

Figure 9.1 Competition (a) and cooperation (b) between the primary and the secondary systems

has been of greater interest to consider practical forms of cooperation where the STs usually act as a simple relay for the PT.

Practical information cooperation between the primary and secondary systems can be realized in different ways. In Reference 6, a spectrum leasing scheme is proposed in which the PU may lease the whole owned bandwidth for a fraction of time exclusively to SUs based on decode-and-forward forwarding scheme and distributed space–time coding. Alternatively, this can be realized via adaptive bandwidth allocation. As illustrated in Figure 9.2(a), in the first half of the time duration (Phase 1), the PT transmits signals over a fraction α ($0 \le \alpha \le 1$) of the whole bandwidth W, and the signals are also received by the SU. In the second half of the time duration (Phase 2), the ST helps relay the signals to the PU. There have also been efforts on simultaneous transmission of the PU and SU data over the same frequency, which simplifies the cooperation protocol design. One such approach employs multiple antennas as depicted in Figure 9.2(b), to eliminate the interference between the primary and secondary systems using zero-forcing (ZF) approach [7] and optimal amplify-and-forward (AF) relaying [8]. Other possible ways in the literature to tackle the primary–secondary interference include transmitting the PU signal on the in-phase (I) channel and the SU signal on the quadrature (Q) channel [9] and exploring dual polarization [10].

9.1.2.2 Energy cooperation

Energy cooperation is a promising solution to transfer energy from energy-rich terminals to help the energy-constrained terminals, and thus improve the overall performance of communications systems, just like information cooperation relaying to repair the channels in outage.

Energy cooperation is a very recent concept and currently there are only limited works available which mainly consider power line systems. For instance, joint communication and energy cooperation between BSs powered by renewable energy is investigated in Reference 11 for the coordinated multi-point downlink cellular networks. The stochastic nature of renewable energy cannot provide reliable energy output, and the cooperation among multiple BSs can greatly increase 'energy diversity'. The proposed joint communication and energy cooperation solution is shown to substantially improve the downlink throughput for EH systems, as compared to

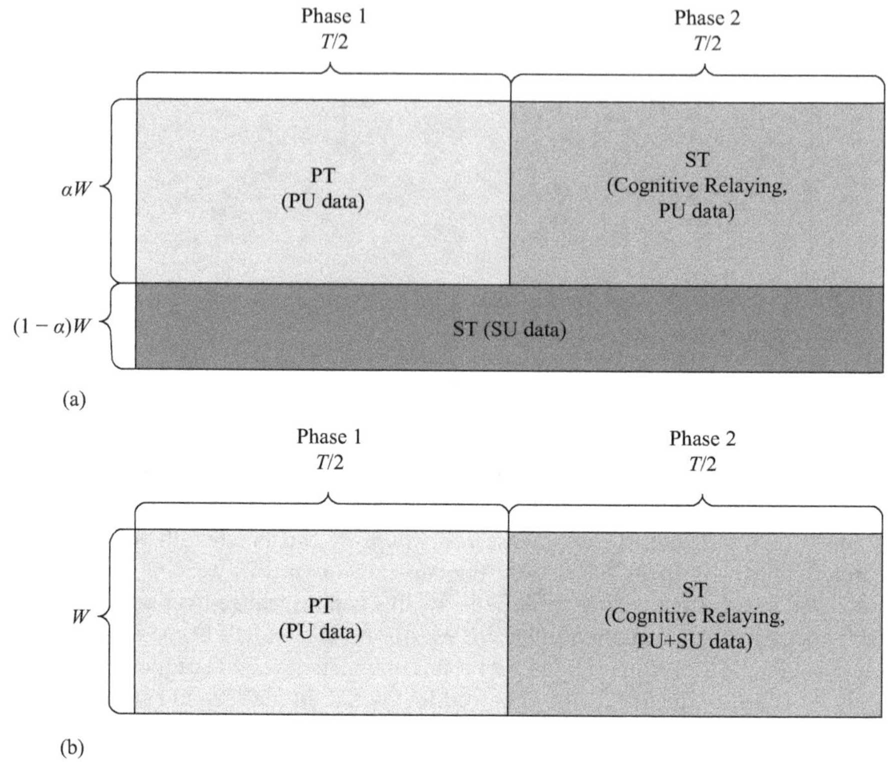

Figure 9.2 Cooperation protocols between the primary and the secondary systems. (a) Adaptive bandwidth and power allocation for a single-antenna ST; (b) Protocol for an multi-antenna ST

the case without energy cooperation. In a similar scenario, the optimal energy cooperation algorithms are designed in Reference 12 for both cases where the renewable energy profile and energy demand profile are deterministic and stochastic.

Wireless energy transfer via the RF-EH technology is a more flexible enabling technology for energy cooperation without the requirements of wiring or close proximity. It has emerged as a new solution where the electromagnetic energy in the environment is captured by the receiver antennas and converted into useful energy. Thanks to recent advances in antenna and rectenna circuit design, there has been great progress towards improving the efficiency of wireless energy transfer. For instance, Powerharvester receivers provided by Powercast can achieve a conversion efficiency as high as 70% in some scenarios [13]. A sensor node powered by a cellular base transceiver station (BTS) at a distance of 200 m from the BTS was reported in Reference 14.

RF-EH technique also enables simultaneous transfer of information and energy using RF signals. This fundamental concept of simultaneous wireless transmission of energy and information is studied in References 15 and 16 from an information-theoretic viewpoint. However, although theoretically appealing, it is impractical to harvest energy and decode information from the same RF signals at the same time. Two practical receiver structures to decode information and EH termed 'time switching' and 'power splitting' are proposed in Reference 2 to separate the signals for energy and information transfer, respectively. 'Power splitting' divides the received signal into two parts, one for harvesting energy and the other for information decoding, which will be adopted in our proposed cooperation. 'Time switching' uses dedicated time slots for harvesting energy and the rest for data transmission. Dynamic switching between information decoding and RF-EH is proposed in Reference 17 then further studied in Reference 18 for a cooperative relaying scenario with a discrete-level battery at the RF-EH relay node.

Relevant to the relay-assisted cooperative communications, an energy-constrained relay node has the opportunity to harvest energy from the received RF signal from the source and uses that harvested energy to forward the source information to the destination. This increases the chance of cooperation significantly and may potentially eliminate the need of an external power supply for a relay. Such a cooperative scheme is studied in Reference 19, where both time switching and power switching relaying protocols are studied and the achievable outage capacity and ergodic capacity are also derived. However, there have been very few works that jointly consider the concept of energy cooperation and information cooperation in a cognitive radio network, and this is the purpose of this chapter.

9.1.3 Organization of the chapter

Understandably, earlier results in this context tend to focus on ensuring or enhancing the PU performance but have little to do with the quality-of-service (QoS) of the SUs. Existing literature also puts no emphasis on the energy cooperation aspect that can be brought from the primary–secondary cooperation paradigm. The purpose of this chapter is to introduce cooperation schemes that can uncover the massive potential of this paradigm and show that the benefit can come with the QoS assurance to both the PUs and the SUs, by presenting information level and energy level cooperation between the primary and the secondary systems to achieve better use of the spectrum.

The rest of this chapter is organized as follows. In Section 9.2, the cognitive relaying network model is described and common system parameters are defined. Section 9.3 introduces information cooperation by multiple STs relaying the primary message. The optimal and distributed optimization of beamforming strategies are proposed, with analysis and comparison on signalling overhead. Joint information and energy cooperation is addressed in Section 9.4. Performance evaluation is presented in Section 9.5. Concluding remarks and future directions are given in Section 9.6.

9.1.3.1 Notations

Throughout this chapter, the following notations will be adopted. Vectors and matrices are represented by boldface lowercase and uppercase letters, respectively. $\|\cdot\|$ denotes the Frobenius norm. $(\cdot)^\dagger$ denotes the Hermitian operation of a vector or matrix. $\mathbf{I}$ denotes an identity matrix of appropriate dimension. $\mathbb{E}[\cdot]$ denotes the expectation. $\mathbf{x} \sim \mathbb{CN}(\mathbf{m}, \Phi)$ denotes a vector $\mathbf{x}$ of complex Gaussian elements with a mean vector of $\mathbf{m}$ and a covariance matrix of Φ. diag($\mathbf{v}$) denotes a diagonal matrix with diagonal elements as the elements of $\mathbf{v}$.

9.2 Network model

In this section, we describe the system model of a cooperative cognitive downlink network as shown in Figure 9.3. The primary system consists of a PT communicating to a single-antenna PU, and the secondary system has M distributed STs each with N transmit/receive antennas and serving a single-antenna SU. We assume that ST_m only serves SU_m. We consider the scenario where cooperation between STs and PT is necessary to meet the QoS target of the primary system. To describe our network model, we define the following system parameters:

h_p, the scalar channel vector between the PT and the PU;

$\mathbf{h}_{mp}^\dagger$, the $1 \times N$ channel vector between ST_m and the PU;

$\mathbf{h}_{jm}^\dagger$, the $1 \times N$ channel vector between ST_j and SU_m;

$\mathbf{g}_m$, the channel between the PT and ST_m;

n_1^p, the noise received at the PU during Phase 1 with $n_1^p \sim \mathbb{CN}(0, N_1^p)$;

n_2^p, the noise received at the PU during Phase 2 with $n_2^p \sim \mathbb{CN}(0, N_2^p)$;

$\mathbf{n}_{sm}$, the $N \times 1$ noise vector received at ST_m during Phase 1 with $\mathbf{n}_{sm} \sim \mathbb{CN}(\mathbf{0}, N_{sm}\mathbf{I})$;

n_m, the noise received at SU_m during Phase 2 with $n_m \sim \mathbb{CN}(0, N_m)$;

P_p, the transmit power of the PT;

s_p, the transmit signal for the PU with $s_p \sim \mathbb{CN}(0, P_p)$;

s_m, the transmit signal for SU_m with $s_m \sim \mathbb{CN}(0, 1)$.

It is assumed that all channels and noises are independent of each other, and the STs adopt AF relaying protocol and work in the half-duplex mode. The communication is synchronous and takes place in two phases. In Phase 1, PT broadcasts its message s_p. The received signals at the PU and ST_m are, respectively, given by

$$y_1 = h_p s_p + n_1^p, \tag{9.1}$$

$$\mathbf{r}_m = \mathbf{g}_m s_p + \mathbf{n}_{sm}. \tag{9.2}$$

Then ST_m processes the received signal using an $N \times N$ relaying matrix $\mathbf{A}_m$ to produce $\mathbf{A}_m\mathbf{r}_m = \mathbf{A}_m\mathbf{g}_m s_p + \mathbf{A}_m\mathbf{n}_{sm}$. In Phase 2, ST_m sends its own message, s_m, using a

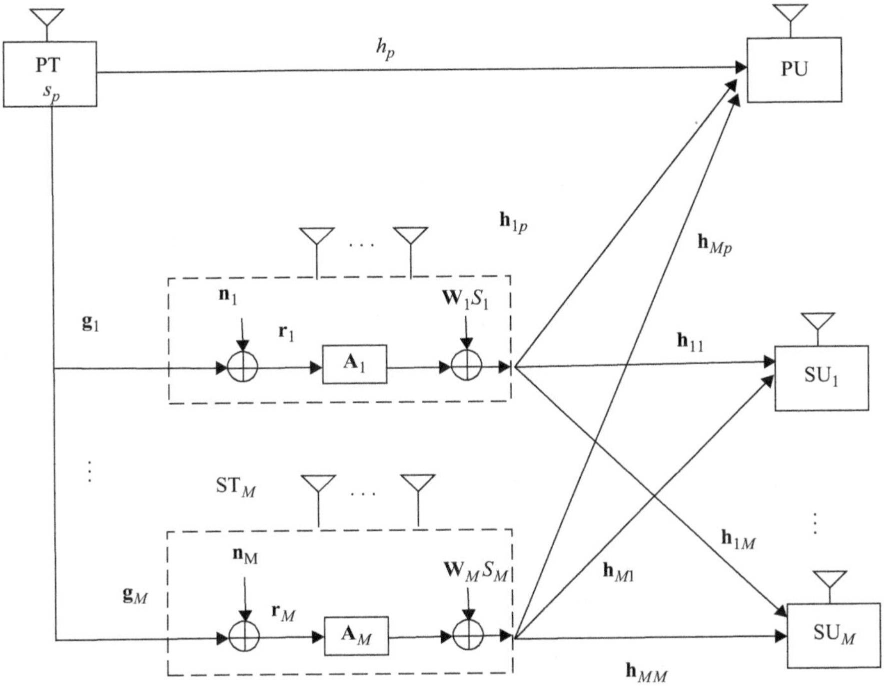

Figure 9.3 System model of a cooperative cognitive downlink network

beamforming vector $\mathbf{w}_m$ together with the processed signals $\mathbf{A}_m\mathbf{r}_m$. We assume that the data symbols (or messages) for the SUs, $\{s_m\}$, and the PU, s_p, are independent of each other. During this period, the PT remains idle and the received signal at SU$_m$ is then given by

$$z_m = \mathbf{h}_{mm}^{\dagger}\left(\mathbf{w}_m s_m + \mathbf{A}_m \mathbf{g}_m s_p + \mathbf{A}_m \mathbf{n}_{sm}\right) + \sum_{\substack{j=1 \\ j\neq m}}^{M} \mathbf{h}_{jm}^{\dagger}\left(\mathbf{w}_j s_j + \mathbf{A}_j \mathbf{g}_j s_p + \mathbf{A}_j \mathbf{n}_{sj}\right) + n_m. \quad (9.3)$$

The received signal-to-interference-plus-noise ratio (SINR) at SU$_m$ is then expressed as

$$\mathsf{SINR}_m = \frac{\left|\mathbf{h}_{mm}^{\dagger}\mathbf{w}_m\right|^2}{\sum_{\substack{j=1 \\ j\neq m}}^{M}\left|\mathbf{h}_{jm}^{\dagger}\mathbf{w}_j\right|^2 + P_p\sum_{j=1}^{M}\left|\mathbf{h}_{jm}^{\dagger}\mathbf{A}_j\mathbf{g}_j\right|^2 + \sum_{j=1}^{M} N_{sj}\left\|\mathbf{h}_{jm}^{\dagger}\mathbf{A}_j\right\|^2 + N_m}. \quad (9.4)$$

Similarly, the received signal at the PU in Phase 2 is given by

$$\begin{aligned} y_2 &= \sum_{j=1}^{M} \mathbf{h}_{jp}^{\dagger} \left(\mathbf{w}_j s_j + \mathbf{A}_j \mathbf{g}_j s_p + \mathbf{A}_j \mathbf{n}_{sj} \right) + n_2^p \\ &= \sum_{j=1}^{M} \mathbf{h}_{jp}^{\dagger} \mathbf{A}_j \mathbf{g}_j s_p + \sum_{j=1}^{M} \mathbf{h}_{jp}^{\dagger} \mathbf{w}_j s_j + \sum_{j=1}^{M} \mathbf{h}_{jp}^{\dagger} \mathbf{A}_j \mathbf{n}_{sj} + n_2^p. \end{aligned} \tag{9.5}$$

Using maximal-ratio combining (MRC) to get an estimate of s_p, the received SINR of the PU becomes

$$\mathsf{SINR}_p = P_P \frac{|h_p|^2}{N_1^p} + \frac{P_P \left| \sum_{j=1}^{M} \mathbf{h}_{jp}^{\dagger} \mathbf{A}_j \mathbf{g}_j \right|^2}{\sum_{j=1}^{M} \left| \mathbf{h}_{jp}^{\dagger} \mathbf{w}_j \right|^2 + \sum_{j=1}^{M} N_{sj} \left\| \mathbf{h}_{jp}^{\dagger} \mathbf{A}_j \right\|^2 + N_2^p}. \tag{9.6}$$

9.3 Information cooperation

9.3.1 Problem formulation

Assuming perfect channel state information (CSI), we aim to minimize the transmit power of STs subject to both PU and SUs' rate constraints r_p and $\{r_m\}$, respectively, by jointly optimizing $\{\mathbf{w}_m\}$ and $\{\mathbf{A}_m\}$. Mathematically, the problem is formulated as

$$\begin{aligned} \min_{\{\mathbf{w}_j, \mathbf{A}_j\}} \quad & \sum_{j=1}^{M} \|\mathbf{w}_j\|^2 + P_P \sum_{j=1}^{M} \|\mathbf{A}_j \mathbf{g}_j\|^2 + \sum_{j=1}^{M} N_{sj} \|\mathbf{A}_j\|^2 \\ \text{s.t.} \quad & \begin{cases} \dfrac{\left| \sum_{j=1}^{M} \mathbf{h}_{jp}^{\dagger} \mathbf{A}_j \mathbf{g}_j \right|^2}{\sum_{j=1}^{M} \left| \mathbf{h}_{jp}^{\dagger} \mathbf{w}_j \right|^2 + \sum_{j=1}^{M} N_{sj} \left\| \mathbf{h}_{jp}^{\dagger} \mathbf{A}_j \right\|^2 + N_2^p} \ge \gamma_p' \triangleq \dfrac{2^{2r_p} - 1}{P_P} - \dfrac{|h_p|^2}{N_1^p}, \\ \mathsf{SINR}_m \ge \gamma_m \triangleq 2^{2r_m} - 1, \quad \forall m, \end{cases} \end{aligned} \tag{9.7}$$

where the factor of 2 associated with the rate requirements is due to the two channel phases used.

9.3.2 Special structure of relaying matrix and reformulation

The problem (9.7) is a convex optimization problem because the objective function is convex and the constraints can also be made convex by setting the imaginary parts of $\mathbf{h}_{mp}^{\dagger} \mathbf{A}_m \mathbf{g}_m$ and $\mathbf{h}_{mm}^{\dagger} \mathbf{w}_m$ to zero [20] without loss of optimality. As a consequence, (9.7) can be solved by the standard interior-point algorithm [21]. However, there are strong reasons to further study the optimization problem to (i) understand the structure that the optimal solution possesses; (ii) derive more efficient algorithms than the interior-point algorithm for convex problems and (iii) develop distributed implementation. The rest of this section will be devoted to the above-mentioned objectives. We will provide deeper understanding of the optimal solution so that physical insights can be gained, which help to develop more efficient algorithms and distributed solutions.

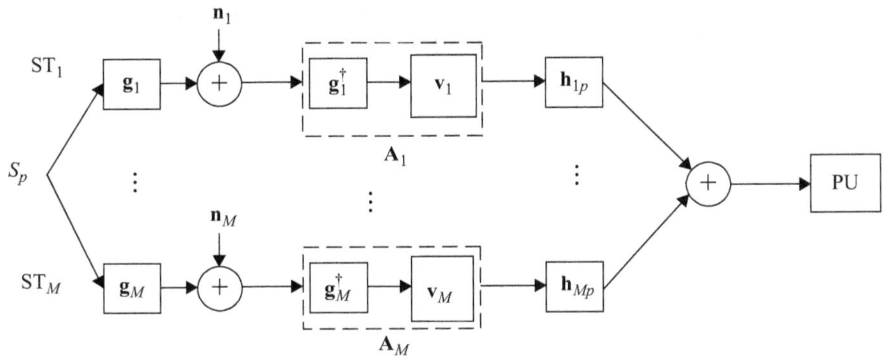

Figure 9.4 Illustration of the relaying processing at the STs

9.3.2.1 Optimal structure of $\{\mathbf{A}_m\}$

Theorem 9.1. *The optimal $\mathbf{A}_m$ has the structure of*

$$\mathbf{A}_m = \mathbf{H}_m \mathbf{a}_m \mathbf{g}_m^\dagger, \quad \forall m, \tag{9.8}$$

where we have defined the composite channel matrix $\mathbf{H}_j \triangleq [\mathbf{h}_{jp}\ \mathbf{h}_{j1} \cdots \mathbf{h}_{jM}]$ from ST_j to the PU and all SUs, and $\mathbf{a}_m$ is a complex parameter vector.

Proof. See Appendix A. □

9.3.2.2 Physical insights and a simplified formulation

Theorem 9.1 states that the optimal structure of $\mathbf{A}_m$ can be divided into two components: $\mathbf{g}_m^\dagger$ and $\mathbf{H}_m\mathbf{a}_m$. This is quite intuitive as there is only one PU message stream and the optimal $\mathbf{A}_m$ is of rank 1. It is observed that each ST first maximizes the received signal-to-noise ratio (SNR) of the PU signal using an MRC receiver, $\mathbf{g}_j^\dagger$, during Phase 1 and then relays the noisy signal using the transmit beamforming vector $\mathbf{H}_j\mathbf{a}_j$ during Phase 2.

This means that during Phase 2, the entire system resembles an interference multiple-input single-output (MISO) channel in which each ST transmits its own message as well as a noisy version of the common PU message using transmit and relay beamforming, respectively. For ST_j, the optimal transmit and relay beamforming vectors are both parameterized as $\mathbf{H}_j\mathbf{a}_j$ where $\mathbf{a}_j$ is the parameter vector to be designed. Defining $\mathbf{v}_j \triangleq \mathbf{H}_j\mathbf{a}_j$, the received PU signal (without noise) via ST_j is expressed as

$$\hat{s}_p = \mathbf{h}_{jp}^\dagger \left[\mathbf{v}_j \mathbf{g}_j^\dagger (\mathbf{g}_j s_p + \mathbf{n}_{sj}) \right] = \|\mathbf{g}_j\|^2 \mathbf{h}_{jp}^\dagger \mathbf{v}_j s_p + \mathbf{h}_{jp}^\dagger \mathbf{v}_j \mathbf{g}_j^\dagger \mathbf{n}_{sj}. \tag{9.9}$$

The overall relaying operation is illustrated in Figure 9.4. Note that all STs should form a collaborative-relaying beam to ensure coherent reception at the PU, i.e., $\mathbf{h}_{jp}^\dagger \mathbf{v}_j$, $\forall j$ is co-phased.

Each $\mathbf{A}_j$ originally is a general $N \times N$ matrix but (9.8) indicates that it is a rank-1 matrix and can be represented by a relaying vector $\mathbf{v}_j$ with dimension N, thus this reduces the dimensionality of optimization variables by a factor of N. As a result, substituting (9.8) into (9.7), we get a new and equivalent formulation

$$\min_{\{\mathbf{w}_j,\mathbf{v}_j\}} \sum_{j=1}^{M} \|\mathbf{w}_j\|^2 + \sum_{j=1}^{M} c_j \|\mathbf{g}_j\|^2 \|\mathbf{v}_j\|^2 \tag{9.10a}$$

$$\text{s.t.} \begin{cases} \dfrac{\left|\mathbf{h}_{mm}^{\dagger}\mathbf{w}_m\right|^2}{\sum_{\substack{j=1\\j\neq m}}^{M}\left|\mathbf{h}_{jm}^{\dagger}\mathbf{w}_j\right|^2 + \sum_{j=1}^{M} c_j\|\mathbf{g}_j\|^2\left|\mathbf{h}_{jm}^{\dagger}\mathbf{v}_j\right|^2 + N_m} \geq \gamma_m, \quad \forall m, \\ \dfrac{\left|\sum_{j=1}^{M}\mathbf{h}_{jp}^{\dagger}\mathbf{v}_j\|\mathbf{g}_j\|^2\right|^2}{\sum_{j=1}^{M}\left|\mathbf{h}_{jp}^{\dagger}\mathbf{w}_j\right|^2 + \sum_{j=1}^{M} N_{sj}\left|\mathbf{h}_{jp}^{\dagger}\mathbf{v}_j\right|^2\left\|\mathbf{g}_j^{\dagger}\right\|^2 + N_2^p} \geq \gamma_p', \end{cases} \tag{9.10b}$$

where $c_j \triangleq (P_P\|\mathbf{g}_j\|^2 + N_{sj})$. For notational convenience, we define the vectors and matrices:

$$\begin{cases} \mathbf{v} \triangleq \mathsf{vec}(\mathbf{v}_1, \ldots, \mathbf{v}_M), \\ \mathbf{h}_p \triangleq \left[\mathbf{h}_{1p}^T\|\mathbf{g}_1\|^2, \ldots, \mathbf{h}_{Mp}^T\|\mathbf{g}_M\|^2\right]^T, \\ \bar{\mathbf{H}}_p \triangleq \mathsf{BlkDiag}\left(\mathsf{vec}\left(\mathbf{h}_{1p}^{\dagger}\sqrt{N_{s1}}\|\mathbf{g}_1\|, \ldots, \mathbf{h}_{Mp}^{\dagger}\sqrt{N_{sM}}\|\mathbf{g}_M\|\right)\right), \\ \bar{\mathbf{H}}_m \triangleq \mathsf{BlkDiag}\left(\mathsf{vec}\left(\mathbf{h}_{1m}^{\dagger}\|\mathbf{g}_1\|\sqrt{c_1}, \ldots, \mathbf{h}_{Mm}^{\dagger}\|\mathbf{g}_M\|\sqrt{c_M}\right)\right), \quad \forall m, \\ \mathbf{D}_v \text{ is a diagonal matrix with its } (n,n)\text{th entry being } c_j\|\mathbf{g}_j\|^2, \quad \text{for } n = (j-1) \\ \qquad M+1, \ldots, jM. \end{cases} \tag{9.11}$$

Then (9.10) can be further rewritten as

$$\mathbb{P} : \min_{\{\mathbf{w}_j,\mathbf{v}\}} \sum_{j=1}^{M} \|\mathbf{w}_j\|^2 + \mathbf{v}^{\dagger}\mathbf{D}_v\mathbf{v}$$

$$\text{s.t.} \begin{cases} \dfrac{\left|\mathbf{h}_p^{\dagger}\mathbf{v}\right|^2}{\sum_{j=1}^{M}\left|\mathbf{h}_{jp}^{\dagger}\mathbf{w}_j\right|^2 + \left\|\bar{\mathbf{H}}_p\mathbf{v}\right\|^2 + N_2^p} \geq \gamma_p', \\ \dfrac{\left|\mathbf{h}_{mm}^{\dagger}\mathbf{w}_m\right|^2}{\sum_{\substack{j=1\\j\neq m}}^{M}\left|\mathbf{h}_{jm}^{\dagger}\mathbf{w}_j\right|^2 + \left\|\bar{\mathbf{H}}_m\mathbf{v}\right\|^2 + N_m} \geq \gamma_m, \quad \forall m. \end{cases} \tag{9.12}$$

Problem (9.12) will help to develop a much more efficient and distributed algorithm than solving the original problem (9.7) directly.

9.3.3 An efficient algorithm and distributed implementation

In this section, we will derive an efficient algorithm that is suitable for distributed implementation without centralized coordination, based on the dual problem formulation that will be developed below.

9.3.3.1 Derivation of the dual problem

We first derive the linear programming (LP) formulation for power optimization. For ease of composition and without loss of optimality, we assume that all noise power is normalize to 1, i.e., $N_m = N_2^p = 1,\ \forall m$.

Define $\mathbf{w}_m = \sqrt{p_m}\tilde{\mathbf{w}}_m, p_j = \|\mathbf{w}_j\|^2, \|\tilde{\mathbf{w}}_j\| = 1, \mathbf{v} = \sqrt{p_0}\tilde{\mathbf{v}}, p_0 = \|\mathbf{v}\|^2, \|\tilde{\mathbf{v}}\| = 1$, and assume $\{\tilde{\mathbf{w}}_j, \tilde{\mathbf{v}}\}$ are fixed, then (9.12) becomes a LP problem in the power vector $\mathbf{p} \triangleq [p_0,\ p_1, \ldots, p_M]^T$:

$$\tilde{\mathbb{P}}(\{\tilde{\mathbf{w}}_j, \tilde{\mathbf{v}}\}) : \min_{\mathbf{p}>0} \sum_{j=1}^{M} p_j + p_p \tilde{\mathbf{v}}^\dagger \mathbf{D}_v \tilde{\mathbf{v}}$$

$$\text{s.t.} \begin{cases} \dfrac{p_m \left|\mathbf{h}_{mm}^\dagger \tilde{\mathbf{w}}_m\right|^2}{\sum_{\substack{j=1\\ j\neq m}}^{M} p_j \left|\mathbf{h}_{jm}^\dagger \tilde{\mathbf{w}}_j\right|^2 + p_p \left\|\bar{\mathbf{H}}_m \tilde{\mathbf{v}}\right\|^2 + 1} \geq \gamma_m, \quad \forall m \geq 1, \\[2ex] \dfrac{p_0 \left|\mathbf{h}_p^\dagger \tilde{\mathbf{v}}\right|^2}{\sum_{j=1}^{M} p_j \left|\mathbf{h}_{jp}^\dagger \tilde{\mathbf{w}}_j\right|^2 + p_p \left\|\bar{\mathbf{H}}_p \tilde{\mathbf{v}}\right\|^2 + 1} \geq \gamma_p'. \end{cases} \tag{9.13}$$

It is well known from the optimization theory [21] that the dual of the LP problem

$$\min_{\mathbf{p}\geq\mathbf{0}} \mathbf{m}^T\mathbf{p} \ \text{ s.t. } \ \mathbf{A}\mathbf{p} \geq \mathbf{n} \tag{9.14}$$

is

$$\max_{\mathbf{q}\geq\mathbf{0}} \mathbf{n}^T\mathbf{q} \ \text{ s.t. } \ \mathbf{A}^T\mathbf{q} \leq \mathbf{m}. \tag{9.15}$$

Therefore, the dual of the (linear) power allocation problem (9.13) is expressed as

$$\tilde{\mathbb{PD}}(\{\tilde{\mathbf{w}}_j, \tilde{\mathbf{v}}\}) : \max_{\mathbf{q}\geq\mathbf{0}} \sum_{j=0}^{M} q_j$$

$$\text{s.t.} \begin{cases} \dfrac{q_m \left|\mathbf{h}_{mm}^\dagger \tilde{\mathbf{w}}_m\right|^2}{q_0 \left|\mathbf{h}_{mp}^\dagger \tilde{\mathbf{w}}_m\right|^2 + \sum\limits_{\substack{n=1\\ n\neq m}}^{M} q_n \left|\mathbf{h}_{mn}^\dagger \tilde{\mathbf{w}}_m\right|^2 + \left\|\tilde{\mathbf{w}}_m^\dagger\right\|^2} \leq \gamma_m, \quad \forall m \geq 1, \\[2ex] \dfrac{q_0 \left|\mathbf{h}_p^\dagger \tilde{\mathbf{v}}\right|^2}{\tilde{\mathbf{v}}^\dagger \mathbf{D}_v \tilde{\mathbf{v}} + \sum_{m=1}^{M} q_m \left\|\bar{\mathbf{H}}_m \tilde{\mathbf{v}}\right\|^2 + q_0 \left\|\bar{\mathbf{H}}_p \tilde{\mathbf{v}}\right\|^2} \leq \gamma_p'. \end{cases} \tag{9.16}$$

A key step is to notice that the above (virtual) problem maximization (9.16) is equivalent to the following power minimization problem, by changing the direction of the constraints:

$$\tilde{\mathbb{Q}}(\{\tilde{\mathbf{w}}_j, \tilde{\mathbf{v}}\}) : \min_{\mathbf{q}\geq 0} \sum_{j=0}^{M} q_j$$

$$\text{s.t.} \begin{cases} \dfrac{q_m \left|\mathbf{h}_{mm}^{\dagger}\tilde{\mathbf{w}}_m\right|^2}{q_0|\mathbf{h}_{m0}^{\dagger}\tilde{\mathbf{w}}_m|^2 + \sum\limits_{\substack{n=1\\n\neq m}}^{M} q_n \left|\mathbf{h}_{mn}^{\dagger}\tilde{\mathbf{w}}_m\right|^2 + \left\|\tilde{\mathbf{w}}_m^{\dagger}\right\|^2} \geq \gamma_m, \quad \forall m \geq 1, \\[2ex] \dfrac{q_0 \left|\mathbf{h}_0^{\dagger}\tilde{\mathbf{v}}\right|^2}{\tilde{\mathbf{v}}^{\dagger}\mathbf{D}_v\tilde{\mathbf{v}} + \sum_{m=1}^{M} q_m \left\|\bar{\mathbf{H}}_m\tilde{\mathbf{v}}\right\|^2 + q_0 \left\|\bar{\mathbf{H}}_0\tilde{\mathbf{v}}\right\|^2} \geq \gamma_p'. \end{cases} \tag{9.17}$$

This is because at the optimum, all constraints in both problems should be satisfied with equality, which uniquely defines the optimal power vector.

Based on the above derivation, we can easily see the following connection, for the power optimization problems:

$$\mathrm{val}\Big(\tilde{\mathbb{P}}\left(\{\tilde{\mathbf{w}}_j, \tilde{\mathbf{v}}\}\right)\Big) = \mathrm{val}\Big(\tilde{\mathbb{P}}\mathbb{D}\left(\{\tilde{\mathbf{w}}_j, \tilde{\mathbf{v}}\}\right)\Big) = \mathrm{val}\Big(\tilde{\mathbb{Q}}\left(\{\tilde{\mathbf{w}}_j, \tilde{\mathbf{v}}\}\right)\Big), \forall\, \|\tilde{\mathbf{w}}_j\| = 1, \quad \|\tilde{\mathbf{v}}\| = 1, \tag{9.18}$$

where 'val' denotes the value of the objective function at the optimum point.

If we combine power optimization and beamforming optimization, we have the following relationship:

$$\mathrm{val}(\mathbb{P}) = \min_{\{\tilde{\mathbf{w}}_j, \tilde{\mathbf{v}}\}} \mathrm{val}\left(\tilde{\mathbb{P}}\left(\{\tilde{\mathbf{w}}_j, \tilde{\mathbf{v}}\}\right)\right) = \min_{\{\tilde{\mathbf{w}}_j, \tilde{\mathbf{v}}\}} \mathrm{val}\left(\tilde{\mathbb{Q}}\left(\{\tilde{\mathbf{w}}_j, \tilde{\mathbf{v}}\}\right)\right). \tag{9.19}$$

This means the optimal objective value of the studied problem (9.12) is equal to that of (9.17) if the beamforming vectors are also optimized, i.e.,

$$\tilde{\mathbb{Q}} : \min_{\mathbf{q}\geq \mathbf{0}} \sum_{j=0}^{M} q_j$$

$$\text{s.t.} \begin{cases} \max\limits_{\|\tilde{\mathbf{w}}_m\|=1} \dfrac{q_m \left|\mathbf{h}_{mm}^{\dagger}\tilde{\mathbf{w}}_m\right|^2}{q_0 \left|\mathbf{h}_{mp}^{\dagger}\tilde{\mathbf{w}}_m\right|^2 + \sum\limits_{\substack{n=1\\n\neq m}}^{M} q_n \left|\mathbf{h}_{mn}^{\dagger}\tilde{\mathbf{w}}_m\right|^2 + \left\|\tilde{\mathbf{w}}_m^{\dagger}\right\|^2} \geq \gamma_m, \quad \forall m \geq 1, \\[2ex] \max\limits_{\|\tilde{\mathbf{v}}\|=1} \dfrac{q_0 \left|\mathbf{h}_p^{\dagger}\tilde{\mathbf{v}}\right|^2}{\tilde{\mathbf{v}}^{\dagger}\mathbf{D}_v\tilde{\mathbf{v}} + \sum_{m=1}^{M} q_m \left\|\bar{\mathbf{H}}_m\tilde{\mathbf{v}}\right\|^2 + q_0 \left\|\bar{\mathbf{H}}_p\tilde{\mathbf{v}}\right\|^2} \geq \gamma_p', \end{cases} \tag{9.20}$$

which in turn can be equivalently written as

$$\mathbb{Q}: \min_{\mathbf{q}\geq\mathbf{0}} \sum_{j=0}^{M} q_j$$

$$\text{s.t.} \begin{cases} q_m = \dfrac{\gamma_m}{\mathbf{h}_{mm}^{\dagger}\left(\mathbf{I}+q_0\mathbf{h}_{mp}\mathbf{h}_{mp}^{\dagger}+\sum_{\substack{n=1\\n\neq m}}^{M} q_n\mathbf{h}_{mn}\mathbf{h}_{mn}^{\dagger}\right)^{-1}\mathbf{h}_{mm}}, & \forall m\geq 1, \\ q_0 = \dfrac{\gamma_p'}{\mathbf{h}_p^{\dagger}\left(\mathbf{D}_v+\sum_{m=1}^{M} q_m\bar{\mathbf{H}}_m^{\dagger}\bar{\mathbf{H}}_m+q_0\bar{\mathbf{H}}_p^{\dagger}\bar{\mathbf{H}}_p\right)^{-1}\mathbf{h}_p}. \end{cases} \tag{9.21}$$

by optimizing the beamforming vectors as follows:

$$\tilde{\mathbf{w}} = \left(\mathbf{I}+q_0\mathbf{h}_{mp}\mathbf{h}_{mp}^{\dagger}+\sum_{\substack{n=1\\n\neq m}}^{M} q_n\mathbf{h}_{mn}\mathbf{h}_{mn}^{\dagger}\right)^{-1}\mathbf{h}_{mm}, \quad \tilde{\mathbf{w}} = \frac{\tilde{\mathbf{w}}}{\|\tilde{\mathbf{w}}\|}, \tag{9.22}$$

$$\tilde{\mathbf{v}} = \left(\mathbf{D}_v+\sum_{m=1}^{M} q_m\bar{\mathbf{H}}_m^{\dagger}\bar{\mathbf{H}}_m+q_0\bar{\mathbf{H}}_p^{\dagger}\bar{\mathbf{H}}_p\right)^{-1}\mathbf{h}_p, \quad \mathbf{v} = \frac{\mathbf{v}}{\|\mathbf{v}\|}. \tag{9.23}$$

The new vector variable $\mathbf{q} = [q_0, \ldots, q_M]^T$ now has an interpretation of dual variable. The procedures on how to recover the solution of the original problem after the dual problem is solved will be discussed in Section 9.4.3. In the following, we will propose a low-complexity algorithm that is suitable for distributed implementation.

9.3.3.2 The proposed algorithm based on fixed-point iteration

The derived problem (9.21) is of great importance in the sense that it not only simplifies the original problem, but also enables a distributed implementation below.

Algorithm 1 Efficient fixed-point iteration to solve (9.21)

---

1) Initialize $\mathbf{q} \triangleq [q_0, q_1, \ldots, q_M]^T$ with arbitrary non-negative vector.
2) Update $\{q_m(\mathbf{q})\}, q_0(\mathbf{q})\}$ as the equality constraints in (9.21), i.e.,

$$q_m(\mathbf{q}) = \frac{\gamma_m}{\mathbf{h}_{mm}^{\dagger}\left(\mathbf{I}+q_0\mathbf{h}_{mp}\mathbf{h}_{mp}^{\dagger}+\sum_{\substack{n=1\\n\neq m}}^{M} q_n\mathbf{h}_{mn}\mathbf{h}_{mn}^{\dagger}\right)^{-1}\mathbf{h}_{mm}}, \quad \forall m\geq 1, \tag{9.24}$$

$$q_0(\mathbf{q}) = \frac{\gamma_p'}{\mathbf{h}_p^{\dagger}\left(\mathbf{D}_v+\sum_{m=1}^{M} q_m\bar{\mathbf{H}}_m^{\dagger}\bar{\mathbf{H}}_m+q_0\bar{\mathbf{H}}_p^{\dagger}\bar{\mathbf{H}}_p\right)^{-1}\mathbf{h}_p}. \tag{9.25}$$

3) Go back to Step (2) until convergence.
4) After obtaining the optimal dual solutions $\{q_m^*(\mathbf{q}), q_0^*(\mathbf{q})\}$, find the primary solutions $\{\tilde{\mathbf{w}}_m^*, \tilde{\mathbf{v}}^*\}$ as derived in (9.22).

Theorem 9.2. *Algorithm 1 finds the optimal solution to (9.21).*

Proof. It is easy to check that the mappings in (9.24) and (9.25) satisfy

- Positivity: $q_m(\mathbf{q}) \geq 0, \ \forall m \geq 0$;
- Monotonicity: If $q \geq q'$, then $q_m(\mathbf{q}) \geq q_m(\mathbf{q}'), \ \forall m \geq 0$;
- Scalability: For all $\alpha > 1, \alpha q_m(\mathbf{q}) > q_m(\alpha \mathbf{q}), \ \forall m \geq 0$.

Therefore, $q_m(\mathbf{q}), \ m = 0, \ldots, M$ are standard interference functions and the optimality of the solution obtained by using Algorithm 1 follows directly from the results in Reference 22, which completes the proof. □

9.3.3.3 Distributed implementation

In practice, distributed implementation is preferred without requiring global CSI or centralized processing. In this section, we will show that the proposed Algorithm 1 can be implemented in a distributed way. To be specific, each SU's power can be updated locally assuming all others are fixed. The details are described as follows.

The optimization of $\mathbf{q}_m$ $(m > 0)$ in (9.24) of Algorithm 1 requires only the CSI from ST_m to all the SUs and the PU, i.e., $\mathbf{H}_m = [\mathbf{h}_{mp} \ \mathbf{h}_{m1} \cdots \mathbf{h}_{mM}]$ defined in Appendix A, and therefore can be implemented at ST_m locally. To obtain $\mathbf{q}$ other than q_m (q_0 will be discussed next), it can be obtained by a small amount of information exchange among all the STs or feedback from the SUs [22]. After the optimal $\mathbf{q}$ is known, the optimization of unit-norm beamforming vector of $\mathbf{w}_m$ ($\tilde{\mathbf{w}}_m$, without power scaling) can be readily found using (9.22) with only local CSI.

The optimization of q_0 in the proposed algorithm (9.25) is less straightforward. At first, it may seem that the update of q_0 involves global CSI, but in fact the matrices $\bar{\mathbf{H}}_m^\dagger \bar{\mathbf{H}}_m$ and $\bar{\mathbf{H}}_p^\dagger \bar{\mathbf{H}}_p$ are both block diagonal, and therefore the right-hand side of (9.25) can be decoupled. After some manipulations, we can get

$$\mathbf{h}_p^\dagger \left(\mathbf{D}_v + \sum_{m=1}^{M} q_m \bar{\mathbf{H}}_m^\dagger \bar{\mathbf{H}}_m + q_0 \bar{\mathbf{H}}_p^\dagger \bar{\mathbf{H}}_p \right)^{-1} \mathbf{h}_p = \sum_{j=1}^{M} \alpha_j, \tag{9.26}$$

where

$$\alpha_j \triangleq \|\mathbf{g}_j\|^2 \mathbf{h}_{jp}^\dagger \mathbf{C}^{-1} \mathbf{h}_{jp}, \tag{9.27}$$

and $\mathbf{C} \triangleq q_0 N_{sj} \mathbf{h}_{jp} \mathbf{h}_{jp}^\dagger + (P_P \|\mathbf{g}_j\|^2 + N_{sj})\mathbf{I} + \sum_{m=1}^{M} q_m (P_P \|\mathbf{g}_j\|^2 + N_{sj}) \mathbf{h}_{jm} \mathbf{h}_{jm}^\dagger$. Note that the SINR is the summation of contributions from all the collaborative STs. For ST_j, α_j can be evaluated using local CSI and $\mathbf{q}$. Thus, the update of q_0 can be done in a distributed manner by

$$q_0 = \frac{\gamma_p'}{\sum_{j=1}^{M} \alpha_j}, \tag{9.28}$$

which requires that the STs exchange information of $\{\alpha_j\}$ with each other, which only incurs a small amount of signalling overhead.

Once the optimal $\boldsymbol{\lambda}$ is found, ST_j calculates the unit-norm relay-beamforming vector by (9.22) which again only relies on local CSI and can be implemented in a distributed manner.

9.3.3.4 Conversion to the original downlink solutions

After obtaining the optimal solution $\mathbf{q}$ to the dual problem (9.21), and the unit-norm beamformning vectors $\{\tilde{\mathbf{w}}_m\}, \tilde{\mathbf{v}}$, in this section, we discuss how to find the optimal primary solutions based on them.

Define

$$\mathbf{D} \in \mathbb{R}^{(M+1)\times(M+1)}: \begin{cases} \mathbf{D}_{1,1} = \dfrac{\gamma_p'}{\left|\mathbf{h}_p^\dagger \tilde{\mathbf{v}}\right|^2}, \\ \mathbf{D}_{n,n} = \dfrac{\gamma_n}{|\mathbf{h}_{nn}\tilde{\mathbf{w}}_n|^2}, \quad \text{for } n = 2, \ldots, M+1, \\ \mathbf{D}_{m,n} = 0, \quad \text{for } m \neq n. \end{cases} \tag{9.29}$$

Having identified $\{\tilde{\mathbf{w}}_m\}$ and $\tilde{\mathbf{v}}$, we can decide on the downlink power based on the fact that at the optimum, all constraints in (9.7) are met with equalities; therefore we have $M+1$ independent linear equations sufficient to solve for the $(M+1)\times 1$ power vector to get

$$\mathbf{p} \triangleq [p_0\, p_1\, \cdots\, p_M]^T = (\mathbf{I} - \mathbf{D}\mathbf{F}^T)^{-1}\mathbf{D}\boldsymbol{\sigma}, \tag{9.30}$$

where

$$\mathbf{F} \in \mathbb{R}^{(M+1)\times(M+1)}: \begin{cases} \mathbf{F}_{1,1} = \left\|\bar{\mathbf{H}}_p\tilde{\mathbf{v}}\right\|^2, \\ \mathbf{F}_{n,n} = 0, \quad \text{for } n = 2, \ldots, M+1, \\ \mathbf{F}_{n+1,1} = |\mathbf{h}_{np}^\dagger\tilde{\mathbf{w}}_n|^2, \quad \text{for } n = 1, \ldots, M, \\ \mathbf{F}_{1,n+1} = \left\|\bar{\mathbf{H}}_m\tilde{\mathbf{v}}\right\|^2, \quad \text{for } n = 1, \ldots, M, \\ \mathbf{F}_{m+1,n+1} = |\mathbf{h}_{nm}^\dagger\tilde{\mathbf{w}}_m|^2, \quad \text{for } m, n = 1, \ldots, M, \text{ and } m \neq n, \end{cases} \tag{9.31}$$

and

$$\boldsymbol{\sigma} \in \mathbb{R}^{(M+1)\times 1}: \begin{cases} \sigma_1 = \tilde{\mathbf{v}}^\dagger \mathbf{D}_v \tilde{\mathbf{v}}, \\ \sigma_n = 1, \quad \text{for } n = 2, \ldots, M+1. \end{cases} \tag{9.32}$$

As a result, the downlink beamforming solution to the original problem (9.12) is given by

$$\begin{cases} \mathbf{v} = \sqrt{p_0}\tilde{\mathbf{v}}, \\ \mathbf{w}_m = \sqrt{p_m}\tilde{\mathbf{w}}_m, \quad \forall m. \end{cases} \tag{9.33}$$

The implementation of the above solution needs to make the matrix $(\mathbf{I} - \mathbf{D}\mathbf{F}^T)^{-1}\mathbf{D}$ known to all the STs, and the associated parameter exchange will be analysed later in this section.

To summarize, the proposed algorithm is desirable for distributed implementation when a central controller is absent to gather global CSI and coordinate the required optimization. With only local CSI at each ST and moderate amount of parameter exchange, the optimal transmit and relaying beamforming can be realized in a distributed manner. Finally, we analyse the signalling requirement of the proposed algorithm.

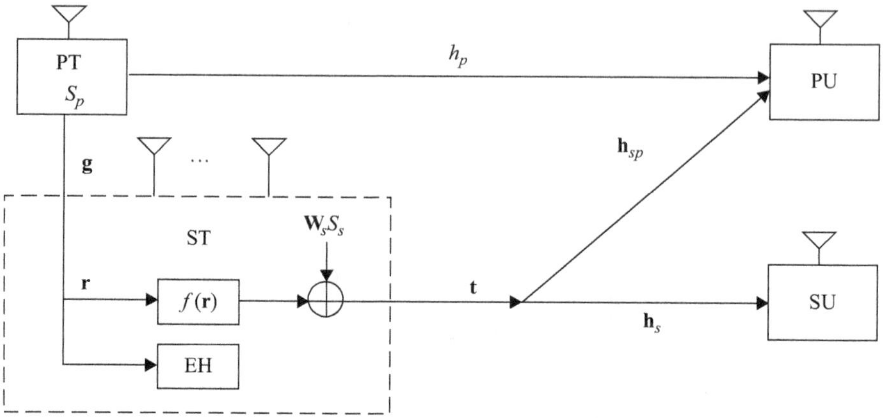

Figure 9.5 Energy and information cooperation in a cognitive radio network

9.3.3.5 Analysis of signalling overhead

The distributed strategy depends largely on the required parameter exchange between the STs in two steps. First, the dual problem (9.21) needs to be solved and at each iteration, the vectors $[q_1, \ldots, q_M]$ and $\boldsymbol{\alpha} = [\alpha_1, \ldots, \alpha_M]$ need to be shared among the STs. The total number of exchanged parameters is $2N_I M$ where N_I denotes the total number of iterations needed for the algorithm to converge. Then for the uplink–downlink conversion, the matrix $(\mathbf{I} - \mathbf{DF}^T)^{-1}\mathbf{D}$ needs to be made available to all the STs, requiring $(M+1)^2$ positive scalars to be shared. The total number of parameter exchange is therefore $2N_I M + (M+1)^2$, regardless of the number of transmit antennas, N. As a comparison, full CSI exchange, which would make possible a centralized optimization, will require $2NM^2$ scalars to be shared where the factor of 2 is due to the complex nature of the parameters. If the same quantization scheme is used, the ratio of the number of exchanged bits in the proposed algorithm over full CSI exchange is

$$\eta = \frac{2N_I M + (M+1)^2}{2NM^2} \approx \frac{N_I + \frac{1}{2}}{NM}. \tag{9.34}$$

When the number of STs, M, is large, a great reduction in backhaul signalling is anticipated.

9.4 Information and energy cooperation

9.4.1 System setting

In this section, we turn to a more sophisticated cooperation between the primary and the secondary systems, which involves energy cooperation in addition to the information cooperation discussed in the previous section. For ease of composition, we consider a simplified network work model in which the secondary system is composed of a single ST sending message to a single SU, as depicted in Figure 9.5.

We adopt slightly different notations of system parameters because of the simplified network model. Specifically, $\mathbf{h}_s$ and $\mathbf{h}_{sp}$ are used to denote the ST-SU and ST-PU channels, respectively, and other parameters are kept the same. The ST itself has an initial total energy of P_{s0} and further receives/harvests energy from the PT. After the ST receives both information and energy from the PT, it processes the primary signal, harvests energy then uses the harvested energy together with its own energy, to serve the SU and relay the signal to the PU. To be consistent with the information cooperation introduced in the previous section, AF relaying protocol is again employed by the ST.

We explicitly consider two noise components received at a terminal: one is the received thermal noise and the other is due to RF to baseband conversion, both are modelled as zero-mean additive white Gaussian noise (AWGN) with variances of N_0 and N_C, respectively. Assuming that they are independent, define the combined received noise power as $\tilde{N}_0 = N_0 + N_C$.

There are different approaches that facilitate the information and energy transfer from the PT to the ST. In the sequel, we will introduce two specific cooperation schemes and find their optimal as well as low-complexity solutions. For fairness, the same amount of energy (for both the PT and the ST) is used to compare all schemes.

9.4.2 The ideal primary-cognitive cooperation

9.4.2.1 System model and problem formulation

We first look into the ideal cooperation between the PT and the ST for information and energy transfer, where the ST has non-causal information about the primary signal and obtains energy from the PT via the PT-ST channel $\mathbf{g}$. Note that although in practice, this cooperation scheme is either too difficult or too costly to implement, it provides a performance upper bound for practical cooperation protocols.

Since the primary message s_p ($\mathrm{E}[|s_p|^2] = 1$) is non-causally known at the ST, it can employ the capacity-achieving technique of DPC to encode the primary signal and superimpose its own secondary signal s_s ($\mathrm{E}[|s_s|^2] = 1$) such that no primary interference is introduced at the SU. The received signal at the PU is

$$y_p = \left(\sqrt{(1-\beta)P_p}h_p + \mathbf{h}_{sp}^{\dagger}\mathbf{w}_p\right)s_p + \mathbf{h}_{sp}^{\dagger}\mathbf{w}_s s_s + n_p, \tag{9.35}$$

where $\mathbf{w}_p$ is the beamforming vector used by the ST to forward the primary signal, and $n_p \in \mathbb{CN}(0, \tilde{N}_0)$ is the combined received noise at the PU. $\beta P_p (0 \le \beta \le 1)$ denotes the amount of energy transferred to the ST and received as $P_p\beta\eta\|\mathbf{g}\|^2$, where η is the efficiency of energy transfer. The ST then has the total power of $P_{s0} + P_p\beta\eta\|\mathbf{g}\|^2$ to serve both the PU and the SU. Due to the use of DPC, the SU receives

$$y_s = \mathbf{h}_s^{\dagger}\mathbf{w}_s s_s + n_s, \tag{9.36}$$

where $n_s \in \mathbb{CN}(0, \tilde{N}_0)$ is the combined received noise at the SU. Then the primary and secondary received SINRs are, respectively,

$$\Gamma_p = \frac{\left|\sqrt{(1-\beta)P_p}h_p + \mathbf{h}_{sp}^{\dagger}\mathbf{w}_p\right|^2}{\left|\mathbf{h}_{sp}^{\dagger}\mathbf{w}_s\right|^2 + \tilde{N}_0} \quad \text{and} \quad \Gamma_s = \frac{\left|\mathbf{h}_s^{\dagger}\mathbf{w}_s\right|^2}{\tilde{N}_0}. \tag{9.37}$$

It is easy to see that the optimal $\mathbf{w}_p$ admits the form $\mathbf{w}_p = \sqrt{q_p}\frac{\mathbf{h}_{sp}}{\|\mathbf{h}_{sp}\|}e^{j\theta}$, where θ is chosen for coherent reception and $q_p \triangleq \|\mathbf{w}_p\|^2$. As a result, the achievable PU rate is

$$R_p = \log_2(1+\gamma_p) = \log_2\left(1 + \frac{\left(\sqrt{(1-\beta)P_p}h_p + \sqrt{q_p}\,\|\mathbf{h}_{sp}\|\right)^2}{\left|\mathbf{h}_{sp}^{\dagger}\mathbf{w}_s\right|^2 + \tilde{N}_0}\right). \tag{9.38}$$

The problem of maximizing the SU rate subject to PU rate and ST power constraints is written as

$$\max_{\mathbf{w}_s, q_p, \beta} \quad \left|\mathbf{h}_s^{\dagger}\mathbf{w}_s\right| \tag{9.39a}$$

$$\text{s.t.} \quad \frac{\left(\sqrt{(1-\beta)P_p}|h_p| + \sqrt{q_p}\|\mathbf{h}_{sp}\|\right)^2}{\left|\mathbf{h}_{sp}^{\dagger}\mathbf{w}_s\right|^2 + \tilde{N}_0} \geq 2^{r_p} - 1, \tag{9.39b}$$

$$\|\mathbf{w}_s\|^2 + q_p \leq P_{s0} + P_p\beta\eta\|\mathbf{g}\|^2, \tag{9.39c}$$

$$0 \leq \beta \leq 1, \quad q_p > 0. \tag{9.39d}$$

9.4.2.2 Feasibility

Before solving (9.39), in Proposition 9.1, we first give the condition under which it is feasible and the proof is given in Appendix B.

Proposition 9.1. *Problem (9.39) is feasible if and only if the PU rate requirement r_p is not larger than*

$$R_{p,\max} \triangleq \begin{cases} \log_2\left(1 + \dfrac{\left(\sqrt{P_p}|h_p| + \sqrt{P_{s0}}\|\mathbf{h}_{sp}\|\right)^2}{\tilde{N}_0}\right), & \text{if } P_p\eta^2\|\mathbf{g}\|^2\|\mathbf{h}_{sp}\|^2 < P_{s0}|h_p|^2; \\ \log_2\left(1 + \dfrac{P_p\eta + P_{s0}}{\eta^2\|\mathbf{g}\|^4\|\mathbf{h}_{sp}\|^2 + \eta\|\mathbf{g}\|^2|h_p|^2}\dfrac{\left(|h_p|^2 + \eta\|\mathbf{g}\|^2\|\mathbf{h}_{sp}\|^2\right)^2}{\tilde{N}_0}\right), & \text{otherwise.} \end{cases} \tag{9.40}$$

According to Proposition 9.1, when $P_p\eta^2\|\mathbf{h}_{sp}\|^2 < P_{s0}|h_p|^2$ which means that the transferred power from the PT to the ST cannot bring sufficient performance gain, no energy transfer is needed. This may happen when primary power is limited, the secondary power is abundant, or the efficiency of the power transfer is too low, etc.

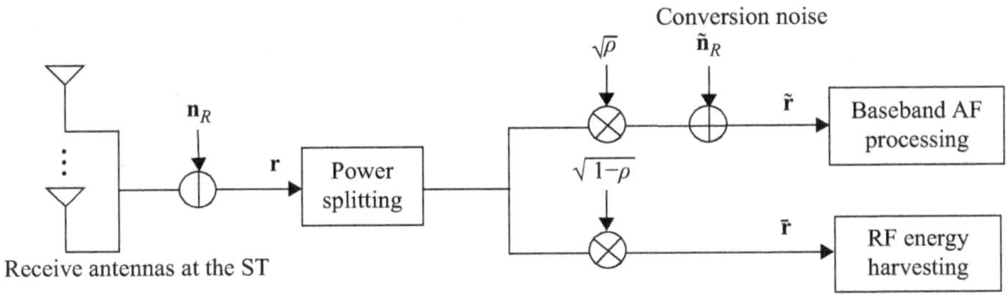

Figure 9.6 The power splitting EH technique at the ST

9.4.2.3 Simplified characterization

Problem (9.39) is equivalent to the following convex problem:

$$\max_{\mathbf{w}_s,\mathbf{v}} \quad Re\left(\mathbf{h}_s^\dagger \mathbf{w}_s\right) \tag{9.41a}$$

$$s.t. \quad \left(\mathbf{f}^T\mathbf{v}\right)^2 \geq (2^{r_p}-1)\left(\left|\mathbf{h}_{sp}^\dagger \mathbf{w}_s\right|^2 + \tilde{N}_0\right), \tag{9.41b}$$

$$\|\mathbf{w}_s\|^2 + \mathbf{v}^T\mathbf{D}\mathbf{v} \leq P_{s0} + P_p\eta\|\mathbf{g}\|^2, \tag{9.41c}$$

$$\mathbf{v} \geq 0, \quad [\mathbf{v}]_1 \leq 1. \tag{9.41d}$$

where $\mathbf{f}^T = \left[\sqrt{P_p}|h_p|, \|\mathbf{h}_{sp}\|\right]$, $\mathbf{D} = \text{diag}(\eta P_p, 1), \mathbf{v} \triangleq \left[\sqrt{1-\beta}, \sqrt{q_p}\right]$. Optimization problem in (9.41) is a second-order cone problem (SOCP) and is convex. It can be solved efficiently using convex optimization techniques.

9.4.3 A practical power-splitting cooperation

9.4.3.1 System model and problem setting

In this section, a realistic energy and information cooperation is studied. We assume that the ST first listens to the primary transmission via the channel $\mathbf{g}$ then forwards it to the PU, therefore two channel phases are required to complete the communications. In Phase 1, the PT broadcasts its data s_p with power $2P_p$ where the factor 2 is because the PT only transmits during the first half duration, then the received signals at the PU and the ST are, respectively,

$$y_{p1} = \sqrt{2P_p}h_p s_p + n_{p1} \quad \text{and} \quad \mathbf{r} = \sqrt{2P_p}\mathbf{g}s_p + \mathbf{n}_R, \tag{9.42}$$

where $n_{p1} \in \mathbb{CN}(0, \tilde{N}_0)$ is the combined noise at the PU while $\mathbf{n}_R \in \mathbb{CN}(0, N_0\mathbf{I})$ is the thermal noise received at ST, respectively.

To forward primary information as well as harvest RF energy at the ST, the practical power splitting technique [2] is used, which is depicted in Figure 9.6 and works as follows. The ST splits the received RF signal from the PT into two portions: one for forwarding to the PU after AF processing and the other for harvesting energy, with relative power of ρ and $1-\rho$, respectively. The signal for AF processing will be converted from the RF to the baseband, and this results in the received signal

$$\tilde{\mathbf{r}} = \sqrt{\rho}\mathbf{r} + \tilde{\mathbf{n}}_R, \tag{9.43}$$

where $\tilde{\mathbf{n}}_R \sim \mathbb{CN}(\mathbf{0}, N_C\mathbf{I})$ is the complex AWGN during the RF to baseband conversion. The ST processes the received signal and produces $f(\mathbf{r}) = \mathbf{A}\tilde{\mathbf{r}}$. Similar to the scenario of information cooperation, it has been shown that the optimal $\mathbf{A}$ has the structure of $\mathbf{A} = \mathbf{w}_p\mathbf{g}^\dagger$ according to Theorem 9.1, where $\mathbf{w}_p$ is a new transmit beamforming vector to be optimized. This is also intuitive because there is a single primary data stream, the best reception strategy for the ST is to use MRC.

The signal for EH is simply

$$\bar{\mathbf{r}} = \sqrt{1-\rho}\mathbf{r} = \sqrt{1-\rho}\left(\mathbf{g}\sqrt{2P_p}s_p + \mathbf{n}_R\right). \tag{9.44}$$

Assuming the energy transfer efficiency of η, the amount of the harvested energy is

$$P_{\text{EH}} = P_p\eta(1-\rho)\|\mathbf{g}\|^2. \tag{9.45}$$

Therefore, the ST will have a total transmit power of $2P_{s0} + 2P_p\eta(1-\rho)\|\mathbf{g}\|^2$ where the factor 2 is due to the fact that the ST only transmits signals in the second half of the communication time.

In Phase 2, the ST superimposes the relaying signal $f(\mathbf{r})$ with its own data s_s using the secondary beamforming vector $\mathbf{w}_s$, then transmits it to both the PU and the SU. Note that DPC is not used at the ST. In this phase, the PT remains idle.

The ST's transmit signal is

$$\begin{aligned}\mathbf{t} &= \mathbf{w}_s s_s + \mathbf{w}_p\mathbf{g}^\dagger\tilde{\mathbf{r}}\\ &= \mathbf{w}_s s_s + \sqrt{2\rho P_p}\mathbf{w}_p\|\mathbf{g}\|^2 s_p + \sqrt{\rho}\mathbf{w}_p\mathbf{g}^\dagger\mathbf{n}_R + \mathbf{w}_p\mathbf{g}^\dagger\tilde{\mathbf{n}}_R,\end{aligned} \tag{9.46}$$

with average power

$$p_R = \mathrm{E}\|\mathbf{t}\|^2 = \|\mathbf{w}_s\|^2 + \left(2P_p\rho\|\mathbf{g}\|^4 + \rho\|\mathbf{g}\|^2 N_0 + \|\mathbf{g}\|^2 N_C\right)\|\mathbf{w}_p\|^2. \tag{9.47}$$

The received signal at the SU is

$$\begin{aligned}y_s &= \mathbf{h}_s^\dagger\mathbf{t} + n_s\\ &= \mathbf{h}_s^\dagger\mathbf{w}_s s_s + \sqrt{\rho}\mathbf{h}_s^\dagger\mathbf{w}_p\|\mathbf{g}\|^2 s_p + \sqrt{\rho}\mathbf{h}_s^\dagger\mathbf{w}_p\mathbf{g}^\dagger\mathbf{n}_R + \mathbf{h}_s^\dagger\mathbf{w}_p\mathbf{g}^\dagger\tilde{\mathbf{n}}_R + n_s,\end{aligned} \tag{9.48}$$

where $n_s \in \mathbb{CN}(0, \tilde{N}_0)$ is the combined noise at the SU. The received SINR at SU is then expressed as

$$\Gamma_s = \frac{\left|\mathbf{h}_s^\dagger\mathbf{w}_s\right|^2}{\left(2P_p\rho\|\mathbf{g}\|^4 + \rho\|\mathbf{g}\|^2 N_0 + \|\mathbf{g}\|^2 N_C\right)\left|\mathbf{h}_s^\dagger\mathbf{w}_p\right|^2 + \tilde{N}_0}, \tag{9.49}$$

and the achievable SU rate is $R_s = \frac{1}{2}\log_2(1+\Gamma_s)$ where the factor $\frac{1}{2}$ arises due to the two orthogonal channel uses. The received signal at the PU is

$$\begin{aligned}y_{p2} &= \mathbf{h}_{sp}^\dagger\mathbf{t} + n_{p2}\\ &= \mathbf{h}_{sp}^\dagger\mathbf{w}_s s_s + \rho\mathbf{h}_{sp}^\dagger\mathbf{w}_p\|\mathbf{g}\|^2 s_p + \rho\mathbf{h}_{sp}^\dagger\mathbf{w}_p\mathbf{g}^\dagger\mathbf{n}_R + \mathbf{h}_{sp}^\dagger\mathbf{w}_p\mathbf{g}^\dagger\tilde{\mathbf{n}}_R + n_{p2},\end{aligned} \tag{9.50}$$

where $n_{p2} \in \mathbb{CN}(0, \tilde{N}_0)$ is the combined noise at the PU during Phase 2.

Applying the MRC strategy to y_{p1} and y_{p2}, the received SINR of the PU is the sum of two channel uses, and consequently, the achievable PU rate is

$$R_p = \frac{1}{2}\log_2\left(1 + \frac{2P_p|h_p|^2}{\tilde{N}_0} + \frac{2P_p\rho\|\mathbf{g}\|^4\left|\mathbf{h}_{sp}^\dagger\mathbf{w}_p\right|^2}{\left|\mathbf{h}_{sp}^\dagger\mathbf{w}_s\right|^2 + (\rho\|\mathbf{g}\|^2N_0 + \|\mathbf{g}\|^2N_C)\left|\mathbf{h}_{sp}^\dagger\mathbf{w}_p\right|^2 + \tilde{N}_0}\right). \quad (9.51)$$

Next we can formulate the problem of maximizing the SU rate R_s subject to the PU's rate constraint r_p and the ST's transmit power constraint $2(P_{s0} + P_{\mathrm{EH}})$, by jointly optimizing the power splitting parameter ρ, the secondary beamforming vector $\mathbf{w}_s$, and the forwarding primary beamforming vector $\mathbf{w}_p$. Using the monotonicity between the received SINR and the achievable rate, the optimization problem can be written as

$$\begin{aligned}
\max_{\mathbf{w}_s,\mathbf{w}_p,\rho}\quad & \frac{\left|\mathbf{h}_s^\dagger\mathbf{w}_s\right|^2}{(2P_p\rho\|\mathbf{g}\|^4 + \rho\|\mathbf{g}\|^2N_0 + \|\mathbf{g}\|^2N_C)\left|\mathbf{h}_s^\dagger\mathbf{w}_p\right|^2 + \tilde{N}_0} \\
\text{s.t.}\quad & \frac{\left|\mathbf{h}_{sp}^\dagger\mathbf{w}_p\right|^2}{\left|\mathbf{h}_{sp}^\dagger\mathbf{w}_s\right|^2 + \tilde{N}_0} \geq \frac{\gamma_p'}{\rho\|\mathbf{g}\|^4 - \gamma_p'(\rho\|\mathbf{g}\|^2N_0 + \|\mathbf{g}\|^2N_C)}, \\
& \|\mathbf{w}_s\|^2 + (2P_p\rho\|\mathbf{g}\|^4 + \rho\|\mathbf{g}\|^2N_0 + \|\mathbf{g}\|^2N_C)\|\mathbf{w}_p\|^2 \leq 2P_{s0} + 2P_p\eta(1-\rho)\|\mathbf{g}\|^2, \\
& 0 \leq \rho \leq 1,
\end{aligned} \quad (9.52)$$

where we have defined $\gamma_p' \triangleq \frac{2^{2r_p}-1}{2P_p} - \frac{|h_p|^2}{\tilde{N}_0}$.

9.4.3.2 The closed-form optimal solution

Assuming problem (9.52) is feasible, we study how to find its optimal solution.

By change of variable $\mathbf{w}_p := \sqrt{2P_p\rho\|\mathbf{g}\|^4 + \rho\|\mathbf{g}\|^2N_0 + \|\mathbf{g}\|^2N_C}\mathbf{w}_p$, we write (9.52) in an equivalent but more compact form as

$$\begin{aligned}
\max_{\mathbf{w}_s,\mathbf{w}_p,\rho}\quad & \frac{\left|\mathbf{h}_s^\dagger\mathbf{w}_s\right|^2}{\left|\mathbf{h}_s^\dagger\mathbf{w}_p\right|^2 + \tilde{N}_0} \\
\text{s.t.}\quad & \frac{\left|\mathbf{h}_{sp}^\dagger\mathbf{w}_p\right|^2}{\left|\mathbf{h}_{sp}^\dagger\mathbf{w}_s\right|^2 + \tilde{N}_0} \geq \gamma_p'', \\
& \|\mathbf{w}_s\|^2 + \|\mathbf{w}_p\|^2 \leq 2P_{s0} + 2P_p\eta(1-\rho)\|\mathbf{g}\|^2, \\
& 0 \leq \rho \leq 1,
\end{aligned} \quad (9.53)$$

where $\gamma_p'' \triangleq \frac{(2P_p\rho\|\mathbf{g}\|^2 + \rho N_0 + N_C)\gamma_p'}{\rho\|\mathbf{g}\|^2 - \gamma_p'(\rho N_0 + N_C)}$. We find the following lemma useful to solve (9.53) and its proof is given in Appendix C.

Lemma 9.1. *Consider a general maximization problem below:*

$$\begin{aligned}
\max_{\mathbf{w}_1,\mathbf{w}_2} \quad & \frac{\left|\mathbf{h}_2^{\dagger}\mathbf{w}_2\right|^2}{\sigma^2+\left|\mathbf{h}_2^{\dagger}\mathbf{w}_1\right|^2} \\
s.t. \quad & \frac{\left|\mathbf{h}_1^{\dagger}\mathbf{w}_1\right|^2}{\sigma^2+\left|\mathbf{h}_1^{\dagger}\mathbf{w}_2\right|^2} \geq \gamma_1 \\
& \|\mathbf{w}_1\|^2+\|\mathbf{w}_2\|^2 \leq P_C,
\end{aligned} \tag{9.54}$$

where $\mathbf{h}_1, \mathbf{h}_2$ *are* $N \times 1$ *vectors and* γ_1, P_C, σ^2 *are positive scalars. Define* $\zeta^2 \triangleq \frac{\left|\mathbf{h}_1^{\dagger}\mathbf{h}_2\right|^2}{\|\mathbf{h}_1\|^2\|\mathbf{h}_2\|^2}$. *Suppose (9.60) is feasible and its optimal objective value is* γ_2, *then* γ_2 *is uniquely determined by the following equation set:*

$$\begin{cases}
\lambda_1 = \dfrac{\gamma_1\sigma^2\left(\sigma^2+\lambda_2\|\mathbf{h}_2\|^2\right)}{\|\mathbf{h}_1\|^2\left(\sigma^2+\lambda_2\|\mathbf{h}_2\|^2\left(1-\zeta^2\right)\right)}, \\
\lambda_2 = \dfrac{\gamma_2\sigma^2\left(\sigma^2+\lambda_1\|\mathbf{h}_1\|^2\right)}{\|\mathbf{h}_2\|^2\left(\sigma^2+\lambda_1\|\mathbf{h}_1\|^2\left(1-\zeta^2\right)\right)}, \\
\lambda_1+\lambda_2 = P_C,
\end{cases} \tag{9.55}$$

where λ_1, λ_2 *are dual variables.*

Given ρ, using Lemma 9.1, the dual variables (λ_1, λ_2) of (9.53) are identified. Then the optimal $\mathbf{w}_s$, $\mathbf{w}_p$ can be expressed as

$$\begin{cases}
\mathbf{w}_1 = \sqrt{p_1}\dfrac{\left(\sigma^2\mathbf{I}+\lambda_2\mathbf{h}_2\mathbf{h}_2^{\dagger}\right)^{-1}\mathbf{h}_1}{\left\|\left(\sigma^2\mathbf{I}+\lambda_2\mathbf{h}_2\mathbf{h}_2^{\dagger}\right)^{-1}\mathbf{h}_1\right\|}, \\
\mathbf{w}_2 = \sqrt{p_2}\dfrac{\left(\sigma^2\mathbf{I}+\lambda_1\mathbf{h}_1\mathbf{h}_1^{\dagger}\right)^{-1}\mathbf{h}_2}{\left\|\left(\sigma^2\mathbf{I}+\lambda_1\mathbf{h}_1\mathbf{h}_1^{\dagger}\right)^{-1}\mathbf{h}_2\right\|},
\end{cases} \tag{9.56}$$

where the downlink power p_1, p_2 can be found using the uplink–downlink duality [23]. Then the optimal solution to (9.53) can be derived by performing one-dimensional optimization of ρ.

9.5 Performance evaluation

Computer simulations are conducted to evaluate the performance of the proposed algorithms. Rician channel model is adopted with the K factor of 5 dB for each channel link. The path loss exponent is chosen to be 3. The distances from the STs to the PT, PU and SUs are all normalized to 1 m while the distance from the PT to the PU is set to 2 m. Thus, the primary channel is much weaker than other links. For convenience, all the SUs have the same target rate r and noise power levels at all terminals are the same. We choose the transmit SNR, which is defined as transmit power normalized by noise power, as the power metric. For each simulation result, 10^3 channel realizations are simulated. Outage occurs when (i) r_p is not supported in the primary system alone for a channel instance and (ii) r_p and r cannot be jointly supported in the case the primary system has cooperation with the STs. Since we consider a slow fading channel, one channel estimation can be utilized for a very long time, and therefore the channel estimation overhead is ignored. Unless otherwise specified, we assume that the secondary system has only one ST and SU pair; and each ST has $N = 4$ transmit antennas and all the other terminals have a single antenna. The PT power is set to be 20 dB and the PU has a rate requirement of $r_p = 3$ bps/Hz.

9.5.1 Benchmark and proposed schemes

We will evaluate the performance of the proposed information and energy cooperation schemes and compare it with benchmark schemes including:

- *No cooperation*. This refers to the case where the PU uses the entire available bandwidth and power for transmission without letting the STs to use part of the bandwidth for cooperation.
- *Information cooperation*. This is what we introduced in Section 9.3 where the STs only relay the PT's message to PU without energy cooperation.
- *Ideal cooperation*. This is the case where the ST has non-causal primary information for energy and information cooperation discussed in Section 9.4.2. It provides a performance upper bound.
- *Proposed cooperation*. This refers to the practical power splitting scheme for energy and information cooperation discussed in Section 9.4.3.

9.5.2 Outage performance due to the information cooperation

First, we examine the effect of the ST power and the number of STs on the PU's outage performance in Figure 9.7 for the information cooperation. We assume that both the PU and the SUs have a minimum rate requirements of 2 bps/Hz. Results for 'No Cooperation' in which the PU uses the entire available bandwidth and power for transmission without letting the STs to use part of the bandwidth for cooperation is provided for comparison. We consider that the PT's transmit SNR is 10 dB. Results reveal that outage occurs with a probability of over 95% for the PU but the situation greatly improves when the PT cooperates with the STs and if the ST transmit power is high. For instance, when the ST power is 20 dB, the outage probability reduces to 5%. At low ST transmit SNR (below 5 dB), cooperation does not pay off (but also does not worsen the PU outage performance) but as the ST transmit power increases beyond 12 dB, the PU's outage probability drops rapidly due to the increased diversity from the STs, showing that cooperation is a promising approach to repair the PU link. When the number of STs increases, the performance due to cooperation in general degrades and this is intuitive since more SUs' rate requirements need to be satisfied.

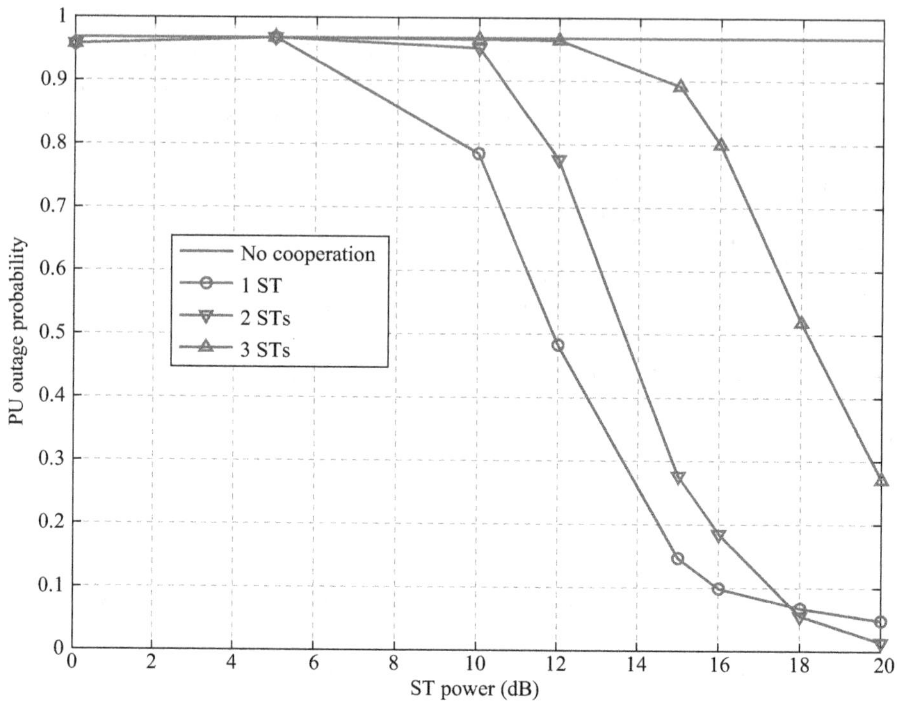

Figure 9.7 The PU outage probability against the ST power

9.5.3 Improved rate region

Next we demonstrate the achievable primary–secondary data rate region because of the cooperation when the ST power is 20 dB. In Figure 9.8, we plot the rate regions for different schemes for a random channel. It is seen that the achievable rate regions are greatly enlarged thanks to the energy cooperation. The proposed energy cooperation doubles the SU rate compared to the information-only cooperation and significantly increases the PU rate. Due to the non-causal information transfer, the ideal information and energy cooperation provides an outer bound for both the practical cooperation scheme and the information-only cooperation, which leaves room for the study of more advanced but still practical cooperation schemes.

9.5.4 Average SU rate

Next we investigate the impact of the ST's self-energy on the achievable average SU rate in Figure 9.9 when the efficiency of energy transfer, η, takes values 0.1, 0.5 and 1. Substantial rate gain is achieved using the proposed schemes compared with the case of information cooperation, especially in low-to-medium ST energy region. Even when the energy conversion efficiency η is as low as 0.1, the rate gain is over an order of magnitude. This is because the energy flow from the primary to the secondary system significantly increases the chance of cooperation when the ST is

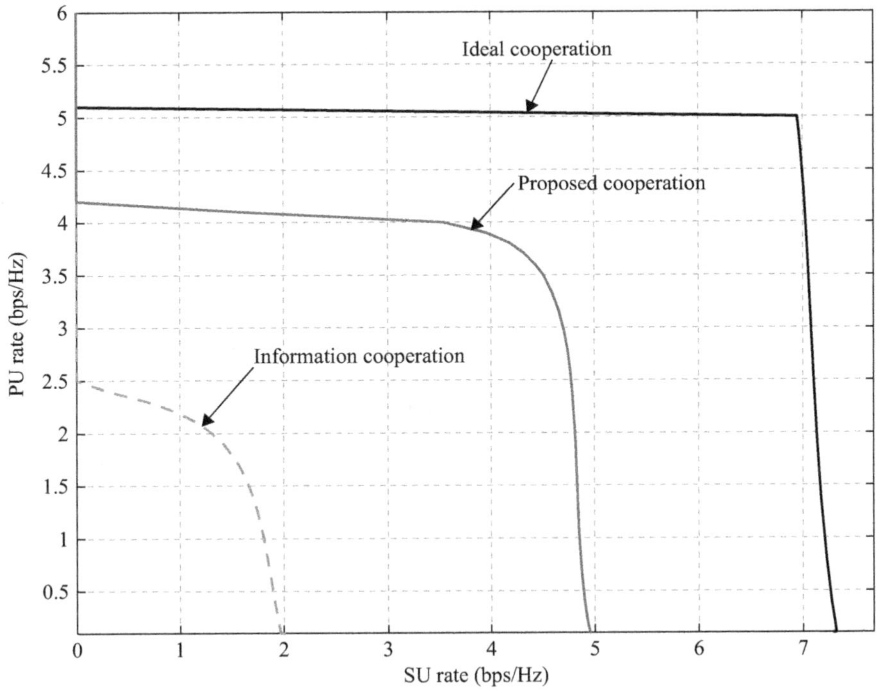

Figure 9.8 The PU-SU rate region due to cooperation. Efficiency of energy transfer $\eta = 0.5$

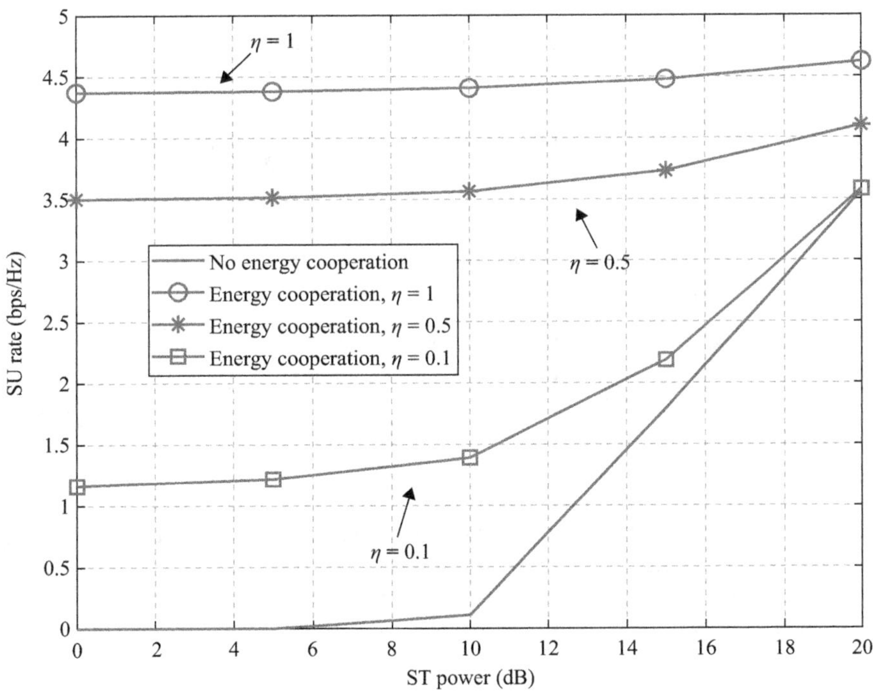

Figure 9.9 The SU rate vs. SU power, $r_p = 3$ bps/Hz

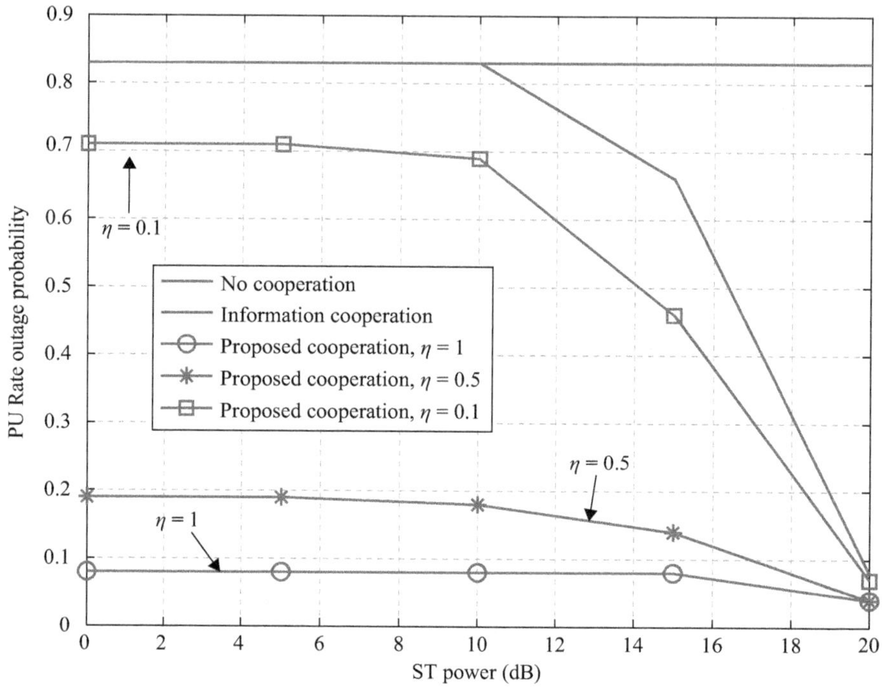

Figure 9.10 The PU rate outage vs. SU power, $r_p = 3$ bps/Hz and $r_s = 4$ bps/Hz

energy constrained. While in the high-energy region and $\eta = 0.1$, the performance of the information only cooperation scheme is close to that of the proposed cooperation based on power splitting. This is because when the ST has sufficient energy and the efficiency of energy transfer is low, harvesting energy from the primary transmission brings little performance gain.

9.5.5 Outage performance due to the additional energy cooperation

We then compare the PU rate outage performance of different schemes when the ST energy varies from 0 to 20 dB in Figure 9.10. We assume that the SU rate requirement is $r_s = 4$ bps/Hz. It is first noted that without any primary–secondary cooperation, the PU experiences rate outage with a high probability of over 80% due to the weak primary channel. If only information cooperation but no energy transfer is used, the outage probability can be reduced only in the high-energy region, which is consistent with the result in Figure 9.7. In contrast to that, when energy cooperation is also used, the outage at low-energy region is reduced tremendously. When $\eta \geq 0.5$ and the ST has 0 dB energy, the proposed cooperation reduces the outage probability to below 20%, which is a substantial improvement.

9.6 Concluding remarks and future directions

In this chapter, we have introduced new cooperation opportunities at both information and energy levels between the primary and secondary systems in cognitive radio networks. Cognitive relaying through the originally interfering links is the key to create incentives for both systems to cooperate and improve the overall utilization of the limited spectrum resources. For the information cooperation, we have discussed the optimal beamforming solution based on AF relaying protocol, and more importantly derived the distributed solution for practical implementation and analysed its reduced overhead. For the joint information and energy cooperation, we have discussed both the ideal cooperation assuming non-causal primary information available at the ST and a practical power-splitting based energy and information transfer. Substantial performance gains have been observed using the proposed joint cooperation in terms of reducing the outage of the primary systems and enlarging the rate region of both the primary and the secondary systems, therefore confirms the mutual benefit of cooperation for future cognitive networks.

We point out a few directions for future work. First, the performance gain brought by the primary–secondary cooperation in a cognitive network comes at the cost of an overhead. Specific to our proposed schemes, the additional overhead includes synchronization between the primary and the secondary systems, as well as acquiring the CSI of the STs-PT and the STs-PU links. The CSI can be estimated at the STs by overhearing the broadcasting information during the channel estimation stage of the primary system. The synchronization and CSI acquisition are also possible via controlling signalling between the two systems since both have strong incentives to cooperate. Second, the discussed relaying protocol is still based on the half-duplex mode, and when the relay is working in the full-duplex mode, it can transmit information while receiving information or harvesting energy at the same time, and this will make the cooperation more effective. Third, the STs may have the capability of harvesting energy from the ambient environment, for instance, wind, solar energy, and even other RF signals, apart from the PT. This will significantly increase the use of green energy and dramatically reduce the energy consumption of the overall system. However, the stochastic nature of the renewable sources will make the cooperation more challenging, and how to stabilize the overall cognitive relaying systems are worth of further study.

Appendix A: Proof of Theorem 9.1

Proof. Without loss of generality, $\mathbf{A}_j$ can be expressed in the form of

$$\mathbf{A}_j = \left[\mathbf{H}_j \ \ \mathbf{H}_j^{\perp}\right]\begin{bmatrix} \mathbf{a} & \mathbf{C} \\ \mathbf{d} & \mathbf{E} \end{bmatrix}\left[\mathbf{g}_j^{\dagger} \ \ (\mathbf{g}_j^{\perp})^{\dagger}\right] = \mathbf{H}_j\mathbf{a}\mathbf{g}_j^{\dagger} + \mathbf{H}_j^{\perp}\mathbf{d}\mathbf{g}_j^{\dagger} + \mathbf{H}_j\mathbf{C}\left(\mathbf{g}_j^{\perp}\right)^{\dagger} + \mathbf{H}_j^{\perp}\mathbf{E}\left(\mathbf{g}_j^{\perp}\right)^{\dagger}, \tag{9.57}$$

where $\mathbf{H}_j \triangleq [\mathbf{h}_{jp} \ \mathbf{h}_{j1} \cdots \mathbf{h}_{jM}]$ and $\mathbf{a} \in \mathbb{C}^{(M+1)\times 1}$, $\mathbf{C} \in \mathbb{C}^{(M+1)\times(N-1)}$, $\mathbf{d} \in \mathbb{C}^{(N-M-1)\times 1}$, $\mathbf{E} \in \mathbb{C}^{(N-M-1)\times(N-1)}$ are parameter vectors and matrices.

A closer observation of (9.7) reveals that the optimization will maximize $\mathbf{h}_{jp}^{\dagger}\mathbf{A}_j\mathbf{g}_j$ (can be assumed real and positive without loss of optimality) while minimizing $\left|\mathbf{h}_{jm}^{\dagger}\mathbf{A}_j\mathbf{g}_j\right|^2$, $\left\|\mathbf{h}_{jm}^{\dagger}\mathbf{A}_j\right\|^2$, $\left\|\mathbf{h}_{jp}^{\dagger}\mathbf{A}_j\right\|^2$, $\left\|\mathbf{A}_j\mathbf{g}_j\right\|^2$ and $\|\mathbf{A}_j\|^2$. It is clearly seen that $\mathbf{d}, \mathbf{E}$ do not affect $\mathbf{h}_{jp}^{\dagger}\mathbf{A}_j\mathbf{g}_j$, $\left|\mathbf{h}_{jm}^{\dagger}\mathbf{A}_j\mathbf{g}_j\right|^2$, $\left\|\mathbf{h}_{jm}^{\dagger}\mathbf{A}_j\right\|^2$ and $\left\|\mathbf{h}_{jp}^{\dagger}\mathbf{A}_j\right\|^2$, due to the appearance of $\mathbf{H}_j^{\perp}$. Setting them to zero also reduces terms $\left\|\mathbf{A}_j\mathbf{g}_j\right\|^2$ and $\|\mathbf{A}_j\|^2$ which are to be minimized. On the other hand, $\mathbf{C}$ has no impact on $\mathbf{h}_{jp}^{\dagger}\mathbf{A}_j\mathbf{g}_j$, $\left|\mathbf{h}_{jm}^{\dagger}\mathbf{A}_j\mathbf{g}_j\right|^2$, and $\left\|\mathbf{A}_j\mathbf{g}_j\right\|^2$, while setting them to zero can reduce $\left\|\mathbf{h}_{jm}^{\dagger}\mathbf{A}_j\right\|^2$, $\left\|\mathbf{h}_{jp}^{\dagger}\mathbf{A}_j\right\|^2$ and $\|\mathbf{A}_j\|^2$, which are to be minimized. As a consequence, without loss of optimality, $\mathbf{C} = \mathbf{0}, \mathbf{d} = \mathbf{0}, \mathbf{E} = \mathbf{0}$ and then we reach the optimal structure in the theorem. □

Appendix B: Proof of Proposition 9.1

The maximum PU rate is achieved when SU rate is zero and $\mathbf{w}_s = \mathbf{0}$. In this case, $q_p = P_{s0} + P_p\beta\eta\|\mathbf{g}\|^2$ and the optimal β that maximizes the PU rate is given by

$$\beta^* = \arg_\beta \max_{0\le\beta\le 1} \left(\sqrt{(1-\beta)P_p}|h_p| + \sqrt{P_{s0} + \beta\eta P_p}\|\mathbf{h}_{sp}\|\right). \tag{9.58}$$

Setting the derivative to zero leads to the unique critical point

$$\bar{\beta} = \frac{P_p\eta^2\|\mathbf{g}\|^4\|\mathbf{h}_{sp}\|^2 - P_{s0}|h_p|^2}{P_p\eta^2\|\mathbf{g}\|^4\|\mathbf{h}_{sp}\|^2 + P_p\eta\|\mathbf{g}\|^2|h_p|^2}. \tag{9.59}$$

Then the optimal β is $\beta^* = \left[\bar{\beta}\right]_0^1$. Accordingly, the maximum PU rate in (9.40) can be derived.

Appendix C: Closed-form solution to Problem (9.60)

Our aim here is to find the closed-form solution to the general maximization problem below:

$$\max_{\mathbf{w}_1,\mathbf{w}_2} \frac{\left|\mathbf{h}_2^{\dagger}\mathbf{w}_2\right|^2}{1+\left|\mathbf{h}_2^{\dagger}\mathbf{w}_1\right|^2}$$

$$\text{s.t.} \quad \frac{\left|\mathbf{h}_1^{\dagger}\mathbf{w}_1\right|^2}{1+\left|\mathbf{h}_1^{\dagger}\mathbf{w}_2\right|^2} \ge \gamma_1, \quad \|\mathbf{w}_1\|^2 + \|\mathbf{w}_2\|^2 \le P_C, \tag{9.60}$$

where $\mathbf{h}_1, \mathbf{h}_2$ are $N \times 1$ vectors and γ_1, P_C are positive scalars. This problem has the following physical meaning. Consider a MISO broadcast system with an N-antenna

BS and two single-antenna users. The channels from the BS to user 1 and user 2 are $\mathbf{h}_1$ and $\mathbf{h}_2$, respectively. The noise powers at users are assumed to be one, otherwise, the channel can be normalized with the noise power. Suppose the BS has a total power constraint P_C and user 1 has a SINR constraint γ_1, then this problem has the interpretation of maximization of user 2's SINR. Suppose its optimal objective value is γ_2^*. To find the optimal solution to (9.60), we first consider the following weighted sum power minimization problem:

$$\min_{\mathbf{w}_1,\mathbf{w}_2} \quad \|\mathbf{w}_1\|^2 + \|\mathbf{w}_2\|^2$$
$$\text{s.t.} \quad \frac{\left|\mathbf{h}_1^\dagger \mathbf{w}_1\right|^2}{1+\left|\mathbf{h}_1^\dagger \mathbf{w}_2\right|^2} \geq \gamma_1, \quad \frac{\left|\mathbf{h}_2^\dagger \mathbf{w}_2\right|^2}{1+\left|\mathbf{h}_2^\dagger \mathbf{w}_1\right|^2} \geq \gamma_2. \tag{9.61}$$

It can be validated that if we set $\gamma_2 = \gamma_2^*$ in (9.61), then its optimal objective value is P_C and vice versa. So we can focus on (9.61) in order to characterize the solution to (9.60). The dual problem of (9.61) can be derived as

$$\max_{\lambda_1,\lambda_2 \geq 0} \quad \lambda_1 + \lambda_2$$
$$\text{s.t.} \quad \mathbf{I} + \lambda_2 \mathbf{h}_2 \mathbf{h}_2^\dagger \succeq \frac{\lambda_1}{\bar{\gamma}_1} \mathbf{h}_1 \mathbf{h}_1^\dagger, \quad \bar{\gamma}_1 \triangleq \gamma_1$$
$$\mathbf{I} + \lambda_1 \mathbf{h}_1 \mathbf{h}_1^\dagger \succeq \frac{\lambda_2}{\gamma_2} \mathbf{h}_1 \mathbf{h}_2^\dagger, \tag{9.62}$$

where λ_1 and λ_2 are dual variables. The two linear matrix inequality constraints uniquely determine λ_1 and λ_2:

$$\lambda_1 = \frac{\bar{\gamma}_1}{\mathbf{h}_1^\dagger \left(\mathbf{I} + \lambda_1 \bar{\mathbf{h}}_2 \mathbf{h}_2^\dagger\right)^{-1} \mathbf{h}_1}, \quad \lambda_2 = \frac{\gamma_2}{\mathbf{h}_2^\dagger \left(\mathbf{I} + \lambda_1 \mathbf{h}_1 \mathbf{h}_1^\dagger\right)^{-1} \mathbf{h}_2}. \tag{9.63}$$

Using matrix inversion lemma and define $\rho^2 \triangleq \frac{\left|\mathbf{h}_1^\dagger \mathbf{h}_2\right|^2}{\|\mathbf{h}_1\|^2 \|\mathbf{h}_2\|^2}$, we have

$$\begin{aligned}
\lambda_1 &= \frac{\bar{\gamma}_1 \left(1 + \lambda_2 \|\mathbf{h}_2\|^2\right)}{\|\mathbf{h}_1\|^2 + \lambda_2 \left(\|\mathbf{h}_2\|^2 \|\mathbf{h}_1\|^2 - \left\|\mathbf{h}_2^\dagger \mathbf{h}_1\right\|^2\right)} \\
&= \frac{\bar{\gamma}_1 \left(1 + \lambda_2 \|\mathbf{h}_2\|^2\right)}{\|\mathbf{h}_1\|^2 \left(1 + \lambda_2 \|\mathbf{h}_2\|^2 (1-\rho^2)\right)}, \\
\lambda_2 &= \frac{\gamma_2 \left(1 + \lambda_1 \|\mathbf{h}_1\|^2\right)}{\|\mathbf{h}_2\|^2 + \lambda_1 \left(\|\mathbf{h}_2\|^2 \|\mathbf{h}_1\|^2 - \left\|\mathbf{h}_2^\dagger \mathbf{h}_1\right\|^2\right)} \\
&= \frac{\gamma_2 \left(1 + \lambda_1 \|\mathbf{h}_1\|^2\right)}{\|\mathbf{h}_2\|^2 \left(1 + \lambda_1 \|\mathbf{h}_1\|^2 (1-\rho^2)\right)}.
\end{aligned} \tag{9.64}$$

Remember we also have a power equation below:

$$\lambda_1 + \lambda_2 = P_C. \tag{9.65}$$

It is observed that $\lambda_1, \lambda_2, \gamma_2$ should satisfy and uniquely determined by the above three equations (9.63–9.65), so the analytical solutions can be found. Define $A = \|\mathbf{h}_1\|^2\|\mathbf{h}_2\|^2(1-\rho^2)$, $B = -\left(\|\mathbf{h}_2\|^2\bar{\gamma}_1 + P_C\|\mathbf{h}_1\|^2\|\mathbf{h}_2\|^2(1-\rho^2) + \|\mathbf{h}_1\|^2\right)$, and $C = \left(P_C\|\mathbf{h}_2\|^2 + 1\right)\bar{\gamma}_1$. Then from (9.64), we have

$$f(\lambda_1) \triangleq A\lambda_1^2 + B\lambda_1 + C = 0. \tag{9.66}$$

Since $A > 0$, $C > 0$, and $B < 0, f(\lambda) = 0$ has positive two roots. Because $f(P_C) < 0$, we know that the optimal λ_1 corresponds to the minimum root. Once λ_1 is found, λ_2 and γ_2 can be easily derived from (9.64) and (9.65).

References

[1] J. Mitola and G. Q. Maguire, "Cognitive radio: Making software radios more personal," *IEEE Pers. Commun.*, vol. 6, no. 6, pp. 13–18, August 1999.

[2] R. Zhang and C. K. Ho, "MIMO broadcasting for simultaneous wireless information and power transfer," *IEEE Trans. Wireless Commun.*, vol. 12, no. 5, pp. 1989–2001, May 2013.

[3] N. Devroye, P. Mitran, and V. Tarokh, "Achievable rates in cognitive radio channels," *IEEE Trans. Inform. Theory*, vol. 52, no. 5, pp. 1813–1827, May 2006.

[4] J. Lv, R. Blasco-Serrano, E. A. Jorswieck, R. Thobaben, and A. Kliks, "Optimal beamforming in MISO cognitive channels with degraded message sets," in *Proc. IEEE Wireless Communications Network Conference (WCNC)*, Paris, France, 1–4 April 2012, pp. 538–543.

[5] J. Lv and E. A. Jorswieck, "Spatial shaping in cognitive system with coded legacy transmission," in *Proc. International ITG Workshop on Smart Antennas (WSA)*, Aachen, Germany, 24–25 February 2011, pp. 1–6.

[6] O. Simeone, I. Stanojev, S. Savazzi, Y. Bar-Ness, U. Spagnolini, and R. Pickholtz, "Spectrum leasing to cooperating secondary ad hoc networks," *IEEE J. Sel. Areas Commun.*, vol. 26, no. 1, pp. 203–213, January 2008.

[7] S. H. Song, M. O. Hasna, and K. B. Letaief, "Prior zero-forcing for cognitive relaying," *IEEE Trans. Wireless Commun.*, vol. 12, no. 2, pp. 938–947, February 2013.

[8] G. Zheng, S. H. Song, K. K. Wong, and B. Ottersten, "Cooperative cognitive networks: Optimal, distributed and low-complexity algorithms," *IEEE Trans. Signal Process.*, vol. 61, no. 11, pp. 2778–2790, June 2013.

[9] B. Cao, L. X. Cai, H. Liang, *et al.*, "Cooperative cognitive radio networking using quadrature signaling," in *Proc. INFOCOM*, Orlando, FL, March 2012.

[10] Q. Zhang, B. Cao, Y. Wang, N. Zhang, X. Lin, and L. Sun, "On exploiting polarization for energy-harvesting enabled cooperative cognitive radio networking," *IEEE Wireless Commun.*, vol. 20, no. 4, pp. 116–124, August 2013.

[11] J. Xu, Y. H. Guo, and R. Zhang, "CoMP meets energy harvesting: A new communication and energy cooperation paradigm," in *IEEE Global Communications Conference (Globecom)*, Atlanta, GA, December 2013, pp. 2508–2513.
[12] Y. K. Chia, S. Sun, and R. Zhang, "Energy cooperation in cellular networks with renewable powered base stations," in *IEEE Wireless Communications and Networking Conference (WCNC)*, Shanghai, China, April 2013, pp. 6996–7010.
[13] POWERCAST Frequently Asked Questions. Available at: *http://www.powercastco.com/frequently-asked-questions/*.
[14] A. Parks, A. Sample, Y. Zhao, and J. R. Smith, "A wireless sensing platform utilizing ambient RF energy," in *IEEE Topical Meeting on Wireless Sensors and Sensor Networks (WiSNET)*, Austin, TX, January 2013, pp. 331–333.
[15] L. R. Varshney, "Transporting information and energy simultaneously," in *Proc. IEEE International Symposium on. Information Theory*, Toronto, Canada, 6–11 July 2008, pp. 1612–1616.
[16] P. Grover and A. Sahai, "Shannon meets Tesla: Wireless information and power transfer," in *Proc. IEEE International Symposium on Information Theory*, Austin, TX, June 2010, pp. 2363–2367.
[17] L. Liu, R. Zhang, and K.-C. Chua, "Wireless information transfer with opportunistic energy harvesting," *IEEE Trans. Wireless Commun.*, vol. 12, no. 1, pp. 288–300, January 2013.
[18] I. Krikidis, S. Timotheou, and S. Sasaki, "RF energy transfer for cooperative networks: Data relaying or energy harvesting?" *IEEE Commun. Lett.*, vol. 16, no. 11, pp. 1772–1775, November 2012.
[19] A. A. Nasir, X. Zhou, S. Durrani, and R. A. Kennedy, "Relaying protocols for wireless energy harvesting and information processing," *IEEE Trans. Wireless Commun.*, vol. 12, no. 7, pp. 3622–3636, July 2013.
[20] M. Bengtsson and B. Ottersten, "Optimal and suboptimal transmit beamforming," in L. C. Godara (Ed.), *Handbook of Antennas in Wireless Communications*, CRC Press, Boca Raton, FL, August 2001.
[21] S. Boyd and L. Vandenberghe, *Convex Optimization*, Cambridge University Press, Cambridge, 2004.
[22] S. Ulukus and R. D. Yates, "Adaptive power control and MMSE interference suppression," *ACM Wireless Net.*, vol. 4, no. 6, pp. 489–496, November 1998.
[23] M. Schubert and H. Boche, "A generic approach to QoS-based transceiver optimization," *IEEE Trans. Commun.*, vol. 55, no. 8, pp. 1557–1566, August 2007.

Chapter 10

Relaying in non-ideal conditions

Yunfei Chen

10.1 Introduction

Wireless relaying has been proposed as a promising technology to achieve either diversity gain via a distributed multiple-input multiple-output setting or extended network coverage without the need for extra infrastructure. In practice, this is often performed under non-ideal channel and signal conditions. In this chapter, the effects of various non-ideal conditions on the performance of wireless relaying will be examined. We focus on amplify-and-forward (AF) relaying, where the relay node amplifies and forwards the signal received from the source node to the destination node without any decoding. The examined non-ideal conditions include feedback delay, link correlation, non-Gaussian interference, and wireless power.

10.2 Relaying with feedback delay

Feedback delay occurs when relaying is performed in a time-division manner. For example, in a relaying system, a pilot frame can be transmitted first to estimate the channel gains of different links. Then, a control frame will be sent to select the best link or links for relaying, based on the estimated channel gains. Finally, a data frame will be sent from the source node to the destination node through the relay node or nodes for information delivery. During this process, however, the channel gain could vary with time in a practical setting such that the channel gain estimated using the pilot frame may not be the same as the actual channel gain during relay selection or signal demodulation. Consequently, feedback delay could degrade the relaying performance by either providing an inaccurate channel gain for relay selection or for signal demodulation.

Thus, this section examines the performances of multi-hop AF relaying with generalized relay selection, where multiple relays are selected for relaying instead of only one relay. The selection criteria will be obtained by extending those proposed in Reference 1. The result will be compared with maximum ratio combining (MRC) where all relays are used in AF relaying. The comparison is conducted in terms of the bit error rate (BER).

Consider a multi-hop AF system with one source node, one destination node, J relaying links, and I hops in each link, where $R_{i,j}$ represents the relay at the ith hop in the jth link, $i = 1, 2, \ldots, I$ index the hops, and $j = 1, 2, \ldots, J$ index the links. No direct link exists between source and destination due to obstacles. Denote the channel gain estimated using the pilot frames as $\tilde{g}_{i,j}$, and the delayed version of $\tilde{g}_{i,j}$ in the signal demodulation with a time delay of τ as $g_{i,j}$. Both are assumed to be complex Gaussian random variables with mean zero and common variance. The relaying transmission happens in I time slots. Thus, the received signal in the jth link of the ith hop is

$$r_{i,j} = g_{i,j}\tilde{\beta}_{i-1,j}r_{i-1,j} + w_{i,j}, \tag{10.1}$$

where $i = 1, 2, \ldots, I$, $r_{0,j} = s$ is the signal from the source, $\tilde{\beta}_{0,j} = \sqrt{E_1}$ is the transmitted signal energy, and $w_{i,j}(t)$ is the additive white Gaussian noise with mean zero and variance $\sigma_{i,j}^2$. In this work, we consider variable gain with $\tilde{\beta}_{i-1,j} = \sqrt{\frac{E_{i,j}}{E_{i-1,j}|\tilde{g}_{i-1,j}|^2+\sigma_{i-1,j}^2}}$ $(i = 2, 3, \ldots, I)$, as well as fixed gain with $\tilde{\beta}_{i-1,j} = 1$, where $E_{i,j}$ is the transmitted energy at the jth relay link of the ith hop. Assume $E\{|\tilde{g}_{i,j}|^2\} = E\{|g_{i,j}|^2\} = \Omega_{i,j}$. Also, let

$$\tilde{\lambda}_{i,j} = \frac{|\tilde{g}_{i,j}|^2 E_{i,j}}{\sigma_{i,j}^2}, \tag{10.2}$$

be the instantaneous signal-to-noise ratio (SNR) used in generalized relay selection. Then, for fixed gain, the instantaneous end-to-end SNR for the jth link can be shown as

$$\tilde{\lambda}_j = E_1\left[\sum_{i=1}^{I}\prod_{n=1}^{i}\left(\frac{E_{i,j}}{\tilde{\lambda}_{i,j}\sigma_{i-1,j}^2}\right)\right]^{-1}, \tag{10.3}$$

and for variable gain, it becomes [2]

$$\tilde{\lambda}_j = \left[\prod_{i=1}^{I}\left(1 + \frac{1}{\tilde{\lambda}_{i,j}}\right) - 1\right]^{-1}. \tag{10.4}$$

In the demodulation, the received signal at the destination will be multiplied with the channel estimate to remove the fading effect as

$$dv = \sum_{m=1}^{U} r_{I,\hat{J}_m} \times \prod_{i=1}^{I} \tilde{g}_{i,\hat{J}_m}^*, \tag{10.5}$$

for fixed gain and

$$dv = \sum_{m=1}^{U} r_{I,\hat{J}_m} \times \prod_{i=1}^{I} \tilde{g}_{i,\hat{J}_m}^* \prod_{i=2}^{I} \tilde{\beta}_{i-1}, \tag{10.6}$$

for variable gain, where dv is the decision variable. The relationship between $g_{i,j}$ and $\tilde{g}_{i,j}$ can be written as

$$g_{i,j} = \tilde{g}_{i,j} \times \rho_{i,j} + g_{i,j}' \times \sqrt{1 - \rho_{i,j}^2}, \tag{10.7}$$

where $g_{i,j}'$ is an independent Gaussian random variable and $\rho_{i,j}$ is the correlation determined by the feedback delay. If the Jakes' model is used, one has $\rho_{i,j} = J_0(2\pi f_d \tau_{i,j})$, where $J_0(\cdot)$ is the zeroth order Bessel function of the first kind, f_d is the normalized

Doppler shift, and $\tau_{i,j}$ is the feedback delay. Finally, assume that $\hat{J}_1, \ldots, \hat{J}_U$ in (10.5) and (10.6) are the links that have been selected for forwarding. The selection criteria will be discussed next.

The selection schemes in Reference 1 only select the best relay in the two-hop system. They can be extended to select the U best relays in an I-hop system. They are discussed as follows. First, one can use the full relay selection scheme that selects U relay links with the highest instantaneous end-to-end SNRs, or

$$\left\{\hat{J}_1, \ldots, \hat{J}_U\right\} = \arg\left\{\max_{j=1,2,\ldots,J}\left\{\tilde{\lambda}_j\right\}\right\}. \tag{10.8}$$

Second, one can use the partial relay selection scheme that selects U relay links with the highest first-hop instantaneous SNRs, or

$$\left\{\hat{J}_1, \ldots, \hat{J}_U\right\} = \arg\left\{\max_{j=1,2,\ldots,J}\left\{\tilde{\lambda}_{1,j}\right\}\right\}. \tag{10.9}$$

Third, one can use the partial relay selection scheme that selects U relay links with the highest ith hop instantaneous SNRs, or

$$\left\{\hat{J}_1, \ldots, \hat{J}_U\right\} = \arg\left\{\max_{j=1,2,\ldots,J}\left\{\tilde{\lambda}_{i,j}\right\}\right\}, \tag{10.10}$$

where $i = 2, 3, \ldots, I$. Finally, one can use the partial relay selection scheme that selects U relay links with the highest received amplitudes instead of the instantaneous SNR, or

$$\left\{\hat{J}_1, \ldots, \hat{J}_U\right\} = \arg\left\{\max_{j=1,2,\ldots,J}\left\{|\tilde{r}_{i,j}|\right\}\right\}. \tag{10.11}$$

In the above equations, $\tilde{\lambda}_j$, $\tilde{\lambda}_{i,j}$ are determined by (10.3) (or (10.4)), and (10.2), respectively, $U \leq J$ and $\tilde{r}_{i,j}$ are the samples used to obtain the channel gain estimate $\tilde{g}_{i,j}$. Denote the scheme in (10.8) as $\max\{\tilde{\lambda}_j\}$, the scheme in (10.9) as $\max\{\tilde{\lambda}_{1,j}\}$, the scheme in (10.10) as $\max\{\tilde{\lambda}_{i,j}\}$, and the scheme in (10.11) as $\max\{|\tilde{r}_{i,j}|\}$. Note also that, in this case, we assume that channel estimation is performed at the same time as relay selection before data transmission. Consequently, the channel gain used in relay selection is assumed to be the same as the estimated channel gain. Thus, we only examine the effect of feedback delay on the signal demodulation. An extension could be made by examining its effect on both relay selection and signal demodulation if one uses another version of the delayed channel gain in (10.3) and (10.4).

One can see that the $\max\{\tilde{\lambda}_j\}$, $\max\{\tilde{\lambda}_{1,j}\}$ and $\max\{\tilde{\lambda}_{i,j}\}$ schemes use the estimated channel gain $\tilde{g}_{i,j}$ or the estimated instantaneous SNR $\tilde{\gamma}_{i,j}$ for relay selection, while the $\max\{|\tilde{r}_{i,j}|\}$ scheme uses the delayed received signal amplitude for relay selection. In the signal demodulation, they also use the outdated channel gain $\tilde{g}_{i,j}$ in (10.5) and (10.6). Moreover, for variable gain relaying, since the amplification factor $\tilde{\beta}_i$ requires the use of the outdated channel gain $\tilde{g}_{i,j}$ due to feedback delay, it is expected to be more vulnerable to feedback delay than fixed gain relaying.

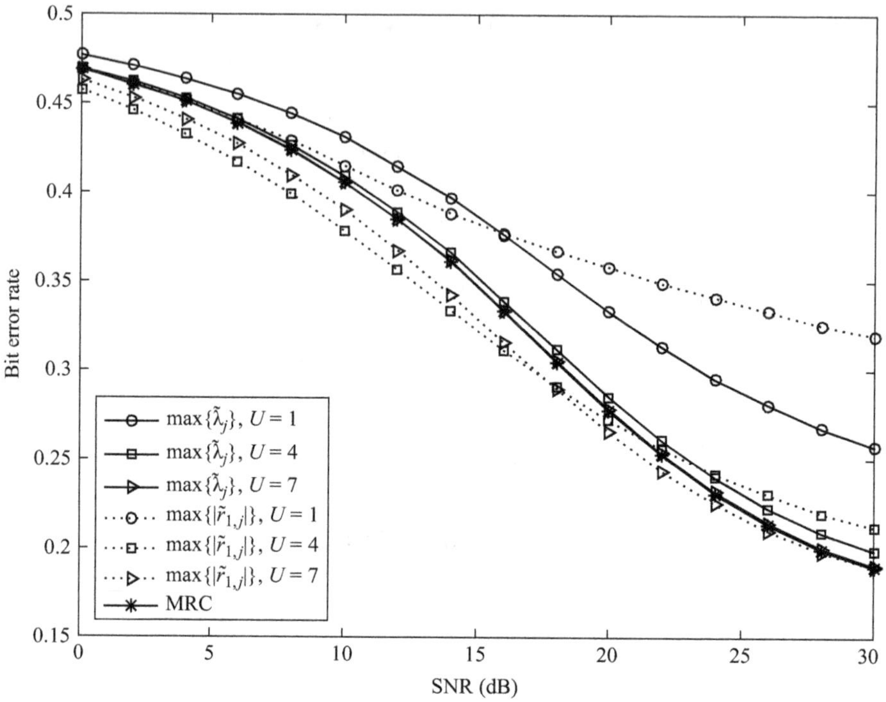

Figure 10.1 Comparison of $\max\{\tilde{\lambda}_j\}$ *and* $\max\{|\tilde{r}_{I,j}|\}$ *schemes when* $ue_1 = ue_2 = ue_3 = 1$, $\rho_1 = 0.6$, $\rho_2 = 0.7$, $\rho_3 = 0.8$, $J = 9$ *in Rayleigh fading channels for a four-hop AF relaying system with fixed gain*

In the following, computer simulation is performed to compare different generalized relay selection schemes when feedback delay occurs. In the simulation, we use binary phase shift keying (BPSK). Also, the parameters are chosen as $\Omega_1 = \Omega_2 = \cdots = \Omega_I = 1, \sigma_1^2 = \sigma_2^2 = \cdots = \sigma_I^2 = 1, E_1 = \frac{E}{\lambda}, E_2 = \frac{E*ue_1}{\lambda}, \ldots, E_I = \frac{E*ue_{I-1}}{\lambda}$, where $\lambda = K(1 + ue_1 + \cdots + ue_{I-1})$, $E = (E_1 + E_2 + \cdots + E_I)J$ is the total energy across all hops and links and $ue_1 = \frac{E_2}{E_1}, ue_2 = \frac{E_3}{E_1}, \ldots, ue_{I-1} = \frac{E_I}{E_1}$.

Figure 10.1 compares the $\max\{\tilde{\lambda}_j\}$ and $\max\{|\tilde{r}_{1,j}|\}$ schemes when the generalized relay selection method chooses different numbers of links for an AF four-hop system with Rayleigh fading. We first examine the case when the amplification factor is fixed to 1. We set $ue_1 = ue_2 = ue_3 = 1, \rho_1 = 0.6, \rho_2 = 0.7, \rho_3 = 0.8, J = 9$. Several observations can be made. First, for $\max\{\tilde{\lambda}_j\}$, one notices that better performances can be achieved with more relays selected. The best performance is obtained when MRC is used that selects all relays. However, this is not the case for the $\max\{|\tilde{r}_{1,j}|\}$ scheme, where more selected relays do not necessarily give better performance. In fact, the best performance is not achieved by MRC at $U = 9$, but at $U = 7$ when only seven relays are used. When no feedback delay occurs in the channel estimation and relay selection, the channel condition is ideal such that the larger the number of selected relays is, the

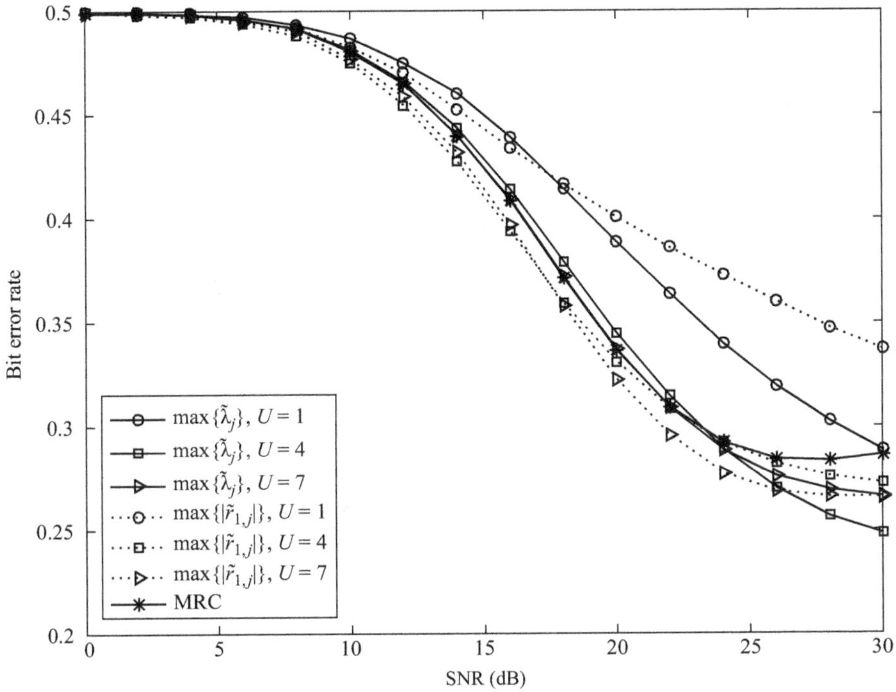

Figure 10.2 Comparison of $\max\{\tilde{\lambda}_j\}$ *and* $\max\{|\tilde{r}_{1,j}|\}$ *schemes when* $ue_1 = ue_2 = ue_3 = 1$, $\rho_1 = 0.6$, $\rho_2 = 0.7$, $\rho_3 = 0.8$, $J = 9$ *in Rayleigh fading channels for a four-hop AF relaying system with variable gain*

more energy the system will collect to achieve better performance. However, when feedback delay occurs in a realistic communication channel, although more selected relays may lead to higher diversity gain, they also suffer more from feedback delay due to an increased number of relays. Thus, the performance degrades when U is too large. For $\max\{|\tilde{r}_{1,j}|\}$, from Figure 10.1, one sees that it only needs to combine the first seven maximums of $|\tilde{r}_{1,j}|$ in order to optimize its BER performance. For $\max\{\tilde{\lambda}_j\}$, choosing the first seven maximums gives a performance very close to MRC but it is not optimum anyway.

Figure 10.2 also compares the $\max\{\tilde{\lambda}_j\}$ and $\max\{|\tilde{r}_{1,j}|\}$ schemes when the generalized relay selection method chooses different numbers of links for an AF four-hop system with Rayleigh fading. However, in Figure 10.2, the amplification factor changes with the estimated channel gain. Similar observations can be made from Figure 10.1. Furthermore, comparing Figure 10.2 with Figure 10.1, the performance gain of the $\max\{|\tilde{r}_{1,j}|\}$ scheme using a combination of seven relays over MRC is even more significant at large SNRs. In this case, even for $\max\{\tilde{\lambda}_j\}$, MRC is not the best scheme. Specifically, $U = 4$ and $U = 7$ perform better, especially for large SNRs. The reason is that variable gain relaying is more sensitive to the feedback delay due to

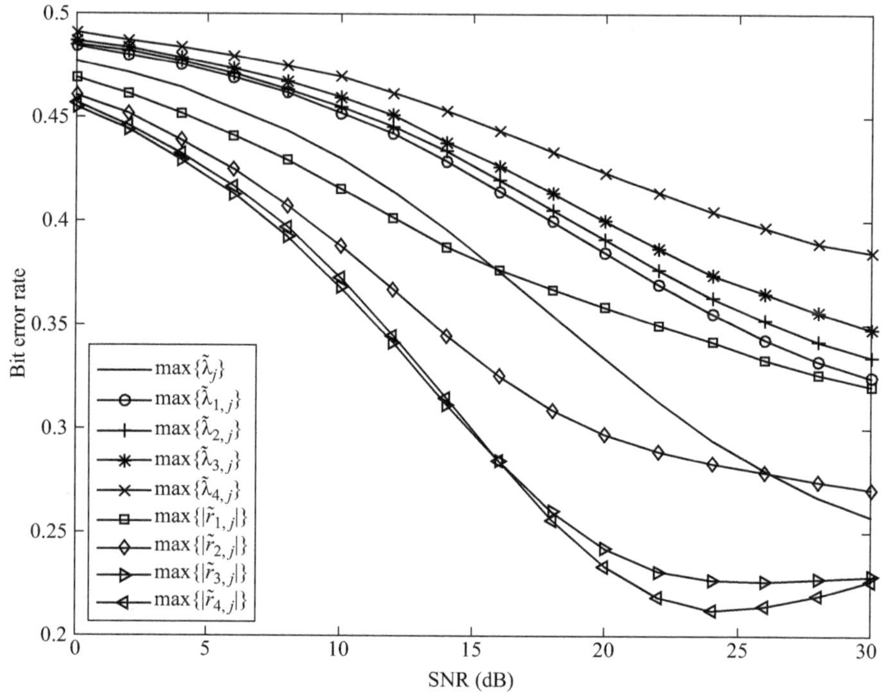

Figure 10.3 Comparison of $\max\{\tilde{\lambda}_{i,j}\}$ *and* $\max\{|\tilde{r}_{i,j}|\}$ *when* $ue_1 = ue_2 = ue_3 = 1$, $\rho_1 = 0.6$, $\rho_2 = 0.7$, $\rho_3 = 0.8$, $J = 9$, $U = 1$ *in Rayleigh fading channels for a four-hop AF relaying system with fixed gain*

the additional use of estimated channel state information in the amplification factor. As a result, it is easier to have more performance losses when using more relays due to the feedback delay, than fixed gain.

Figure 10.3 compares $\max\{\tilde{\lambda}_{i,j}\}$ and $\max\{|\tilde{r}_{i,j}|\}$ for a four-hop AF relaying system with fixed gain in Rayleigh fading channels. From Figure 10.3, it can be seen that the selection scheme that chooses relays using the received signal amplitude outperforms the one using the instantaneous SNR for a multi-hop system, which agrees with what was observed in Reference 1 for a dual-hop system. For example, $\max\{|\tilde{r}_{2,j}|\}$, $\max\{|\tilde{r}_{3,j}|\}$ and $\max\{|\tilde{r}_{4,j}|\}$ outperform the SNR-based selection schemes in most cases, including the selection criterion based on the end-to-end SNR, especially when the operating SNR is small. Another observation is that the relaying performance degrades when the hop index i increases for $\max\{\tilde{\lambda}_{i,j}\}$, while it improves when the hop index i increases for $\max\{|\tilde{r}_{i,j}|\}$. In particular, the performance of $\max\{|\tilde{r}_{1,j}|\}$ scheme is worse than the performance of $\max\{|\tilde{r}_{2,k}|\}$, which is worse than the performance of $\max\{|\tilde{r}_{3,j}|\}$ and so on, while the performance of $\max\{\tilde{\lambda}_{1,j}\}$ is better than the performance of $\max\{\tilde{\lambda}_{2,j}\}$, which is better than the performance of $\max\{\tilde{\lambda}_{3,j}\}$. Among them, the $\max\{|\tilde{r}_{4,k}|\}$ and $\max\{|\tilde{r}_{3,k}|\}$ schemes have the best performances in

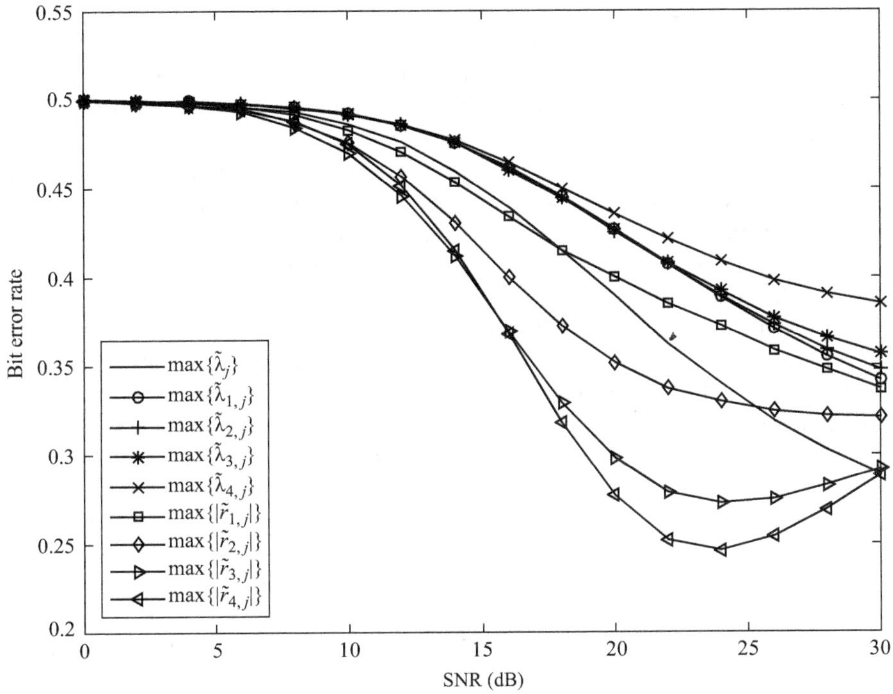

Figure 10.4 Comparison of $\max\{\tilde{\lambda}_{i,j}\}$ *and* $\max\{|\tilde{r}_{i,j}|\}$ *at* $ue_1 = ue_2 = ue_3 = 1$, $\rho_1 = 0.6$, $\rho_2 = 0.7$, $\rho_3 = 0.8$, $J = 10$, $U = 1$ *in Rayleigh fading channels for a four-hop AF relaying system with variable gain*

all cases. Figure 10.4 compares the performances for variable gain under the same conditions. Again, the selection schemes based on the received amplitude have large performance gains over that based on the SNR.

In summary, this section has examined the performance of multi-hop AF relaying using generalized relay selection by considering the effect of feedback delay. Extensive computer simulations have shown that the full relay selection based on the end-to-end SNR is not always the best, MRC is not always the better than relay selection and SNR-based selection is not always better than amplitude-based selection. It has also been shown that variable gain relaying is more sensitive to feedback delay than fixed gain relaying.

10.3 Relaying with link correlation

In an ideal channel condition, the link between the source and the relay, the link between the relay and the destination or the direct link between the source and the destination are often assumed to be independent. In practice, however, these links may actually be correlated due to some common scatterers or due to a specific geometric structure of the links. For example, experiments performed in Reference 3 showed that, if a relaying system is set up in a corridor, the source-to-relay link and the relay-to-destination link will be correlated as the signal goes along the corridor. The same observation was made in Reference 4, where the source-to-destination link and the relay-to-destination link may be correlated due to the same scatters affecting the destination, or the source-to-relay link and the source-to-destination link may be correlated due to common scatterers affecting the source. This correlation will degrade the relaying performance, as the full diversity gain of relaying is only achieved when the links are independent.

This section examines the effect of link correlation on the AF relaying performance. To do this, exact analytical expressions for the moment-generating functions (MGFs) of the end-to-end SNR in a two-hop three-terminal AF relaying system with correlated links will be derived for Nakagami-*m* fading channels, assuming that each node has a single antenna and that different links have the same integer value of the Nakagami-*m* parameter. Using these MGFs, the error rate performance of AF relaying can be easily obtained. Both fixed gain relaying and variable gain relaying are studied.

Consider a three-terminal relaying system where the source sends its signal to the destination directly as well as through a relay. For simplicity, only a two-hop system with a single relay is used. Each terminal has a single antenna. There have been works on terminals using multiple antennas, including correlated antennas. However, this is beyond the scope of this chapter, although extension to the case of multiple antennas is also possible. All the links suffer from Nakagami-*m* fading. It can be easily shown that the combined end-to-end SNR for this three-terminal relaying system is given by [5]

$$\lambda_T = \lambda_{sd} + \lambda_r, \tag{10.12}$$

where λ_{sd} is the SNR for the direct link between source and destination and λ_r is the end-to-end SNR of the relaying link from source to destination through the relay. The value of λ_r depends on the relaying protocol used. If variable gain relaying is used, one has [6]

$$\lambda_r = \frac{\lambda_{sr}\lambda_{rd}}{\lambda_{sr} + \lambda_{rd} + 1}, \tag{10.13}$$

where λ_{sr} is the SNR for the source-to-relay link and λ_{rd} is the SNR for the relay-to-destination link. In many works, however, due to the complexity of this expression, the exact SNR is often approximated as [5]

$$\lambda_r \approx \frac{\lambda_{sr}\lambda_{rd}}{\lambda_{sr} + \lambda_{rd}}, \tag{10.14}$$

which is tight when the link SNR is large. If fixed gain relaying is used, one has [7]

$$\lambda_r = \frac{\lambda_{sr}\lambda_{rd}}{\lambda_{rd} + c}, \tag{10.15}$$

where $c = \frac{P_r}{\beta^2 N_r}$, P_r is the transmitted signal power of the relay, N_r is the noise power at the relay, and β is the amplification factor. For fixed gain relaying, β is an arbitrary constant such that c can also take an arbitrary value [7–9]. Further, one has the probability density function (PDF) as

$$f(\lambda_i) = \frac{\lambda_i^{m_i-1}}{\Gamma(m_i)\omega_i^{m_i}} e^{-\frac{\lambda_i}{\omega_i}}, \tag{10.16}$$

where m_i is the Nakagami-m parameter, ω_i is the average SNR, $i = sd$ for the source-to-destination link, $i = sr$ for the source-to-relay link, and $i = rd$ for the relay-to-destination link. It was also derived in Reference 10 that the bivariate Gamma PDF is written as

$$\begin{aligned} f(\lambda_i, \lambda_j) &= \frac{e^{-\frac{\lambda_i}{(1-\rho)\omega_i} - \frac{\lambda_j}{(1-\rho)\omega_j}}}{\Gamma(m)\omega_i\omega_j(1-\rho)\left(\sqrt{\omega_i\omega_j\rho}\right)^{m-1}} \\ &\quad \times I_{m-1}\left(\frac{2\sqrt{\rho}\sqrt{\lambda_i\lambda_j}}{\sqrt{\omega_i\omega_j}(1-\rho)}\right)(\lambda_i\lambda_j)^{\frac{m-1}{2}}, \end{aligned} \tag{10.17}$$

where ρ represents the correlation between two Gamma variables, $I_{m-1}(\cdot)$ is the modified Bessel function of the first kind of order $m-1$, m is the common parameter for i and j, and $i \neq j = sd, sr, rd$. The value of m is assumed to be an integer. Using these expressions, the MGF in three different cases of correlated links can be derived.

The first case occurs when the source-to-destination link and the source-to-relay link are correlated, both of which are independent of the relay-to-destination link. In other words, λ_{sd} and λ_{sr} are correlated. In this case, for variable gain relaying in (10.14), the MGF is calculated as

$$G(x) = \int_0^\infty \int_0^\infty \int_0^\infty e^{-x\lambda_{sd} - x\frac{\lambda_{sr}\lambda_{rd}}{\lambda_{sr}+\lambda_{rd}}} f(\lambda_{sd}, \lambda_{sr}) f(\lambda_{rd})\, d\lambda_{sd}\, d\lambda_{sr}\, d\lambda_{rd}.$$

By expanding the Bessel function using its series expression in Reference 11, (8.445) and solving the integration over λ_{sd} first, one has

$$\begin{aligned} G(x) &= \sum_{i=0}^{\infty} \frac{\rho^i (i+m-1)!}{i!\Gamma(m+i)\Gamma(m)\Gamma(m_2)(\omega_0\omega_1)^{m+i}} \\ &\quad \times \frac{1}{(1-\rho)^{m+2i}\omega_2^{m_2}} \frac{1}{\left(x + \frac{1}{(1-\rho\omega_0)}\right)^{i+m}} \\ &\quad \times \Omega_1(x \mid i+m, m_2, (1-\rho)\omega_1, \omega_2), \end{aligned} \tag{10.18}$$

where $m_0 = m_1 = m$ is an integer and

$$\Omega_1(x \mid p_1, p_2, q_1, q_2) = \int_0^\infty \int_0^\infty e^{-x\frac{x_1x_2}{x_1+x_2}} x_1^{p_1-1} e^{-\frac{x_1}{q_1}} x_2^{p_2-1} e^{-\frac{x_2}{q_2}}\, dx_1 dx_2,$$

has to be solved. To do this, we notice that, for two independent Gamma variables X_1 and X_2, the cumulative distribution function (CDF) of $Y = \frac{X_1 X_2}{X_1+X_2}$ is given by

$$F_Y(y) = 1 - \sum_{j=0}^{p_1-1} \sum_{n=0}^{p_2+j-1} \frac{2\binom{p_2+j-1}{n} q_2^{\frac{n-j+1}{2}} y^{p_2+j}}{j!\Gamma(p_2) q_2^{p_2} q_1^{\frac{n+j+1}{2}}} e^{-y\left(\frac{1}{q_1}+\frac{1}{q_2}\right)} K_{n-j+1}\left(\frac{2y}{\sqrt{q_1 q_2}}\right),$$

using [11, (8.445)], [11, (3.351.1)], [11, 8.352.1], and [11, (3.471.9)]. Using the CDF, its MGF can be calculated as $G_Y(x) = x\int_0^\infty e^{-xy} F_Y(y) dy$ and using the PDF, the MGF can be calculated as

$$G_Y(x) = \int_0^\infty \int_0^\infty e^{-x\frac{x_1 x_2}{x_1+x_2}} \frac{x_1^{p_1-1} e^{-\frac{x_1}{q_1}}}{\Gamma(p_1) q_1^{p_1}} \frac{x_2^{p_2-1} e^{-\frac{x_2}{q_2}}}{\Gamma(p_2) q_2^{p_2}} dx_1\, dx_2.$$

Equating the two different expressions of MGF, one can solve Ω_1 as

$$\Omega_1(x \mid p_1, p_2, q_1, q_2) = \Gamma(p_1)\Gamma(p_2) q_1^{p_1} q_2^{p_2} - \Gamma(p_1) q_1^{p_1} 2x$$

$$\times \sum_{j=0}^{p_1-1} \sum_{n=0}^{p_2+j-1} \frac{\binom{p_2+j-1}{n}}{j!} \frac{2^{2n-2j+2}\sqrt{\pi}}{q_1^{n+1}} \frac{\Gamma(p_2+n+2)\Gamma(p_2+2j-n)}{\Gamma\left(p_2+j+\frac{3}{2}\right)}$$

$$\times \frac{F\left(p_2+n+2, n-j+\frac{3}{2}; p_2+j+\frac{3}{2}; \frac{x+\left(\frac{1}{\sqrt{q_1}}-\frac{1}{\sqrt{q_2}}\right)^2}{x+\left(\frac{1}{\sqrt{q_1}}+\frac{1}{\sqrt{q_2}}\right)^2}\right)}{\left(x+\left(\frac{1}{\sqrt{q_1}}+\frac{1}{\sqrt{q_2}}\right)^2\right)^{p_2+n+2}},$$

where $\Gamma(\cdot)$ is the Gamma function and $F(\cdot,\cdot;\cdot;\cdot)$ is the hypergeometric function. For fixed gain relaying, following a similar method, the MGF is derived as

$$G(x) = \sum_{i=0}^{\infty} \frac{\rho^i (i+m-1)!}{i!\Gamma(m+i)\Gamma(m)\Gamma(m_2)(\omega_0\omega_1)^{m+i}}$$

$$\times \frac{1}{(1-\rho)^{m+2i}\omega_2^{m_2}\left(x+\frac{1}{(1-\rho\omega_0)}\right)^{i+m}}$$

$$\times \Omega_2\left(x \mid i+m, m_2, (1-\rho)\omega_1, \omega_2\right), \tag{10.19}$$

where

$$\Omega_2(x \mid p_1, p_2, q_1, q_2) = \Gamma(p_1)\Gamma(p_2)q_1^{p_1}q_2^{p_2} - (p_1 - 1)!x \sum_{j=0}^{p_1-1}\sum_{n=0}^{j}\binom{j}{n}$$
$$\times \frac{\Gamma(p_2+1+n)c^{\frac{j+p_2-n-1}{2}} e^{\frac{c}{2q_1q_2\left(x+\frac{1}{q_1}\right)}}}{q_1^{\frac{j+p_2+n-1-2p_1}{2}} q_2^{\frac{j-p_2-n-1}{2}} \left(x+\frac{1}{q_1}\right)^{\frac{j+p_2+n+1}{2}}}$$
$$\times W_{-\frac{j+p_2+n+1}{2}, \frac{j-p_2-n}{2}}\left(\frac{c}{q_1q_2\left(x+\frac{1}{q_1}\right)}\right),$$

and $W(\cdot)$ is the Whittaker-W function, using Reference 11, (6.643.3) by defining $Z = \frac{X_1X_2}{X_2+c}$ and finding its MGF expressions using PDF and CDF.

Using the same method, other cases can be studied as well. In the second case when the source-to-destination link and the relay-to-destination link are correlated, it can be shown that the MGF for variable gain relaying is

$$G(x) = \sum_{i=0}^{\infty} \frac{\rho^i(i+m-1)!}{i!\Gamma(m+i)\Gamma(m)\Gamma(m_1)(\omega_0\omega_2)^{m+i}}$$
$$\times \frac{1}{\left(x+\frac{1}{(1-\rho)\omega_0}\right)^{i+m}(1-\rho)^{m+2i}\omega_1^{m_1}}$$
$$\times \Omega_1(x \mid m_1, i+m, \omega_1, (1-\rho)\omega_2), \tag{10.20}$$

and the MGF for fixed gain relaying is

$$G(x) = \sum_{i=0}^{\infty} \frac{\rho^i(i+m-1)!}{i!\Gamma(m+i)\Gamma(m)\Gamma(m_1)(\omega_0\omega_2)^{m+i}}$$
$$\times \frac{1}{\left(x+\frac{1}{(1-\rho)\omega_0}\right)^{i+m}(1-\rho)^{m+2i}\omega_1^{m_1}}$$
$$\times \Omega_2(x \mid m_1, i+m, \omega_1, (1-\rho)\omega_2). \tag{10.21}$$

In the third case when the source-to-relay link and the relay-to-destination link are correlated, one also has the MGF for variable gain relaying as

$$G(x) = \frac{1}{(1+\omega_0 x)^{m_0}}$$
$$\times \sum_{i=0}^{\infty} \frac{\rho^i}{i!\Gamma(m+i)\Gamma(m)(\omega_1\omega_2)^{m+i}(1-\rho)^{m+2i}}$$
$$\times \Omega_1(x \mid i+m, i+m, (1-\rho)\omega_1, (1-\rho)\omega_2), \tag{10.22}$$

and the MGF for fixed gain relaying as

$$G(x) = \frac{1}{(1+\omega_0 x)^{m_0}} \times \sum_{i=0}^{\infty} \frac{\rho^i}{i!\Gamma(m+i)\Gamma(m)(\omega_1\omega_2)^{m+i}(1-\rho)^{m+2i}} \times \Omega_2(x \mid i+m, i+m, (1-\rho)\omega_1, (1-\rho)\omega_2). \tag{10.23}$$

Finally, using these derived MGFs, the error rate expressions can be obtained. For example, for non-coherent detection of binary frequency shift keying (BFSK) signals, the BER is given by [12]

$$P_b = \frac{1}{2}M\left(-\frac{1}{2}\right). \tag{10.24}$$

For differential detection of binary phase shift keying (DBPSK) signals, the BER is given by [12]

$$P_b = \frac{1}{2}M(-1). \tag{10.25}$$

For coherent detection, the BER using MGF will have an extra integration using the Craig's formula.

Next, some numerical examples are shown to examine the relaying performance with correlated links. In these examples, DBPSK is considered. Also, for fixed gain relaying, $c = 1$.

Figures 10.5 and 10.6 give the error rates using variable gain and fixed gain, respectively, when the source-to-relay and source-to-destination links are correlated. It can be seen that in this case the link correlation has considerable effect on the error rate performance of relaying. As expected, the performance degrades when the correlation increases, due to reduced diversity gain. Also, fixed gain relaying is more sensitive to the correlation but has a better overall performance than variable gain relaying. On the other hand, one can ignore the link correlation when $\rho < 0.5$.

Figures 10.7 and 10.8 give the error rates using variable gain and fixed gain relaying, respectively, when the relay-to-destination and source-to-destination links are correlated. Unlike in Figures 10.5 and 10.6, in this case, the link correlation has negligible effect on fixed gain relaying but still has considerable effect on variable gain relaying. Explanation to this observation is provided as follows. From (10.15), one sees that the end-to-end SNR for fixed gain is only determined by λ_{sr} when λ_{rd} approaches infinity or is very large. Consequently, the link correlation between the relay-to-destination and source-to-destination links will not affect the end-to-end SNR, when the average SNR is large. On the other hand, for variable gain, from (10.14), the end-to-end SNR is always determined by both λ_{sr} and λ_{rd}, when the average SNR is large. As a result, it is more sensitive to the link correlation between λ_{sd} and λ_{rd}.

Figures 10.9 and 10.10 give the error rates using variable gain and fixed gain, respectively, when the source-to-relay and relay-to-destination links are correlated.

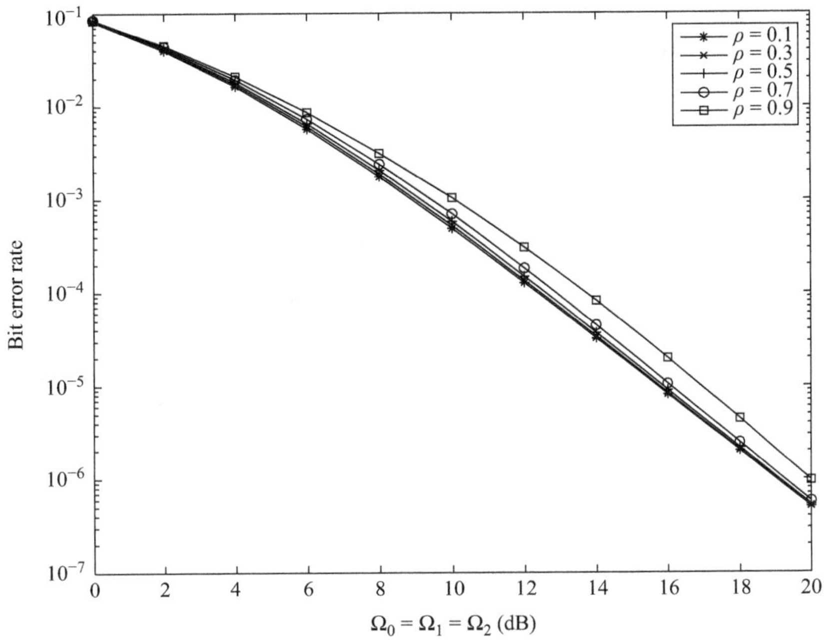

Figure 10.5 BER of DBPSK using variable gain when the source-to-relay and source-to-destination links are correlated for $m_0 = m_1 = 2$, $m_2 = 1$

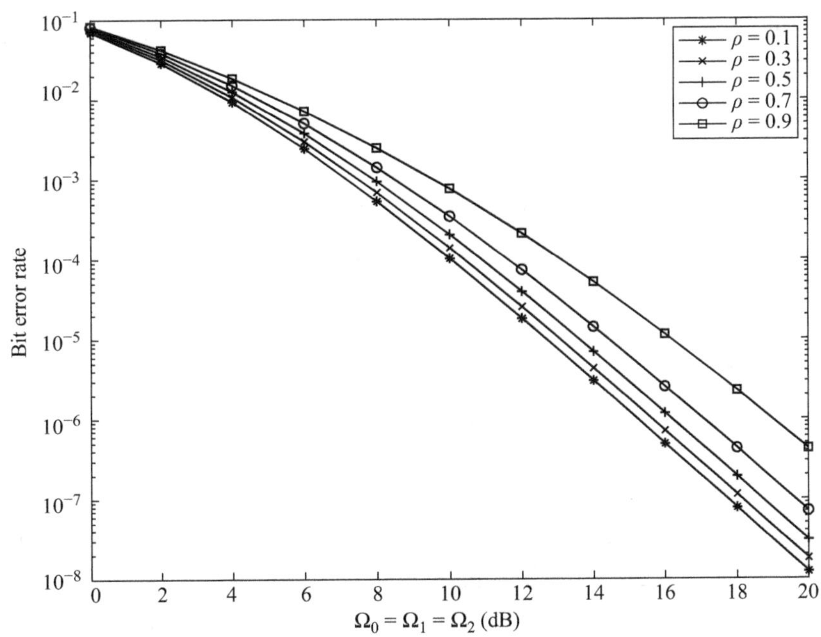

Figure 10.6 BER of DBPSK using fixed gain when the source-to-relay and source-to-destination links are correlated for $m_0 = m_1 = 2$, $m_2 = 1$

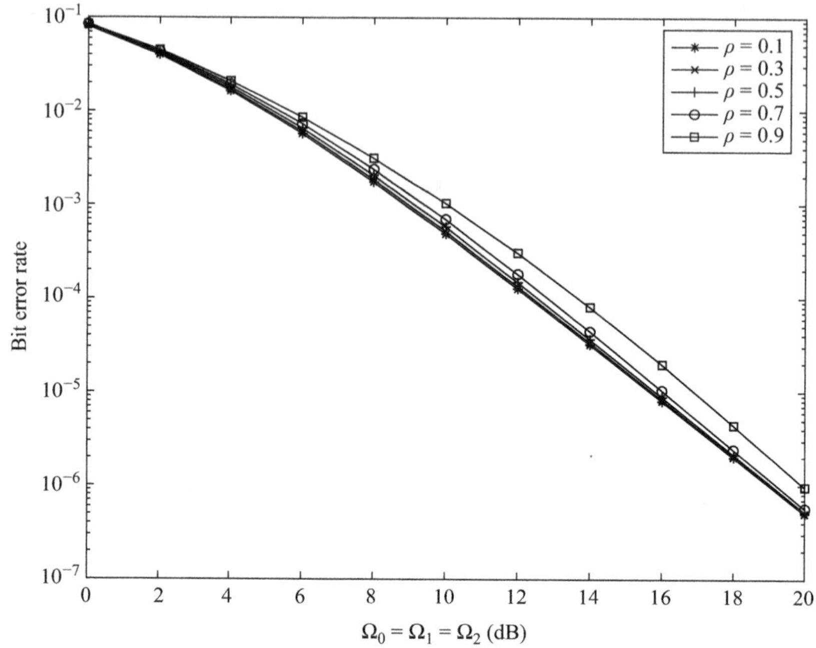

Figure 10.7 BER of DBPSK using variable gain when the relay-to-destination and source-to-destination links are correlated for $m_0 = m_1 = 2$, $m_2 = 1$

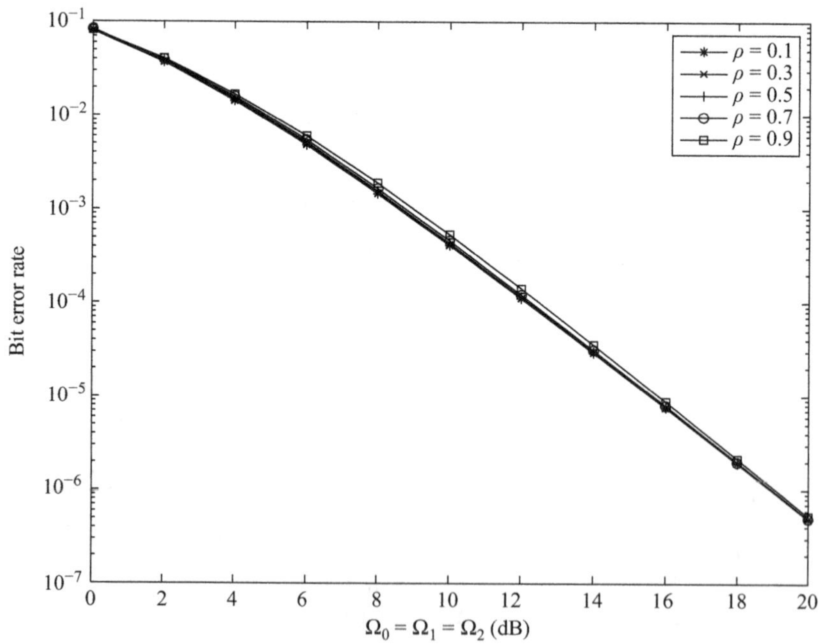

Figure 10.8 BER of DBPSK using fixed gain when the relay-to-destination and source-to-destination links are correlated for $m_0 = m_1 = 2$, $m_2 = 1$

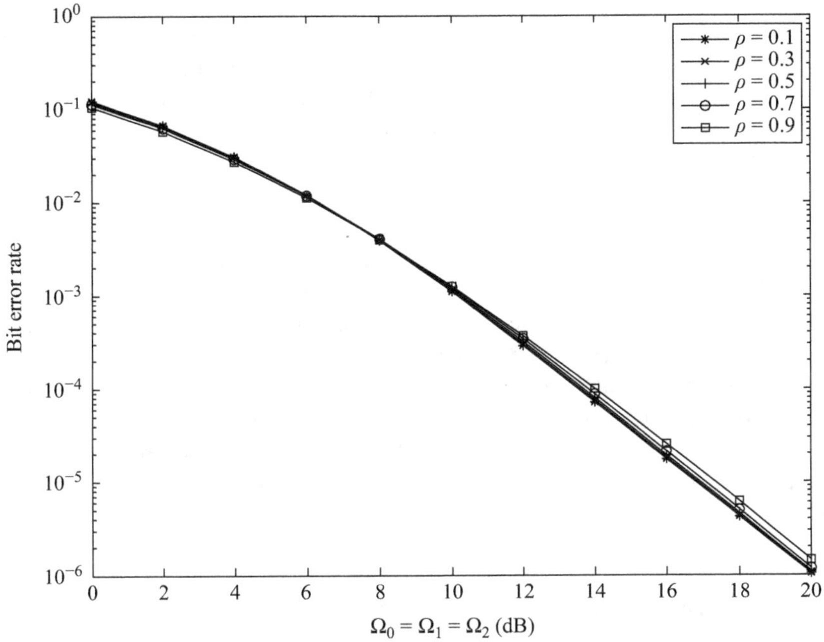

Figure 10.9 BER of DBPSK using variable gain when the source-to-relay and relay-to-destination links are correlated for $m_0 = m_1 = 2$, $m_2 = 1$

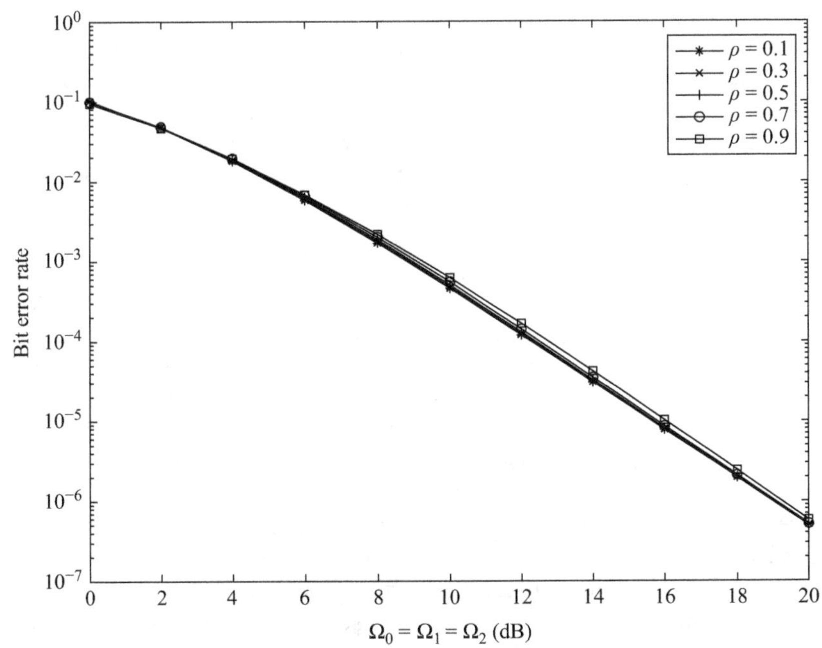

Figure 10.10 BER of DBPSK using fixed gain when the source-to-relay and relay-to-destination links are correlated for $m_0 = m_1 = 2$, $m_2 = 1$

In this case, the link correlation has minimum effect on either fixed gain or variable gain. This can be explained as follows. From (10.12), the diversity gain is achieved by combining signals from both the direct link and the relay link. In this case, only the source-to-relay and relay-to-destination links are correlated, which means that the direct link is still independent of the relay link. As a result, the diversity gain suffers very little from this correlation between two hops of the relay link, as it mainly comes from the independence between direct link and relay link.

In summary, this section has examined the performances of AF relaying for different cases when the links in the systems are correlated. The MGF method has been used and provides accurate analysis. The numerical examples have shown that the effect of link correlation varies for different relaying schemes. Specifically, fixed gain relaying has considerable performance degradation when the source-to-relay and source-to-destination links are correlated, while variable gain relaying has considerable performance degradation when the source-to-relay and source-to-destination links or relay-to-destination and source-to-destination links are correlated. The value of the correlation coefficient larger than 0.5 causes more performance degradation than that smaller than 0.5.

10.4 Relaying with non-Gaussian interference

Interference is another mitigating factor in wireless relaying. There are different sources of interferences, such as inter-symbol interference, inter-carrier interference, and multi-user interference. For a relaying network with multiple nodes, one of the main interfering sources is co-channel/multi-user interference, caused by concurrent transmissions from other nodes that share the same channel. In many works, for simplicity, this interference is ideally assumed to be a Gaussian random variable, when the central limit theorem applies. However, the central limit theorem is not always valid. In particular, if the variance is not finite, the sum of a number of random variables does not approach a Gaussian distribution any more. In this case, the interference will follow a non-Gaussian distribution.

In particular, several works have reported that in a large-scale wireless network there is a considerable amount of co-channel interference due to cell frequency reuse or multiple access [13–17]. Instead of the Gaussian model, they have shown and proved that the symmetric alpha stable (SαS) model is a better model for this co-channel interference, when a Poisson field of interfering sources are considered. For example, in Reference 13, Sousa used a SαS model for the interference caused by a Poisson field of interferers. Similar studies in References 14, 15 show that the co-channel interference in a spectrum sharing system follows the SαS model. Recently, many other works, such as References 16 and 17, also found that co-channel interference in cellular, sensor, and ad hoc networks can be well described by the SαS model.

Motivated by these observations, this section aims to propose some new detectors for AF relaying in SαS interference, a practical scenario in wireless relaying systems suffering from co-channel interference modeled in [13–17]. These detectors will be used for coherent detection of BPSK signals. Specifically, the proposed work uses a mixture of Cauchy and Gaussian approximation, Cauchy approximation and Myriad approximation to the SαS model, under the assumption that the co-channel interference dominates the channel distortion to simplify the designs of the proposed detectors.

Again, consider a three-terminal AF relaying system with one source node, one destination node, and one relay node. The relaying transmission happens in two phases. In the broadcasting phase, the signal is transmitted from the source to the relay. In the relaying phase, the received signal at the relay is forwarded to the destination. There is no direct link between the source and the destination due to obstacles between them, and hence the purpose of relaying. Then, the received signal at the destination can be written as

$$r = gs + I_1 + I_2, \tag{10.26}$$

where $g = g_{sr}g_{rd}\beta$ is the cascaded channel gain, $I_1 = g_{rd}\beta I_{sr}$ and $I_2 = I_{rd}$ are the interference components, g_{sr} is the channel coefficient of the source-to-relay link, s is the signal transmitted by the source, I_{sr} is the co-channel interference at the relay, g_{rd} is the channel coefficient of the relay-to-destination link, β is the amplification factor, and I_{rd} is the co-channel interference at the destination. In this section, BPSK signals are assumed such that $s = +1$ and $s = -1$ with equal probabilities. Also, g_{sr} and g_{rd} are assumed to be Gaussian random variables with mean zero and variance ω_{sr}^2 and ω_{rd}^2, respectively. This is the worst fading condition, worse than Rayleigh fading. The co-channel interferences I_{sr} and I_{rd} follow a SαS distribution, which will be discussed in more details later. The amplification factor β depends on the actual protocol. For variable gain relaying, we set $\beta = \frac{1}{\sqrt{|g_{sr}|^2}}$ and for fixed gain relaying, we set $\beta = \frac{1}{\omega_{sr}}$.

The detection of the signal s depends on the sample distribution. The conventional Gaussian detector using I received samples assumes a Gaussian distribution for all samples and therefore, its detection metric can be given as [6]

$$V_C = \sum_{i=1}^{I} r^{(i)} g_{sr} g_{rd} \beta, \tag{10.27}$$

where $r^{(i)}$ is the ith sample of the received signal at the destination. The decision is $+1$ if V_C is larger than 0 and the decision is -1 if V_C is smaller than 0. Effectively, (10.27) calculates the sample mean of the offset received signal and uses this sample mean for detection. This is based on the assumption that the interference is Gaussian. However, this is not the practical case in some wireless networks. In these networks, the SαS model is more suitable. The usefulness of the SαS distribution comes from the fact that it is the limiting distribution of a sum of random variables with infinite variances based on a generalized central limit theorem [18]. It is more impulsive than the Gaussian distribution and hence has a longer tail. The characteristic function (CF) of SαS is given by [18]

$$\Phi(jv) = e^{-|\gamma v|^\alpha + j\delta v}, \tag{10.28}$$

where γ is the scale parameter with $\gamma > 0$, δ is the location parameter with $-\infty < \delta < \infty$, and α is the shape parameter with $0 < \alpha < 2$. Then, the PDF of SαS does not have a closed-form expression and has to be calculated using an integral as

$$f_\alpha(x) = \frac{1}{2\pi} \int_{-\infty}^{\infty} \Phi(jv) e^{-jvx}\, dv, \tag{10.29}$$

for $-\infty < x < \infty$. The lack of a closed-form expression for the PDF has caused a lot of problems when using this distribution. Even worse, the integral in (10.29) converges very slowly, especially for small values of α. One important property of the SαS distribution in the above is that its order of moment only exists when the order is between -1 and α. This means that all orders of moment larger than α are infinite. Consequently, if $\alpha < 2$, the variance of SαS is infinite and when $\alpha < 1$, the mean of SαS is also infinite. In this section, I_{sr} and I_{rd} are assumed to follow the SαS distributions with shape parameter α_{sr}, scale parameter γ_{sr}, location parameter δ_{sr} and shape parameter α_{rd}, scale parameter γ_{rd}, location parameter δ_{rd}, respectively. Furthermore, assume that I_{sr} and I_{rd} are independent. In the following, consider a homogeneous relaying system with $\alpha_{sr} = \alpha_{rd} = \alpha$ and the location parameters are assumed to be zero such that $\delta_{sr} = \delta_{rd} = 0$.

In a wireless system, $\alpha = \frac{2}{m}$ and the value of the path loss exponent m is often larger than 2. Thus, the shape parameters of I_{sr} and I_{rd} are smaller than 1. In this case, the means of I_{sr} and I_{rd} are infinite. This is the main reason why the performance of the conventional Gaussian detector in (10.27) using the sample mean will not perform well, which will be shown by numerical examples.

To derive the detectors, one has the CFs of I_{sr} and I_{rd} as

$$\Phi_{I_{sr}}(jv) = e^{-|\gamma_{sr} v|^\alpha}, \tag{10.30}$$

and

$$\Phi_{I_{rd}}(jv) = e^{-|\gamma_{rd} v|^\alpha}, \tag{10.31}$$

respectively. Using (10.30) and (10.31), the CF of the received sample r can be derived as

$$\Phi_r(jv) = e^{jvg_{sr}g_{rd}\beta s - \left(|g_{rd}\beta\gamma_{sr}|^\alpha + \gamma_{rd}^\alpha\right)|v|^\alpha}. \tag{10.32}$$

Using (10.32), the PDF of the received sample r is calculated as

$$f_r(x) = \frac{1}{\pi} \int_0^{\infty} e^{-\left(|g_{rd}\beta\gamma_{sr}|^\alpha + \gamma_{rd}^\alpha\right)v^\alpha} \cos\left(v\left(x - g_{sr}g_{rd}\beta s\right)\right) dv. \tag{10.33}$$

Thus, using I independent samples, the optimum maximum likelihood detector is derived as

$$V_O = \sum_{i=1}^{I} \log\left(f_r(r_i|s=+1)\right) - \sum_{i=1}^{I} \log\left(f_r(r_i|s=-1)\right), \tag{10.34}$$

where r_i is the ith sample of the received signal. Since $f_r(x)$ does not have a closed-form expression, (10.34) has to be calculated numerically, which is too complicated. Next, simpler suboptimum detectors will be derived.

It was shown in Reference 19 that a mixture of Cauchy and Gaussian can approximate the SαS well as

$$f_\alpha(x) \approx \frac{(1-\epsilon)}{2\sqrt{\pi\psi^2}} e^{-\frac{x^2}{4\psi^2}} + \frac{\epsilon\psi}{\pi(x^2+\psi^2)}, \tag{10.35}$$

where $\epsilon = 2 - \alpha$ and $\psi = \left(|g_{rd}\beta\gamma_{sr}|^\alpha + \gamma_{rd}^\alpha\right)^{\frac{1}{\alpha}}$ is the scale parameter of the Cauchy distribution. In the frequency domain, this is equivalent to

$$\Phi(jv) = e^{-|\psi v|^\alpha} \approx (1-\epsilon)e^{-\psi^2|v|^2} + \epsilon e^{-\psi|v|}. \tag{10.36}$$

Using (10.35) in (10.34), one has

$$V_{\text{Mix}} = \sum_{i=1}^{I} \log\left(\frac{e^{-\frac{(r_i - g_{sr}g_{rd}\beta)^2}{4\psi^2}} + \frac{2\epsilon\psi^2}{(1-\epsilon)\sqrt{\pi}\left[(r_i - g_{sr}g_{rd}\beta)^2 + \psi^2\right]}}{e^{-\frac{(r_i + g_{sr}g_{rd}\beta)^2}{4\psi^2}} + \frac{2\epsilon\psi^2}{(1-\epsilon)\sqrt{\pi}\left[(r_i + g_{sr}g_{rd}\beta)^2 + \psi^2\right]}}\right). \tag{10.37}$$

Compared with (10.34), (10.37) does not require any numerical integration in the calculation of the detection metric. Thus, it is much simpler than the optimum detector.

It was also discussed in Reference 20 that the Cauchy PDF could be a good approximation to the SαS PDF. Thus, using the Cauchy approximation, one has

$$f_\alpha(x) \approx \frac{\chi}{\pi(x^2+\chi^2)}, \tag{10.38}$$

where $\chi = |g_{rd}\beta\gamma_{sr}| + \gamma_{rd}$. Equivalently, one has

$$\Phi(jv) \approx e^{-\chi|v|}. \tag{10.39}$$

Using (10.38) in (10.34), one has the decision variable for the detector using the Cauchy approximation as

$$V_{\text{Cau}} = \sum_{i=1}^{I} \log\left(\frac{(r_i + g_{sr}g_{rd}\beta)^2 + \chi^2}{(r_i - g_{sr}g_{rd}\beta)^2 + \chi^2}\right). \tag{10.40}$$

This detection metric has similar complexity to that in (10.37). Both are simpler than the optimum detector.

The Cauchy-approximation-based detector in (10.40) does not rely on the shape parameter α. To improve this approximation, one may also use the Myriad filter for a pure SαS interference [21]. Following this method, one has the detection metric using the Myriad approximation as

$$V_{\text{Myr}} = \sum_{i=1}^{I} \log\left(\frac{(r_i + g_{sr}g_{rd}\beta)^2 + K^2}{(r_i - g_{sr}g_{rd}\beta)^2 + K^2}\right), \tag{10.41}$$

where χ in (10.40) has been replaced by $K = \sqrt{\frac{\alpha}{2-\alpha}}\chi$. The derived new detectors have similar complexities.

Next, the performances of these new detectors will be compared with the conventional Gaussian detector using some numerical examples. In this comparison, the Gaussian noise is added to the samples to emulate the practical scenario, although the derivation above ignores the noise for simplicity. The SNR is defined as $\text{SNR} = \frac{I\omega_{sr}^2}{\sigma_{sr}^2} = \frac{I\omega_{rd}^2}{\sigma_{rd}^2}$, where σ_{sr}^2 and σ_{rd}^2 are the noise variances at the relay and destination, respectively. The SIR is defined as $\text{SIR} = \frac{I\omega_{sr}^2}{\gamma_{sr}^{\alpha}} = \frac{I\omega_{rd}^2}{\gamma_{rd}^{\alpha}}$. Also, set $\omega_{sr}^2 = \omega_{rd}^2 = 1$, while the values of $\gamma_{sr} = \gamma_{rd}$ change with the examined SIR and the values of $\sigma_{sr}^2 = \sigma_{rd}^2$ change with the examined SNR.

Figures 10.11 and 10.12 compare the approximate PDFs using a mixture of Cauchy and Gaussian, using Cauchy and using Myriad with the exact PDF for $\alpha = 0.3$ and $\alpha = 0.7$, respectively. One sees that the exact PDF becomes more impulsive at the center when the value of α decreases. The Myriad approximation overestimates the PDF near the center, i.e., when $-2 < x < 2$, and underestimates the PDF near the tail when $x > 2$ or $x < -2$. The mixture and Cauchy approximations overestimate the PDF near the center, i.e., $-7 < x < 7$, and underestimates the PDF near the tail when $x > 7$ or $x < -7$. Also, the Myriad approximation tracks the mode of the PDF well at $\alpha = 0.7$, but when α decreases, all the approximations lose track of the mode. The smaller the value of α is, the larger the approximation error seems to be from Figures 10.11 and 10.12. At $\alpha = 0.3$, the PDF is so impulsive that none of these approximations can track the exact PDF closely. A good approximation to the exact PDF will certainly give good detector performance. Figures 10.11 and 10.12 imply that there may still be room to improve these approximations and therefore to improve the detector performance.

Figures 10.13–10.16 compare the performances of the detectors with the conventional Gaussian detector. One sees that the new approximation-based detectors outperform the conventional Gaussian detector with significant performance gains in all cases. Also, the new approximation-based detectors have almost identical performances. In many cases, the detectors based on the Cauchy approximation and the Myriad approximation are better than the detector based on the mixture approximation. The performances of the detectors change little when the value of α decreases.

In summary, this section has proposed several new detectors for AF relaying signals in SαS interference by using a mixture of Cauchy and Gaussian, Cauchy and Myriad approximations to the exact PDF of the SαS distribution. The derivation has been based on the assumption of dominant interference but the simulation has taken the Gaussian noise into account as well. Numerical results have shown that the derived detectors have much better performances than the conventional Gaussian detector that assumes a Gaussian interference, which has been widely used in the literature for AF relaying. This allows us to save tens of dB in effective SNR to achieve the same BER performance as the Gaussian detector. Thus, the new detectors are very useful for highly impulsive environments where the Gaussian detector fails due to a fundamental flaw.

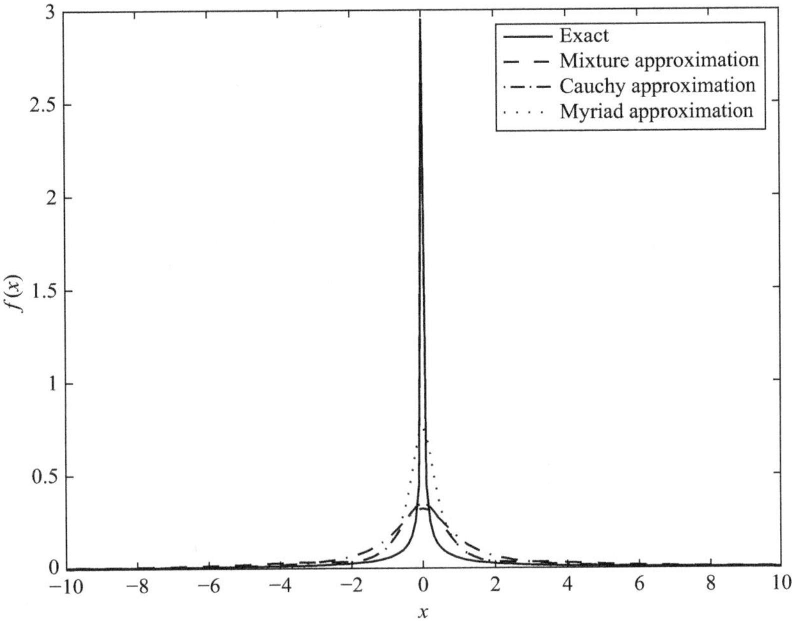

Figure 10.11 *Comparison of different approximations to the PDF of a standard SαS random variable at α = 0.3*

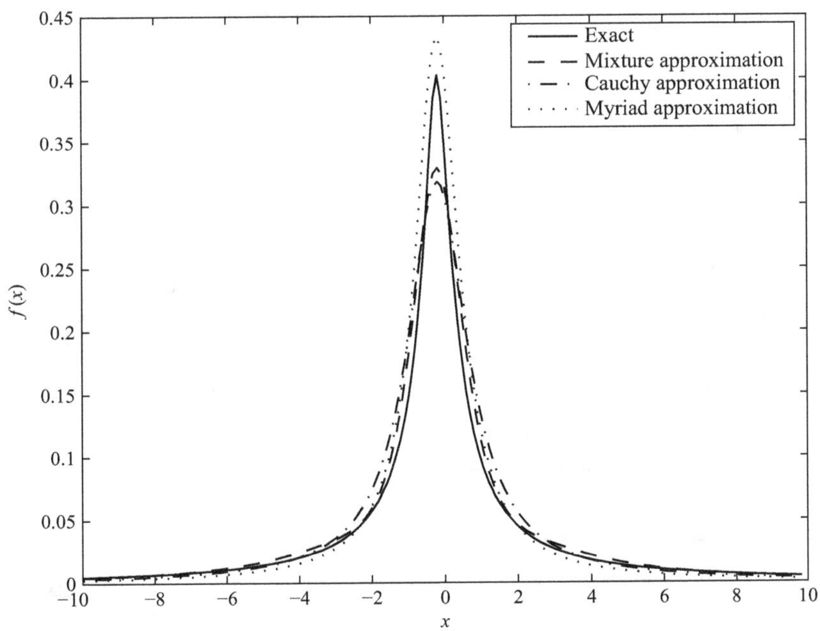

Figure 10.12 *Comparison of different approximations to the PDF of a standard SαS random variable at α = 0.7*

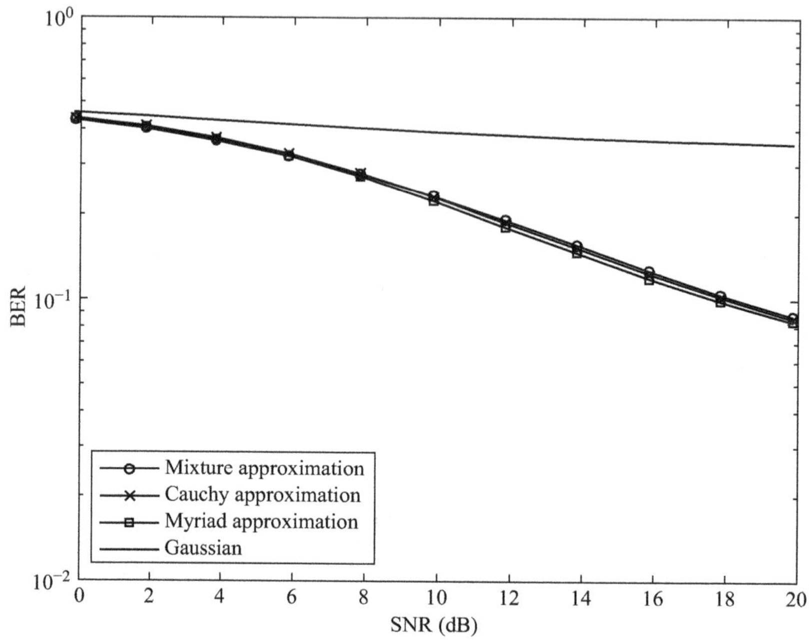

Figure 10.13 *Comparison of different detector performances at $\alpha = 0.3$, SIR = 10 dB, and N = 4 for fixed gain relaying*

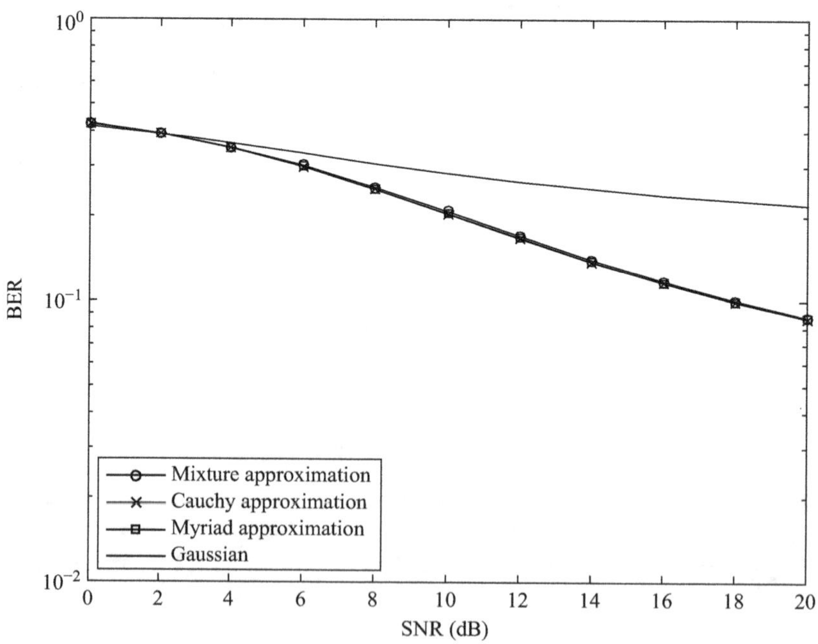

Figure 10.14 *Comparison of different detector performances at $\alpha = 0.7$, SIR = 10 dB, and N = 4 for fixed gain relaying*

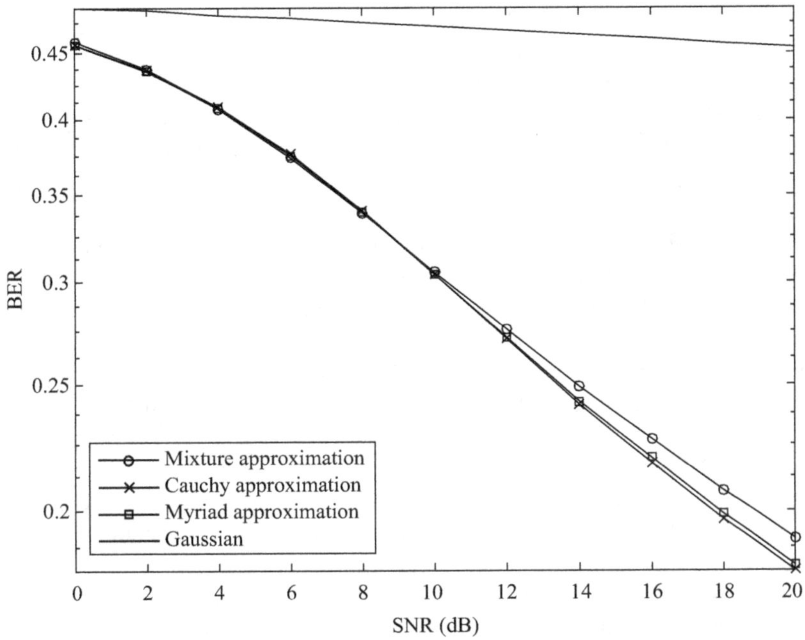

Figure 10.15 Comparison of different detector performances at $\alpha = 0.5$, $SIR = 10\,dB$, and $N = 10$ for fixed gain relaying

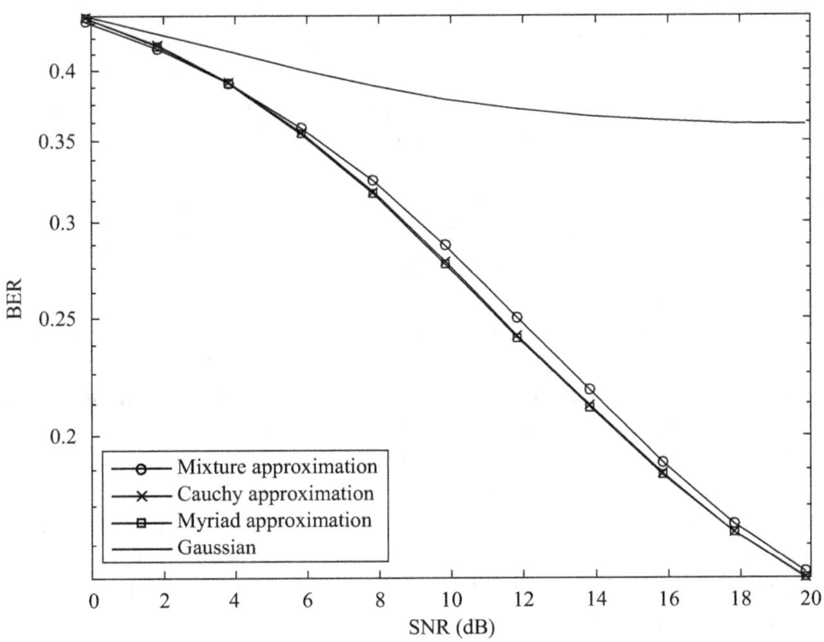

Figure 10.16 Comparison of different detector performances at $\alpha = 0.5$, $SIR = 10\,dB$ and $N = 4$ for variable gain relaying

10.5 Relaying with wireless power

In the conventional relaying system with ideal operating conditions, the relay node can rely on a fixed power source through a wired connection to the power grid. In this case, the relay node may not need to worry about its energy consumption. In practical situations, however, the relay node may not have such a reliable power source or may not want to utilize its own power source, for relaying other nodes' information. In these situations, wireless power could be a viable solution, where the relay node harvests energy from either the source node or some external renewable source, such as the sun or wind, and only uses the harvested wireless power for relaying transmission.

In this section, we focus on relaying using wireless power. Relaying relies on idle nodes in the wireless network to forward information. However, such operation, costs the idle node extra energy and therefore, may prevent them from taking part in relaying, especially for those mobile nodes that operate on batteries and therefore have a limited lifetime. Wireless power relieves this worry by allowing the relay node to harvest energy from either the source node or external ambient sources for sustainable operation. This can greatly reduce the use of the relay node's own battery and therefore, encourages idle nodes to participate in relaying.

If the relay node harvests energy from the source node, relaying with wireless power actually becomes a simultaneous wireless information and power transfer (SWIPT) system, as the relay node receives both information and energy from the source node. Similar to other SWIPT systems, since the relay node cannot perform information reception and energy harvesting at the same time due to the hardware limitation, information transmission and power transfer have to be separated [22].

In the first method, time switching (TS) can be used, where an extra energy harvesting phase is added to the conventional relaying process that includes broadcasting phase and relaying phase. Specifically, assuming that α is the TS coefficient with $0 \leq \alpha \leq 1$, in the first phase, the source node transfers a fixed amount of energy to the relay node for αT seconds. In the second phase, the source node transmits its information to the relay node for $(1-\alpha)T/2$ seconds. In the third phase, the relay node uses the harvested energy in the first phase to forward the information received in the second phase to the destination node for $(1-\alpha)T/2$ seconds, where T is the total duration of transmission.

In the second method, power splitting (PS) can be used, where a portion of the signal received from the source node is harvested as energy by the relay node. This eliminates the need for a dedicated harvesting time. In this method, wireless-powered relaying is similar to conventional relaying. In the broadcasting phase, the source node transmits the signal to the relay node for $T/2$ seconds, part of which is harvested for energy and part of which is used for information decoding. In the relaying phase, the relay node forwards the received information signal to the destination node for $T/2$ seconds using the harvested energy. Assuming that the PS factor is ρ with $0 \leq \rho \leq 1$, a portion ρ of the received power from the source is used for energy while the remaining $1-\rho$ is used for information.

Comparing TS with PS, it is noted that both require a separate energy harvester in addition to the information transceiver at the relay. However, PS is slightly more complicated due to the non-trivial power splitter required in hardware implementation. On the other hand, TS requires a dedicated harvesting time, which adds complexity to synchronization and also reduces the throughput of the system, which essentially trades time for complexity. Due to the extra harvesting time, under similar conditions, TS often has a smaller throughput than PS. An issue with PS is that it often has a smaller transmission range than TS in the AF protocol, as if one uses a large portion of the signal for energy, it will give a weak signal for forwarding while if one uses a small portion of the signal for energy, it will lead to a small transmission power in the relaying phase, both of which reduces the quality of the received signal at the destination.

Works on relaying with wireless power have mainly focused on the analysis of the relaying performance and therefore, the optimization of system parameters, such as α and ρ. For TS, when α is large, more energy can be harvested by the relay node such that higher end-to-end SNR will be achieved with larger throughput but on the other hand, the effective throughput decreases due to the increased harvesting time for large α. For PS, when ρ is large, more energy can be harvested but a weaker data signal will be forwarded by the relay node. Thus, optimum values of α and ρ exist. In fading channels, two performance measures are widely used, delay-limited throughput and delay-tolerant throughput. For delay-limited throughput, the source node transmits at a constant rate such that one has to derive the average outage probability of the end-to-end SNR, based on which the throughput can be calculated by excluding outage events. For delay-tolerant throughput, one calculates the ergodic capacity.

In the following, we give an example for the analysis of delay-limited throughput for AF relaying using wireless power. In particular, if TS is used, based on the same system model as used in Reference 22, the end-to-end SNR can be obtained as

$$\lambda = \frac{P_r\beta^2|g_{rd}|^2P_s|g_{sr}|^2}{P_r\beta^2|g_{rd}|^2\left(\sigma_{ra}^2+\sigma_{rb}^2\right)+\sigma_{da}^2+\sigma_{db}^2}, \tag{10.42}$$

where β is the amplification factor, P_s is the transmission power of the source, g_{sr} is the channel gain of the source-to-relay link, g_{rd} is the channel gain in the relay-to-destination link, σ_{ra}^2 and σ_{rb}^2 are the variances of the noise at the relay from the antenna and baseband conversion, respectively, σ_{da}^2 and σ_{db}^2 are the variances of the noise at the destination from the antenna and baseband conversion, respectively, and P_r is the transmission power of the relay.

All the noise are additive white Gaussian noise with mean zero. Define $\lambda_{sr} = \frac{P_s|g_{sr}|^2}{\sigma_{ra}^2+\sigma_{rb}^2}$ and $\lambda_{rd} = \frac{|g_{rd}|^2}{\sigma_{da}^2+\sigma_{db}^2}$. For Rayleigh fading channels, they are exponentially distributed random variables with

$$f_{\lambda_{sr}}(x) = \frac{1}{\omega_{sr}}e^{-\frac{x}{\omega_{sr}}}, \quad x > 0, \tag{10.43}$$

$$f_{\lambda_{rd}}(x) = \frac{1}{\omega_{sr}}e^{-\frac{x}{\omega_{rd}}}, \quad x > 0, \tag{10.44}$$

respectively, where $\omega_{sr} = \frac{P_s E\{|g_{sr}|^2\}}{\sigma_{ra}^2+\sigma_{rb}^2}$ and $\omega_{rd} = \frac{E\{|g_{rd}|^2\}}{\sigma_{da}^2+\sigma_{db}^2}$. Also, the transmission power of the relay is given by [22]

$$P_r = \frac{E_h}{(1-\alpha)\frac{T}{2}}, \tag{10.45}$$

where E_h is the harvested energy from the source signal, α is the TS coefficient, and T is the total transmission time.

For variable gain, one has $\beta = \frac{1}{\sqrt{P_s|g_{sr}|^2+\sigma_{ra}^2+\sigma_{rb}^2}}$. Thus, from (10.42), one further has

$$\lambda_{\text{TS-VG}} = \frac{\lambda_{sr}\lambda_{rd}}{\lambda_{rd} + \dfrac{(1+\lambda_{rd})}{P_r}}. \tag{10.46}$$

For fixed gain relaying, in this section, we assume that $\beta = \frac{1}{\sqrt{P_s E\{|g_{sr}|^2\}+\sigma_{ra}^2+\sigma_{rb}^2}}$. Thus, using (10.42), one also has

$$\lambda_{\text{TS-FG}} = \frac{\lambda_{sr}\lambda_{rd}}{\lambda_{rd} + \dfrac{1+\omega_{sr}}{P_r}}. \tag{10.47}$$

If the conversion efficiency of the energy harvester is a constant, independent of the input power, one has $E_h = \eta P_s|g_{sr}|^2\alpha T$, where η is the conversion efficiency of the energy harvester. Using this in (10.46) and (10.47), one has

$$\lambda_{\text{TS-VG}}^{\text{Cons}} = \frac{\lambda_{sr}\lambda_{rd}}{\lambda_{rd} + \dfrac{(1-\alpha)(1+\lambda_{sr})}{2\alpha\eta\lambda_{sr}\left(\sigma_{ra}^2+\sigma_{rb}^2\right)}}, \tag{10.48}$$

and

$$\lambda_{\text{TS-FG}}^{\text{Cons}} = \frac{\lambda_{sr}\lambda_{rd}}{\lambda_{rd} + \dfrac{(1-\alpha)(1+\omega_{sr})}{2\alpha\eta\lambda_{sr}\left(\sigma_{ra}^2+\sigma_{rb}^2\right)}}. \tag{10.49}$$

If the conversion efficiency is dependent of the input power, one has $E_h = \eta\left[P_s|g_{sr}|^2\right]P_s|g_{sr}|^2\alpha T$. Using this in (10.46) and (10.47), one has

$$\lambda_{\text{TS-VG}}^{\text{NonCons}} = \frac{\lambda_{sr}\lambda_{rd}}{\lambda_{rd} + \dfrac{(1-\alpha)(1+\lambda_{sr})}{2\alpha\eta\left[\lambda_{sr}\left(\sigma_{ra}^2+\sigma_{rb}^2\right)\right]\lambda_{sr}\left(\sigma_{ra}^2+\sigma_{rb}^2\right)}}, \tag{10.50}$$

and

$$\lambda_{\text{TS-FG}}^{\text{NonCons}} = \frac{\lambda_{sr}\lambda_{rd}}{\lambda_{rd} + \dfrac{(1-\alpha)(1+\omega_{sr})}{2\alpha\eta\left[\lambda_{sr}\left(\sigma_{ra}^2+\sigma_{rb}^2\right)\right]\lambda_{sr}\left(\sigma_{ra}^2+\sigma_{rb}^2\right)}}. \tag{10.51}$$

Using (10.50) and (10.51), their outage probabilities can be derived as

$$P_{\text{out}-\text{TSVG}}^{\text{NonCons}} = 1 - \frac{e^{-\frac{D}{\omega_{sr}}}}{\omega_{sr}} \int_0^\infty e^{-\frac{t}{\omega_{sr}} - \frac{D(1-\alpha)(t+D+1)}{\omega_{rd}(2\alpha)\eta\left[(t+D)\left(\sigma_{ra}^2+\sigma_{rb}^2\right)\right](t+D)\left(\sigma_{ra}^2+\sigma_{rb}^2\right)t}} dt, \tag{10.52}$$

for variable gain relaying and

$$P_{\text{out}-\text{TSFG}}^{\text{NonCons}} = 1 - \frac{e^{-\frac{D}{\omega_{sr}}}}{\omega_{sr}} \int_0^\infty e^{-\frac{t}{\omega_{sr}} - \frac{D(1-\alpha)(\omega_{sr}+1)}{\omega_{rd}(2\alpha)\eta\left[(t+D)\left(\sigma_{ra}^2+\sigma_{rb}^2\right)\right](t+D)\left(\sigma_{ra}^2+\sigma_{rb}^2\right)t}} dt, \tag{10.53}$$

for fixed gain relaying, where $D = 2_0^R - 1$ is the predetermined threshold of SNR and R_0 is the desired information rate imposed at the source. These expressions can be easily solved using standard mathematical software. Finally, the delay-limited throughput in this case is derived as [22]

$$\tau_{\text{TSVG}}^{\text{NonCons}} = \frac{R_0}{2}(1-\alpha)\left(1 - P_{\text{out}-\text{TSVG}}^{\text{NonCons}}\right), \tag{10.54}$$

and

$$\tau_{\text{TSFG}}^{\text{NonCons}} = \frac{R_0}{2}(1-\alpha)\left(1 - P_{\text{out}-\text{TSFG}}^{\text{NonCons}}\right). \tag{10.55}$$

The above results are for TS. If PS is used instead, one has the end-to-end SNR as

$$\lambda = \frac{(1-\rho)P_r\beta^2|g_{rd}|^2 P_s|g_{sr}|^2}{P_r\beta^2|g_{rd}|^2\left((1-\rho)\sigma_{ra}^2 + \sigma_{rb}^2\right) + \sigma_{da}^2 + \sigma_{db}^2}, \tag{10.56}$$

where ρ is the PS factor in PS which measures how much portion is used for energy harvesting. Then, if a constant conversion efficiency is assumed, one has $E_h = \eta\rho P_s|g_{sr}|^2 T/2$ such that

$$\lambda_{\text{PS}-\text{VG}}^{\text{Cons}} = \frac{\lambda'_{sr}\lambda_{rd}}{\lambda_{rd} + \dfrac{\left(1+\lambda'_{sr}\right)}{\eta\rho\lambda'_{sr}\left(\sigma_{ra}^2 + \dfrac{\sigma_{rb}^2}{1-\rho}\right)}}, \tag{10.57}$$

and

$$\lambda_{\text{PS}-\text{FG}}^{\text{Cons}} = \frac{\lambda'_1\lambda_2}{\lambda_2 + \dfrac{\left(1+\omega'_{sr}\right)}{\eta\rho\lambda'_{sr}\left(\sigma_{ra}^2 + \dfrac{\sigma_{rb}^2}{1-\rho}\right)}}, \tag{10.58}$$

where $\beta = \frac{1}{\sqrt{(1-\rho)P_s|g_{sr}|^2 + (1-\rho)\sigma_{ra}^2 + \sigma_{rb}^2}}$ and $\beta = \frac{1}{\sqrt{(1-\rho)P_s E\left\{|g_{sr}|^2\right\} + (1-\rho)\sigma_{ra}^2 + \sigma_{rb}^2}}$ are for variable gain and fixed gain, respectively, $\lambda'_{sr} = \frac{P_s|g_{sr}|^2}{\sigma_{ra}^2 + \frac{\sigma_{rb}^2}{1-\rho}}$ is an exponential random variable with PDF

$$f_{\lambda'_{sr}}(x) = \frac{1}{\omega'_{sr}} e^{-\frac{x}{\omega'_{sr}}}, \quad x > 0, \tag{10.59}$$

with $\omega'_{sr} = \frac{P_s E\{|g_{sr}|^2\}}{\sigma_{ra}^2 + \frac{\sigma_{rb}^2}{1-\rho}}$. If the conversion efficiency η is a function of the input power, one also has $E_h = \eta[\rho P_s |g_{sr}|^2]\rho P_s |g_{sr}|^2 T/2$, giving

$$\lambda_{\text{PS-VG}}^{\text{NonCons}} = \frac{\lambda'_{sr}\lambda_{rd}}{\lambda_{rd} + \frac{(1+\lambda'_{sr})}{\eta\left[\rho\lambda'_{sr}\left(\sigma_{ra}^2 + \frac{\sigma_{rb}^2}{1-\rho}\right)\right]\rho\lambda'_{sr}\left(\sigma_{ra}^2 + \frac{\sigma_{rb}^2}{1-\rho}\right)}}, \tag{10.60}$$

and

$$\lambda_{\text{PS-FG}}^{\text{NonCons}} = \frac{\lambda'_{sr}\lambda_{rd}}{\lambda_{rd} + \frac{(1+\omega'_{sr})}{\eta\left[\rho\lambda'_{sr}\left(\sigma_{ra}^2 + \frac{\sigma_{rb}^2}{1-\rho}\right)\right]\rho\lambda'_{sr}\left(\sigma_{ra}^2 + \frac{\sigma_{rb}^2}{1-\rho}\right)}}. \tag{10.61}$$

Using (10.60) and (10.61), their outage probabilities are obtained by

$$P_{\text{out-PSVG}}^{\text{NonCons}} = 1 - \frac{e^{-\frac{D}{\omega'_{sr}}}}{\omega'_{sr}} \int_0^\infty e^{-\frac{t}{\omega'_{sr}} - \frac{D(t+D+1)}{\omega_{rd}\eta\left[\rho(t+D)\left(\sigma_{ra}^2 + \frac{\sigma_{rb}^2}{1-\rho}\right)\right]\rho(t+D)\left(\sigma_{ra}^2 + \frac{\sigma_{rb}^2}{1-\rho}\right)t}} dt, \tag{10.62}$$

for variable gain and

$$P_{\text{out-PSFG}}^{\text{NonCons}} = 1 - \frac{e^{-\frac{D}{\omega'_{sr}}}}{\omega'_{sr}} \int_0^\infty e^{-\frac{t}{\omega'_{sr}} - \frac{D\left(\omega'_{sr}+1\right)}{\omega_{rd}\eta\left[\rho(t+D)\left(\sigma_{ra}^2 + \frac{\sigma_{rc}^2}{1-\rho}\right)\right]\rho(t+D)\left(\sigma_{ra}^2 + \frac{\sigma_{rc}^2}{1-\rho}\right)t}} dt, \tag{10.63}$$

for fixed gain. Then, the throughput is given by [22]

$$\tau_{\text{PSVG}}^{\text{NonCons}} = \frac{R_0}{2}\left(1 - P_{\text{out-PSVG}}^{\text{NonCons}}\right), \tag{10.64}$$

and

$$\tau_{\text{PSFG}}^{\text{NonCons}} = \frac{R_0}{2}\left(1 - P_{\text{out-PSFG}}^{\text{NonCons}}\right). \tag{10.65}$$

Next, these performances are calculated in the numerical examples. In the calculation, we set $P_s = 1$, $R_0 = 3$, $E\{|g_{sr}|^2\} = E\{|g_{rd}|^2\} = 1$, and $\sigma_{ra}^2 = \sigma_{rb}^2 = \sigma_{da}^2 = \sigma_{db}^2 = \sigma^2$. Also, for the constant conversion efficiency, we use $\eta = 0.5$. For the varying efficiency, we use the harvesters in Reference 23 by Le *et al.* and in Reference 24 by Stoopman *et al.*

Figure 10.17 gives the delay-limited throughput using TS for different conversion efficiencies. One sees that there exists a maximum throughput in all the cases, although the optimal values of α are considerably different for different curves. Also, fixed gain relaying and variable gain relaying have different performances. In particular, from Figure 10.17, the maximum throughput for fixed gain relaying is larger than that for variable gain relaying. In all the considered cases, the constant conversion efficiency has larger throughput or overestimates the realistic throughput.

Figure 10.18 gives the delay-limited throughput using PS for different conversion efficiencies. In Figure 10.18, the throughput using the harvester in Reference 24

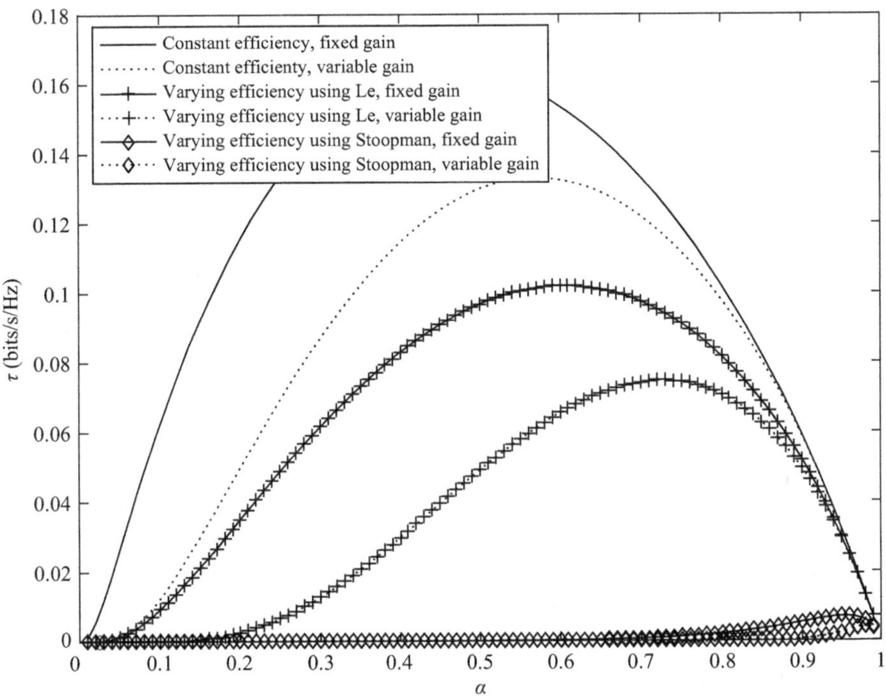

Figure 10.17 τ versus α for the constant and varying efficiencies using AF relaying and TS when $\sigma^2 = 0.05$

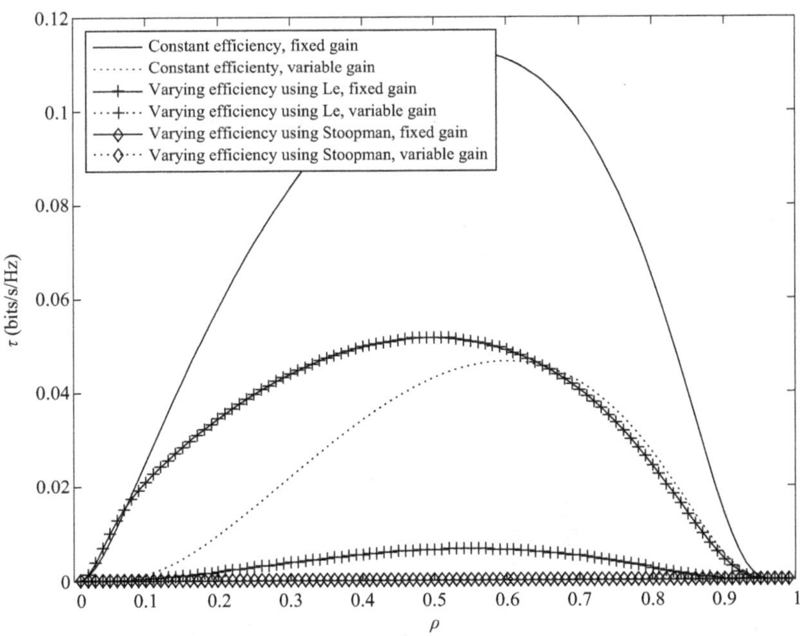

Figure 10.18 τ versus ρ for the constant and varying efficiencies using AF relaying and PS when $\sigma^2 = 0.05$

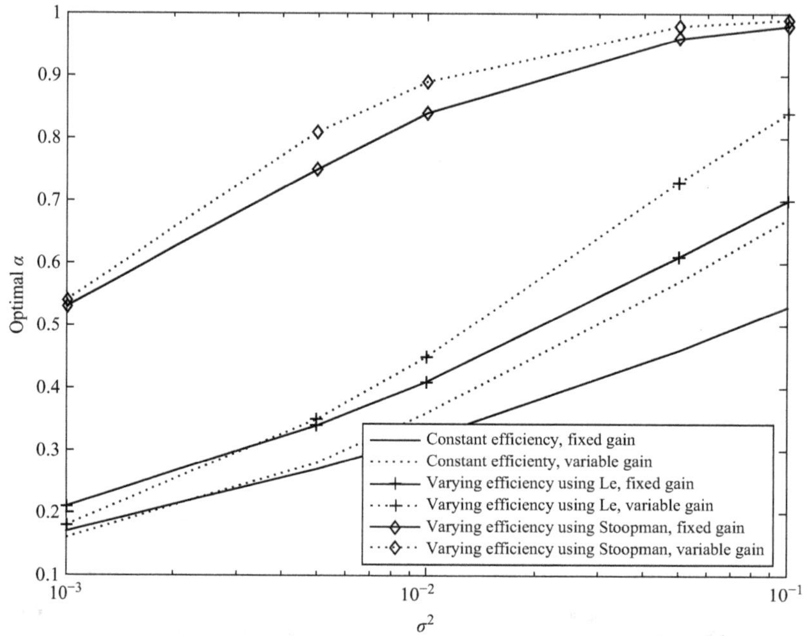

Figure 10.19 Optimal α versus σ^2 for the constant and varying efficiencies using AF relaying and TS

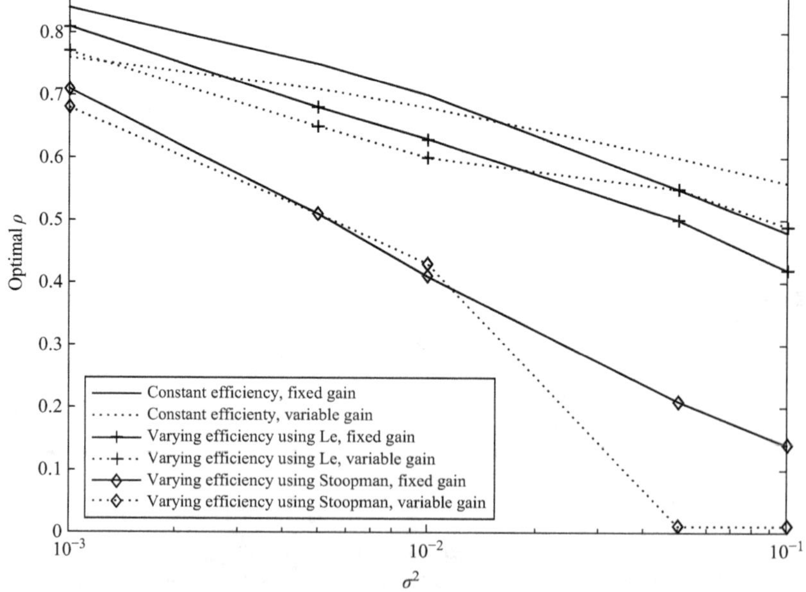

Figure 10.20 Optimal ρ versus σ^2 for the constant and varying efficiencies using AF relaying and PS

is close to zero in most cases. Again, a maximum throughput exists in all the curves. The optimal value of the PS factor for the harvester in Reference 23 is quite different from that assumes a constant conversion efficiency. Specifically, the curves for References 23 and 24 have smaller maximum throughput and smaller optimal PS factor than those for the constant conversion efficiency.

Figures 10.19 and 10.20 give the optimal values of α and ρ, respectively, for different values of σ^2 that determine the SNR. The optimal value of α increases with σ^2. Also, the optimal value of α using the constant efficiency is much smaller than that using the varying efficiency. On the other hand, the optimal value of ρ decreases with σ^2. The optimal value of ρ for the constant efficiency is always larger than that using the varying efficiency or the realistic harvester.

The above deals with the SWIPT-based relaying with wireless power. The other type of relaying with wireless power harvests energy from external renewable sources, such as radio frequency energy and interference, but not from the source node. Thus, they do not assume a SWIPT structure, as their energy and information comes from different places. Owing to this, scheduling becomes a very important issue for this type of wireless-powered relaying, as energy and information may not arrive at the same time such that one needs to ensure enough energy before conducting any information transmission. This system is not discussed here due to space limitations.

References

[1] Y. Chen, C.-X. Wang, H. Xiao, D. Yuan, "Novel partial relay selection schemes for AF relaying in Nakagami-*m* fading channels," *IEEE Transactions on Vehicular Technology*. 2011; 60(7): 3497–3503.

[2] M.O. Hasna, M.-S. Alouini, "Outage probability of multihop transmission over Nakagami fading channels," *IEEE Communications Letters*. 2003; 7(5): 216–218.

[3] P. Wang, Y. Zhou, X. Yin, X. Huang, F. Liu, "Measurement and analysis of relay channels in indoor environments," *Proceedings of 2010 Sixth International Conference on Wireless Communications Networking and Mobile Computing*; Chengdu, China, September 2010.

[4] Q.H. Chu, J.-M. Conrat, J. Cousin, "Experimental characterization and modeling of shadow fading correlation for relaying systems," *Proceedings of 2011 IEEE Vehicular Technology Conference*; San Francisco, CA, USA, September 2011.

[5] P.A. Anghel, M. Kaveh, "Exact symbol error probability of a cooperative network in a Rayleigh-fading environment," *IEEE Transactions on Wireless Communications*. 2004; 3(5): 1416–1421.

[6] J.N. Laneman, G.W. Wornell, "Energy-efficient antenna sharing and relaying for wireless networks," *Proceedings of 2000 IEEE Wireless Communications and Networking Conference*; Chicago, IL, March 2000.

[7] M.O. Hasna, M.-S. Alouini, "A performance study of dual-hop transmissions with fixed gain relays," *IEEE Transactions on Wireless Communications*. 2004; 3(6): 1963–1968.

[8] G.K. Karagiannidis, "Performance bounds of multihop wireless communications with blind relays over generalized fading channels," *IEEE Transactions on Wireless Communications*. 2006; 5(3): 498–503.

[9] M. Di Renzo, F. Graziosi, F. Santucci, "A comprehensive framework for performance analysis of dual-hop cooperative wireless systems with fixed-gain relays over generalized fading channels," *IEEE Transactions on Wireless Communications*. 2009; 8(10): 5060–5074.

[10] J. Reig, L. Rubio, "Bivariate Nakagami-*m* distribution with arbitrary fading parameters," *Electronics Letters*. 2002; 38(25): 1715–1717.

[11] I.S. Gradshteyn, I.M. Ryzhik. *Table of Integrals, Series, and Products*. 6th edn. New York, NY: Academic Press; 2000.

[12] M.K. Simon, M.-S. Alouini. *Digital Communication over Fading Channels*. 2nd edn. Hoboken, NJ: John Wiley & Sons; 2005.

[13] E. Sousa, "Performance of a spread spectrum packet radio network link in a Poisson field of interferences," *IEEE Transactions on Information Theory*. 1992; 38(6): 1743–1754.

[14] J. Ilow, D. Hatzinakos, A.N. Venetsanopoulos, "Performance of FH SS radio networks with interference modeled as a mixture of Gaussian and alpha-stable noise," *IEEE Transactions on Communications*. 1998; 46(3): 509–520.

[15] J. Ilow, D. Hatzinakos, "Analytic alpha-stable noise modeling in a Poisson field of interferers or scatterers," *IEEE Transactions on Signal Processing*. 1998; 46(6): 1601–1611.

[16] M. Haenggi, R.K. Ganti, "Interference in large wireless networks," in *Foundations and Trends in Networking*. New York, NY: Now Publishers; 2008, pp. 127–248.

[17] M.Z. Win, P.C. Pinto, L.A. Shepp, "A mathematical theory of network interference and its applications," *Proceedings of IEEE*. 2009; 97(2): 205–230.

[18] G. Samorodnitsky, M.S. Taqqu. *Stable Non-Gaussian Random Processes: Stochastic Models with Infinite Variance*. New York, NY: Chapman & Hall; 1994.

[19] J.H. McCulloch, "Numerical approximation of the symmetric stable distribution and density," in *A Practical Guide to Heavy Tails: Statistical Techniques and Applications*. Boston, MA: Birkhauser; 1998.

[20] G.A. Tsihrintzis, C.L. Nikias, "Performance of optimum and suboptimum receivers in the presence of impulsive noise modeled as an alpha-stable process," *IEEE Transactions on Communications*. 1995; 43(2–4): 904–914.

[21] J.G. Gonzalez, G.R. Arce, "Optimality of the myriad filter in practical impulsive-noise environments," *IEEE Transactions on Signal Processing*. 2001; 49(2): 438–441.

[22] A.A. Nasir, X. Zhou, S. Durrani, R.A. Kennedy, "Relaying protocols for wireless energy harvesting and information processing," *IEEE Transactions on Wireless Communications*. 2013; 12(7): 3622–3636.

[23] T. Le, K. Mayaram, T. Fiez, "Efficient far-field radio frequency energy harvesting for passively powered sensor networks," *IEEE Journal of Solid-State Circuits*. 2008; 43(5): 1287–1302.

[24] M. Stoopman, S. Keyrouz, H.J. Visser, K. Philips, W.A. Serdijn, "A self-calibrating RF energy harvester generating 1 V at −26.3 dBm," *Proceedings of 2013 Symposium on VLSI Circuits*; Kyoto, Japan, June 2013.

Chapter 11
Relaying and physical layer security

Raef Bassily and Sennur Ulukus

11.1 Introduction

Modern communication systems are designed and implemented with both reliability and security of communication in mind. Consequently, the aspect of security of communications has been subjected to careful theoretical study and investigation over the last few decades. As cryptography provides us with efficient and practical solutions for the security problem that are acceptable under reasonable conjectures within the framework of the theory of computational complexity, the underlying physical model of the communication process is not exploited in the cryptographic approach. On the other hand, the study of the physical layer offers a new interesting set of ideas and methods that takes into account the aspects of how communication takes place and exploits them to achieve unconditional security. The theoretical framework of this area is rooted in information theory and is referred to as *information-theoretic security*. Information-theoretic security not only provides us with fundamental limits of secure communication, but it can also provide us with both insights and practical techniques to achieve a robust form of security.

The first rigorous information-theoretic treatment of the security aspect of communication was introduced by Shannon [1] in 1949. In his model, a private, authenticated, and noiseless channel is assumed to be available between the legitimate communicating pair. This channel is used by the communicating pair to agree on a key which is later used by one of them to encrypt its confidential message to the other on a public noiseless channel where any transmission could be perfectly intercepted by an eavesdropper with unbounded computational power. Shannon showed that to achieve perfect security in this model the length of the key must be at least as long as the length of the confidential message. Later, in 1975, Wyner was the first to introduce the notion of information-theoretic security to channels with imperfections in his seminal work [2]. In his *wiretap channel*, the sender and the receiver of a confidential message are connected by only one imperfect communication channel which is wiretapped by a passive but informed eavesdropper with unbounded computational power. Wyner showed that one can indeed exploit the channel randomness to pay for the extra randomness required to encrypt the message and achieve secure communication without the help of a private channel between the legitimate pair. In particular, Wyner introduced a measure for security called the *equivocation* which is defined as the conditional entropy of the message given the eavesdropper's observation normalized by the length of the transmission duration. Accordingly, for an information rate to be secure, the normalized mutual information between the message and the eavesdropper's observation must go to zero as the length of the transmission duration goes to infinity. Wyner obtained an expression for the supremum of the set of achievable secure information rates known as the *secrecy capacity* of the wiretap channel. However, in Wyner's model, it was assumed that the received signal by the eavesdropper is a *degraded* version of the signal received by the legitimate receiver. This constraint was later removed by Csiszar and Korner in Reference 3 where they obtained the secrecy capacity in the general case. For a model well suited to the wireless channels, Leung-Yan-Cheong and Hellman obtained the secrecy capacity of the Gaussian wiretap channel in Reference 4. Starting from Wyner's work, the basic underlying idea of these works is that it is possible to exploit the characteristics of the communication channel to achieve information-theoretic security. When communication takes place over a wireless channel, the properties of such a channel, e.g., its broadcast nature, can be exploited to enable secure communication at high rates. The broadcast nature of the wireless channel gives rise to two relevant concepts, namely, interference and cooperation. These two concepts are shown to be useful in the context of secure communication. In particular, in a cooperative wireless network, a trusted node can help to increase the secure communication rate of the legitimate source–destination pair either by introducing a useful interference to confuse the eavesdropper and hence limit its ability to obtain any information about the transmitted message, i.e., by *passive (deaf) cooperation*, or by listening to the sender's transmission and accordingly helping communicate the sender's message to the receiver, i.e., by *active cooperation*.

In this chapter, we investigate the problem of physical-layer security in wireless networks with cooperative relays. In Sections 11.2 and 11.3, we study the notion of *deaf cooperation* to reinforce security of transmission in wireless relay networks. We distinguish between two main approaches to deaf cooperation, namely, the *cooperative jamming* (CJ) approach and the *noise forwarding* (NF) approach. In a CJ scheme, a helping interferer transmits Gaussian noise when it can hurt the eavesdropper more than it can hurt the legitimate receiver, and hence improves the achievable secure information rate. The idea of introducing artificial noise in a Gaussian wiretap channel by a helper node was introduced in References 5–7. In relay networks with secrecy constraints, the role of CJ was further investigated, e.g., in References 8–10. References 11–13 proposed CJ strategies for multiple-antenna relay networks. On the other hand, in the NF scheme which was introduced in Reference 14, the relay node sends a dummy (context-free) codeword drawn at random from a codebook that is known to both the legitimate receiver and the eavesdropper to introduce helpful interference that would hurt the eavesdropper more than the legitimate receiver.

In Section 11.2, we investigate the role of a deaf helper in improving the achievable secrecy rates of a Gaussian wiretap channel by using either the CJ mode or the NF mode of deaf cooperation. We derive the conditions under which each mode of deaf cooperation improves over the secrecy capacity of the original wiretap channel and show that a helping node can be either a useful cooperative jammer or a useful noise forwarder but not both at the same time. We discuss some key features in the optimal power allocation for both the source and the helping node for each of the two modes of deaf cooperation. Next, we consider the deaf helper selection problem where there are N relays present in the system and it is required to select the best K deaf helpers, $K \geq 1$, that yield the maximum possible achievable secrecy rate with deaf cooperation using K relays. In the special case of $K = 1$, i.e., when selecting a single deaf helper, the optimal strategy is straightforward and requires $O(N)$ computations. For the general case, i.e., when $K > 1$, the computational complexity of the optimal selection is prohibitive. We propose a suboptimal computationally efficient strategy for the selection problem in the general case.

In Section 11.3, we study the CJ and the NF modes of deaf cooperation when the helper node is equipped with multiple antennas. We regard the channel from the helper to the eavesdropper as being composed of two orthogonal components: one that is aligned with the channel between the helper and the legitimate receiver (direct component) and another that is orthogonal to the channel between the helper and the legitimate receiver (orthogonal component). We propose a strategy in which the helper uses the orthogonal component to transmit pure Gaussian noise as in the CJ strategy while it uses the direct component for either CJ or NF depending on the given channel conditions. We discuss the optimal power control policy for this strategy and show the channel conditions where CJ is better than NF for signaling over the direct component and vice versa. Next, we consider the reversely degraded multiple-antenna relay–eavesdropper channel. We show that a simple strategy in which the relay jams with full power along the orthogonal component and transmits nothing in the direct component achieves a secrecy rate that approaches the secrecy capacity of this channel as the relay's average power goes to infinity.

We then turn our attention to the role of active cooperation for secrecy in wireless relay networks. In active cooperation, the relay listens to the source transmissions and uses its observation to improve the achievable secrecy rate. This mode is based on well-known strategies, e.g., decode-and-forward (DF), compress-and-forward (CF), and amplify-and-forward (AF) strategies, devised originally for cooperative models with no secrecy constraint. These strategies were first introduced in Reference 15 for the basic relay channel. In the context of multiple relay networks with no secrecy constraints, References 16 and 17 proposed multi-hop DF strategies. In Reference 14, the basic relay–eavesdropper channel was introduced and the basic strategies of Reference 15 were extended to the secrecy context. In this chapter, we focus on DF-based strategies for multiple relay networks with secrecy constraint. In References 18 and 19, two-stage (half-duplex) cooperative secrecy protocols were proposed in which a set of multiple relays decode the source's message in the first stage, then the relays forward the source's message to the destination using beamforming. Both references investigated the role of the beamforming relays in improving secrecy.

In Section 11.4, we investigate full-duplex relaying strategies for secrecy in cooperative multiple relay networks. We first study the DF strategy for secrecy in a single relay channel with an eavesdropper. We propose a suboptimal decode-and-forward with zero-forcing (DF/ZF) strategy. Next, we consider the general problem with multiple relays. We propose three strategies based on the DF/ZF technique. In the first strategy, all the relays decode the source message at the same time, and then perform beamforming. A major disadvantage in this strategy is that the relays far from the source create a bottleneck and limit the achievable rate. In the second strategy, the relays are ordered with respect to their distance from the source and they perform DF in a multi-hop fashion. This strategy is shown to overcome the main drawback of the first strategy. We discuss the ZF technique in the second strategy and show that only half of the relays' signals can be eliminated from the eavesdropper's observation. Hence, we propose a third strategy which is also a multi-hop DF strategy where it is possible to fully eliminate all the relays' signals from the eavesdropper's observation.

11.2 Deaf cooperation in multiple relay networks

Deaf cooperation refers to a helper (relay) transmitting signals that are independent of the message signals of the source node. There are essentially two known ways of doing this: CJ [6, 7] and NF [14]. While the original CJ in References 6 and 7 was done by using independent identically distributed (i.i.d.) Gaussian signals over a multiple access channel, the concept of CJ is much more general and in fact, has become an integral part of achievable schemes in many multi-user extensions of the wiretap channel. There have been several extended CJ mechanisms which do not use i.i.d. signaling and/or Gaussian signals, see for example, References 20–25, however, in this chapter, we will focus on the original form of CJ using i.i.d. Gaussian signals. In the sequel, unless stated otherwise, we will use the term CJ to refer to this particular form of jamming. In NF [14], on the other hand, a helper relay terminal transmits additional randomness in the form of randomly chosen (non-information carrying) codewords from a known codebook instead of transmitting i.i.d. Gaussian noise signals. The major difference between CJ with Gaussian noise and NF with a Gaussian codebook is that, in the latter, by choosing the rates appropriately, the legitimate user can be enabled to decode the helper (added confusion) signal, hence receiving a clean information-carrying signal whereas the eavesdropper's channel remains jammed; while in the former, both legitimate and eavesdropping links are jammed simultaneously. These strategies can outperform one another depending on the channel conditions.

Both strategies have been used to extend the concept of multi-user secrecy to networks of relays in Reference 26. We discuss here some of the main ideas and techniques for incorporating both strategies to ensure physical-layer security in a multiple relay network. We consider the communication scenario illustrated in Figure 11.1. A source, s, sends a confidential message to a destination, d, over an additive white Gaussian noise (AWGN) channel in the presence of an informed eavesdropper, e. The communication occurs in the presence of a set of N nodes, $\mathcal{N} = \{r_1, \ldots, r_N\}$, from which K nodes are selected to help improving the achievable secure communication rate between s and d via deaf cooperation, i.e., via CJ or NF, or both. Let $\mathcal{K}$ denotes the set of selected helpers. The signals observed by the destination d and the eavesdropper e can be expressed as

$$Y = \sqrt{\gamma_{s,d}}X_s + \sum_{r\in\mathcal{K}} \sqrt{\gamma_{r,d}}X_r + \tilde{N} \tag{11.1}$$

$$Z = \sqrt{\gamma_{s,e}}X_s + \sum_{r\in\mathcal{K}} \sqrt{\gamma_{r,e}}\tilde{X}_r + \tilde{N}' \tag{11.2}$$

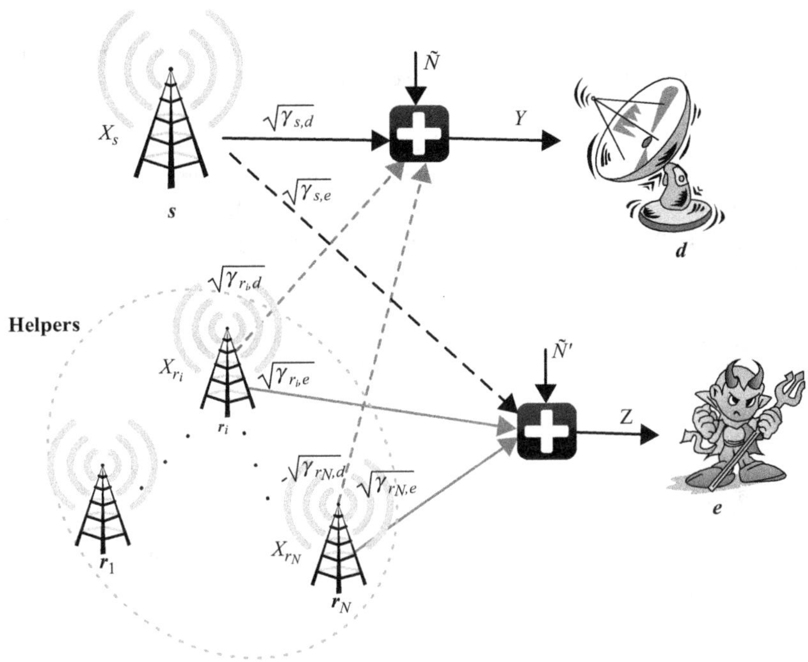

Figure 11.1 A network with multiple deaf helpers

where $\gamma_{i,j}$ is the channel gain between nodes i and j, X_i is the channel input at node i, and $\tilde{N}$, $\tilde{N}'$ are real-valued zero mean, unit variance AWGN at the destination and the eavesdropper, respectively. The channel inputs satisfy the following average power constraints $\mathbb{E}\,[X_i^2] \leq \bar{p}_i$, $i = s, r \in \mathcal{K}$.

11.2.1 CJ versus NF

Let us consider the case of a *single* deaf helper, i.e., $|\mathcal{K}| = K = 1$. We will compare the CJ and NF strategies in this setting. Generally speaking, it is not useful to perform CJ when the helper is closer to the destination than to the eavesdropper; on the other hand, one can still introduce helpful interference in this case by transmitting a dummy codeword from a codebook that is known to the destination and the eavesdropper. The transmission of dummy codewords refers to Wyner's idea of stochastic encoding for secrecy [2] where multiple codewords are associated with a single message. Since the cost of these dummy codewords is a decrease in the transmitter's rate, if the helper takes the responsibility of sending these dummy codewords, then the secrecy rate of the transmitter may improve [27].

Now, let us formalize these observations. In both CJ and NF schemes, the channel input at the source X_s in (11.1) and (11.2) is a symbol of the codeword that represents the encoded confidential message. Such codeword is drawn from an i.i.d. Gaussian codebook, i.e., X_s is a Gaussian random variable with zero mean and variance P_s where

$P_s \leq \bar{\rho}_s$. In both schemes, the channel input at the deaf helper X_r is also Gaussian with zero mean and variance P_r where $P_r \leq \bar{\rho}_r$. However, the difference between the two schemes comes from the origin of X_r. In the CJ scheme, X_r is white Gaussian noise that plays the same role as the background noise at the destination and the eavesdropper except for the fact that it is generated artificially. On the other hand, in the NF scheme, X_r is a symbol of a dummy (context-free) codeword drawn from a Gaussian codebook that is assumed to be available at both the destination and the eavesdropper. Accordingly, for given power values P_s and P_r, the secrecy rate achievable by the CJ scheme [7], R^{CJ}, is given by

$$R^{\mathrm{CJ}}(P_s, P_r) = \frac{1}{2}\log\left(\frac{(1+\gamma_{s,d}P_s+\gamma_{r,d}P_r)(1+\gamma_{r,e}P_r)}{(1+\gamma_{s,e}P_s+\gamma_{r,e}P_r)(1+\gamma_{r,d}P_r)}\right) \tag{11.3}$$

Whereas the secrecy rate achievable by the NF scheme [14], R^{NF}, is given by

$$R^{\mathrm{NF}}(P_s, P_r) = \min\left\{\frac{1}{2}\log\left(\frac{(1+\gamma_{s,d}P_s)(1+\gamma_{r,e}P_r)}{1+\gamma_{s,e}P_s+\gamma_{r,e}P_r}\right), \frac{1}{2}\log\left(\frac{1+\gamma_{s,d}P_s+\gamma_{r,d}P_r}{1+\gamma_{s,e}P_s+\gamma_{r,e}P_r}\right)\right\} \tag{11.4}$$

On the other hand, when no helper node is involved, the secrecy capacity of the original Gaussian wiretap channel [4] for a given power value P_s is given by

$$C^{\mathrm{GWT}}(P_s) = \left(\frac{1}{2}\log\left(\frac{1+\gamma_{s,d}P_s}{1+\gamma_{s,e}P_s}\right)\right)^+ \tag{11.5}$$

where $(x)^+ = \max(0, x)$.

Using these expressions, it is not hard to derive conditions under which one strategy outperforms the other in terms of the achievable secrecy rate. The reader may refer to Reference 26 for more details. However, one important observation to state here is that $\gamma_{r,e}/\gamma_{r,d} \geq \max(\gamma_{s,e}/\gamma_{s,d}, 1)$ is a necessary condition for the CJ scheme to achieve higher secrecy rate than the secrecy capacity of the original Gaussian wiretap channel. On the other hand, $\gamma_{r,e}/\gamma_{r,d} \leq 1$ is a necessary condition for the NF scheme to achieve higher secrecy rate than the secrecy capacity of the original Gaussian wiretap channel. In Reference 26, the optimal power control policies are derived for (11.3) and (11.4). Based on such policies, one can reach slightly stronger necessary conditions for achieving higher rate than the capacity of the Gaussian wiretap channel via deaf cooperation. Namely, one can show that if the optimal value of (11.3), denoted as $\bar{R}^{\mathrm{CJ}}$, is strictly larger than C^{GWT}, then we must have $\gamma_{r,e}/\gamma_{r,d} > \max(\gamma_{s,e}/\gamma_{s,d}, 1)$. On the other hand, if the optimal value of (11.4), denoted as $\bar{R}^{\mathrm{NF}}$, is strictly larger than C^{GWT}, then we must have $\gamma_{r,e}/\gamma_{r,d} < \min\left(1, \frac{1+\gamma_{s,e}\bar{\rho}_s}{1+\gamma_{s,d}\bar{\rho}_s}\right)$. We note that these conditions imply that a useful deaf helper can either be a cooperative jammer or a noise forwarder but not both.

11.2.2 Deaf helpers selection

We now discuss strategies for selecting relays that act as deaf helpers for the purpose of boosting the secure communication rate between the legitimate pair. We will first consider the case where a single relay is to be selected out of the set $\mathcal{N}$ of available relays, then we will discuss the general case where $K > 1$ relays are to be selected.

11.2.2.1 Single deaf helper selection strategy

Using the conditions discussed above, one can find two disjoint subsets $\mathcal{N}^{\mathrm{CJ}}$ and $\mathcal{N}^{\mathrm{NF}}$ of $\mathcal{N}$ where

$$\mathcal{N}^{\mathrm{CJ}} \triangleq \left\{ r_j \in \mathcal{N} : \frac{\gamma_{r_j,e}}{\gamma_{r_j,d}} > \max\left(1, \frac{\gamma_{s,e}}{\gamma_{s,d}}\right)\right\} \tag{11.6}$$

is the set of potential cooperative jammers, and

$$\mathcal{N}^{\mathrm{NF}} \triangleq \left\{ r_j \in \mathcal{N} : \frac{\gamma_{r_j,e}}{\gamma_{r_j,d}} < \min\left(1, \frac{1+\gamma_{s,e}\bar{\rho}_s}{1+\gamma_{s,d}\bar{\rho}_s}\right)\right\} \tag{11.7}$$

is the set of potential noise forwarders.

For each $r \in \mathcal{N}$, using its knowledge of its own channel gains, r identifies which mode of cooperation (CJ or NF) based on the conditions in (11.6) and (11.7). Accordingly, r computes one of the two rates $\bar{R}^{\mathrm{CJ}}$ and $\bar{R}^{\mathrm{NF}}$ depending on its target mode of cooperation where the rate is computed using the optimal power control policy. Then r sends this information to s. Upon receiving such information from all $r \in \mathcal{N}$, s identifies the relay r^* with the maximum rate R^* and knows its mode of cooperation. Consequently, s notifies r^* that it has been selected as the optimal deaf helper which in turn notifies d of the former's selection. It is assumed that this information is also intercepted by e. By executing this strategy, the optimal single deaf helper r^* is identified together with its mode of deaf cooperation.

11.2.2.2 Multiple deaf helpers selection strategy

Suppose that the system permits us to involve at most K relays, $1 \leq K \leq N$, as deaf helpers. Each relay can be either a cooperative jammer or a noise forwarder. Let $\mathcal{K}^{\mathrm{CJ}} \subseteq \mathcal{N}^{\mathrm{CJ}}$ denotes the set of the selected cooperative jammers and $\mathcal{K}^{\mathrm{NF}} \subseteq \mathcal{N}^{\mathrm{NF}}$ denotes the set of the selected noise forwarders where $|\mathcal{K}^{\mathrm{CJ}} \bigcup \mathcal{K}^{\mathrm{NF}}| \leq K$ (see Figure 11.2). The achievable secrecy rate in this case for fixed power values ρ_s, ρ_r, $r \in \mathcal{K}^{\mathrm{CJ}} \bigcup \mathcal{K}^{\mathrm{NF}}$ is given as a function of $(\mathcal{K}^{\mathrm{CJ}}, \mathcal{K}^{\mathrm{NF}})$ by

$$\begin{aligned} R(\mathcal{K}^{\mathrm{CJ}}, \mathcal{K}^{\mathrm{NF}}) = \min_{\mathcal{M}\subseteq\mathcal{K}^{\mathrm{NF}}} & \left\{ \frac{1}{2}\log\left(\frac{1+\gamma_{s,d}P_s+\sum_{r\in\mathcal{M}}\gamma_{r,d}P_r}{1+\sum_{r\in\mathcal{K}^{\mathrm{CJ}}}\gamma_{r,d}P_r}\right)\right. \\ & \left. -\frac{1}{2}\log\left(\frac{1+\gamma_{s,e}P_s+\sum_{r\in\mathcal{M}}\gamma_{r,e}P_r}{1+\sum_{r\in\mathcal{K}^{\mathrm{CJ}}}\gamma_{r,e}P_r+\sum_{r\in\mathcal{K}^{\mathrm{NF}}\setminus\mathcal{M}}\gamma_{r,e}P_r}\right)\right\} \end{aligned} \tag{11.8}$$

When there are more than one deaf helper, the problem of finding an optimal power control policy for (11.8) becomes analytically intractable and no closed-form

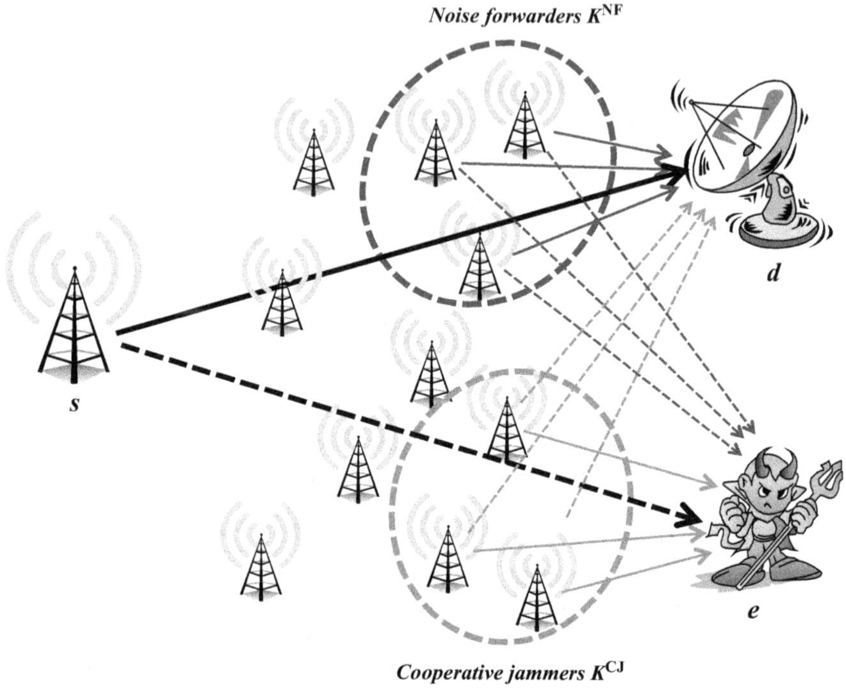

Figure 11.2 Multiple deaf helpers selection: typically yields a set of cooperative jammers $\mathcal{K}^{CJ}$ and a set of noise forwarders $\mathcal{K}^{NF}$

solution could be found. One could possibly resort in this case to numerical algorithms. However, numerical algorithms usually have large running time and their convergence to the global optimum is not guaranteed. Hence, using one such algorithm inside a selection strategy will slow it down and substantially increase the total number of computations carried out through the selection strategy. Thus, in the context of the selection problem, using numerical methods for finding the optimal power allocations is not an efficient option. Therefore, when the number of deaf helpers to be selected is greater than 1, we will use a fixed power allocation policy. That is, when $K > 1$, we set $P_s = \bar{\rho}_s$ and $P_r = \bar{\rho}_r$, $r \in \mathcal{N}$.

Now, for the selection problem, it is clear that in an optimal selection strategy, which picks the best K deaf helpers out of the N available relays, one would possibly have to compute the achievable rate for every possible disjoint pair of subsets $\mathcal{K}^{CJ}$, $\mathcal{K}^{NF} \subseteq \mathcal{N}$ such that $|\mathcal{K}^{CJ} \bigcup \mathcal{K}^{NF}| \leq K$, and then picks the pair that achieves the maximum. Unfortunately, this strategy is computationally expensive especially for large values of N and K. Thus, we describe below a suboptimal strategy that requires significantly less computations. This strategy builds upon the single deaf helper selection strategy discussed above. In particular, the strategy follows a *greedy* algorithm that runs for at most K iterations where in each iteration the *best single deaf helper* is selected.

We now describe the steps of the *multiple deaf helpers selection (MDHS)* strategy. First, fix the power allocation policy as $P_s = \bar{\rho}_s$ and $P_r = \bar{\rho}_r$, $r \in \mathcal{N}$, then do the following:

1. Find the best deaf helper, its mode of cooperation, and the corresponding achievable rate as in the *single deaf helper selection strategy*.
2. While the number of selected deaf helpers is less than K, do the following:
   a. Every *unselected* node, in the set available relays $\mathcal{N}$, computes the achievable secrecy rate twice: once when it adds itself to the set of already selected cooperative jammers, and another time when it adds itself to the set of already selected noise forwarders. This is done using (11.8). Then, it finds the maximum of the two rates, identifies the corresponding mode of deaf cooperation (whether it is CJ or NF), and sends the resulting rate and the mode of cooperation to the source.
   b. The source finds the maximum of all the values it receives from all the relays involved in Step 2a. This is the value of the achievable rate of the current selection stage. The source also identifies the corresponding relay whose rate is the maximum among all the rates it received in Step 2a and identifies its mode of cooperation (CJ or NF). Then, the source compares the current rate with the rate value obtained in the previous selection stage.
      i. **If** the rate obtained in the previous selection stage is greater than or equal to the current rate value: **stop** and **output** the selected helpers (the set of the selected cooperative jammers and the set of the selected noise forwarders up to the previous selection stage), and the achievable rate obtained in the previous selection stage as the resulting achievable rate.
      ii. **Otherwise**, the source **updates** the value of the achievable rate with the current value it obtained in Step 2b, **adds** the corresponding relay either to the set of the selected cooperative jammers or the set of the selected noise forwarders depending on its mode of deaf cooperation identified in Step 2b above.

When K is a fixed constant (i.e., it does not depend on N), the number of computations required in the above strategy can be shown to be $O(N)$ while the number of computations in the optimal selection strategy is $\Omega(N^K)$. When K is allowed to grow with N, in particular, when $K/N = \alpha$ for some fixed $\alpha \in (0, 1]$, the number computations required in the above strategy is $O(N^2 2^{\alpha N})$ while the number of computations required in the optimal strategy is $\Omega(N 2^{\alpha N \log(\frac{3}{\alpha})})$. This shows that the complexity reduction achieved by the selection strategy proposed above compared to the optimal selection strategy is of order $O(N^{K-1})$ when K is a fixed constant, and is of exponential order, namely, $O\left(\frac{2^{\alpha N \log(\frac{3}{2\alpha})}}{N}\right)$ when $K = \alpha N$ for some $\alpha \in (0, 1]$. This is a substantial reduction in complexity leading to a faster and more efficient selection. For example, when $N = 50$ and $K = 25$, the reduction factor is approximately 2^{55}.

11.2.3 Empirical evaluation

First, let us consider the single deaf helper case. For simpler notation, we will normalize the gains of the source–eavesdropper and the relay–eavesdropper channels by the respective channel gains to the destination. In other words, we will be dealing with relative channel gains rather than the absolute channel gains. We define the relative channel gain of the source by $h_s \triangleq \gamma_{s,e}/\gamma_{s,d}$ and that of the eavesdropper by $h_r \triangleq \gamma_{r,e}/\gamma_{r,d}$. We compare the two modes of deaf cooperation by plotting the optimal secrecy rate achievable by each of CJ and NF modes against the relative channel gain between the deaf helper and the eavesdropper, h_r. We also normalize the power constraints $\bar{\rho}_s$ and $\bar{\rho}_r$ of the source and the relay by $\gamma_{s,d}$ and $\gamma_{r,d}$, respectively.

In Figure 11.3, we set the normalized power constraints of the source and the deaf helper as $\bar{\rho}_s = \bar{\rho}_r = 5$. We consider two cases. In the first case, we choose $h_s < 1$, namely, we set $h_s = 0.75$. In the second case, we choose $h_s > 1$, namely, $h_s = 1.25$. For each case, we plot $\bar{R}^{\mathrm{CJ}}$ and $\bar{R}^{\mathrm{NF}}$ versus the relative channel gain h_r. We observe that $\bar{R}^{\mathrm{CJ}} = C^{\mathrm{GWT}}$ when $h_r \leq \max(1, h_s)$ and $\bar{R}^{\mathrm{CJ}} > C^{\mathrm{GWT}}$ otherwise. One can also see that $\lim_{h_r \to \infty} \bar{R}^{\mathrm{CJ}}(h_r) = C^{\mathrm{G}}$ where C^{G} is the capacity of the Gaussian channel between the source and the destination when no secrecy constraint is imposed, i.e., when the eavesdropper is not present. On the other hand, we observe that $\bar{R}^{\mathrm{NF}} = C^{\mathrm{GWT}}$ when $h_r \geq \min\left(1, \frac{1+h_s\bar{\rho}_s}{1+\bar{\rho}_s}\right)$ whereas $\bar{R}^{\mathrm{NF}} > C^{\mathrm{GWT}}$ otherwise.

Next, we consider the multiple deaf helper case. Consider a disk of radius 1 km where the source is located at the center, both the destination and the eavesdropper are located at some fixed points on the circumference. Consider N relays whose locations are chosen randomly and uniformly in this disk. Each channel gain is generated according to the formula: $\gamma = \frac{SV}{d^{\zeta}}$ where γ is the channel gain, S is a lognormal random variable to account for shadowing, and V is a Rayleigh random variable for fading, d is the distance, and ζ is the path loss [28]. We assume that the underlying Gaussian random variables from which S and V are generated are independent, zero mean, and unit variance Gaussian random variables. We also take $\zeta = 3$. We set $\bar{\rho}_s = 10$ and $\bar{\rho}_r = 1$, $\forall r \in \mathcal{N}$.

In Figure 11.4, we plot the achievable secrecy rate against the maximum allowed number of helpers, K, for $N = 25$ and 50, in three different cases. In the first case, the secrecy rate is obtained using the *MDHS* strategy described in the previous section. In the second case, we only consider CJ as the only deaf cooperation mode, i.e., ignore all the relays that could be useful noise forwarders and use the MDHS strategy only for useful cooperative jammers. In the third case, we consider only NF as the only mode available for deaf cooperation. It is clear from Figure 11.4 that making use of the two modes (CJ/NF) together in the system could significantly increase the achievable secrecy rates. Also, we notice that one could benefit from considering a larger set of relays, i.e., larger N, as this may lead to a better selected set of helpers.

In Figure 11.5, the achievable secrecy rate, R^*, is plotted against the maximum allowed number of helpers, K, for three different realizations of the relays where $N = 50$. It can be seen that the selected helpers could be cooperative jammers (CJ)

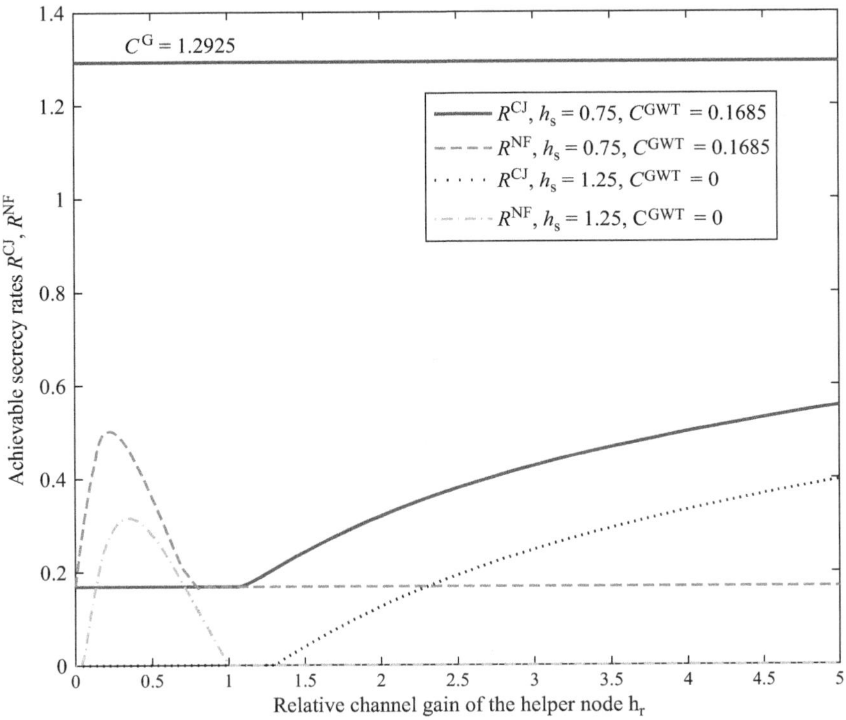

Figure 11.3 *The optimal achievable rates by the two modes of deaf cooperation, $\bar{R}^{CJ}$ and $\bar{R}^{NF}$ as functions of the relative channel gain from the deaf helper to the eavesdropper h_r, plotted for two cases of the relative channel gain from the source to the eavesdropper h_s*

or noise forwarders (NF), or both, and that one can improve the achievable rate by selecting more than one helper. One can also see that the number of selected helpers could be less than K. Specifically, for the realizations considered here, the numbers of selected helpers are 2, 4, and 6.

11.3 Deaf cooperation with multiple antennas

In this section, we consider the role of a deaf helper with multiple antennas in improving secure communication rates over the Gaussian wiretap channel. Interestingly, this extension leads to a new set of results that were not available in the single antenna case. In particular, having multiple antennas at the relay allows us to decompose the relay–eavesdropper channel into two orthogonal components, one in the direction of the relay–destination channel (direct component) and the other in the orthogonal direction to the relay–destination channel (orthogonal component). Accordingly, we

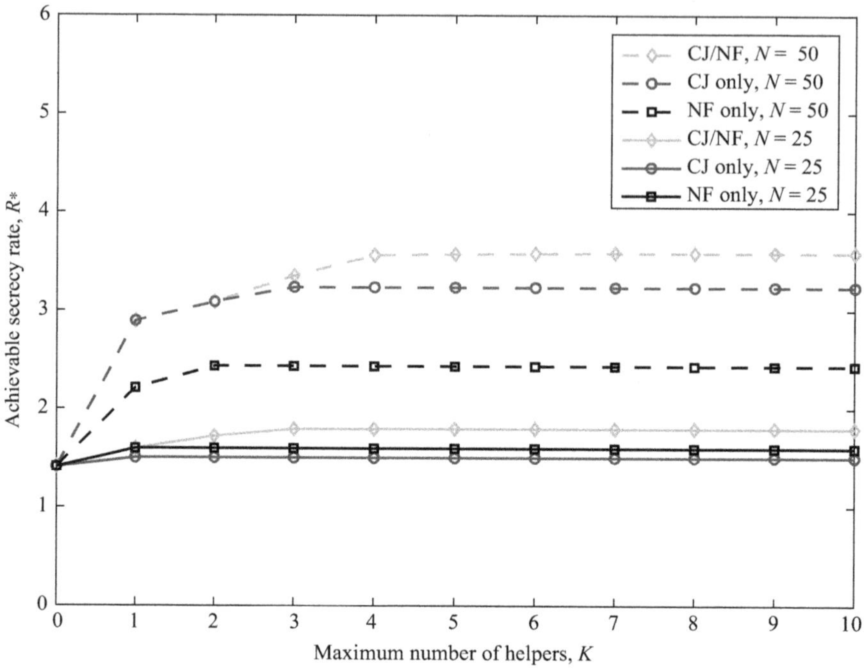

Figure 11.4 The achievable secrecy rate, R^*, versus the maximum allowed number of deaf helpers, K, for three cases: CJ/NF, NF only, and CJ only. This is done for $N = 25$ and 50

can consider using each of these components for deaf cooperation (CJ or NF). It is intuitive that the orthogonal component should be used for CJ. However, it is not clear what strategy should be used along the direct component. It is not also clear how the relay should distribute its power on these two components.

In this section, we highlight the answers to these questions. We discuss the necessary conditions for each of the CJ and the NF strategy to be useful when employed along the direct component, i.e., to improve over the best secrecy rate achievable when the transmission from the relay is constrained to the orthogonal component only. In particular, it is shown that along the direct component of the channel either CJ is useful or NF is useful but not both. Moreover, there are some cases in which neither CJ nor NF is useful along the direct component. The reader may refer to Reference 29 for more details.

Next, we turn our attention to a certain class of the multiple-antenna relay–eavesdropper channels, namely, the reversely degraded channel. The strategy in which the relay jams with full power along the orthogonal component of the channel and transmits nothing in the direct component is shown to be optimal when the relay's average power goes to infinity. In particular, the secrecy rate achieved by this strategy approaches the capacity of the reversely degraded multiple-antenna relay channel as the relay's average power increases, and hence this strategy achieves the optimal

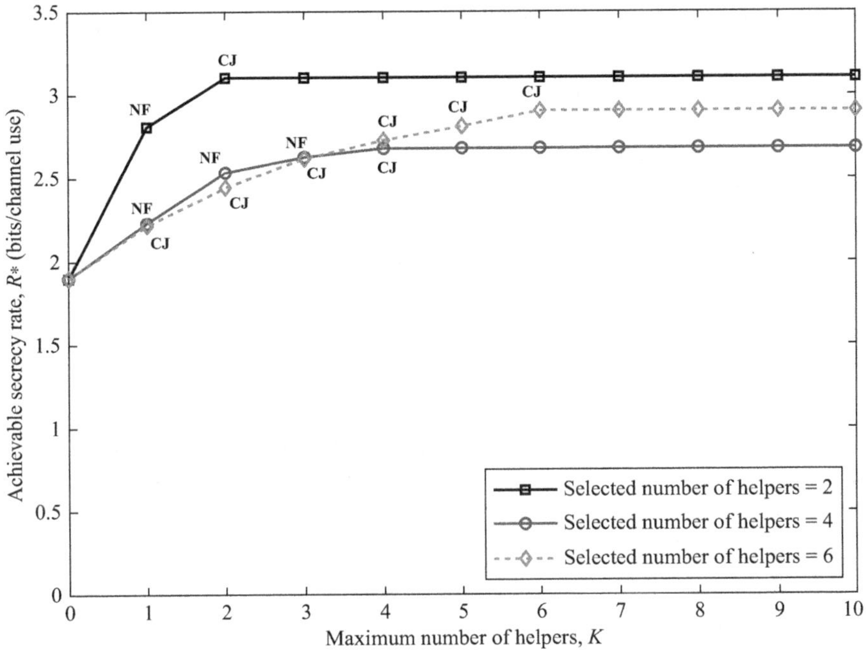

Figure 11.5 The achievable secrecy rate versus the maximum allowed number of helpers, K, for three different realizations of relays locations, for N = 50

secure degrees of freedom (DoF) of the reversely degraded multiple-antenna relay–eavesdropper channels.

The system can be described by the following equations:

$$Y = X_s + \mathbf{h}_r^T \mathbf{X}_r + \tilde{N} \tag{11.9}$$

$$Z = \sqrt{g_s} X_s + \mathbf{g}_r^T \mathbf{X}_r + \tilde{N}' \tag{11.10}$$

where $\mathbf{h}_r \in \mathbb{R}^K$ is the vector of channel gains between the helper r and the destination d, and $g_s \in \mathbb{R}$ and $\mathbf{g}_r \in \mathbb{R}^K$ are the scalar channel gain and the vector of channel gains from the source s and the helper r to the eavesdropper, respectively. The channel inputs are subjected to the following average power constraints:

$$E\left[|X_s|^2\right] \leq \bar{\rho}_s \quad \text{and} \quad E\left[\|\mathbf{X}_r\|^2\right] \leq \bar{\rho}_r \tag{11.11}$$

The relay–eavesdropper channel vector $\mathbf{g}_r$ can be written as $\mathbf{g}_r = \sqrt{\alpha}\mathbf{h}_r + \mathbf{u}_r$ where $\mathbf{h}_r^T \mathbf{u}_r = 0$, i.e., we decompose $\mathbf{g}_r$ into two orthogonal components: $\sqrt{\alpha}\mathbf{h}_r$ (in the direction of the legitimate channel $\mathbf{h}_r$) and $\mathbf{u}_r$ (orthogonal to $\mathbf{h}_r$). Hence, one can write $\mathbf{X}_r$ in (11.9) and (11.10) as the sum of two orthogonal components: $\mathbf{X}_{r0}$ (in the direction of $\mathbf{h}_r$) and $\mathbf{X}_{r1}$ (in the direction of $\mathbf{u}_r$). That is, $\mathbf{X}_r = \mathbf{X}_{r0} + \mathbf{X}_{r1}$ where

$$\mathbf{X}_{r0} = X_{r0}\mathbf{h}_r = \frac{\mathbf{h}_r^T \mathbf{X}_r^T}{\gamma_{r0}}\mathbf{h}_r \tag{11.12}$$

$$\mathbf{X}_{r1} = X_{r1}\mathbf{u}_r = \frac{\mathbf{u}_r^T \mathbf{X}_r}{\gamma_{r1}}\mathbf{u}_r \tag{11.13}$$

where $\gamma_{r0} = \|\mathbf{h}_r\|^2$ and $\gamma_{r1} = \|\mathbf{u}_r\|^2$. Thus, we can write (11.9) and (11.10) as

$$Y = X_s + X_{r0} + \tilde{N} \tag{11.14}$$

$$Z = \sqrt{g_s}X_s + \sqrt{\alpha}X_{r0} + X_{r1} + \tilde{N}' \tag{11.15}$$

Clearly, X_{r0} and X_{r1} are the scaled projections of $\mathbf{X}_r$ in the direction of $\mathbf{h}_r$ and $\mathbf{u}_r$, respectively. We will assume that X_{r0} and X_{r1} are uncorrelated. We denote the power of the direct component by $Q_{r0} \triangleq E[X_{r0}^2]$ and the power of the orthogonal component by $Q_{r1} \triangleq E[X_{r1}^2]$. Hence, the second constraint in (11.11) is equivalent to

$$\frac{Q_{r0}}{\gamma_{r0}} + \frac{Q_{r1}}{\gamma_{r1}} \leq \bar{\rho}_r \tag{11.16}$$

11.3.1 CJ versus NF in multi-antenna deaf cooperation

Now, we consider the possible signaling strategy for X_{r0} and X_{r1}, the two signaling strategies CJ and NF. Clearly, if the CJ strategy is used for X_{r1}, the eavesdropper is the only one who is possibly harmed by the resulting noise, not the destination. Hence, we assume that the helper will use the orthogonal component X_{r1} for CJ, i.e., X_{r1} is a Gaussian random variable with zero mean and variance Q_{r1}. Hence, we distinguish between two possible strategies depending on whether the helper uses the direct component X_{r1} for CJ or NF. In both strategies, the channel input at the source X_s is a symbol of the codeword that represents the encoded confidential message. Such codeword is drawn from an i.i.d. Gaussian codebook, i.e., X_s is a Gaussian random variable with zero mean and variance P_s where $P_s \leq \bar{\rho}_s$. Also, in both strategies, the direct component of the channel input at the helper X_{r0} is a Gaussian random variable with zero mean and variance Q_{r0}. The difference between the two strategies comes from the origin of X_{r0}. In the CJ strategy, X_{r0} is a Gaussian random variable that plays the role of background noise at both the destination and the eavesdropper except for the fact that it is generated artificially. On the other hand, in the NF strategy, X_{r0} is a symbol of a dummy (context-free) codeword drawn from an i.i.d. Gaussian codebook that is assumed to be available at both the destination and the eavesdropper.

If X_{r0} is used for CJ, the achievable secrecy rate [7], denoted as R^{CJ}, is given by

$$R^{\mathrm{CJ}} = \frac{1}{2}\log\left(\frac{(1+P_s+Q_{r0})(1+\alpha Q_{r0}+Q_{r1})}{(1+g_sP_s+\alpha Q_{r0}+Q_{r1})(1+Q_{r0})}\right) \tag{11.17}$$

On the other hand, if X_{r0} is used for NF, the achievable secrecy rate [14], denoted as R^{NF}, is given by

$$R^{\text{NF}} = \min\left\{\frac{1}{2}\log\left(\frac{(1+P_s)(1+\alpha Q_{r0}+Q_{r1})}{1+g_sP_s+\alpha Q_{r0}+Q_{r1}}\right), \frac{1}{2}\log\left(\frac{(1+P_s+Q_{r0})(1+Q_{r1})}{1+g_sP_s+\alpha Q_{r0}+Q_{r1}}\right)\right\} \tag{11.18}$$

For the sake of comparison, let R_o denotes the optimal secrecy rate achievable when no transmission is carried out along the direct component of the channel, i.e., when the transmission is constrained only to the orthogonal component of the channel. Hence, R_o is given by

$$R_o = \left(\frac{1}{2}\log\left(\frac{(1+\bar{\rho}_s)(1+\gamma_{r1}\bar{\rho}_r)}{1+g_s\bar{\rho}_s+\gamma_{r1}\bar{\rho}_r}\right)\right)^+ \tag{11.19}$$

In Reference 29, a detailed discussion is given for the optimal power control policies for (11.17) and (11.18). One important consequence of these policies is deriving conditions that tell us when it is better to use CJ rather than NF along the direct component X_{r0} and vice versa. We summarize these conditions here.

Let $\tilde{\alpha} = \frac{\alpha - \frac{\gamma_{r1}}{\gamma_{r0}}}{1+\gamma_{r1}\bar{\rho}_r}$, $\tilde{g}_s = \frac{g_s}{1+\gamma_{r1}\bar{\rho}_r}$, and $\beta = \frac{\gamma_{r1}}{\gamma_{r0}+\gamma_{r0}\gamma_{r1}\bar{\rho}_r}$. For CJ along the direct channel component to be useful, it is necessary to have $\tilde{\alpha} > \max(1, \tilde{g}_s)$. On the other hand, for NF along the direct channel component to be useful, it is necessary to have $0 < \tilde{\alpha} < 1 - \beta$. In other words, if $\bar{R}^{\text{CJ}} > R_o$ then $\tilde{\alpha} > \max(1, \tilde{g}_s)$, whereas, if $\bar{R}^{\text{NF}} > R_o$ then $0 < \tilde{\alpha} < 1 - \beta$, where $\bar{R}^{\text{CJ}}$ and $\bar{R}^{\text{NF}}$ are the optimal values for (11.17) and (11.18), respectively, computed using the optimal power policies in Reference 29. Hence, if $\tilde{\alpha} \leq 0$ or $1 - \beta \leq \tilde{\alpha} \leq \max(1, \tilde{g}_s)$, neither CJ nor NF along the direct component is useful.

11.3.2 The reversely degraded multi-antenna relay–eavesdropper channel

We now turn our attention to a special case of the model above. Here, we assume that the relay receives a vector $\mathbf{Y}_r$ which is a noisy version of the source transmission and hence the relay can use this observation in one way or another to help increasing the achievable secrecy rate. We also assume that, given the relay's channel input $\mathbf{X}_r$, the relay's observation is a degraded version of the destination's observation. In particular, the relay's observation, $\mathbf{Y}_r \in \mathbb{R}^K$, is given by

$$\mathbf{Y}_r = \boldsymbol{\eta} Y + \boldsymbol{\Theta}\mathbf{X}_r + \tilde{\mathbf{N}}_r \tag{11.20}$$

where $\boldsymbol{\eta} \in \mathbb{R}^K$ is the vector of channel coefficients from the destination to the relay, $\boldsymbol{\Theta} \in \mathbb{R}^{K\times K}$ is the matrix of channel coefficients from the relay's input to the relay's ouUput, and $\tilde{\mathbf{N}}_r \in \mathbb{R}^K$ is AWGN vector of zero mean and identity covariance matrix and is independent of $(X_s, \mathbf{X}_r, \tilde{N}, \tilde{N}')$. Accordingly, we have the following Markov chain $X_s \to (Y, \mathbf{X}_r) \to \mathbf{Y}_r$. We further assume that the channel's orthogonal component is non-zero, i.e., $\mathbf{u}_r \neq \mathbf{0}$; hence, given the source's input X_s, neither the destination's observation Y nor the eavesdropper's observation Z is a degraded version of one another.

In this special case of the model, it can be shown that CJ strategy over the orthogonal component X_{r1} and no signaling over the direct component X_{r0} yields a secrecy rate that approaches the secrecy capacity of this channel as $\bar{\rho}_r \to \infty$. As a consequence, this strategy also achieves the optimal secure DoF for this model (see Reference 29 for details). This is formally stated as follows. Let $C_s(\bar{\rho}_r)$ be the secrecy capacity of the reversely degraded relay–eavesdropper channel described above for a given value of the relay's average power constraint $\bar{\rho}_r$. Suppose that $\mathbf{u}_r \neq \mathbf{0}$. Recall that $R_o(\bar{\rho}_r)$, defined in (11.19), is the secrecy rate achievable by using the total source's power $\bar{\rho}_s$ for information transmission and using the total relay's power $\bar{\rho}_r$ for CJ along the orthogonal component of the relay–eavesdropper channel. Then, for every $\varepsilon > 0$, there is a sufficiently large value $\bar{\rho}_r$ such that $R_o(\bar{\rho}_r) > C_s(\bar{\rho}_r) - \varepsilon$. In particular, $\lim_{\bar{\rho}_r \to \infty} R_o(\bar{\rho}_r) = C^{\mathrm{G}}$ where $C^{\mathrm{G}} = \frac{1}{2}\log(1 + \bar{\rho}_s)$ is the capacity of the Gaussian channel between the source and the destination when there is no eavesdropper in the system.

Clearly, C^{G} is an upper bound on the secrecy capacity of the reversely degraded relay–eavesdropper channel since the relay in this case cannot increase the reliable information rate from the source to the destination and hence the capacity of the relay channel with no secrecy constraints is indeed C^{G}.

One can even make a stronger statement than the one above. Even in the case where the relay–eavesdropper channel $\mathbf{g}_r$ is unknown to all the nodes, if we allow the relay to choose a random signaling direction $\mathbf{s}_r \in \mathbb{R}^K$ for jamming in the subspace orthogonal to $\mathbf{h}_r$, then the secrecy rate achievable by this strategy is $R_o(\bar{\rho}_r) = \frac{1}{2}\log(1 + \bar{\rho}_s) - \frac{1}{2}\log\left(\frac{1+\mathbf{g}_r^T\mathbf{s}_r\bar{\rho}_r + g_s\bar{\rho}_s}{1+\mathbf{g}_r^T\mathbf{s}_r\bar{\rho}_r}\right)$. It is clear that $\mathbf{g}_r^T\mathbf{s}_r \neq 0$ with probability 1. Hence, $R_o(\bar{\rho}_r) \to C^{\mathrm{G}}$ almost surely as $\bar{\rho}_r \to \infty$. Thus, even if the relay–eavesdropper's channel $\mathbf{g}_r$ is unknown, the above result still holds with probability 1.

11.3.3 Empirical evaluation

We first compare the optimal secrecy rates R^{CJ} and R^{NF} achievable by the CJ and NF strategies proposed in Section 11.3.1 with the optimal secrecy rate R_o achievable by the strategy that uses only the orthogonal component of the channel for CJ. We also compare these rates to the secrecy capacity C^{GWT} of the original Gaussian wiretap channel with no relay. In Figure 11.6, we set $\bar{\rho}_s = 5$, $\bar{\rho}_r = 2$, $g_s = 0.85$, $\gamma_{r0} = 2$, and $\gamma_{r1} = 1$. We plot R^{CJ}, R^{NF}, R_o, and C^{GWT} versus $\sqrt{\alpha}$, $0 \leq \sqrt{\alpha} \leq 4$, where $\sqrt{\alpha}$ is defined as $\frac{\mathbf{g}_r^T\mathbf{h}_r}{\gamma_{r0}}$. It is clear from Figure 11.6 that the necessary conditions for $R^{\mathrm{CJ}} > R_o$ and $R^{\mathrm{NF}} > R_o$ that we discussed above are satisfied here. Note that the necessary condition for $R^{\mathrm{CJ}} > R_o$ is equivalent to $\alpha > \frac{\gamma_{r1}}{\gamma_{r0}} + \max(g_s, 1 + \gamma_{r1}\bar{P}_r)$, i.e., $\alpha > 3.5$. Note also that the necessary condition for $R^{\mathrm{NF}} > R_o$ is equivalent to $\frac{\gamma_{r1}}{\gamma_{r0}} < \alpha < 1 + \gamma_{r1}\bar{P}_r$, i.e., $0.5 < \alpha < 3$. It is clear that, in general, the CJ and NF strategies discussed in this section yield greater secrecy rates than R_o and C^{GWT}.

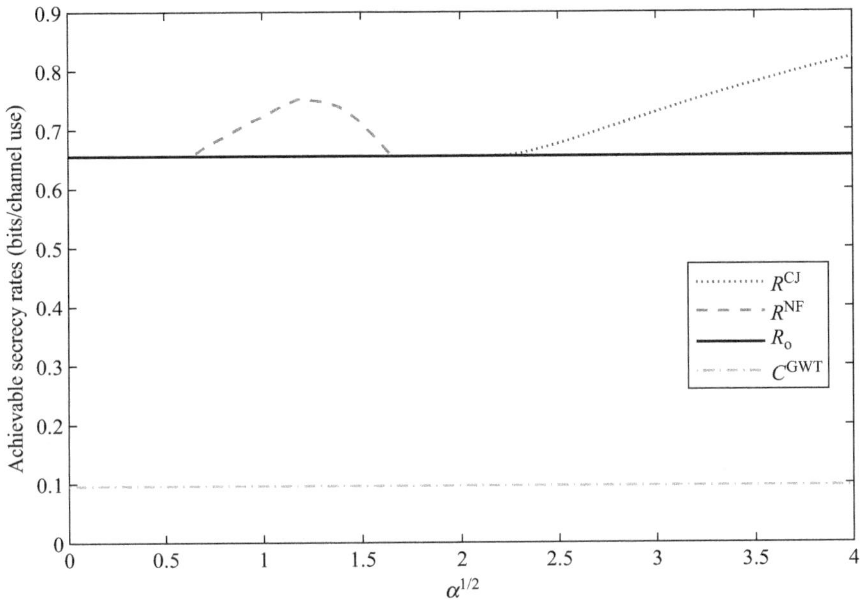

Figure 11.6 The optimal achievable secrecy rates by the CJ and NF strategies, R^{CJ} and R^{NF}, the optimal achievable secrecy rate with no transmission along the direct channel component, R_o, and the secrecy capacity of the original Gaussian wiretap channel, C^{GWT}, as functions $\sqrt{\alpha}$ where $\sqrt{\alpha} = \frac{g_r^T h_r}{\gamma_{r0}}$

Next, we consider the case where the relay is constrained to use only one of the two modes (CJ or NF) over all the channel components, i.e., the relay cannot split its power between CJ and NF. We denote the secrecy rate achievable in this case by either $R^{\mathrm{SM-CJ}}$ or $R^{\mathrm{SM-NF}}$ depending on the single mode of deaf cooperation that the relay is using. It is clear that $R^{\mathrm{SM-CJ}} = R^{\mathrm{CJ}}$ where R^{CJ} is the optimal secrecy rate achieved by the CJ strategy since in this strategy the relay jams over the two orthogonal components of the channel and hence it is indeed a single-mode strategy. However, in the NF strategy the relay uses the orthogonal component for CJ whereas it uses the direct component for NF. Therefore, intuitively, we must have $R^{\mathrm{NF}} \geq R^{\mathrm{SM-NF}}$ in general. To illustrate this, in Figure 11.7, we plot R^{NF}, $R^{\mathrm{SM-CJ}}$, $R^{\mathrm{NF-SM}}$, and C^{GWT} versus $\sqrt{\alpha}$, $0 \leq \sqrt{\alpha} \leq 2$. The values of $\bar{\rho}_s$, $\bar{\rho}_r$, g_s, γ_{r0}, and γ_{r1} are fixed and chosen as in the previous example.

Finally, we consider the reversely degraded relay–eavesdropper channel with multiple antennas at the relay as described in Section 11.3.2. In Figure 11.8, we illustrate the results we discussed above for this model. We fix $\bar{\rho}_s = 5, \gamma_{r1} = 1$. We plot the achievable secrecy rate R_o as a function of $\bar{\rho}_r$ for three different values of the channel gain g_s, namely, $g_s = 0.25$, 0.75, and 1.5. In this example, the capacity of the Gaussian channel between the source and the destination without secrecy

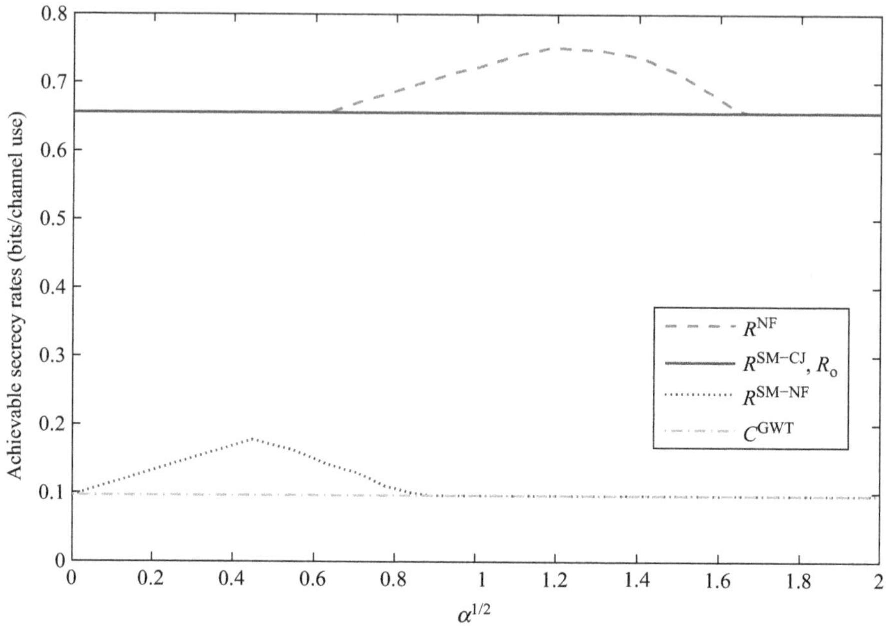

Figure 11.7 The optimal achievable secrecy rate by the NF strategy, R^{NF}, the optimal achievable secrecy rate by the single-mode CJ strategy, R^{SM-CJ}, the optimal achievable secrecy rate by the single-mode NF strategy, R^{SM-NF}, and the secrecy capacity of the original Gaussian wiretap channel, C^{GWT}, as functions $\sqrt{\alpha}$

constraints is $C^{\mathrm{G}} = \frac{1}{2}\log(1+\bar{\rho}_s) = 1.292$ bits/channel use. It is clear from Figure 11.8 that $R_o(\bar{\rho}_r)$ converges to C^{G} as $\bar{\rho}_r$ increases and the rate of convergence increases as g_s decreases.

11.4 Active cooperation in multiple relay networks

The second mode of cooperation for secrecy we consider is the active mode of cooperation in which a relay listens to the source transmissions and uses its observation to improve the achievable secrecy rate. This model with a single relay is developed in Reference 14. In multiple relay networks, the role of active cooperation has been investigated in some recent works. For example, the role of active cooperation in improving secrecy via beamforming relays is investigated in References 18 and 19. In both References 18 and 19, a two-stage cooperative secrecy protocol is proposed in which a set of multiple relays decode the source's message in the first stage, then the relays forward the source's message to the destination using beamforming. Reference 18 proposes an iterative strategy, when the global channel state information (CSI) is perfectly available, to design the beamforming coefficients either to maximize the

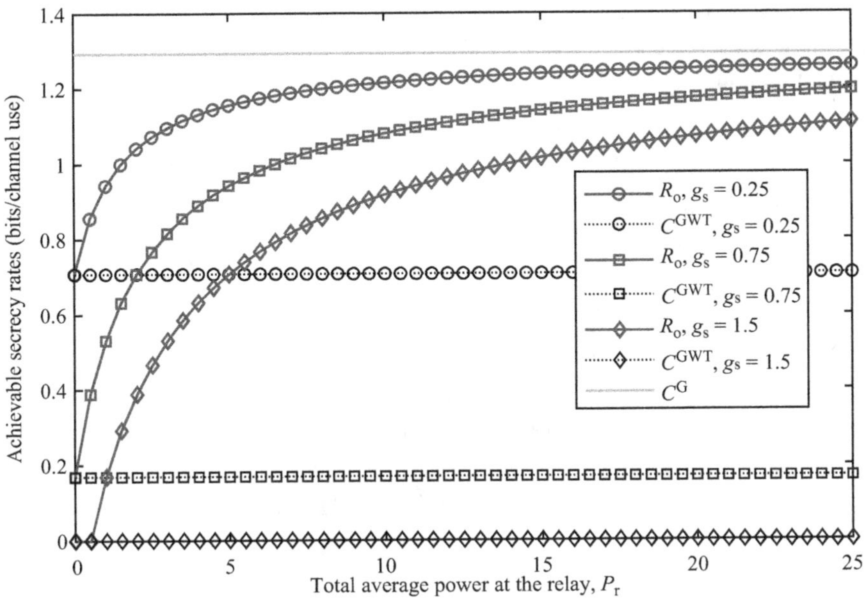

Figure 11.8 The achievable secrecy rate of the reversely degraded multiple-antenna relay channel as a function of $\bar{\rho}_r$, $R_o(\bar{\rho}_r)$, the capacity of the Gaussian channel between the source and the destination without secrecy constraints, C^G, and the secrecy capacity of the original Gaussian wiretap channel, C^{GWT}

secrecy rate for a fixed transmit power or to minimize the transmit power for a fixed secrecy rate. The same reference proposes a suboptimal ZF strategy in which an additional constraint of canceling out the signals from the eavesdropper's observation is imposed. In Reference 19, the problem of maximizing the secrecy rate achieved by the collaborative beamforming of the relays when the global CSI is perfectly available is investigated. Both References 18 and 19 assume that the communication occurs in two stages where in the first stage (source to relays) neither the destination nor the eavesdropper can hear the source and hence no secrecy requirement is involved in this stage, whereas in the second stage only the relays (but not the source) send the source's message by beamforming to the destination and hence their model becomes similar to a multiple-input single-output (MISO) wiretap channel [30–33]. In Reference 34, this assumption is removed, namely, any node in the system can hear any other transmitting node(s) at any time during the message is being communicated. The same reference proposes three different strategies based on DF/ZF.

11.4.1 DF with multiple relays

Let $\mathcal{T} = \{1, \ldots, T\}$ denotes the set of relays. For simplicity of notation, we will denote the source s as node 0, the destination d as node $T + 1$, and the eavesdropper as node $T+2$. Also, for the sake of generality, signals and channel coefficients will be assumed to take complex values. We can express the outputs at the relays, the destination, and the eavesdropper as

$$Y_i = h_{0i}X_0 + \sum_{j\in\mathcal{N}\backslash\{i\}} h_{ji}X_j + \tilde{N}_i, \quad i \in \mathcal{T} \tag{11.21}$$

$$Y_{T+1} = X_0 + \sum_{i\in\mathcal{T}} X_i + \tilde{N}_{T+1} \tag{11.22}$$

$$Y_{T+2} = h_{0,T+2}X_0 + \sum_{i\in\mathcal{T}} h_{i,T+2}X_i + \tilde{N}_{T+2} \tag{11.23}$$

where for $i, j \in \{0, 1, \ldots, T+2\}$, h_{ij} is the complex channel gain from node i to node j, X_i is the channel input at node i, and $\tilde{N}_i$ is the complex circularly symmetric zero mean unit variance Gaussian noise at node i. We assume perfect knowledge of all channel gains at all the nodes. The average power constraints are given by

$$E[|X_0|^2] \triangleq P_0 \le \bar{\rho}_s \quad \text{and} \quad E[|X_i|^2] \triangleq P_i \le \bar{\rho}_r, \quad i \in \mathcal{T} \tag{11.24}$$

where we assume that all the relays have equal power constraint $\bar{\rho}_r$.

11.4.1.1 A single-hop strategy

In the first strategy, all the relays decode the source message at the same time, then perform beamforming by transmitting scaled versions of the same signal to the destination, i.e., in this strategy each message block is transmitted to the destination in a single hop (see Figure 11.9). Moreover, all the relays' signal components can be eliminated from the eavesdropper's observation, i.e., *full* ZF can be achieved.

The strategy requires that all the relays perform signal beamforming as they forward the source message to the destination. In particular, we choose $X_0 = \tilde{X}_0 + \alpha_0 X_r$ and $X_i = \alpha_i X_r$, $i \in \mathcal{T}$ where $\tilde{X}_0$, X_r are independent circularly symmetric complex Gaussian random variables with zero mean and variances $\tilde{P}_0$ and P_r, respectively, and α_0, α_i, $i \in \mathcal{T}$ are some complex numbers. From (11.24), we must have $\tilde{P}_0 + |\alpha_0|^2 P_r \le \bar{\rho}_s$ and $|\alpha_i|^2 P_r \le \bar{\rho}_r$, $i \in \mathcal{T}$.

The value of α_0 is chosen to force the term of the eavesdropper's observation that depends on X_r to zero. This goal can be attained by choosing $\alpha_0 = \alpha^{\mathrm{ZF}} \triangleq -\frac{\sum_{j\in\mathcal{T}} \alpha_j h_{j,T+2}}{h_{0,T+2}}$. Hence, the achievable rate becomes

$$R^{\mathrm{DF/ZF}} = \min\left\{ \log\left(\frac{1+|h_{0i^\star}|^2\tilde{P}_0}{1+|h_{0,T+2}|^2\tilde{P}_0}\right), \log\left(\frac{1+\tilde{P}_0+\left|\sum_{j\in\mathcal{T}}\alpha_j\left(1-\frac{h_{j,T+2}}{h_{0,T+2}}\right)\right|^2 P_r}{1+|h_{0,T+2}|^2\tilde{P}_0}\right)\right\} \tag{11.25}$$

where $i^\star = \arg\min_{i\in\mathcal{T}} |h_{0i}|$. For a fixed value of $\tilde{P}_0$, the rate in (11.25) is maximized by choosing $\alpha_j = \frac{\left(1-\frac{h_{j,T+2}}{h_{0,T+2}}\right)^*}{\left|1-\frac{h_{j,T+2}}{h_{0,T+2}}\right|}$, $\forall j \in \mathcal{T}$, where x^* denotes the complex conjugate of

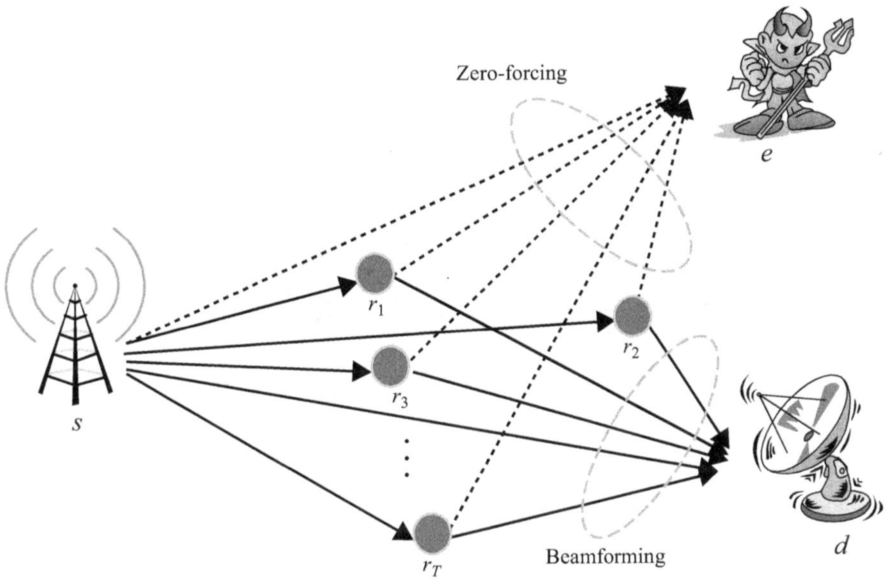

Figure 11.9 Multiple relay single-hop DF strategy with ZF

the complex number x. Hence, the achievable secrecy rate becomes

$$R^{\mathrm{DF/ZF}} = \min\left\{ \log\left(\frac{1+|h_{0i^*}|^2\tilde{P}_0}{1+|h_{0,T+2}|^2\tilde{P}_0}\right), \log\left(\frac{1+\tilde{P}_0+\left(\sum_{j\in\mathcal{T}}\left|1-\frac{h_{j,T+2}}{h_{0,T+2}}\right|\right)^2 P_r}{1+|h_{0,T+2}|^2\tilde{P}_0}\right)\right\} \tag{11.26}$$

where $\tilde{P}_0+|\alpha^{\mathrm{ZF}}|^2P_r \le \bar{\rho}_s$ and $P_r \le \bar{\rho}_r$ $\left(\text{where } \alpha^{\mathrm{ZF}} = -\sum_{j\in\mathcal{T}}\frac{h_{j,T+2}}{h_{0,T+2}}\frac{\left(1-\frac{h_{j,T+2}}{h_{0,T+2}}\right)^*}{\left|1-\frac{h_{j,T+2}}{h_{0,T+2}}\right|}\right)$.

Although this strategy is simple and allows for full ZF, it has an obvious drawback. From (11.26), note that for this strategy to be useful we need that $\min_{i\in\mathcal{T}}|h_{0,i}| > \max\{1,|h_{0,T+2}|\}$. Hence, the relays which are far from the source could possibly create a bottleneck that limits the achievable rate.

11.4.1.2 A multiple hop strategy

The goal of this strategy is to overcome the drawback of the above strategy, i.e., to eliminate the requirement that all relays must decode the source message at the same time, which creates a bottleneck. The strategy here is inspired by References 16 and 17 that considered the model with no secrecy constraints. In this strategy, the relays are ordered with respect to their distance from the source and they perform DF in a multi-hop fashion, i.e., the closest relay decodes the source message first, forwards it (with the help of the source) to the second closest relay and till it reaches the destination. Thus, if the total number of the relays is T, then the transmission

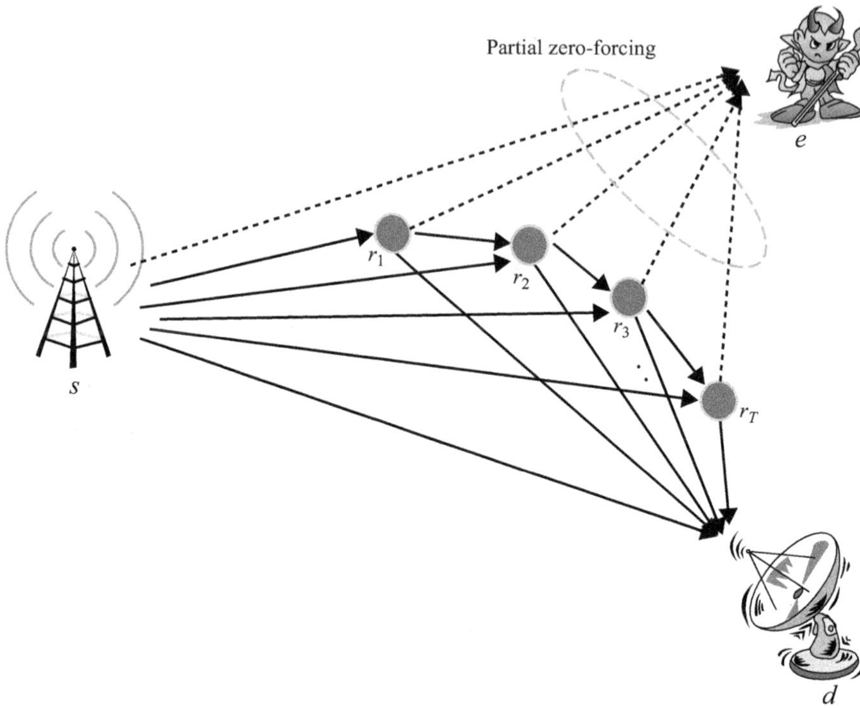

Figure 11.10 Multiple relay T-hop DF strategy with partial ZF

of each message block is done in T hops. More precisely, the transmission of each message block occurs over T hops as follows. In any given transmission block b of the source message, the first relay decodes the current message block and forwards it to the second relay in the transmission block $b+1$ and so on till the last relay decodes the source message block and forwards it to the destination in the transmission block $b+T$. Since the multi-hop transmission is pipelined, we only have an overhead delay of T blocks before the first message block reaches the destination (see Figure 11.10). Under the usual assumption that the source message is composed of sufficiently large number of blocks $B \gg T$, the achievable rate loss due to such overhead is negligible.

In this strategy, we choose the channel inputs as follows: $X_i = \tilde{X}_i + \alpha_i X_{i+1}$, $i = 0, \ldots, T-1$ and $X_T = \tilde{X}_T$ where all $\tilde{X}_i, i = 0, \ldots, T$ are independent circularly symmetric complex Gaussian random variables with zero mean and variances $\tilde{P}_i$, $i = 0, \ldots, T$, respectively, and $\alpha_i, i = 0, \ldots, T-1$ are some complex numbers. Equivalently, we have $X_i = \tilde{X}_i + \sum_{j=i+1}^{T} \prod_{\ell=i}^{j-1} \alpha_\ell X_j$, $i = 0, \ldots, T-1$ and $X_T = \tilde{X}_T$. The resulting expressions for the achievable rate and the associated power constraints are more complicated than those of the first strategy and we refer the reader to Reference 34 for full details. However, one important remark to make here is that this strategy eliminates the bottleneck created by the far-off relays. In particular, a necessary condition for this strategy to be useful is to have $\max_{i \in \mathcal{T}} |h_{0i}| > \max\{1, |h_{0,T+2}|\}$,

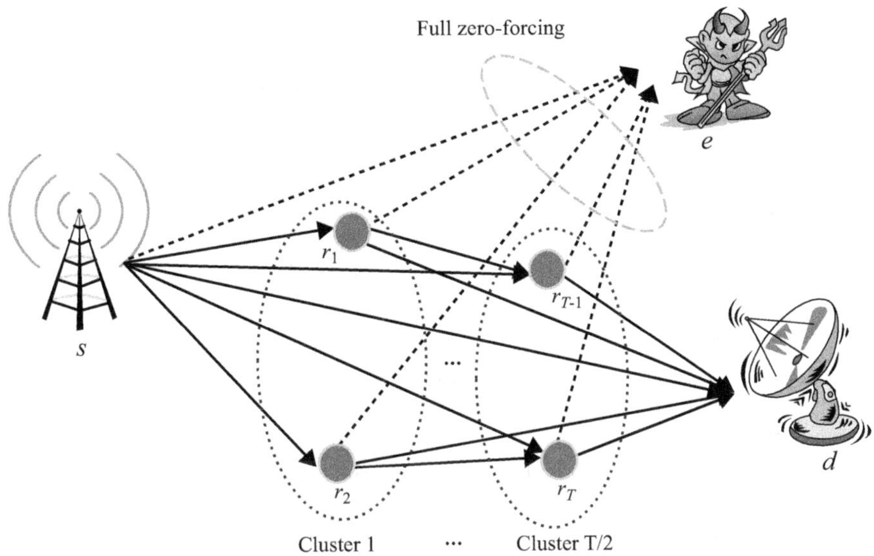

Figure 11.11 Multiple relay $\frac{T}{2}$-hop DF strategy with full ZF

which shows that the relays far from the source do not necessarily limit the achievable rate as in the single-hop strategy.

In this strategy, however, since all the relays transmit fresh information in every transmission block, only half of the relays' signal components can be zero-forced in the eavesdropper's observation. More precisely, we can only eliminate the signals of either the odd (or the even) relays in the multi-hop ordering from the eavesdropper's observation but not both. Hence, this gives us a multi-hop strategy with *partial* ZF.

11.4.1.3 A multiple hop strategy with full ZF

We now describe a $\frac{T}{2}$-hop strategy that aims to combine the advantages of the two aforementioned strategies in an efficient way. That is, in this strategy, the achievable rate is not limited by the worst source–relay channel as in the first strategy, yet one can eliminate all the relays' signals from the eavesdropper's observation.

In this strategy, the relays are ordered with respect to their distance from the source and then grouped into clusters of two relays per cluster. The source transmits the message to the relays in the first cluster (closest to the source) which decode the message and forwards it (with the help of the source) to the relays in the second cluster and so on till the message is forwarded to the destination (see Figure 11.11). That is, the transmission of each message block takes place over $T/2$ hops (assuming T is even without loss of generality) as follows. In any given transmission block b of the source message, the closest pair of relays to the source decodes the bth message block transmitted by the source and forwards it to the second closest pair of relays in the transmission block $b + 1$ and so on till the furthest pair of relays from the source decodes the bth message block and forwards it to the destination in the transmission block $b + \frac{T}{2}$. As before, since the transmission is pipelined, the overhead delay of this strategy is only $\frac{T}{2}$ blocks.

In this strategy, we choose the channel inputs as follows:

$$X_0 = \tilde{X}_0 + \alpha_0 X_1 \tag{11.27}$$

$$X_{2i-1} = \tilde{X}_{2i-1} + \alpha_{2i-1,2i} X_{2i+1} \tag{11.28}$$

$$X_{2i} = \beta_{2i-1,2i} X_{2i-1} \tag{11.29}$$

for $i = 1, \ldots, T/2$, where $\tilde{X}_0$ and $\tilde{X}_{2i-1},\ i = 1, \ldots, T/2$ are independent circularly symmetric complex Gaussian random variables with zero mean and variances $\tilde{P}_0$ and $\tilde{P}_{2i-1},\ i = 1, \ldots, T/2$, respectively, and $\alpha_0,\ \alpha_{2i-1,2i},\ \beta_{2i-1,2i},\ i = 1, \ldots, T/2$ are complex numbers that represent the design parameters whose values are chosen to maximize the achievable secrecy rate. In Reference 34, it is shown that by properly adjusting these parameters, we can zero-force all the relays' signals at the eavesdropper. Hence, in typical situations, this strategy provides a good trade-off between the first two strategies.

11.4.2 Empirical evaluation

We give here simulation results for the following experiment. Consider a two-dimensional coordinate system where the source (node 0) is located at the origin. The channel gain $h_{\ell k}$ between any two nodes ℓ and k is given by $h_{\ell k} = d_{\ell k}^{-\zeta} e^{j\theta_{\ell k}}$ where $d_{\ell k}$ is the distance between ℓ and k, $\zeta > 1$ is the path loss coefficient, and $\theta_{\ell k}$ accounts for independent phase fading and is uniformly and independently distributed over $(0, 2\pi)$ for all ℓ, k. We choose $d_{0,T+1} = d_{0,T+2} = 1$ km and take $\zeta = 3$. We use a constant power allocation policy at all the relays where the transmit powers of all the relays are set to $\bar{\rho}_r = 10$ and accordingly power is allocated at the source to maximize the achievable rate where the total average power at the source is set to $\bar{\rho}_s = 50$. All the channel gains are assumed to be fixed for the whole transmission duration and assumed to be known at all the nodes. We consider two scenarios. In the first scenario, all the T relays are uniformly spread over a disc of radius 0.75 km centered at the source. In the second scenario, all the T relays are at the same distance of 0.5 km from the source.

In Figure 11.12, for $T = 1, \ldots, 10$, we plot the achievable secrecy rate by each of the multiple relay strategies discussed above: the first is the *multiple relay single-hop decode-and-forward with zero-forcing (MRSH-DF/ZF)* strategy, the second is the *multiple relay multiple hop decode-and-forward with partial zero-forcing (MRMH-DF/PZF)* strategy where the number of hops is T, and the third is the *multiple relay multiple hop decode-and-forward with full zero-forcing (MRMH-DF/FZF)* strategy where the number of hops is $T/2$.

Figure 11.12 shows that the MRMH-DF/PZF strategy usually achieves higher rates than the MRSH-DF/ZF strategy when there is a noticeable variation in the magnitudes of the channel gains $h_{0,k},\ k \in \mathcal{T}$ between the source and the relays which is the case captured by the first scenario. However, since in the MRMH-DF/PZF strategy,

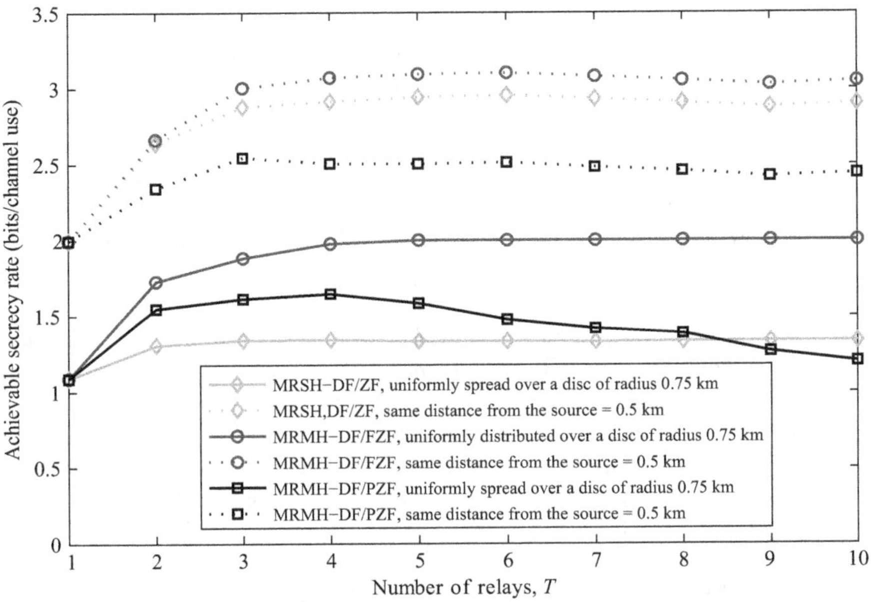

Figure 11.12 *The achievable secrecy rate, $R^{DF/ZF}$, by the MRSH-DF/ZF, the MRMH-DF/PZF, and the MRMH-DF/FZF strategies versus the number of relays, T, for two cases: when the relays are uniformly spread over a disc of radius 0.75 km centered at the source, and when all the relays are the same distance (0.5 km) from the source*

we can eliminate only half of the signal terms from the eavesdropper's observation, as T increases, the MRMH-DF/PZF strategy becomes less efficient due to the increase in the number of signal components observed at the eavesdropper. One can also see that the MRSH-DF/ZF strategy is usually better than the MRMH-DF/PZF strategy when the amount of variation in the magnitudes of the channel gains between the source and the relays is small. This is clearly captured by the second scenario, where all such channel gains have the same magnitude. On the other hand, one can see that the rate achieved by the MRMH-DF/FZF strategy is better in both examples. This is due to the fact that the MRMH-DF/FZF strategy enjoys the advantages of the two previous strategies with insignificant loss in the achievable rate in typical situations.

11.5 Conclusions

In this chapter, we discussed several physical-layer techniques to reinforce and imrove security of communications over wireless relay networks. These techniques are based on various cooperation strategies among network nodes.

In Section 11.2, we investigated two modes of deaf cooperation for secrecy, namely, CJ and NF. We first showed that a helping node can either be a useful cooperative jammer or a useful noise forwarder but not both. Next, when multiple nodes are available for deaf cooperation, we gave a simple strategy based on a greedy algorithm that selects a set of deaf helpers to boost the achievable secrecy rate between the source and the destination. Such strategy requires significantly less computations than the computationally inefficient optimal strategy. Our strategy is optimal when there is a single helper to be selected; however, it is suboptimal in the general case. Empirically, this strategy yields significant gains in terms of the achievable secrecy rates.

In Section 11.3, we extended the idea of deaf cooperation to the case where the helper is equipped with multiple antennas. We showed that the multiple spatial dimensions can be exploited in the deaf cooperation paradigm to improve secrecy rates by decomposing the helper–eavesdropper channel into a *direct* component aligned with the helper–eavesdropper channel and an *orthogonal* component. We discussed a deaf cooperation technique in which the orthogonal component is used for CJ whereas the direct component is used either for CJ or NF depending on the relative channel conditions. We showed both analytically and empirically the secrecy gains achieved by this technique. Moreover, in the case where the helper channel is reversely degraded, we discussed a simple CJ strategy that yields a secrecy rate that approaches the secrecy capacity as the helper's power increases.

Finally, in Section 11.4, we investigated the role of active cooperation for secrecy in multiple relay networks via DF strategies. We discussed three strategies based on DF with ZF. In the first strategy, which is a single-hop strategy, we showed that all the relays' signals can be eliminated at the eavesdropper (full ZF), however, the rate achieved by this strategy is limited by the worst source–relay channel. The second strategy is a multiple hop strategy that was shown to overcome the drawback of the first strategy, however, with the disadvantage of enabling only partial ZF. We then discussed a third strategy, which is also a multiple hop strategy, that offers a reasonable trade-off between the first two strategies. For this strategy, we showed that full ZF is possible and the rate achieved does not suffer from the drawback of the first strategy.

References

[1] C. E. Shannon. Communication theory of secrecy systems. *Bell System Technical Journal*, 28:656–715, 1949.

[2] A. Wyner. The wire-tap channel. *Bell System Technical Journal*, 54(8): 1355–1387, Jan. 1975.

[3] I. Csiszar and J. Korner. Broadcast channels with confidential messages. *IEEE Transactions on Information Theory*, 24(3):339–348, May 1978.

[4] S. K. Leung-Yan-Cheong and M. E. Hellman. The Gaussian wire-tap channel. *IEEE Transactions on Information Theory*, 24(4):451–456, Jul. 1978.

[5] R. Negi and S. Goel. Guaranteeing secrecy using artificial noise. *IEEE Transactions on Wireless Communications*, 7(6):2180–2189, Jun. 2008.

[6] E. Tekin and A. Yener. The Gaussian multiple access wiretap channel. *IEEE Transactions on Information Theory*, 54(12):5747–5755, Dec. 2008.

[7] E. Tekin and A. Yener. The general Gaussian multiple-access and two-way wiretap channels: Achievable rates and cooperative jamming. *IEEE Transactions on Information Theory*, 54(6):2735–2751, Jun. 2008.

[8] I. Krikidis, J. Thompson, and S. McLaughlin. Relay selection for secure cooperative networks with jamming. *IEEE Transactions on Wireless Communications*, 8(10):5003–5011, Oct. 2009.

[9] J. Chen, R. Zhang, L. Song, Z. Han, and B. Jiao. Joint relay and jammer selection for secure two-way relay networks. In *IEEE ICC 2011*, Kyoto, Japan, pages 1–5, Jun. 2011.

[10] J. P. Vilela, M. Bloch, J. Barros, and S. W. McLaughlin. Friendly jamming for wireless secrecy. In *IEEE ICC 2010*, Capetown, South Africa, pages 1–6, May 2010.

[11] L. Dong, Z. Han, A. P. Petropulu, and H. V. Poor. Cooperative jamming for wireless physical layer security. In *15th IEEE Workshop on Statistical Signal Processing*, Cardiff, United Kingdom, pages 417–420, Sep. 2009.

[12] J. Huang and A. L. Swindlehurst. Cooperation strategies for secrecy in MIMO relay networks with unknown eavesdropper CSI. In *ICASSP 2011*, Prague, Czech Republic, pages 3424–3427, May 2011.

[13] J. Huang and A. L. Swindlehurst. Secure communications via cooperative jamming in two-hop relay systems. In *IEEE GLOBECOM 2010*, Miami, FL, USA, pages 1–5, Dec. 2010.

[14] L. Lai and H. El Gamal. The relay-eavesdropper channel: Cooperation for secrecy. *IEEE Transactions on Information Theory*, 54(9):4005–4019, Sep. 2008.

[15] T. Cover and A. El Gamal. Capacity theorems for the relay channel. *IEEE Transactions on Information Theory*, 25:572–584, Sep. 1979.

[16] P. Gupta and P. R. Kumar. Towards an information theory of large networks: An achievable rate region. *IEEE Transactions on Information Theory*, 49(8): 1877–1894, Aug. 2003.

[17] G. Kramer, M. Gastpar, and P. Gupta. Cooperative strategies and capacity theorems for relay networks. *IEEE Transactions on Information Theory*, 51(9):3037–3063, Sep. 2005.

[18] L. Dong, Z. Han, A. P. Petropulu, and H. V. Poor. Secure wireless communications via cooperation. In *46th Annual Allerton Conference on Communications, Control and Computing*, University of Illinois at Urbana-Champaign, Champaign, IL, pages 1132–1138, Sep. 2008.

[19] J. Zhang and M. C. Gursoy. Collaborative relay beamforming for secrecy. *EURASIP Journal on Advances in Signal Processing, IEEE ICC 2010*, Capetown, South Africa, pages 1–5, May 2010.

[20] X. He and A. Yener. Providing secrecy with structured codes: Two-user Gaussian channels. *IEEE Transactions on Information Theory*, 60(4):2121–2138, Apr. 2014.

[21] R. Bassily and S. Ulukus. Ergodic secret alignment. *IEEE Transactions on Information Theory*, 58(3):1594–1611, Mar. 2012.

[22] G. Bagherikaram, A. S. Motahari, and A. K. Khandani. On the source DoF of the single-antenna MAC, *IEEE ISIT 2010*, Austin, TX, pages 2588–2592, June 2010.

[23] J. Xie and S. Ulukus. Secure degrees of freedom of one-hop wireless networks. *IEEE Transactions on Information Theory*, 60(6):3359–3378, Jun. 2014.

[24] J. Xie and S. Ulukus. Secure degrees of freedom of *K*-user Gaussian interference channels: A unified view. *IEEE Transactions on Information Theory*, 61(5):2647–2661, May 2015.

[25] J. Xie and S. Ulukus. Secure degrees of freedom of the Gaussian wiretap channel with helpers and no eavesdropper CSI: Blind cooperative jamming. In *Conference on Information Sciences and Systems*, John Hopkins University, Baltimore, MD, pages 1–5, Mar. 2013.

[26] R. Bassily and S. Ulukus. Deaf cooperation and relay selection strategies for secure communication in multiple relay networks. *IEEE Transactions on Signal Processing*, 61(6):1544–1554, Mar. 2013.

[27] E. Ekrem and S. Ulukus. Cooperative secrecy in wireless communications. In W. Trappe and R. Liu (Eds.), *Securing Wireless Communications at the Physical Layer*. Springer-Verlag, Berlin, 2009.

[28] D. Tse and P. Viswanath. *Fundamentals of Wireless Communication*. Cambridge University Press, Cambridge, 2005.

[29] R. Bassily and S. Ulukus. Deaf cooperation for secrecy with multiple antennas at the helper. *IEEE Transactions on Information Forensics and Security*, 7(6):1855–1864, Dec. 2012.

[30] S. Shafiee, N. Liu, and S. Ulukus. Towards the secrecy capacity of the Gaussian MIMO wire-tap channel: The 2-2-1 channel. *IEEE Transactions on Information Theory*, 55(9):4033–4039, Sep. 2009.

[31] A. Khisti and G. W. Wornell. Secure transmission with multiple antennas – Part II: The MIMOME wiretap channel. *IEEE Transactions on Information Theory*, 56(11):5515–5532, Nov. 2010.

[32] F. Oggier and B. Hassibi. The secrecy capacity of the MIMO wiretap channel. *IEEE Transactions on Information Theory*, 57(8):4961–4972, Aug. 2011.

[33] T. Liu and S. Shamai. A note on the secrecy capacity of the multiple-antenna wiretap channel. *IEEE Transactions on Information Theory*, 55(6):2547–2553, Jun. 2009.

[34] R. Bassily and S. Ulukus. Secure communication in multiple relay networks through decode-and-forward strategies. *Journal of Communications and Networks*, 14(4):352–363, Aug. 2012.

Chapter 12
Relaying technologies for smart grid

Jing Jiang and Hongjian Sun

12.1 Introduction

Smart grid (SG) is a modernized electrical grid that uses information and communications technology (ICT) to gather and act on information for improving the efficiency, reliability, economics, and sustainability of the electricity distribution and management. Communication technologies can support a broad range of SG applications, including advanced metering infrastructure (AMI) and demand response (DR), and thus are the key to achieve the potential advantages of SGs. However, there are many formidable challenges when communication technologies are applied to the SG, such as the trade-offs between communication coverage and capacity, the high-reliability requirement for communication and limited spectral resources. Relaying has emerged as one of the most promising candidate solutions for addressing these issues. In this chapter, an introduction of communication networks and relaying strategies for SGs is presented, together with a discussion of how to improve spectral efficiency and coverage in relay-based ICT infrastructure for SG applications. Considering communication channel medium, there are two sets of networks in SGs, i.e., wired and wireless communication networks. Special attention is paid to the use of various relaying strategies in power line communication (PLC) and wireless communication networks.

There is a widely recognized need to upgrade existing electricity grids in order to improve power delivery, reduce operating costs, and to support renewable energy sources. Due to the dependence of these goals on the data acquisition and control, such SGs must combine existing electricity grids with advanced ICT infrastructure. A mature SG will consist of a number of applications, e.g., supervisory control and data acquisition (SCADA), AMI, and DR. As different applications require distinct degrees of coverage, capacity, reliability, security, and latency, the implementation of ICT infrastructure for SG networks raises many challenging design issues [1, 2].

Most SG applications, e.g., SCADA and AMI, should exhibit high reliability, large coverage, and high security, while requiring different scales of latency and data rates, e.g., 0.1–1 s latency and 100 Kbps data rate for SCADA, and 10–20 s latency and 1 Mbps data rate for AMI. Considering different communication channel media, both wireline and wireless technologies can be used to meet these requirements. In the former case, PLC is a natural solution [2]. Using PLC, relatively small equipment investment is needed because it uses existing power lines as the data transmission medium. In the latter case, either ZigBee or Wi-Fi can be employed to enable AMI and DR applications due to their good capacity and low transmit power [3, 4].

12.1.1 PLCs for SGs

PLC enables a number of SG applications, including home energy management, distributed generation management, demand side management, and plug-in electric vehicle management [5]. Further, PLC has been proposed as the most natural technology for providing broadband Internet access to domestic or commercial premises, because it uses existing power lines with only small equipment investment required. In local area networks, PLC could support a series of indoor services, such as video streaming, online gaming, as well as complement wireless local area networks by providing a link through walls [6].

Using the power grid as a communication medium for transmitting signals, however, presents many formidable challenges for telecommunications engineers, because the power lines were not initially designed for communication purposes. First, the PLC channel exhibits strong frequency-selective fading with deep notches at some frequencies. Second, noise and interference can be coupled to power lines, leading to high power, time- and frequency-varying background noise. Third, the transmit power of PLC is capped due to electromagnetic compatibility considerations. Finally, adding more bandwidth yields only marginal improvements because the signal attenuation over power lines is proportional to the transmission frequency. As a result, the transmission capability of PLC is highly restricted, i.e., it has a small transmission range and low capacity. On the other hand, there is an increasing demand for the high capacity and large bandwidths for the indoor PLC network. For example, high-definition television (HDTV) users may require 15–20 Mbps bandwidth when using MPEG-2 compression, in addition to the stringent transmission delay constraints.

To improve the PLC capacity, an opportunistic relaying strategy was proposed where the relay (repeater) is opportunistically used only if it can provide higher capacity than the direct transmission [7]. It has been shown that the opportunistic relaying strategy performs very well for in-home PLC networks. Previous studies have also revealed that the transmission delay over multihop PLC networks can be significantly reduced by adopting distributed space–time block codes (DSTBCs) at the relays [8]. The advantages of the DSTBC-based retransmission scheme lie in the fact that the signals from different relays are orthogonal and can be easily decoded at the destination.

12.1.2 Wireless mesh network for SGs

Either ZigBee or Wi-Fi can be employed to enable AMI and DR applications in SG. However, because of their low transmit power levels, both technologies have limited coverage. A wireless network that consists of various nodes (e.g., Wi-Fi and ZigBee) organized in a mesh topology can enhance the coverage [9–12] in SGs. In addition, wireless mesh networks are inherently more reliable as they can take advantage of self-forming and self-healing network concepts.

Wireless mesh architectures are usually implemented at the network layer or the data link layer. Relevant research focuses on protocol design for transferring data between network entities [10–12]. The performance of a wireless mesh network depends on the quality, reliability, and efficiency of communications between different nodes in the network. Taking advantage of spatial diversity, relaying technologies can improve the performance of wireless links between neighboring nodes to meet the communication quality requirements of a wireless mesh network. However, conventional relaying technologies, e.g., amplify-and-forward (AF) and decode-and-forward (DF), enhance the capacity at the expense of consuming resources, e.g., radio frequency (RF) spectrum. Inefficient use of these resources could lead to low spectral efficiency [1]. Thus, it is critical to study advanced relaying technologies to improve the spectral efficiency while retaining the advantages of relaying transmission.

12.1.3 Chapter overview

In this chapter, we focus on relaying technologies in SG communication networks. We first present an introduction of communication technologies for SG applications and analyze the potential challenges. Then in the context of the indoor PLC network, we provide an example of using PLC for building an indoor entertainment system, and analyze some relaying transmission strategies. The spectral efficiency of relaying strategies is discussed and demonstrated. Next, we focus on wireless communications for SGs, and study some advanced relaying strategies including unidirectional two-relay strategy with collaborative beamforming, bidirectional relaying for information exchange, and virtual multiple-input multiple-output (MIMO) strategy. The aim of this study is to identify relaying strategies that can achieve greater spectral efficiency, extended transmission range, and improved reliability for SG applications.

12.2 Communication networks for SGs

12.2.1 An overview of SG communication architectures

SG is a term which refers to a modernized electrical grid that uses ICT to gather and act on information for improving the efficiency, reliability, economics, and sustainability of the electricity distribution and management [13]. It is important to note that the design and implementation of an effective, secure, scalable, and reliable communication network is the key to achieve the potential advantages of the SG [14]. A SG is a complex system with different viewpoints consisting of electrical grid and several information management viewpoints. From the viewpoint of an electrical processes, a SG is an integration of power generation, transmission (representing the infrastructure which transports electricity over long distance), distribution (representing the infrastructure which distributes electricity to customers), and power consumption (i.e., end users of electricity).

From the viewpoint of communication networks, a SG consists of three main segments, as shown in Figure 12.1:

- Home area network (HAN): It gathers data from a variety of smart devices within the home area, and it is responsible for delivering the control commands (from utility companies) to the smart devices for enabling smart energy management. Typically, HAN needs to cover areas up to 200 m^2 and support data rate from 10 to 100 kb/s [13].
- Neighbor area network (NAN): It is deployed within the distribution domain and is responsible for communications of metering data enabling utility companies to interact with meters through monitoring and control. NAN endpoints are smart meters (SMs) that provide real-time or near-real-time energy consumption as feedback corresponding to the demand to/from utility companies. A Communication network is required to form a framework for real-time bidirectional information transmission and exchange in SG NANs [15].
- Wide area network (WAN): It gathers data from multiple NANs (from multiple data concentrators) and is responsible to transfer data to utility companies' private backhaul networks. It also enables two-way information flow needed for substation communications, distribution automation, power quality monitoring, and so on. WANs usually need to cover areas of several thousands of kilometers and may aggregate thousands of supported devices that require data rate from 10 to 100 Mb/s [13].

12.2.2 Communication technologies for SGs

Various advanced communication technologies have been considered for their applications in different domains of SG networks. Considering communication channel medium, there are two sets of networks, i.e., wired and wireless communication networks.

Wired communication is a mature and stable communication technology. For example, fiber-optic can be used in the private backhaul networks in SGs to provide high speed and reliable communications; but it is very expensive. PLC allows exploiting the existing electrical infrastructure and greatly saves the deployment cost for an extra communication channel [16]. It enables a natural upgrade from simple electricity grid to SG that has hybrid electricity and data communication solutions. But the transmission capability of PLC is highly restricted due to limited bandwidth, low transmit power, and the harsh noise conditions over power lines. Transmission range extension and capacity enhancement are both urgent challenges that telecommunication engineers face before a mass deployment of PLC devices is feasible.

Compared to wired solutions, wireless technologies offer benefits such as lower cost of equipment and installation, quick deployment, widespread access and greater flexibility [17]. ZigBee, Bluetooth, and Wi-Fi can be used to enable wireless communications into HAN devices. IEEE has set standards for these three technologies where physical layer and medium access control layer have been defined: IEEE 802.15.4b for ZigBee, IEEE 802.15.1a for Bluetooth, and IEEE 802.11a/b/g/n for Wi-Fi. WiMAX

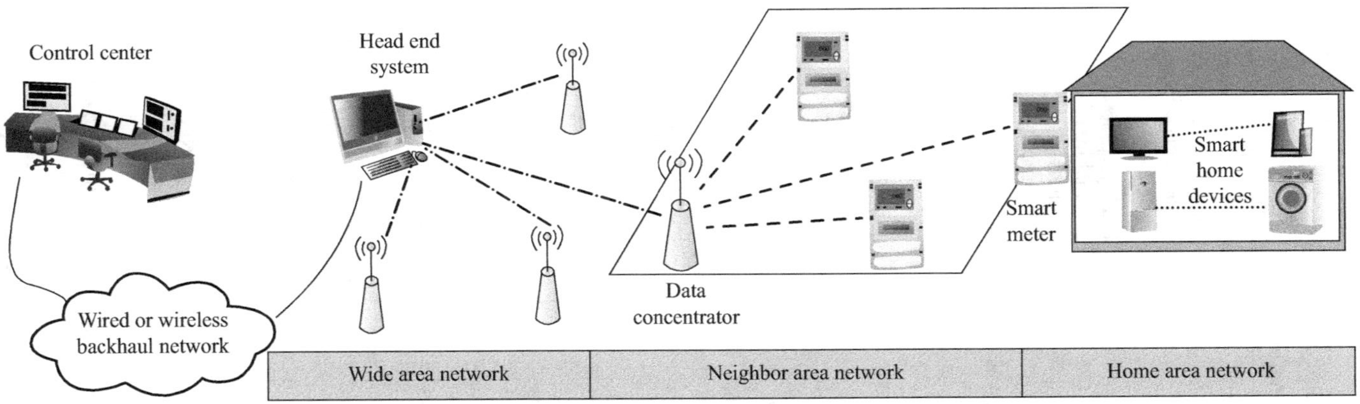

Figure 12.1 The communication architecture of SG

Table 12.1 Comparison of communication technologies for SGs

Technology	Data rate and domains	Benefits	Drawbacks
ZigBee (IEEE 802.15.4)	250 kb/s; suitable for HAN	Low cost; low power usage; a large number of nodes; secure and reliable	Low data rates; very short range; does not penetrate structures well
Wi-Fi (IEEE 802.11)	11–300 Mb/s; suitable for HAN and NAN	High speed; low cost; mature standards; widespread use	Security issues; does not penetrate concrete buildings
Bluetooth (IEEE 802.15.1)	Up to 1 Mb/s; suitable for HAN	Ease of access; low power consumption; widespread use	Very short range; limited number of nodes; strong noise issues
WiMAX (IEEE 802.16)	289 Mb/s for download; suitable for NAN and WAN	Simple and scalable; low latency; faster speed than 3G cellular; proper security	User shared bandwidth; Limited access to spectrum licenses or spectrum leasing required
Cellular network (3G cellular or 4G LTE)	More than 2 Mb/s for 3G; 300 Mb/s download for 4G; suitable for NAN and WAN	Widely adopted; stable and well standardized; low latency, and high data rate (especially 4G)	Utility must rent the infrastructure from a cellular carrier; in the transition phase from 3G to 4G deployment [13]
PLC	More than 100 Mb/s for broadband; 234 kb/s for narrowband; suitable for HAN and NAN	Allow exploiting the existing electrical infrastructure and greatly save the deployment cost	Low transmit power; low data rate; harsh noise; short transmission range

and cellular network communications provide wireless communication solutions for NANs.

Table 12.1 shows a comparison among various wired and wireless communication technologies that are commonly used to enable SG communications. Their technical specifications, benefits, and drawbacks are also discussed.

12.2.3 Challenges of communication technologies in SGs

SG communication infrastructure should allow utilities to interact with their electrical devices as well as with the customers on a near real-time basis. However, as shown in Table 12.1, there are several challenges that still need to be addressed before their deployments for SGs. These challenges are described in the following subsections.

12.2.3.1 Coverage and capacity trade-off

Some SG applications, e.g., SCADA and AMI, require the access network to cover a large area. Unfortunately, since the interference level increases as the number of nodes increases, the coverage of a communication network has an inverse relationship with the channel capacity; therefore, there is a trade-off between coverage and capacity in conventional direct transmission systems [1].

One potential solution is to use relaying technologies, whereby one long communication link is broken into two or more shorter, lower power links. Due to the inherent broadcast nature of the communications from the source, it may be possible for one or more nodes receiving strong signals from the source to forward them to the destination. Therefore, relaying transmission is an important technique to widen the coverage and enhance the capacity [6].

12.2.3.2 Reliability

The reliability of a network can be defined in terms of its robustness, survivability, and sufficiency of its connectivity to support a prescribed level of performance. Most SG applications, e.g., AMI, require reliable communications paths from the customers back to the high-speed core network. However, radio propagation in wireless communications is affected by several factors, e.g., multipath fading which may result in temporary failure of wireless communication due to a severe drop in the received signal-to-noise ratio (SNR). Furthermore, wireless networks may suffer disruptions caused by adverse weather conditions (e.g., thunderstorms) that could attenuate the transmission ability of the wireless network [1].

Using relaying technologies, the transmit signal can be passed through both the direct communication channel and the relay channel(s). With the aid of a receiver combining strategy, the multipath fading effects can be averaged or even removed. Under adverse weather conditions, direct transmission link may be blocked. It is feasible to build alternate links using the rest of the nodes, therefore offering capabilities of self-forming and self-healing to the ICT infrastructure.

12.2.3.3 Spectrum issues

RF spectrum is the lifeblood of wireless communication systems. However, current communications for SGs have access to limited numbers of frequencies, which are primarily designated for SCADA and AMI. Using current transmission techniques, it is very challenging to support additional demands from certain new SG applications, such as video surveillance. The efficient use of RF spectrum is, therefore, a critical issue that needs to be addressed before the deployment of extensive wireless networking in SGs.

The spectral utilization efficiency is often measured by spectral efficiency, which is defined as the number of bits that can be communicated over a given bandwidth within a unit of time (in bits/s/Hz). Due to the contributions of relay channels which can boost the signal strength at the destination, relaying technologies can achieve higher spectral efficiency than direct transmission.

12.2.4 Relaying strategies in SG communication networks

Based on the above discussions, there exists a common need for applying relaying technologies in SG networks. Using relaying technology, one or more relays are deployed and used to assist transmit signals to the destination. In a traditional relay-based system, the data communication can be divided into two time slots. This is required due to the half-duplex constraint, which means that the relays are unable to receive and transmit data simultaneously. In the first time slot, the source broadcasts its information to the destination and one or more relays. In the second time slot, the relays forward the received data to the destination. By using relays, the destination could achieve much higher reliability in decoding the information from the source by taking advantage of spatial diversity.

The main challenge in the relay-based system is how to use the relays efficiently which requires study of how to use the relay(s) and also how many relays are needed. When we consider a single-relay system, the relaying protocol at the relay could significantly affect the system performance. We allow the relay to decode, re-encode and re-transmit the message, or to amplify its received signal subject to its power constraint, or to forward a quantized and compressed version of the message employing a source coding method. We refer to these three protocols as AF, DF, and compress-and-forward (CF), respectively [18]. Compared to AF, the CF protocol can provide better performance, but it is more complicated as the relay needs to quantize the signal before forwarding. The performance of DF heavily relies on whether the relay can successfully decode the transmitted signals. The DF protocol outperforms CF when the relay is closer to the source, but performs worse when the relay is closer to the destination.

12.3 Relaying technologies for indoor PLC networks

In this section, we focus on the application of different relaying strategies in the context of indoor PLC networks. We first provide a simple example of using PLC for building an indoor entertainment system, and then introduce some relaying transmission strategies. Further, as frequency spectrum is a precious resource in PLC systems, the spectral efficiency of relaying strategies is also discussed in this section. The aim of this study is to identify relaying strategies that can achieve improved spectral efficiency and extend transmission range for indoor PLC networks.

12.3.1 Indoor PLC network: example, advantages, and challenges

We start with describing an application example of using PLC technologies for building an indoor entertainment system. As shown in Figure 12.2, we assume that a router is connected to the Internet through a broadband connection (e.g., DSL, cable, and

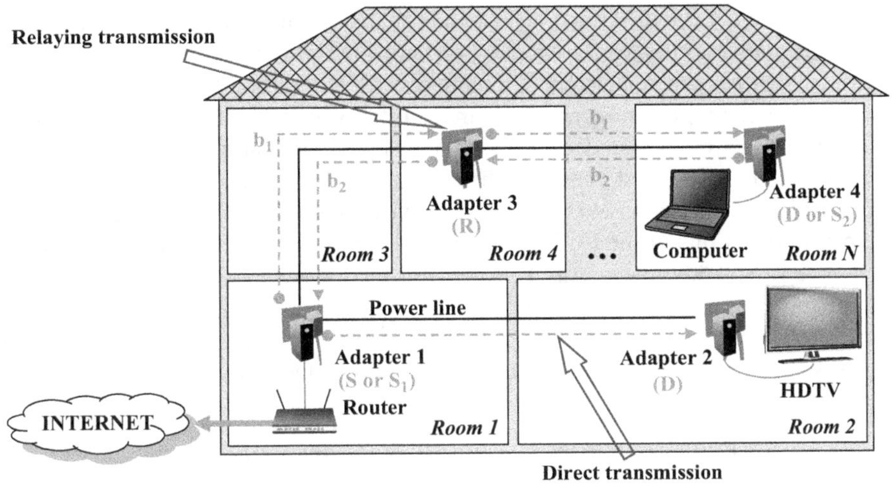

Figure 12.2 Practical relaying transmission in the indoor PLC network; S, source; R, relay; and D, destination

fiber). In such a scenario, an indoor PLC network can be established using the existing electrical wiring and some power line adapters whose function is to enable digital communication over power lines. First, we plug an adapter into a power socket, and then connect it to the router by using a category 5 Ethernet cable. After switching on the adapter, the Internet signals can be communicated over power lines. Some other adapters may be plugged in at any other power socket in the building. Once the power lines in the building are connected, the Internet signals can be transmitted to all adapters. Therefore, high-speed Internet access can be provided to the indoor entertainment devices via the adapters. Typical entertainment devices include: computers, HDTV, Internet video boxes, Blu-ray players, and game consoles.

There are several advantages of using PLC for building such an indoor entertainment system. A key advantage is that it allows entertainment devices to access the high-speed Internet with relatively small equipment investment as only the adapters are required. The entertainment device uses existing power sockets in the building, making it easy to set up. The entertainment device can be connected in a "plug-and-play" fashion. Further, there is no need for complicated wiring in such a PLC network. By comparison, the cost of deploying wires for Ethernet in many buildings can be prohibitive. Finally, but not the least, as power lines are ubiquitously available in the home or office, PLC can cover areas that are hard to reach by wireless connections due to propagation losses in the building materials.

Despite the advantages of PLC, it faces a number of challenges. One of the challenges is the standardization of PLC for the successful implementation in the local area network. Without standardization, PLC devices may conflict with each other, leading to a very poor performance. There are several groups and organizations addressing PLC standards for home or office applications. Such standardization efforts include IEEE P1901 and ITU-T G.hn [19]. By using sophisticated modulation techniques such as orthogonal frequency-division multiplexing (OFDM), IEEE 1901-compliant devices can achieve bit rate more than 100 Mbps at the physical layer [20]. Another challenge is RF interference from the PLC system to other electrical devices. As power lines are not shielded, they can act as large antennas, potentially causing interference. The PLC system shares the same frequency bands used by a number of radio services such as amateur radio, short-waves and medium-waves radio services. The communications of the PLC system may therefore cause interference to nearby radio receivers. Thus, the transmit power of a PLC system is limited according to federal communications commission (FCC) part 15 rules. Various sources of noise and disturbances can be coupled to power lines. As a result, direct transmission in PLC system often results in short transmission ranges and low capacity.

12.3.2 Relaying strategies for indoor PLC networks

Data transmission in PLC typically involves the information being transferred directly from a source (S) to the destination (D) via power lines, e.g., from adapter 1 (source) to adapter 2 (destination) as shown in Figure 12.2. Based on the discussions in the above section, the received signals at the destination may be too weak to be decoded successfully when the transmission distance is too long, e.g., from the router to a computer in Figure 12.2. Due to the inherent broadcast nature of the communications from the source, it may be possible for one or more adapters connected to the power lines (e.g., adapter 3 in Figure 12.2) to receive the transmitted signals and help to forward them to the destination. Therefore, relaying transmission is an important technique to widen the coverage and enhance the capacity of PLC. We consider such a relay-based PLC system, where one or more relays working in the half-duplex mode are used to retransmit the signals to the destination. It is assumed that the destination cannot receive the signals directly transmitted from the source due to the long transmission distance. In this PLC system, the data communication can be divided into two time slots: the source broadcasts its information through power lines to one or more relays in the first time slot, and then the relays forward the received data to the destination in the second time slot. In the following subsection, we will start with traditional relaying strategies where only one relay is considered, and then investigate two potential strategies that are more practical for the indoor PLC network.

12.3.2.1 Unidirectional single-relay system

When we consider a single-relay PLC system, the relaying protocol at the relay could significantly affect the system performance. Here we consider popular relaying protocols, i.e., AF and DF, as demonstrated in Figure 12.3. For the AF relaying protocol, as illustrated in Figure 12.3(a), it allows the relay to amplify and retransmit these noisy data to the destination. In Reference 21, Laneman *et al.* showed that for a wireless scenario the AF protocol could perform efficiently at high SNRs. But in low-SNR scenarios, it is argued that the performance of this protocol will be degraded.

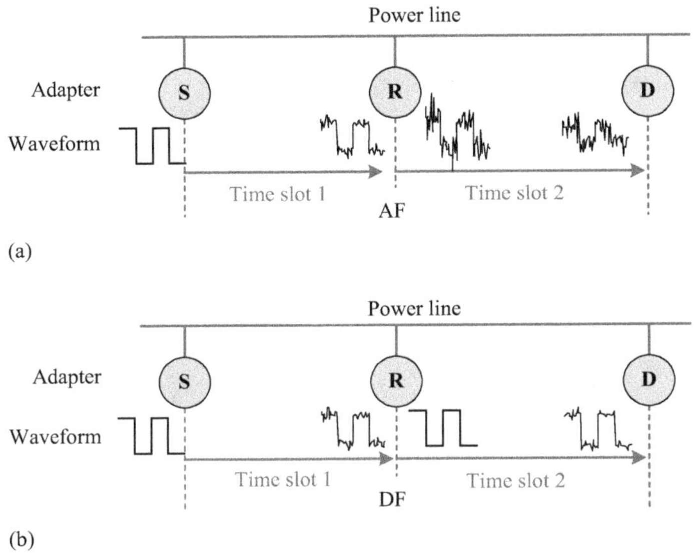

Figure 12.3 Demonstration of relaying protocols in the PLC network: (a) amplify-and-forward (AF) and (b) decode-and-forward (DF)

This is because the noise is also amplified in addition to the signal amplification at the relay.

For the DF relaying protocol, as shown in Figure 12.3(b), it allows the relay to decode the received signals from the source, and then re-encode and forward them to the destination. The performance of DF heavily relies on whether the relay can successfully decode the transmitted signals. If the relay fails to decode the signal due to high signal attenuation over power lines, these errors will be propagated to the destination and lead to even worse performance than for direct transmission. On the other hand, if the signal is correctly decoded at the relay, the destination will receive a stronger signal and thus obtain improved performance.

It is noteworthy that in PLC the high-frequency band (i.e., 3–30 MHz) has very high path loss, e.g., approximately 150 dB/km on low-voltage (LV) power lines. The spectrum in PLC is therefore a very valuable resource as only low-frequency band can propagate efficiently over long distances. Unfortunately, as we will see, both AF and DF boost the signals at the destination at the expense of losing resources, e.g., the time and power allocated to the relay. The spectral efficiency is defined as the number of bits that can be communicated over a given bandwidth within a unit of time (often measured in bits/s/Hz). Inefficient use of these resources leads to a low spectral efficiency. The spectral efficiency loss in a multiple-relay system could be even worse if either multiple time slots or frequency bands are exclusively allocated to different relays. Some strategies can be used to improve the spectral efficiency while retaining the advantages of relaying transmission. In the next two subsections, we will

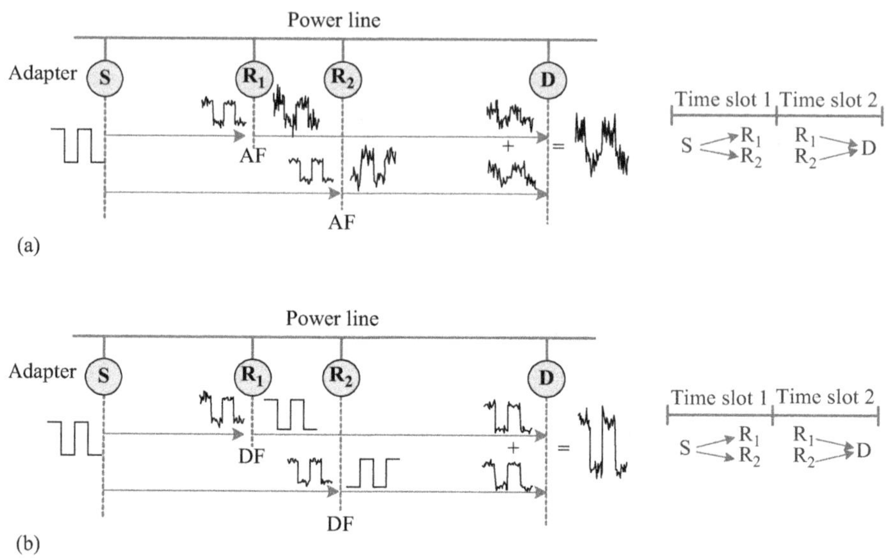

Figure 12.4 Relaying protocols combined with two-relay beamforming strategy: (a) unidirectional AF relaying protocol plus beamforming and (b) unidirectional DF relaying protocol plus beamforming

investigate two potential strategies that are practical for the indoor PLC network, i.e., a two-relay system using beamforming concepts, and a bidirectional relaying strategy for a two-way information-exchange system.

12.3.2.2 Unidirectional two-relay system with beamforming

When the destination cannot receive information directly from the source, all the available adapters between the source and destination can be considered as potential relays. In an indoor PLC environment, power sockets often appear in pairs. It means that we can often find two adapters (which are closely located) to serve as two relays. Due to the physical distribution of potential relays, a two-relay system is reasonable and feasible, and thus discussed in this subsection.

In this system, the two relays forward certain versions of the received signals to the destination, as shown in Figure 12.4(a) for AF and Figure 12.4(b) for DF. No matter which relaying protocol is used, AF or DF, the two-relay system faces a challenge. Because of different channel phases at the two relays, the correlation properties of the received signals at the destination will be distorted. That means, a superposition of the signals at the destination will not necessarily strengthen the intended signals. Thus, beamforming is introduced to adaptively adjust the transmit signal phases at the two relays. For example, as shown in Figure 12.4, the signal phase at R_2 is adjusted. Taking advantage of beamforming, the received signals at the destination can be constructively added at the destination to improve SNR.

To enable beamforming, the relays are assumed to know the channel state information (CSI) for the relay–destination links. Due to the fact that PLC channel is stationary for a long time (often a few seconds to tens of seconds), CSI can be estimated accurately at the destination and communicated back to the relays. Moreover, realizing beamforming also requires synchronization of the carriers in terms of phase and frequency at the relays. As the two relays are closely located, this requirement could be satisfied by exchanging timing messages. Hence, the two-relay system with beamforming is a feasible way to improve information transmission in the indoor PLC network.

12.3.2.3 Bidirectional relaying for information-exchange system

In the indoor PLC network, two nodes may require to exchange information. For example, as shown in Figure 12.2, the router (connected to adapter 1, i.e., S_1) sends data b_1 to the computer (connected to adapter 4, i.e., S_2), while the computer at the same time is uploading data b_2 to the router. Using conventional unidirectional relaying strategies as mentioned above, four time slots are needed to accomplish the information exchange process as shown in Figure 12.5(a). In contrast, the bidirectional relaying strategy requires only two time slots to complete an information exchange process as represented in Figure 12.5(b). In the first time slot, two end nodes (i.e., S_1 and S_2) send their information to the relay using the same frequency band, leading to a superposition of the received signals at the relay. Using the AF relaying protocol, the relay directly forwards its received signals to both end nodes. Because these two end nodes have their own copies of transmitted signals, they can subtract out their own signals and obtain the information transmitted from the other node. In this way, a higher spectral efficiency can be achieved, since the information at the two nodes is exchanged using fewer time slots.

If the bidirectional relaying strategy is applied together with the DF relaying protocol, the relay needs to decode the superposed signals. Optimal performance can be achieved using a maximum likelihood (ML) detection algorithm. Using ML, the detection error can be minimized but at the expense of high complexity. With a lower computational cost, other approaches can obtain almost ML performance. One approach is the so-called vertical-Bell laboratories layered space time (V-BLAST) detection algorithm [22]. The advantage of the V-BLAST algorithm is that its computational complexity is low and is fixed for the whole range of SNRs, while still achieving a good detection performance. We adopt V-BLAST detection algorithm at the relay for the bidirectional DF relaying system studied in this section.

12.3.3 Numerical results and discussions

In this subsection, simulation results are presented to show the advantages and drawbacks of the different relaying strategies in indoor PLC networks. Comparisons of relaying transmission versus direct transmission, beamforming versus non-beamforming, unidirectional versus bidirectional relaying strategies are also illustrated.

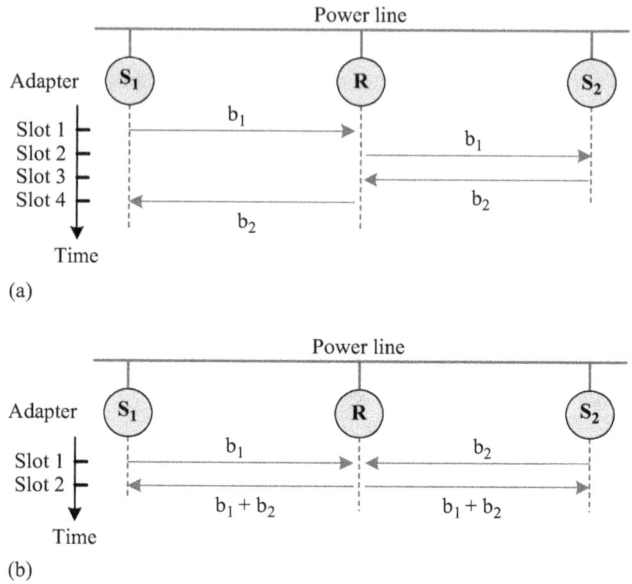

Figure 12.5 Comparison of conventional unidirectional relaying strategy with bidirectional relaying strategy when two end nodes exchange information: (a) conventional unidirectional relaying that uses four time slots and (b) bidirectional relaying strategy that uses two time slots

The simulations are carried out based on a framework presented in Reference 23, where the polyvinyl chloride (PVC)-insulated cables are used to emulate the power line network. We note that the PVC-insulated cables are widely used in the LV power distribution grid. The OFDM frame structure is adopted: one super frame, which has a length of 1.92 s, consists of 64 equal-length frames, while each frame is composed of 256 OFDM symbols. Moreover, both frequency-selective channels and colored background noise are used to make the simulations more realistic. Channel transfer functions are generated by assuming a group of randomly located branches with each branch having a random length, and using the multiconductor transmission line (MTL) model [24]. As the PLC channel can be stationary for a few seconds in practice, one generated channel is used over a whole super frame, and will be regenerated for a new super frame. To model the colored background noise, a synthesis process in Reference 25 that passes white Gaussian noise (WGN) through a colored filter is used. The noise power spectral density (PSD) is assumed to be unchanged over a super frame at one node. When we consider two relay nodes, we assume that they are closely located and share the same noise PSD. Furthermore, it is assumed that each node in the PLC system can transmit signals at the same power level with an equal power allocation on each sub-carrier. The PSD of the transmitted signals is −60 dBm/Hz, which is chosen based on FCC regulations part 15.

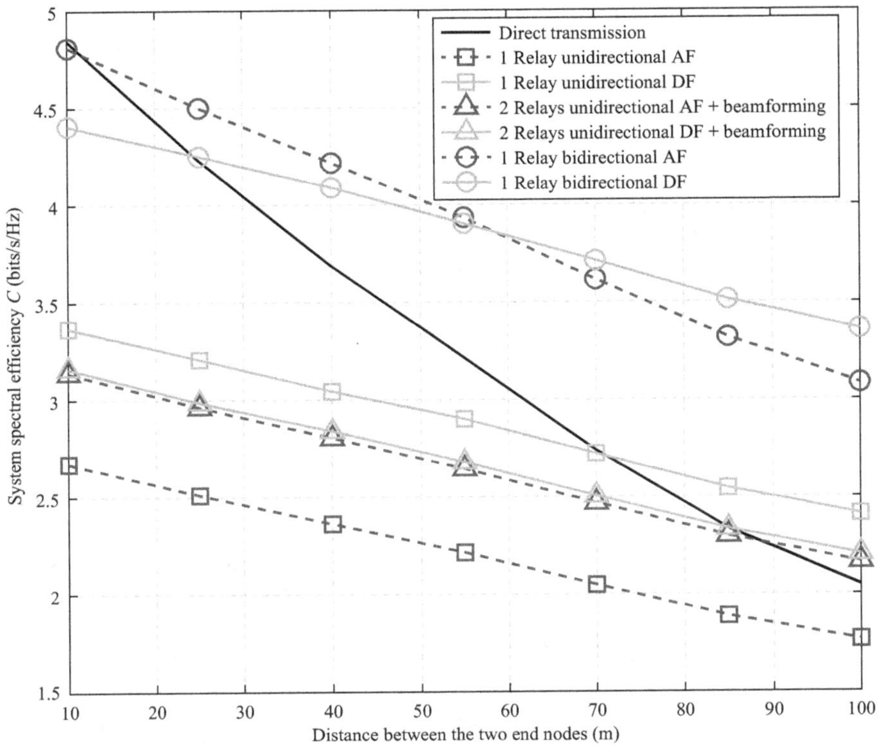

Figure 12.6 Spectral efficiency when the distance between the two end nodes varies

The spectral efficiencies of various relaying strategies are compared in Figure 12.6, where the x-axis denotes the distance between two end nodes and the y-axis denotes the system spectral efficiency. It can be seen that the single-relay unidirectional AF and DF strategies are not necessary if the distance between the two end nodes is short. This is because both unidirectional relaying strategies use two time slots. Even though they can help the system to obtain a higher information rate, the inefficient use of time resources leads overall to a lower spectral efficiency compared to direct transmission. However, it should be noted that direct transmission is more vulnerable to an increase of transmission distance (i.e., higher channel attenuation), and thus has a smaller transmission coverage than the single-relay unidirectional DF strategy. For example, if a spectral efficiency of 2.5 bits/s/Hz is required, the direct transmission scheme has a transmission range of 80 m, less than 90 m obtained by using the single-relay DF strategy. As to the single-relay unidirectional AF strategy, its performance is always worse than the direct transmission. It is reasonable, not only because of a higher time expense, but also due to the amplified noise over the power lines when the AF protocol is used.

From Figure 12.6, it is also observed that adding more relays while performing beamforming is not very helpful in the indoor PLC system. Beamforming in the unidirectional two-relay system has enabled only a 0.5 bits/s/Hz spectral efficiency gain compared to the single-relay AF strategy, but has a 0.3 bits/s/Hz loss compared to the single-relay DF strategy. The loss in spectral efficiency of the two-relay DF strategy is due to the fact that beamforming only improves the information rate for the second time slot. The overall information rate of the system is, in fact, limited by the channel conditions between the source S and the two relays R_1 and R_2 in the first time slot (as shown in Figure 12.4). One poorly conditioned channel from the source to one of the relays will eventually impair the overall spectral efficiency of the two-relay DF strategy. As for the two-relay AF strategy, it only offers a marginal improvement in performance compared to the single-relay AF case, while adding an extra relay node also increases the system complexity.

In contrast to beamforming, it is encouraging to note that bidirectional relaying strategies offer much higher spectral efficiencies. Compared to the direct transmission, the bidirectional relaying strategies can achieve a much larger transmission range. For example, given a specific spectral efficiency of 3.5 bits/s/Hz, the transmission range of the bidirectional DF strategy is 90 m, almost double the 46 m obtained by the direction transmission. For short-range transmission, the performance of the bidirectional DF strategy is impaired by using the sub-optimal V-BLAST detection algorithm, while AF provides better performance. As the transmission distance increases, the impact of amplified noise becomes noticeable for the bidirectional AF strategy, and its spectral efficiency finally drops below the bidirectional DF case.

The empirical cumulative distribution function (CDF) of spectral efficiency is shown in Figure 12.7, where the x-axis denotes the spectral efficiency X and the y-axis denotes the empirical CDF $F(X)$. The spectral efficiency CDF $F(X)$ is defined as the percentage of systems having a spectral efficiency less than or equal to X. The simulation results are based on an assumption that the distance between the two end nodes is 55 m. It should be emphasized that the CDF shows results for 10,000 different indoor deployments of the PLC system of a given distance between the two end nodes. From Figure 12.7, we can see that the CDF curves shift to the right when bidirectional relaying strategies are adopted. It implies that bidirectional relaying strategies have relatively high spectral efficiencies. In addition, we note that the slope of the CDF curve of the bidirectional DF strategy is steeper than the bidirectional AF case. It means that the bidirectional DF strategy is more robust against channel condition changes. In other words, the bidirectional AF strategy and direct transmission are more sensitive to channel and noise conditions.

12.4 Relaying technologies for SG wireless communications

As we discussed in the above sections, there is a common need for applying relaying technologies in SG communications. In Section 12.3, we have demonstrated the implementation of various relaying strategies in the context of the indoor-wired PLC network. In this section, we will focus on wireless communication networks, and will study some potential relaying technologies that can be applied to HAN and/or NAN.

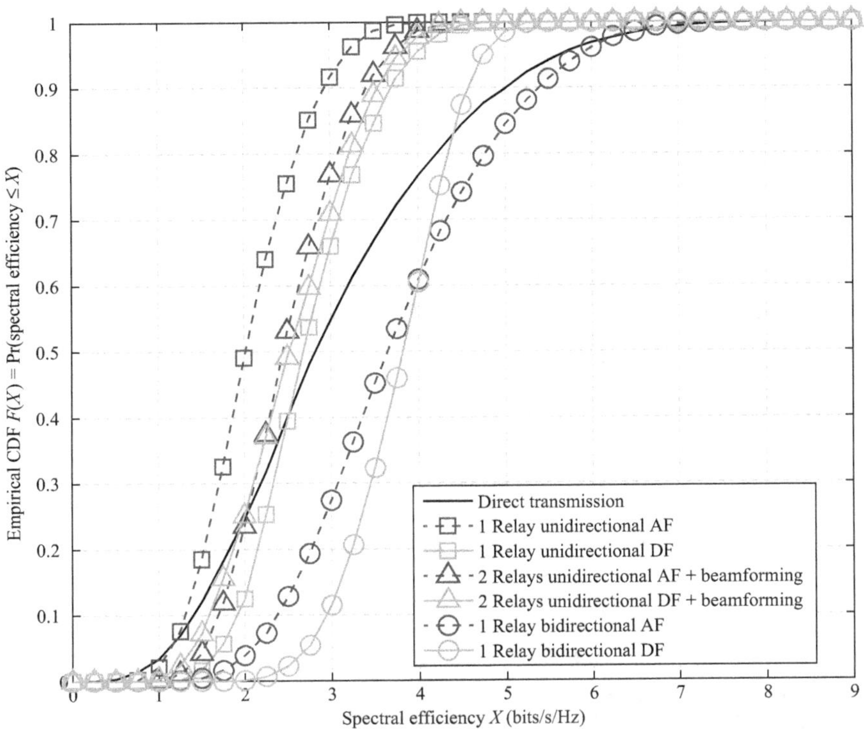

Figure 12.7 Empirical spectral efficiency CDF when the distance between the two end nodes is 55 m

We consider AMI, a typical SG application, with the aim to upload energy consumption data to the utility for DR applications. An example of a wireless network for implementing AMI is shown in Figure 12.8. There are two scenarios in which relaying technologies can be used. In NAN, customers 2 and 3 could act as relays that help customer 1 to transmit data to the advanced metering regional collector (AMRC). Further, AMRCs can help each other to transmit data to the utility's head-end system (UHES).

As we discussed in Section 12.3, conventional unidirectional single-relay technologies (no matter using AF or DF) boost the signals at the destination at the expense of consuming extra resources (e.g., the time and spectrum allocated to the relay) which leads to low system spectral efficiency. In order to improve the spectral efficiency while retaining the advantages of relaying transmission, we studied two advanced relaying technologies, i.e., unidirectional two-relay system with beamforming and bidirectional relaying for information-exchange system. Thus, in this section, we start with these two relaying strategies and investigate whether they can improve system spectral efficiency for SG wireless communication networks. Another advanced relaying technology, virtual-MIMO, will be studied later in this section.

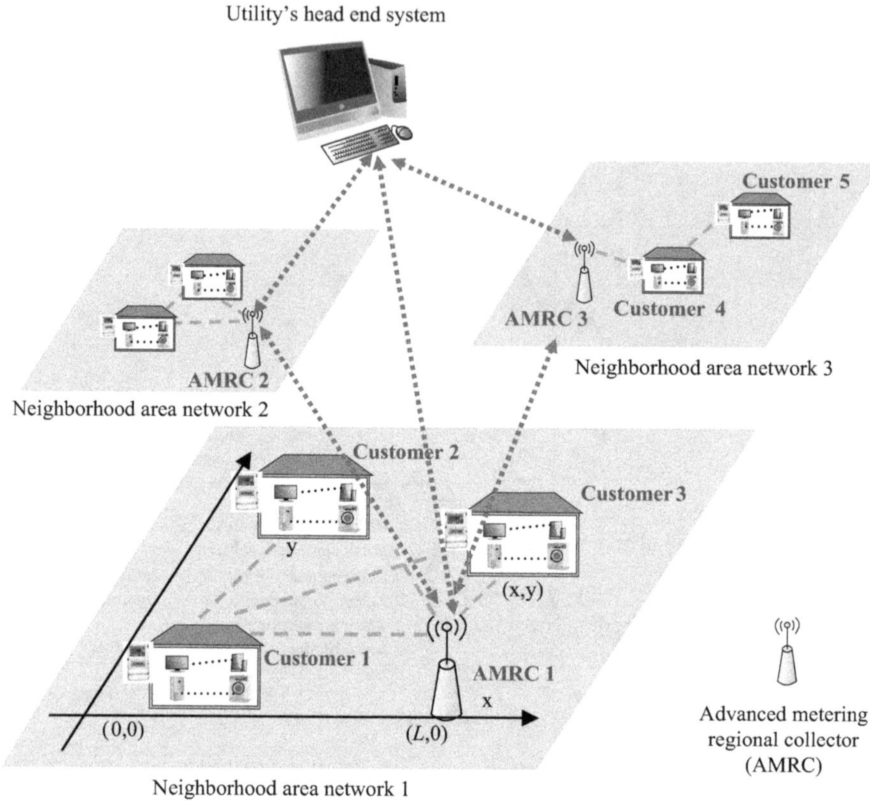

Figure 12.8 An example of relaying technology implementation in the SG: relaying technology-based AMI

12.4.1 Unidirectional two-relay system with collaborative beamforming

Suppose that NAN is implemented using wireless technology based on the ZigBee standard, and the channel center frequency is chosen to be $2405 + 5(k - 11)$ MHz, where the channel index k is a random integer, $k \in [11, 26]$. The bandwidth is assumed to be 2 MHz, and the channels between the relay and the end nodes are assumed to suffer frequency-flat Rayleigh fading. Considering the path loss, we adopt the International Telecommunication Union (ITU) indoor propagation model [26], in which the distance power loss coefficient is set to be 28. The transmit power is set to be 0 dBm, and the antenna gain is 2.5 dB. Additive white Gaussian noise (AWGN) is added to the communication channels with the power level of -110 dBm. Without significant loss of generality, all customers are assumed to be located on the x–y plane. Customer 1 is located at the origin (0,0), and the AMRC is at the coordinates $(L, 0)$. The coordinates of the relay (either customer 2 or customer 3) are denoted by (x, y), where x and y are uniformly distributed values with ranges $x \in [0, L]$

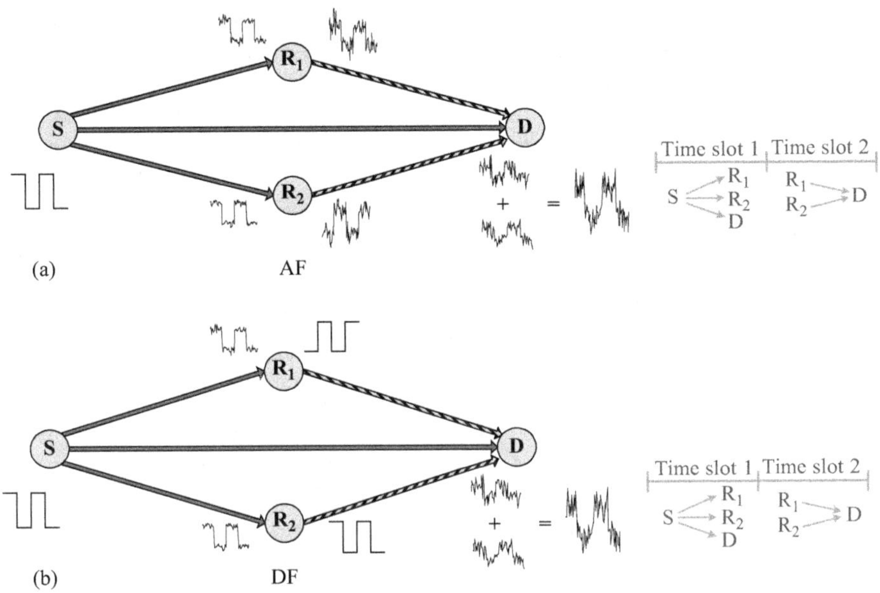

Figure 12.9 Relaying protocols combined with two-relay beamforming strategy: (a) unidirectional AF relaying protocol plus beamforming and (b) unidirectional DF relaying protocol plus beamforming

and $y \in \left[-\frac{L}{2}, \frac{L}{2}\right]$. To emulate the interference from other unlicensed spectrum users (e.g., Wi-Fi or ZigBee), we assume that 1–3 users (with 3 dBm transmit power) are using the same frequency band, with random locations in the ranges $x \in [0, L]$ and $y \in \left[-\frac{L}{2}, \frac{L}{2}\right]$ [1].

Suppose that one SM of customer 1 (source) is uploading the data to the AMRC (destination) while two neighboring customers (relays) could assist the data transmission procedure. Using conventional relaying strategies, two relays forward certain versions of the received signals to the destination. Because of different channel phases at the two relays, the correlation properties of the received signals at the destination will be distorted. We use collaborative beamforming [27] to adaptively adjust the transmit signal phases and amplitudes at the two relays. For example, as shown in Figure 12.9, the signal phase at the relay 2 is adjusted. Taking advantage of collaborative beamforming, the received signals at the destination can be constructively added at the destination to improve SNR [1]. To enable collaborative beamforming, the relays are assumed to be synchronized by the use of reference signals from a positioning system such as the global positioning system (GPS). In addition, we assume that the relays are sufficiently separated so that any mutual coupling effects among their antennas are negligible.

The spectral efficiencies of different relaying strategies are compared in Figure 12.10(a) and (b), where the x-axis denotes the distance between two end nodes and the y-axis denotes the system spectral efficiency. As shown in Figure 12.10(a), when direct transmission is available, relaying strategies yield spectral efficiency (also coverage) gain over direct transmission for any transmission distance. Beamforming in the unidirectional two-relay AF system results in marginal improvement compared to the single-relay AF strategy, while adding an extra relay node also increases the system complexity. Meanwhile, the performance of the two-relay DF system with collaborative beamforming is similar to the single-relay DF strategy. This is because the overall information rate of the system is, in fact, limited by the channel conditions between the source and the two relays in the first time slot. One poorly conditioned channel from the source to one of the relays will eventually impair the overall spectral efficiency of the two-relay DF strategy. In Figure 12.10(b), we can see that, if direct transmission is blocked, relaying technologies can still achieve satisfactory spectral efficiency.

The empirical CDF of spectral efficiency is shown in Figure 12.11(a) and (b), where the x-axis denotes the spectral efficiency X and the y-axis denotes the empirical CDF $F(X)$. The spectral efficiency CDF $F(X)$ is defined as the percentage of systems having a spectral efficiency less than or equal to X. The simulation results are based on an assumption that the distance between the two end nodes is 70 m. The CDF shows results for 10,000 different channel conditions for a given distance between the two end nodes. It should be emphasized that the steeper the curve is, the more robust the system can be. In addition, the curve shifting to the right implies that it can obtain higher spectral efficiency. In Figure 12.11(a) and (b), we can see that all relaying strategies are more reliable than direct transmission.

12.4.2 Bidirectional relaying for information-exchange system

DR in SGs requires high data-rate two-way communications between the customers and the UHES. Using conventional unidirectional relaying strategies in NAN, four time slots are needed to accomplish the information exchange process, leading to low spectral efficiency, as illustrated in Figure 12.5(a). The bidirectional relaying strategy requires only two time slots to complete an information exchange process. In the first time slot, two end nodes (e.g., customer 5 and the AMRC 3 in Figure 12.8) send their information to the relay (e.g., customer 4) using the same frequency band, leading to a superposition of the received signals at the relay. Note that because we assume that all nodes are working in the half-duplex mode, two end nodes cannot decode the signals in the first time slot. Using the AF relaying protocol, the relay directly forwards its received signals to both end nodes. Because these two end nodes have their own copies of transmitted signals, they can subtract their own signals and obtain the information transmitted from the other node. We note that the CSI between the relay and two end nodes can be estimated by using a channel estimation scheme, e.g., pilot symbol insertion or training bits [28, 29]. In this way, a higher spectral efficiency can be achieved, since the information at the two nodes is exchanged using fewer time slots.

If the bidirectional relaying strategy is applied together with the DF relaying protocol, the relay needs to decode the superposed signals. As discussed in Section 12.3.2, optimal performance can be achieved using an ML detection algorithm; near-optimal

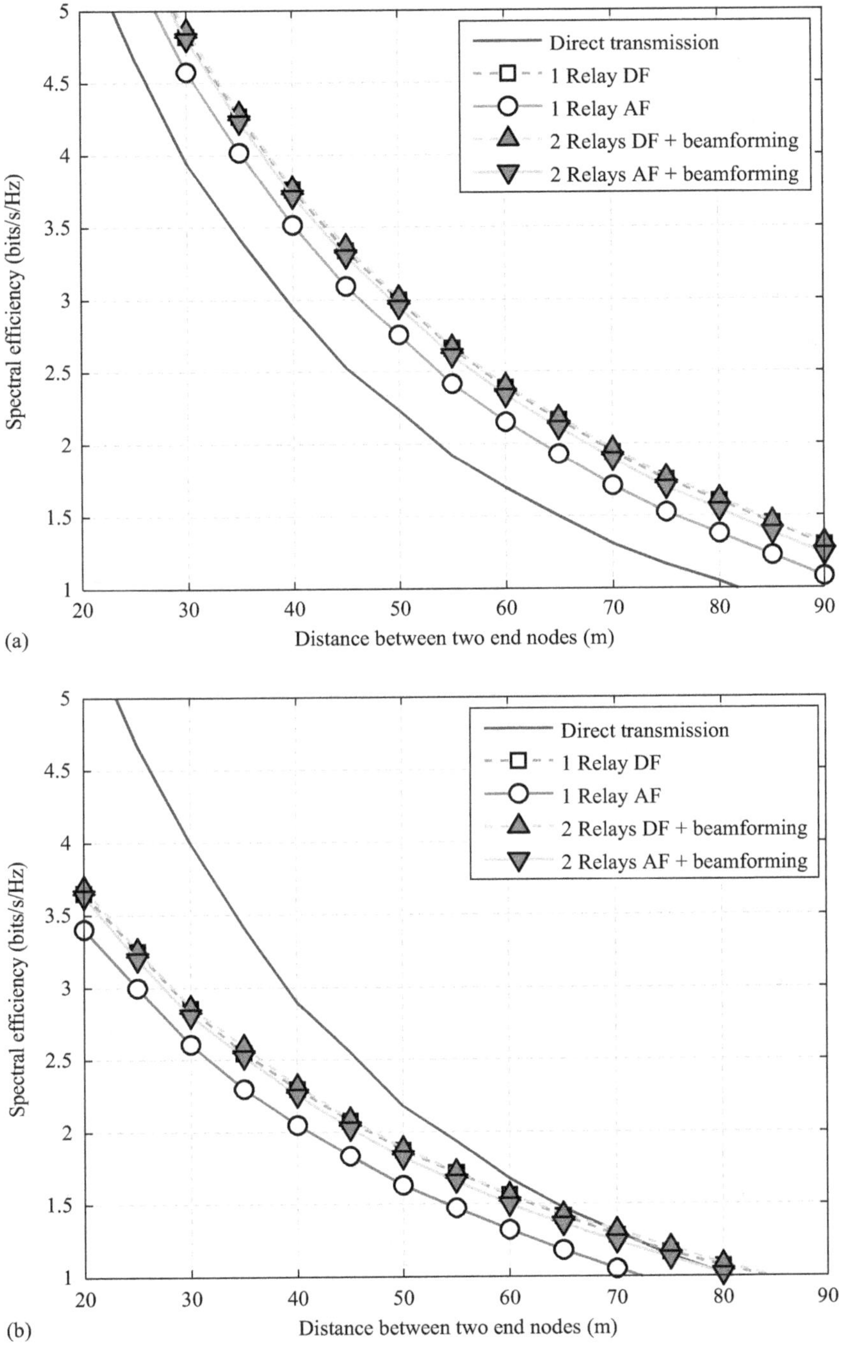

Figure 12.10 Comparisons of relaying technologies with direct transmission: (a) spectral efficiency when direct transmission is available and (b) spectral efficiency when direct transmission is blocked

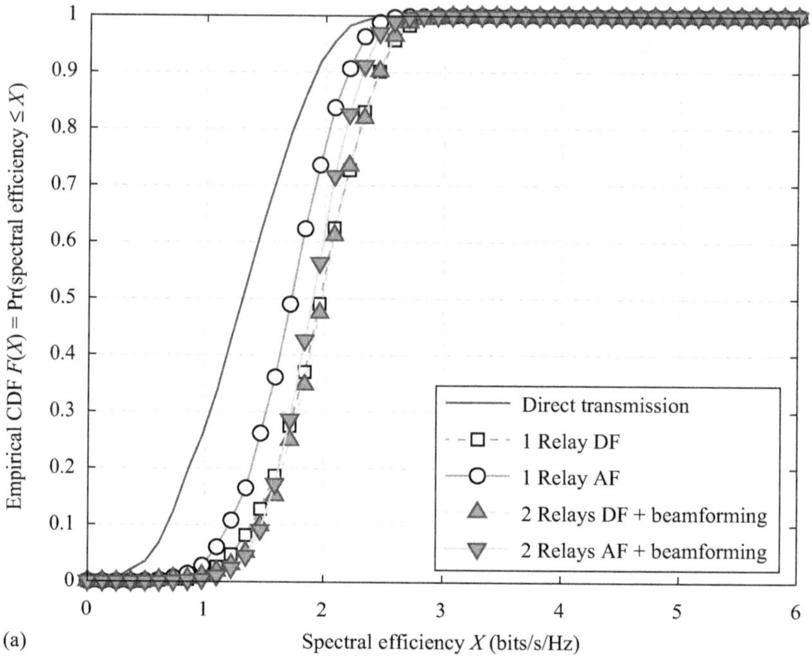

(a)

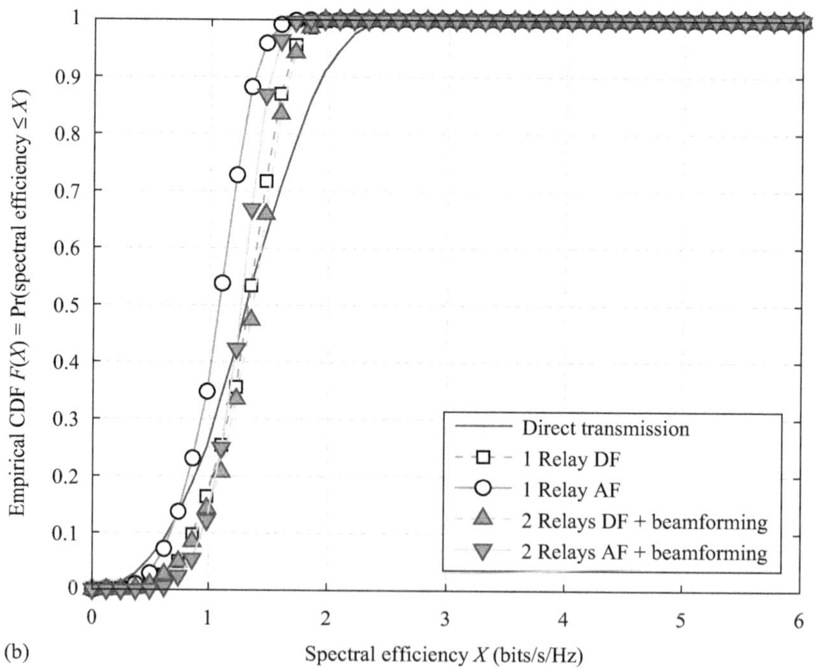

(b)

Figure 12.11 Comparisons of relaying technologies with direct transmission: (a) empirical spectral efficiency CDF when direct transmission is available and (b) empirical spectral efficiency CDF when direct transmission is blocked

performance can be achieved by using V-BLAST detection algorithm [30] with lower computational cost. We thus consider the V-BLAST detection algorithm at the relay for the bidirectional DF relaying system.

We consider the same simulation settings as in Section 12.4.1, and obtained the results as shown in Figures 12.12 and 12.13. It can be seen from Figure 12.12(a) and (b) that, despite the blocked direct transmission, the bidirectional relay strategies are always superior to direct transmission when the transmission distance is relatively long. This phenomenon arises from the facts that the bidirectional relay strategies exchange information using fewer time slots when compared to the unidirectional relay strategies, and that the bidirectional relay strategies do not use the direct link for transmitting information as the two end nodes are working in the half-duplex mode. The bidirectional DF is inferior to direct transmission when the transmission distance L is less than 35 m. This is because, when using V-BLAST, we treat the signal from another node as interference (it becomes stronger as the transmission distance becomes shorter), thereby decreasing the information rate of the whole system. Furthermore, we can see from Figure 12.13(a) and (b) that even though the bidirectional relay strategies are not as reliable as the conventional relaying strategies, higher spectral efficiency can be achieved with them regardless of the availability of direct transmission link.

12.4.3 Virtual-MIMO in SG wireless communications

MIMO systems have recently emerged as one of the most significant wireless techniques, as they can greatly improve spectral efficiency, channel capacity and link reliability of wireless communications [31, 32]. These benefits have encouraged extensive research on a virtual-MIMO system where the transmitter has multiple antennas and single-antenna receivers can work together to form a virtual antenna array and reap some performance benefits of MIMO systems [33, 34].

With the customer-side local cooperation, virtual-MIMO is another advanced relaying technology which can be used in SG wireless communications to improve system spectral efficiency and reliability. For example, suppose in an NAN, an AMRC has multiple antennas and a SM has single antenna, as illustrated in Figure 12.14. There is another device closed to the SM which can serve as a relay. The relay could be a computer, customer device or another SM nearby. Since the distance between the SM device and the relay device is generally much shorter than that from the AMRC, the devices could cooperate through their short-range Wi-Fi, or Bluetooth communications links. With such cooperation, the customer could expect traditional MIMO benefits as if the single-antenna SM had multiple antennas. Note that customer-side communication used for cooperation is realized via an orthogonal channel and allows much higher transmission rate and frequency reuse compared with the long-distance data communication. As for the cooperation protocol, since the relays get closer to the destination (i.e., the SM) in our scenario, it has been shown that, compared with AF and DF, the CF protocol provides superior performance [35, 36] and therefore serves as the best candidate for this system.

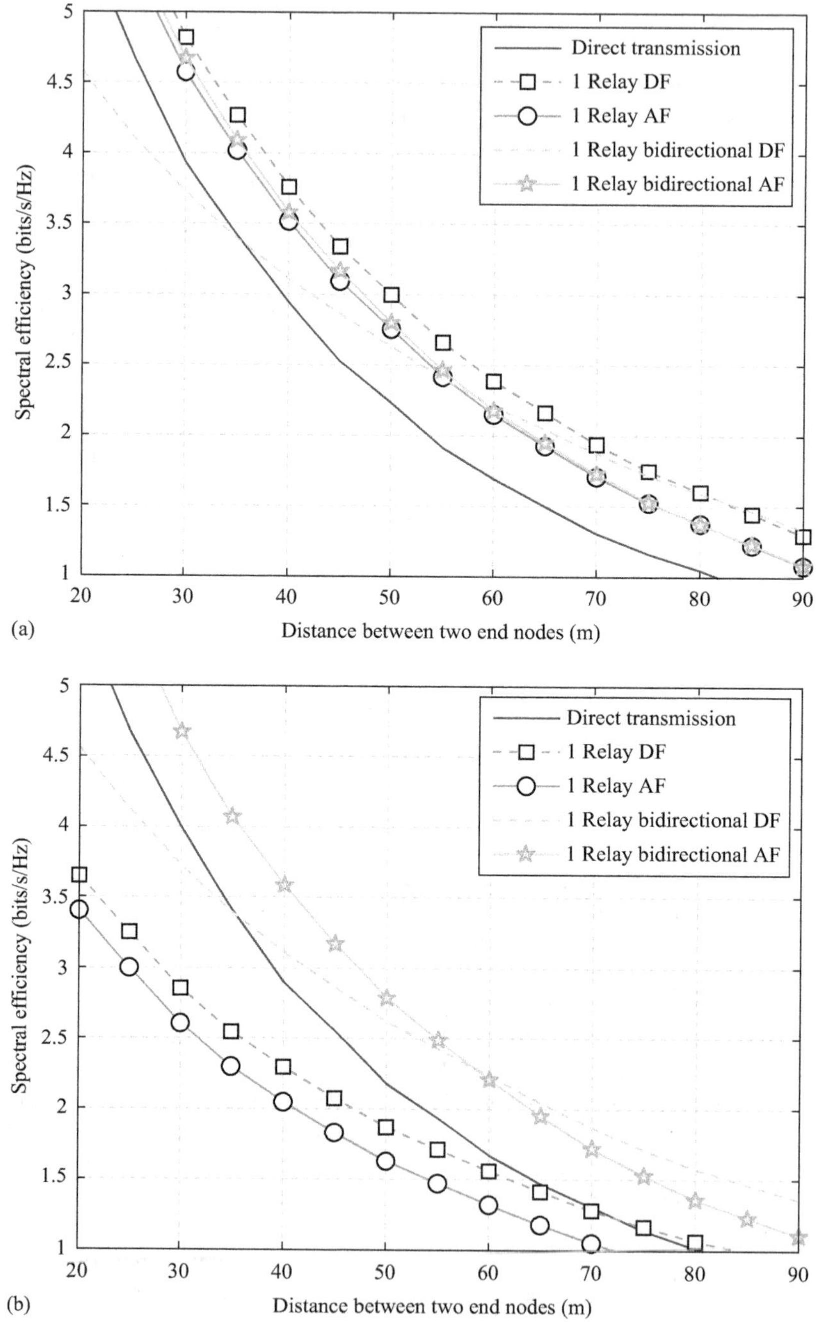

Figure 12.12 Comparisons of relaying technologies with direct transmission: (a) spectral efficiency when direct transmission is available and (b) spectral efficiency when direct transmission is blocked

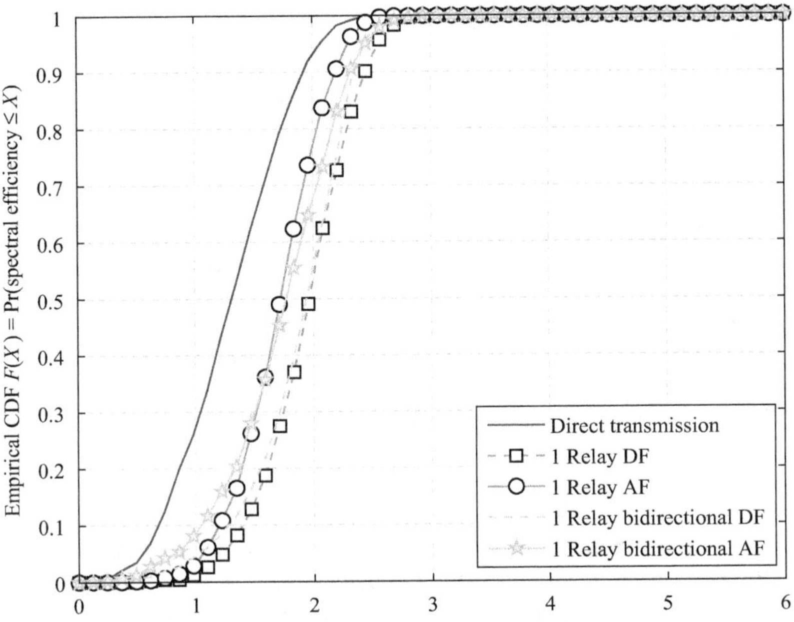

(a)

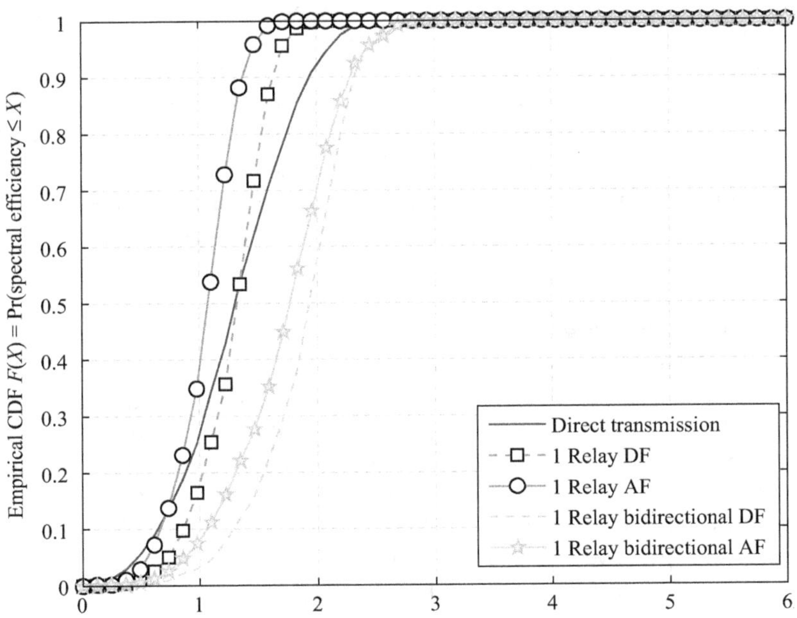

(b)

Figure 12.13 Comparisons of relaying technologies with direct transmission: (a) empirical spectral efficiency CDF when direct transmission is available and (b) empirical spectral efficiency CDF when direct transmission is blocked

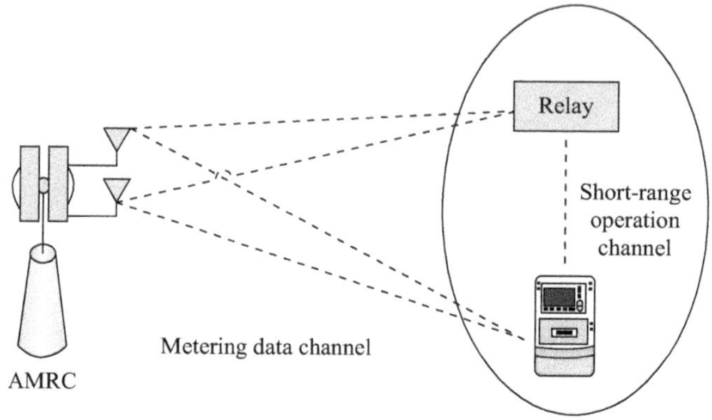

Figure 12.14 An example of virtual-MIMO technology implementation in the SG

We thus consider such a virtual-MIMO relaying system, with one remote multiple antenna AMRC communicating with closely spaced single-antenna devices. Since the AMRC is far away from the customer devices, a block fading channel model is assumed here. Fading is flat and constant on each block, but independently Rayleigh distributed on different blocks. As the customer devices are closely spaced, it is reasonable to expect that the communication between the devices is much better and more stable than that between the AMRC and devices. It is more likely that we could achieve high channel capacity with high reliability on this short-range link. Hence, as also considered in References 35, 37, and 38, we assume customer devices cooperate by way of an error-free conference link, with capacity C. The impact of a non-ideal cooperation link has been also illustrated in Reference 35. It has been shown that, considering the short-range communication and using a channel-aware adaptive CF scheme, the impact of the non-ideal cooperation link is too slight to impair the system performance significantly.

12.4.3.1 The CF protocol

This virtual-MIMO operation is performed by the customer-side cooperation on the conference link. In order to avoid interfering with the metering data channel, the conference link is realized via an orthogonal channel to the metering data channel. That is, two different frequency bands are used for transmitting and receiving data at the relay. Compared with the long-range metering data channel, the orthogonal conference link is much short range with low transmission power, and could be reused many times over the coverage area of the long-range link. Thus, cooperation will not affect the overall system throughput. The relay may help the virtual-MIMO system to achieve almost ideal MIMO performance.

As for cooperation protocol, CF is chosen here, so that a compressed version of the data will be sent over the cooperation or conference link. To enable CF, a standard source coding technique is employed for practical considerations. We do not employ the Wyner–Ziv coding technique, as it does not improve the performance significantly

[35, 39], but introduces extra complexity. We refer the readers to Reference for more details on the comparison of standard source coding and Wyner–Ziv coding techniques. The relay is equipped with a vector quantizer to enable source coding for CF. The key tasks of the relay thus include constructing a good codebook and quantizing the received data. We assume a fixed quantization rate (i.e. source coding rate), denoted by R_c bits/sample. Since the two receivers cooperate by way of an error-free conference link, the fixed source coding rate R_c could be smaller than but arbitrarily close to the link capacity.

12.4.3.2 Numerical results and discussions

To focus on the performance and practical implementation of the CF cooperation, we simplify the system to the 2-by-2 antennas case. That is, AMRC is equipped with two antennas and each customer device has a single antenna. Further performance improvements are expected for the case of more cooperating terminals equipped with larger numbers of antennas. We implement the bit-interleaved coded modulation technique [40] to provide forward error correction at AMRC. Bit-interleaved coded modulation separates the aspects of coding and modulation by bit-interleaving, so that a symbol error at the decoder does not cause consecutive error bits in the codeword, which improves system performance. In addition, at AMRC, a binary convolutional code is assumed with the generator polynomials $[133, 171]_{\text{octal}}$. Gray-labeled quadrature phase shift keying (QPSK) and 16QAM (quadrature amplitude modulation) modulations are considered. The channels between the transmitter and receivers are assumed to be independent identically distributed normalized block Rayleigh fading, with 10^5 fading blocks and each block has 200 consecutive symbol periods. Furthermore, the Voronoi VQ is implemented at the relay to enable CF cooperation. The simulation results are obtained using the Monte Carlo method, and we plot error ratios against the information bit SNR.

The block error rate (BLER) performance of the virtual-MIMO with QPSK or 16QAM mapping, under various fixed source coding rates, is shown in Figure 12.15. Because of the error-free conference link, the link capacity C equals the source coding rate. The BLERs are compared against the lower bound of the corresponding MIMO system, and the non-cooperative MISO system. Figure 12.15(a) shows that, with help from the relay, the system always performs better than the MISO system, as the CF protocol always provides a gain over direct transmission. As C increases beyond 3 bits/s, the simulation results of virtual-MIMO decrease and approach the ideal MIMO system performance. A similar trend can be seen in Figure 12.15(b) for the 16QAM mapping. As its constellation size is larger than that of QPSK, higher source coding rates are considered. It is obvious that, when $C = 4$ bits/s for QPSK and $C = 7$ bits/s for 16QAM, the BLER of CF cooperation performs quite close to the ideal MIMO system. This phenomenon can also been seen from Figure 12.16 where we demonstrate the system throughput using numerical integration. With sufficient C, the helping relay enables the proposed single-antenna user device to achieve MIMO performance.

As to cooperation protocols, the Alamouti space–time block coding (STBC) based DF scheme [41] could be implemented, with two signals x_1 and x_2 transmitted in one

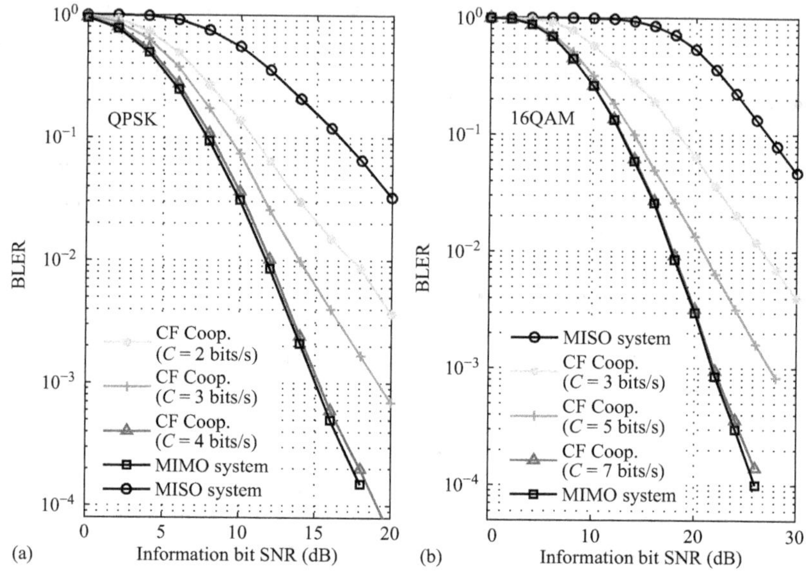

Figure 12.15 BLER of the cooperative virtual-MIMO system over block Rayleigh fading channels (QPSK mapping is considered in (a), while 16QAM in (b))

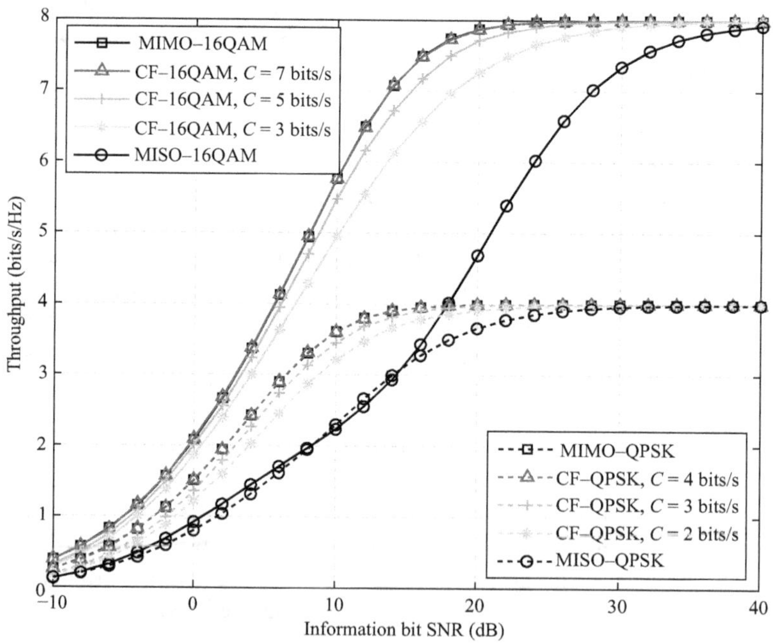

Figure 12.16 Throughput of the cooperative virtual-MIMO system over block Rayleigh fading channels. (Solid curves correspond to 16QAM mapping, while dash-dotted curves correspond to QPSK mapping.)

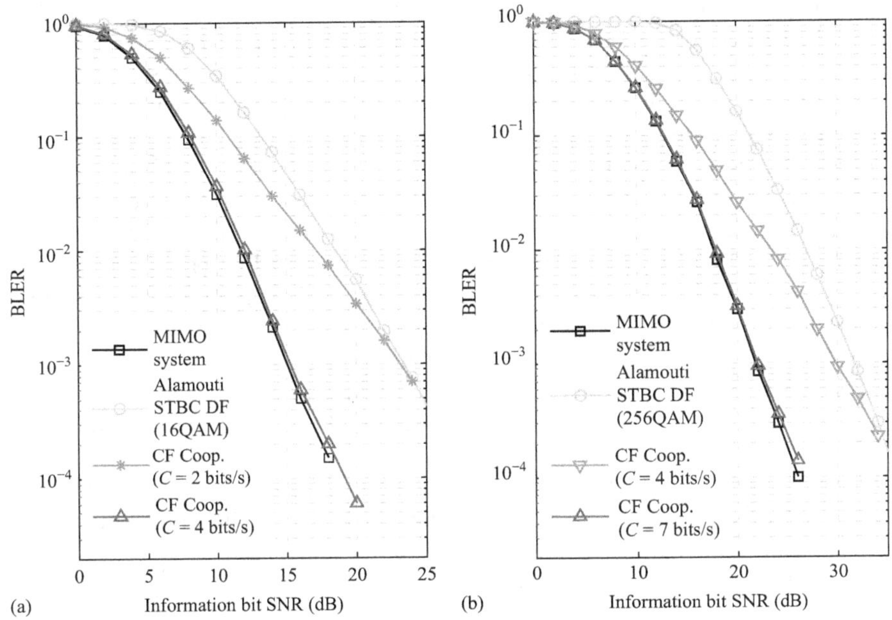

Figure 12.17 BLER comparison for same spectral efficiency between the CF cooperation scheme and the Alamouti STBC-based DF scheme. (CF with QPSK and Alamouti DF with 16QAM are compared in (a), while CF with 16QAM and Alamouti DF with 256QAM are considered in (b).)

symbol period, and $-x_2^*$ and x_1^* originated from x_1 and x_2, sent in the next symbol period. But the spectral efficiency of the Alamouti DF scheme is only one-half of that with CF or AF protocols. To illustrate the benefit of CF cooperation in this virtual-MIMO system, we compare it with the Alamouti STBC-based DF scheme in Figure 12.17. To make a fair comparison, we consider 16QAM modulation for the Alamouti DF scheme, compared to QPSK modulation for CF, since both of them allow a spectral efficiency of 2 bits/s/Hz. Also, the CF scheme with 16QAM and the Alamouti DF with 256QAM are compared. As shown in Figure 12.17, only at high-information bit SNR, the Alamouti DF can achieve similar error ratios comparable to the CF scheme which uses $C = 2$ bits/s for QPSK and $C = 4$ bits/s for 16QAM. Specifically, for a target BLER of 10^{-2}, the CF scheme using a reasonable C, i.e., $C \geq 2$ bits/s for QPSK and $C \geq 4$ bits/s for 16QAM, is expected to perform much better than the Alamouti DF scheme.

12.5 Conclusions

In this chapter, we have discussed the challenges of communication technologies when applied to SGs and the need for implementing relaying technologies in SG communications. In order to improve the coverage, the spectral efficiency, and the reliability of SG communications, we have investigated several potential relaying strategies, e.g., collaborative beamforming for multiple-relay systems, bidirectional relaying for information-exchange systems, and virtual-MIMO systems. It has been shown that all relaying strategies can improve the reliability of SG communications thanks to the spatial diversity, regardless of indoor PLC network or wireless communication networks. In the context of the indoor PLC network, unidirectional relaying strategies are not necessary when the transmission distance is short, and the two-relay system with beamforming cannot give significant performance improvements over the single-relay case. In contrast, bidirectional relaying strategies are helpful even if the transmission distance is as short as 25 m in the indoor PLC network. For wireless communication networks, single-direction two-relay systems can only achieve marginal improvement over a single-relay system at the expense of increased implementation complexity. Regardless of whether direct transmission is available, the spectral efficiency and the coverage of SG communications can be improved by using bidirectional relaying strategies. In addition, for the proposed virtual-MIMO system, with the CF protocol, the relay device could enable the single-antenna SM device to expect MIMO benefits as if it had multiple antennas.

References

[1] H. Sun, A. Nallanathan, B. Tan, J. Thompson, J. Jiang, and H. Poor, "Relaying technologies for smart grid communications," *IEEE Wireless Communications*, vol. 19, no. 6, pp. 52–59, Dec. 2012.

[2] Trilliant Incorporated, "Wireless wan for the smart grid: Distribution networking for today's (and tomorrow's) smart grid communications infrastructure," Redwood City, CA, Technical Report, March 2010.

[3] P. Parikh, M. Kanabar, and T. Sidhu, "Opportunities and challenges of wireless communication technologies for smart grid applications," in *Proc. 2010 IEEE Power and Energy Society General Meeting*, Minneapolis, Minnesota, USA, July 2010, pp. 1–7.

[4] J. Wang and V. Leung, "Comparisons of home area network connection alternatives for multifamily dwelling units," in *Proc. 2011 Fourth IFIP International Conference on New Technologies, Mobility and Security (NTMS)*, Paris, France, February 2011, pp. 1–5.

[5] G. Bumiller, L. Lampe, and H. Hrasnica, "Power line communication networks for large-scale control and automation systems," *IEEE Communications Magazine*, vol. 48, no. 4, pp. 106–113, Apr. 2010.

[6] B. Tan and J. Thompson, "Relay transmission protocols for in-door powerline communications networks," in *Proc. Communications Workshops (ICC), 2011 IEEE International Conference on*, Kyoto, Japan, June 2011, pp. 1–5.

[7] A. Tonello, F. Versolatto, and S. D'Alessandro, "Opportunistic relaying in in-home PLC networks," in *Proc. 2010 IEEE Global Telecommunications Conference (GLOBECOM 2010)*, Miami, Florida, USA, December 2010, pp. 1–5.

[8] L. Lampe, R. Schober, and S. Yiu, "Distributed space-time coding for multi-hop transmission in power line communication networks," *IEEE Journal on Selected Areas in Communications*, vol. 24, no. 7, pp. 1389–1400, Jul. 2006.

[9] V. Gungor, B. Lu, and G. Hancke, "Opportunities and challenges of wireless sensor networks in smart grid," *IEEE Transactions on Industrial Electronics*, vol. 57, no. 10, pp. 3557–3564, Oct. 2010.

[10] B. Lichtensteiger, B. Bjelajac, C. Mller, and C. Wietfeld, "Rf mesh systems for smart metering: System architecture and performance," in *Proc. 2010 First IEEE International Conference on Smart Grid Communications (SmartGrid-Comm)*, Gaithersburg, Maryland, USA, October 2010, pp. 379–384.

[11] G. Iyer, P. Agrawal, E. Monnerie, and R. Cardozo, "Performance analysis of wireless mesh routing protocols for smart utility networks," in *Proc. 2011 IEEE International Conference on Smart Grid Communications (SmartGridComm)*, Brussels, Belgium, October 2011, pp. 114–119.

[12] T. Iwao, K. Yamada, M. Yura, *et al.*, "Dynamic data forwarding in wireless mesh networks," in *Proc. 2010 First IEEE International Conference on Smart Grid Communications (SmartGridComm)*, Gaithersburg, Maryland, USA, October 2010, pp. 385–390.

[13] Q.-D. Ho, Y. Gao, and T. Le-Ngoc, "Challenges and research opportunities in wireless communication networks for smart grid," *IEEE Wireless Communications*, vol. 20, no. 3, pp. 89–95, Jun. 2013.

[14] Y. Yan, Y. Qian, H. Sharif, and D. Tipper, "A survey on smart grid communication infrastructures: Motivations, requirements and challenges," *IEEE Communications Surveys Tutorials*, vol. 15, no. 1, pp. 5–20, First 2013.

[15] W. Meng, R. Ma, and H.-H. Chen, "Smart grid neighborhood area networks: a survey," *IEEE Network*, vol. 28, no. 1, pp. 24–32, Jan. 2014.

[16] S. Galli and T. Lys, "Next generation narrowband (under 500 kHz) power line communications (PLC) standards," *China Communications*, vol. 12, no. 3, pp. 1–8, Mar. 2015.

[17] A. Mahmood, N. Javaid, and S. Razzaq, "A review of wireless communications for smart grid," *Renewable and Sustainable Energy Reviews*, vol. 41, pp. 248–260, 2015 [Online]. Available: http://www.sciencedirect.com/science/article/pii/S1364032114007126

[18] G. Kramer, M. Gastpar, and P. Gupta, "Cooperative strategies and capacity theorems for relay networks," *IEEE Transactions on Information Theory*, vol. 51, no. 9, pp. 3037–3063, Sep. 2005.

[19] M. Rahman, C. S. Hong, S. Lee, J. Lee, M. Razzaque, and J. H. Kim, "Medium access control for power line communications: An overview of the IEEE 1901 and ITU-T G.hn standards," *IEEE Communications Magazine*, vol. 49, no. 6, pp. 183–191, Jun. 2011.

[20] *IEEE Standard for Broadband over Power Line Networks: Medium Access Control and Physical Layer Specifications*, IEEE Std 1901-2010 Std., Dec. 2010.

[21] J. Laneman, D. Tse, and G. W. Wornell, "Cooperative diversity in wireless networks: Efficient protocols and outage behavior," *IEEE Transactions on Information Theory*, vol. 50, no. 12, pp. 3062–3080, Dec. 2004.

[22] G. Golden, C. Foschini, R. Valenzuela, and P. Wolniansky, "Detection algorithm and initial laboratory results using V-BLAST space-time communication architecture," *Electronics Letters*, vol. 35, no. 1, pp. 14–16, Jan. 1999.

[23] M. Zimmermann and K. Dostert, "A multipath model for the powerline channel," *IEEE Transactions on Communications*, vol. 50, no. 4, pp. 553–559, Apr. 2002.

[24] S. Galli and T. Banwell, "A deterministic frequency-domain model for the indoor power line transfer function," *IEEE Journal Selected Areas in Communications*, vol. 24, no. 7, pp. 1304–1316, Jul. 2006.

[25] D. Benyoucef, "A new statistical model of the noise power density spectrum for powerline communication," in *Proceedings of the International Symposium on Power Line Communications and Its Applications*, Kyoto, Japan, March 2003, pp. 136–141.

[26] *Propagation Data and Prediction Methods for the Planning of Indoor Radio Communication Systems and the Radio Local Area Networks in the Frequency Range 900 MHz to 100 GHz*, ITU-R Recommendations Std., 2001.

[27] H. Ochiai, P. Mitran, H. Poor, and V. Tarokh, "Collaborative beamforming for distributed wireless ad hoc sensor networks," *IEEE Transactions on Signal Processing*, vol. 53, no. 11, pp. 4110–4124, Nov. 2005.

[28] B. Jiang, F. Gao, X. Gao, and A. Nallanathan, "Channel estimation and training design for two-way relay networks with power allocation," *IEEE Transactions on Wireless Communications*, vol. 9, no. 6, pp. 2022–2032, Jun. 2010.

[29] T.-H. Pham, Y.-C. Liang, A. Nallanathan, and H. K. Garg, "Optimal training sequences for channel estimation in bi-directional relay networks with multiple antennas," *IEEE Transactions on Communications*, vol. 58, no. 2, pp. 474–479, Feb. 2010.

[30] C. Hassell, J. Thompson, B. Mulgrew, and P. M. Grant, "A comparison of detection algorithms including BLAST for wireless communication using multiple antennas," in *Proc. The 11th IEEE International Symposium on Personal, Indoor and Mobile Radio Communications, 2000. PIMRC 2000*, London, UK, vol. 1, 2000, pp. 698–703.

[31] D. Gesbert, M. Shafi, D.-S. Shiu, P. J. Smith, and A. Naguib, "From theory to practice: An overview of MIMO space-time coded wireless systems," *IEEE Journal on Selected Areas in Communications*, vol. 21, no. 3, pp. 281–302, Apr. 2003.

[32] J. Jiang, J. Thompson, H. Sun, and P. Grant, "Practical analysis of codebook design and frequency offset estimation for virtual-multiple-input-multiple-output systems," *IET Communications*, vol. 7, no. 6, pp. 585–594, Apr. 2013.

[33] C. T. K. Ng, N. Jindal, A. J. Goldsmith, and U. Mitra, "Capacity gain from two-transmitter and two-receiver cooperation," *IEEE Transactions on Information Theory*, vol. 53, no. 10, pp. 3822–3827, Oct. 2007.

[34] J. Jiang, J. S. Thompson, and H. Sun, "A singular-value-based adaptive modulation and cooperation scheme for virtual-MIMO systems," *IEEE Transactions on Vehicular Technology*, vol. 60, no. 6, pp. 2495–2504, Jul. 2011.

[35] J. Jiang, J. Thompson, H. Sun, and P. Grant, "Performance assessment of virtual multiple-input multiple-output systems with compress-and-forward cooperation," *IET Communications*, vol. 6, no. 11, pp. 1456–1465, Jul. 2012.
[36] A. Host-Madsen and J. Zhang, "Capacity bounds and power allocation for wireless relay channels," *IEEE Transactions on Information Theory*, vol. 51, no. 6, pp. 2020–2040, Jun. 2005.
[37] R. Dabora and S. D. Servetto, "On the role of estimate-and-forward with time sharing in cooperative communication," *IEEE Transactions on Information Theory*, vol. 54, no. 10, pp. 4409–4431, Oct. 2008.
[38] C. T. K. Ng, I. Maric, A. J. Goldsmith, S. Shamai, and R. D. Yates, "Iterative and one-shot conferencing in relay channels," in *Proc. IEEE Information Theory Workshop*, Punta del Este, Uruguay, March 2006, pp. 193–197.
[39] M. Kuhn, J. Wagner, and A. Wittneben, "Cooperative processing for the WLAN uplink," in *IEEE WCNC*, Las Vegas, NV, March 2008, pp. 1294–1299.
[40] G. Caire, G. Taricco, and E. Biglieri, "Bit-interleaved coded modulation," *IEEE Transactions on Information Theory*, vol. 44, no. 3, pp. 927–946, 1998.
[41] Z. Bai, D. Yuan, and K. Kwak, "Performance evaluation of STBC-based cooperative systems over slow Rayleigh fading channel," *Computer Communications*, vol. 31, pp. 4206–4211, Nov. 2008.

Chapter 13

Simultaneous wireless information and power transfer in relay interference channels

He (Henry) Chen and Yonghui Li

13.1 Introduction

In energy-constrained wireless networks (e.g., wireless sensor networks), devices are typically powered by batteries with limited operation duration, which significantly confines the network performance. As a sustainable approach to prolonging the network's lifetime, the energy harvesting technique has recently drawn significant interest since it allows terminals to replenish their batteries from external energy sources in the surrounding environment. Initial efforts on integrating energy harvesting techniques into wireless communication systems have mainly focused on renewable energy sources, such as solar, wind, thermal, and vibration (see References 1–3 and references therein). However, the intermittent and unpredictable nature of these renewable energy sources could make the energy harvesting technique inapplicable for many practical applications with minimum quality-of-service requirements. In this context, the radio-frequency (RF) energy transfer technique, which uses RF signals as energy sources, provides an attractive solution [4]. Since RF signals can be controlled, the RF energy transfer technique can provide much more reliable energy than conventional energy harvesting techniques that rely on the uncontrollable external natural resources.

On the other hand, RF signals have been widely used for information transmission in wireless communications. In this regard, simultaneous wireless information and power transfer (SWIPT) [5] becomes attractive since it enables the dual utilization of same RF signals for the delivery of both information and energy at the same time. The basic concept of SWIPT was first proposed in Reference 5. A comprehensive receiver architecture and the corresponding rate-energy trade-off were developed in Reference 6. Inspired by these two seminal results, the designs of SWIPT schemes for wireless networks have attracted a lot of attention recently. SWIPT schemes for multiple-input multiple-output (MIMO) broadcasting channels, multiple-input single-output (MISO) interference channels, and MIMO interference channels were designed and evaluated in References 7–9, respectively. The resource allocation algorithms for SWIPT in broadband wireless systems were investigated in Reference 10, while an energy-efficient resource allocation algorithm was developed in Reference 11 for SWIPT in orthogonal frequency division multiple access (OFDMA) systems.

In Reference 12, Nasir *et al.* first merged the idea of SWIPT into a classical three-node relay network, where the energy-constrained relay node harvests energy from the source signal to enable forwarding the received signal. Two practical relaying protocols, namely time switching-based relaying (TSR) and power splitting-based relaying (PSR), were proposed for the considered relay network. Analytical expressions for the outage probability and the ergodic capacity of the proposed protocols were derived for delay-limited and delay-tolerant modes, respectively. This seminal work of SWIPT in relay networks has been extended in various ways recently. More specifically, SWIPT schemes were designed and analyzed for two-way relaying channels in References 13 and 14, for two-hop full-duplex relaying channels in References 15 and 16, for two-hop relaying channels with a multi-antenna relay in References 17–19, and for two-hop non-regenerative MIMO-OFDM relay networks in Reference 20, respectively.

Besides the aforementioned works that considered the classical three-node relay networks, there have been some works that designed SWIPT in larger relay networks [21–25]. In Reference 21, Ding *et al.* studied the setup where multiple source–destination pairs communicate with each other via a common energy harvesting relay. Specifically, several power allocation schemes were proposed to efficiently distribute the power harvested at the relay among multiple pairs. A multiple-relay scenario was investigated in Reference 22, where different strategies to use the available energy harvesting relays and their impact on the system performance were characterized by applying the advanced stochastic geometry theory. In addition, Krikidis [23] also adopted stochastic geometry to analyze the outage performance of SWIPT in large-scale networks with/without relaying, where the transmitters and the relays (if they exist) are assumed to be connected to constant power supplies, while the receivers harvest the energy from the signals received from the source and relay based on a power splitting technique [6]. Besides, in our previous works [24, 25], we first characterized the design of SWIPT in relay interference channels, which have many practical applications, such as cellular networks, wireless sensor networks, wireless local area networks (WLANs), etc. [26–29]. Particularly, we considered the scenario that multiple source–destination pairs communicate simultaneously with the help of their dedicated energy harvesting relays, which do not have their own power supply and need to harvest energy from the source signal before forwarding the information signal. power splitting framework for the SWIPT in relay interference channels. In particular, each source–relay–destination link in the relay interference channels is modeled as a strategic player who chooses its dedicated relay's power splitting ratio to maximize its individual rate.

In this chapter, we extend our previous works [24, 25] to a more general scenario of relay interference channels with energy harvesting relays equipped with multiple antennas. It is worth pointing out that implementing multiple antennas at the relays is particularly appealing in practice since it can not only boost the amount of harvesting energy at the relay in the first hop but also enhance the strength of the received signal at the destination by exploiting the transmit diversity technique in the second hop. The power splitting technique is assumed to implement at relays. Specifically, each relay node splits the signal received from all sources into two parts according to a power splitting ratio: one part is sent to the information processing unit, and the rest is used to harvest energy for forwarding the received information in the second time slot. We consider that each link's performance is characterized by its achievable rate and thus regard the sum-rate of all links as a network-wide performance metric. However, the network-wide sum-rate maximization problem of the considered system is shown to be analytically non-tractable due to its complexity and non-convexity. Motivated by this, we propose a heuristic two-stage network optimization approach. Specifically, considering the energy harvesting feature of the relays, we first figure out that the maximum ratio combining/maximal ratio transmission (MRC/MRT) scheme would be a particularly appropriate signal processing technique for the relays in practice owing to its low complexity and light requirement of channel state information (CSI). In the second stage, we optimize the power splitting ratios of the relays with the selected MRC/MRT technique.

Unfortunately, the simplified sum-rate maximization problem regarding power splitting ratios with the selected MRC/MRT scheme is still difficult to resolve. This is because the power splitting ratio of each link not only affects the performance of this link, but also affects the performance of other links due to mutual interference between different links. This means that the optimization of each ratio depends on all other ratios and they are tangled together. Moreover, the maximization of the sum-rate is shown to be also non-convex even though the beamforming vectors at relays have been determined. The global optimal power splitting ratios cannot be efficiently achieved even in a centralized fashion with a heavy signaling overhead. To tackle this problem, we develop a distributed power splitting framework for the considered network by resorting to the non-cooperative game theory. In particular, we formulate a non-cooperative game by modeling each source–relay–destination link in the relay interference channels as a strategic player who controls its dedicated relay's power splitting ratio to maximize its individual rate. We then prove the existence and uniqueness of the Nash equilibrium (NE) for the formulated game. Moreover, a distributed algorithm is proposed with provable convergence to achieve the NE. Extensive numerical simulations are finally performed to validate all theoretical analyses and demonstrate the impacts of various system parameters on the network performance.

The remainder of this chapter is organized as follows. We first introduce some preliminaries of SWIPT and non-cooperative game theory in Section 13.2 to make this chapter sufficiently self-contained. Section 13.3 describes the system model and problem formulation. In Section 13.4, we elaborate the proposed game-theoretic power splitting framework for the considered network. Numerical results are provided in Section 13.5 and conclusions are drawn in Section 13.6.

13.2 Preliminaries of SWIPT and game theory

To make this chapter sufficiently self-contained, in this section, we introduce some basic concepts of SWIPT and non-cooperative game theory.

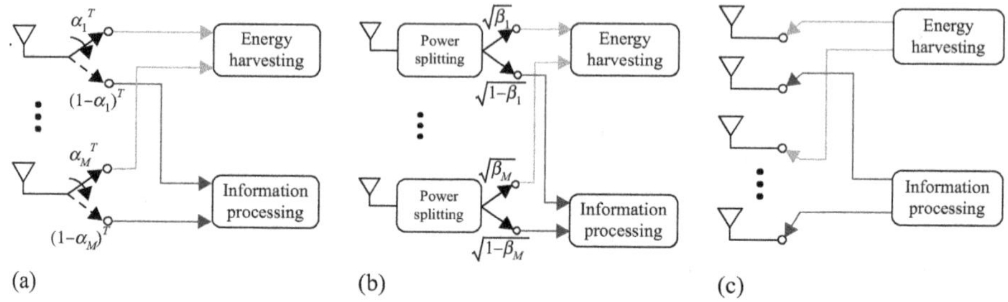

Figure 13.1 Three basic structures for a multiple-antenna SWIPT receiver, where α_i denotes the time switching factor of the ith antenna, β_i denotes the power splitting factor of the ith antenna, and T denotes the duration of each transmission block. (a) Time switching; (b) power splitting; and (c) antenna splitting

13.2.1 Basic receiver structures of SWIPT

An ideal SWIPT receiver is assumed to be able to harvest energy and decode information from the same signal [5]. However, this could not be achieved by practical circuits currently. To realize SWIPT practically, the received signal should be split in to two distinct streams, one for energy harvesting and the other for information decoding. Some practical receiver structures for SWIPT have been proposed in the literature. In the following, we briefly introduce three basic structures for a multiple-antenna SWIPT receiver [30, 31], namely time switching, power splitting, and antenna switching, as shown in Figure 13.1.

Time switching receiver: In this receiver structure, as depicted in Figure 13.1(a), each antenna can switch between the energy harvesting unit and information processing unit periodically based on a pre-determined time switching factor. In this case, the received signals are split in the time domain such that the signals can only be used for either energy harvesting or information processing at a certain instant. This structure has a low-complexity hardware implementation but needs stringent information/energy scheduling and time synchronization to achieve good performance.

Power splitting receiver: This structure shown in Figure 13.1(b) implements SWIPT by splitting the received signal at each antenna into two power streams based on a certain power splitting ratio before performing any analog/digital signal processing. Then one stream is sent to the energy harvesting unit and the other one is connected to the information processing unit. Compare with the time splitting structure, the power splitting technique requires a higher receiver complexity since it needs to employ extra power splitting units. However, this structure can realize the instantaneous SWIPT as the received signals are used for both energy harvesting and information processing at the same time.

Antenna switching receiver: This receiver structure achieves SWIPT by splitting the received signals in the antenna domain. Specifically, the antennas at the receiver are divided into two groups, where one group is used by the energy harvesting unit and the other group is connected to the information processing unit (see Figure 13.1c). From a theoretical perspective, antenna switching could be considered as a special case of power splitting with binary power splitting factors for each antenna.

Among the described receiver structures, it is shown in Reference 31 that the optimized power splitting receiver outperforms the other two receivers at the expense of higher hardware complexity and computational burden for resource allocation.

13.2.2 Basic concepts of non-cooperative game theory

Game theory, which was originally developed to model problems in the area of economics, has been widely adopted to solve the competitive resource allocation problems in wireless communications recently [32]. As one of the most important branches of game theory, non-cooperative game theory mainly aims at the study and analysis of competitive decision-making processes involving multiple players. It provides analytical tools to characterize the strategic interactions between the rational players. The players normally have partially or totally conflicting interests, which are determined by their joint actions.

A non-cooperative game generally consists of three basic components:

- ***Players***: The players are the decision-makers of a non-cooperative game, denoted by a finite set $\mathcal{N} = \{1, 2, \ldots, N\}$.
- ***Strategies***: The strategies are the actions that the players can take during the interactive decision-making process. Let $s_i \in \mathcal{S}_i$ denotes the strategy selected by player i, where $\mathcal{S}_i$ is the non-empty strategy set of player i. Then, $\mathbf{s} = (s_1, s_2, \ldots, s_N)$ is called a strategy profile that consists of all players' strategies. Moreover, it is obvious that $\mathbf{s} \in \mathcal{S}$ with $\mathcal{S} = \mathcal{S}_1 \times \mathcal{S}_2 \times \cdots \times \mathcal{S}_N$ denoting the Cartesian product of the strategy sets.
- ***Utilities***: The utility of player i, denoted by $u_i : \mathcal{S} \rightarrow \mathbb{R}$, is a measurement function used to represent the player's evaluation of all possible outcomes. It assigns a number to each possible outcome, with higher utilities representing more desirable outcomes.

For convenience, we normally use the triple $\langle \mathcal{N}, \{\mathcal{S}_i\}, \{u_i(s_i, \mathbf{s}_{-i})\} \rangle$ to denote a non-cooperative game. In non-cooperative games, the players are considered to be rational and selfish. Specifically, each player only aims to maximize its individual utility, irrespective of other players' utilities and the overall system performance. For the sake of notations, the vector $\mathbf{s}_{-i}$ is normally defined to denote the strategy profile excluding s_i. Then, we have $\mathbf{s} = (s_i, \mathbf{s}_{-i})$. Now we can say that player i prefers the strategy s_i to s_i' if $u_i(s_i, \mathbf{s}_{-i}) > u_i(s_i', \mathbf{s}_{-i})$. For a given strategy profile $\mathbf{s}_{-i}$ (i.e., other players' strategies are fixed), player i can select a strategy to maximize its utility function, which is called a best response of player i.

The most well-known solution to the non-cooperative games is the (pure strategy) NE [33]. A NE of a given non-cooperative game $\langle \mathcal{N}, \{\mathcal{S}_i\}, \{u_i(s_i, \mathbf{s}_{-i})\} \rangle$ is a feasible point $\mathbf{s}^*$ such that $\forall i \in \mathcal{N}$, we have the following:

$$u_i\left(s_i^*, \mathbf{s}_{-n}^*\right) \geq u_i\left(s_i, \mathbf{s}_{-i}^*\right), \quad \forall s_i \in \mathcal{S}_i \tag{13.1}$$

In other words, a NE is a feasible strategy profile with the property that no single player can increase the utility by deviating from the strategy corresponding to the equilibrium, given the strategies of the other players. The following theorem proposed in Reference 34 is usually adopted to verify the existence of the NE.

Theorem 13.1. *A NE* exists *in the game* $\langle \mathcal{N}, \{\mathcal{S}_i\}, \{u_i(s_i, \mathbf{s}_{-i})\}\rangle$ *if* $\forall i \in \mathcal{N}$, $\mathcal{S}_i$ *is a compact and convex set;* $u_i(\mathbf{s})$ *is continuous in* $\mathbf{s}$ *and quasi-concave in* s_i, *where* $\mathbf{s} = (s_i, \mathbf{s}_{-i})$.

13.3 System model and problem formulation

13.3.1 System model

In this chapter, we consider the design of SWIPT in relay interference channels, as depicted in Figure 13.2. The system consists of N source–relay–destination (S–R–D) links and the set of these links is denoted as $\mathcal{N} = \{1, \ldots, N\}$. Particularly, within the link $S_i \to R_i \to D_i$, $i \in \mathcal{N}$, the source S_i communicates with its corresponding destination D_i, assisted by a dedicated relay R_i. The relay nodes are assumed to employ the decode-and-forward relaying protocol to forward the received signals [35]. We also assume that the direct source–destination channels are neglected due to a high path loss and shadowing attenuation. Since these two-hop links share the same spectrum, they interfere with each other over the dual hops.

We consider that sources and destinations are equipped with only one antenna, while the relays are equipped with multiple antennas. Moreover, all nodes operate in a half-duplex mode. The relay nodes do not have their own power supply and need to harvest energy from the received signal in order to forward the received signal to the destinations. It is assumed that the energy harvesting and information transmission are implemented for every received message block. For the purpose of exposition, we further assume that the processing power consumed by the transmit/receive circuitry at the relay nodes is negligible compared to the power used for signal forwarding [12]. Besides, we consider that all links experience slow and frequency-flat fading, where the channel gains remain constant during each transmission block but may change from one block to another.

We use M_i to denote the number of antennas equipped at the ith relay. Let $\mathbf{g}_{ij} \in \mathbb{C}^{M_j \times 1}$ and $\mathbf{h}_{ij} \in \mathbb{C}^{M_i \times 1}$ denote the channel gain from S_i to R_j in the first hop and that from R_i to D_j in the second hop, respectively. In the first time slot, all sources transmit simultaneously and the signal received by the relay R_i can be written as

$$\mathbf{y}_{R_i} = \sqrt{P_i}\mathbf{g}_{ii}x_i + \sum_{j=1, j\neq i}^{N} \sqrt{P_j}\mathbf{g}_{ji}x_j + \mathbf{n}_i^a \tag{13.2}$$

where P_i and x_i are the fixed transmit power and the transmitted information of the source S_i with $\mathbb{E}\{|x_i|^2\} = 1$, $\mathbf{n}_i^a$ is a $M_i \times 1$ vector and denotes the additive noise introduced by the receiver antenna at the relay R_i with $\mathbb{E}\{\mathbf{n}_i^a(\mathbf{n}_i^a)^\dagger\} = \sigma_{i,a}^2\mathbf{I}$.

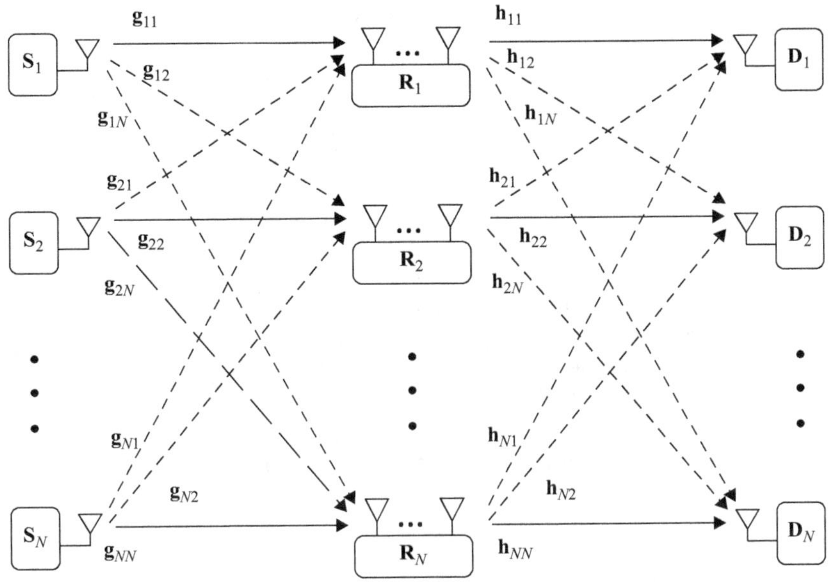

Figure 13.2 System model for relay interference channels with multi-antenna relays

In this chapter, we consider that the relays adopt the power splitting receiver structure due to its superior performance. To reduce the complexity, the power splitting ratios for different antennas of a certain relay are set to be the same. More specifically, the received signal at each antenna of the relay R_i is split into two streams according to a power splitting ratio ρ_i. The fraction $\sqrt{\rho_i}$ of the received signal at each antenna is used for energy harvesting, while the remaining one is sent to the information processing unit. In practice, the antenna noise has a negligible impact on both the information receiving and energy harvesting, since $\sigma_{i,a}^2$ is generally much smaller than the noise power introduced by the baseband processing circuit, and thus even lower than the average power of the received signal [36]. For simplicity, we ignore the noise term $\mathbf{n}_i^a$ in the following analysis, i.e., setting $\sigma_{i,a}^2 = 0$ [36]. For the sake of simplicity, we assume a normalized transmission time for each hop (i.e., the transmission duration of each hop is equal to one). Then, the terms "energy" and "power" can be used interchangeably. In this case, the energy harvested by the relay R_i at the end of the first hop can thus be expressed as [19]

$$Q_i = \eta\rho_i\left(P_i\|\mathbf{g}_{ii}\|^2 + \sum_{j=1,j\neq i}^{N} P_j\|\mathbf{g}_{ji}\|^2\right) \tag{13.3}$$

where $0 < \eta \leq 1$ is the energy conversion efficiency that depends on the rectification process and the energy harvesting circuit, and $\|\cdot\|$ denotes the Euclidean norm. Note that the second term inside the bracket of (13.3) comes from the interference generated by other links. Meanwhile, the information signal received by the information processing unit at the relay R_i is given by

$$\begin{aligned} \mathbf{y}_{R_i}^I &= \sqrt{1-\rho_i}\mathbf{y}_{R_i} + \mathbf{n}_i^b \\ &= \sqrt{1-\rho_i}\sqrt{P_i}\mathbf{g}_{ii}x_i + \sqrt{1-\rho_i}\sum_{j=1,j\neq i}^{N}\sqrt{P_j}\mathbf{g}_{ji}x_j + \mathbf{n}_i^b \end{aligned} \tag{13.4}$$

where n_i^b is $M_i \times 1$ vector and denotes the additive white Gaussian noise (AWGN) with $\mathbb{E}\{\mathbf{n}_i^b(\mathbf{n}_i^b)^\dagger\} = \sigma_{R_i}^2\mathbf{I}$, which is introduced by the signal processing circuit from passband to baseband at the relay R_i. Subsequently, the relay node will first decode the information based on the received information signal $\mathbf{y}_{R_i}^I$ given in (13.4) before forwarding it using the harvested energy. Let $\mathbf{w}_i \in \mathbb{C}^{M_i\times 1}$ denotes the receive beamforming vector adopted at the relay $R_i\|\mathbf{w}\|^2 = 1$. In this case, the received signal-to-interference-plus-noise ratio (SINR) can thus be written as

$$\gamma_{i,1} = \frac{(1-\rho_i)P_i|\mathbf{w}_i^\dagger\mathbf{g}_{ii}|^2}{(1-\rho_i)\sum_{j=1,j\neq i}^{N}P_j|\mathbf{w}_i^\dagger\mathbf{g}_{ji}|^2 + \sigma_{R_i}^2\|\mathbf{w}_i\|^2} \tag{13.5}$$

In the second time slot, the relay nodes forward the decoded information to their corresponding destinations using the energy harvested in the first time slot. Here, we consider the case that the relays do not perform power control and exhausts all the harvested energy for information forwarding. Then, the received signal at D_i is given by

$$y_{D_i} = \sqrt{Q_i}\mathbf{h}_{ii}^\dagger\mathbf{v}_i x_i + \sum_{j=1,j\neq i}^{N}\sqrt{Q_j}\mathbf{h}_{ji}^\dagger\mathbf{v}_j x_j + n_{D_i} \tag{13.6}$$

where $\mathbf{v}_i \in \mathbb{C}^{M_i\times 1}$ denotes the transmit beamforming vector at the relay R_i with $\|\mathbf{v}_i\|^2 = 1$ and n_{D_i} is the AWGN at the destination D_i with $\mathbb{E}\{n_{D_i}^* n_{D_i}\} = \sigma_{D_i}^2$. The received SINR at the destination D_i can thus be written as

$$\gamma_{i,2} = \frac{\rho_i\eta\left(\sum_{n=1}^{N}P_n\|\mathbf{g}_{ni}\|^2\right)|\mathbf{h}_{ii}^\dagger\mathbf{v}_i|^2}{\sum_{j=1,j\neq i}^{N}\rho_j\eta\left(\sum_{n=1}^{N}P_n\|\mathbf{g}_{nj}\|^2\right)|\mathbf{h}_{ji}^\dagger\mathbf{v}_j|^2 + \sigma_{D_i}^2} \tag{13.7}$$

The achievable rate of the ith link in this case is thus given by

$$\begin{aligned} \mathcal{R}_i &= \frac{1}{2}\min\left(\log_2(1+\gamma_{i,1}), \log_2(1+\gamma_{i,2})\right) \\ &= \frac{1}{2}\log_2(1+\gamma_i) \end{aligned} \tag{13.8}$$

where

$$\gamma_i = \min(\gamma_{i,1}, \gamma_{i,2}) \tag{13.9}$$

can be regarded as the end-to-end SINR of the ith link. Without loss of generality, we hereafter assume that $\sigma_{R_i}^2 = \sigma_{D_i}^2 = \sigma^2$, for any $i \in \mathcal{N}$.

13.3.2 Problem formulation

In this chapter, we consider that each link's performance is characterized by its achievable rate and thus regard the sum-rate of all links as a network-wide performance metric. Now we are ready to formulate the following network-wide sum-rate maximization problem for the considered network:

$$\begin{aligned} &\max_{\{\rho_i, \mathbf{w}_i, \mathbf{v}_i\}_{i=1}^N} \sum_{i=1}^N \mathcal{R}_i \\ &\text{s.t. } 0 \le \rho_i \le 1, \quad \forall i \in \mathcal{N}, \\ &\|\mathbf{v}_i\| = 1, \quad \forall i \in \mathcal{N} \end{aligned} \tag{13.10}$$

Unfortunately, the optimal solution to problem (13.10) does not seem to be analytically tractable due to its complicated and non-convex nature. Motivated by this, we choose to apply a heuristic two-stage optimization approach. Specifically, we first design the receive and transmit beamforming vectors (i.e., $\mathbf{w}_i$s and $\mathbf{v}_i$s) by integrating the inherent features of the considered system and then optimize the power splitting ratios (i.e., ρ_is) with the selected beamforming vectors in the second stage.

Various linear processing schemes have been proposed for multi-antenna relays to deal with the information relaying in presence of co-channel interference in open literature. Different schemes require different amount of CSI and computation complexity. Generally, with larger amount of available CSI, more sophisticated signal processing schemes with higher complexity could be designed to improve the transmission efficiency and to achieve a better end-to-end performance. For the considered system, due to the energy harvesting feature, it would be challenging for the relays to maintain the full CSI of both the first and second hops. Thus, it is sufficiently reasonable to assume that each relay only grasps the CSI of inner-link channels (i.e., $\mathbf{g}_{ii}$ and $\mathbf{h}_{ii}$ for the relay R_i). In this situation, the MRC/MRT scheme would be a particularly appropriate signal processing technique for the energy harvesting relays owing to its low-complexity light requirement of CSI. In particular, the relays adopt the MRC technique to combine the signals received by multiple antennas in the first hop and the MRT technique to forward the received signal to their associated destinations in the second hop. Mathematically, we have $\mathbf{w}_i = \frac{\mathbf{g}_{ii}}{\|\mathbf{g}_{ii}\|}$ and $\mathbf{v}_i = \frac{\mathbf{h}_{ii}}{\|\mathbf{h}_{ii}\|}$. We now can degrade the formulated problem (13.10) to the following optimization problem in terms of all power splitting ratios:

$$\begin{aligned} &\max_{\boldsymbol{\rho}} \sum_{i=1}^N \mathcal{R}_i(\boldsymbol{\rho}) \\ &\text{s.t. } \boldsymbol{\rho} \in \mathcal{A}, \\ &\mathbf{w}_i = \frac{\mathbf{g}_{ii}}{\|\mathbf{g}_{ii}\|}, \ \mathbf{v}_i = \frac{\mathbf{h}_{ii}}{\|\mathbf{h}_{ii}\|}, \quad \forall i \in \mathcal{N} \end{aligned} \tag{13.11}$$

where $\boldsymbol{\rho} = [\rho_1, \rho_2, \ldots, \rho_N]^T$ denotes the vector made up of the power splitting ratios of all links and $\mathcal{A} = \{\boldsymbol{\rho} | 0 \leq \rho_i \leq 1, \ \forall i \in \mathcal{N}\}$ is the feasible set of $\boldsymbol{\rho}$.

Even though the receive and transmit beamforming vectors have been determined, the simplified problem (13.11) regarding power splitting ratios only is still very difficult to solve because it is not only non-convex but also non-differentiable due to the min operator inside the rate expression given in (13.8). This means that the globally optimal power splitting profile (i.e., the optimal solution to problem (13.11)) cannot be efficiently calculated even in a centralized fashion regardless of the heavy signaling overhead required by the centralized method. To tackle this issue, we resort to the well-established game theory and develop a distributed power splitting approach specified in the following section.

13.4 Distributed power splitting via game theory

In this section, we propose a game-theoretic approach to solve the optimization problem (13.11) regarding the power splitting ratios of all links. To this end, we develop a distributed framework by considering that all links are strategic and they only aim to maximize their individual achievable rates by controlling their own power splitting ratios. Mathematically, this will involve the ith link solving the following optimization problem:

$$\begin{aligned} \max_{\rho_i} \quad & \mathcal{R}_i(\rho_i, \boldsymbol{\rho}_{-i}) \\ \text{s.t.} \quad & \rho_i \in \mathcal{A}_i, \\ \mathbf{w}_i = & \frac{\mathbf{g}_{ii}}{\|\mathbf{g}_{ii}\|}, \quad \mathbf{v}_i = \frac{\mathbf{h}_{ii}}{\|\mathbf{h}_{ii}\|} \end{aligned} \tag{13.12}$$

where $\boldsymbol{\rho}_{-i} = [\rho_1, \ldots, \rho_{i-1}, \rho_{i+1}, \ldots, \rho_N]^T$ denotes the vector of all links' power splitting ratios, except the ith one, and $\mathcal{A}_i = \{\rho_i | 0 \leq \rho_i \leq 1\}$ is the feasible set of the ith link's power splitting ratio.

13.4.1 Non-cooperative game formulation

We can observe from (13.12) that the optimization problem to be solved by each link is tangled together because the power splitting ratio of each link not only affects the rate of this link but also affects the rate of other links due to mutual interference between links. To solve this problem, we model the considered power splitting problem as a non-cooperative game in game theory [33]. Particularly, the considered power splitting problem for the considered network can be characterized by the following non-cooperative game:

- *Players*: The N S–R–D links.
- *Strategies*: Each link determines its power splitting ratio $\rho_i \in \mathcal{A}_i$ to maximize the achievable rate for its own link.
- *Utilities*: The achievable rate $\mathcal{R}_i(\rho_i, \boldsymbol{\rho}_{-i})$ with the selected beamforming vectors.

For convenience, we denote the formulated non-cooperative game as

$$\mathcal{G} = \langle \mathcal{N}, \{\mathcal{A}_i\}, \{\mathcal{R}_i(\rho_i, \boldsymbol{\rho}_{-i})\}\rangle \tag{13.13}$$

Note that we treat each link consisting of three nodes as a "virtual" single player for the sake of presentation. In practice, each player is supposed to be one node of each link (e.g., relay) that acts as the coordinator of each link. Recall that the most well-known solution to non-cooperative games is the NE. In the following two subsections, we first analyze the existence and uniqueness for the NE of the formulated game $\mathcal{G}$, and then develop a distributed algorithm to achieve the NE.

13.4.2 Existence and uniqueness of the NE

We first evaluate the existence of NE by following the method specified in Theorem 13.1. After investigating the properties of the action sets and the utility functions for the formulated game $\mathcal{G}$, we have the following proposition regarding the existence of the NE.

Proposition 13.1. *The formulated power splitting game $\mathcal{G}$ with the selected beamforming vectors for the considered network possesses at least one NE.*

Proof. It is obvious that the strategy set $\mathcal{A}_i$ of the formulated game is convex and compact and the utility $\mathcal{R}_i(\rho_i, \boldsymbol{\rho}_{-i})$ is continuous in $\boldsymbol{\rho}$ for any $i \in \mathcal{N}$. In addition, it is straightforward to check that both $\gamma_{i,1}$ and $\gamma_{i,2}$ with the selected beamforming vectors are concave functions of ρ_i, which means that the utility $\mathcal{R}_i(\rho_i, \boldsymbol{\rho}_{-i})$ is a concave (also quasi-concave) function of the strategy ρ_i. Thus, with reference to Theorem 13.1, we can claim that the formulated game admits at least one NE. □

Once the NE is proved to exist, the following natural question that arises is whether it is unique. This is important not only for predicting the state of the network but also crucial for convergence issues. In principle, the uniqueness of the NE can be analyzed by several methods, which has been summarized in Reference 37. However, due to the non-differentiability of the utility function (the min operator), most of the methodologies cannot be applied to the formulated game except the standard function approach [38] because, as shown below, it only requires that the best response function satisfies certain properties. To proceed, we first figure out the best response functions of the links (players) given in the following lemma.

Lemma 13.1. *Given a power splitting strategy profile $\boldsymbol{\rho}$ and the selected beamforming vectors, the best response function of the link $S_i \to R_i \to D_i$ in the game $\mathcal{G}$ can be expressed as*

$$\mathcal{B}_i(\boldsymbol{\rho}) = \frac{(X_iW_i + X_i + Y_iZ_i + Z_i) - \sqrt{(X_iW_i + X_i - Y_iZ_i + Z_i)^2 + 4Y_iZ_i^2}}{2Y_iZ_i} \tag{13.14}$$

where

$$X_i = \frac{P_i \|\mathbf{g}_{ii}\|^2}{\sigma^2} \tag{13.15a}$$

$$Y_i = \sum_{j=1, j\neq i}^{N} \frac{P_j |\mathbf{g}_{ii}^{\dagger}\mathbf{g}_{ji}|^2}{\sigma^2 \|\mathbf{g}_{ii}\|^2} \tag{13.15b}$$

$$Z_i = \frac{\eta \left(\sum_{n=1}^{N} P_n \|\mathbf{g}_{ni}\|^2 \right) \|\mathbf{h}_{ii}\|^2}{\sigma^2} \tag{13.15c}$$

$$W_i = \sum_{j=1, j\neq i}^{N} \frac{\rho_j \eta \left(\sum_{n=1}^{N} P_n \|\mathbf{g}_{nj}\|^2 \right) |\mathbf{h}_{ji}^{\dagger}\mathbf{h}_{jj}|^2}{\|\mathbf{h}_{jj}\|^2 \sigma^2} \tag{13.15d}$$

are defined for the simplicity of notations.

Proof. Let ρ_i^* denotes the best response of the ith link corresponding to a given power splitting strategy profile $\boldsymbol{\rho}$ and the selected beamforming vectors. That is, ρ_i^* is the optimal solution to problem (13.12). It can be easily observed that $\gamma_{i,1}$ in (13.5) is monotonically decreasing in ρ_i, while $\gamma_{i,2}$ in (13.7) is monotonically increasing in ρ_i. Thus, ρ_i^* must satisfy the following condition:

$$\gamma_{i,1}(\rho_i^*) = \gamma_{i,2}(\rho_i^*) \tag{13.16}$$

Substituting the selected beamforming vectors into the expressions of $\gamma_{i,1}$ and $\gamma_{i,2}$, we have

$$\frac{(1-\rho_i^*) X_i}{(1-\rho_i^*) Y_i + 1} = \frac{\rho_i^* Z_i}{W_i + 1} \tag{13.17}$$

where X_i, Y_i, Z_i, and W_i are defined in (13.15). After rearranging (13.17), we obtain

$$Y_i Z_i (\rho_i^*)^2 - [X_i (W_i + 1) + Y_i Z_i + Z_i] \rho_i^* + X_i (W_i + 1) = \ell_i (\rho_i^*) = 0 \tag{13.18}$$

Note that (13.18) is a quadratic equality of ρ_i^*, denoted by $\ell_i(\rho_i^*) = 0$. Thus, the value of ρ_i^* can be obtained by figuring out the roots of the quadratic equality in (13.18).

We note that

$$Y_i Z_i > 0 \tag{13.19a}$$

$$\ell_i(0) = X_i (W_i + 1) > 0 \tag{13.19b}$$

$$\ell_i(1) = -Z_i < 0 \tag{13.19c}$$

Based on (13.19), we can deduce that the quadratic equality $\ell_i(\rho_i^*) = 0$ admits two roots lying in the intervals (0, 1) and $(1, +\infty)$, respectively. Since the feasible set of the power splitting ratio is [0, 1], the valid value of ρ_i^* can only be the smaller root of the quadratic equality $\ell_i(\rho_i^*) = 0$. Mathematically, we have,

$$\rho_i^* = \frac{(X_iW_i + X_i + Y_iZ_i + Z_i) - \sqrt{(X_iW_i + X_i - Y_iZ_i + Z_i)^2 + 4Y_iZ_i^2}}{2Y_iZ_i} = \mathcal{B}_i(\boldsymbol{\rho}) \tag{13.20}$$

which completes the proof. □

We now define the vector best response function $\boldsymbol{\mathcal{B}}(\boldsymbol{\rho}) = [\mathcal{B}_1(\boldsymbol{\rho}), \ldots, \mathcal{B}_N(\boldsymbol{\rho})]^T$. Then, with reference to the well-known fixed point theorem [37], the strategy profile $\boldsymbol{\rho}^*$ is a NE of the formulated game $\mathcal{G}$ *if and only if* it is a fixed point of the function $\boldsymbol{\mathcal{B}}(\boldsymbol{\rho})$. Mathematically, we should have $\boldsymbol{\mathcal{B}}(\boldsymbol{\rho}^*) = \boldsymbol{\rho}^*$. Hence, the uniqueness evaluation for the NE of the formulated game is equivalent to that for the fixed point of the function $\boldsymbol{\mathcal{B}}(\boldsymbol{\rho})$. Furthermore, it is proved in Reference 38 that the fixed point of the function $\boldsymbol{\mathcal{B}}(\boldsymbol{\rho})$ is unique if $\boldsymbol{\mathcal{B}}(\boldsymbol{\rho})$ is a *standard* function. To be more precise, we give the definition of the standard function as follows:

Definition 13.1. *A function* $\mathbf{f}(\mathbf{x})$ *is said to be standard if it satisfies the following properties for all* $\mathbf{x} \geq \mathbf{0}$*:*

- Positivity*:* $\mathbf{f}(\mathbf{x}) > \mathbf{0}$.
- Monotonicity*: If* $\mathbf{x} \geq \mathbf{x}'$*, then* $\mathbf{f}(\mathbf{x}) \geq \mathbf{f}(\mathbf{x}')$.
- Scalability*: For all* $\alpha > 1$, $\alpha\mathbf{f}(\mathbf{x}) > \mathbf{f}(\alpha\mathbf{x})$.

Here, all the inequalities are componentwise.

After investigating the properties of the best response functions given in Lemma 13.1, we have the following proposition regarding the uniqueness of the NE.

Proposition 13.2. *The formulated game* $\mathcal{G}$ *for the considered system with the selected beamforming vectors always possesses a unique NE.*

Proof. To prove this proposition, we need to show that the best response vector function $\boldsymbol{\mathcal{B}}(\boldsymbol{\rho})$ of the formulated game is standard. To this end, in the following, we show that the function $\boldsymbol{\mathcal{B}}(\boldsymbol{\rho})$ satisfies the three properties of the standard function.

(1) Positivity: As shown in the proof of Lemma 13.1, for any player i and any strategy profile $\boldsymbol{\rho}$, the best response function $\mathcal{B}_i(\boldsymbol{\rho})$ is always larger than zero, which guarantees the positivity of the function $\boldsymbol{\mathcal{B}}(\boldsymbol{\rho})$.

(2) Monotonicity: Suppose $\boldsymbol{\rho}$ and $\boldsymbol{\rho}'$ are two different strategy profiles and $\boldsymbol{\rho} \geq \boldsymbol{\rho}'$. Then, the corresponding best response functions of any player i can be written as

$$\mathcal{B}_i(\boldsymbol{\rho}) = \frac{(X_iW_i + X_i + Y_iZ_i + Z_i) - \sqrt{(X_iW_i + X_i - Y_iZ_i + Z_i)^2 + 4Y_iZ_i^2}}{2Y_iZ_i} \tag{13.21}$$

and

$$\mathcal{B}_i(\boldsymbol{\rho}') = \frac{(X_iW_i' + X_i + Y_iZ_i + Z_i) - \sqrt{(X_iW_i' + X_i - Y_iZ_i + Z_i)^2 + 4Y_iZ_i^2}}{2Y_iZ_i} \tag{13.22}$$

where the terms X_i, Y_i, Z_i, and W_i are defined in (13.15), and

$$W_i' = \sum_{j=1, j\neq i}^{N} \frac{\rho_j'\eta\left(\sum_{n=1}^{N} P_n\left\|\mathbf{g}_{nj}\right\|^2\right)\left|\mathbf{h}_{ji}^{\dagger}\mathbf{h}_{jj}\right|^2}{\left\|\mathbf{h}_{jj}\right\|^2\sigma^2}$$

After a careful comparison of (13.21) and (13.22), we can see that all the terms in $\mathcal{B}_i(\boldsymbol{\rho})$ and $\mathcal{B}_i(\boldsymbol{\rho}')$ are the same except the terms W_i and W_i'. Hence, the inequity $\mathcal{B}_i(\boldsymbol{\rho}) \geq \mathcal{B}_i(\boldsymbol{\rho}')$ holds if we can prove $\mathcal{B}_i(W_i) \geq \mathcal{B}_i(W_i')$. Moreover, we have $W_i \geq W_i'$ since $\boldsymbol{\rho} \geq \boldsymbol{\rho}'$. Thus, the proof of $\mathcal{B}_i(\boldsymbol{\rho}) \geq \mathcal{B}_i(\boldsymbol{\rho}')$ is equivalent to proving that the $\mathcal{B}_i(W_i)$ is a non-decreasing function of W_i. That is, $\frac{\partial \mathcal{B}_i(W_i)}{W_i} \geq 0$. Expanding $\frac{\partial \mathcal{B}_i(W_i)}{W_i} \geq 0$, we have

$$\frac{\partial \mathcal{B}_i(W_i)}{W_i} = \frac{X_i}{2Y_iZ_i}\left[1 - \frac{X_iW_i + X_i - Y_iZ_i + Z_i}{\sqrt{(X_iW_i + X_i - Y_iZ_i + Z_i)^2 + 4Y_iZ_i^2}}\right] \tag{13.23}$$

Since the term $X_i/(2Y_iZ_i) > 0$ and the term in the square bracket of (13.23) is always large than zero, we can claim that $\frac{\partial \mathcal{B}_i(W_i)}{W_i} > 0$, which complete the proof of monotonicity.

(3) Scalability: For any $\alpha > 1$, we define the function $\mathcal{F}_i(\alpha, \boldsymbol{\rho}) = \alpha\mathcal{B}_i(\boldsymbol{\rho}) - \mathcal{B}_i(\alpha\boldsymbol{\rho})$. Then, the proof of the scalability is equivalent to proving that $\mathcal{F}_i(\alpha, \boldsymbol{\rho}) > 0$ for any $\alpha > 1$. First, it is obvious that $\mathcal{F}_i(1, \boldsymbol{\rho}) = 0$. Thus, a sufficient condition for $\mathcal{F}_i(\alpha, \boldsymbol{\rho}) > 0$ is that $\mathcal{F}_i(\alpha, \boldsymbol{\rho})$ is an increasing function of α, i.e., $\frac{\partial \mathcal{F}(\alpha, \boldsymbol{\rho})}{\partial \alpha} > 0$.

To proceed, we first derive the first-order and second-order partial derivatives of $\mathcal{F}_i(\alpha, \boldsymbol{\rho})$ with respect to α and obtain

$$\begin{aligned}\frac{\partial \mathcal{F}_i(\alpha, \boldsymbol{\rho})}{\partial \alpha} &= \frac{1}{2Y_iZ_i}\left\{X_i + Y_iZ_i + Z_i\right. \\ &\quad + \frac{(\alpha X_iW_i + X_i - Y_iZ_i + Z_i)X_iW_i}{\sqrt{(\alpha X_iW_i + X_i - Y_iZ_i + Z_i)^2 + 4Y_iZ_i^2}} \\ &\quad \left. - \sqrt{(X_iW_i + X_i - Y_iZ_i + Z_i)^2 + 4Y_iZ_i^2}\right\}\end{aligned} \tag{13.24}$$

$$\frac{\partial^2 \mathcal{F}_i(\alpha, \boldsymbol{\rho})}{\partial \alpha^2} = \frac{2Z_i(X_iW_i)^2}{\left[(\alpha X_iW_i + X_i - Y_iZ_i + Z_i)^2 + 4Y_iZ_i^2\right]^{3/2}} \tag{13.25}$$

From (13.25), we can see that $\frac{\partial^2 \mathcal{F}_i(\alpha,\boldsymbol{\rho})}{\partial \alpha^2}$ is always larger than zero, which indicates that $\frac{\partial \mathcal{F}_i(\alpha,\boldsymbol{\rho})}{\partial \alpha}$ is increasing in α. Thus, a sufficient condition for $\mathcal{F}_i(\alpha, \boldsymbol{\rho}) > 0$ can now be simplified as $\frac{\partial \mathcal{F}_i(\alpha,\boldsymbol{\rho})}{\partial \alpha}|_{\alpha=1} > 0$. Substituting $\alpha = 1$ into (13.24), we get

$$\begin{aligned}\left.\frac{\partial \mathcal{F}_i(\alpha, \boldsymbol{\rho})}{\partial \alpha}\right|_{\alpha=1} &= \frac{1}{2Y_iZ_i}\left\{X_i + Y_iZ_i + Z_i\right. \\ &\quad + \frac{(X_iW_i + X_i - Y_iZ_i + Z_i)X_iW_i}{\sqrt{(X_iW_i + X_i - Y_iZ_i + Z_i)^2 + 4Y_iZ_i^2}} \\ &\quad \left. - \sqrt{(X_iW_i + X_i - Y_iZ_i + Z_i)^2 + 4Y_iZ_i^2}\right\}\end{aligned} \tag{13.26}$$

To proceed, we derive the first-order derivative for the right-hand side of (13.26) with respect to W_i. After some algebraic manipulations, we have

$$\partial\left(\left.\frac{\partial \mathcal{F}_i(\alpha,\boldsymbol{\rho})}{\partial \alpha}\right|_{\alpha=1}\right) \Big/ \partial W_i = \frac{2Z_iW_iX_i^2}{\left[(X_iW_i + X_i - Y_iZ_i + Z_i)^2 + 4Y_iZ_i^2\right]^{3/2}}$$

which is shown to be always positive. Thus, $\frac{\partial \mathcal{F}_i(\alpha,\boldsymbol{\rho})}{\partial \alpha}|_{\alpha=1}$ is an increasing function in W_i. Since $W_i > 0$, we further have

$$\begin{aligned}\left.\frac{\partial F_i(\alpha, \boldsymbol{\rho})}{\partial \alpha}\right|_{\alpha=1} &> \left.\frac{\partial F_i(\alpha, \boldsymbol{\rho})}{\partial \alpha}\right|_{\alpha=1, W_i=0} \\ &= \frac{1}{2Y_iZ_i}\left\{X_i + Y_iZ_i + Z_i - \sqrt{(X_i + Y_iZ_i + Z_i)^2 - 4X_iY_iZ_i}\right\} > 0\end{aligned} \tag{13.27}$$

Therefore, we can claim that $\alpha\mathcal{B}_i(\boldsymbol{\rho}) > \mathcal{B}_i(\alpha\boldsymbol{\rho})$, which completes the whole proof of this proposition. □

13.4.3 Distributed algorithm

So far, we have proved that the formulated game $\mathcal{G}$ with the selected beamforming vectors for the considered system always has a unique NE for any system profile (channel realization, transmit powers of the source, and noise power). However, the unique equilibrium is meaningful in practice only if an algorithm that is able to achieve such an equilibrium from any non-equilibrium state can be developed. To this end, in this subsection, we propose an iterative algorithm with provable convergence to achieve the NE, in which the links update their own power splitting ratios in a distributed manner. In addition, the practical implementation of the proposed algorithm is also discussed.

Algorithm 1 Distributed power splitting algorithm

1: Set $t = 0$ and each player (link) $i \in \mathcal{N}$ chooses a random power splitting ratio $\rho_i(0)$ from the feasible set $\mathcal{A}_i$.
2: If a suitable termination criterion is satisfied: STOP.
3: Each link $i \in \mathcal{N}$ updates the power splitting ratio via executing

$$\rho_i(t+1) = \mathcal{B}_i\left(\boldsymbol{\rho}(t)\right) \tag{13.28}$$

4: $t \leftarrow t + 1$; go to STEP 2.

13.4.3.1 Algorithm description

Different types of distributed algorithms have been developed to find the NEs of non-cooperative games in open literature (see Reference 39 and the references therein). Here, we have more interests in best response-based algorithms since we have derived best response functions of the formulated game. Based on the derived best response functions given in (13.14), we develop a distributed power splitting algorithm for the considered system with the selected beamforming vectors. In the proposed algorithm, each player iteratively updates its power splitting ratio using the best response function, given its opponent's strategies fixed. The iterative algorithm is formally described in **Algorithm 1**. In terms of the convergence of Algorithm 1, we have the following proposition:

Proposition 13.3. *From any initial point, Algorithm 1 always converges to the unique NE of the formulated games $\mathcal{G}$ both synchronously and asynchronously.*

Proof. Since the best response vector function $\boldsymbol{\mathcal{B}}(\boldsymbol{\rho})$ is always standard (see proof of Proposition 13.2), the proof of this proposition follows with reference to Theorem 13.2 in Reference 38. □

Note that Proportion 13.3 guarantees that the iterative procedure in Algorithm 1 is globally convergent with no extra requirement on the initial point, system profile (channel realization, transmit powers of sources, and noise power), and updating manners (i.e., synchronous rule or asynchronous rule). In particular, thanks to the asynchronous updating capability of the proposed algorithm, the links are allowed to adjust their power splitting ratios in different updating frequencies. In other word, the algorithm can still achieve the NE even some links perform updates based on outdated information of its suffered inter-link interference. This indicates that the proposed algorithm can actually cope with lack of synchronization in wireless communications, which may be caused by many practical issues, e.g., sampling errors, measurement imperfections, and information aging.

13.4.3.2 Implementation discussion

It is worth emphasizing that the distributed nature of the proposed algorithm is on the basis of treating each link as a single player. However, the source, relay, and destination that constitute each link (player) are geographically separated in practical scenarios. In this regard, an efficient implementation of the proposed algorithm with the minimum information sharing is of great importance in practice.

In our design, the relay node is appointed as the link coordinator that perform the information collection and the best response computation. Here, we assume that the energy consumed for the algorithm computations at the relay nodes are negligible compared with the energy used for information forwarding since these computations are not complex and energy-consuming at all. This assumption can be further supported by the rapid development of the low-power chips. We subsequently identify the information that is needed to collect or exchange for the implementation of the proposed algorithm. According to the best response functions given in (13.14), the relay R_i needs to know the values of the terms X_i, Y_i, Z_i, and W_i defined in (13.15). To this end, the relay R_i should execute the following steps:

- *Step 1*: Measure the inner channel gains (i.e., $\mathbf{g}_{ii}$ and $\mathbf{h}_{ii}$) and implement the selected transmit and receive beamforming vectors for information combining and forwarding.
- *Step 2*: Acquire the transmit power of its source P_i as well as the noise power σ^2, and then calculate the value of X_i based on these information.
- *Step 3*: Measure[1] the power of its received signal, acquire the power of the received signal at its associated destination D_i, and then calculate the values of Y_i, Z_i, and W_i.
- *Step 4*: Calculate the best response power splitting ratio based on (13.14).

From the above description, we can observe that the overheads are required in acquiring the following three kinds of information for each link in the proposed algorithm: (1) pilots for estimating the CSI from source to its dedicated relay and

[1] The measurement of the signal power can be performed by the radio scene analyzer [40].

CSI from relay to destination, (2) the value of transmit power from the source to relay, and (3) the power of received signal at the destination, which needs to be sent from the destination to its dedicated relay. The first two kinds of overheads are only needed once for each channel realization, while the third one is required in each iteration of the proposed algorithm. From the above discussion, we can see that in the proposed algorithm, only some local information needs to be exchanged within each link and no information needs to be exchanged among different links. Finally, it is worth mentioning that a possible termination criterion for the proposed Algorithm 1 could be $[\rho_i(t+1) - \rho_i(t)]/\rho_i(t+1) \leq \zeta$, where ζ is a sufficiently small constant.

13.5 Numerical results

In this section, some numerical results are presented to illustrate and validate the theoretical analysis in the previous section. For simplicity, we restrict our attention to a linear topology for each link. More specifically, we consider that S_i–R_i–D_i forms a straight line with unit length, i.e., $d_{S_iD_i} = d_{S_iR_i} + d_{R_iD_i} = 1, \ \forall i \in \mathcal{N}$, with d_{XY} denoting the distance between nodes X and Y. The channels between all transceiver pairs are assumed to be subject to mutually independent Rayleigh fading. To take into account the impact of path loss, we adopt the channel model that $\mathbb{E}\{\mathbf{g}_{ij}\mathbf{g}_{ij}^{\dagger}\} = (d_{S_iR_j})^{-\tau}\mathbf{I}$ and $\mathbb{E}\{\mathbf{h}_{ij}\mathbf{h}_{ij}^{\dagger}\} = (d_{R_iD_j})^{-\tau}\mathbf{I}$, where $\tau \in [2, 5]$ is the path loss factor [41]. In all the simulations, without loss of generality, we set the path loss factor $\alpha = 3$, the noise power $\sigma^2 = 1$, the energy conversion efficiency $\eta = 0.5$ and all relays are equipped with the same number of antennas, i.e., $M_1 = \cdots = M_N = M$.

13.5.1 *Verification of best response function and algorithm convergence*

To verify the correctness of the derived best response function and the convergence of the proposed algorithm, we consider a simple network consisting of two links, where each relay is equipped with three antennas. In Figure 13.3(a), we first plot the best response functions (i.e., $\rho_1\,(\rho_2)$ and $\rho_2\,(\rho_1)$) of the formulated game with a set of randomly generated parameters, respectively. With reference to Reference 37, the intersection points of the best response functions are actually the NEs of the corresponding game. From Figure 13.3(a), we can see that the two curves only admit one intersection point, which indicates that the formulated game in the considered two-link network possess a unique NE. In Figure 13.3(b), the proposed algorithm is executed to achieve the corresponding NE from a randomly generated initial point in both synchronous and asynchronous updating manners. With regard to the asynchronous update, we let link 1 update three times more often than link 2. It can be seen from Figure 13.3(b) that the proposed algorithms can converge to the corresponding NE obtained in Figure 13.3(a) with both synchronous and asynchronous updates. We thus can claim that the observations from Figure 13.3(a) and 13.3(b) validate our theoretical analysis regarding the best response function and the global convergence of the proposed algorithm. In all the following results, we adopt the synchronous update in the proposed algorithm.

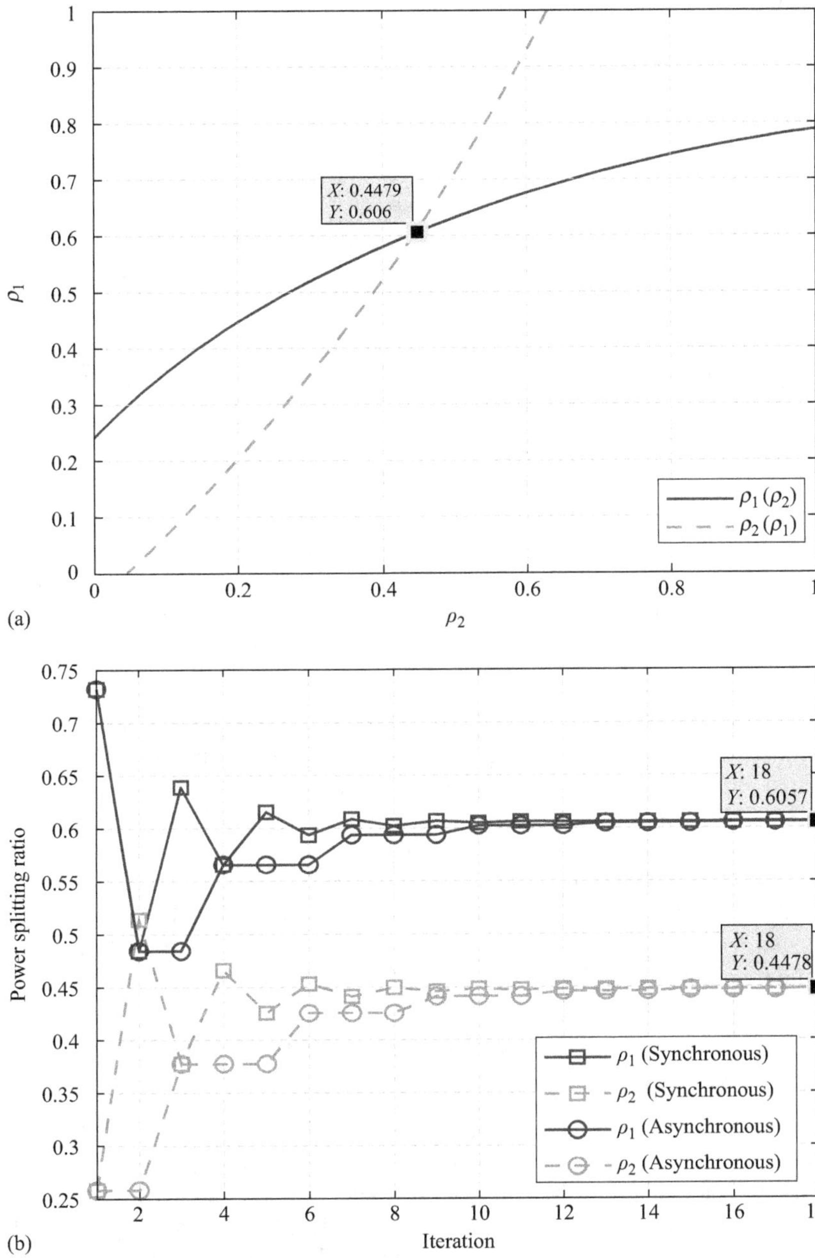

Figure 13.3 The best response functions of the formulated game and the convergence of Algorithm 1 in a two-link network with a set of randomly generated parameters. (a) Best responses of the formulated game and (b) convergence of Algorithm 1

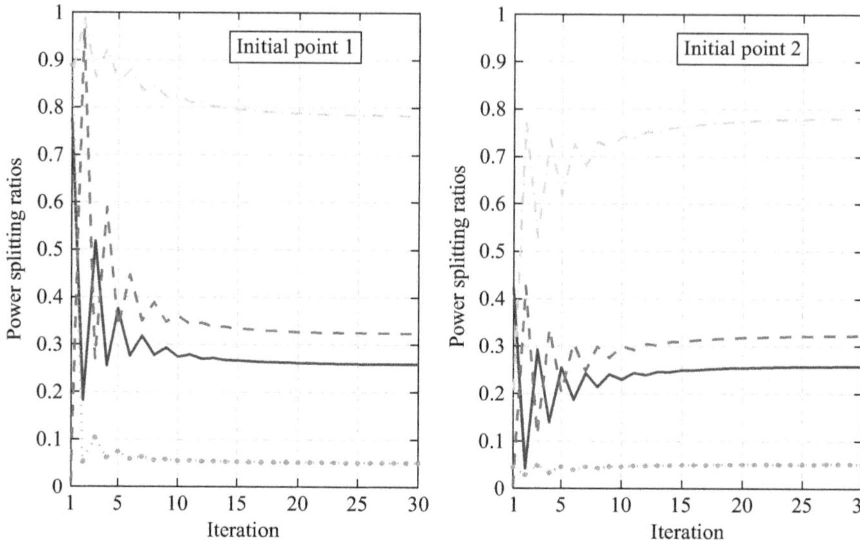

Figure 13.4 Convergence of the proposed algorithm for a randomly generated four-link network with two different initial points (point 1 – left and point 2 – right), where $M = 3$ and $P_i = 15$ dB, $\forall i \in \{1, 2, 3, 4\}$

To demonstrate that the proposed algorithm can also converge to the NE in multi-link scenarios, we show its convergence performance for an example of a randomly generated four-link network in Figure 13.4. As can be observed from Figure 13.4, the proposed algorithm can converge to the same values (i.e., the NE) from two different initial points, which further validates the theoretical analysis. Note that due to the space limitation, we only show results in Figures 13.3 and 13.4 for some specific network setups with one random realization of channel gains and/or transmit powers of the sources, although similar results can also be shown for other network setups and random realizations.

13.5.2 System average performance and effects of system parameters

Next, we simulate the average sum-rate of the relay interference channels with SWIPT that implement the selected beamforming vectors and the proposed power splitting scheme. We characterize the performance of both two-link and multi-link scenarios. For simplification and illustrative purpose, we consider that all links in the considered networks are mutually parallel. Thus, the two-link setting is characterized by the interlink distance, denoted by d_L. In the multi-link setting, we assume that the distances between adjacent links are equal. Hence, this scenario is characterized by the number of links N and the distance between the farthest two links, denoted by $d_{\max}$. In the following figures, each curve is plotted by averaging over the results from 5000 independent channel realizations.

Figure 13.5 illustrates the average sum-rate of a two-link network, where the performances are compared across the optimal sum-rate obtained by the centralized optimization problem (13.11) solved via exhaustive search, the proposed game-theoretic approach, the random scheme in which the power splitting ratios are randomly generated over [0, 1]. Both symmetric and asymmetric topologies with different number of antennas at relays are simulated. We can observe from Figure 13.5 that the proposed game-theoretic method always outperforms the random scheme in both network topologies. In addition, the performance gap between these two schemes becomes larger when the inter-link distance d_L increases. Moreover, it can be observed from Figure 13.5 that the proposed game-theoretic approach suffers performance loss compared to the optimal scheme when the inter-link distance is very small, i.e., in high interference scenarios. However, as the inter-link distance increases, it quickly coincides with the centralized optimal scheme with the selected beamforming vectors in all simulated cases. Therefore, we can claim that the proposed game-theoretic approach can achieve a near-optimal performance on average, especially for the scenarios with medium and large interlink distance (i.e., relatively low and moderate interference). Finally, Figure 13.5 also shows that the average performance of all three schemes improves as the number of antennas equipped at relays increases. The reason behind this phenomenon is that the relays can not only harvest larger amount of energy with multiple antennas during the first time slot but also utilize the multiple antennas to implement the MRT scheme to enhance the signal power received at their corresponding destinations during the second time slot.

Figures 13.6 and 13.7, respectively, demonstrate the impacts of the number of links and the number of antennas equipped at relays on the average sum-rate of relay interference channels with SWIPT in the multi-link scenario, where all relays are assumed in middle positions (i.e., $d_{S_iR_i} = d_{R_iD_i} = 0.5,\ \forall i$). Note that the performance of the centralized optimal scheme is omitted in these two figures because the corresponding optimization problem is non-convex and thus cannot be efficiently solved in a multi-link scenario. In Figure 13.6, we draw the average sum-rate curves achieved by the proposed game-theoretic scheme and the random scheme versus the number of links with $d_{\max} = 5$ and $P_i = 10$ dB, $\forall i$. From this figure, we can see that the average sum-rates of both schemes first increase and then keep decreasing with the growth of the number of links. This observation is understandable. Specifically, the initial sum-rate increase is actually a multiplexing gain as more links share the same spectrum with relatively low mutual interference. Nevertheless, with a further increase in the number of links, the interlink interference becomes stronger, which leads to a monotonically decreasing sum-rate. We can also observe from Figure 13.6 that the proposed game-theoretic approach always outweighs the random method in all simulated cases. The performance gap between these two schemes keeps almost the same with the increasing of the number of link. Furthermore, similar to an observation in Figure 13.5, higher sum-rate can be achieved when more antennas are equipped at energy harvesting relays.

The impact of the number of antennas equipped at relays on the average sum-rate of the considered system is shown in Figure 13.7. As can be observed from this figure, the sum-rates of both schemes increase monotonically with the rising of the antenna

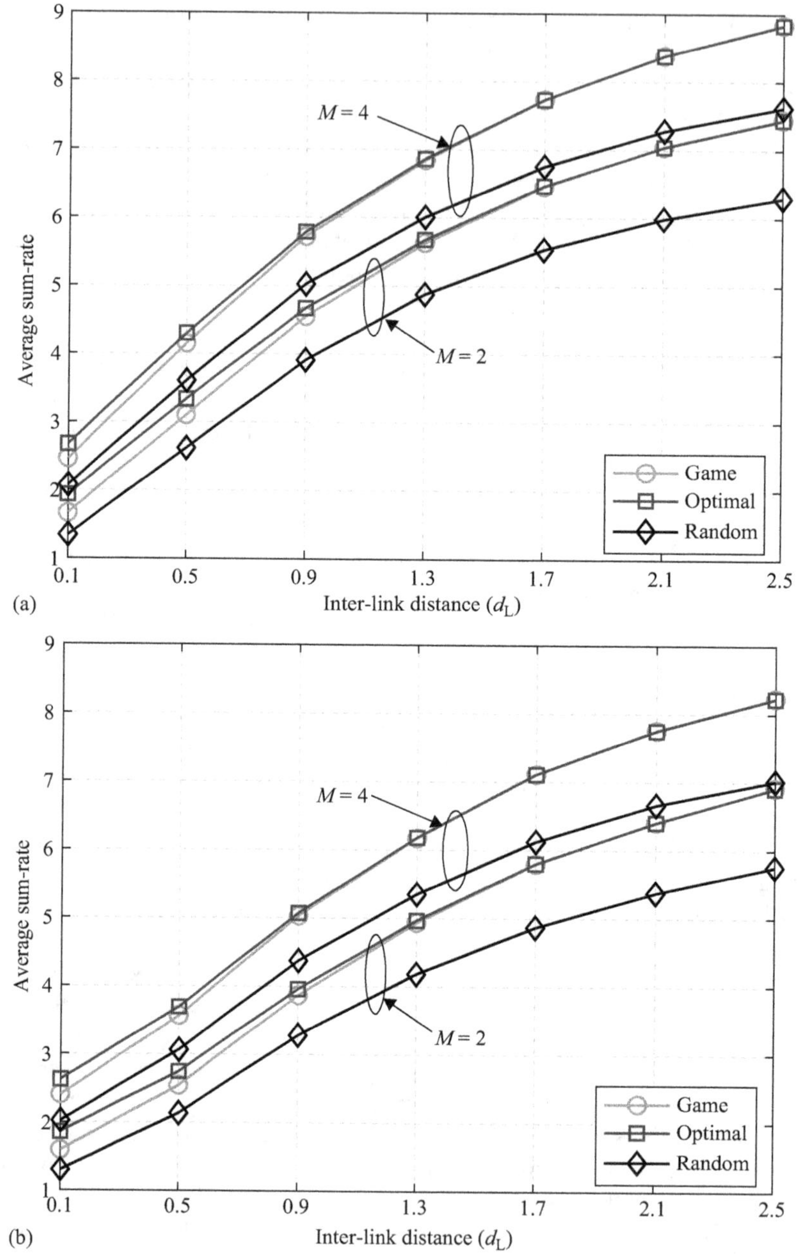

Figure 13.5 The average sum-rates of two-link network with different number of relay antennas for (a) symmetric network with $d_{S_iR_i} = d_{R_iD_i} = 0.5$ and $P_i = 15\,dB$, $\forall i = 1,2$, (b) asymmetric network with $d_{S_1R_1} = d_{R_2D_2} = 0.25$ and $P_i = 15\,dB$, $\forall i = 1,2$

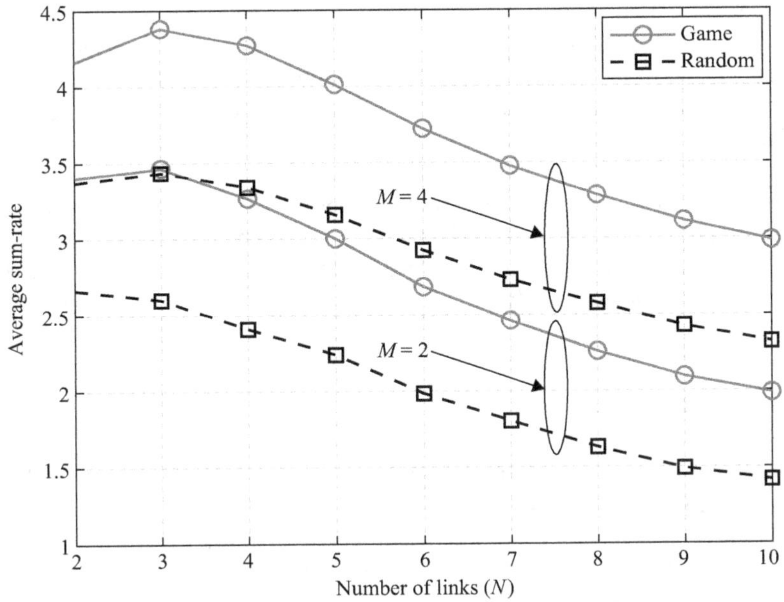

Figure 13.6 The impact of link numbers on average sum-rate of multi-link networks with different number of relay antennas, where $d_{\max} = 5$ and $P_i = 10\,dB$, $\forall i$.

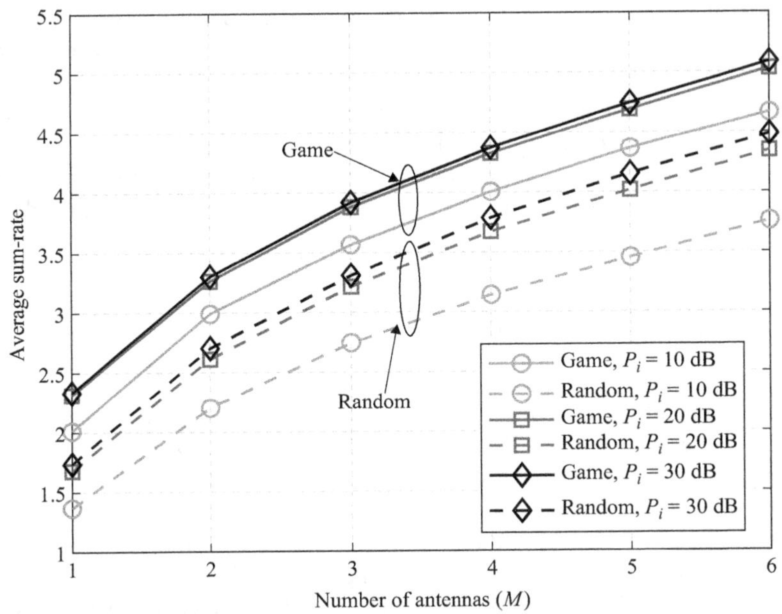

Figure 13.7 The impact of the relay antenna number on average sum-rate of a multi-link network with different source transmit powers, where $d_{\max} = 5$ and $N = 5$.

number, which is consistent to the observations in previous two figures. Once again, the proposed game-theoretic approach is always superior to the random method in all cases. Furthermore, we can observe from Figure 13.7 that increasing the source transmit powers can also improve the system average sum-rate to a certain extent, especially from low to moderate SNRs (e.g., 10–20 dB). However, the average sum-rate achieved by the proposed game-theoretic approach tends to get saturated at high SNR as the performance gap between 20 dB and 30 dB is shown to be very small. This indicates that at high SNR, the power control at the sources should be jointly designed with the power splitting at the relays to further improve the sum-rate of relay interference channels with SWIPT.

13.6 Conclusions

In this chapter, we designed SWIPT in relay interference channels with energy harvesting relays equipped with multiple antennas. We resolved the sum-rate maximization problem of the considered system by proposing a heuristic two-stage optimization approach. More specifically, we first identified that the MRC/MRT scheme would be a particularly appropriate signal processing technique for the energy harvesting relays in practice owing to its low complexity and light requirement of CSI. With this practical signal processing scheme, in the second stage we developed a distributed framework to optimize power splitting ratios at relays by formulating a non-cooperative game. We proved that the formulated game always admits a unique NE. Best response-based distributed algorithms with provable convergence were also developed to achieve the NE. The numerical results showed that the proposed algorithm can converge to the NE from different starting points, and the developed game-theoretic approach can achieve a near-optimal network-wide performance on average, especially for the scenarios with relatively low and moderate interference. Furthermore, to implement the proposed algorithm with the MRC/MRT scheme, only some local information needs to be exchanged within each link and no information needs to be exchanged among different links.

References

[1] B. Medepally and N. Mehta, "Voluntary energy harvesting relays and selection in cooperative wireless networks," *IEEE Trans. Wireless Commun.*, vol. 9, no. 11, pp. 3543–3553, Nov. 2010.

[2] O. Ozel, K. Tutuncuoglu, J. Yang, S. Ulukus, and A. Yener, "Transmission with energy harvesting nodes in fading wireless channels: Optimal policies," *IEEE J. Sel. Areas Commun.*, vol. 29, no. 8, pp. 1732–1743, Sep. 2011.

[3] J. Xu and R. Zhang, "Throughput optimal policies for energy harvesting wireless transmitters with non-ideal circuit power," *IEEE J. Sel. Areas Commun.*, vol. 32, no. 2, pp. 322–332, Feb. 2014.

[4] H. Visser and R. Vullers, "RF energy harvesting and transport for wireless sensor network applications: Principles and requirements," *Proc. IEEE*, vol. 101, no. 6, pp. 1410–1423, Jun. 2013.

[5] L. Varshney, "Transporting information and energy simultaneously," in *IEEE International Symposium on Information Theory (ISIT)*, 2008, Toronto, Canada, pp. 1612–1616.

[6] X. Zhou, R. Zhang, and C. K. Ho, "Wireless information and power transfer: Architecture design and rate-energy tradeoff," *IEEE Trans. Commun.*, vol. 61, no. 11, pp. 4754–4767, 2013.

[7] R. Zhang and C. K. Ho, "MIMO broadcasting for simultaneous wireless information and power transfer," *IEEE Trans. Wireless Commun.*, vol. 12, no. 5, pp. 1989–2001, 2013.

[8] S. Timotheou, I. Krikidis, G. Zheng, and B. Ottersten, "Beamforming for MISO interference channels with QoS and RF energy transfer," *IEEE Trans. Wireless Commun.*, vol. 13, no. 5, pp. 2646–2658, May 2014.

[9] J. Park and B. Clerckx, "Joint wireless information and energy transfer in a K-user mimo interference channel," *IEEE Trans. Wireless Commun.*, vol. 13, no. 10, pp. 5781–5796, Oct. 2014.

[10] K. Huang and E. Larsson, "Simultaneous information and power transfer for broadband wireless systems," *IEEE Trans. Signal Process.*, vol. 61, no. 23, pp. 5972–5986, 2013.

[11] D. W. K. Ng, E. S. Lo, and R. Schober, "Wireless information and power transfer: Energy efficiency optimization in OFDMA systems," *IEEE Trans. Wireless Commun.*, vol. 12, no. 12, pp. 6352–6370, 2013.

[12] A. A. Nasir, X. Zhou, S. Durrani, and R. A. Kennedy, "Relaying protocols for wireless energy harvesting and information processing," *IEEE Trans. Wireless Commun.*, vol. 12, no. 7, pp. 3622–3636, 2013.

[13] Z. Chen, B. Xia, and H. Liu, "Wireless information and power transfer in two-way amplify-and-forward relaying channels," in *2014 IEEE Global Conference on Signal and Information Processing (GlobalSIP)*, Atlanta, GA, USA, December 2014, pp. 168–172.

[14] Z. Fang, X. Yuan, and X. Wang, "Distributed energy beamforming for simultaneous wireless information and power transfer in the two-way relay channel," *IEEE Signal Process. Lett.*, vol. 22, no. 6, pp. 656–660, Jun. 2015.

[15] C. Zhong, H. Suraweera, G. Zheng, I. Krikidis, and Z. Zhang, "Wireless information and power transfer with full duplex relaying," *IEEE Trans. Commun.*, vol. 62, no. 10, pp. 3447–3461, Oct. 2014.

[16] Y. Zeng and R. Zhang, "Full-duplex wireless-powered relay with self-energy recycling," *IEEE Wireless Commun. Lett.*, vol. 4, no. 2, pp. 201–204, Apr. 2015.

[17] I. Krikidis, S. Sasaki, S. Timotheou, and Z. Ding, "A low complexity antenna switching for joint wireless information and energy transfer in MIMO relay channels," *IEEE Trans. Commun.*, vol. 62, no. 5, pp. 1577–1587, May 2014.

[18] Z. Zhou, M. Peng, Z. Zhao, and Y. Li, "Joint power splitting and antenna selection in energy harvesting relay channels," *IEEE Signal Process. Lett.*, vol. 22, no. 7, pp. 823–827, Jul. 2015.

[19] G. Zhu, C. Zhong, H. Suraweera, G. Karagiannidis, Z. Zhang, and T. Tsiftsis, "Wireless information and power transfer in relay systems with multiple antennas and interference," *IEEE Trans. Commun.*, vol. 63, no. 4, pp. 1400–1418, Apr. 2015.

[20] K. Xiong, P. Fan, C. Zhang, and K. Letaief, "Wireless information and energy transfer for two-hop non-regenerative MIMO-OFDM relay networks," *IEEE J. Sel. Areas Commun.*, vol. 33, no. 8, pp. 1595–1611, July 2015.

[21] Z. Ding, S. M. Perlaza, I. Esnaola, and H. V. Poor, "Power allocation strategies in energy harvesting wireless cooperative networks," *IEEE Trans. Wireless Commun.*, vol. 13, no. 2, pp. 846–860, 2014.

[22] Z. Ding, I. Krikidis, B. Sharif, and H. Poor, "Wireless information and power transfer in cooperative networks with spatially random relays," *IEEE Trans. Wireless Commun.*, vol. 13, no. 8, pp. 4440–4453, Aug. 2014.

[23] I. Krikidis, "Simultaneous information and energy transfer in large-scale networks with/without relaying," *IEEE Trans. Commun.*, vol. 62, no. 3, pp. 900–912, Mar. 2014.

[24] H. Chen, Y. Jiang, Y. Li, Y. Ma, and B. Vucetic, "A game-theoretical model for wireless information and power transfer in relay interference channels," in *IEEE ISIT'14*, July 2014, pp. 1161–1165.

[25] H. Chen, Y. Li, Y. Jiang, Y. Ma, and B. Vucetic, "Distributed power splitting for SWIPT in relay interference channels using game theory," *IEEE Trans. Wireless Commun.*, vol. 14, no. 1, pp. 410–420, Jan. 2015.

[26] O. Simeone, O. Somekh, Y. Bar-Ness, H. V. Poor, and S. Shamai, "Capacity of linear two-hop mesh networks with rate splitting, decode-and-forward relaying and cooperation," in *Proceedings of the 45th Annual Allerton Conference on Communication, Control and Computing*, Monticello, Illinois, USA, 2007.

[27] P. Thejaswi, A. Bennatan, J. Zhang, R. Calderbank, and D. Cochran, "Rate-achievability strategies for two-hop interference flows," in *Proceedings of 46th Annual Allerton Conference on Communication, Control, and Computing*, Illinois, USA, 2008, pp. 1432–1439.

[28] Y. Cao and B. Chen, "Capacity bounds for two-hop interference networks," in *Proceedings of 47th Annual Allerton Conference on Communication, Control, and Computing*, Monticello, Illinois, USA, 2009, pp. 272–279.

[29] K. T. Truong and W. Robert Jr., "A distributed algorithm using interference pricing for relay interference channels," *EURASIP J. Adv. Signal Process.*, vol. 2013, no. 1, pp. 1–16, 2013.

[30] I. Krikidis, S. Timotheou, S. Nikolaou, G. Zheng, D. W. K. Ng, and R. Schober, "Simultaneous wireless information and power transfer in modern communication systems," *IEEE Commun. Mag.*, vol. 52, no. 11, pp. 104–110, Nov. 2014.

[31] Z. Ding, C. Zhong, D. W. K. Ng, *et al.*, "Application of smart antenna technologies in simultaneous wireless information and power transfer," *IEEE Commun. Mag.*, vol. 53, no. 4, pp. 86–93, Apr. 2015.

[32] Z. Han, D. Niyato, W. Saad, T. Basar, and A. Hjorungnes, *Game Theory in Wireless and Communication Networks: Theory, Models, and Applications*. Cambridge University Press, Cambridge, 2012.

[33] D. Fudenberg and J. Tirole, *Game Theory*. MIT Press, Cambridge, MA, 1992.
[34] I. L. Glicksberg, "A further generalization of the Kakutani fixed point theorem with application to Nash equilibrium points," *Proc. Am. Math. Soc.*, vol. 3, no. 1, pp. 170–174, 1952.
[35] A. Nosratinia, T. E. Hunter, and A. Hedayat, "Cooperative communication in wireless networks," *IEEE Commun. Mag.*, vol. 42, no. 10, pp. 74–80, 2004.
[36] L. Liu, R. Zhang, and K.-C. Chua, "Wireless information and power transfer: A dynamic power splitting approach," *IEEE Trans. Commun.*, vol. 61, no. 9, pp. 3990–4001, 2013.
[37] S. Lasaulce, M. Debbah, and E. Altman, "Methodologies for analyzing equilibria in wireless games," *IEEE Signal Process. Mag.*, vol. 26, no. 5, pp. 41–52, 2009.
[38] R. Yates, "A framework for uplink power control in cellular radio systems," *IEEE J. Sel. Areas Commun.*, vol. 13, no. 7, pp. 1341–1347, 1995.
[39] S. Lasaulce and H. Tembine, *Game Theory and Learning for Wireless Networks: Fundamentals and Applications*. Elsevier, Amsterdam, 2011.
[40] Y. Shi, J. Wang, K. Letaief, and R. Mallik, "A game-theoretic approach for distributed power control in interference relay channels," *IEEE Trans. Wireless Commun.*, vol. 8, no. 6, pp. 3151–3161, 2009.
[41] H. Chen, J. Liu, L. Zheng, C. Zhai, and Y. Zhou, "Approximate SEP analysis for DF cooperative networks with opportunistic relaying," *IEEE Signal Process. Lett.*, vol. 17, no. 9, pp. 779–782, 2010.

Chapter 14
Relaying in optical wireless communication

Majid Safari, Wasiu Popoola and Sinan Sinanovic

14.1 Introduction

Optical wireless communication (OWC) refers to the transmission of data using light propagating through the atmosphere over ranges of few metres to few kilometres (terrestrial) to thousands of kilometres (space/satellite). This technology has recently attracted a renewed interest although its roots can be traced back to Alexander Graham Bell's photophone, which transmitted human conversation carried by a beam of light over a distance of about 200 m [1]. OWC technology provides an inherently secure, license-free, and high capacity alternative to radio frequency (RF) wireless communication and a cost-effective and easy-to-install alternative to fibre optics. These advantages make this technology an attractive solution for intermediate, last mile, or access connectivity in the future congested networks by bridging the gap between the end user and the fibre-optic infrastructure already in place.

Long-range OWC systems are based on the transmission of infrared light over the atmosphere and are commonly referred as free-space optical (FSO) communication [2] while most of short-range applications use visible light communication (VLC) [3]. In FSO systems, infrared coherent sources (e.g., lasers) are typically used to generate highly directed beams that can be transmitted over long line-of-sight links. On the other hand, VLC systems employ incoherent sources [e.g., light emitting diodes (LEDs)] operating at the visible range of the light spectrum in order to integrate both high-speed data access and lighting in an indoor environment.

Unlike RF wireless technology, OWC systems are mostly based on intensity modulation and direct detection (IM/DD) allowing only for non-negative real signalling thereby significantly simplifying the transceiver design by avoiding the estimation and processing of the phase of the received signal. Furthermore, because of much smaller wavelength of light, the channel modelling (whether for indoor/short range or outdoor/long-range applications) and the characteristics of transmit and receive components are fundamentally different from the RF wireless systems [4].

The optical beam propagated over the atmosphere undergoes different impairments caused by scattering, absorption, diffraction, and atmospheric turbulence. The OWC channel influenced by these effects experiences random performance fluctuations in adverse weather and/or strong turbulence conditions. These fluctuations (known as fading) can cause long outages in the order of millisecond, which is long at the scale of optical data rate. Spatial diversity techniques have been proposed to improve the reliability of turbulent FSO channels through the insertion of redundancy. In such systems, multiple transmit/receive apertures are employed to establish a multiple-input multiple-output (MIMO) FSO system with several parallel diversity paths through which copies of data can be sent simultaneously [5, 6].

Co-located MIMO systems might not be realisable for all applications and this encourages the implementation of distributed MIMO systems with widely separated nodes cooperating to provide spatial diversity. Such systems, which are also known as cooperative or relay-assisted communication, have been extensively investigated in the context of RF wireless communication. Multi-hop transmission is an alternative relay-assisted scheme, which employs multiple relays in serial configuration. RF multi-hop systems do not offer any diversity gain but they are typically employed to extend the coverage of communication networks. This chapter will investigate the application of distributed MIMO and relaying strategies [7–18] as efficient solutions to overcome atmospheric degradations in OWC channels.

14.2 The optical wireless channel

Coherent OWC, although possible, is rarely employed in current systems due to technical difficulties and high cost. Therefore, this chapter will focus on IM/DD communication systems where the data are modulated on transmitted optical power (rather than in phase or frequency) using IM techniques such as on–off keying (OOK), pulse position modulation (PPM), pulse amplitude modulation (PAM), and optical orthogonal frequency division multiplexing (O-OFDM).

The main impairments of the IM/DD OWC channel include (1) optical power attenuation caused by absorption, scattering, and diffraction, (2) signal distortions caused by noise, (3) fading caused by atmospheric turbulence, and (4) optical component imperfections such as non-linear effects. The light beam travelling through the atmosphere expands as it propagates and is scattered and absorbed by aerosols and molecules. In addition, the air thermal inhomogeneities cause random fluctuation of refractive index, known as atmospheric turbulence, which degrades the reliability of OWC systems. Finally, the received signal is distorted by non-linear effects and different sources of noise such as background, thermal, and shot noise. In this section, we briefly discuss a theoretical OWC channel model that can capture these important distortions.

Using the complex quasi-monochromatic notation [19], the optical field modulated by data can be described as a function of time and space as

$$f(t,\vec{\rho}) = \left(\sqrt{I(t)}\xi(\vec{\rho})\right) e^{j2\pi ft} \tag{14.1}$$

where $I(t)$ is the intensity modulated signal, $\xi(\vec{\rho})$ is the spatial beam pattern, and f is the carrier frequency. Assuming that the beam pattern $\xi(\vec{\rho})$ is transmitted from an aperture R located at $z=$ plane, the received beam pattern $\xi'(\vec{\rho}')$ collected within aperture R' at $z=L$ plane can be described by the extended Huygens–Fresnel principle as

$$\xi'(\vec{\rho}') = \int_R \xi(\vec{\rho})h(\vec{\rho},\vec{\rho}')e^{-aL/2}d\vec{\rho} \tag{14.2}$$

where a is the extinction coefficient which models the loss due to absorption and scattering and $h(\vec{\rho}, \vec{\rho}')$ denotes the paraxial Green's function for atmospheric propagation and is given by [19]

$$h(\vec{\rho}, \vec{\rho}') = \frac{e^{jkL + jk|\vec{\rho} - \vec{\rho}'|^2/2L}}{j\lambda L} e^{\chi(\vec{\rho}, \vec{\rho}') + j\phi(\vec{\rho}, \vec{\rho}')} \tag{14.3}$$

where λ is the wavelength, $k = 2\pi/\lambda$ is the wave number, and $\chi(\vec{\rho}, \vec{\rho}')$ and $\phi(\vec{\rho}, \vec{\rho}')$ are, respectively, the turbulence-induced random log-amplitude and phase fluctuation terms. Assuming normalised transmit beam pattern, the optical signal power collected by the receive aperture can be obtained as

$$P_s = P \int_{R'} |\xi'(\vec{\rho}')|^2 \, d\vec{\rho}' \tag{14.4}$$

where $P = \frac{1}{T}\int_T I(t)dt$ is the average transmitted optical power and T is the symbol duration. In a far-field scenario where the receive and transmit aperture diameters are much smaller than the link length and are small compared to the spatial coherence length of the turbulence, the average received optical power in (14.4) reduces to [9]

$$P_s = \ell h P \tag{14.5}$$

where $h = |e^{\chi + j\phi}|^2 = e^{2\chi}$ denotes the turbulence-induced fading coefficient and ℓ is the path loss of the optical link including the atmospheric attenuation caused by diffraction, absorption, and scattering given by

$$\ell(L) = e^{-aL} \frac{A_{TX} A_{RX}}{(\lambda L)^2} \tag{14.6}$$

where A_{TX} and A_{RX} are respectively, the transmit and receive aperture areas.

14.2.1 Fading statistics

Long-range atmospheric optical communication systems are susceptible to turbulence-induced fading. Turbulence is caused by temperature variations in air generating a chaotic state of atmospheric flows. According to Rytov's theory [20], under weak turbulence regime, the turbulence-induced fading coefficient h can be described by lognormal distribution with log-amplitude mean and variance of (μ_x, σ_x^2) and the probability density function (pdf) given by

$$f(h) = \frac{1}{h\sqrt{8\pi\sigma_x^2}} \exp\left(-\frac{(\log(h) - 2\mu_x)^2}{8\sigma_x^2} \right) \tag{14.7}$$

In order to ensure that the fading does not attenuate or amplify the average power, the fading coefficient is normalised such that $E[h] = 1$ implying $\mu_x = -\sigma_x^2$. Assuming spherical wave propagation through a horizontal atmospheric path, the log-amplitude variance σ_χ^2 can be expressed in terms of wave number k, refractive index structure constant C_n^2, and the distance between transmitter and receiver L as [20]

$$\sigma_\chi^2 = 0.124 k^{7/6} C_n^2 L^{11/6} \tag{14.8}$$

For moderate and strong turbulence conditions, gamma-gamma distribution is a commonly used statistical model to describe the turbulence-induced fading [20] and is given by

$$f(h) = \frac{2\alpha\beta^{(\alpha+\beta)/2}}{\Gamma(\alpha)\Gamma(\beta)} h^{(\alpha+\beta)/2-1} K_{\alpha-\beta}\left(2\sqrt{\alpha\beta h}\right) \tag{14.9}$$

where $K_v(\cdot)$ is the modified Bessel function of order v, $\Gamma(\cdot)$ is the Gamma function, and $\alpha > 0$ and $\beta > 0$ are the turbulence parameters which depend on the link length, refractive index structure constant, and wave number.

Unlike RF communications, the strength of the turbulence-induced fading (e.g., in (14.9)) is a function of the link length and this factor significantly affects the performance of OWC systems. As reflected by the performance analysis of optical relaying systems in the next sections, this inherent characteristic allows serial (multi-hop) relaying techniques bring performance improvements against the degrading effects of turbulence-induced fading.

14.2.2 Optical detection and receiver noise

In OWC systems, information waveforms are modulated onto an optical carrier and the generated optical field is radiated through the atmosphere towards destination. At the receiver, the optical field collected at the receive aperture is focused onto a photodetector located at the focal plane of an optical lens. The field produced in the focal plane is commonly known as 'Airy pattern' and occupies a width of about 2λ in the absence of turbulence. The photodetector is a *square-law* detector and measures the optical power of the incident field on its photosensitive area and converts it to electrical current. The receiver then processes the detected electrical current to recover the original transmitted information [4].

Note that the small wavelength of light allows for the spatial resolvability of Airy patterns received from sufficiently separated sources. Therefore, the photodetector outputs the incoherent superposition of the interfering optical beams (i.e., sum of individual optical power) since the generated Airy patterns are resolvable (i.e., spatially orthogonal) at the detector plane providing sufficient separation of the relays.

In optical wireless channels, the receiver apertures are commonly exposed to strong background light along with the desired signal power. The amount of background noise collected by the receiver is proportional to the receiver field of view (FOV) and the receiver aperture area. The photodetector estimates the received optical power (including signal and background light) by measuring the incident photon count. This estimate is distorted by shot noise and thermal noise. The shot noise is caused by the random arrival of photons at the detector plane. The photon count distorted by shot noise is thus governed by a Poisson statistics. At high received photon rates, the Poisson distributed shot noise converges to a Gaussian noise [4]. This occurs in background-limited regime where background noise is strong and dominant compared to the other sources of noise. The additive white Gaussian noise (AWGN) model is valid for both background-limited and thermal-limited regimes (i.e., when

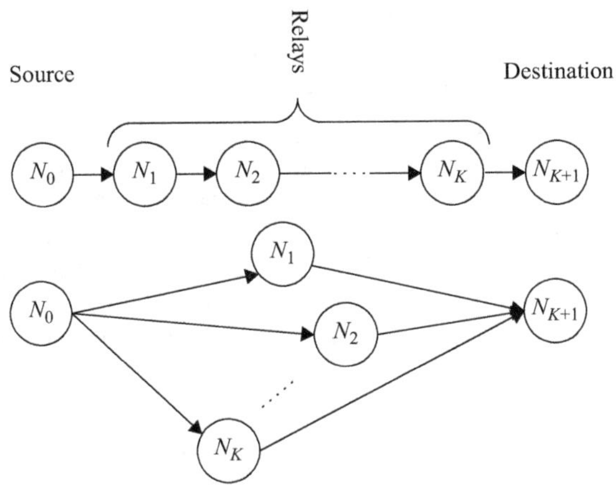

Figure 14.1 Serial (top) and parallel (bottom) relaying configurations

electrical thermal noise is dominant). Considering the AWGN model, the received electrical signal can be expressed as

$$r(t) = RP_b + R\ell hI(t) + n(t) \tag{14.10}$$

where $R = \eta q/\hbar f$ is the responsivity of the detector, q is the electron charge, η is the quantum efficiency of the photodetector, $\hbar$ is the Planck constant, and $n(t)$ is the zero-mean white Gaussian noise with power spectral density of N_0. The received background light generates the DC current RP_b and the corresponding shot noise, which contributes to $n(t)$. Assuming that the receiver integrates the received electrical signal over the symbol duration T, the output of the optical receiver can be expressed as

$$r = RTP_b + RTP\ell h + n \tag{14.11}$$

14.3 Relaying configurations

In this section, a number of geometrical configurations and signalling schemes of the relay-assisted OWC will be discussed. Serial, parallel, and multipath relaying configurations are defined and different signalling schemes such as decode-and-forward (DF) and amplify-and-forward (AF) are discussed.

Figure 14.1 illustrates the serial and parallel relaying schemes typically considered for relay-assisted OWC systems [9]. In serial relaying which is also known as multi-hop relaying, the source directs the modulated light beam towards a single relay node. Based on the signalling schemes considered, the relay performs proper signal processing either by amplifying or by regenerating the received signal and retransmits it to the next relay. This continues until the source's data arrives at the destination. In RF wireless communication, the serial relaying configuration does not provide any diversity gain but can enhance the performance of long links by limiting attenuation (i.e., providing power gain).

The parallel relaying configuration illustrated in Figure 14.1 comprises a single-input multiple-output (SIMO) link connecting the sources to multiple (K) relays and a multiple-input single-output (MISO) link connecting the relays to the destination. The source node transmits the same signal to the K relays and the relays either amplify or regenerate the received signal and synchronously forward it to the destination based on AF or DF relaying schemes, which will be discussed next. In RF wireless communication, parallel relaying is applied to provide spatial diversity gain for communication systems where the employment of multiple antennas is not possible. However, unlike RF systems, broadcasting is not possible in long-range FSO communication systems since directed light beams are used to establish line of site (LOS) optical links with minimum path loss. Therefore, in FSO parallel relaying systems, the source must be equipped with a multi-laser transmitter with each of the lasers pointing towards its corresponding relay node. In effect, the total transmit power P_t is determined by the sum of transmit powers at $2K$ lasers (K lasers at the source and one laser at each relay) in parallel relaying while it is described by the sum of transmit powers in $K+1$ lasers (one laser at the source and each relay) in serial relaying. Therefore, assuming equal power allocation, the transmit power at each node is given by $P=P_t/N$ where $N=2K$ for parallel relaying and $N=K+1$ for serial relaying. To simplify the analysis, this chapter focuses on equal power allocation among the nodes (relays and source). Nevertheless, the problem of optimal power allocation for an optical wireless relaying system has been investigated in the literature. For example, Safari *et al.* [10] have proposed optimal power control laws for OWC relay-assisted systems assuming different levels of available channel state information (CSI) under short (over channel coherence time) and long average power constraints.

Note that, in OWC parallel relaying, a simple repetition code is employed rather than a distributed space–time coding scheme, which is required in RF parallel relaying systems to ensure constructive superposition of the signals and achieving spatial diversity gain. This significantly simplifies the signalling scheme as the same signal is repeated across relays at both SIMO and MISO links of the parallel configuration. At the destination, an optical receiver with a wide FOV is employed where the optical beams received from separated relaying nodes can be orthogonally detected by the photodetector. As mentioned in Section 14.2.2, the photodetector measures the sum of optical powers collected from individual light beams and inherently ensures the constructive superposition of the signals received from multiple relays and thus the spatial diversity gain. In fact, Safari and Uysal [21] show that the employment of space–time coding not only increases the complexity of the relaying system but also deteriorates the performance of OWC system.

Multipath relaying is a more complex relaying configuration [12], which combines the two configurations of serial and parallel relaying. In this relaying configuration, there are parallel paths between source and destination and each path may include single or multiple relays arranged in serial. This chapter focuses on the serial and parallel relaying configurations, illustrated in Figure 14.1, to simplify the analysis and provide important conclusions regarding the efficiency of these two distinct relaying approaches. A number of these conclusions can be readily extended to OWC multipath relaying as well [12].

14.3.1 Relaying signaling schemes

The optical relay can perform different electrical and/or optical signal processing on the received signal before forwarding it to the destination node. This section discusses different signalling schemes for optical wireless relaying.

DF relaying: The relay decodes the signal after direct detection, and remodulates, recodes, and retransmits it to the next relay or to destination. In order to avoid error propagation and provide reliable communication, the DF relay only forwards the optical signal if the received signal-to-noise ratio (SNR) exceeds a given decoding SNR threshold $\gamma_{th} = C^{-1}(R_0)$ where $C(\cdot)$ denotes the instantaneous capacity of the OWC link which is an increasing function of γ and R is the data rate. If SNR exceeds γ_{th} (i.e., $R_0 \leq C$), no outage happens and signal can be decoded with arbitrarily low error probability at the receiver. In Reference 13, the application of a convolutional forward-error-correction code in DF relaying was investigated and the performance of the coded relaying system was compared against uncoded relaying communication. In the uncoded relaying scheme, which is called detect-and-forward relaying, the relay directly detects the binary signal modulated over the optical beam and remodulates its laser beam with the detected signal without any further coding or signal processing.

AF relaying: The relay can operate either in electrical or optical mode as illustrated in Figure 14.2. Before forwarding the signal to the next relay or destination, the relay boosts the received signal by performing a background light removal operation and amplification. In electrical AF relaying, the optical signal is first converted to an electrical signal using photodetector. To improve the power efficiency of the relaying system, the background light DC current (RTP_b in (14.11)), which contains no information, is mostly removed through a debiasing operation [9]. To achieve maximum power efficiency, the debiasing operation reduces the DC level of the signal to the lowest level that still preserves the non-negativity of the output electrical signal. The output electrical signal can then directly modulates the intensity of transmit laser

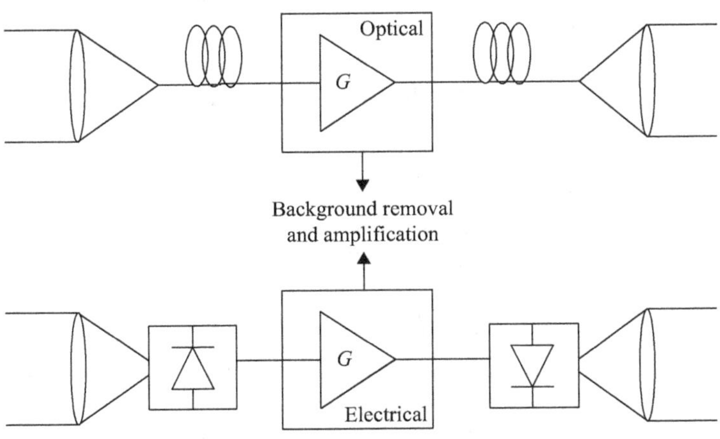

Figure 14.2 Optical (top) and electrical (bottom) AF relays

beam after scaling by an amplifier. The amplification factor can be a fixed value (i.e., fixed-gain relaying) selected based on the channel statistics such that it keeps the average optical transmit power constant. If instantaneous CSI is available, the amplifier gain can be adaptively changed (i.e., adaptive-gain relaying) based on the turbulence condition of the previous link to enhance the performance of the AF relaying.

In optical AF relaying (also known as all-optical relaying) [14–17], the relay focuses the collected optical signal into an optical fibre, optically amplifies it [e.g., using an erbium-doped fibre amplifier (EDFA)] and then forwards it to the next relay or destination without conversion to electrical signal. To avoid the accumulation of the background noise over multiple hops, the relay should efficiently reduce the level of background light collected along with the optical signal before optical amplification. In Reference 16, a background light reduction technique is proposed using highly non-linear fibre and optical filters. The gain of optical amplifier can be either fixed or adaptive depending on the availability of CSI and the complexity of the relay.

14.4 Performance of relay-assisted OWC systems

This section studies the performance of relay-assisted OWC systems in serial or parallel configurations whether employing DF or AF relaying schemes. The performance of the relaying systems is evaluated based on different performance metrics such as outage probability and diversity gains. The performance analysis provides interesting conclusions highlighting the crucial difference between the underlying OWC channel and the more familiar RF wireless channel.

Turbulence-induced fading, which is the major impairment of long atmospheric optical channels, is a slowly varying phenomenon with channel coherence time of about 1–10 ms. At typical transmission rates of optical communication, the fading remains constant over hundreds of thousands up to millions of consecutive bits resulting in a quasi-static channel. The errors caused by this fading occur in burst and are no longer independent. Therefore, outage probability is an appropriate metric to evaluate the performance of OWC relaying systems operating over such channels. The outage probability is defined as the probability of outage events in which the communication system does not work reliably (i.e., the instantaneous capacity of the channel C falls below the transmission rate R_0) and is given by

$$P_{\text{out}} \triangleq \Pr(C(\gamma) < R_0) = \Pr(\gamma < \gamma_{\text{th}}) \tag{14.12}$$

where γ is the instantaneous SNR and the second equality comes from the definition of the decoding threshold SNR $\gamma_{\text{th}} = C^{-1}(R_0)$.

14.4.1 Performance of DF relaying systems

The end-to-end outage probability performance of DF relaying systems can be described as a function of the outage probability of intermediate links since the individual link operates independently. Therefore, we first calculate the outage probability of an intermediate single-input single-output (SISO) link as the building block of both serial and parallel relaying configurations. Noting that the background light DC level in (14.11) can be estimated and removed before decoding, the received electrical SNR for an intermediate SISO link connecting ith and jth nodes ($\forall i,j \in \{0, 1, \ldots, K+1\}$) is given by

$$\gamma = \frac{(RTP\ell_{i,j}h_{i,j})^2}{N_0} \tag{14.13}$$

where indices $i=0$ and $i=K+1$ refer to the source and destination, respectively, and indices $i \in \{1, 2, \ldots, K\}$ refer to the relay nodes. Inserting (14.13) in (14.12), the outage probability of the SISO link is

$$P_{\text{out-SISO}} = \Pr\left(h_{i,j} < \frac{N}{P_M \ell_{i,j}}\right) \tag{14.14}$$

where P_M denotes power margin [9] and is defined as $P_M = P_t/P_{\text{th}}$, P_{th} is a threshold transmit power that guarantees outage-free transmission over the OWC channel connecting source and destination in the absence of turbulence. The power margin can be thus written as

$$P_M = \frac{RTP_t}{\sqrt{N_0 \gamma_{\text{th}}}} \tag{14.15}$$

Considering the lognormal model for the fading coefficient $h_{i,j}$ as described in Section 14.2.1, the outage probability can be written using the cumulative distribution function (CDF) of the lognormal distribution as

$$P_{\text{out-SISO}}(L_{i,j}) = Q\left(\frac{\log(\ell_{i,j}P_M/N) + 2\mu_\chi(L_{i,j})}{2\sigma_\chi(L_{i,j})}\right) \tag{14.16}$$

where $L_{i,j}$ is the length of link between ith and jth nodes. Another building block of the relaying configurations illustrated in Figure 14.1 is the MISO link connecting K relays to the destination in parallel relaying. As mentioned in Section 14.2.1, the photodetector at the destination simply outputs the sum of individual optical signals received from K relays. Furthermore, the receiver at the destination needs to widen its FOV to see all the K relays. Therefore, considering background-noise-limited regime, the noise power at this receiver is assumed to be K times the noise power at the relays. Hence, the received SNR for the MISO link can be written as

$$\gamma = \frac{\left(RTP\sum_{i\in\mathbf{D}} \ell_{i,K+1}h_{i,k+1}\right)}{KN_0} \tag{14.17}$$

where $\mathbf{D}$ is the set of all relays that successfully decoded the source signal. Note that, in DF parallel relaying, only these nodes forward their signals to the destination. Inserting (14.17) in (14.12), the outage probability of the MISO link is obtained as

$$P_{\text{out-MISO}}(\mathbf{D}) = \Pr\left(\sum_{i\in\mathbf{D}} \ell_{i,K+1}h_{i,K+1} < \tfrac{2K^2}{P_M}\right) \tag{14.18}$$

In Reference 9, a closed-form expression of the same form as (14.16) is provided for (14.18) by approximating the weighted sum of lognormal random variables as a single lognormal random variable using moment matching method [9].

In serial relaying, if any of the intermediate SISO links (i.e., hops) fails an end-to-end outage occurs. Hence, the corresponding outage probability can be obtained as

$$\begin{aligned} P_{\text{out}} &= \Pr\left(\bigcup_{i=0}^{K}\{\gamma_i < \gamma_{\text{th}}\}\right) = 1 - \Pr\left(\bigcap_{i=0}^{K}\{\gamma_i > \gamma_{\text{th}}\}\right) \\ &= 1 - \prod_{i=0}^{K}\left(1 - P_{\text{out-SISO}}(L_{i,i+1})\right) \end{aligned} \tag{14.19}$$

where γ_i is the SNR of the ith hop. In parallel relaying, the existence of multiple paths allows for reliable operation even when some of the intermediate links are in outage. A parallel relaying outage occurs if either the decoded set $\mathbf{D}$ is empty or the MISO link connecting the successful relays to the destination fails. There are 2^K distinct possible decoded sets. Let $\mathbf{S}(i)$ and $P_{\mathbf{S}(i)}$, respectively, denote the ith possible set and the probability of the event $\{\mathbf{D} = \mathbf{S}(i)\}$, the outage probability for parallel relaying scheme can be expressed as

$$\begin{aligned} P_{\text{out}} &= \sum_{i=1}^{2^K} P_{\mathbf{S}(i)} P_{\text{out-MISO}}(\mathbf{S}(i)) \\ &= \sum_{i=1}^{2^K}\left(\prod_{j\in\mathbf{S}(i)}\left(1 - P_{\text{out-SISO}}(L_{0,j})\right)\prod_{j\notin\mathbf{S}(i)} P_{\text{out-SISO}}(L_{0,j})\right) P_{\text{out-MISO}}(\mathbf{S}(i)) \end{aligned} \tag{14.20}$$

Figure 14.3 demonstrates the end-to-end outage probability of an optical DF relaying system with serial and parallel configurations assuming $\lambda = 1550\,\text{nm}$, $a \approx 0.1$ (i.e., atmospheric attenuation of 0.43 dB/km), turbulence structure constant of $C_n^2 = 1 \times 10^{-14}\,\text{m}^{-2/3}$, and the link range (i.e., distance between source and destination) of $L_{0,K+1} = 5\,\text{km}$. For serial relaying, the consecutive nodes are equidistant along the path from the source to the destination and for parallel relaying, the relays are located at about the halfway point. The results are obtained through analytical derivations and were validated by Monte Carlo simulation with identical match. As a benchmark, outage probability of the direct transmission is also included.

Figure 14.3 shows an impressive performance gain for serial relaying against both direct transmission and parallel relaying. Particularly, for a target outage probability of 10^{-6}, we observe performance improvements of 18.5, 25.4, and 29.2 dB for $K = 1, 2,$ and 3 with respect to direct transmission. Furthermore, for $K = 2$ and 3, respectively, the serial relaying outperforms the parallel relaying with the same number of relays by 5.5 and 10 dB. This observation shows that it would be more efficient to use relays in a serial (multi-hop) rather than parallel configuration. This result is quite different from RF wireless systems where only parallel relaying can provide diversity gain and is thus superior at high SNR.

Unlike RF wireless systems, the strength of fading is distance dependent in OWC channels (see (14.8)). This allows serial relaying to provide fading mitigation capability by reducing the length (and strength of fading) of intermediate hops through insertion of relays. Therefore, the OWC serial relaying provides diversity gain without

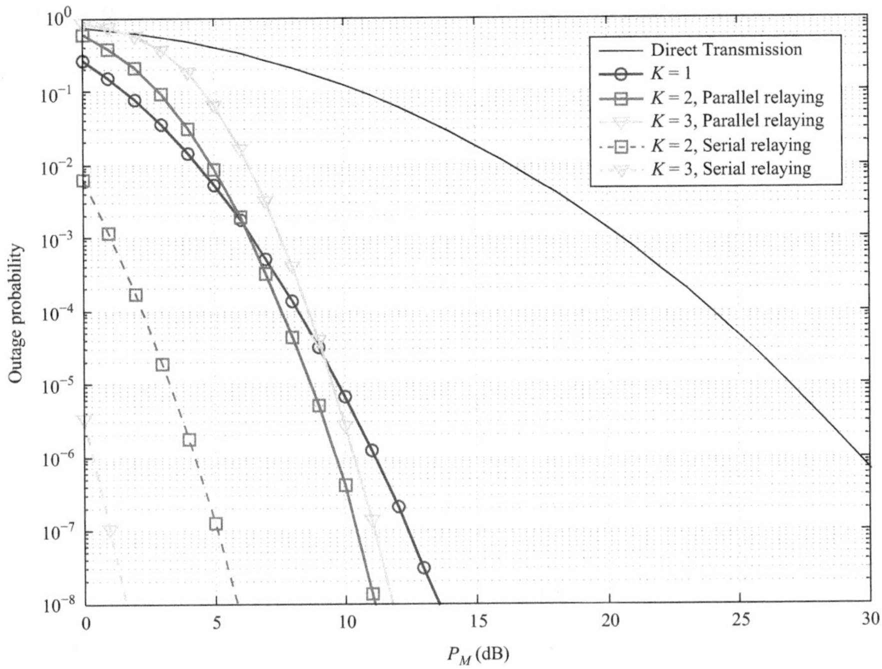

Figure 14.3 The outage probability of DF relaying with serial and parallel configurations and different number of relays

requiring multiple paths between source and destination. This diversity gain can be observed by comparing the slope of serial relaying curves at high SNR for different number of relays.

For $K = 1$, serial and parallel relaying configurations perform identically since they are equivalent. However, as K increases, parallel relaying is outperformed by serial relaying since parallel relaying (with only two hops) exploits distance-dependency of fading variance to a lesser extent. Note that, within the range of the figure, the parallel relaying with $K = 2$ performs better than that with $K = 3$ despite lower diversity gain. This can be explained by the low-power efficiency of parallel relaying with high number of relays, which consumes more power at source (with multi-laser transmitter) and receives more background noise at destination (with wide FOV).

14.4.2 Performance of AF relaying systems

Unlike DF relaying, the outage probability of AF relaying systems is determined by the total received SNR at the destination (rather than SNRs of intermediate links) since intermediate relay nodes do not decode their received signals. In order to simplify the analysis and provide a lower complexity relaying solution, a fixed-gain electrical AF relaying system is considered where the ith relay amplifies the received signal by a fixed factor of a_i after a debiasing operation. Safari and Uysal [9] show that the amplification factor can be well approximated by the reciprocal of the path loss of the preceding hop (i.e., $a_i \approx 1/\ell_{i-1,i}$) in order to keep the average transmit power fixed. The received signal at the destination includes the effect of aggregated channel gains of multiple hops on the signal as well as accumulation of noise over different hops. Moreover, the debiasing operations add a deterministic term (given CSI is available at the destination) to the received signal at the destination. For serial and parallel relaying, respectively, the received SNR at the destination can be obtained using similar approach as in Reference 9 as

$$\gamma = \frac{\left(RTP\ell_{0,1}h_{0,1}\prod_{i=1}^{K} a_i\ell_{i,i+1}h_{i,i+1}\right)^2}{N_0\left(\sum_{j=1}^{K}\prod_{i=1}^{K} a_i^2\ell_{i,i+1}^2h_{i,i+1}^2 + 1\right)} \triangleq \frac{\left(RTP\tilde{h}_{\text{ser}}\right)^2}{N_0} \tag{14.21}$$

$$\gamma = \frac{\left(RTP\sum_{i=1}^{K} a_i\ell_{0,i}\ell_{i,K+1}h_{0,i}h_{i,K+1}\right)^2}{N_0\left(\sum_{i=1}^{K} a_i^2\ell_{i,K+1}^2h_{i,K+1}^2 + K\right)} \triangleq \frac{\left(RTP\tilde{h}_{\text{par}}\right)^2}{N_0} \tag{14.22}$$

where the value K appears in the denominator of (14.22) because of assuming the receiver FOV being K times wider than the relays' FOVs for parallel relaying. Furthermore, the debiasing term does not appear in the SNR expressions since it can be estimated and removed from the signal assuming CSI is available at the receiver. For PPM modulation, the effect of debiasing term is simply cancelled out without requiring CSI [9]. Letting $\tilde{h} = \tilde{h}_{\text{ser}}$ and $\tilde{h} = \tilde{h}_{\text{per}}$ as defined in (14.21) and (14.22), the outage probability of AF serial and parallel relaying are, respectively, given by

$$P_{\text{out}} = \Pr\left(\tilde{h} < \frac{N}{P_M}\right) \tag{14.23}$$

Noting that the product of independent lognormal random variables is also lognormal and using the approximation of weighted sum of lognormal random variables as a single lognormal random variable, approximated analytical expressions can be obtained for (14.23) as in Reference 9.

Figure 14.4 demonstrates the end-to-end outage probability of relay-assisted OWC system with electrical AF relaying and serial and parallel configurations assuming similar channel parameters as in Figure 14.3. Compared to direct transmission, the AF relaying system reveals significant performance improvements. However, the performance gains are less than those observed in DF relaying. Moreover, in AF relaying, the performance gain of serial configuration over parallel configuration is significantly reduced. On the other hand, AF relays enjoy a lower complexity in comparison with DF counterparts since they simply amplify the signal with a fixed gain without requiring any decoding process.

The performance of optical AF relaying (all-optical relaying) over atmospheric optical channels has been investigated in the literature and is compared with electrical

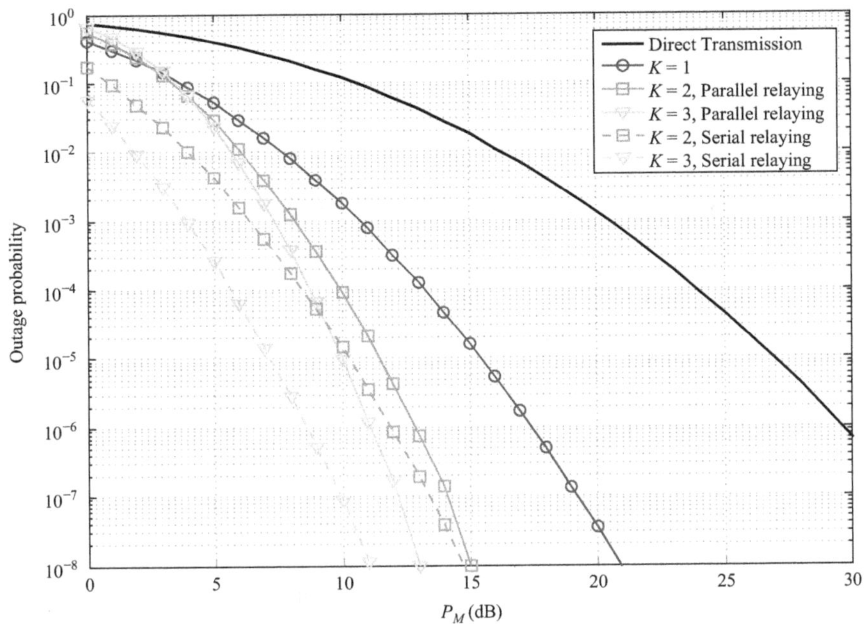

Figure 14.4 The outage probability of electrical AF relaying with serial and parallel configurations and different number of relays

AF relaying schemes [14–17]. To analyse the performance of optical AF relaying systems, additional impairments such as EDFA's amplified spontaneous emission noise and the fibre insertion loss needs to be carefully considered. Comparing all-optical and electrical AF relaying, Kashani *et al.* [14] show that all-optical relaying presents a favourable trade-off between complexity and performance.

14.4.3 Diversity gain analysis

Diversity gain is commonly defined as the negative asymptotic slope of the performance curves versus SNR on a log-log scale. The bit-error rate (BER) or outage probability performance curves of communication systems operating over lognormal fading channels at high SNR decay with (log SNR)^2 rather than log SNR. Therefore, the conventional definition of diversity gain does not converge for lognormal fading channels whether for RF [22] or optical wireless systems [11]. In Reference 22, a relative measure of diversity gain was proposed to address this issue and is given by

$$D_{\mathrm{R}} = \lim_{\mathrm{SNR}\to\infty} \frac{\partial \log P_{\mathrm{out}} / \partial \log \mathrm{SNR}}{\partial \log P_{\mathrm{out0}} / \partial \log \mathrm{SNR}} = \lim_{P_M\to\infty} \frac{\partial \log P_{\mathrm{out}} / \partial \log P_M}{\partial \log P_{\mathrm{out0}} / \partial \log P_M} \tag{14.24}$$

which calculates the asymptotic ratio of the decaying slope of outage probability for the system under consideration (P_{out}) with respect to the slope of outage probability for a benchmark scheme (P_{out0}). If direct transmission (i.e., SISO link) is considered as the benchmark scheme, the definition of relative diversity gain in (14.24) yields the same value as the conventional definition of diversity gain for Rayleigh fading channel [22]. This is because the conventional diversity gain of SISO direct transmission is unity over Rayleigh fading. It is also easy to see that SNR and power margin P_M are interchangeable in (14.24). In the following, we provide the relative diversity gain of DF relaying with serial and parallel configurations over lognormal fading for which there exist exact or approximate analytical outage probability expressions as discussed in Section 14.4.1

Serial DF relaying: The relative diversity gain of serial DF relaying can be obtained by inserting (14.19) in (14.24) and considering direct transmission as the benchmark scheme. Assuming that the nodes in serial configuration are equidistant and using the Chernoff bound on Q function, the diversity gain can be derived after some manipulations as the ratio of the fading variance of the direct link with respect to the fading variance of each hop as [11,23]

$$D_{\mathrm{R}} = \frac{\sigma_\chi^2(L)}{\sigma_\chi^2(L/(K+1))} = (K+1)^{11/6} \tag{14.25}$$

From (14.25), it is interesting to note that, unlike RF wireless systems, serial relaying or multi-hop transmission provides diversity gain in optical wireless systems. In fact, the distance-dependency of the fading variance in (14.8) is the crucial factor in making this difference by allowing multi-hop optical transmission to exploit the shorter distance in the resulting hops with much weaker turbulence effect.

Parallel DF relaying: The relative diversity gain of parallel DF relaying can be obtained by inserting (14.9) in (14.24). Assuming that relays in parallel configuration are located around the halfway point from the source to destination and using the Chernoff bound on Q function, the diversity gain can be approximated after some manipulations as [11,23]

$$D_{\mathrm{R}} = 2^{11/6}K \tag{14.26}$$

Comparing (14.25) and (14.26), the diversity gain of OWC DF relaying with serial and parallel configurations can be expressed in a general form as

$$D_{\mathrm{R}} = N_{\mathrm{H}}^{11/6} N_{\mathrm{D}} \tag{14.27}$$

where N_{H} and N_{D} are respectively the number of hops and the number of diversity paths in the relaying configuration. It is observed that the number of hops N_{H} (raised to the power of $11/6 \approx 2$), which can be seen as the serial relaying factor, has a stronger impact on the diversity gain compared to the number of diversity paths as the parallel relaying factor. Therefore, it is concluded that a parallel relaying scheme with $N_{\mathrm{D}} \geq N_{\mathrm{H}}$ takes advantage of the distance-dependency of turbulence-induced fading variance less than a serial relaying scheme ($N_{\mathrm{H}} \geq N_{\mathrm{D}}$). In effect, serial relaying outperforms parallel relaying as the number of relays increases. This result confirms our earlier observations on the superiority of serial relaying in Figure 14.3.

14.5 Relay-assisted quantum communication

Cryptosystems are an important part of today's communication networks protecting privacy of data transmission. However, the majority of available cryptosystems are only able to offer computational security that are potentially at risk considering the availability of current supercomputers and will be of no use by the realisation of future quantum computers. Quantum cryptography is a radically different approach and is based on the well-known rule of quantum mechanics that measuring a quantum system will perturb it if the measurement is not compatible with the quantum state. Quantum cryptography guarantees unconditionally secure data communication between a transmitting party (Alice) and a receiving party (Bob) using quantum key distribution (QKD). Any intervention from a potential eavesdropper (Eve) in QKD between Alice and Bob introduces perturbations in the quantum system and reveals her presence [24].

14.5.1 Relay-assisted QKD

The first proposed QKD protocol commonly known as BB84 [25] uses elementary quantum systems such as polarised photons to transfer secret keys between Alice and Bob. As illustrated in Figure 14.5, for each quantum bit (qubit), Alice randomly selects a polarisation basis from two known bases (e.g., 0°/90° and −45°/+45°) and sends a single photon modulated with a bit value of 0 or 1 using the randomly selected basis. At the receiver side, Bob measures the quantum state of the incoming photon again based on a randomly selected polarisation basis. To avoid using uncorrelated measurement results caused by incompatible bases separately selected by Alice and Bob, they construct the secret key only using the qubits received at the *sift* events. Sift events, where both Alice and Bob have chosen the same basis, can be identified through exchange of information in a public (classical) channel. Finally, they perform some error correction and privacy amplification procedures respectively, to improve the reliability of the generated secret key and to prevent Eve from keeping useful information about the key [24].

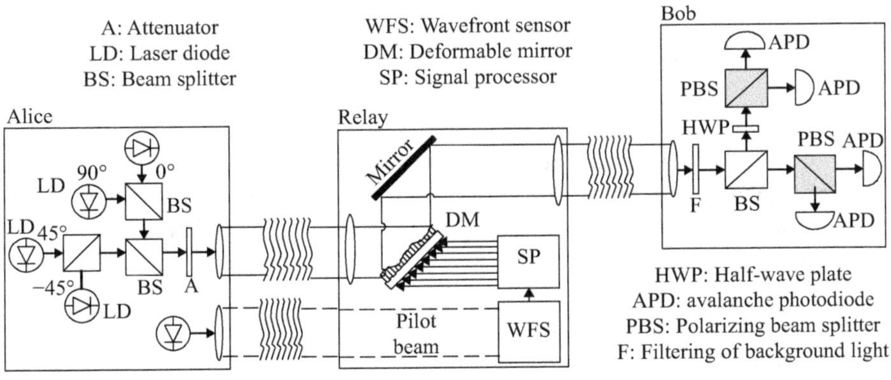

Figure 14.5 The schematic diagram of the QKD with adaptive optics-based relay

A major drawback of QKD systems either operating through optical fibre or atmospheric links is the range limitation. *Quantum repeaters* have been originally proposed in Reference 26 for fibre-optic systems by creating perfect entangled pairs and transfer them through repeaters by entanglement swapping. However, this technique requires quantum memory and entanglement purification, which are currently difficult to realise. Furthermore, a class of quantum relays called *trusted relays* have been introduced in References 27 and 28 which assume the relays can be trusted to have partial or perfect knowledge of the distributing keys.

Here, we focus on the analysis of an alternative classical relaying scheme for free-space QKD that does not require complex quantum signal processing or compromising security by trusting relays. In this scheme, the qubits generated by Alice travel through relay nodes in a serial relaying configuration before detection by Bob. The conventional AF and DF relaying schemes are of no use for QKD systems since both relaying procedures perturb the quantum state [24] and this might be confused as Eve's intervention. As illustrated in Figure 14.5, the relays passively receive the transmitted optical beam and reconstruct its turbulence-degraded wavefront using adaptive optics (AO) before pointing it to the next relay or Bob. The AO system [29] uses a wave-front sensor that measures the turbulence distortions of high-intensity pilot beams transmitted at the beginning of each block of key transmission. This information feeds into a deformable mirror that corrects the atmospheric wavefront distortions of quantum beams.

14.5.2 Qubit error rate performance analysis

In the proposed relay-assisted QKD system illustrated in Figure 14.5, Alice generates a polarised optical pulse for each qubit with an average photon number of n_s and sends it to the relay. Due to the atmospheric channel impairments, the ith relay node ($i = 1, 2 \ldots, K$) collects only a random fraction γ_i of the transmitted photons in the ith hop. At the destination side, Bob also collects only a fraction γ_{K+1} of the transmitted photons from the last relay node and an overall fraction $\gamma = \gamma_1\gamma_2 \cdots \gamma_{K+1}$ of the transmitted photons from Alice. The fraction parameter of the ith hop γ_i ($i = 1, 2, \ldots, K+1$) is expressed in terms of ith transmit and receive beam patterns as

$$\gamma_i = \frac{\int_{R_i'} |\xi_i'(\vec{\rho}')|^2 d\vec{\rho}'}{\int_{R_i} |\xi_i(\vec{\rho})|^2 d\vec{\rho}} \tag{14.28}$$

Assuming that the relays and the destination collect an average number of n_{B0} background photons per polarisation and taking into account the background photons redirected by the relays, the total average number of background photons collected per polarisation by Bob can be written as

$$n_B = n_{B0}\left(1 + \sum_{i=2}^{K+1}\prod_{j=i}^{K+1}\gamma_j\right) \tag{14.29}$$

Each of the single-photon detectors (Geiger-mode avalanche photodiodes) at Bob's receiver further registers an average number of n_D dark counts. The performance of the underlying relay-assisted QKD system can be evaluated through the derivation of an upper bound on qubit error rate (QBER) which is given by [24]

$$\mathrm{QBER} = \frac{\Pr(\mathrm{error})}{\Pr(\mathrm{sift})} \tag{14.30}$$

where Pr(sift) and Pr(error) are probabilities of sift and error events, respectively. The conditional sift and error probabilities of the relay-assisted system given $\{\gamma_1, \gamma_2, \ldots, \gamma_{K+1}\}$ is obtained as [30]

$$\Pr(\mathrm{sift}|\boldsymbol{\gamma}) = \eta(n_S\gamma/2 + 2n_N)\exp[-\eta(n_S\gamma + 4n_N)] \tag{14.31}$$

$$\Pr(\mathrm{error}|\boldsymbol{\gamma}) = \eta n_N \exp[-\eta(n_S\gamma + 4n_N)] \tag{14.32}$$

where η and $n_N \triangleq n_\mathrm{B}/2 + n_\mathrm{D}$ are respectively the quantum efficiency and the average number of noise counts at each of Bob's detectors. To obtain unconditional probabilities of sift and error, statistical characterisation of fraction parameters is required. Using singular value decomposition of the channel gains $h_i(\vec{\rho}, \vec{\rho}')$ and considering the most significant eigenvalues μ_i of the corresponding hops with index $i = 1, 2, \ldots, K+1$, theoretical bounds can be found (see Reference [18] for details) on the average optical power transfer and sift and error probabilities at each hop. In fact, the exact statistical description of μ_i is not available for near-field propagation in the literature. However, an upper bound on $E\{\mu_i\}$ can be simply obtained as [30]

$$E\{\mu_i\} \geq \hat{\mu}_i \equiv \frac{8\sqrt{D_f}}{\pi} \int_0^1 e^{-\frac{D_i(d_i x)}{2}} \left(\cos^{-1}(x) - x\sqrt{1-x^2}\right) J_1\left(4x\sqrt{D_f}\right) dx \tag{14.33}$$

where $J_1(\cdot)$ is the first-order Bessel function of the first kind, $D_i(\nu) = 1.09k^2C_n^2L_{i-1,i}\nu^{5/3}$ is the spherical-wave wave structure function of ith hop, and D_f is the Fresnel number product of ith hop. Letting $\hat{\mu} = \hat{\mu}_1\hat{\mu}_2\cdots\hat{\mu}_{K+1}$ and $L = L_{0,1} + L_{1,2} + \cdots + L_{K,K+1}$ and using the bounds on sift and error probabilities, the QBER can bounded as [18]

$$\mathrm{QBER} \leq \frac{2\hat{n}_N e^{-4\eta\hat{n}_N}[(1-\hat{\mu}) + e^{-\eta n_S e^{-aL}}\hat{\mu}]}{(2n_{\mathrm{B}0} + 4n_\mathrm{D})e^{-\eta(2n_{\mathrm{B}0}+4n_\mathrm{D})}(1-\hat{y}) + (n_S e^{-aL} + 4\hat{n}_N)e^{-\eta(n_S e^{-aL}+4\hat{n}_N)}\hat{y}} \tag{14.34}$$

where $\hat{y}$ and $\hat{n}_N$ are, respectively, defined as

$$\hat{y} \triangleq \frac{n_S e^{-aL}\prod_{j=1}^{K+1}\hat{\mu}_j + 2n_{B0}\left(\sum_{i=2}^{K+1}\prod_{j=i}^{K+1} e^{-aL_{j-1,j}}\hat{\mu}_j\right)}{n_S e^{-aL} + 2n_{B0}\sum_{i=2}^{K+1}\prod_{j=i}^{K+1} e^{-aL_{j-1,j}}} \tag{14.35}$$

$$\hat{n}_N = \frac{n_{B0}}{2}\left(1 + \sum_{i=2}^{K+1}\prod_{j=i}^{K+1} e^{-aL_{j-1,j}}\right) + n_D \tag{14.36}$$

which the latter is a pessimistic measure of noise count at each of Bob's detectors. Finally, the exact optimum QBER of a free-space QKD system operating over non-turbulent air can be obtained as a benchmark as [30]

$$\mathrm{QBER} = \frac{2n_N}{n_S\mu_0 e^{-aL_{\mathrm{AB}}} + 4n_N} \tag{14.37}$$

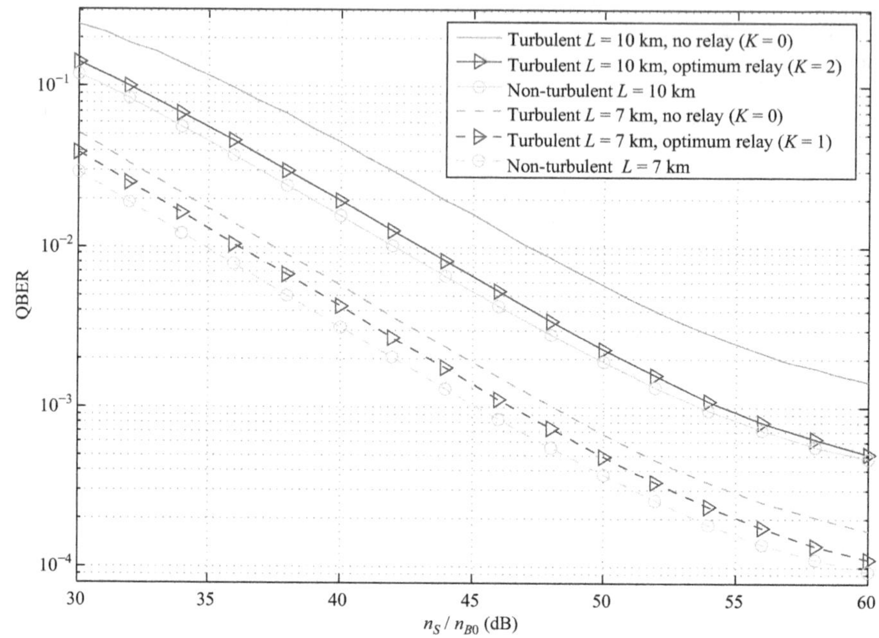

Figure 14.6 The QBER of relay-assisted quantum communication over different link lengths

where L_{AB} is the length of the line-of-sight link connecting Alice and Bob and μ_0 is the largest eigenvalue of the singular value decomposition of vacuum-propagation Green's function given in Reference 31.

Figure 14.6 demonstrates the QBER performance of the proposed relay-assisted QKD scheme operating over turbulent atmosphere in comparison with direct transmission as well as with the scheme operating over non-turbulent channel. The channel parameters are assumed as $\lambda = 0.77\ \mu\text{m}$, $C_n^2 = 1 \times 10^{-15}\ \text{m}^{-2/3}$ with absorption and scattering loss of 2 dB/km. The receive and transmit aperture diameters are 10 cm and the relay nodes are located equidistant along the path connecting Alice and Bob, i.e., $L_{\mathrm{AB}} = L$. Furthermore, Alice transmits at the average number of transmitted photons of $n_s = 1$ and Bob's detectors are operated at the detection efficiency of $\eta = 0.5$ with average dark count of $n_{\mathrm{D}} = 10^{-6}$ and varying average number of background count, n_{B0}, depending on the complexity of the temporal, spectral, and spatial filters employed.

Figure 14.6 shows that the proposed relay-assisted QKD system significantly outperforms the direct transmission in different ranges of received background noise and both link lengths of $L = 7$ km and $L = 10$ km. This gain is more significant in the longer link length where the relayed transmission almost eliminates the degrading effects of turbulence at high SNR and converges to the transmission over non-turbulent atmosphere. It is also observed that the optimum number of relays for $L = 10$ km is

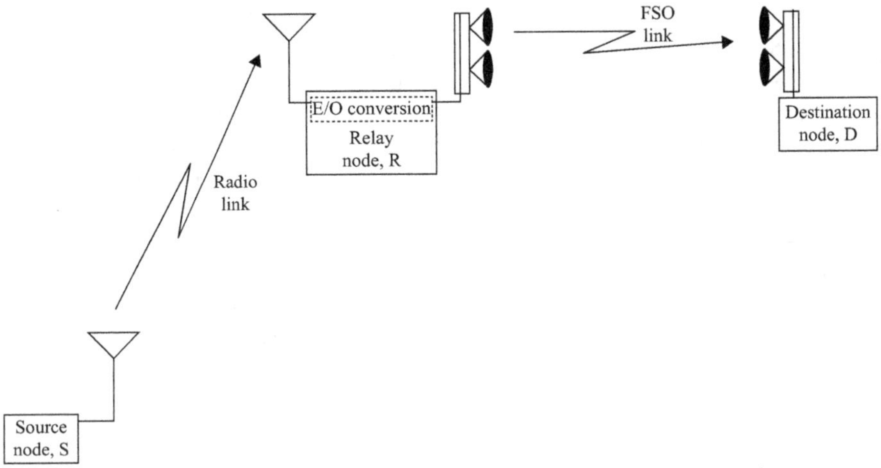

Figure 14.7 An illustration of mixed RF-FSO dual-hop relay network

two relays while it is one relay for $L = 7$ km. Finally, note that the result shows the raw values of QBER calculated without considering error correction procedures that needs to be performed by Alice and Bob.

14.6 Mixed RF-FSO relay transmission

The previous section describes a symmetrical FSO relay system in which the fading distribution is the same on both links of the relay system. In several practical instances, however, each relay link is likely to encounter a different type of fading. One of such situation is the mixed RF/FSO communication link depicted in Figure 14.7. In this kind of asymmetric dual-hop system, the advantages of an FSO link are being used to extend the reach of the RF link.

In the asymmetric RF/FSO relay transmission, the RF link experiences the Rayleigh fading while the FSO link suffers from atmospheric turbulence induced fading. The fading statistics on the FSO link is modelled as a gamma-gamma distribution – a widely used model for fading due to atmospheric turbulence. The relay node amplifies the incoming RF signal and converts it to an optical signal for onward transmission. This kind of relaying is often referred to as AF relay system with fixed gain. At the relay node, the conversion of the RF signal into its optical equivalent is done by the employing subcarrier intensity modulation (SIM) scheme. The basic concept of SIM is further discussed in Section 14.6.1.

At the relay node, the parameters relating to the S–R link are given subscript '1' while those of the R–D link are given subscript '2'. Hence, the signal r_1 reaching the relay node from the source can be denoted as [32]

$$r_1 = \alpha_1 s + n_1 \tag{14.38}$$

where α_1 and n_1 represent the fading amplitude of the Rayleigh fading channel and AWGN with variance N_{01} for the S–R link, respectively. The data carrying RF signal itself is denoted as s. The relay node amplifies the received signal with a fixed gain G. Thereafter, the SIM technique of Section 14.6.1 is used to impress the amplified RF signal onto the intensity of an optical carrier (typically a laser) for onward transmission to the destination node, D. The retransmitted optical signal at the relay is hence given by the following equation:

$$s_{\text{opt}} = G(1 + \eta r_1) \tag{14.39}$$

In the above equation, the electrical-to-optical conversion coefficient of the laser driver is represented as η. By filtering out all DC components, the signal received at the destination node can be modelled as [32]

$$r_2 = IG\eta(\alpha_1 s + n_1) + n_2 \tag{14.40}$$

where n_2 is the AWGN on the R–D link with variance N_{02}. The overall SNR of the full asymmetric RF/FSO is as given by the following equation:

$$\gamma = \frac{\gamma_1\gamma_2}{\gamma_2 + C} \tag{14.41}$$

In the above equation, $C = (G^2N_{01})^{-1}$ and $\gamma_1 = \alpha_1^2 P_1/N_{01}$ represent the SNR on the S–R hop, with P_1 being the transmitted power at the source node. This γ_1 (S–R SNR) suffers from Rayleigh fading whose pdf is given by the following equation:

$$f_{\gamma_1}(\gamma_1) = \frac{1}{\bar{\gamma}_1} e^{-\gamma_1/\bar{\gamma}_1} \tag{14.42}$$

where $\bar{\gamma}_1$ represents the average SNR on the S–R link.

Similarly, $\gamma_2 = \eta^2 I^2/N_{02}$ is the SNR on the R–D link and experiences gamma-gamma fading. According to Lee *et al.* [32], the pdf of γ_2 under the influence of gamma-gamma fading is as given by the following equation:

$$f_{\gamma_2}(\gamma_2) = \frac{(\alpha\beta)^{\frac{(\alpha+\beta)}{2}} \gamma_2^{\frac{(\alpha+\beta)}{4}-1}}{\Gamma(\alpha)\Gamma(\beta)\bar{\gamma}_2^{\frac{(\alpha+\beta)}{4}}} K_{\alpha-\beta}\left(2\sqrt{\alpha\beta\sqrt{\frac{\gamma_2}{\bar{\gamma}_2}}}\right) \tag{14.43}$$

Again $\bar{\gamma}_2$ is the average SNR of the R–D link while $K_a(\cdot)$ is the modified Bessel function of the second kind of order a. The parameters α and β are related to the strength of the atmospheric conditions.

The outage probability performance metric, P_{out}, is defined as the probability that the system SNR is lower than a predefined threshold value γ_{th}. Hence,

$$P_{\text{out}} = \Pr[\gamma < \gamma_{\text{th}}] = \Pr\left[\frac{\gamma_1\gamma_2}{\gamma_2 + C} < \gamma_{\text{th}}\right] \tag{14.44}$$

From the foregoing, the end-to-end outage probability of the relay system can thus be derived as follows [32]:

$$\begin{aligned} P_{\text{out}} &= \int_0^\infty \Pr\left[\frac{\gamma_1\gamma_2}{\gamma_2 + C} < \gamma_{\text{th}} | \gamma_2\right] f_{\gamma_2}(\gamma_2) d\gamma_2 \\ &= 1 - e^{-\gamma_{\text{th}}/\bar{\gamma}_1} \frac{(\alpha\beta)^{\frac{(\alpha+\beta)}{2}}}{\Gamma(\alpha)\Gamma(\beta)\bar{\gamma}_2^{\frac{(\alpha+\beta)}{4}}} E(\gamma_{\text{th}}) \end{aligned} \tag{14.45}$$

The integral function $E(\gamma_{\text{th}})$ in the above equation is defined as:

$$E(\gamma_{\text{th}}) = \int_0^\infty \gamma_2^{\frac{(\alpha+\beta)}{4}-1} e^{(-\gamma_{\text{th}}/\bar{\gamma}_1)(C/\gamma_2)} K_{\alpha-\beta}\left(2\sqrt{\alpha\beta\sqrt{\frac{\gamma_2}{\bar{\gamma}_2}}}\right) \mathrm{d}\gamma_2 \tag{14.46}$$

14.6.1 Subcarrier intensity modulation

The SIM technique is a scheme in which a time-continuous electrical signal at a much lower frequency than the optical carrier is used to modulate the intensity of an optical carrier. This electrical signal could comprise of a single subcarrier or a number of subcarriers aggregated using frequency division multiplexing [33]. In essence, the SIM technique is the IM equivalent of the multi-carrier modulation approach in RF communications that is already deployed in applications such as digital television, local area networks, asymmetric digital subscriber line, 4G communication system [34,35]. It has equally found applications in optical fibre communications. For example, the SIM techniques have been commercially adopted in transmitting cable television signals over optical fibres and have also been used in conjunction with wavelength division multiplexing [36]. For the seamless integration of FSO systems into the present and future networks, which already harbour subcarrier modulated (or multiple carrier) signals, the understanding of subcarrier modulated FSO is thus imperative. The following are some of the benefits of using SIM in an FSO system.

- It benefits from already developed and evolved RF communication components such as stable oscillators and narrow filters.
- It avoids the need for an adaptive threshold required by optimum performing OOK modulated FSO [37].
- It can be used to increase capacity by accommodating data from different users on different subcarriers.
- It has comparatively lower bandwidth requirement than the PPM for example.

Some possible challenges in implementing SIM in FSO systems include:

- Relatively high average transmitted power due to:
  (i) the optical source being ON during the transmission of both binary digits ‘1’ and ‘0’, unlike in OOK where the source is ON during the transmission of bit ‘1’ only.

(ii) the multiple subcarrier composite electrical signal being the sum of the modulated sinusoids (i.e., dealing with both negative and positive values) requires a DC bias. This is to ensure that this composite electrical signal, that will eventually modulate the laser irradiance, is never negative. Increasing the number of subcarriers leads to increased average transmitted power, because the minimum value of the composite electrical signal decreases (becomes more negative) and the required DC bias therefore increases [38]. This factor results in poor power efficiency and places a bound on the number of subcarriers that can be accommodated when using multiple SIM.

- The possibility of signal distortions due to inherent laser non-linearity and signal clipping due to over-modulation.
- Stringent synchronisation requirements at the receiver side.

In view of the foregoing, it is important to highlight that the use of multiple subcarriers in SIM should only be considered when the need for higher capacity/throughput far outweighs the highlighted challenges or where an FSO system is to be integrated into an existing network that already contain multiple RF carriers. Several methods have been reported in the literature, including References 38–40, to improve the poor power efficiency of SIM but these will not be discussed here.

In SIM modulated FSO links, an RF subcarrier signal, $m(t)$, is pre-modulated with the source data, $d(t)$. The electrical signal $m(t)$ is then used to modulate the intensity, P_T, of the optical source – typically a continuous wave laser diode. Figure 14.8 illustrates the system block diagram of a SIM-FSO with N subcarriers. The serial-to-parallel converter distributes the incoming data across the N subcarriers. Each subcarrier carries a reduced symbol rate but the aggregate must be equal to the symbol rate of $d(t)$.

Another obvious possibility, not shown in Figure 14.8, is to have different users occupying the N different subcarriers. For the M-PSK subcarrier modulation shown in Figure 14.8(a), the encoder maps each subcarrier symbol onto the symbol amplitude $\{a_{ic}, a_{is}\}_{i=1}^{N}$ that corresponds to the constellation in use. Since the subcarrier signal, $m(t)$, is bipolar (i.e., having both positive and negative values), a DC level b_0 is usually added to $m(t)$ before it is used to directly drive the laser diode. This is to avoid clipping of the negative part of $m(t)$ signal since the optical intensity can only be positive.

The general expression for $m(t)$ in an N-subcarrier SIM-FSO system is given as:

$$m(t) = \sum_{i=1}^{N} m_i(t) \tag{14.47}$$

During a symbol duration, each RF subcarrier signal in $m(t)$ is of the following form:

$$m_i(t) = g(t)a_{ic}\cos(\omega_{ci}t + \varphi_i) + g(t)a_{is}\sin(\omega_{ci}t + \varphi_i) \tag{14.48}$$

where $g(t)$ is the pulse shaping function. The subcarrier angular frequency and phase are represented by $[\omega_{ci}, \varphi_i]_{i=1}^{N}$. It follows therefore that each subcarrier $m_i(t)$ can be modulated by any standard RF digital/analogue modulation technique, such as QAM,

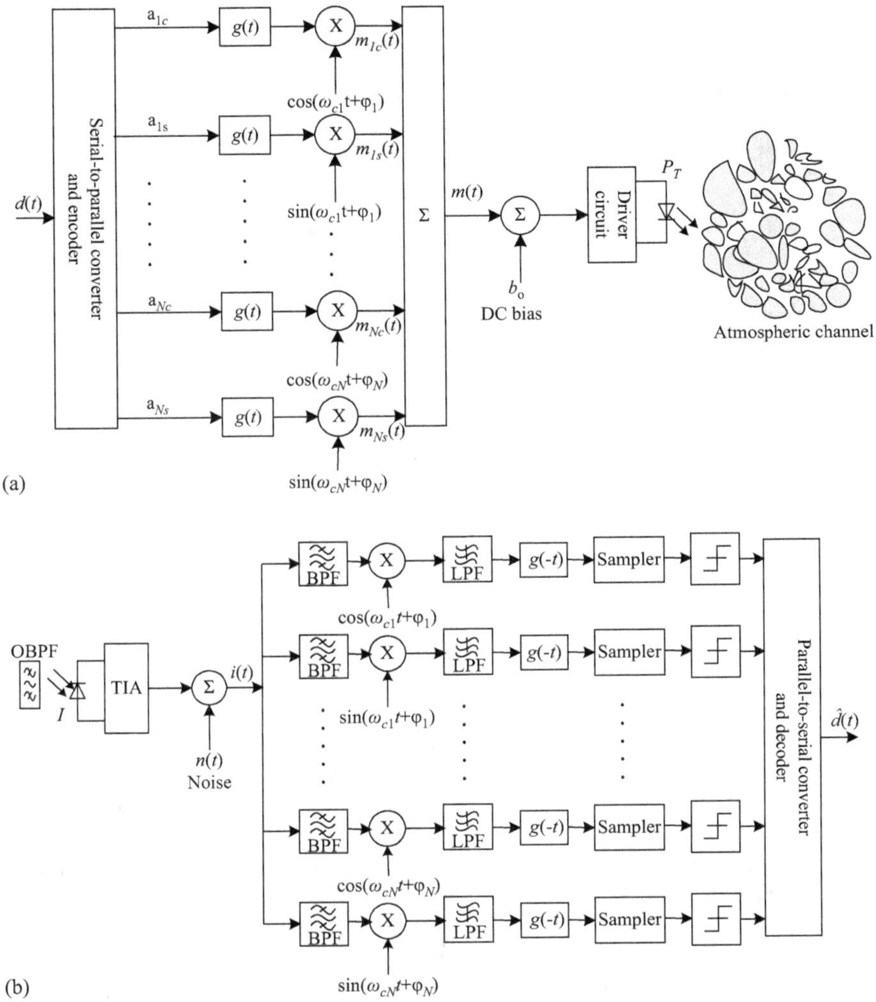

Figure 14.8 A block diagram illustrating the SIM-FSO concept: (a) transmitter and (b) receiver. TIA – trans-impedance, OBPF – optical band pass filter

M-PSK, M-FSK, and M-ASK. At the receiver, the incoming optical radiation, P_R, is converted into an electrical signal, $i(t)$ via direct current. This is then followed by a standard RF demodulator to recover the transmitted symbol as shown in Figure 14.8(b). By normalising the receiver area to unity and representing the received power by irradiance, I, the received signal can be modelled as:

$$i(t) = \Re I[1 + \xi m(t)] + n(t) \tag{14.49}$$

where the optical modulation index $\xi = \left| \frac{\mathrm{m(t)}}{i_B - i_{Th}} \right|$ (see Figure 14.9).

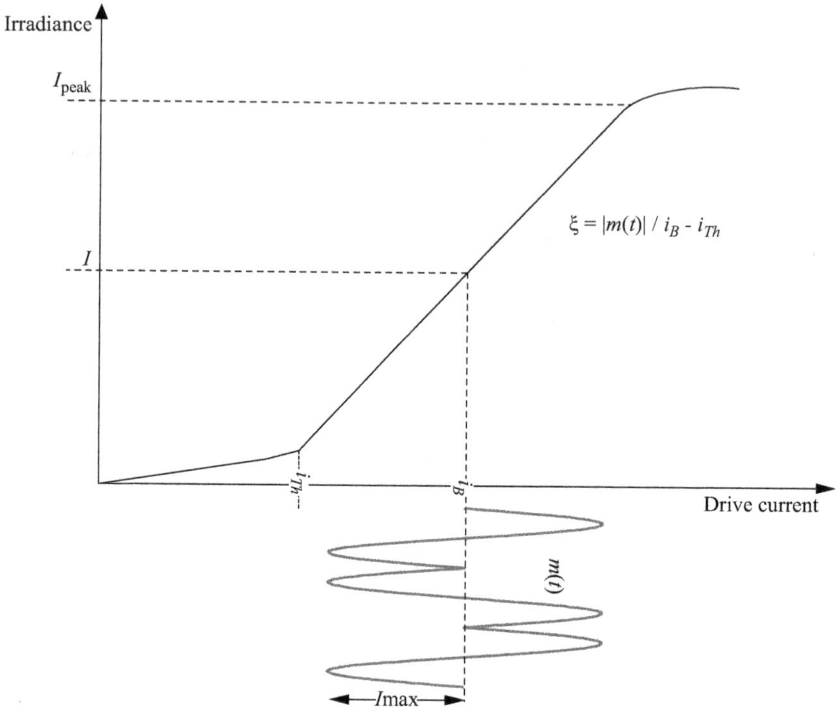

Figure 14.9 Output characteristic of an optical source driven by a subcarrier signal showing optical modulation index

The electrical band pass filter (BPF) with a minimum bandwidth of $2R_b$ performs the following functions:

1. selection of the individual subcarrier for demodulation,
2. reduction of the noise power, and
3. suppress any slow varying $\Re I$ component present in the received signal.

For a subcarrier at ω_{ci}, the received signal is thus:

$$i(t) = I_{\text{comp}} + Q_{\text{comp}} \tag{14.50}$$

where

$$I_{\text{comp}} = \Re I \xi g(t) a_{ic} \cos(\omega_{ci} t + \varphi_i) + n_I(t) \tag{14.51}$$

$$Q_{\text{comp}} = -\Re I \xi g(t) a_{is} \sin(\omega_{ci} t + \varphi_i) + n_Q(t) \tag{14.52}$$

where $n_1(t)$ and $n_Q(t)$ are the independent AWGN each with a zero mean and a variance denoted by σ^2. The quadrature components I_{comp} and Q_{comp} are further down converted by the reference signals cos $\omega_c t$ and sin $\omega_c t$, respectively, and applied to the standard receiver architecture. The electrical low-pass filters, which are part of the standard

RF receiver, remove any out of band (unwanted) signals from the down converted signal and then pass it onto the decision circuit.

14.7 Optical orthogonal frequency multiplexing and effects of signal clipping

OWC system conveys the information modulating the signal intensity of the optical emitter, such as laser or LED in infrared or visible part of the spectrum, as opposed to a RF system where the data are used to modulate the electromagnetic field radiated by an antenna. This means that only signal intensity can be detected and, hence, it is not possible to use all digital modulation techniques known in RF communications without modification. Unipolar techniques from the pulsed modulation schemes, such as PAM, PPM, and pulse width modulation (PWM)can be adopted in a straightforward way and are used in OWC systems with IM/DD [41]. However, the challenging, dispersive nature of the optical wireless channel presents a major throughput limiting factor for these modulation schemes because of the severe inter-symbol interference (ISI) at high data rates. For this reason, a more robust technique, such as orthogonal frequency division multiplexing (OFDM), is preferred choice. Some of advantages of OFDM include equalisation and adaptive loading. Equalisation can be performed with single-tap equalisers in the frequency domain which reduces design complexity while frequency subcarriers can be adaptively loaded according to the channel transfer function. The adaptive loading therefore improves the use of the channel when attenuation or interference is significant in particular frequency bands.

However, conventional RF OFDM signals are bipolar and complex valued which imply that they have to be adjusted to be usable in IM/DD systems which require signals to be real and unipolar. O-OFDM uses the time-domain signal envelope to modulate the intensity of the LED [42]. The signal needs to be real valued in the time domain and this goal is achieved when Hermitian symmetry is imposed on the OFDM subcarriers [43]. However, Hermitian symmetry effectively sacrifices one half of the original OFDM spectral efficiency in order to obtain real signal.

There are a number of approaches to deal with the issue of bipolarity in OFDM signals which have been proposed in the literature. Well-known approaches which ensure unipolarity include direct-current biased optical OFDM (DCO-OFDM) [42], asymmetrically clipped optical OFDM (ACO-OFDM) [44], pulse-amplitude-modulated discrete multitone modulation (PAM-DMT) [45], and unipolar orthogonal frequency division multiplexing (U-OFDM) [46]. DCO-OFDM generates a unipolar signal by introducing a DC bias. In ACO-OFDM, only the odd subcarriers are modulated whereby the additional 50% loss in spectral efficiency is sacrificed for unipolarity. In PAM-DMT, the frequency subcarriers in an OFDM frame are modulated with imaginary symbols from the M-PAM modulation scheme so that distortion of removing the negative part of the signal is transformed only into a real-valued signal, which is orthogonal to the information-carrying signal. In U-OFDM, all possible subcarriers in the frequency domain are modulated as in DCO-OFDM, and after the time-domain signal is obtained, it is divided into a positive block and a negative block. The positive block is a copy of the original signal frame with all negative samples set to zero. The negative block is a copy of the original signal frame with all samples inverted, and then, the negative samples are set to zero.

An OFDM system employs the inverse fast Fourier transform (IFFT) to generate the multiplexed signal at the transmitter. Since IFFT is a summing operation, the non-distorted time-domain signal follows a close to normal distribution for large IFFT sizes due to the central limit theorem (CLT) [47]. A total subcarrier number as small as 64 is usually considered to be sufficient to provide the normal distribution [48]. Optical front-ends contain numerous imperfections which result in a non-linear distortion of the transmitted signal such as signal clipping or, more generally, non-linear transfer effects. However, these non-linear distortions can be modelled by means of the Bussgang theorem as an attenuation of the data-carrying signal plus a non-Gaussian uncorrelated clipping noise component [49]. More specifically, if X is a zero-mean normally distributed random variable and $d(X)$ is an arbitrary memoryless distortion on X, then, following the exposition of Bussgang theorem in Reference 50:

$$d(X) = \alpha X + Y \tag{14.53}$$

$$E[XY] = 0 \tag{14.54}$$

where α is a constant, $E[\cdot]$ represents expectation and Y is a noise component which is uncorrelated with X. Using (14.53) and (14.54), α can be derived as

$$\alpha = E[X\ d(X)]/\sigma_x^2 \tag{14.55}$$

where α_x^2 represents the variance of X. The uncorrelated noise Y can then be computed as follows:

$$E[Y^2] = E[d^2(X)] - \alpha^2\sigma_x^2 \tag{14.56}$$

$$E[Y] = E[d(X)] \tag{14.57}$$

$$\sigma_Y^2 = E[Y^2] - E[Y]^2 \tag{14.58}$$

where α_Y^2 stands for the variance of Y.

At the receiver, the fast Fourier transform (FFT) is used for demultiplexing. Due to summing operation in FFT, the CLT can be applied again and the clipping noise can be represented as a Gaussian process. This approach is used in Reference 51 where symmetric signal clipping in O-OFDM is studied following similar work where the non-linear transfer effects of high-power amplifiers (HPAs) in OFDM-based RF systems. DC biasing of a *symmetrically* clipped signal is studied in Reference 52. However, in DCO-OFDM, the time-domain signal is likely to be *asymmetrically* clipped double sidedly because of insufficient forward biasing and the maximum power limit of the transmitter front-end. Furthermore, ACO-OFDM also requires investigation of *asymmetric* clipping due to the nature of its asymmetric signal. For these reasons, double-sided time-domain signal clipping at the transmitter is simulated and analysed for DCO-OFDM and ACO-OFDM in Reference 43. The attenuation factor and the variance of the complex-valued normally distributed clipping noise are determined in closed form using the Bussgang theorem approach from (14.53) to (14.58). These variables are then included in the derivation of the electrical SNR which is confirmed via BER Monte Carlo simulation of clipped DCO-OFDM and ACO-OFDM. The DCO-OFDM and ACO-OFDM systems are analysed at different clipping levels and it is demonstrated that ACO-OFDM is more robust to the clipping than DCO-OFDM for the same size modulation order at the expense of a 50% reduction in spectral efficiency. Moreover, a comparison between different orders of modulation in ACO-OFDM case suggests that higher order modulation is more vulnerable to signal clipping. This effect is seen because the clipping noise at the information-carrying subcarriers is added independently of the modulation order.

14.7.1 Signal shaping

We have seen in the previous subsection that various non-linearities of the optical front-ends due to the use of non-optimised components result in a limited linear dynamic range of radiated optical power. In practical terms, that fact implies that the transmitted signal is constrained between levels of minimum and maximum optical power if one is to avoid or reduce the amount of signal clipping. The average optical power level could be further constrained by the eye safety regulations and the design requirements. Therefore, in order to condition the signal in accord with these constraints, careful signal scaling and DC biasing is required in the digital signal processing chip and the analogue circuit, respectively. In particular, due to the normal distribution of the OFDM signals, DCO-OFDM and ACO-OFDM require optimisation of biasing and scaling by using the results obtained by using the fact that OFDM signal is normally distributed and then applying the Bussgang theorem.

Since the PPM and PAM signals have pdfs with finite supports, they can satisfy the constraints without signal clipping. However, clipping represents only one type of non-linearity. Other non-linearities, such as those due to the relationship between the voltage and current in the LED or due to quantisation also have to be accounted for. The fully general approach, using the Bussgang theorem, for modelling various non-linearities, such as quantisation and clipping, for DCO-OFDM, ACO-OFDM, PAM-DMT, and U-OFDM, is developed in Reference 53.

In Reference 54, the aforementioned analysis is used to find the optimum signal scaling and DC-biasing to minimise the electrical SNR per bit which is required for a target BER. While in VLC systems, the DC bias power can be used for illumination as a primary functionality in indoor deployments, in infrared communication systems, the DC bias power is constrained by the eye safety regulations. Therefore, DC bias is generally included in the electrical SNR calculation in the infrared case. Assuming unlimited non-negative range, PPM, PAM, and a variant of DCO-OFD M are compared in a flat fading channel in terms of optical power requirement and spectral efficiency. The non-negative signal is therefore scaled down to accommodate the large peak-to-average-power ratio (PAPR), resulting in an increased optical power requirement. Such approach enabled a comparison of single-carrier and OFDM-based schemes in terms of spectral efficiency and electrical SNR in a dispersive optical wireless channel. It also enabled optimisation of signal shaping for a dynamic range of the transmitter whether a DC bias power is included or excluded in the computation of the SNR. In particular, for the normally distributed O-OFDM signals, as optimal strategy, the required electrical SNR is minimised. The PPM, PAM, and O-OFDM can be compared in terms of electrical SNR requirement and spectral efficiency in the dispersive optical wireless channel.

It is shown in Reference 54 that when the additional DC bias is not considered, DCO-OFDM and PAM show the greatest spectral efficiency for a flat fading channel in the mid-to-high SNR region while DCO-OFDM demonstrates a superior spectral efficiency over PAM for low SNR due to its robustness to channel dispersion and bit loading. However, when the DC bias power is included in the electrical signal power, DCO-OFDM and ACO-OFDM suffer a greater SNR penalty due to the DC bias as compared to PAM and PPM. Still, the developed optimum signal shaping framework enables O-OFDM schemes to significantly decrease the penalty to single-carrier transmission. In a related work in Reference 55, it is shown that by applying a minimum mean squared estimator linear approximation to the biasing and clipping process, the non-linear process can be modelled as a virtual relay channel. Then, using the developed SNR formula, an optimal biasing point that maximises the effective SNR can be determined with the proposed fast searching algorithm.

14.7.2 PAPR reduction

High PAPR can cause undesirable clipping which leads to poor performance of the OFDM systems. One way to reduce the PAPR can be achieved by rotating the phase of the data symbols with several iterations of pilot symbol sequences [56]. By repeating the process of randomly generated pilot symbol sequence, the phase rotation is achieved which is followed by IFFT operation and the PAPR evaluation. Each iteration leads to change in the statistical behaviour of the data symbols. The goal is to generate pilot-assisted (PA) O-OFDM (PAO-OFDM) signal which reduces coherent addition of OFDM subcarriers as much as possible. Hence, the PAPR which gives the minimum value after several iterations is selected as the best PAO-OFDM signal to be transmitted. Using this approach, PAPR can be reduced by several decibels, depending on number of pilots, number of subcarriers, and the threshold for PAPR which in turn improves overall performance of the system. Figure 14.10 shows typical O-OFDM deployment in OWC with PA addition.

In Reference 57, Ogunkoya *et al.* derive the closed-form approximation of the PAPR distributions of the PAO-OFDM signal using the maximum and minimum order statistics and the distribution of real-valued OFDM [58, 59]. To obtain the PAPR of a continuous time-domain OFDM signal, digital-to-analogue conversion process is followed by the IFFT operation. However, the PAPR of four times oversampled discrete time-domain signal has been generally accepted to approximate the continuous time-domain OFDM signal. Using the *maximum* order statistics, the CDF of the PAPR for O-OFDM frame can be derived. Since several iterations of pilot symbol sequence on a single PAO-OFDM frame generate several peaks, the *minimum* order statistics is then used to describe the complementary CDF (CCDF) of the PAO-OFDM frame which gives the smallest PAPR after rotating the phase data symbols with several pilots. Finally, using the metric of optical modulation index, it is shown that the results

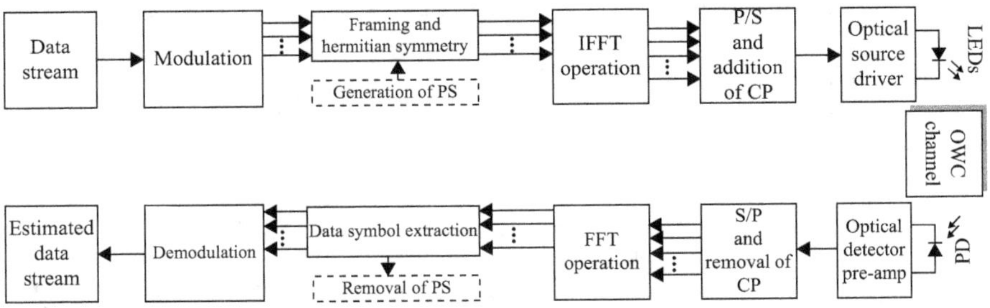

Figure 14.10 Block diagram illustrating the PAO-OFDM communication system; PS, pilot symbol; CP, cyclic prefix; P/S, parallel-to-serial conversion; S/P, serial-to-parallel conversion; and OWC, optical wireless channel

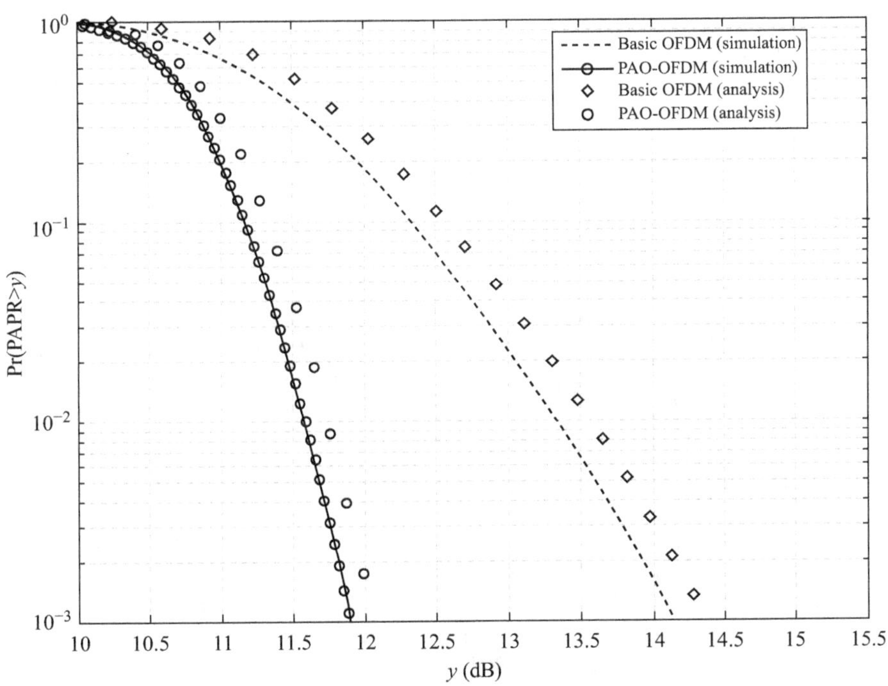

Figure 14.11 CCDF plot for basic and PAO-OFDM signal using simulation and analysis

obtained from the implementation of PAO-OFDM closely match theoretical prediction [57]. Figure 14.11 shows reduction in CCDF when PAO-OFDM is implemented as opposed to the basic OFDM. Even larger gains are possible for larger number of pilots.

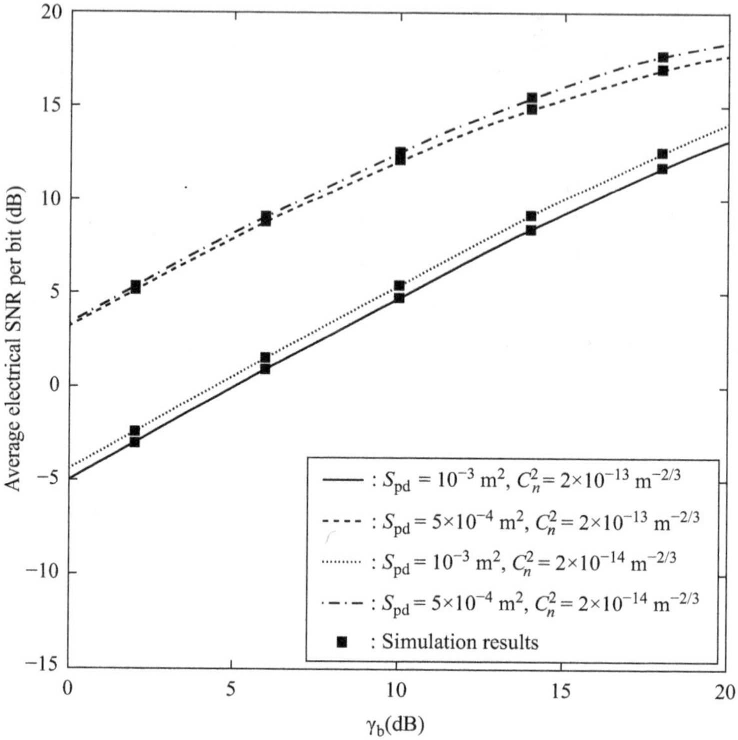

Figure 14.12 Average electrical SNR per bit is shown as a function of undistorted electrical SNR per bit, γ_b, combining the signal clipping effect and moderate ($C_n^2 = 2 \times 10^{-14}$ $m^{-2/3}$) or strong ($C_n^2 = 2 \times 10^{-13}$ $m^{-2/3}$) turbulence conditions. S_{pd} stands for the photosensitive area while C_n^2 is the refractive index structure parameter. Reproduced by permission of the Institution of Engineering and Technology. Full acknowledgment to the Nistazakis H. E., Stassinakis A.N., Sinanovic S., Popoola W. O., Tombras G. S. 'Performance of quadrature amplitude modulation orthogonal frequency division multiplexing-based free space optical links with non-linear clipping effect over gamma–gamma modelled turbulence channels.' IET Optoelectronics, 2015

14.7.3 Combined effects of non-linearities and turbulence in FSO communication

FSO communications offer the promise of achieving high-performance characteristics, such as achieving high data rate in the unregulated part of the spectrum while having relatively low cost of installation and operation. However, FSO links suffer from the challenging atmospheric channel which can severely affect their performance. In particular, scintillation effect, resulting from the atmospheric turbulence, causes the irradiance of the optical signal at the receiver to significantly fluctuate. Irradiance fluctuation depends on the atmospheric turbulence strength which is modelled by gamma-gamma distribution due to its ability to accurately represent a range of turbulence conditions shown in (14.9). In order to achieve the high rate, FSO link may use O-OFDM (FSO-OFDM) due to its well-known properties of having orthogonal subcarriers and its robustness to ISI. However, FSO-OFDM has its own significant challenges which affect its performance.

Beside the influence of the propagation channel in optical communication systems, signal distortions and clippings, due to the limited dynamic range and non-linearities in the circuitry as seen in earlier sections, also occur. Specifically, O-OFDM the signal can have very large PAPR resulting in further clipping at the optical front-end, similar to the other OFDM variants. The normally distributed OFDM signal and the fact that its non-linear distortion can be quantified by applying the Bussgang theorem imply that FSO-OFDM signal can be analysed as previously indicated in expressions (14.43)–(14.48). Although individual impairments such as FSO channel turbulence and clipping have been investigated separately, it is beneficial to have a unified analytical framework that combines these effects. It then becomes possible to investigate the *joint* effects of atmospheric turbulence on such OFDM-FSO links and the clipping due to non-linearities. This approach has been developed in Reference 60.

More specifically, accurate closed-form approximation for evaluating the performance of FSO-OFDM FSO in the presence of turbulence *and* clipping has been achieved by deriving SNR formula which has been used in outage and BER performance metrics. The theoretical results have been validated via simulation. Figure 14.12 shows how average electrical SNR per bit depends on undistorted electrical SNR per bit. The key ingredient in successfully combining the effects of non-linearities and turbulence in unified analytical framework has been Meijer-G function. Further results revealed a close match between derived outage and BER formulas and the simulation for various turbulence amounts, photodiode areas, and clipping conditions [60]. This approach can then easily be extended for the relaying deployments in order to fully account for non-linearities in such scenarios.

References

[1] A. G. Bell, "Selenium and the photophone," *The Electrician*. 1880; vol. 5, p. 214.

[2] X. Zhu, J. M. Kahn, "Free-space optical communication through atmospheric turbulence channels," *IEEE Transactions on Communications*. 2002; vol. 50, no. 8, pp. 1293–1300.

[3] H. Elgala, R. Mesleh, H. Haas, "Indoor optical wireless communication: potential and state-of-the-art," *IEEE Communications Magazine*. 2011; vol. 49, p. 56.

[4] R. M. Gagliardi, S. Karp, *Optical Communications*, John Wiley & Sons, New York, NY, 1995.

[5] S. M. Navidpour, M. Uysal, M. Kavehrad, "BER performance of free-space optical transmission with spatial diversity," *IEEE Transactions on Wireless Communications*. 2007; vol. 6, no. 8, pp. 2813–2819.

[6] M. Safari, S. Hranilovic, "Diversity and multiplexing for near-field atmospheric optical communications," *IEEE Transactions on Communications*. 2013; vol. 61, p. 1988.

[7] T. A. Tsiftsis, H. G. Sandalidis, G. K. Karagiannidis, N. C. Sagias, "Multihop free-space optical communications over strong turbulence channels," In *Proceedings of IEEE International Conference on Communications 2006 (ICC'06)*, Istanbul, Turkey.

[8] G. K. Karagiannidis, T. A. Tsiftsis, H. G. Sandalidis, "Outage probability of relayed free space optical communication systems," *Electronics Letters*. Aug. 2006; vol. 42, pp. 994–995.

[9] M. Safari, M. Uysal, "Relay-assisted free-space optical communication," *IEEE Transactions on Wireless Communications*. 2008; vol. 7, p. 5441.

[10] M. Safari, M. Rad, M. Uysal, "Multi-hop relaying over turbulent atmospheric Poisson channel: Outage analysis and optimization," *IEEE Transactions on Communications*. 2012; vol. 60, p. 817.

[11] M.A. Kashani, M. Safari, M. Uysal, "Optimal relay placement and diversity analysis of relay-assisted free-space optical communication systems," *Journal of Optical Communications and Networking*. 2013; vol. 5, p. 37.

[12] M.A. Kashani, M. Uysal, "Outage performance and diversity gain analysis of free-space optical multi-hop parallel relaying," *Journal of Optical Communications and Networking*. 2013; vol. 5, pp. 901–909.

[13] M. Karimi, M. Nasiri-kenari, "BER analysis of cooperative systems in free-space optical networks," *Journal of Lightwave Technology*. 2009; 27, pp. 5639–5647.

[14] M.A. Kashani, M. Rad, M. Safari, M. Uysal, "All-optical amplify-and-forward relaying system for atmospheric channels," *IEEE Communications Letters*. 2012; vol. 16, p. 1684.

[15] M. Karimi, M. Nasiri-Kenari, "Free space optical communications via optical amplify-and-forward relaying," *IEEE/OSA Journal of Lightwave Technology*. 2011; vol. 29, pp. 242–248.

[16] S. Kazemlou, S. Hranilovic, S. Kumar, "All-optical multihop free-space optical communication systems," *Journal of Lightwave Technology*. 2011; vol. 29, no. 18, pp. 2663–2669.

[17] E. Bayaki, E. Michalopoulos, R. Schober, "EDFA-based all-optical relaying in free-space optical systems," *IEEE Transactions on Communications*. 2012; vol. 60, no. 12, pp. 3797–3807.

[18] M. Safari, M. Uysal, "Relay-assisted quantum-key distribution over long atmospheric channels," *IEEE/OSA Journal of Lightwave Technology*. 2009; vol. 27, p. 4508.

[19] J. H. Shapiro, "Imaging and optical communication through atmospheric turbulence," In *Laser Beam Propagation in the Atmosphere*, J. W. Strohbehn (Ed.), Berlin: Springer-Verlag, 1978, chapter: 6.E.

[20] L. C. Andrews, R. L. Phillips, C. Y. Hopen, *Laser Beam Scintillation with Applications*, SPIE Press, Bellingham, WA, 2001.

[21] M. Safari, M. Uysal, "Do we really need OSTBCs for free-space optical communication with direct detection?" *IEEE Transactions on Wireless Communications*. 2008; vol. 7, p. 4445.

[22] M. Safari, M. Uysal, "Cooperative diversity over log-normal fading channels: Performance analysis and optimization," *IEEE Transactions on Wireless Communications*. 2008; vol. 7, p. 1963.

[23] M. Safari, M. Uysal, "Diversity gain analysis of free-space optical communication systems," In *Proceedings of Canadian Conference on Electrical and Computer Engineering (CCECE'08)*, Niagara Falls, Canada, 2008; pp. 1239–1244.

[24] N. Gisin, G. Ribordy, W. Tittel, H. Zbinden, "Quantum cryptography," *Reviews of Modern Physics*. 2002; vol. 74, p. 145.

[25] C. H. Bennett, G. Brassard, "Quantum cryptography: Public key distribution and coin tossing," In *Proceedings of the IEEE International Conference on Computer Systems and Signal Processing*, Bangalore, India, 1984; pp. 175–179.

[26] H. J. Briegel, W. Dür, J. I. Cirac, P. Zoller, "Quantum repeaters: The role of imperfect local operations in quantum communication," *Physical Review Letters*. 1998; vol. 81, pp. 5932–5935.

[27] H. Bechmann-Pasquinucci, A. Pasquinucci, "Quantum key distribution with trusted quantum relay," 2005; eprint: quant-ph/0505089.

[28] M. Aspelmeyer, T. Jennewein, M. Pfenningbauer, W. R. Leeb, A. Zeilinger, "Long-distance quantum communication with entangled photons using satellites," *IEEE Journal of Selected Topics in Quantum Electronics*. 2003; vol. 9, pp. 1541–1551.

[29] J. W. Hardy, *Adaptive Optics for Astronomical Telescopes*, Oxford University Press, Oxford, 1998.

[30] J. H. Shapiro, "Near-field turbulence effects on quantum-key distribution," *Physical Review A*. 2003; vol. 67, no. 2, p. 022309.

[31] D. Slepian, "Analytic solution of two apodization problems," *Journal of Optical Society of America*. 1965; vol. 55, pp. 1110–1115.

[32] E. Lee, L. Park, D. Han, G. Yoon, "Performance analysis of the asymmetric dual-hop relay transmission with mixed RF/FSO links," *IEEE Photonics Technology Letters*. 2011; vol. 23, no. 21, pp. 1642–1644.

[33] Z. Ghassemlooy, W. Popoola, S. Rajbhandari, *Optical Wireless Communications: System and Channel Modelling with MATLAB*. Boca Raton, FL: CRC Press, ISBN-10: 1439851883, August 2012.

[34] T. Ohtsuki, "Multiple-subcarrier modulation in optical wireless communications," *IEEE Communications Magazine*. 2003; vol. 41, no. 3, pp. 74–79.

[35] I. B. Djordjevic, B. Vasic, "100-Gb/s transmission using orthogonal frequency division multiplexing," *IEEE Photonics Technology Letters*. 2006, vol. 18, no. 15, pp. 1576–1578.

[36] G. P. Agrawal, *Fiber-Optic Communication Systems*, 3rd ed. New York, NY: Wiley-Interscience, 2002.

[37] W. O. Popoola, "Subcarrier intensity modulated free-space optical communication systems," Ph.D. Thesis, 2009.

[38] R. You, J. M. Kahn, "Average power reduction techniques for multiple-subcarrier intensity-modulated optical signals," *IEEE Transaction on Communications*. 2001, vol. 49, no. 12, pp. 2164–2171.

[39] S. Teramoto, T. Ohtsuki, "Multiple-subcarrier optical communication systems with peak reduction carriers," In *Proceedings of IEEE Global Telecommunications Conference (GLOBECOM)*, San Francisco, USA, 2003; pp. 3274–3278.

[40] S. Teramoto, T. Ohtsuki, "Multiple-subcarrier optical communication systems with subcarrier signal-point sequence," *IEEE Transactions on Communications*. 2005; vol. 53, no. 10, pp. 1738–1743.

[41] J. M. Kahn, J. R. Barry, "Wireless infrared communications," *Proceedings of IEEE*. 1997; vol. 85, no. 2, pp. 265–298.

[42] J. B. Carruthers, J. M. Kahn, "Multiple-subcarrier modulation for nondirected wireless infrared communication," *IEEE Journal of Selected Areas of Communication*. 1996; vol. 14, no. 3, pp. 538–546.

[43] S. Dimitrov, S. Sinanovic, H. Haas, "Clipping noise in OFDM based optical wireless communication systems," *IEEE Transactions on Communications*. 2012; vol. 60, no. 4, pp. 1072–1081.

[44] J. Armstrong, A. Lowery, "Power efficient optical OFDM," *Electronic Letters*. 2006; vol. 42, no. 6, pp. 370–372.

[45] S. C. J. Lee, S. Randel, F. Breyer, A. M. J. Koonen, "PAM-DMT for intensity-modulated and direct-detection optical communication systems," *IEEE Photonics Technology Letters*. 2009; vol. 21, no. 23, pp. 1749–1751.

[46] D. Tsonev, S. Sinanovic, H. Haas, "Novel unipolar orthogonal frequency division multiplexing (U-OFDM) for optical wireless," *Proceedings of IEEE Vehicular Technology Conference*, Yokohama, Japan, May 2012. pp. 1–5.

[47] J. Rice, *Mathematical Statistics and Data Analysis*, 2nd ed. Pacific Grove, CA: Duxbury Press; 1995.

[48] D. Dardari, V. Tralli, A. Vaccari, "A theoretical characterization of nonlinear distortion effects in OFDM systems," *IEEE Transactions on Communications*. 2000; vol. 48, no. 10, pp. 1755–1764.

[49] J. Bussgang, "Cross correlation function of amplitude-distorted Gaussian signals," Research Laboratory for Electronics, Massachusetts Institute of Technology, Cambridge, MA, Technical Report 216, March 1952.

[50] H. E. Rowe, "Memoryless nonlinearities with Gaussian inputs: Elementary results," *Bell System Technical Journal*. 1982; vol. 61, no. 7, pp. 1519–1525.

[51] S. Randel, F. Breyer, S. C. J. Lee, J. W. Walewski, "Advanced modulation schemes for short-range optical communications," *IEEE Journal of Selected Topics in Quantum Electronics*. 2010; vol. 16, no. 5, pp. 1280–1290.

[52] S. C. J. Lee, F. Breyer, S. Randel, H. P. A. van der Boom, A. M. J. Koonen, "High-speed transmission over multimode fiber using discrete multitone modulation," *Journal of Optical Networking*. 2008; vol. 7, no. 2, pp. 183–196.

[53] D. Tsonev, S. Sinanovic, H. Haas, "Complete modeling of nonlinear distortion in OFDM-based optical wireless communication," *Journal of Lightwave Technology*. 2013; vol. 31, no. 18, pp. 3064–3076.

[54] S. Dimitrov, S. Sinanović, H. Haas, "Signal shaping and modulation for optical wireless communication," *Journal of Lightwave Technology*; 2012. vol. 30, no. 9, pp. 1319–1328.

[55] L. Chen, B. Krongold, J. Evans, "Theoretical characterization of nonlinear clipping effects in IM/DD optical OFDM systems," *IEEE Transactions on Communications*. 2012; vol. 60, no. 8. pp. 2304–2312.

[56] W. O. Popoola, Z. Ghassemlooy, B. G. Stewart, "Pilot-assisted PAPR reduction technique for optical OFDM communication systems," *Journal of Lightwave Technology*. 2014; vol. 32, no. 7, pp. 1374–1382.

[57] F. B. Ogunkoya, W. O. Popoola, A. Shahrabi, S. Sinanović, "Performance evaluation of pilot-assisted PAPR reduction technique in optical OFDM systems," *IEEE Photonics Technology Letters*. 2015; vol. 27, no. 10, pp. 1088–1091.

[58] G. Casella, R. L. Berger, *Statistical Inference*, 2nd ed. Pacific Grove, CA: Duxbury Press; 2002.

[59] H. Yu, M. Chen, G. Wei, "Distribution of PAR in DMT systems," *Electronic Letters*. 2003; vol. 39, no. 10, pp. 799–801.

[60] H. E. Nistazakis, A.N. Stassinakis, S. Sinanovic, W. O. Popoola, G. S. Tombras, "Performance of quadrature amplitude modulation orthogonal frequency division multiplexing-based free space optical links with non-linear clipping effect over gamma–gamma modelled turbulence channels," *Optoelectronics, IET*, vol. 9, no. 5, pp. 269–274, October 2015.

Chapter 15

Multibeam joint processing satellites: cooperative relays, high above

Dimitrios Christopoulos, Stefano Andrenacci, Danilo Spano, Symeon Chatzinotas, Jens Krause and Björn Ottersten

15.1 An introduction to multibeam satellite networks

In satellite networks, information is transmitted either from the GateWay (GW) to the users, i.e., the Forward Link (FL), or from the users back to the GW, i.e., the Return Link (RL). Each of these paths is comprised of two wireless links, namely the uplink, connecting a ground segment (GW or user terminal) with the satellite and the downlink, vice versa. A diagram of a typical satellite network is given in Figure 15.1, where the wireless links of issue are illustrated. The notion of a feedback link is also given, which can be established either over the satellite RL or over other networks (e.g., terrestrial networks can offer low-rate links).

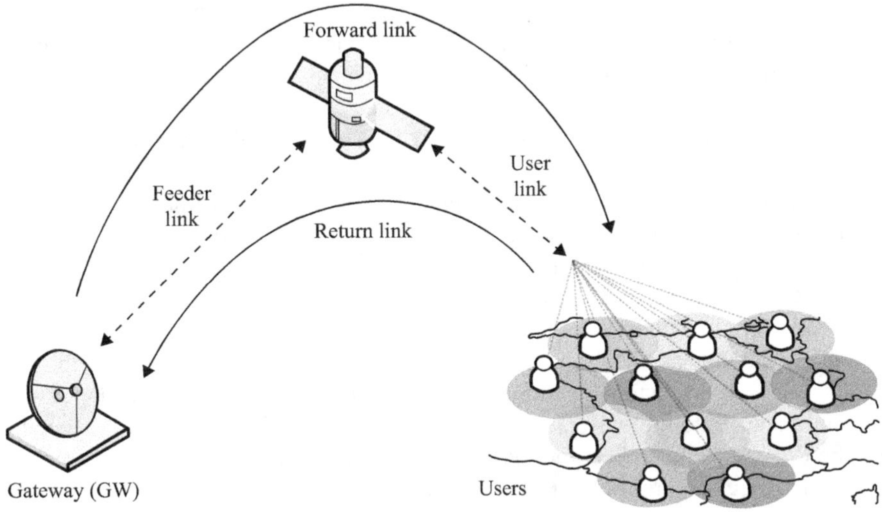

Figure 15.1 Multibeam satellite system

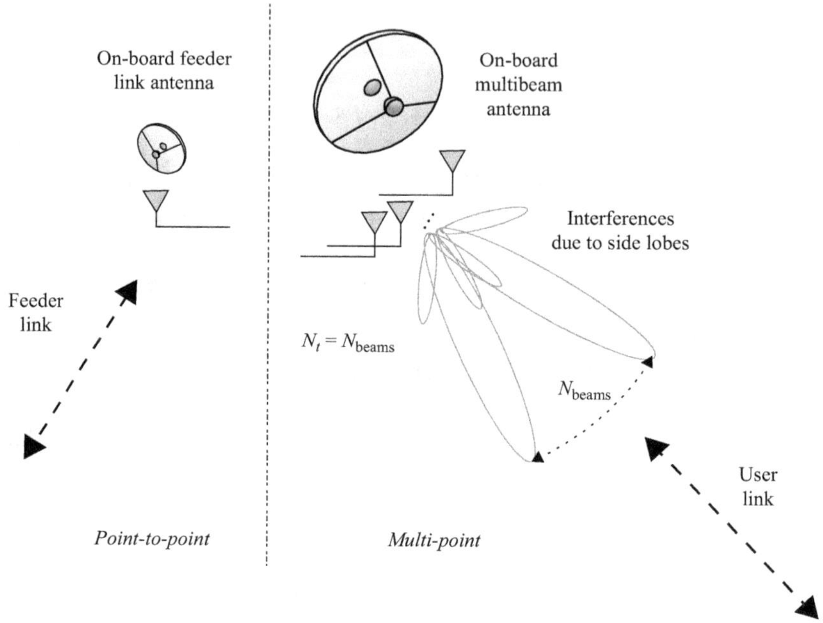

Figure 15.2 Multibeam antenna

The focus of this chapter is on increasing the offered throughput of the satellite network. The bottleneck of this network lies in the user link where a large number of users are competing over a limited amount of resources. A first step to reduce this obstacle has been the introduction of multibeam antennas that facilitate frequency reuse, by spatially separating co-channel users. Due to the antenna imperfections, however, residue interference still limits the system. A simple example of an on-board multibeam antenna is presented in Figure 15.2. Predominantly in the higher frequency bands, a large number of recent satellite procurements have clearly confirmed the multibeam trend as a broadband reference system architecture. Examples include systems such as Wildblue-1 [1] and Anik F2 [2] (66Ka-band spot beams) covering parts of America and many others.

In a multibeam system, the fundamental attribute of the user link is the existence of multiple antennas at the satellite that are simultaneously serving multiple users over the coverage. Therefore, a multiuser multiple antenna channel is realized. In the RL, this can be viewed as a multiuser single input (single antenna users are assumed) multiple output (the multibeam antenna is the receiver) (SIMO) channel. In the FL, equivalently as a Multiple Input Single Output (MISO) channel. Since the total capacity of the relay channel is limited by the channel hop with the worst conditions, in a multibeam satellite system, the feeder link (i.e., the link between the satellite and the GW) is can also become the bottleneck. Since the feeder link is a point-to-point link, frequency multiplexing is necessary. Subsequently, an increase in the user sum rate requires proportional increase in the bandwidth of the feeder link. Higher frequency bands (e.g., Q/V bands) for this case are being considered in the design of the future terabit satellite [3–5]. Nevertheless, to provide extremely high throughput in the user link (i.e., the link between the satellite and the users), novel techniques for the feeder link need to be developed. In this direction, the deployment of multiple GWs to serve a single satellite is being considered [6]. In the same direction, multiple gateway (multi-GW) scenarios are also discussed in this chapter.

The key advantage of applying multiple antenna signal processing in satcoms lies in the inherent nature of multibeam satellite networks: a large number of antennas are illuminating a vast coverage area, while all RL signals are processed in one or more central locations, namely the GW. A transparent multibeam satellite illuminating central Europe with 12 beams is already shown Figure 15.1. For a typical multibeam satellite covering Europe, more than 200 beams are expected to be deployed. The SoA techniques developed and evaluated in this chapter are motivated by the inherent characteristics of satellite communications which impose specific constraints. An overview of these constraints is given in the following sections.

15.2 Practical constraints

Albeit the throughput enhancement provided by cooperative techniques in satellite networks, as in detail presented in the following, several issues arise with the adoption of these techniques in satcoms and need to be addressed. These practical constraints are identified in the following paragraphs.

15.2.1 Channel state information acquisition

Channel state information (CSI) is one of the most important enablers for the application of the multibeam joint processing techniques. In the context of satcoms, frequency division multiplexing prevents any reciprocity between the FL and the RL. Therefore, channel knowledge can only be acquired with dedicated pilots that are received by the end-user ground terminals. Based on the specific precoded and un-precoded sequences, as it will be detailed in the following, the channel is estimated at the terminals and then fed back to the GW. In more detail, CSI is acquired by broadcasting pilot signals through the FL to all terminals which in turn measure them and feed the quantized measurements back to the GW through the RL.

Furthermore, the CSI acquisition process in satcoms introduces a long delay which may result in outdated CSI. This complication is especially acute for the FL where CSI is needed before transmission in order to calculate the precoding vectors. As a direct consequence, in the following sections, we will only focus on fixed terminals for the FL. This slow fading channel enables the practical implementation of joint processing since CSI is easier to acquire.

15.2.2 Frequency reuse

In general, the aggressive reuse of the PHY resources is bound to increase the on-board payload requirements. Conventional payloads typically employ Travelling Wave Tube Amplifiers (TWTAs) as on-board High Power Amplifiers (HPAs). Despite its reliability, this equipment significantly burdens the multibeam satellite in terms of weight and power requirements. To optimize the payloads, multiple beams are amplified by a single HPA. In more detail, a four-color frequency reuse scheme is commonly realized by splitting the available frequency/polarization resources into four colors: half the available bandwidth and one circular polarization (right hand – RHCP or left hand – LHCP) define one color. Then adjacent beams are allocated a different color (i.e., polarization and frequency combination), so that an adequate signal-to-interference-plus-noise ratio (SINR) can be attained at the receive side. As a result, an x number of beams will reuse $x/4$ times the available PHY resources. In this configuration, since half of the total user link bandwidth is used in each beam, a single TWTA can be dedicated to every two beams. By advancing to more aggressive reuse schemes, e.g., two-color reuse, some of the adjacent beams will be operating over the same resources, thus greatly increasing the intra-system interference. In this case, x beams will provide $x/2$ times the PHY resources at the expense of requiring two TWTA per beam, one for each polarization. By pushing the reuse to the extreme, a full frequency reuse scheme provides the highest possible linear increase of the PHY resources with the number of antennas, since x beams will reuse all resources x times. Nevertheless, all adjacent beams will be co-channel and interference will tremendously increase. Also, two TWTA per beam will be necessary. The only plausible realization of such scenarios is via joint multibeam signal processing as it will be shown in the following.

Since signal processing methods will be designed to mitigate the increased interference, the second fundamental limitation for the application of aggressive frequency reuse, namely the added on-board complexity, needs to be addressed. The heavy TWTAs hinder the deployment of full frequency reuse payloads that accommodate a large number of antennas. Nevertheless, future multibeam payloads compatible with full frequency are enabled by the already developed low cost, efficient, reliable, and low-weight Solid State Power Amplifiers (SSPAs). The payload complexity increase is illustrated in Figure 15.3. In this figure, the total power consumption of each payload

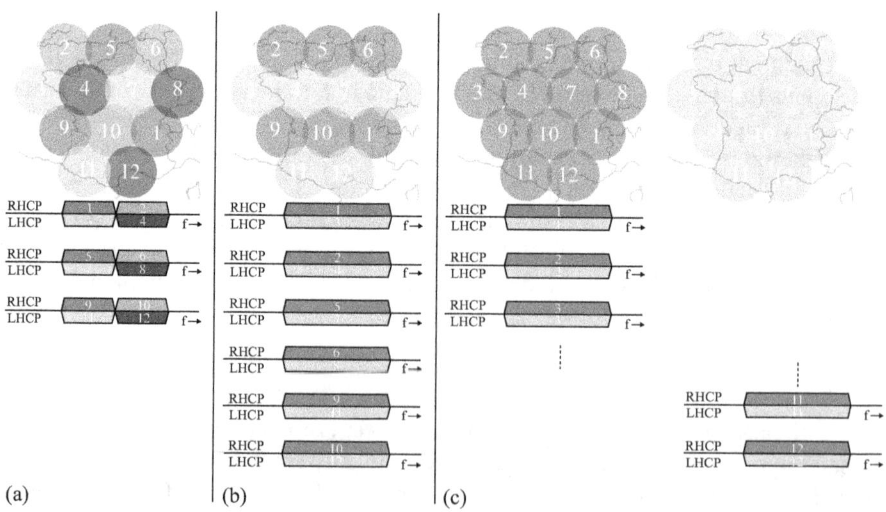

Figure 15.3 Payload complexity increase due to aggressive frequency reuse

architecture is not shown. For a fair comparison, the TWTAs of scenario (c) need to be operating at 4 times lower power than the ones of scenario (a).

15.2.3 Feeder link

An increase in the capacity of the user link of the multibeam system can only be enabled by a proportional increase in the capacity of the feeder link. Therefore, the need to deploy multiple GW to cover the feeder link requirements is emanated. High-throughput satellite systems have to be supported by multiple GW. Due to feeder link limitations, one single GW cannot accommodate the total number of employed beams. In this chapter, to perform multibeam joint processing, centralized processing is commonly assumed, unless stated otherwise. This theoretical assumption can be supported by a real system implementation via two approaches. One solution would be the exploitation of higher frequency bands for the feeder link (optical feeder links), assuming that such a system can be practically employed. As a result, a single GW could serve the multibeam satellite system. Alternatively, another approach is the full interconnection amongst the multiple GWs so that they all share the same information (CSI and data). The second approach is easier to implement if we consider the bandwidth capabilities of broadband cable networks. Of course, the added delay is an issue to be considered, especially in the satcoms context where delay is already a major issue. Subsequently, both approaches lead to the verification of the simplistic assumption of cooperative system that utilizes a central precoder/decoder. Additionally, this chapter also touches upon the subject of decentralized precoding. Works in the existing literature examine the performance degradation effects of the adoption of decentralized precoder designs, for the case where full GW interconnection cannot be assumed. An example of such an approach for multibeam satellite systems can be found in Reference 6 where the level of cooperation amongst GW is examined and the most promising technique is shown to be partial data and CSI exchange amongst the interconnected GW.

15.2.4 Framing

The rigid framing structure of satcoms standards introduces constraints in the application of signal processing in satcoms. The PHY design of satcoms standards encompasses long forward error correction (FEC) codes to cope with noise limited channels and long propagation delays. Besides block coding, synchronization issues and fade mitigation techniques that rely on an adaptive link layer design (Adaptive Coding and Modulation – ACM) further necessitate this structure. More specifically, DVBS2 is the latest generation standard for satcoms enabling broadband and interactive services via satellite. It has been designed for broadcasting services (standard and high-definition TV), Internet, and professional services such as TV contribution links and digital satellite news gathering. During the formulation of DVBS2, three main concepts were carefully considered: (a) best transmission performance approaching the Shannon limit, (b) total flexibility, and (c) reasonable receiver complexity. High performance and low-complexity iterative decoding schemes like LDPC codes along with high-order APSK modulations were adopted for efficient operation over the non-linear satellite channel in the quasi-error-free region. Compared to previous standards, the second generation standard attains 20%–35% capacity increase or alternatively 2–2.5 dB more robust reception for the same spectrum efficiency by virtue of the advanced waveforms. Furthermore, to facilitate the provision of interactive services, the standard features operation under a great range of ACM parameters. When used for interactive services, ACM allows optimization of the transmission parameters adaptive to varying path conditions. The optimization of the transmission parameters is also possible in combination with TDM data for multiple receivers. Transmission parameters can change instantaneously one a frame-by-frame basis. Hence, resources are optimally exploited, since operation under a constant fading margin according to a worst-case scenario design is no longer necessary.

Focusing on the most recent extensions of the second generation digital video broadcasting for satellite standard, DVB-S2X [7], specific modifications allow for the application of precoding. The latest evolution of satcoms standards suggests PHY frames that can be perfectly aligned in time over the beams. Consequently, precoding can be applied over this optimized PHY design.[1] Nevertheless, the unequal data payloads that correspond to each user, along with the variable number of users per frame, hinder the calculation of a precoding matrix on a user-by-user basis. Therefore, despite the optimality, in the channel capacity sense, of channel-by-channel precoding, practical system implementations impose a frame-based precoding design.

15.2.5 The non-linear satellite channel

This chapter also discusses the non-linear dependence of the input versus the output power of the efficient and reliable amplifiers that drive each satellite antenna. The non-linear amplification carried out over the satellite is emanated from a power and cost efficiency perspective. On-board power is a scarce resource and therefore operating the amplifiers close to saturation is of utmost importance. Notwithstanding the fact that a great deal of investment is being put to equipment linearization, one has always to keep two points in mind. First, the precoded waveforms have a significantly different structure and as a result, even the slightest non-ideality can lead to unexpected behavior. Second, the perfect linearization always comes at a cost. As a result, the development of cost efficient methods with significant performance over non-linearized amplifiers can prove valuable. Subsequently, this chapter aims at providing a first evaluation of the effects of a practical satellite channel on generic linearly precoded signals.

15.3 System model

The focus is on a single broadband multibeam satellite relaying signals to multiple single antenna users. Let N_t denote the number of transmitting elements, which for the

[1] It should be noted that the PHY design is optimized to cope with the inherent attributes of the satellite channel such as long propagation delay and low signal power and is therefore challenging to modify it in favor of the signal processing methods.

purposes of the present work, are considered equal to the number of beams (one feed per beam assumption) and N_u the total number of users simultaneously served. The received signal at the ith user will read as $y_i = \mathbf{h}_i^\dagger \mathbf{x} + n_i$, where $\mathbf{h}_i^\dagger$ is a $1 \times N_t$ vector composed of the channel coefficients (i.e., channel gains and phases) between the ith user and the N_t antennas of the transmitter, $\mathbf{x}$ is the $N_t \times 1$ vector of the transmitted symbols and n_i is the complex circular symmetric (c.c.s.) independent identically distributed (i.i.d.) zero mean additive white Gaussian noise (AWGN), measured at the ith user's receiver. Herein, for simplicity, the noise will be normalized to one and the impact of noise at the receiver side will be incorporated in the channel coefficients, as will be shown in the following section (see (15.4)).

15.3.1 Multicast model

Let us assume that a total of N_t multicast groups are realized where $\mathcal{I} = \{\mathcal{G}_1, \mathcal{G}_2, \ldots, \mathcal{G}_{N_t}\}$ the collection of index sets and $\mathcal{G}_k$ the set of users that belong to the kth multicast group, $k \in \{1, \ldots, N_t\}$. Each user belongs to only one frame (i.e., group), thus $\mathcal{G}_i \cap \mathcal{G}_j = \emptyset$, $\forall i, j \in \{1, \ldots, N_t\}$, while $\rho = N_u/N_t$ denotes the number of users per group. Let $\mathbf{w}_k \in \mathbb{C}^{N_t \times 1}$ denotes the precoding weight vector applied to the transmit antennas to beamform towards the kth group of users. By collecting all user channels in one channel matrix, the general linear signal model in vector form reads as $\mathbf{y} = \mathbf{H}\mathbf{x} + \mathbf{n} = \mathbf{H}\mathbf{W}\mathbf{s} + \mathbf{n}$, where $\mathbf{y}$ and $\mathbf{n} \in \mathbb{C}^{N_u}$, $\mathbf{x} \in \mathbb{C}^{N_t}$ and $\mathbf{H} \in \mathbb{C}^{N_u \times N_t}$. Since, the frame-based precoding imposes a single precoding vector for multiple users, the matrix will include as many precoding vectors (i.e., columns) as the number of multicast groups. This is the number of transmit antennas, since one frame per antenna is assumed. Also, the symbol vector includes a single equivalent symbol for each frame, i.e., $\mathbf{s} \in \mathbb{C}^{N_t}$, in line with the multicast assumptions. Consequently, a square precoding matrix is realized, i.e. $\mathbf{W} \in \mathbb{C}^{N_t \times N_t}$. The assumption of independent information transmitted to different frames implies that the symbol streams $\{s_k\}_{k=1}^{N_t}$ are mutually uncorrelated. Also, the average power of the transmitted symbols is assumed normalized to one. Therefore, the total power radiated from the antenna array is equal to

$$P_{tot} = \sum_{k=1}^{N_t} \mathbf{w}_k^\dagger \mathbf{w}_k = \text{Trace}\left(\mathbf{W}\mathbf{W}^\dagger\right) \tag{15.1}$$

where $\mathbf{W} = [\mathbf{w}_1, \mathbf{w}_2, \ldots, \mathbf{w}_{N_t}]$. The power radiated by each antenna element is a linear combination of all precoders and reads as [8]

$$P_n = \left[\sum_{k=1}^{N_t} \mathbf{w}_k \mathbf{w}_k^\dagger\right]_{nn} = \left[\mathbf{W}\mathbf{W}^\dagger\right]_{nn} \tag{15.2}$$

where $n \in \{1, \ldots, N_t\}$ is the antenna index. The fundamental difference between the sum power constraint of (15.1) and the proposed per-antenna constraint (PAC) is clear in (15.2), where instead of one, N_t constraints are realized, each one involving all the precoding vectors.

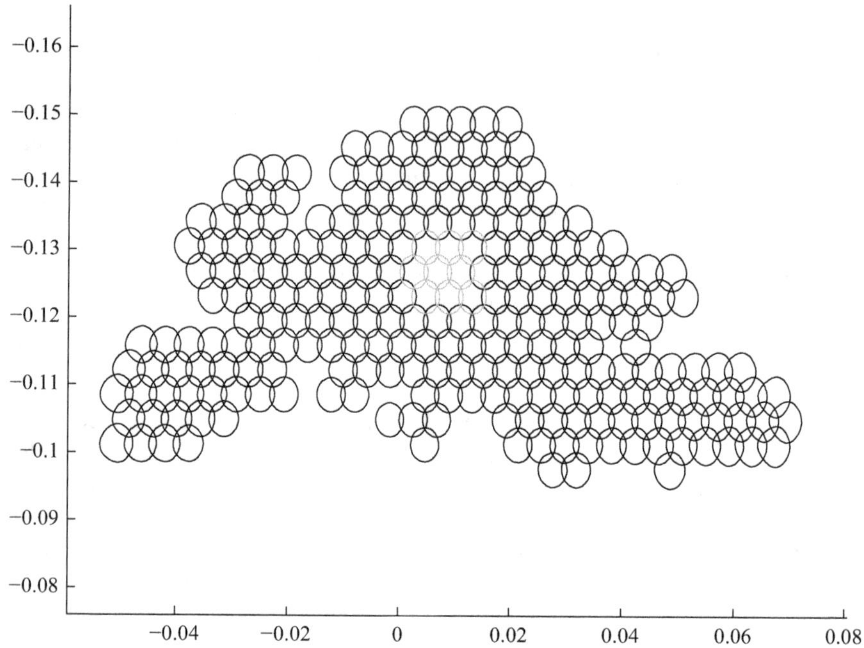

Figure 15.4 Beam pattern covering Europe, with the nine beams considered herein highlighted

15.3.2 Multibeam satellite channel

The above general system model is applied over a multibeam satellite channel explicitly defined as follows. A 245 beam pattern that covers Europe is employed [12]. For the purposes of the present work, only a subset of the 245 beams will be considered, as presented in Figure 15.4. Such a consideration is in line with the multi-GW assumptions of large multibeam systems. The effects of interference from adjacent clusters will also be briefly discussed. A complex channel matrix that models the link budget of each user as well as the phase rotations induced by the signal propagation is employed [12].

In more detail, the total channel matrix $\mathbf{H} \in \mathbb{C}^{N_u \times N_t}$ is generated as

$$\mathbf{H} = \mathbf{\Phi B} \tag{15.3}$$

and includes the multibeam antenna pattern (matrix $\mathbf{B}$) and the signal phase due to different propagation paths between the users (matrix $\mathbf{\Phi}$). The real matrix $\mathbf{B} \in \mathbb{R}^{N_u \times N_t}$ models the satellite antenna radiation pattern, the path loss, the receive antenna gain, and the noise power. Its i, jth entry is given by Reference 12:

$$b_{ij} = \left(\frac{\sqrt{G_{\mathrm{R}} G_{ij}}}{4\pi \left(d_k \cdot \lambda^{-1}\right) \sqrt{\kappa T_{\mathrm{cs}} B_u}} \right) \tag{15.4}$$

with d_k the distance between the ith user and the satellite (slant range), λ the wavelength, κ the Boltzmann constant, T_{cs} the clear sky noise temperature of the receiver, B_u the user link bandwidth, G_{R} the receiver antenna gain, and G_{ij} the multibeam antenna gain between the ith single antenna user and the jth on-board antenna (= feed). Hence, the beam gain for each satellite antenna–user pair depends on the antenna pattern and on the user position.

An inherent characteristic of the multibeam satellite channel is the high correlation of signals at the satellite side. Thus, a common assumption in multibeam channel models is that each user will have the same phase between all transmit antennas due to the long propagation path. The identical phase assumption between one user and all transmit feeds is supported by the relatively small distances between the transmit antennas and the long propagation distance of all signals to a specific receiver. Hence, in (15.3), the diagonal square matrix $\mathbf{\Phi}$ is generated as $[\mathbf{\Phi}]_{xx} = e^{\text{j}\phi_x}$, $\forall x = 1, \dots, N_u$ where ϕ_x is a uniform random variable in $[2\pi, 0)$ and $[\mathbf{\Phi}]_{xy} = 0$, $\forall x \neq y$.

15.3.3 Average user throughput

Based on the above link budget considerations, the achievable average user throughput is normalized over the number of beams, in order to provide a metric comparable with multibeam systems of any size. Therefore, the average user throughput, R_{avg} as will be hereafter referred to, is given as

$$R_{\text{avg}} = \frac{2B_u}{1+\alpha} \frac{1}{N_t} \sum_{k=1}^{N_t} f_{\text{DVB-S2X}} \left(\min_{i \in \mathcal{G}_k} \{\text{SINR}_i\}, \mathbf{t} \right) \tag{15.5}$$

in Gbps/beam, where all parameters are defined in Table 15.1. In (15.5), the spectral efficiency function $f_{\text{DVB-S2X}}$ receives as input each users SINR as well as a threshold vector $\mathbf{t}$. Then, $f_{\text{DVB-S2X}}$ performs a rounding of the input SINR to the closest lower floor given by the threshold vector $\mathbf{t}$ and outputs the corresponding spectral efficiency in bps/Hz. This operation is denoted as $\lfloor \cdot \rfloor_{\mathbf{t}}$. The mapping of receive SINR regions to a spectral efficiency achieved by a respective modulation and coding (MODCOD) scheme is explicitly defined in the latest evolution of the SatComs standards [7]. It should also be noted that the conventional four-color frequency reuse calculations are based on the exact same formula, with the only modifications being the input SINR, calculated under conventional four-color reuse pattern and with the pre-log factor reduced by four times, equal to the conventional fractional frequency reuse.

15.4 Sum rate and throughput maximization

For the precoding design, optimal multigroup multicast precoders under per-antenna constraints are proposed to maximize the throughput of the multibeam satellite system. The design of throughput maximizing optimal precoders is a complicated problem without an explicit solution even for the unicasting case [9]. When advancing to multicasting assumptions, the structure of the problem becomes even more

Table 15.1 Link budget parameters

Parameter	Value
Frequency band	Ka (20 GHz)
User terminal clear sky temperature, T_{cs}	235.3 K
User link bandwidth, B_u	500 MHz
Output back off, OBO	5 dB
On-board power, P_{tot}	50 dBW
Roll off, α	0.20
User terminal antenna gain, G_R	40.7 dBi
Multibeam antenna gain, G_{ij}	Reference 12

involved, as already explained [10]. Consequently, heuristic methods for the sum-rate maximization have been developed in References 11 and 12.

Since a multigroup multicasting scenario entails the flexibility to maximize the total system rate by providing different service levels amongst groups, the multigroup multicast max SR optimization aims at increasing the minimum SINR within each group while in parallel maximizing the sum of the rates of all groups. Intuitively, this can be accomplished by reducing the SINR of users with better conditions than the worst user of their group. Also, groups that contain compromised users might need to be turned off and their users driven to service unavailability to save power resources and degrees of freedom. Subsequently, power is not consumed for the mitigation of poor channel conditions. Any remaining power budget is then reallocated to well conditioned and balanced in terms of performance groups.

15.4.1 Per-antenna power constrained sum-rate maximization

This section focuses on the per-antenna power constrained max SR problem, formally defined as

$$\mathcal{SR}: \max_{\{\mathbf{w}_k\}_{k=1}^{N_t}} \sum_{i=1}^{N_u} \log_2(1+\gamma_i)$$

$$\text{subject to: } \gamma_i = \min_{m\in\mathcal{G}_k} \frac{\left|\mathbf{w}_k^\dagger \mathbf{h}_m\right|^2}{\sum_{l\neq k}^{N_t}\left|\mathbf{w}_l^\dagger \mathbf{h}_m\right|^2 + \sigma_m^2}, \tag{15.6}$$

$$\forall i \in \mathcal{G}_k, k, l \in \{1,\ldots,N_t\}$$

$$\text{and to: } \left[\sum_{k=1}^{N_t} \mathbf{w}_k \mathbf{w}_k^\dagger\right]_{nn} \leq P_n, \tag{15.7}$$

$$\forall n \in \{1,\ldots,N_t\}$$

Problem $\mathcal{SR}$ receives as input the channel matrices as well as the per-antenna power constraint vector $\mathbf{p}_{\text{ant}} = [P_1, P_2, \ldots, P_{N_t}]$. Following the notation of Reference 13 for ease of reference, the optimal objective value of $\mathcal{SR}$ will be denoted as $c^* = \mathcal{SR}(\mathbf{p}_{\text{ant}})$ and the associated optimal point as $\{\mathbf{w}_k^{\mathcal{SR}}\}_{k=1}^{N_t}$.

15.4.2 Throughput maximization under availability constraints

Constraints inspired by the inherent nature of satellite communications emanate the definition of novel optimization problems. This paragraph focuses on enabling demanding in terms of availability satellite services. Increased skepticism over spectrally efficient, aggressive frequency reuse multibeam satellites stems from the effects of such configurations on the SINR distribution across the coverage. In full frequency reuse scenarios, the useful signal power at the receiver is greatly reduced due to the intra-system interferences. Despite the throughput gains due to the increased user link bandwidth and the adequate management of interferences by linear precoding, the mean and variance of the SINR distribution over the coverage area is generally reduced. This is the price paid for increasing the frequency reuse. Naturally, this reduction in the average SINR will lead to a higher utilization of lower MODCODs and increase the probability of service unavailability over the coverage (outage probability). Retransmissions that incur in these outage instances are bound to burden the system in terms of efficiency. What is more, by acknowledging the multiuser satellite environment, these outage periods can potentially become comparable to the inherent long propagation delay of SatComs. Such a case will render the overall delay experienced by the user, unacceptable. As a result, the probability of compromised users to experience long outage periods needs to be considered. In this work, the introduction of minimum rate constraints (MRCs) over the entire coverage is proposed, as a means to guarantee the stringent availability requirements typically accustomed in SatComs. The guarantee of a minimum level of service availability is introduced for the first time in a max SR multigroup multicast optimization.

15.4.2.1 Sum-rate maximization under MRCs

To the end of providing high service availability, the gains of the sum rate optimization can be traded-off in favor of a minimum guaranteed rate across the coverage. This trade-off mostly depends on the minimum MODCOD supported by the ACM.[2]

Since an intermediate solution between the fairness and the sum rate optimizations is of high engineering interest, a novel optimization problem, namely the throughput maximization under availability constraints, is proposed. The innovation,

[2] For instance in DVB-S2X under normal operation over a linearized channel, the most robust modulation and coding rate can provide quasi-error-free communications (frame error probability lower than 10^{-5}) for as low as -2.85 dB of user SINR, thus achieving a minimum spectral efficiency of 0.4348 bps/Hz [7]. Beyond this value, a service outage occurs.

aspired by operational requirements, lies in the incorporation of MRCs in the PAC sum-rate maximizing problem (equivalently minimum SINR constraints). Formally, the new optimization problem is defined as

$$\mathcal{SRA}: \max_{\{\mathbf{w}_k\}_{k=1}^{N_t}} \sum_{i=1}^{N_u} \log_2(1+\gamma_i)$$

$$\text{s.t.: } \gamma_i = \min_{m \in \mathcal{G}_k} \frac{\left|\mathbf{w}_k^{\dagger}\mathbf{h}_m\right|^2}{\sum_{l \neq k}^{N_t} \left|\mathbf{w}_l^{\dagger}\mathbf{h}_m\right|^2 + \sigma_m^2}, \quad \forall i \in \mathcal{G}_k, k, l \in \{1, \ldots, N_t\} \tag{15.8}$$

$$\text{and to: } \left[\sum_{k=1}^{N_t} \mathbf{w}_k\mathbf{w}_k^{\dagger}\right]_{nn} \leq P_n, \quad \forall n \in \{1, \ldots, N_t\}, \tag{15.9}$$

$$\text{and to: } \gamma_i \geq \gamma_{\min}, \ \forall i \in \{1, \ldots, N_u\} \tag{15.10}$$

In $\mathcal{SRA}$, the power allocation needs to account for the MRCs, i.e., (15.10), as discussed in more detail in Reference 12.

As intuitively expected, it is shown in Reference 12 that the introduction of MRCs is bound to decrease the system throughput performance. However, this trade-off can be leveraged towards more favorable conditions, by considering other system aspects, as will be discussed in the following.

15.4.2.2 Throughput maximization

A modulation constrained practical system employs higher order modulations to increase its rate with respect to the useful signal power. The strictly increasing logarithmic cost functions describe communications based on Gaussian alphabets and provide the Shannon upper bound of the system spectral efficiency. Therefore, the sum-rate maximization problems solved hitherto fail to account for the modulation constrained throughput performance of practical systems. The complication lies in the analytically intractable, at least by the methods considered herein, nature of a step cost function. In this section, an attempt to leverage this cost function in favor of the system throughput performance is presented. In more detail, benefiting from the finite granularity of the rate function (15.5) over the achieved SINR, an extra system level optimization can be defined as

$$\mathcal{SRM}: \max_{\{\mathbf{w}_k\}_{k=1}^{N_t}} \sum_{i=1}^{N_u} f_{\text{DVB-S2X}}(\gamma_i, \mathbf{t})$$

$$\text{s.t.: } \gamma_i = \min_{m \in \mathcal{G}_k} \frac{\left|\mathbf{w}_k^\dagger \mathbf{h}_m\right|^2}{\sum_{l \neq k}^{N_t} \left|\mathbf{w}_l^\dagger \mathbf{h}_m\right|^2 + \sigma_m^2}, \quad (15.11)$$

$$\forall i \in \mathcal{G}_k, k, l \in \{1, \ldots, N_t\}$$

$$\text{and to: } \left[\sum_{k=1}^{N_t} \mathbf{w}_k \mathbf{w}_k^\dagger\right]_{nn} \leq P_n, \quad (15.12)$$

$$\forall n \in \{1, \ldots, N_t\},$$

$$\text{and to: } \gamma_i \geq \gamma_{\min}, \ \forall i \in \{1, \ldots, N_u\} \quad (15.13)$$

where $f_{\text{DVB-S2X}}(\cdot, \cdot)$ is the finite granularity step function defined in (15.5). The realization of a non-strictly increasing cost function inhibits the application of gradient-based solutions and necessitates a different solution process, as in detail provided in Reference 12.

15.4.3 Performance evaluation

The validity of the heuristic sum-rate maximization algorithm is established by comparing the performance of the herein proposed precoders with the optimal in a max–min fair sense, solutions of Reference 14. The throughput versus availability trade-off between the two formulations is also be exhibited in the following paragraphs.

In Figure 15.5, the average user throughput of the considered multibeam satellite is plotted versus an increasing total on-board available power in Gbps/beam. Two users per frame are considered, i.e., $\rho = 2$. Clearly, the proposed precoding designs outperform existing approaches. The $\mathcal{SR}$ problem achieves more than 30% gains over the max–min fair solutions of References 14 and 15. These gains are reduced when the max SR under MRCs is considered, i.e., $\mathcal{SRA}$. This is the price paid for guaranteeing service availability over the coverage. Finally, the maximum gains are observed when the modulation aware max SR precoding, i.e., $\mathcal{SRM}$ is employed, which also guarantees service availability. Consequently, the best performance is noted for $\mathcal{SRM}$ with more than 30% of gains over the max–min fair formulation of Reference 14 and as much as 100% gains over conventional systems in the high-power region, for 2 users per frame.

For the same simulation setting, the cumulative distribution functions (CDFs) of the SINRs over the coverage area are given in Figure 15.6. Clearly, conventional systems achieve higher SINRs by the means of the fractional frequency reuse. This value is around 17 dB, in line with the results of Reference 12. However, this does not necessarily translate to system throughput performance. To guarantee increased SINRs,

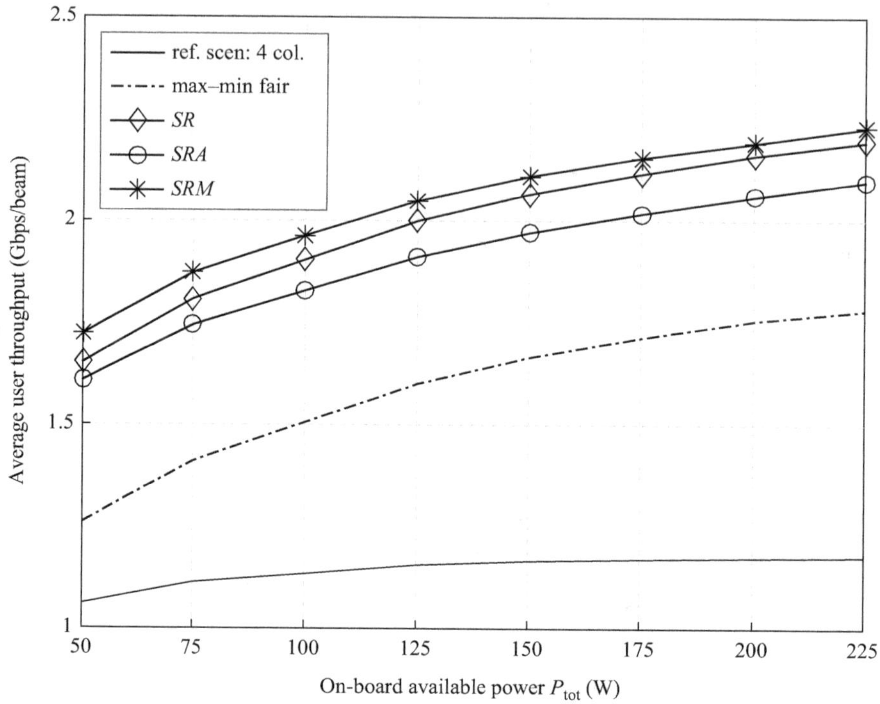

Figure 15.5 Average user throughput versus on-board available transmit power, for 2 users per frame

the frequency allocated per user is four times reduced. On the other hand, aggressive frequency reuse reduces the average SINR values and increases its variance, as seen in Figure 15.6. This, however, allows for more efficient resource utilization and consequently higher throughput, as seen before in Figure 15.5. Moreover, the superiority of the max SR techniques proposed herein, over the fair solutions is also evident. Amongst these methods, the best one is $\mathcal{SRM}$ as already shown.

The benefits of $\mathcal{SRA}$ over $\mathcal{SR}$ are clear in Figure 15.6, where the SINR CDF of all methods is presented. Clearly, $\mathcal{SRA}$ guarantees a minimum SINR of −2.85 dB but attains SINRs higher than 2 dB with less probability than $\mathcal{SR}$. Furthermore, $\mathcal{SRA}$ can be regarded as a middle step towards advancing to the more elaborate, $\mathcal{SRM}$ algorithm. Since $\mathcal{SRM}$ includes the same availability constraints as $\mathcal{SRA}$, identical availability gains are noted for both methods. However, $\mathcal{SRM}$ exploits the granular nature of the spectral efficiency function towards achieving SINRs higher than $\mathcal{SR}$. In Figure 15.6, it is clear that the proposed optimization manages to adapt each user's SINR to the throughput function, since the SINR distribution follows the granular spectral efficiency function. Users have SINR values in between the DVB-S2X thresholds with very low probability. This insightful result justifies the increased gains of $\mathcal{SRM}$, even for guaranteed availability. Also, according to Figure 15.6, 40% of the users operate utilizing the first four available MODCODs.

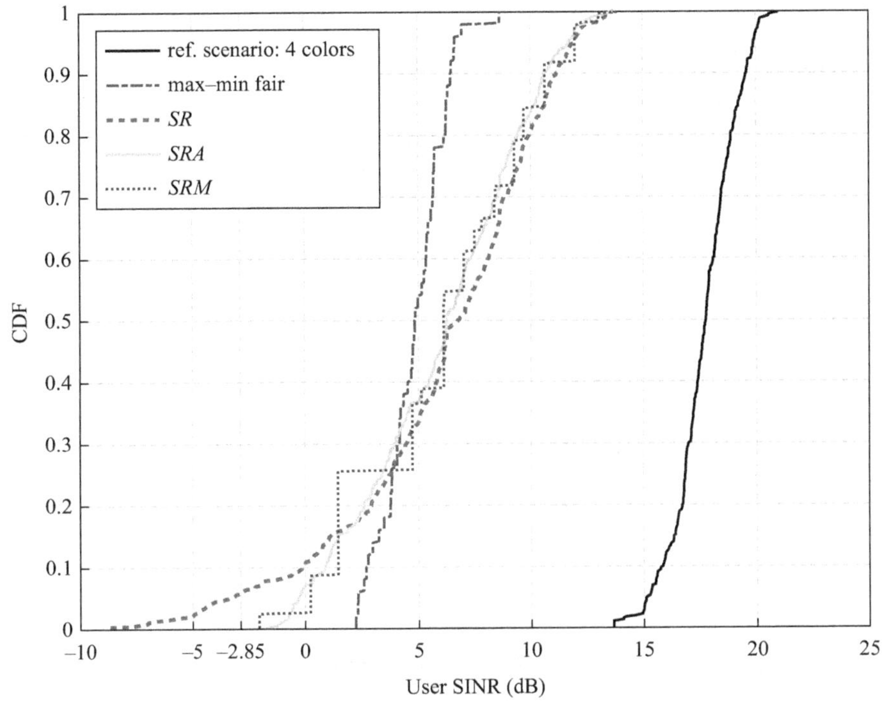

Figure 15.6 CDF of user SINR over the coverage, for 2 users per frame

Moreover, Figure 15.7 provides the rate of CDFs of the conventional and the max–min fair systems and exhibits the very low variance of their receive SINR. On the contrary, $\mathcal{SR}$ not only achieves very high rates but also drives some users to the unavailability region. A 5% outage probability is noted for this precoding scheme. However, this is not the case for the $\mathcal{SRA}$ and $\mathcal{SRM}$ problems, which guarantee at least 0.3 Gbps to all users.

An important issue is the performance of the developed methods with respect to an increasing number of users per frame. As presented in Figure 15.8, $\mathcal{SRM}$ manages to provide more than 30% of gains for $\rho = 3$ users per frame. Both the conventional and the proposed systems suffer from an increase in the number of users per frame, since the worst user defines the MODCOD for all users. For conventional systems, this degradation is negligible when compared to the frame-based precoding systems. The performance degradation, when a precoding vector is matched to more than one channels, is expected. As initially proven in Reference 10, when advancing from unicasting to multicasting, the precoding problem becomes NP-hard. Added to that, when more users are grouped together, then the chances are that one of them will be compromised and thus constrain the performance of all other users. This observation further justifies the results of Figure 15.8. Nevertheless, in the same figure, positive gains over the conventional systems are reported even for 6 users per frame unlike all other state of the art techniques. These results are given for a

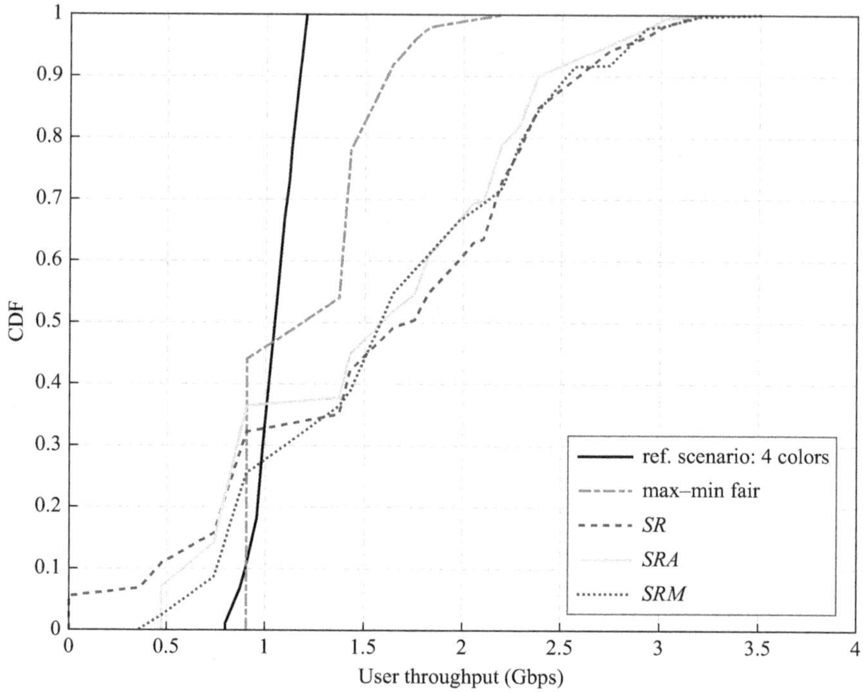

Figure 15.7 Per-user throughput CDF, for 2 users per frame

nominal on-board available power of 50 W. This performance is compromised by the random user scheduling since users with very different SINRs are co-scheduled and thus constrained by the performance of the worst user.

All results presented hitherto, ignore any interference originating from adjacent clusters of beams, when multi-GW scenarios are considered. A first evaluation of the performance degradation due to this interference can be found in Figures 15.9 and 15.10. More specifically, Figure 15.9 presents the per beam average throughput for 5 users per frame, when interference from adjacent beams is accounted for. By comparing these results with the results of Figure 15.8, for an identical operating point of 50 dBW/beam, it is clear that the interferences are significantly impacting the systems in terms of total throughput. Nevertheless, interference still plays a crucial role as revealed in Figure 15.10. In this figure, the SINR CDF is given in a logarithmic scale to focus on the effects of interference on the very low SINR values. In this regime, notably total system availability cannot be guaranteed. Even when the availability-constrained algorithms are employed (e.g., SRA at −2.85 dB), signal processing cannot manage interference from adjacent clusters and therefore the edge users will experience a much lower value of SINR than the one aimed. More specifically, SRM without min SINR constraints exhibits a 96% availability, for a minimum supported MODCOD of around −3 dB or lower. On the other hand, SRA attains 99.9% availability, at the expense of throughput gains, if the system can still operate below −7 dB. Based on these results, distributed precoding methods are of

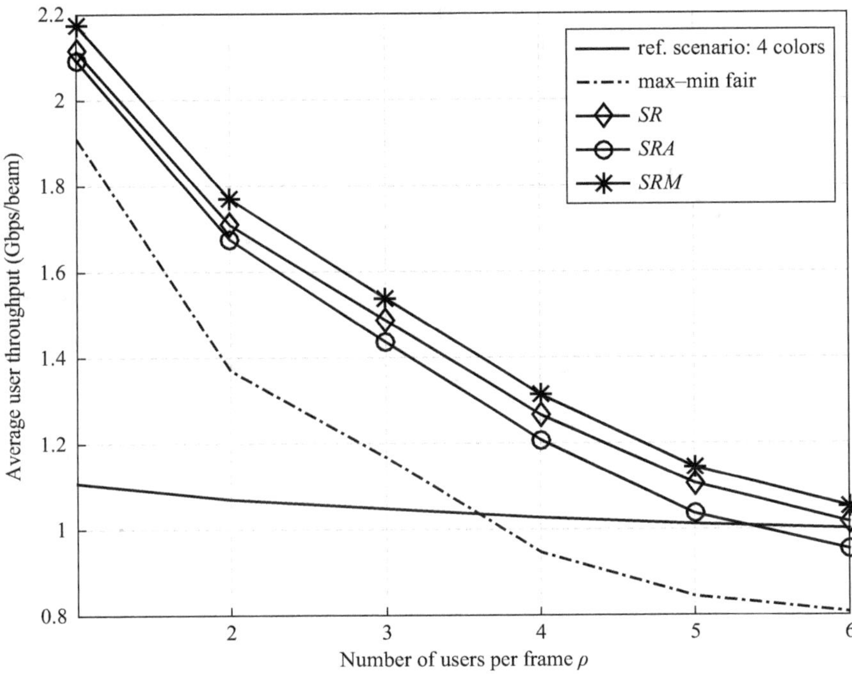

Figure 15.8 Average user throughput versus the number of users per frame

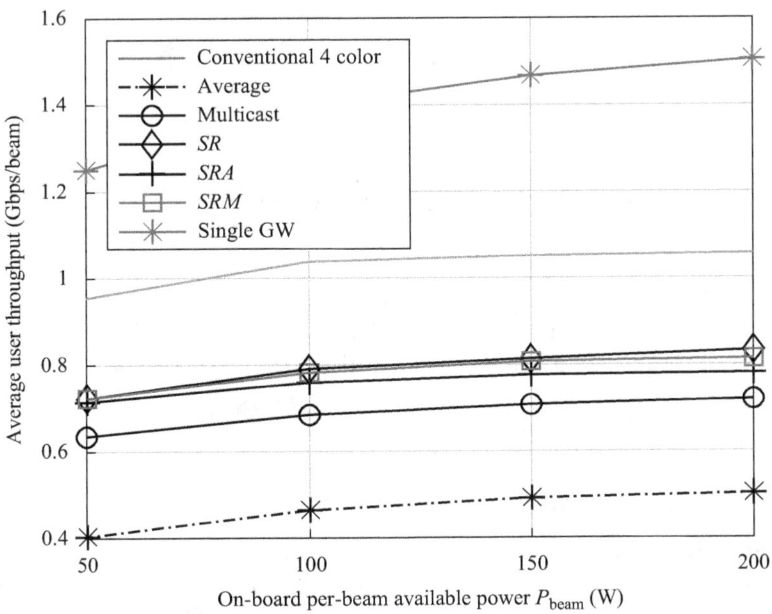

Figure 15.9 Average user throughput versus increasing per beam power budget, for a multi-GW scenario with 5 users per frame and random user scheduling

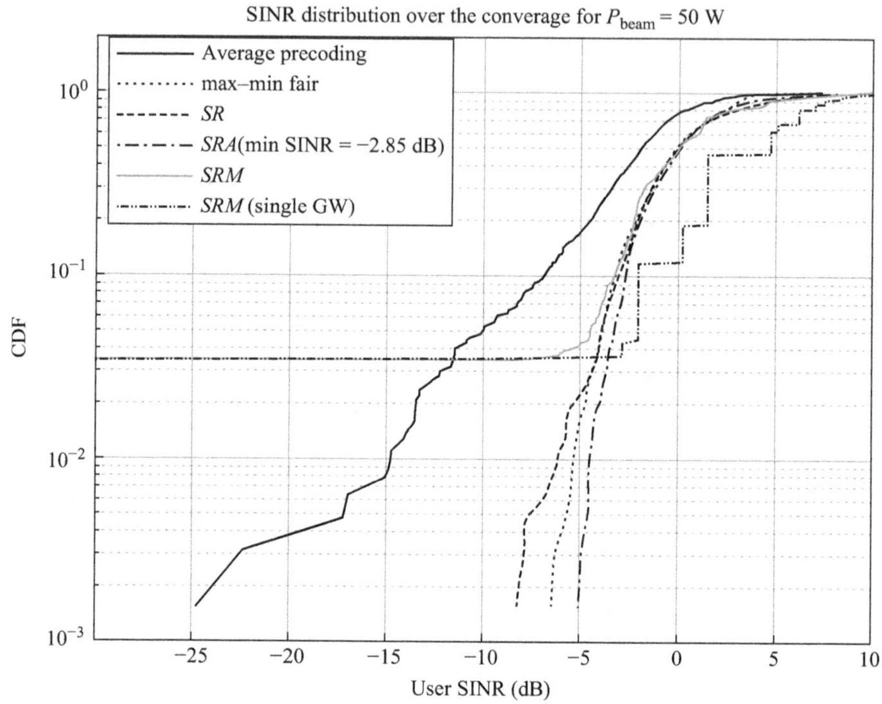

Figure 15.10 SINR distribution in multi-GW scenarios where no cooperation between GWs is assumed

high importance towards guaranteeing the operation of the multibeam satellite under acceptable outage probabilities.

15.5 Linear precoding over non-linear channels

The multibeam satellite is a power limited relay, dependent on battery and solar energy. As a result, the on-board high-power amplifiers need to be operated at their saturation limits. On the other hand, the input output characteristic of cost efficient and reliable TWTAs is not a linear function throughout the operational regime. Consequently, the practical implementation of precoding over multibeam satellite systems needs to consider this non-ideal amplification stage.

15.5.1 The non-linear multibeam channel

The non-linearity of the satellite channel, which determines a deterioration of the transmitted precoded signals, is introduced by the non-ideal on-board amplification stage. Indeed, the signals feeding the on-board TWTAs are subject to a non-linear effect, which affects both the amplitude and the phase. Considering the

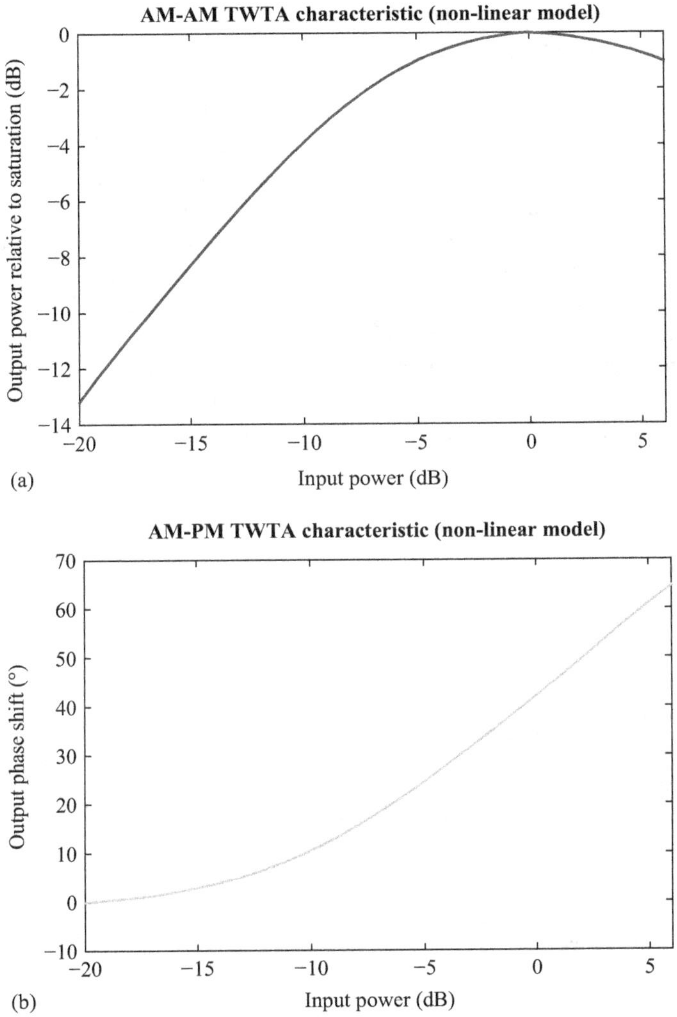

Figure 15.11 Input–output characteristics of the on-board TWTA ((a) AM-AM and (b) AM-PM)

non-linearized TWTA model defined in Reference 7, the corresponding amplitude-to-amplitude (AM-AM) and amplitude-to-phase (AM-PM) characteristics are shown in Figure 15.11. These curves represent relationship between the input and the output power of the amplifier, and between the input power and the output phase shift, respectively. The on-board TWTA needs to be operated as close as possible to saturation, to efficiently exploit the scarce available power. As clear from the curves, this drives the operating point to a region where the instantaneous output power is a non-linear function of the input power. This non-linear amplification, in particular when combined with channels that introduce memory, leads to signal distortion.

In the remainder of this work, the effects of non-linear amplification over precoded waveforms are investigated, via simulations. To this end, a relevant parameter that is analyzed is the peak-to-average power ratio (PAPR) of the transmitted signals.

For the sake of clarity, in Figure 15.12, the overall block scheme of the simulated satellite FL is reported. As shown in the scheme, in the ground segment, after the mapping of each bit stream into a symbol stream according to a specific modulation, a pilot sequence (P2) is attached to the payload. The P2 pilot sequence, present in the DVB-S2X frame structure [7], is a sequence of 180 symbols, known at the receiver side, which is modulated and precoded as the information sequence. Such characteristics enable a receive equalization scheme, which will be described in the following. After the precoding module, each branch contains a square-root-raised-cosine (SRRC) filter for the shaping and the transmission of the signals. Moreover, the space segment, besides the described non-linear TWTAs, includes the blocks corresponding to the input multiplexer (IMUX) and the output multiplexer (OMUX). Further on, an

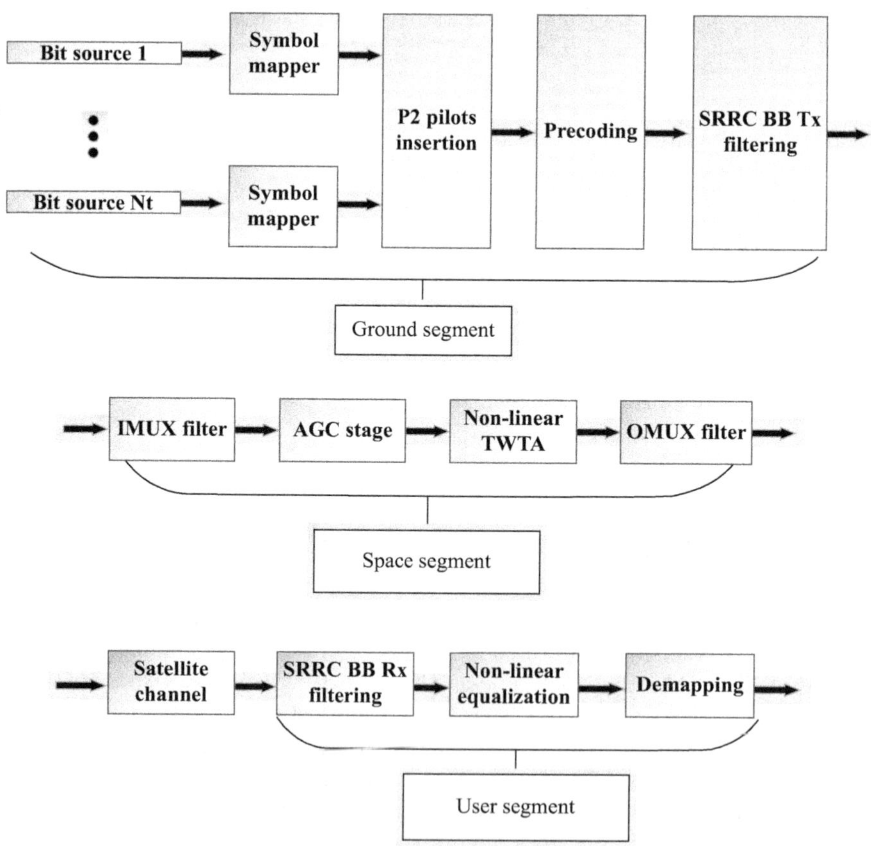

Figure 15.12 Block scheme of the considered satellite forward link

automatic gain control (AGC) scheme is considered before the amplification stage. Such component, fundamental for handling the non-linearity issue, will be described in the remaining of this section. Finally, after the satellite channel, the user segment branches include the mentioned equalization block and the demapper, which returns the information symbols.

15.5.2 Effects of non-linear amplification

In order to analyze the effects induced by the non-linearities of the satellite channel onto the precoded waveforms, it is worth considering the scatter plot representations of the transmitted symbols as resulting by simulations, according to the block scheme of Figure 15.12. However, at first, the blocks referring to the AGC stage and to the non-linear equalization, which will be characterized in the remainder of this section, are not considered.

The following scatter plots are obtained by assuming a 32-APSK modulation scheme and a E_s/N_0 value (ratio of the average energy per symbol over the noise power spectral density) of 25 dB, on a seven-beam system using ZF precoding.

15.5.2.1 PAPR analysis

A first evaluation of the effect of the channel non-linearities onto the precoded waveforms can be given by studying the PAPR of the transmitted signals. Considering ZF precoding, Figure 15.13 shows how the PAPR increases with the number of

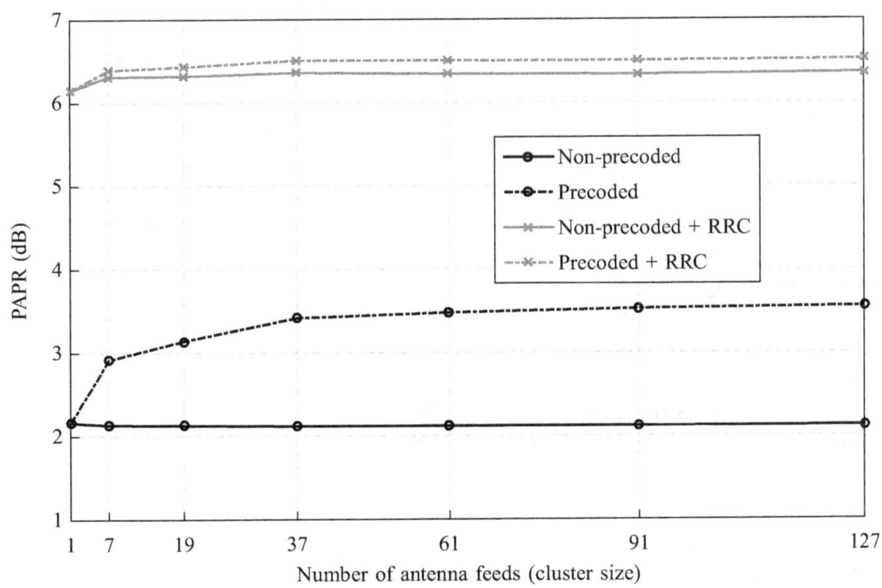

Figure 15.13 PAPR of 32-APSK modulated signals versus number of transmitting elements, with and without precoding

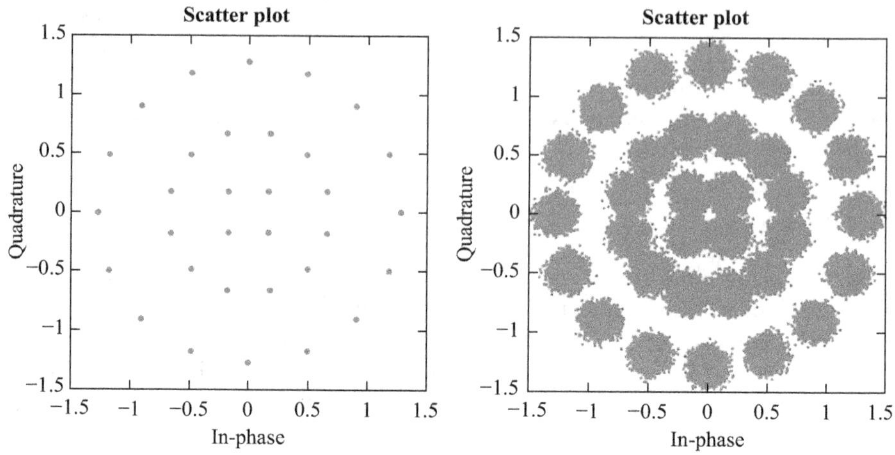

Figure 15.14 Tx constellations before (left) and after (right) precoding

feeds of the multibeam antenna, both for precoded and non-precoded waveforms. Moreover, the PAPR curves are shown in two cases, one considering a SRRC filtering operation and the other without. Therefore, the curves including filtering evaluate the signal in the sample domain, while the un-filtered in the symbol domain. The modulation scheme considered for these curves is 32-APSK. Analogous curves, here not reported for the sake of brevity, can be obtained considering different modulation orders. As one can easily envisage, the PAPR increases with the modulation order.

From such result, it can be seen that the precoded waveforms show a higher PAPR with respect to the non-precoded ones. As a consequence, the distortion induced on the signals by the TWTA will be stronger when precoding is applied. Nevertheless, the percentage of the increase of PAPR in the sample domain is lower. Thus, we can deduce that the effect of precoding is not as detrimental if one considers a more accurate modeling of the transmission chain.

In addition to the increase of the PAPR for the precoded waveforms, it should be kept in mind that the channel distortion also affects the performance of precoding, leading to an imperfect compensation of the Multiuser Interference. Also for this reason, the signal corruption due to the channel non-linearities is expected to be stronger when precoding is applied.

15.5.2.2 Scatterplot analysis

Figure 15.14 shows the transmitted symbols, for one of the data streams, as they are before the precoding operation (left side) and after it (right side). It is visible how the precoding operation changes the constellations, correlating the different symbol streams and therefore increasing the PAPR, as already discussed. Then, in Figure 15.15, the scatter plots of the received symbols are depicted, compared with the

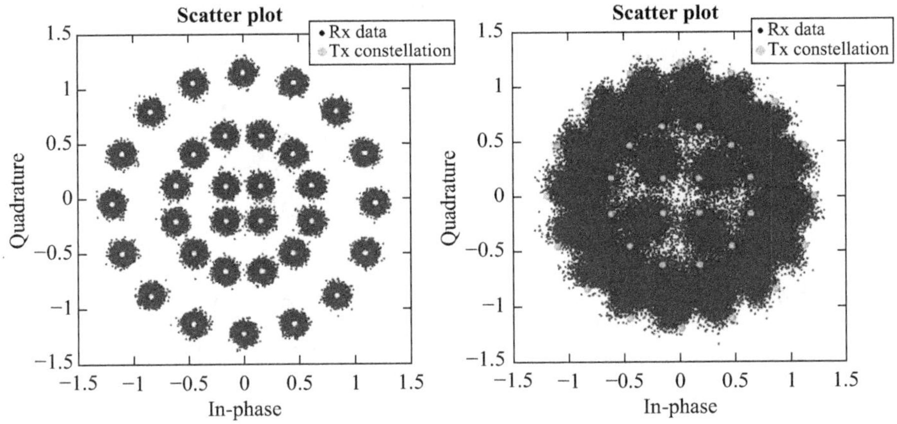

Figure 15.15 (Left) scatterplots of the Tx and Rx constellations, with a linear AWGN channel; (right) same constellations with a non-linear satellite channel

transmitted ones, both in the case of a linear AWGN channel, hence without the non-linear TWTA block (left side), and in the case including the channel non-linearities (right side). The deterioration of the transmitted precoded signals is highly visible in the second case. More specifically, we can observe two effects of the channel non-linearities on the received constellations [16]:

- a warping effect on the constellation centroids, due to the non-linear characteristics of the TWTA, which consists in a displacement of the centroids both in amplitude and in phase;
- a clustering effect on the received symbols, due to the inter-symbol interference (ISI), which is enhanced by the TWTA non-linearities.

In the following section, we focus on two solutions aiming at counteracting the signal corruption due to the channel non-linearities.

15.5.3 Counteracting non-linearities

In the following paragraphs, we focus on two general approaches aiming at counteracting the signal corruption due to the channel non-linearities [17], for point-to-point communications:

1. a proper AGC stage, applying a power back-off to the transmitted signals so to reduce the distortion;
2. a non-linear equalization module at the receiver, which aims at compensating the distortion induced by the TWTAs. This compensation would be based on the transmission of a known sequence, i.e., a pilot sequence or a training sequence.

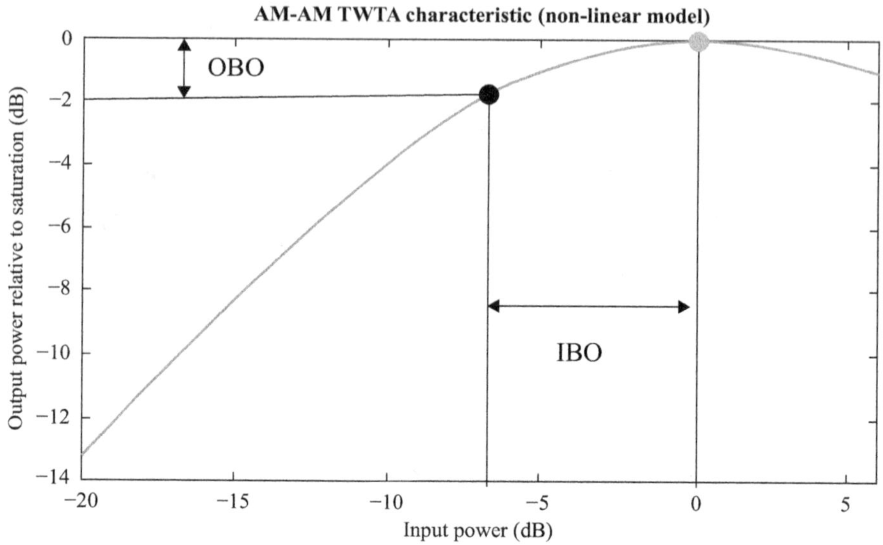

Figure 15.16 Back-off operation onto the transmitted useful power

The AGC stage applies to the signals an input back-off (IBO), which can be defined, in dB, as:

$$\mathrm{IBO} = 10\log(P_{\mathrm{SAT}}/P_{\mathrm{AV}}) \tag{15.14}$$

where P_{SAT} represents the saturation power of the amplifier, while P_{AV} denotes the average signal power after the back-off operation. Such operation is depicted in Figure 15.16, where the lighter mark represents the saturation point of the TWTA and the black mark is the operational point. Its clear how, at the cost of a reduced transmitted power, the distortion is reduced.

Concerning the second approach, namely the non-linear equalization performed in the user segment, pilot sequences need to be exploited.[3] An example of the equalization performed at the receiver is shown in Figure 15.17. In the left-hand side, the received pilot sequence is shown together with the 32-APSK constellation used at the transmitter; in the right, the equalized constellation derived for decoding is shown. For determining the new constellation, the rings' amplitudes, as well as the relative phase shifts, are obtained using the received distorted version of the pilot symbols and exploiting the knowledge of their original transmitted version.

[3] Herein, we use the P2 pilot sequence attached to the payload in the DVB-S2X frame structure [7]. The P2 pilot is a sequence of 180 symbols, and it is modulated and precoded as the information sequence. Such characteristics make it suitable to be exploited in an equalization scheme. In particular, the received P2 sequence is leveraged by the receiver towards estimating the distortion suffered by the signals. Consequently, the design of the constellation to be used for the demapping of the received symbols is made possible at the receiver side.

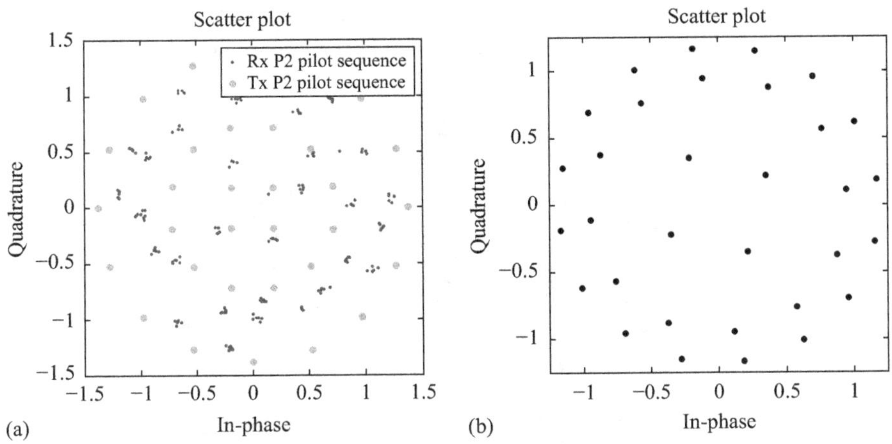

Figure 15.17 Design of a new constellation for demapping, based on the received P2 pilot. (a) Tx and Rx P2 pilot and (b) equalize constellation

Table 15.2 Simulation parameters for the non-linear performance evaluation

Parameter	**Value**
Symbol rate	25 MBaud
Number of beams	7
Number of Tx symbols	10^5
Modulation	32-APSK
Roll-off	0.3
IBO	5 dB
$\frac{E_s}{N_0}$	25 dB

In order to show how the proposed chain of Figure 15.12 behaves, with the inclusion of the AGC and non-linear equalization stages, some simulation results are reported hereafter. The results are obtained using the parameters listed in Table 15.2.

Figure 15.18 shows the scatter plot of the received constellation after demapping, obtained applying the AGC stage with an IBO of 5 dB. The black constellation represents the constellation used for demapping the received symbols, designed in the equalization stage. It is apparent how this constellation matches the received clouds better than the original one. The distortion on the symbols is still clearly visible in the clustering and warping effects. Nevertheless, they appear strongly mitigated with respect to the scatter plot of Figure 15.15 (right), resulting in a lower symbol-error-rate (SER).

Concerning the SER obtained by applying the proposed solutions, Figure 15.19 (left) shows how it varies with the E_s/N_0 value, comparing the multibeam case, where

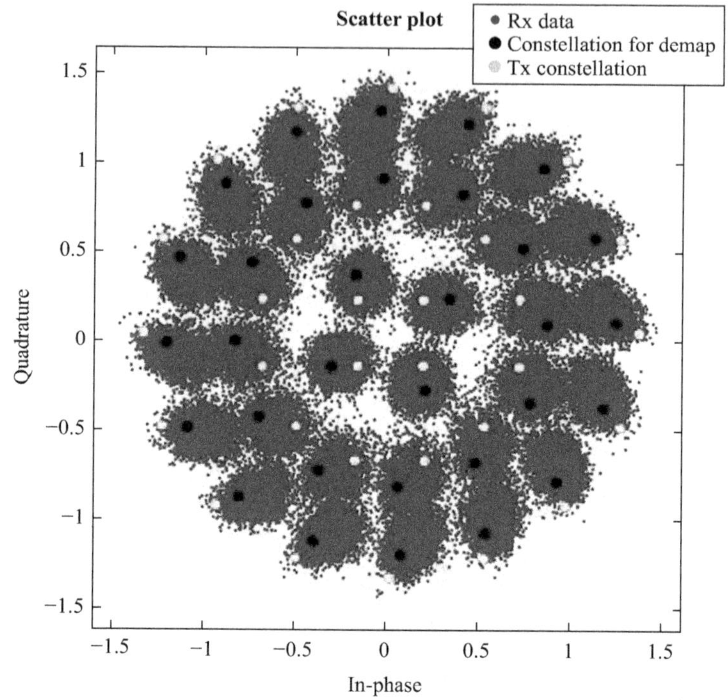

Figure 15.18 Scatterplots of the Tx and Rx constellations, with an IBO of 5 dB and equalization at the Rx. The black constellation is available at the output of the equalization block

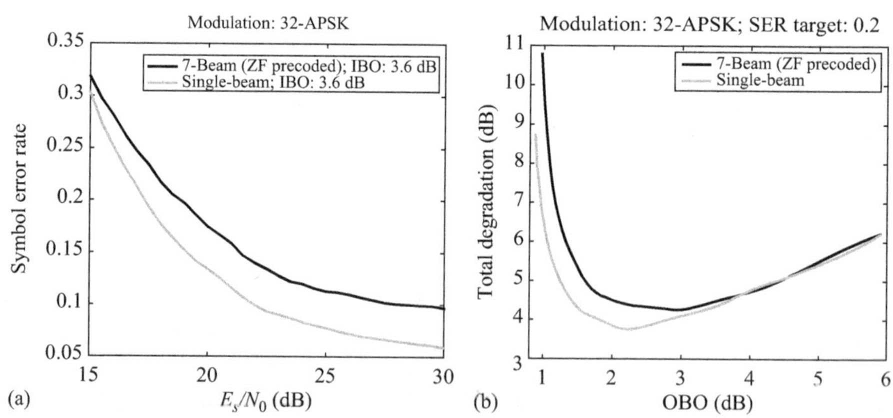

Figure 15.19 (Left) SER versus E_s/N_0 curves, for the precoded and non-precoded cases; (right) TD versus OBO curves, comparing the same cases

ZF precoding is applied, with the single-beam case. The result confirms how, in the precoded case, the system suffers more the effects of the channel non-linearities.

In Figure 15.19 (right), another important performance metric is considered for the system at hand. Such metric is given by the total degradation (TD) of the link, defined as [16] (all the quantities are in dB):

$$\mathrm{TD} = \left[\frac{E_s}{N_0}\right]_{\mathrm{req}}^{\mathrm{NL}} - \left[\frac{E_s}{N_0}\right]_{\mathrm{req}}^{\mathrm{AWGN}} + \mathrm{OBO} \tag{15.15}$$

where $[E_s/N_0]_{\mathrm{req}}^{\mathrm{NL}}$ is the E_s/N_0 value required to achieve a predetermined target SER with the non-linear channel, $[E_s/N_0]_{\mathrm{req}}^{\mathrm{AWGN}}$ is the same quantity considering a linear AWGN channel, and OBO is the output power back-off. The OBO of the system has to be set in order to minimize the resulting TD. In Figure 15.19 (right), the TD curves, plotting the TD as a function of the OBO, are depicted, comparing the multibeam precoded case with the single-beam case. The TD curves are obtained assuming a target SER of 0.2, which once using also some typical FEC techniques, can lead to a quasi-error-free transmission. The optimum back-off corresponds to the minimum of the TD curves. As visible, also this result shows that in the multibeam precoded cases, the degradation of the link is higher with respect to the single-beam case. In addition to this, the optimum back-off, which corresponds to the minimum of the TD curves, is about 0.7 dB higher in the multibeam case. Ultimately, we can identify a trade-off between precoding, which allows for more spectra per user link in multibeam systems, and the applied power back-off, which reduces the useful transmitted power.

15.6 Receiver implementation

In this section, receiver implementation aspects related to precoding operations are addressed. The calculation of the CSI to be fed back at the GW is one of the most challenging part. Thanks to the superframe structure specified in Reference 7, a DVB-S2X-related synchronization procedure is derived in the following. The received signal, $y(t)$, at the generic kth User Terminal (UT) can be modeled as

$$y_k(t) = \sum_{n=1}^{N} r_{kn}(t) + n_k(t) \tag{15.16}$$

where

$$r_{kn}(t) = \hat{h}_{kn}(t)\,x_n(t) \tag{15.17}$$

is the waveform received by the kth terminal from the nth beam with $\hat{h}_{kn}(t)$ the channel coefficient containing all of the impairments, $x_n(t)$ the transmitted information signal, and $n_k(t)$ the AWGN noise at terminal k. In Reference 12, 18–20, various precoding techniques for multibeam systems have been proposed. Accordingly, precoding is applied at the GW and, in order to construct the precoding matrix, it is foreseen that each terminal provides estimates of the channel parameters for all of the detectable received signals, i.e., the terminal specific waveform plus all of the detectable interferers. This requires that a terminal synchronizes not only onto its own signal, but also onto the detectable interferers. The precoding techniques have been developed with reference to the DVB-S2X signal model and scenarios. As a consequence, in order to design and assess the required terminal synchronization procedures, we also considered the DVB-S2X superframe structure as defined in Reference 7. In the following section, the main characteristics of the superframe are overviewed in order to ease the understanding of the proposed synchronization procedures. For more details on the superframe features, the interested reader is referred to Reference 7.

Before entering into the details of the study, it is worthwhile noting that the objective of designing the synchronization chain for such system is far from being a trivial application of already known procedure. The system is in fact highly interference and noise limited by design. More importantly, precoding helps to reduce the interference on the reference beam signal. This means reducing power of the interfering signals. Nevertheless, these waveforms still need to be estimated by the receiver in order to feedback the exact vector channel. This emanates a framing design, that includes not-precoded data such as the pilot symbols, on which estimation and synchronization is preformed. As it will be clear from the following discussion, the synchronization task is extremely challenging because it requires to detect and estimate signals with carrier to interference ratio as low as -16 or -20 dB.

15.6.1 DVB-S2X superframe structure

The superframing structure is introduced for multiple purposes, in particular to increase resilience to co-channel interference by other beams thanks to the use of different scrambling sequences and to support synchronization algorithms thanks to the fixed framing structure. In the following, the main and general features of the superframe structure related to the synchronization procedures are listed:

- the length of the superframe is fixed and equal to 612,540 symbols;
- pilot fields and pilot structure are different for each superframe format; they are specified in the Super Frame Format Indicator (SFFI) field and are fixed and known for a specific frame format;
- the Start of Super Frame (SoSF) field length is equal to 270 symbols while SFFI field length is equal to 450 symbols;
- in order to take advantage from orthogonality properties in case of perfect alignment, Walsh–Hadamard sequences are used for SoSF and pilot fields;
- the SoSF field is composed by the concatenation of two Walsh–Hadamard sequences, respectively, of length 256 and 14 symbols;
- pilot field is composed by the concatenation of two Walsh–Hadamard sequences, respectively, of 32 and four symbols for the long Physical Layer Frame (PLFrame), and 32 and 16 symbols for the short PLFrame;
- the superframe is scrambled using two different scrambling sequences: the first sequence, so-called reference data scrambler, is the same for all beams and it overlays SoSF and pilot fields. The second one, the so-called payload data scrambler, is beam dependent and provides resilience to co-channel interference.

- two types of pilot fields are present: the superframe modulated and, possibly, precoded pilot symbols, called P2, that can be used for precoded data demodulation, and the unprecoded pilots, P, that can be used for the estimation of the CSI to be sent back to the GW.

The SFFI field specifies the format of the superframe. For precoding purposes, the format specifications to be considered are:

- Bundled PLFrames (64,800 payload size) with SF-Pilots
- Bundled PLFrames (16,200 payload size) with SF-Pilots

For the two specified formats, different pilot structures are defined. In the following, a brief description of the structures is reported.

15.6.1.1 Bundled PL-Frames – long frame

In Figure 15.20, the structure of the Bundled long PL-Frame is shown. The main characteristics to keep in mind for synchronization purposes are:

- pilot field length is 36 symbols
- the number of pilot fields in one superframe is 639
- the length of the superframe precoded pilot field P2 is 180 symbols
- the distance between two consecutive pilot fields is 956 symbols (including 36 pilot symbols)
- the Physical Layer Header (PLH) field is composed by six Physical Layer Signaling (PLS) code replica
- the superframe consists of nine Bundled PLFrames of length 64,800 symbols
- the distance between two consecutive PLH fields is 67,920 symbols

15.6.1.2 Bundled PL-Frames – short frame

In Figure 15.20, the structure of the Bundled short PL-Frame is shown. The main characteristics to keep in mind for synchronization purposes are:

- pilot field length is 48 symbols
- the number of pilot fields in one superframe is 324
- the length of the superframe precoded pilot field P2 is 96 symbols
- the distance between two consecutive pilot fields is 1887 symbols (including 48 pilot symbols)
- the PLH field is composed by four PLS code replica

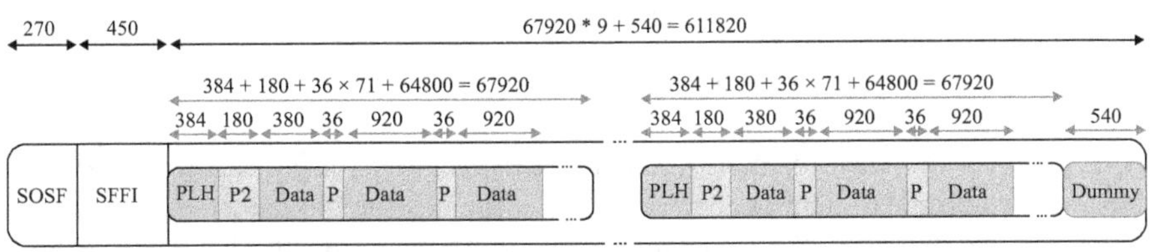

Figure 15.20 Structure of the Bundled PLFrame with precoded and not precoded pilots – long frame [7]

- the superframe consists of 36 Bundled PLFrames of length 16,200 symbols
- the distance between two consecutive PLH fields is 16,984 symbols

For the sake of the synchronization and estimation techniques that will be discussed in the next section, it is fundamental to note also that:

- SoSF and SF-Pilots are not precoded;
- SoSF and SF-Pilots consist of beam-specific orthogonal Walsh–Hadamard sequences;
- two different scrambling sequences are applied to the superframe: the first sequence, the so-called reference data scrambler, is the same for all beams, it overlays only the SoSF and SF-Pilots, and is restarted at each start of superframe. The second scrambling sequence, the so-called payload data scrambler, is beam dependent, overlays the data payload, and provides resilience to co-channel interference.
- the superposition of the beam-specific Walsh–Hadamard sequences in the SoSF and SF-Pilots and the common reference data scrambler yields a unique beam-specific signature that can be used for waveform/beam identification.

15.6.2 Synchronization and channel acquisition

Before entering into the details of the synchronization strategy, it is worthwhile noting that the objective of designing the synchronization chain for a multicast precoded system is far from being a trivial application of already known procedures. On the one hand, multicast precoding is in fact designed to reduce, ideally remove, the interference generated by the co-channel beam waveforms, thus making auxiliary parameters estimation almost impossible on the precoded interferer waveforms. On the other hand, non-precoded SOSF and SF-Pilots are subject to extreme high interference levels, much higher than that experienced in non-precoded systems. As a matter of fact, if that would not be the case, precoding would not be necessary in the system itself. Therefore, we can safely state that synchronization and estimation in precoded High Throughput Satellite (HTS) systems are a formidable challenge by design. The composite received signal, $y_k[n]$, at the kth UT consists of the superposition of waveforms transmitted through B different interfering beams and it can be modeled as:

$$y_k[m] = \left(\sum_{n=1}^{N} h_{kn}[m] x_n[m - \tau_{kn} - \tau_d m] \mathrm{e}^{-j(2\pi \Delta f_{kn} mT + \theta_{kn} + \phi_{kn}^{\mathrm{Pay}}[m])}\right) \times \mathrm{e}^{-j\left(2\pi \left(f_o + \frac{1}{2} f_d m\right) mT + \phi[m]\right)} + n_k \tag{15.18}$$

where $x_n[m]$ is the waveform component received from the nth antenna feed, and $h_{kn}[m]$ is the complex channel coefficient from the nth feed to the kth user, τ_{kn}, Δf_{kn} and θ_{kn} are, respectively, the time, frequency, and phase offsets of the nth received waveform at the kth terminal. Moreover, due to the characteristics of the receiver, the composite signal is affected by the following common impairments which are time and frequency drift τ_{dk}, f_{dk}, frequency offset f_o and phase noise $\phi[n]$. In addition, $\phi_{\text{Pay}}[m]$ is the differential phase noise given by the payload.

Let us consider an indexing function $i(n)$ taking values in the set $S = \{0, 1, 2, \ldots, N\}$, i.e., $i(n) \in S = \{0, 1, 2, \ldots, N\}$, $n = 1, 2, \ldots, N$, where $i(n) = 0$ means that the nth waveform component is not considered in the processing. Each UT shall proceed with the following general synchronization/estimation operations:

a. identify the frame boundary, i.e., frame synchronization, for the $i(n)$th waveform component, $n = 1, \ldots, N$;
b. synchronize frequency, phase, and time for the $i(n)$th waveform component, $n = 1, \ldots, N$;
c. perform channel estimation for the $i(n)$th waveform component, $n = 1, \ldots, N$ to be sent back to the GW.

To this aim, the kth UT shall:

- perform and apply a coarse frequency estimation by means of a non-data aided estimator on $y_k(t)$, e.g., through the algorithm proposed in Reference 21;
- for each waveform $i(n), n = 1, 2, \ldots, N$, perform frame synchronization, e.g., through algorithms described in Reference 22 if pilot fields or unique words are present in the frame format to identify frame boundaries. Non-coherent post-detection integration can be applied to cope with the residual frequency uncertainty and the time varying phase impairment. For those waveform components for which frame synchronization is not successfully achieved $i(n)$ is set to zero so as to exclude them from the subsequent processing;
- for each waveform $i(n)$, $n = 1, 2, \ldots, N$ for which frame synchronization is successfully achieved, perform fine time tracking, phase and frequency tracking, and channel estimation.

A high-level block diagram of the synchronization procedures described above is depicted in Figure 15.21.

It is worth mentioning that time tracking, phase and frequency tracking as well as CSI estimation procedure are performed using pilot-assisted algorithms, which can take advantage of the orthogonal Walsh–Hadamard pilot sequences specified in Annex E of the DVB-S2X standard [7].

For the superframe structure described therein, a possible procedures of the UT synchronization for a system employing precoding is hereafter described with reference to Figure 15.21. It is worthwhile underlining that the following procedure is referred to general scenarios and they are not developed ad hoc with respect to considerations on the possible waveforms misalignment.

15.6.2.1 Synchronization procedure

1. Coarse frequency acquisition.
2. Matched filtering.
3. Frame synchronization.

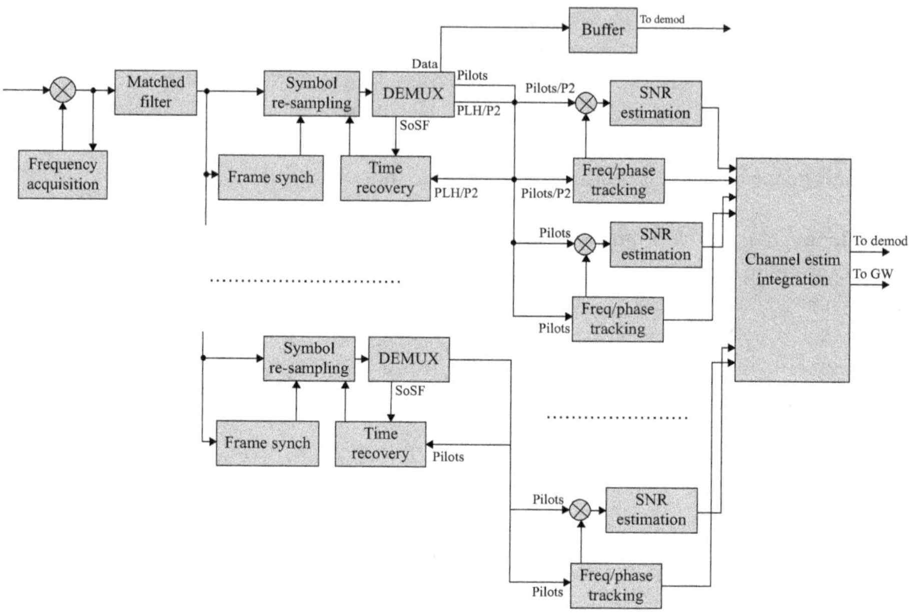

Figure 15.21 Block diagram of synchronization procedure taking place at the UT in a system employing precoding

4. Demultiplexing to separate SoSF, the pilots, the non-precoded pilots (PLH and P2), and data (according to the SF structure contained in the SFFI field).
5. Data-aided time tracking and resampling on the basis of the SoSF and non-precoded pilots (for every signal).
6. Frequency and phase tracking on non-precoded pilots (for every signal).
7. Perform channel estimation to be sent back to the GW on the basis of the non-precoded pilots for each signal.

Regarding Steps 5 and 6, since the precoding operation aims at reducing the interference on the reference signal caused by the remaining signals, precoded pilots cannot be reliably used for synchronization and estimation purposes other than for the reference user. Steps 3–7 can be performed for each waveform component, $i(n)$, $n = 1, 2, \ldots, N$, sequentially or in parallel.

Finally, it is worthwhile noting that, assuming the same baud rate for all of the received waveforms, two possible implementations of the general procedure described above can be devised, depending on the relative delays of the received waveforms. One implementation shall address the case of low-data rates for which the received waveforms can be considered "quasi-synchronous", i.e., the relative time delays τ_{kb} are within one symbol time. The second implementation deals with the complementary case of "non-synchronous" waveforms, i.e., the relative time delays τ_{kb} may exceed one symbol time.

15.7 Conclusions

The main goal of this chapter has been to introduce the reader to the most recent research developments on joint processing multibeam satellite communications. Since a joint processing enabled multibeam satellite constitutes a cooperative amplify-and-forward relay, these developments are closely related to the cooperative relaying literature. Nevertheless, as in detail described herein, the inherent nature of the satellite channel emanates fundamental differentiations in the proposed signal processing methods. Three main practical limitations have been overviewed. First, the need to jointly process signals over a fixed framing structure, coined as framed-based precoding. Second, the proper consideration of the non-idealities of real amplifiers and the effect they can have on precoded signals. Finally, but equally important, the detailed design of receivers that can operate in an interference-limited environment and adequately estimate the channel state of each user, so that precoding at the transmitter can be ultimately performed. If a single take home lesson needs to be gained from this overview would be that cooperative satellite relaying from high above is a mature technology from a research perspective and has reached the doorstep of practical demonstration and implementation. More details on this subject can be found in Reference 22 and 23.

References

[1] [Online]. Available: http://sslmda.com

[2] [Online]. Available: http://www.telesat.com

[3] P.-D. Arapoglou, B. Shankar, A. Panagopoulos, and B. Ottersten, "Gateway diversity strategies in Q/V band feeder links," in *17th Ka and Broadband Communications Conference*, Palermo, Italy, Oct. 2011.

[4] A. Gharanjik, B. S. M. R. Rao, P.-D. Arapoglou, and B. Ottersten, "Large scale transmit diversity in Q/V band feeder link with multiple gateways," in *Personal Indoor and Mobile Radio Communications (PIMRC), 2013 IEEE 24th International Symposium on*, London, Sep. 2013, pp. 766–770.

[5] A. Gharanjik, B. Rao, P.-D. Arapoglou, and B. Ottersten, "Gateway switching in Q/V band satellite feeder links," *Communications Letters, IEEE*, vol. 17, no. 7, pp. 1384–1387, Jul. 2013.

[6] G. Zheng, S. Chatzinotas, and B. Ottersten, "Multi-gateway cooperation in multibeam satellite systems," in *Proceedings of 23rd IEEE Symposium on Personal, Indoor and Mobile Radio Communications, (PIMRC)*, Sydney, 2012, pp. 1360–1364.

[7] ETSI EN 302 307-2 V1.1.1 (2015-02): "Digital Video Broadcasting (DVB); Second generation framing structure, channel coding and modulation systems for broadcasting, interactive services, news gathering and other broadband satellite applications; part 2: DVB-S2-extensions (DVB-S2X)", March 2014.

[8] W. Yu and T. Lan, "Transmitter optimization for the multi-antenna downlink with per-antenna power constraints," *IEEE Trans. Signal Process.*, vol. 55, no. 6, pp. 2646–2660, Jun. 2007.

[9] E. Bjornson, M. Bengtsson, and B. Ottersten, "Optimal multi-user transmit beamforming: Difficult problem with a simple solution structure," *IEEE Signal Processing. Mag.*, vol. 31, no. 4, pp. 142–148, Jul. 2014.

[10] N. Sidiropoulos, T. Davidson, and Z.-Q. Luo, "Transmit beamforming for physical-layer multicasting," *IEEE Trans. Signal Process.*, vol. 54, no. 6, pp. 2239–2251, 2006.

[11] D. Christopoulos, S. Chatzinotas, and B. Ottersten, "Sum rate maximizing multigroup multicast beamforming under per-antenna power constraints," in *Proceedings of IEEE Global Communications Conference*, Austin,TX, Dec. 2014, *preprint:* arXiv:1407.0005 [cs.IT].

[12] D. Christopoulos, S. Chatzinotas, and B. Ottersten, "Multicast multigroup precoding and user scheduling for frame-based satellite communications," *IEEE Trans. Wireless Commun.*, vol. 14, no. 9, pp. 4695–4707, Sept. 2015, doi: 10.1109/TWC.2015.2424961.

[13] E. Karipidis, N. Sidiropoulos, and Z.-Q. Luo, "Quality of service and max-min fair transmit beamforming to multiple cochannel multicast groups," *IEEE Trans. Signal Process.*, vol. 56, no. 3, pp. 1268–1279, 2008.

[14] D. Christopoulos, S. Chatzinotas, and B. Ottersten, "Weighted fair multicast multigroup beamforming under per-antenna power constraints," *IEEE Trans. Signal Process.*, vol. 62, no. 19, pp. 5132–5142, Oct. 2014.

[15] D. Christopoulos, S. Chatzinotas, and B. Ottersten, "Multicast multigroup beamforming under per-antenna power constraints," in *Proceedings of IEEE International Communications Conference (ICC)*, Sydney, Australia, Jul. 2014, *preprint:* arXiv:1407.0004 [cs.IT].

[16] E. Casini, R. D. Gaudenzi, and A. Ginesi, "DVB-S2 modem algorithms design and performance over typical satellite channels," *International Journal of Satellite Communications and Networking*, vol. 22, no. 3, pp. 281–318, 2004. [Online]. Available: http://dx.doi.org/10.1002/sat.791

[17] D. Spano, D. Christopoulos, S. Andrenacci, S. Chatzinotas, J. Krause, and B. Ottersten, "Total degradation analysis of precoded signals onto non-linear satellite channels," in *21st Ka and Broadband Communications Conference*, Bologna, Italy, Oct. 2015.

[18] P.-D. Arapoglou, A. Ginesi, S. Cioni, *et al.*, "DVB-S2x enabled precoding for high throughput satellite Systems," *International Journal of Satellite Communication and Networking (JSCSN)*, 2016.

[19] G. Taricco, "Linear precoding methods for multi-beam broadband satellite systems," in *European Wireless 2014; 20th European Wireless Conference; Proceedings of*, Barcelona, May 2014, pp. 1–6.

[20] D. Christopoulos, S. Chatzinotas, and B. Ottersten, "Frame based precoding in satellite communications: A multicast approach," in *Proceedings of IEEE Advanced Satellite Multimedia Systems Conference (ASMS)*, Livorno, Italy, Sep. 2014, *preprint:* arXiv:1406.6852 [cs.IT].

[21] A.N. D'Andrea and U. Mengali, "Design of quadricorrelators for automatic frequency control systems," *IEEE Trans. Commun.*, vol. 41, no. 6, pp. 988–997, Jun. 1993.

[22] P. Kim, R. Pedone, M. Villanti, *et al.*, "Robust frame synchronization for the DVB-S2 system with large frequency offsets," *IEEE Trans. Commun.*, vol. 58, no. 4, pp. 1235–1246, Apr. 2010.

[23] D. Christopoulos, S. Chatzinotas, G. Taricco, *et al.*, *Cooperative and Cognitive satellite systems*. Elsevier, Amsterdam, 2014, chapter: Multibeam Joint Precoding: Frame Based Design.

[24] D. Christopoulos, S. Chatzinotas, and B. Ottersten, *Cooperative and Cognitive satellite systems*. Elsevier, Amsterdam, 2014, chapter: User Scheduling in Cooperative Satellite Systems.

Index

www.ingramcontent.com/pod-product-compliance
Lightning Source LLC
LaVergne TN
LVHW011857200926
848484LV00002B/471